The Greening of Pharmaceutical Engineering Volume 1

Scrivener Publishing
100 Cummings Center, Suite 541J
Beverly, MA 01915-6106

Publishers at Scrivener
Martin Scrivener(martin@scrivenerpublishing.com)
Phillip Carmical (pcarmical@scrivenerpublishing.com)

The Greening of Pharmaceutical Engineering Volume 1

Practice, Analysis, and Methodology

M.R. Islam
Trans Canada Training, London, Ontario, Canada

J.S. Islam
Dalhousie University, Halifax, Nova Scotia, Canada

G.M. Zatzman
EEC Research Org., Halifax, Nova Scotia, Canada

M.A.H. Mughal
Bethesda 365 Inc., Mississauga, Ontario, Canada

and

M. Safiur Rahman
Bangladesh Atomic Energy Commission

Scrivener Publishing

WILEY

Co-published by John Wiley & Sons, Inc. Hoboken, New Jersey, and Scrivener Publishing LLC, Salem, Massachusetts.
Published simultaneously in Canada.

For general information on our other products and services or for technical support, please contact our Customer Care Department within the United States at (800) 762-2974, outside the United States at (317) 572-3993 or fax (317) 572-4002.

Wiley also publishes its books in a variety of electronic formats. Some content that appears in print may not be available in electronic formats. For more information about Wiley products, visit our web site at www.wiley.com.

For more information about Scrivener products please visit www.scrivenerpublishing.com.

Cover design by Kris Hackerott

Library of Congress Cataloging-in-Publication Data:

ISBN 978-0-470-62603-0

Printed in the United States of America

10 9 8 7 6 5 4 3 2 1

Contents

Preface

In the double barreled sense of the ancient Chinese proverb about "living in interesting times", the epoch we moderns inhabit is "interesting" indeed. Everything jumps out in the form of contradiction upon contradiction.

We are witness to unprecedented progress in technological development. Meanwhile, no less a personage than Nobel Chemistry Laureate Robert Curl can suppress mentioning that our era is a 'technological disaster' pregnant with as-yet-unknown portents.

Acutely attuned to signs of open savagery and other lesser levels of indifference and lack of empathy, we proudly assert our embrace of modern-day Enlightenment on an ever-more-encompassing scale. That said, whence the dissonant drumbeat emanating simultaneously from some of the most ardent supporters of modern European capitalism? Pope Francis insists with the greatest indignation that unfair economic structures today are devastating entire continents (like South America, his home continent), compounding huge inequalities within and between countries. As His Holiness himself acknowledged, these inequalitie themselves constitute acts of the most unbridled terrorism, threatening all while apparently not actually aimed at or singling out any one country or people over others.

Those ruling us from on high deceive others, if not themselves, that we have reached the pinnacle of human equality and democratic values. Meanwhile, cognitive dissonance is fuelled by such Nobel Peace laureates as U.S. President Obama and the Egyptian politician Mohamed el-Baradei accepting without elaboration that such rights in practice belong only to a small group. This is no longer in the realm of Noam Chomsky's "manufactured consent". This is deliberately-fostered disinformation trundled out in a form created and purveyed in Hollywood movies populated by some of the most accomplished actors of our time such as Jack Nicholson's portrayal of the eternally victimized Josie Wales in Cool Hand Luke being casually yet viciously brutalized by the Sherriff, his tormentor, who "explains" to the audience that "What we have here is a failure to communicate."

The realities rendered utterly incoherent by such twisting and reversal of roles include the news of the July 2013 military coup led by General

Sisi, the US War College graduate and treacherous Defence Minister in the cabinet of Egyptian President Mohammed Morsi. President Morsi stands indicted before the entire world, and sentenced to be hanged as the first ever democratically elected president of Egypt. The US celebration of this latest outrage is foisted on the world in an utterly disinformed and disinforming headline: "Kerry Lauds Egypt Military for 'Restoring Democracy'".

We continuously hear about the threats to human existence that loom, while the effort to find serious possible solutions is accompanied by extremely toxic slogans developed over the last 20 years such as: 'save the planet and decrease population', and 'save the planet and plant trees', etc. While the intention may seem pro-Humanity, the complete absence of anything scientific in the foundations of these propositions ... ... that reduce the serious search for a truly balanced application of actual natural law to nothing but sloganeering.

Meanwhile, the sphere of human-set judicial law, outside the realm of formal science, is attacked via the same disinforming methodology. Thus, for example, revising our laws, claiming we are reaching pinnacle of justice, only to read (with a gasp) in the *Washington Post* that "US Court System Criminally Unjust".[1]

- We brag about how we have so many breakthrough technologies in medical science, resulting in record breaking profit for every new medicine, yet we discover there is no cure to a single medicine. We brag about saving children at birth and prolonging life expectancy only to discover record breaking rate of abortion and suicide (1.5 million per year in USA alone).

When it comes to the driver of modern civilization, viz, Economics, we have confronted or sided with either of two arguments that seem to have derived opposite conclusions from the same initial premises. On the one hand, Karl Marx — whose productive period extends from the 1840s to the 1880s — believed capitalism was a mug's game that must reach a day of reckoning between the wage-enslaved workforce and the owner-employers

[1] Meanwhile, prison systems generate brisk profits, and no wonder: with numerous privately-owned and -operated "detention centres" sprinkled along the US-Mexican border from California to Texas populated mostly with "illegal immigrants" branded as such by immigration judges who send them there while sitting on or "advising" the boards of these corporations. The authors became aware of this level of decay from the personal experience of the life-partner of the eldest granddaughter of one of our wives.

who hoarded to themselves all the value added by the workers' collective labor even as the average rate of profit continued to fall. On the other hand, Simon Kuznets in the middle of the 20th century, leader of an entire line of Nobel-Prize economists, made the case for inequality gaps growing smaller as we become more sophisticated as a human race, siding essentially with the Grand Equilibrator, extremely successful stock-market pirate, and policy god of post-World War Two Western welfare economics John Lord Maynard Keynes. As the Keynesian consensus broke down at the US-led World Bank in the 1970s, between these poles the ultimate rebalancing role would be played by Prof. Joseph Stiglitz — the profiteer-turned-'prophet', Nobel laureate economist and World Bank employee who rose to broader public fame castigating the International Monetary Fund (IMF) for offering 'remedies' that made things worse in the so-called developing economies of the Third World/Global South. Amid a steady and massive output of all manner of prescriptive policy studies and experiments, the true expertise of the Stiglitz camp and its following seemed to be the turning of slowdowns into recessions and recessions into depression.

Today, there is no consensus as to who is the Prophet and who the Profiteer. Nevertheless there are these facts to that tell us whatever we need to hear about the state of the world today.

- Data from tax returns show that the top one percent of households in the United States received 8.9 percent of all pre-tax income in 1976. In 2012, the top one percent share had more than doubled to 22.46 percent.
- The total inflation-adjusted net worth of the Forbes 400, an annual listing of America's richest individuals, rose from \$507 billion in 1995 to \$1.62 trillion in 2007, before increasing again to \$2 trillion in 2012.
- Estimates from the Credit Suisse Research Institute, released in October 2010, show that the richest 0.5 percent of global adults hold well over a third of the world's wealth.
- Approximately one third of annual deaths in the United States, epidemiological researchers believe, can be credited to the nation's excessive inequality.
- Approximately one third of annual deaths in the United States, epidemiological researchers believe, can be credited to the nation's excessive inequality.
- In USA, non-white families make in average 65% of the white families (as reported in 2010). In the mean time,

white families claim about six times the net worth of non-white families, a gap that has changed little over the past generation.

- In terms of global disparity, world's wealth owned by the richest 1% increased from 44% in 2009, slated to be over 50% of the world's wealth by 2016.

What exactly is the mechanism that has made it possible for the capitalist world to create such an imbalance? It is increasingly clear that the machineries of war, coupled with 'health care,' have become fuel for the ultimate train wreck that has become the economy. Even in the United States, the question being asked is: have escalating health care costs contributed to the huge economic gap between America's rich and the rest?

This matter is considered debatable. However, what is not debatable is the fact that, from 1999 to 2009, rising health care spending—both on premiums and out-of-pocket costs—totally erased wage gains for a typical family. By 2012, more than one in four families experienced financial burdens of medical care. In 2013, per capita national health expenditure was $9255 (17.4% of GDP). In the meantime, families with incomes at or below 250% of the federal poverty level (FPL) were more likely to experience financial burdens of medical care than families with incomes above 250% of the FPL. Families with children aged 17 years and under were more likely than families without children to experience financial burdens of medical care.

This makes the most promising section of the society the most vulnerable and the worst victim of profiteering of the pharmaceutical industry. Not surprisingly, the rates of childhood asthma, overweight/obesity, learning disability, ADHD, food allergies, seizures, autism, and overall chronic illnesses all skyrocketed as well.

In the meantime, the cost of pharmaceutical stocks have been on fire for years now. Some of the biggest income-earners are generated by drugs that supposedly help manage 'incurable' conditions. With that comes the increasing numbers of profit margin, some (Pfizer) raking in as much as 43% profit. In 2014, the average return on pharmaceutical company stocks 32% in USA as the 'addiction' to pharmaceutical drugs has surpassed that of illicit drug. Today, United States represents 5% of the world's population yet consumes 50% of the worlds' prescription polls and over 80% of the world's prescription narcotics. Then there is prescription drug abuse with reported 2 million Americans that abuse prescription pills every year. There has been 80% increase in doctors prescribing illegal drugs. This is not a coincidence and this process has a new co-conspirator, the FDA.

This new scheme began after PDUFA Prescription Drug User Fee Act was passed in 1992. This act allows FDA to collect funds from pharmaceutical companies to fund new drug approval process. Today, that bill stands at an average of $350,000 per new drug application. This process makes it impossible to carry out independent research on the long-term impact of any drug.

The society has moved from addiction to drugs to addiction to prescription drug and the entire world is paying the price. The drug business globally has raked in $731 billion in 2007, $967 billion in 2012, and projected to rake in some $1.2 trillion in 2017. Today, the United States is the sickly society and tomorrow it would be the entire world.

So, how do we clean up this mess? One thing we cannot do is sanction another epidemiological study or another twist to the current closed loop of fear → sale → Security → fear. Every time there is an epidemiological study, soon appears another study that nullifies it as long as the first study gives the slightest hint that a particular medicine or a new science technology or a diagnostic tool may be the one causing the disease or at least making it worse, e.g. birth defect with antidepressant, vaccine with Autism, ADHD with artificial, microwave, the list is endless.

Also, we cannot agree on disagreeing and settling instead for the so-called 'lesser evil'. History tells us, both of these evils are necessary for creating spiralling down our civilization into the sewage of obscene disparity. This book deconstructs all the dubious yet unchallenged bits and pieces of the most prominent theories, be they from social science or the so-called "hard" sciences. We do so by uncovering and exposing how absurd the fundamental premises of these theories are. It is shown in clearest terms how these theories are the reasons we couldn't get out of the Einstein box and the entire civilization was inescapably trapped. In this way, all paradoxes turn out to be either apparent or outrightly false, and can then be safely removed. This lays the foundation for a set of correct theories with fundamentally sound premises. (That begins in Volume 2.)

1

Introduction

1.1 Opening Remarks[1]

A professor of medicine in Canada was asked if there is any cure in modern era of any contemporary disease. After some reflection, he named penicillin as the only medicine that cures a disease. "Why then do physicians routinely ask if a patient has ever taken penicillin in his lifetime?", he was asked. This time the professor was quick in replying, "Oh, that's because today's penicillin is synthetic (artificial)". Of course, that poses the pointed question of whether there is any medicine today that is not artificial. The same question was posed five years later, this time to an American professor of medicine. He couldn't come up with any medicine. When the name penicillin was mentioned to him, he quipped, "Oh, penicillin is the proof that modern era has no medicine that cures; it only delays the symptoms".

This book is about finding cause of diseases and proposing cures. Taking the time to look into the real science and moving away from dogma science that has incapacitated European education system for many centuries, this book invokes a paradigm shift.

The starting point of this book is: Nature is perfect. This is not a new theme *per se*. It is a new, indeed revolutionary, theme that is

[1]The authors use this chapter both to summarize the forthcoming chapters and to lay out their own opinions on these matters.

fervently disallowed with a near-fanatical obsession in our current era. The ancient Greeks had this theme, as did the ancient Chinese, Indians, and Mayans. This theme reached its culmination in the works of Avicenna, the father of modern medicine, whose most famous medical work, out of nearly 450 treatises, includes the 14-volume *Canon of Medicine*. The *Canon of Medicine* covered every single subject of medicine, all of the known diseases and treatments, methods for testing new treatments, diagnosis, etc. It was a standard full-version medical textbook, used for centuries all over the known world including Europe until the 17th century. Avicenna looked at the theoretical aspect, covering fields of cosmology, the temperaments and humours, etc. He then examined basic anatomy, physiology and phycology. In the second part, he examined health disorders. There, he covered definitions, causes of disease, dietetics, pulse, urine analysis and so on. Finally, he looked at treatment of disease and preservation of health.

Today, meanwhile, we don't know the root cause of a single disease, let alone knowing the cure.[2]

Today's medical science does not include determining cause of a disease, let alone suggesting a cure. Every theory of diagnosis, prediction, and remedy involves dogmatic premises that go unchallenged for centuries. Basically, the modern medicinal system has four major shortcomings. They are:

- the cause of deadliest diseases are unknown;
- when the origin of a disease is believed to be bacterial, all antibiotics used are synthetic;
- vaccines are introduced without knowing the cause of a disease that they are supposed to immunize against; and
- the role of lifestyle on human health is not understood.

In terms of science, it comes from (all these are "New Science"-related):

- conflation of human traits with Godly traits;
- incorrect and illogical narration of origin of universe, universal order, and life on earth;

[2]This book offers theories that describe root causes of disorders, both physical mental. In the second volume of the present work, cures of diseases will be presented in a handbook format.

- incorrect and illogical characterization of nature and sustainability criterion;
- disconnection of conscience from humanity;
- lack of characterization of human thought material (HTM);
- disconnection between matter and energy;
- absence of scientific criteria for defining disease and disorder;
- disconnection between mental and physical disorder; and
- theories of illogical and incoherent fundamental premises.

The above shortcomings have rendered modern science incapable of cognizing in the right direction, the direction in which cognitive processing of data is completed to some explicit point. Everything is done to justify the conclusion in an overt display of preposterous cognition — which is more illogical than dogma. In social science (as well as medical science), the crisis is most acute. We don't even know what makes a human. If one goes by the definition of 'selfishness defining humanity' and/or 'making tools is the sign of human intelligence', violent chimps that make tools to hurt others are humans (Viegas, 2015), along with feuding Neanderthals in Europe (National Geography, 2008). If we have to go by DNA similarity or complexity, plants would be more human than humans. In defining humanity, everything is on the table — except... conscience! Scientifically, conscience should be the criterion that defines humanity, but Europe disconnected conscience from Humanity soon after dogma became enthroned. This is just the beginning of cognitive malfunction that created today's technological disaster and infinite social injustice. Masked for centuries in the form of generalized Eurocentric bias valorizing any scientific finding in Europe or America ahead of science produced outside Europe, this has been difficult to isolate, challenge or eliminate.

A component piece of the theoretical outlook framed by the present work is the notion that anyone other than the human being has no need of empathy in order to act conscientiously. On the contrary: they are part of universal order and hardwired to be conscientious — as part of the universal order. Humans need to be conscious about their intention and have the ability to act on self-interest in the short term or self-interest in the long term (this is acting on conscience). This notion in itself is *not* Eurocentric: indeed it is captured in the concept of *karma* and is very old. New Science — which is Eurocentric at its core —

disconnected this process and everything became about self-interest in the short term. Dogma, meanwhile, was useful for the anti-social purposes of unjust rulers in the East no less than in the West. The bifurcation between science and nonsense became possible and actual to the extent that dogma and Eurocentrism shared a common aim.

This understanding of the matter helps illuminate why it is no surprise that today we cannot explain any phenomenon without resorting to dogmatic assertions and spurious justifications. This ranges from diagnosis of a disease to prediction, from remediation to prevention, and from simplest of disorders to deadliest of diseases. We do not know what causes addiction, light pollution, noise pollution, and numerous other insults to the environment and human thought material, let alone attempting to remedy them. In the meantime, drugs, vaccines, and perpetual therapies and 'management tactics' continue to sell at an unprecedented pace and economy. The driver of this runaway train bound for 'technological disaster' continues to flourish, keeping all concerned content with the status quo.

Another element that frequently goes missing in medical scientific research is what the authors call 'metadata'.

In this book, the authors set up logical theories that address the above shortcomings and properly defines what constitutes disorder, both mental and physical. It then presents a guideline for preventing and correcting disorder, both cognitive and physiological.

1.2 Are we Trained to Develop Contempt for Conscience and Addiction to Selfishness?

Plato said, "Strange times are these in which we live when old and young are taught falsehoods. And the one man that dares to tell the truth is called at once a lunatic and fool." Few question the notion that this 'strange times' is now when it comes to politics. However, fewer understand the science behind this 'strange times', even fewer appreciate how this 'strange times' have pervaded all aspects of our civilization, and practically no one sees this as a problem in the science and technology development sector. Many dislike the current system but few see the big picture and the direction that our civilization is moving and none can tell us how to fix the system.

According to former U.S. President George W. Bush, the Department of Homeland Security in Washington, DC and many others, the maintenance of beliefs by any individual that counter officially accepted views is a personality disorder of such toxicity as to mandate deployment of an entire system for attacking the psyche of such individuals until they "crack" or are destroyed. As a 2006 article in the Sunday *New York Times Magazine* disclosed, this was indeed the object of an elaborate and carefully-planned program of government-funded research. As part of this research, an entire regime of randomized psychological "torture-testing" of people was launched and justified as an effort to catch lies and liars in general on the basis of refining and overcoming the defects of polygraph technology in particular. To grasp the decadence implicit in this proposition, consider the underlying logic of this matter launched during the Bush Administration and continuing to date:

- Either you are with us or you are with the terrorists (MAJOR PREMISE)
- Those who are with us never lie (MINOR PREMISE); therefore
- All liars must be terrorists and all terrorists must be liars (CONCLUSION)

The authors have been striving more than a decade to further develop a genuine paradigm shift emerging in the education system that teaches real science and avoids dogma 'science' (Islam et al., 2013). Well aware, meanwhile, of the fact that 'paradigm shift' is a popular buzzword, we have anticipated readers' skepticism about such a claim and now hoist the following flag from our mast.

1.3 Metadata

It is to be expected that Humankind's consciously-collected and recorded collective medical knowledge would predate the rise of European civilization. Relative to the length of time humans have been present on the Earth, on the other hand, this development is not so far back as many might think. Early in the 20th century, records of such knowledge were known — from findings of collected by archaeological expeditions organized from the UK and Germany in the late 19th century — to date back to at least the 3rd millennium BCE.[3] More recently,

[3]According to Sir William Osler's 1913 Silliman Foundation lecture at Yale University, Egypt was the locale of this breakthrough

even earlier records of acupuncture and other Chinese medical practices have been dated as far back as the 5th millennium BCE.[4]

With these earliest instances, however, we generally lack significant information about the depth of theoretical understanding attached to these evidences back in their own time. As a consequence, systemic explanations of these medical practices could not be provided. At the same time, the importance of immediate specifics of, and conditions surrounding/attending, what was recorded loom correspondingly large.

The possible scientific value or significance of the information content of such specifics is thus entirely tangential. At the same time, until more information is available to fill the void, these specifics are not without some potential probative value. Their value is very much like that of the metadata generated with computerized logging of data being generated by or within any activity that has been placed under some kind of organized observation. Generically speaking: in this sense, such pieces of associated information — data falling outside the immediate focus of interest — can be considered to constitute the metadata of the phenomenon.

The idea of metadata has become widely popularized in connection with Edward Snowden's disclosures of the methods and procedures of intelligence-gathering efforts applied by the National Security Agency on the global public, not excluding citizens and residents of the United States itself. The impression has been left (without being explicitly affirmed), that metadata are a phenomenon peculiar and specific to large-scale computerized data collection.

Such a conclusion is a hasty and extremely important unexplored assumption. **All real, meaningful data of human experience have metadata associated with them**. At the same time, in the conventional scientific knowledge-gathering model followed by millions of scientific workers every minute of every day, the value of a wide range of metadata is largely discounted, and not entirely without good reason. For example: even if a completely-made-up story about a young Isaac Newton being inspired to cogitate on the notion of gravity and free-fall acceleration while sitting under an apple tree were factually true, its probative value and significance for theoretical physics would remain extremely moot (some would say: completely irrelevant). Gravitational forces are not in themselves sense-data, but rather a product of cognitive processing of certain sense-data. But then, even if we qualify it as

[4]Visit http://decodedpast.com/traditional-chinese-medicine-earliest-written-records/1159

metadata, it is useless and irrelevant for theorizing about physical principles, and any its possible significance appears to have nothing to do with the actual aim of any given scientific investigation. But now consider the weighing of evidence in the justice system. Anyone working in that system, from the most senior of judges to the lowliest police detective, will admit that some metadata possess enormous forensic value, to the extent that they serve to throw any light on the who, where, what and why of crimes and crime-related events, making it possible to include or exclude certain guesses about the possible interconnected significance of the time, place or order of those events.

It is the authors' contention that similar value should be attached to the metadata generated by any studies of investigations of the human organism. This is where any unstated conventional assumption that all scientifically gathered data be pure, unalloyed and unbiased could prove highly misleading.

In a certain sense the authors have been here before. In our last volume (*Reconstituting the Curriculum* [Wiley 2013]), several key theses to illuminate the notion of "human thought material" (or HTM for short) were introduced. The book itself, which at one level exhaustively catalogued issues arising in the field of curriculum development in public primary and secondary education, became an extended gloss and investigation of many aspects of the HTM concept. Unexpected sources — such as the Holy Qur'an, the standard collections of hadiths about the life and conduct of Prophet Mohammed and capsule histories of the most significant work and contributions of rightly-guided caliphs over the last 1400 years — were mined for surprising confirmations of the essence of the HTM concept. The key point is: accomplishing this aim could not begin to be accomplished adequately at less than book length. Similarly, the role of metadata in everything — from profiling to establishing orders of relevance of very different sources of nonetheless scientifically useful information from the medical sciences — needs to be explored at book's length. This is that book.

1.4 INTRODUCING CHAPTER TWO: What's Behind Giving Up Honey and Promoting Aspartame as the Cure, i.e., Dumping the Natural Option in Favour of the Artificial One?

Chapter 2 presents the current state of the world and how we got here. In no uncertain terms the direction of our civilization is shown through the following quotes:

An Arab is no better than a non-Arab, and a non-Arab is no better than an Arab; a red man is no better than a black man and a black man is no better than a red man - except if it is in terms of piety. — Prophet Muhammad, 632

I, as much as any other man, stand in favor of having the superior position assigned to the white race... I have no purpose to introduce political and social equality between the white and the black races. — Abraham Lincoln (still a backwoods Illinois politician and lawyer but married to Mary Todd, daughter of a wealthy & politically highly-connected family of slaveowners), 1858

Abraham Lincoln's statement creates an apparent paradox not unlike the election of President Barack Hussein Obama[5] to the same office centuries later. These contradictions make dogmatic cognition look logical, especially one is faced with calling this system 'the best we can have'. If this paradox is not corrected by changing the first premise, other paradoxes arise[6]. Of course, observing and writing *post* 9-11, 'Islamic terrorism' emerged as the most overused term in the media. If 'Islamic terrorism' is not recognized as anything but an oxymoron, then terrorism becomes equated with 'opposing the Establishment' and any possibility for invoking a paradigm shift is disrupted if not completely blocked.

---

[5]On April 11, 2015, at the Summit of Americas convened in Panama with the Cuban government represented for the first time following 54 years of exclusion by unilateral order of the US government, President Obama remarked that "there are dark chapters in our own history in which we have not observed the principles and ideals upon which the country was founded... America never makes claim about being perfect, we do make claim about being open to change."

[6]In this instance there were two first principles overturned — the notion that the US does not tolerate social discrimination against its own people on the basis of racial origin, and the notion of excluding or treating as a pariah countries that govern by means of a political system that is neither congressional, parliamentary nor installed at US gunpoint.

Picture 1.1 Lifestyle of two icons of two contrasting civilizations: the top photo depicts the White House during Lincoln era. The bottom photo depicts a replica of Prophet Muhammad's residence. Once these icons are set as standard, divergence occurs and nature of the divergence cannot be objectively analyzed unless one is prepared to look at the first premises that these icons happen to reference. Any paradigm shift must include a fundamental change — a change in the first premise.

Picture 1.1 shows the inside of the White House during Lincoln's presidency, contrasted with the replica of Prophet Muhammad's house in 7th Century. It was the same time that the Persian and Roman empires were erecting some of the most spectacular icons of lavishness that dwarf today's most extravagant lifestyles. Now, if the picture at the bottom is considered to be savage, then we have indeed progressed throughout centuries. If that is true, we have indeed turned the picture of social progress upside down, showing all decline as progress and *vice versa*. After that, everything is paradox. Prophet Muhammad was also

Picture 1.2 depicts how our intention has defined our perception, then created a science that only confirms our intended conclusion. It has turned the big picture upside down that sees real progress as regression and real regression as progress. In effect, each of our conclusions has become a function of our intention and New Science has made it impossible to cognize objectively.

the same man that said, "Every illness has a cure, and when the proper cure is applied to the disease, it heals by Allah's Will", setting stage for natural cure for everything. All of a sudden natural solutions become matters of terrorism and subversion of the State.

Chapter 2 also highlights a number of paradoxes characterizing modern society. The current dismal state of the world is due to the environmental insult, starting with water pollution and overall deviation from natural state. It is for the same reason that we have an onset of ailment. It is shown that the lifestyle is the driver whereas physical conditions and ailments are manifestation of the lifestyle.

Neither is there a single medicine that cures disease, nor any product of Big Pharma that doesn't make the symptoms worse over the long term. Thus for example we have Prozac consumption increasing suicidal behavior. Parenthetically we can add: there isn't one pharmaceutical remediation of symptoms without unanticipated horrific consequences, e.g., thalidomide and its various substitutes. Nor is there any therapy or procedure that cannot worsen the diseased condition it is intended to 'cure', e.g., chemotherapy, radiotherapy, gene therapy, lobotomy in its various forms). Similarly, there is no diagnostic tool that does not become implicated in helping spread the disease it is supposed to flag, e.g., mammography, MRI, ultrasound, CT scans. Nor are there "alternatives" that are not potentially or actually worse than what they replace, e.g., the e-cigarette).

One cannot help wondering: how is it possible that we continue to sell these as technologies worth spending money and time? It turns out this question can be answered only after reviewing the history in each case objectively (what the authors elsewhere have discussed extensively as "delinearized history"[7]).

1.4.1 Physics as Mastermind of Aphenomenality[8]

The word 'physics' has the root meaning of 'science of nature'. In our own time, however, it has come about that the word 'physics' has come to apply to the science of artificial. For example: all recent Nobel prizes in Physics deal with theories that applicable to engineering processes. The underlying assumption of applying engineering theories to describe natural processes is that nature behaves the same way as a human-engineered process. Based on this premise, physicists have freely delved into describing natural objects, ranging from tiny quarks to large quasars. Since Newton, few scientists have attempted to explain creation (or the emergence) of the universe from nothing. Recent work meanwhile by Stephen Hawking — holder of the chair once occupied by Newton himself at Cambridge —has opened a new line of discussion in which old questions that apparently eluded modern scientists have resurfaced.

Once again, age-old questions such as the ones listed below are being asked (Hawking, 2010, Table 1.1).

1. What is the purpose of our (humans) existence?
2. What is the purpose of the existence of the universe?
3. Is time a creation, if so, when was created?
4. When did the universe begin?
5. How was the universe created?
6. How can we understand nature?

---

[7]For several extensive discussions of aphenomenal reasoning and its consequences, see (Zatzman & Islam 2007). For a detailed examination of the disastrous consequences proceeding from environmental regulations formulated according to the conclusions generated from unchecked aphenomenal reasoning, see especially the Document entitled "Semi-Official View of US Petroleum Refinery Waste and Waste Regulation Practices", placed at the end of Chapter 4 of Gary M. Zatzman *Sustainable Resource Development* (Wiley 2012) pp 213-292

[8]The *absence* of a real question, or of a cognition process supported by the above-mentioned five steps, is what the authors have deemed "aphenomenal" (Zatzman and Islam, 2007). At best, an aphenomenon by itself is irrelevant.

While asking questions is necessary, questions themselves are not sufficient for assuring the utility let alone the accuracy of the answers. Unfortunately, post-Newton Europe has lost the scientific cognition process that would allow these questions to be addressed properly instead of re-stating the false premises (Khan and Islam, 2012). Islam et al. (2013) identified necessary and sufficient conditions for such cognition. They are:

1. Clearly identified purpose;
2. Clearly identified first premise;
3. Ability to discard a first premise and restart the process if contradictions arise;

Table 1.1 List of Aphenomenal Questions asked by Hawking (2010).

Questions of Hawking (2010)	Embedded contradictory premises
How can we understand the world in which we find ourselves?	We find ourselves; it disconnects from any notion that creation may not be a random act that is self created;
Did the Universe need a creator?	The need of creation can create a 'creator'; the notion of creation is dependent on 'need'.
Why is there something rather than nothing?	We have access to the 'intention' of the creator.
Why do we exist?	Our existence has a purpose.
Why this particular set of laws and not some other?	We have access to the knowledge of something that doesn't exist
What is the origin of the laws of nature?	The laws of nature have been originated by an entity
Is there only one set of possible laws?	There is no comprehensive and unique set of laws of nature
What is reality?	Reality is subjective and non-unique
Is there a miracle?	Exceptions and magic are not miracles
What is the speed of light in Maxwell's equations measured relative to?	Maxwell's equation is universal; electromagnetic field is comprised of rigid balls.

4. Existence of real questions;
5. Conclusions are supported by guiding but open questions that precede the conclusion.

Real questions are crucial; the lack thereof would only support the premise that the cognition is based on. A real question is a question that does not have a premise attached to it. In the following, we list a few questions posed by Hawking (2010) and demonstrate how these questions have premises attached to them, the exclusion of which would expose the question as aphenomenal. These are all "why are you lying?" type questions. The idea is to change these questions into real questions in order to find answers. Table 1.2 shows how all New science scholars failed to pose real questions.

Overall, physics as a discipline has traveled from flat earth theory to flat universe theory.[9] (All the parallel transitions cited in the footnote just referenced have the same driver that once controlled the Roman Catholic Church, viz., Money.) This is true in describing the cosmos as well as describing the smallest entities that we know.

The case in point is the article by Lewis (2013). This one discusses a small problem, relying on electron beam rather than natural-light beam in the best electron microscopes today, the article referenced mentions how electron beam destroys certain life-forms at near-atomic microscopic level. Apart from that [sic], it's marvelous technology! A gem of aphenomenal reasoning, in which death appears as an unfortunate side-effect.

1.4.2 False Promises

Corporatized economics, the driver of the modern era, has become anything but economizing. In a manner similar to how physics has been turned into 'science of the artificial' diverging ever further away from its root meaning of 'science of nature', every new product rollout consciously disguises or conceals the extent to which today's "new and improved" versions are either potentially or actually more wasteful than what they are replacing. The plain fact is that "the whole truth and

[9]Consider the following brief catalogue of contemporaneous transitions: from Trinity to infinite god (desire being the god), from God, Son, and Holy ghost to Money, Sex, and Control, from Church, Monarch, and Feudal lords to Corporation, Church, and Government. Constantly, absurdities have been introduced as 'science'.

Table 1.2 Various theories and fundamental premises behind them (conclusions are not necessarily that of the scientist that posited the premise).

Name of the scientist	Premise	Conclusion
Ancient Greek philosophers	1. All matters originate from the Creator and to the creator all matters return.	Universe ruled by unique set of laws; purpose of creation is to have these laws executed; Humans ultimately accountable for their obedience of that unique set of laws; Humans responsible for humanization of the environment; Humans judged by their intentions*; Humans are representative of the Creator*; Earthly life is a test*; Eternal life is the ultimate reality
Ancient Indian philosophers		
Ancient Chinese philosophers		
Middle eastern prophets (Abraham, Moses, Jesus, Muhammad)		
Aristotle	2. A 'substance' that filled all the universe	Absolute speed relative to that 'substance' is infinity.
	3. We see matter because something emerges from our eyes for us to be able to see.	
	4. Everything is either A or not A.	A human (mortal) cannot be god (immortal)
Thomas Aquinas	5. God, son of God, Holly spirit, all can exist in one	Any exception is possible, if deemed convenient

(Continued)

Table 1.2 cont.

Name of the scientist	Premise	Conclusion
	6. Everything is governed by intention of the creator.	There is an intelligent person
	7. Time is a property of matter	
	8. All knowledge comes from the bible	Authority has access to knowledge
Averroes	8. All true knowledge comes from the Qur'an	A good first premise produces correct cognition and vice versa
		Every person has access to knowledge

(Continued)

Table 1.2 cont.

Name of the scientist	Premise	Conclusion
Ibn Haitham	8. Everything about the creator is unique (Only Creator can be infinity and external to the universe).	No creation can have a speed of infinity
		No creation can have constant speed
		Every creation is internal and connected to each other
		Thoughts are internal
	9. Everything, including time, matter, and thought, originates from and controlled by the Creator	Universal order is unique and absolute
	10. Humans have unique purpose and are judged based on their intention	Accountability is based on intention that has no bearing on universal order
	11. Source leaves a signature on whatever emerges from it	Light source affects the quality of light
		A good source emits good light and vice versa

(*Continued*)

Table 1.2 cont.

Name of the scientist	Premise	Conclusion
Newton	12. There is a steady state	First law of motion
	13. There is a state of uniform velocity	First law of motion
	14. There is an external force	Second law of motion
	15. Light travels in waveform	Newton's wave theory
	16. God interferes with universal order	
Dalton	17. All matter comprised of solid spherical, rigid balls	Atomism is true
Maxwell	18. All energy forms comprised of solid, spherical, rigid balls	Photons are uniform and independent of the light source
	19. Luminiferous, all pervasive 'substance' exists (see Premise 2)	Speed of light relative to this 'substance' is variable
Lord Kelvin	Ether (all pervasive 'substance', see Premise 2) exists	
	Flying is an absurd concept	
	The universe is evolving toward 'heat death'	Degree of chaos increasing

(*Continued*)

Table 1.2 cont.

Name of the scientist	Premise	Conclusion
Einstein	20. God doesn't play dice	The time function is deterministic and exact
	21. Light has constant velocity	Light is uniform and static
	22. Energy comprised of solid spherical balls	Energy is disconnected from mass
	23. Time is a perception of individuals	Humans control time
Feynman	24. Every matter has numerous historical path	Creation from nothing is a continuous process
		Reality is a chaotic process
		Reality is subjective
	25. Observation affects the history	Past can be affected by the present
Stephen Hawking	26. Creation of everything from nothing through Big Bang of an infinite mass and zero volume	The universe is expanding with slowing rate of expansion
Saul Perimutter and Brian Schmidt (2011 Nobel Laureates)	27. Creation of everything from nothing through Big Bang of an infinite mass and zero volume	The universe is expanding with accelerating rate of expansion

(*Continued*)

Table 1.2 cont.

Name of the scientist	**Premise**	**Conclusion**
James Quach	27. Creation of everything from nothing (amorphous state) through crystallization (Big Chill)	
Dmitri Krioukov	28. Creation of everything from nothing through Big Bang with further expansion of brain-like fractals	Universe is a super-intelligent design

*Derived from the Holy Qur'an and Book of Hadiths

nothing but" doesn't necessarily sell, so falsehoods to some greater or lesser degree must be promoted. These falsehoods are frequently disguised as what the courts and regulatory agencies euphemistically call "exaggerated claims".[10] Table 2.2 shows a list of promises and how they all imploded with time.

Here are a few samples:

1. In 1960, when birth control pills were first introduced, each pill contained 10 times more male hormone than necessary to abort the egg. The promise behind this was the Liberation of women. Soon after, the anti-nausea drug thalidomide was introduced for pregnant women. The promise here was that women could have easy pregnancies by removing nausea. In reality, 20% of babies who mothers were on the drug became severely deformed. This drug was banned in 1962 but now it is making a comeback. Today, even a 12 year old can get prescribed for birth control pills (at least in Canada) and the same industry is busy producing 'correction pills' that would 'eliminate' the inherent injustice of woman's biology by stopping menstruation altogether.
2. In the 1940's, baby disposable diapers were introduced. The inventor, Marion Donovan, noticed that her babies would 'nearly instantaneously' wet their cloth diapers as soon as they were changed. In 1946, she introduced the 'breakthrough' technology of the disposable [sic] waterproof diaper. Did the habit of 'nearly instantaneously' wetting the diaper go away? Of course not. In fact, the first name of these diapers was 'the boat', indicating it was meant to keep babies afloat on their own urine! However we are convinced that disposable diapers are synonymous with keeping the babies dry and civilized. Cotton nappies are expensive, and are today even considered germ carriers.
3. Thalidomide was first marketed in 1958 in Germany by Chemie Grünenthal. Its origins with scientists from Nazi-era chemical monopolies,were carefully covered up to

---

[10]Nay more: matters get this far only if and when an "unsatisfied customer" musters the means to file a class action.

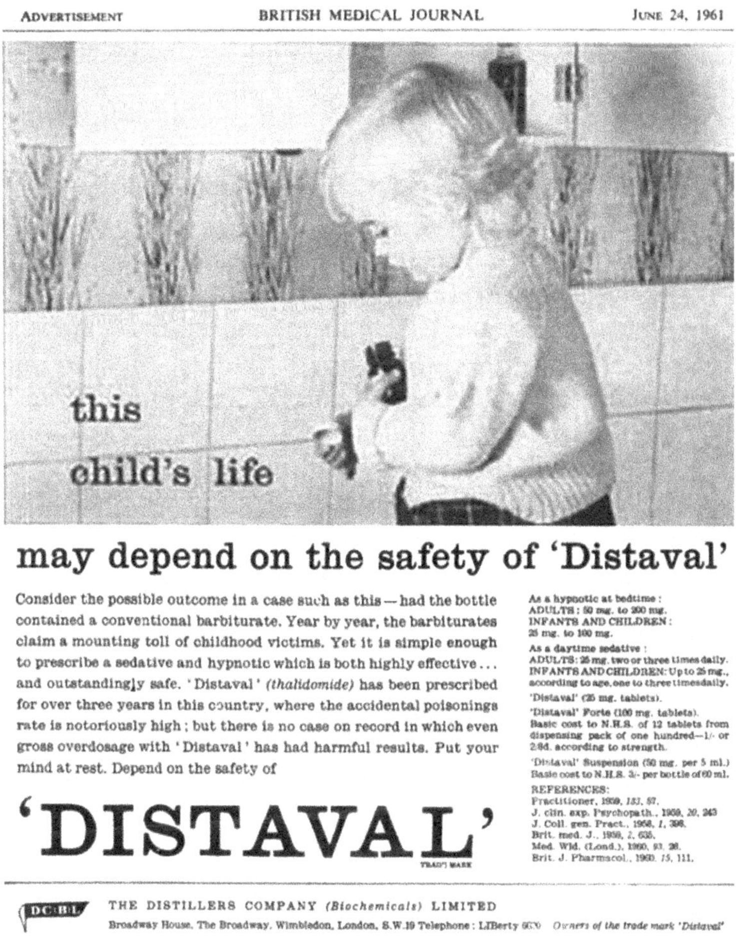

Picture 1.3 False advertisement is in the core of all modern medicine.

facilitate its marketing less than 15 years after the end of World War Two as a treatment for seizures, and later as an anti-nausea and sedative. At the time, barbiturates were frequently used as sedatives. Furthermore, compared to barbiturates, which were highly toxic in overdose, thalidomide was "well tolerated", even in overdose. Based on its apparent safety, no prescription was required. Eventually (see Picture 1.3) its attractiveness as an anti-nausea led to its use in pregnancy for morning sickness.

It wasn't known at the time, but fetal exposure to thalidomide between days 35 and 48 was causing severe limb and organ defects in 20-30% of children.[11]

In the 1950's it wasn't even recognized that drugs could cross the placenta and cause adverse effects to the fetus. Thalidomide hadn't been tested on pregnant animals prior to marketing for use in pregnancy. Regulators didn't require it. And the consequences were horrific.

Thalidomide became a popular drug because of its apparent safety and effectiveness, and it was marketed alone and in combination with other drugs in Germany, the UK, Canada, and other countries whose pharmacological marketplace was cartelized by European-based monopolies. But within a few years, babies started being born with characteristic limb and organ deformities (Picture 1.4). In 1961, two independent researchers identified thalidomide as the likely causal agent. The manufacturer sought to undermine and discredit the findings, but it was clear - the drug had caused catastrophic harm to thousands of fetuses. Subsequent animal testing confirmed this.

The USA, in which the European pharmaceutical cartel was excluded by the American giants (such as Johnson and Johnson *et al.*), largely escaped the thalidomide baby tragedy. But with the expiration of the European cartel's thalidomide patents in the 1970s, US-based research independently uncovered cancer-killing side effects of thalidomide administration. Thus repurposed, the drug has been enjoying a modest comeback for treating chemotherapeutically-induced nausea in post-menopausal patients.[12].

Overall, all that is "modern" and acclaimed as social progress in contemporary social-economic development is the unprecedented scale of the greed manifesting itself so intensively among the ruling caste and their tools — that notorious "one per cent of the One Percent". Nature, on the other hand, operates on the basis of need and therefore there is

[11]A major social development of the 1950s was that drugs could help women surmount ever kind of unhappiness in their lives. Tranqulizer consumption heavily concentrated among married women reached epidemic proportions, reflecting the deep unhappiness that was real and material among this social stratum. This mass *ennui* followed from large numbers of these women having been declared surplus to the needs of the economy once the soldiers returned home from the Second World War, after experiencing the reality of being socially included outside the home in a major way for the first time to supply the labor very much needed in the production processes of the wartime economy. Meanwhile, the explosion in drugs use that became general starting in the 1950s went far beyond bored housewives, into the ranks of the youth. In itself, the initiation of a generalized drugs culture aimed primarliy at destroying the social participation of

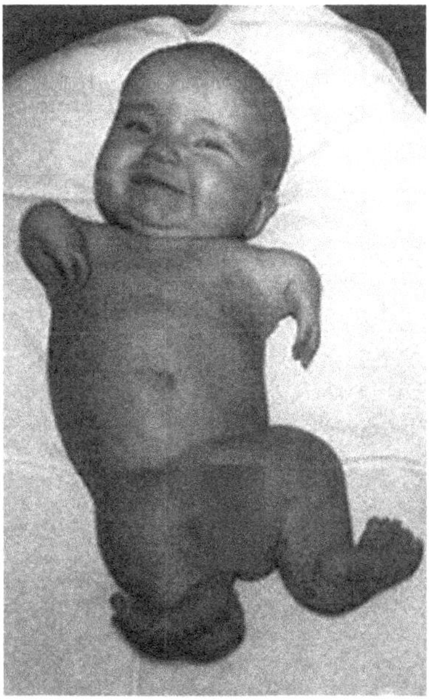

Picture 1.4 Side effects of morning sickness pills.

no need to make false promises or to institute opacity if one wishes to introduce pro-nature development. Indeed: the current 'technological disaster' can be seen and best understood as a predictable result of such greed-driven "social development". One highly-visible marker of these processes is the contemporaneous decay observable in the quality of human health and societally-organized health care. Even if the rates of decay in the British "national health" model or the Canadian-style government-backed institution-delivered forms of care or the emerging Obamacare rejigging of health insurance across the United States may

women posed no a threat to the established order. However, the rising up of youth was indeed another matter entirely, as they needed to be pacified iunto accepting being sent hither and thither as cannon fodder for America's wars on the peoples of southeast Asia. Everything was done to prop up the drugs culture among the youth, setting them up to be attacked and blamed for the so called drugs culture. But the fundamental truth of matters is that, in the contemporary social-economic order, *there is no god but monopoly and maximum is its profit*. Blame for the drugs culture was assigned entirely to the rebellious sections of the youth across North America, South America and Europe, because rebellion against the ruling economic order of monopolies, oligopolies and cartels had to be smashed at all costs.

[12]As a headline in April 2013 put it: "FDA Approves Morning Sickness Drug Once Feared Unsafe". Sources inside the US Government confirmed that from 2010-2013, the

Picture 1.5 Chapter 3 describes how our current civilization has made no progress from the era that we call savage (be it in criminal justice, top picture or human rights, bottom picture).

differ in the details, the common trend of unrelenting decay is unmistakeable. For instance, in last 50 years, there has been an increase of 50 times per capita in the use of sugar ('refined', externally processed, carbohydrate) plastic ('wrinkle free' leather or fabric, 'durable' wood, cheap water container), fertilizer ('refined' biomass), spirit ('refined' alcohol), cigarettes ('refined' tobacco), chemicals ('preservatives', Pasteurization, antibiotics), and 'remediative' surgery, while the 'life' expectancy has increased somewhat. In the words of Albert Einstein, this 'life' isn't

FDA pocketed over $2.5 billion in direct payments from major drug companies, including Bristol-Myers, Merck, Johnson & Johnson, Pfizer, Eli Lilly and even one of the EU-based giants GlaxoSmithKline.

worth living. Unfortunately, this 'life' is being promoted as the only life human beings should live for.

1.5 INTRODUCING CHAPTER THREE: Are the Premises of New Science Sufficient for Uncovering or Establishing The Cause of Anything?

Chapter 3 describes overall decline first in cognition, then in products that were created through 'corporate research' in a continuous move from real to artificial. The present authors, who have been addressing this phenomenon over the last decade in a number of works, have described this trend thus: Honey → Sugar → Saccharine → Aspartame (HSSA). This trend of degradation is seen in every aspect of modern life, from education systems to the policies applied I the fields of energy production, distribution and management. Chapter 3 deconstructs all major theories of physics as well as social science to show how rooted such a generalized trend of self-destruction has become as a result of deepgoing transformations in the theory and applied practices of modern economics during the bipolar division of the world that set in following the end of the Second World War and has continued as a US-dominated project since the collapse of the Soviet bloc a generation ago. Of course, a great deal of this is very easily correlated with developments in international politics but not so easily identified in the very fundaions of the globalized economic order which we currently inhabit. What our investigations have identified that other investigators either overlooked or considered as trivial or beneath further comment or investigation is — for example — how it turns out that, as a product becomes more artificial, the corporate profit margin goes up. In effect: profiteering has become the overall *modus operandi* while the general public being victimized and disenfranchised on a massive scale is seriously challenged when it comes time to resist by the very fact that the trusted professionals, such as engineers and doctors, are acting in fact as agents of The Corporation, the invisible hand of rapacious corporate interests.

Centuries ago, a typical research question of Harvard College would investigate "how many angels can dance on a head of a pin. Today, some of the research questions involve: "is pedophilia natural?" with questions that are motivated by the following 'scholarly' premise:

"Paedophilic interest is natural and normal for human males… At least a sizeable minority of normal males would like to have sex with children … Normal males are aroused by children."

The ground for such an obscene premise was prepared by none other than the evolutionary 'scientist' Richard Dawkins. His previous work on 'mild (*sic*) pedophilia' encountered minimal resistance and the doctrine of the so-called 'selfish gene' faced no criticism. New Science has become recycled dogma, only packaged as 'science'.[13]

New Science has not been able to answer simplest of questions of substance — not, at least, without resorting to dogma and exceptionalism. We do not know what is nature or what is natural. Then, we don't know what a human is. What qualities are unique to the human animal? Researchers routinely spew out conclusions and throw the readership to all directions. Only recently, researchers at a meeting said they had found the oldest tools made by human ancestors—stone flakes dated to 3.3 million years ago. This is 700,000 years older than the previously oldest-known tools. The implicit assumption is that toolmaking and usage is something that definitively distinguishes *Homo sapiens* from the rest of the animal kingdom. However, another unstated assumption lies behind this assumption. According that further assumption, whatever is truly definitive of and unique to the human being must be entirely tangible. The unstated understanding that informs the approach taken by New Science to answering this question is that the explanation / answer will be neither complete nor truly scientific without establishing this tangible existence.[15] Similarly, propensity to violence and aggression is equated in part — as an allegedly tangible expression with humanity— to humanity to large brain sizes. The domain is diverse and the conclusions are absurd and illogical at best. Here is a sampler on the use of milk in Europe (Curry, 2013):

"During the most recent ice age, milk was essentially a toxin to adults because — unlike children — they could not produce the lactase enzyme required to break down lactose, the main sugar in milk. But as farming started to replace hunting and gathering in the Middle East around 11,000 years ago, cattle herders learned how to reduce lactose in dairy products to tolerable levels by fermenting milk to make cheese or yogurt. Several thousand years later, a genetic mutation spread through Europe that gave people the ability to produce lactase — and drink

---

[13]Henceforth the authors apply this terminology ironically

[14]Similarly, psychological states in New Science do not exist without a tangible form, viz., behavior.

milk — throughout their lives. That adaptation opened up a rich new source of nutrition that could have sustained communities when harvests failed. If correct, the new evidence could confirm disputed claims for very early tool use, and it suggests that ancient australopithecines like the famed "Lucy" may have fashioned stone tools, too."

"Lucy", of course, is the invention of paleoanthropologist Donald Johanson who discovered a fossil skeleton, called her a female, and named 'her' after Beatle's song "Lucy in the sky" in 1974. All of a sudden, an African baboon became Lucy the person – all personified by a Beatle's song. So, where is the science in it? Some century ago, a fossilized skull (perceived to be 2.8 million years old) was discovered by anatomist, Raymond Dart, who called the skull a child, and wrote:

"I knew at a glance that what lay in my hands was no ordinary anthropoidal brain. Here in lime-consolidated sand was the replica of a brain three times as large as that of a baboon and considerably bigger than that of an adult chimpanzee..." The Taung Child's teeth were more like a human child's than an ape's. Dart also concluded that it could walk upright, like humans, because the part of the skull where the spinal cord meets the brain was human-like."

When such conjectures and illogical premises packaged with most hyperbolic conclusions are left unchallenged, we arm ourselves with no defense against the likes of Phillipe Rushtons that write 'scholarly' papers on the correlation of body parts with intelligence and equally obscene sophistry.

This mockery of science has taken the most preposterous root in the field of medical science. Everyday, there are publications on the merit of certain drug, or the branding of certain condition as a 'disorder' only to be followed by publications that refute those findings and glorify another round of science. This is particularly obtuse when it comes to lifestyle. As recently as April 10, 2015, Gallagher (2015) wrote: Being overweight 'reduces dementia risk'. This would counter Jha's headline (2011): Obesity in middle age increases risk of dementia, to which it was also added: Diseases such as Alzheimer's almost four times as likely to affect people who are obese in middle age, new study shows. So, what to make out of this lunacy? We don't know but what we know for sure is: being clever and being able to question is not good for our health. David Robson (2015)'s headline reads: The surprising downsides of being clever.

1.6 INTRODUCING CHAPTER FOUR: For How Long We Have Been Lied To?

It will be an understatement to say New Science is grossly inadequate for conducting scientific investigation. Chapter 4 shows how desperate the situation is when it comes to New Science. As cases studies, the chapter analyses Colony Collapse disorder (CCD) along with several human disorders whose causes are unknown. Why CCD? Because honey bees represent the most sustainable unit of nature that behave in exemplary form. CCD, on the other hand represents the current disastrous state of human health and environment. Human health disorders, on the other hand represent what's wrong with our society in general. After all, a human body is the representative of individual organism, be it a community, a state, a continent, or the entire globe.

Ancient people knew 'truth unravels itself', therefore, they tried their best to uphold the truth. Even Abraham Lincoln knew:

"You can fool all the people some of the time, and some of the people all the time, but you cannot fool all the people all the time."

However, nothing stopped the orchestrators of the aphenomenal model, once known as dogma, from trying to control access to information in order to control the past. In George Orwell's word, "He who controls the past controls the future. He who controls the present controls the past." Orwell would know how modern era, obsessed with tangibles and externals operates. The lust for control, sex, and money has replaced the original Trinity. All we have is falsehood after falsehood so the first premises are never questioned, be it in science, political science, or individual or corporate intentions. Of course, real truth can set you free, but for a culture disconnected from Conscience, deceit holds the key to money - money for more money, sex, and control. "History is written by the victor" has become governing principle of our knowledge that must seek guidance from history. A corrupt history is the essence of misguidance and gateway to ignorance. Each falsehood has enormous ripple effects and cannot be corrected without addressing the source, which is never touched in modern day cognition. Some fantasize that we have become enlightened and no longer need resort to dogma. Yet, as Chapter 3 demonstrates, we have become progressively more dogmatic in our cognition, arriving eventually at the HSSA degradation detectable universally throughout every pore of contemporary society.

At no time in known history was it actually recognized that history is not immutable. Both in science (quantum theories of the origin of universe) and social science now openly accept the notion of multiple histories as fact. Today, if someone doesn't like the past, all he need do is re-write it and spend some money selling it as fact. In the old days, someone actually had to disprove the factual errors of history. Dogma has erased this by invoking Authority of God, albeit without any logic or justification. Nowadays, the dispute is in what constitutes 'truth'. After the rise of pragmatism as the new dogma, 'truth' has become anything that can be 'believed in'. The newer version of pragmatism adds 'scientific' flavor to the same nonsense and makes it more difficult to counter with logical discourse. Such a collapse of science has profound implications going forward.

1.7 INTRODUCING CHAPTER FIVE: A Starting-Point for Society-Wide Corruption

Mark Twain said, "The lack of money is the root of all evil." This has become the first premise of today's religions. Ever since the yoke and diktat of the Roman Catholic Church was put in place, 'religion' has been formally conflated with the Money god that remains today as the undisputed almighty. Ramified with the moral compass of the Aquinian bible and its Church-approved Aristotelian conception of the natural order, the Eurocentric view makes room for both the notion of truth as a spectrum and the notion of knowledge as an amalgam of truths and falsehoods.

These are not ordinary blends of truth and falsehood. Rather, this is falsehood with a purpose. That purpose is to maximize profit and other short-time gains for a tiny minority. This would manifest throughout history, albeit being under disguise of secularism at later stages. Furthermore, there is no separation of 'church' and 'state' when it comes to policy and money laundering schemes. Every policy, political or 'religious', is rooted in the dogma: There is no god but Money, Maximum is the Profit. It is not a coincidence that Roman Catholic Church is dubbed as the World Capital of Money laundering. Manhattan (1983) made the point with the following comment:

"The Vatican has large investments with the Rothschilds of Britain, France and America, with the Hambros Bank, with the Credit Suisse in London and Zurich. In the United States it has large investments with the Morgan Bank, the Chase-Manhattan Bank, the First National Bank of New York, the Bankers Trust Company, and others.

The Vatican has billions of shares in the most powerful international corporations such as Gulf Oil, Shell, General Motors, Bethlehem Steel, General Electric, International Business Machines, T.W.A., etc.

Some idea of the real estate and other forms of wealth controlled by the Catholic church may be gathered by the remark of a member of the New York Catholic Conference, namely 'that his church probably ranks second only to the United States Government in total annual purchase.'

The scheme of calling Saul of Tarsus 'the best apostle', creating December 25 the birthday of the 'son of God', 'Good Friday' the death anniversary, "Easter" the rebirth day, developing the notion of original sin all created Money for the Roman Catholic church and anyone that followed that model. In Chapter 5 we demystify much of this process.[15]

People are oblivious to vast majority of the chemicals and potent drugs that have become part of the lifestyle. At every stage, the HSSA degradation is evident but ignored. Chapter 5 presents the science behind the disinformation that has taken over modern society in all walks of life. They are called 'advertisements' and yet each piece is a testimony of falsehood and deliberate lies planted to maximize profit at the expense of public health and quality of life (e.g. Picture 1.6) Follow-up chapters establish how every artificial chemical is addictive and there is no such thing as 'safe dose' for these chemicals. The fact that they are not illegal is a testimony of a corrupt system that is focused on maximizing profit at the expense of public health. There is a direct correlation between profit margin and addiction level of these drugs.

It is in every chemical, be it smoke, liquid or solid (Pictures 1.8 - 1.10). Scandals abound but only represent tiny percentage of the bigger picture. Chapter 5 discusses how this has been systematically introduced in our lifestyle.

1.8 INTRODUCING CHAPTER SIX: Deconstruction of the Foundations of Aphenomenal Science in the Works of Newton *(Part A)* and Einstein *(Part B)*

Hart (1978) ranked Isaac Newton at number-2, ahead of Jesus Christ who is ranked number-3 as the most influential humans of history.

[15]They were clear falsehoods, deliberately perpetrated with the motive of amassing wealth. The establishment of the United States itself is an important case in point discussed at length in Appendix 2 at the end of the current chapter.

Picture 1.6 False advertisement itself has taken turn to the worse, following HSSA degradation.

With discoveries in optics, motion and mathematics, Newton developed the principles of modern physics. In 1687, he published his most acclaimed work, *Philosophiae Naturalis Principia Mathematica* (Mathematical Principles of Natural Philosophy), which has been called the single most influential book on physics. Newton died in London on March 31, 1727 and got King's honour at his funeral. He died a wealthy

Picture 1.7 There is no such thing as false advertisement. In modern era, every advertisement is a deliberate falsehood, bar none.

man with the fame of a great Monarch. His kingly status was matched with his divine status. Alexander Pope wrote:

God said, "Let Newton be!" and all was light.

What Jesus Christ stood for as regards the Roman Catholic Church, Newton stood for in New Science. Everything after Newton had the premise embedded in it. Newton's laws are a form of absolute truth and applicable not just to certain limited finite state(s) of matter.

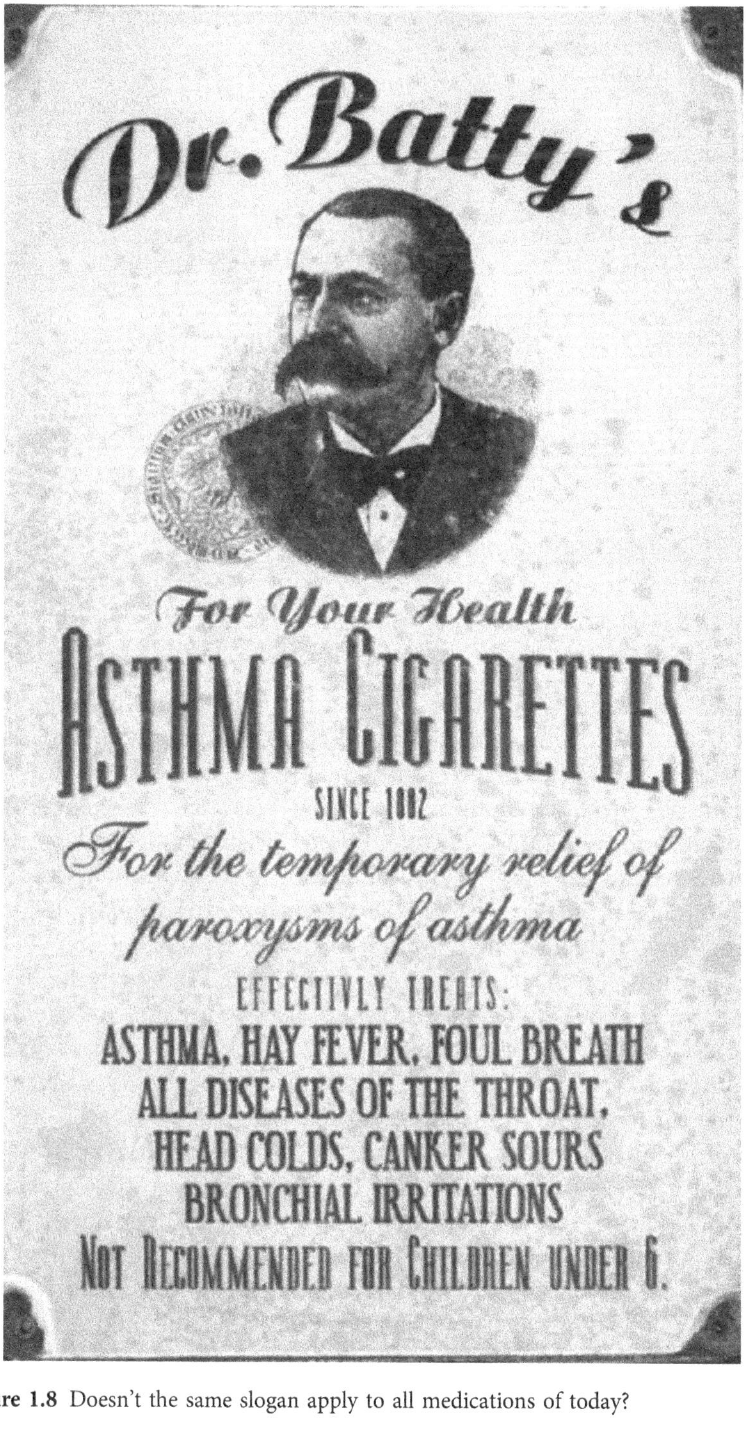

Picture 1.8 Doesn't the same slogan apply to all medications of today?

Picture 1.9 Liquid toxin is ubiquitous, that we have to pay for it.

Blake's depiction of Newton's persona is symbolic and non-literal, yet closer to the real Newton than any other artistic rendition. This is the Newton symbolizing the New Trinity of Money, Corporation and Government, all attributed nominally to what was called 'secularism' but manipulated conventionally like any other Christian religion.

New Science — including the foundation of social science provided by economics — has come to be based on the extraordinary contributions of Newton such as his (Trinitarian) three laws of motion (including the principles of inertia, force, action and reaction), the law of gravitation, and his discoveries in optics, astronomy, and mathematics. Newton's laws enabled measurements of actual distances, speeds, and weights to be calculated, laying the foundation of modern inventions from the steam engine to the space rocket.

If we are to give credit to Newton for his contribution to New Science, however, we must also admit that his greatest promoters of the present day bears responsibility for the technological disaster of today. Accordingly, in the first part of Chapter 6, Newton is re-examined with a view to deconstructing his fundamental premises. The facts we work on include both those describing his scientific-academic and public governmental work (including his status as the man that sent money counterfeiters to the guillotine and his high standing with the Stuart

Picture 1.10 Solid toxins only change name and more toxic replacements are introduced in the market.

Picture 1.11 In his painting 'Newton', the British poet and painter, William Blake, represents Newton as a divine geometer. He is sitting naked on a rock at the bottom of the ocean leaning over a scroll, and measuring the symbol of the Trinity [ex (Jones, 2014)].

Restoration of the English monarchy)[16], as well as on the still somewhat hidden private side of the man.[17]

Newton's father, Isaac Newton, Sr. was described as a "wild and extravagant man." He died before Newton was born. When Newton was three, his mother remarried and went to live with her new husband, the Reverend Barnabus Smith, leaving her son in the care of his maternal grandmother, Margery Ayscough. The young Isaac disliked his stepfather and held some enmity towards his mother for marrying him, as revealed by this entry in a list of sins committed up to the age of 19:

[16]This was initially reestablished in 1660 with the return from years of exile in France of Charles II, the senior Stuart family member properly in line to the throne. It was consolidated after the 'Glorious Revolution' of 1689 that placed James II on the English throne, on a foundation that explicitly rejected every single genuinely progressive idea identified with Oliver Cromwell's Commonwealth of 1649-1660 that followed the English Civil War.

[17]The authors draw readers attention here to Newton's unpublished alchemical work on how to convert lead into gold, and the equally remarkable effort by the prominent 20th century British economist John Maynard Lord Keynes who purchased these private papers at public auction. Concerned about controlling any shock to Newton's

"Threatening my father and mother Smith to burn them and the house over them." Facts of this kind fill out a profile of Newton as quite an angry man, alien to the disposition normally associated with prophets or other divinely-inspired individuals.

Newton's biography is a catalog of the symptoms of bipolar (or manic depressive) disorder, an illness he suffered from most of his life. Romantic writers, enamored with any white man perceived as divine, often called manic depression 'a disease of men of genius', while others considered it an essential element for creativity. In a society that accepted Newton as the closest thing to god, Newton could have nothing negative about him. It was argued that depression made one a perfectionist and mania led to intense periods of productivity, faith in ones own talent, and the need to prove oneself right. Similar statements have been made about Abraham Lincoln and his mental state.

Newton exhibited signs of bipolar disorder early in life. He was a solitary child who didn't engage in games with other children. He spent most of his time alone, building miniature mills, machines, carts, and other inventions. He was high-strung, egotistical, and domineering. He experienced attacks of rage, which he directed toward his friends and family. He later recalled 'threatening my father and mother to burn them and the house over them.' Newton also had intense moments of remorse, when he made long lists of his 'sins' or wrongdoings. His list recorded 'striking many', 'punching my sister', 'peevishness with my mother'. His violent temper made him unpopular and his peers and the servants rejoiced when Newton left home for Cambridge. At times of depression, Newton hallucinated and had conversations with absent people. He became obsessed with religion and immersed himself in alchemy. While at a certain level a similar 'obsession' among polymaths of the Islamic golden era cold be observed, these individuals were not hallucinating, their god was not a Money god and their alchemy had nothing to do with amassing wealth.

Like many who suffer from bipolar disorder, Newton developed grandiose delusions. In his notes on alchemy and religion he wrote that he was appointed by God to bring His truth to the world. This is similar to the statements made by none other than George W Bush. Here is where it becomes challenging, to say the least, to seriously attempt to maintain the illegitimacy of drawing any connection between politics and science through the aphenomenon of profiteering.

---

established reputation, Lord Keynes opened them to no-one outside a tiny circle of extremely close personal associates.

Newton neither married nor established serious long-lasting friendships. In his later years, a combination of pride, insecurity and side trips on peculiar scientific inquiries led even some of his few friends to worry about his mental stability, even at the height of his career.

If we are to give credit to Newton for his contribution to New Science, we must be able to say he is the orchestrator of the technological disaster of today. Newton wasn't above criticism or critiques, but no one considered connecting Newton's 'laws' to their absurd premises or their applications to some of the most extremist 'unintended outcomes' of capitalism. Chapter 6 does both and shows how Newton must be deconstructed and his absurd premises exposed before a paradigm shift can be implemented.

If we are to give credit to Newton for his contribution to New Science, we must be able to admit that he orchestrated the technological disaster of today. Newton wasn't above criticism or critiques, but no one considered connecting Newton's 'laws' to their absurd premises or their applications to some of the most extremist 'unintended outcomes' of capitalism. Chapter 6 does both, disclosing how Newton must be deconstructed and his absurd premises exposed before a paradigm shift can be implemented.

If Newton was the man that turned everything intangible into tangible pointing toward interest of the Establishment, Einstein is the man that oriented every tangible to Self-Interest in the shortest term. He deified Self. Thanks to Einstein's quantum theory, today, the so-called "immortality gene" (Weller, 2013) is a 'reality' whereas the so-called "selfish gene" is... a dogma.

Einstein's early work on the theory of relativity (1905) dealt only with systems or observers in uniform (unaccelerated) motion with respect to one another and is referred to as the special theory of relativity; among other results, it demonstrated that two observers moving at great speed with respect to each other will disagree about measurements of length and time intervals made in each other's systems, that the speed of light is the limiting speed of all bodies having mass, and that mass and energy are interchangeable. This is profoundly important in falsifying anything that would have any connection to science.

In 1911 he asserted the equivalence of gravitation and inertia, and in 1916 he completed his mathematical formulation of a general theory of relativity that included gravitation as a determiner of the curvature of a space-time continuum. Even though he merely rehashed Maxwell's theory, which was another rehash of Atomism, he added time, as a dimension, albeit making it in wrong and illogical sequence.

In addition to the theory of relativity, Einstein is also known for his contributions to the development of the quantum theory. He postulated (1905) light quanta (photons), upon which he based his explanation of the Photoelectric Effect, and he developed the quantum theory of specific heat. He reserved his main efforts for his unified field theory, feeling that when it was completed the quantization of energy and charge would be found to be a consequence of it. As a result, Newton's mechanical materialism got transferred from mass to energy. Ever since, we lost the means of distinguishing natural energy from artificial energy. With it we lost the ability to assess the effect of artificial on human body.

Some consider him the most creative genius of modern era. In 1905, sometimes referred to as his *annus mirabilis* (wonderful year), and while he was still working in the Patent Office, the young 26-year-old Einstein completed his PhD (with a thesis on "A new determination of molecular dimensions") and had no less than four important papers published in the "Annalen der Physik", the leading German physics journal:

- a paper on the particulate nature of light, in which he explained the "photoelectric effect" and certain other experimental results by proposing that light interacts with matter as discrete "packets" or quanta of energy, rather than as a wave (an idea first suggested by Max Planck as a purely mathematical manipulation), which in itself is based on Atomic theories;
- a paper explaining Brownian motion (the seemingly random movement of particles suspended in a fluid) as direct evidence of molecular action, thus supporting the atomic theory (that all matter is made up of tiny atoms and molecules);
- a paper, which has become known as the Special Theory of Relativity, on the electrodynamics of moving bodies, which showed that the speed of light is independent of the observer's state of motion, and introduced the idea that the space-time frame of a moving body could slow down and contract in the direction of motion relative to the frame of the observer. This contradicts Einstein's premise that time itself is a function of perception and can be explained dogmatically by disconnecting time from matter; and

- a paper on mass-energy equivalence, in which he deduced the famous equation $E = mc^2$ from his special relativity equations, suggesting that tiny amounts of mass could be converted into huge amounts of energy (which presaged the later development of nuclear power). This one includes a premise that speed of light could be constant in a 'vacuum' (a vacuum cannot exist anywhere in actual space-time).

All these theories are taken later on as the most revolutionary findings of New Science since Newton. So, what's the background of this creative genius? Young Albert is labeled dull, dyslexic, even autistic or schizophrenic, by a considerable number of today's experts and interested parties. Who dares to determine, post facto, whether Einstein's genius is a result of autistic traits or of schizophrenic features? As long as the experts base their judgments on outright erroneous assertions about his childhood deficiencies, on misunderstandings regarding his performance at school, or on trivia of the kind of "he let his hair grow long and did not comb it" or "he wore old clothes and did not care about style", those judgments can hardly pass for reliable scientific expertise. As long as the same symptom is cited as an evidence of schizoid traits by one and as proof of being an autism spectrum disorder by another expert, one ought rather to trust a third expert who frankly admits that while a pre-mortem diagnosis of a disorder with no known biologic markers would seem difficult enough, definitive post-mortem diagnoses are clearly impossible.

Einstein was no doubt an exceptional individual. He was highly gifted and acquired early in his life the ability to exploit his talents. The stimulating milieu of his childhood, an ambitious mother who supported the son's self-reliance, and a counterbalancing and comforting father provided the environment where the child could develop his own personality. The "Creator and Rebel" eventually found a way of reaching self-actualization in the framework of his society.

The second section of Chapter 6 deals with deconstructing Einstein's theories at their root premises. This is necessary for correcting misconceptions regarding both physical and mental effects of matter and energy.

1.9 INTRODUCING VOLUME TWO: The Way Out

If the Information age has given us anything, it is access to facts, information, misinformation, disinformation, and most importantly conclusions of the "pundits". As long as these pundits are not motivated by 'finding the truth', which would require them to be open to changing their first premise (including intention) to avoid dogmatic thinking, no guidance is provided by these conclusions. Similarly, information has no meaning as long as the readership has already decided which conclusion he desires. This book is a guide for those that wish to make up their minds only after going through the process of cognition. An open mind is the key requirement and Volume 2 of this book series is all about guiding a readership with an open mind.

Everyday, we have breaking news in all aspects of health and environment. What can be learned from them? Take for instance, those incessant streams of DNA articles. In an article by Reynolds (2015), we read: "Identical twins in Finland who shared the same sports and other physical activities as youngsters but different exercise habits as adults soon developed quite different bodies and brains, according to a fascinating new study that highlights the extent to which exercise shapes our health, even in people who have identical genes and nurturing." This study purports to show how that our environment and physical activity shape our development, but it clouds the issue of what can be said to be inborn. Can this also be the answer, then, to those that say our sexual preference is innate?

The facts of this paper are:

1. Twins with identical DNA will have different physique depending on physical regimen followed
2. They will have different brain structure depending on physical regimen followed.

If the first premise is either that "DNA is the controller of behavior" or that "DNA is the fundamental 'particle' that defines humanity", then this paper creates dogmatic contradiction. People on both sides can then be expected to argue forever reframing the old and tired "Christian vs Atheist" debates. They are both wrong, because of their first premise. On the other hand, if the first premise is: Brain (psyche) and physique (body) work like yin and yang with both controlled by the heart where freedom of intention rules, then this article makes perfect sense and there is no contradiction in interpretation of facts presented here.

Scientists are continually coming up with contradictory reports/studies. Any conclusion that comes out matches precisely the first premise that was carefully selected in order to align the "evidence" with the desired outcome. The moment such a conclusion is drawn, others prepare the counter-argument in support of conclusions of their own that come from another premise or premises. It is therefore no surprise that such nonsense as the proposed "God gene", "Selfish gene", "Poverty gene" and "Gay gene" were allowed to enter modern-day science. The only remaining unanswered question would be: when may we expect the inevitable new studies proposing an 'incest gene', a 'pedophilia gene', and-or other equally dubious concoctions of similarly 'scientific' merit?[18]

If previous chapters have taught us anything, it is: you cannot trust anything that comes from someone that has an agenda other than disseminating the truth. In fact, if we have to describe the reason behind the current dismal state of the world, we have to point finger at all the dogmas that we have believed in. So, how do we know whom to trust? How about not trusting those that are addicted to Trinity (the old religious one or the new one, i.e., Money, Sex, and Control). Information age makes it easy for us to find out the background of the world leaders today and with the correct attitude we can sort out prophets from profiteers. Table 1.3. shows the list of some of world leaders of modern era and how they got their inspirations. Of course, this list is not comprehensive and serves only as a guideline.

Figure 1.1 below shows how opium cultivation in Afghanistan has varied over the years. With the exception of the Central region, all regions experienced an increase in poppy-cultivation levels in 2014, with the largest relative increase being in the Northeastern region (79%; mainly driven by increases in Badakhshan), followed by the Eastern (9%), Western (8%) and Northeastern regions (4%). However, the vast

[18]In the meantime, the work of lowering the legal age of consent to legalize the predatory behavior against minors is already in the works. Of course, this is lowering the legal age of consent and not for marriage purposes, just for sex — as though that makes it "better". Thus, a 50-year-old man can have sex with an 11-year-old boy because... the latter has allegedly given his consent! As for the overwhelmingly unequal power relationship involved ... where?!? What power relationship? Who said that?!! "Everyone knows" it's just God versus Satan with Satan occasionally scoring a point here and there at God's expense... Furthermore — according to the likes of Tom Flanagan, a retired University of Calgary professor — consumption and redistribution of "kiddie porn" (of the men-sexually-abusing-children variety) by itself is innocent activity unworthy of police intervention! (Flanagan remains highly connected, having advised current

Table 1.3 Creative genius and its sources.

Creative Genius	The person	The persona
Steve Jobs	Steve Jobs is a visionary known around the world for his contributions to Apple,	Jobs credits a lot of his innovation to experimenting with LSD at a younger age. According to sources, Jobs believed that experimenting with LSD in the 1960s was "one of the two or three most important things he had done in his life."
Sigmund Freud	Known as world's greatest psychoanalyst. The man that introduced Sex as the sole motivator of human cognition.	He is also known for his admiration of cocaine and his frequent use of the drug. He spoke publicly about the benefits of moderate cocaine usage. In a letter to his fiancée, Martha, Freud wrote that "if all goes well, I will write an essay [on cocaine] and I expect it will win its place in therapeutics by the side of morphine and superior to it...I take very small doses of it regularly against depression and with the most brilliant of success."
Bill Gates	World's richest man ($80 billion asset in 2014) and the most famous college dropout.	Was experimenting with LSD. In an interview with Playboy, Gates touched on his experimentation with the drug, saying that "there were things I did under the age of 25 that I ended up not doing subsequently."

(*Continued*)

Table 1.3 cont.

Creative Genius	**The person**	**The persona**
Carl Sagan	Renowned astrophysicist and cosmologist, the guy that wrote the introduction of the book: A Brief History of Time by Stephen Hawking.	He openly discussed his marijuana habits in addition to advocating for its recreational use to stimulate intellectual endeavors. He wrote a 1971 essay called "Marijuana Reconsidered," in which he talked about the benefits of the drug.
Richard Feynman	Feynman was a celebrated physicist known for his groundbreaking work in quantum mechanics. He is a Nobel Laureate in Physics.	He dabbled in LSD, marijuana and ketamine. He stopped, however, when he worried he might become addicted. He once wrote that he got "such fun out of thinking that I don't want to destroy this most pleasant machine that makes life such a big kick."
Francis Crick	Crick was part of the team Watson, Crick and Franklin, who together discovered the all-important DNA-structure.	LSD was his motivator. Crick told many people about his LSD experimentation and how he used the drug while working on the molecular structure experiments.

(Continued)

Table 1.3 cont.

Creative Genius	The person	The persona
Thomas Edison	The iconic American inventor, who 'invented' the light bulb.	Regularly consumed a cocaine-laced drink of a Bordeaux wine treated with coca leaves (the main ingredient of which is cocaine). In some recipes, the ethanol in the wine would extract cocaine from the coca leaves in concentrations exceeding 7mg per fluid ounce.
John C. Lilly	Lilly is remembered for his groundbreaking work in the area of electronic brain stimulation. He was the first person to ever map pain and pleasure pathways in the brain.	He toyed around with mind-altering drugs such as LSD and ketamine.
Paul Erdos	Erdos was a Hungarian mathematician, remembered for his eccentric lifestyle and his influential contributions to 20th century academia. He was one of the most prolific mathematicians who ever lived.	Largely motivated by amphetamines. In fact, he once entered a bet with a friend that he could live without amphetamines for an entire month. He won the bet, but reported that he couldn't do any math during that month-long period. He resumed popping pills and math was forever changed.

(Continued)

Table 1.3 cont.

Creative Genius	The person	The persona
Friedrich Nietzsche	Nietzsche is synonymous with brilliance today, but he was 'crazy' that needed to keep his brain under control with drugs.	He was addicted to opium, taking huge doses at times where his work would have definitely be influenced. He reportedly wrote "The Genealogy of Morals" in just two weeks during an opium binge.
Pablo Picasso	Some call Picasso the most important artist of the 20th century, citing his innovative techniques such as Cubism as evidence.	Cubism could have been a result of the artist's drug use, as Picasso dabbled in psychotropic drugs during his lifetime. His use of opium, morphine and hashish leads some historians to believe that a movement as influential as Cubism was created under the influence of something else entirely.
Dr. William Stewart Halsted	Halsted is credited as inventing the mastectomy, a groundbreaking surgical procedure.	Halsted not only used cocaine as an anesthetic for patients, but he experimented with it himself.
Howard Hughes	Hughes was a businessman turned engineer billionaire.	He would pop everything from codeine to valium pills.

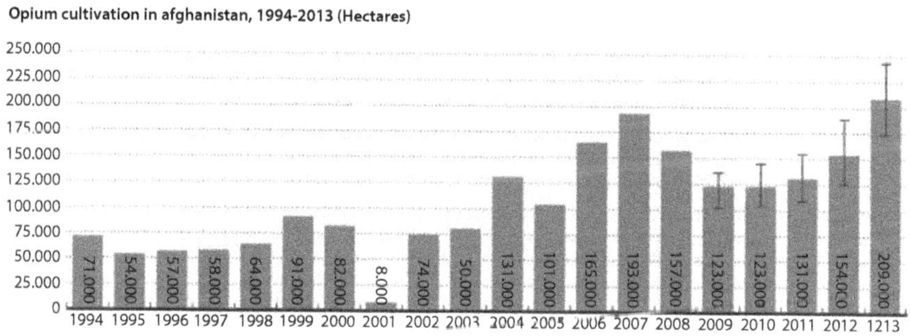

Figure 1.1 Opium cultivation in Afghanistan, 1994-2014 (Hectares) Sources: UNODC and UNODC/MCN opium surveys 1994-2014. The high-low lines represent the upper and lower bounds of the 95% confidence interval.

majority (89%) of total opium cultivation took place in nine provinces in Afghanistan's Southern and Western regions. Opium cultivation increased in most of the main opium poppy-growing provinces, including Badakhshan, Badghis, Nangarhar, Kandahar, Farah, Hilmand, whereas it decreased in Nimroz and Uruzgan provinces.

Volume 2 describes what constitutes natural mass and energy, as well as natural cognition in terms of human thought material (HTM). The model used is yin yang that shows the balance of tangible and intangible at all levels. At the end, the readership is empowered with the process that distinguishes between natural and unnatural, between truth and falsehood, between cure and insult. Picture 1.13 Shows fungus and charcoal on organic lemon and handmade bread, respectively. It shows anecdotally how the pathway followed can determine effect of such chemicals. New science doesn't have a criterion to distinguish between the two pathways. Volume 2 shows how pathway analysis can lead to determining the final state of all matter, including mass, energy and thought material. This discussion is of paramount importance for determining causes of any disease as well as the remediation of environmental health. Once we understand how natural mass and energy are produced and how to detect natural processes one must have a natural cognition tool, it becomes easy to characterize matters and rank them

Canadian Prime Minster Stephen Harper on a wide range of policy issues over a couple of decades. Today reportedly not so much in favour, Flanagan was front-and-centre amid Harper's first battles to lead a "new" Conservative Party in the mid-1990s through the winning election campaigns that secured Harper two consecutive minority governments between 2006 and 2011.)

Picture 1.12 *Organic orange peel with organic olive oil produces real light. Thanks to Newton and Einstein, we have lost the ability to discern any difference between artificial light that gives cancer and natural light that prevents it.*

in terms of beneficial and harmful. Once characterized, the way to determining cause of a disease is paved.

1.10 INTRODUCING CHAPTER SEVEN: Concluding the Discussion-So-Far

We started off this chapter with an anecdote involving two professors of medicine telling us there is no cure to any disease – at least a cure that is available at the pharmacy self. We discussed the hopeless of diagnostic tools and the fact that we don't know the cause of all 'incurable diseases'. We also discussed the prophets of doom – the kind that have been scaring people with fears – ranging from being pushed off the

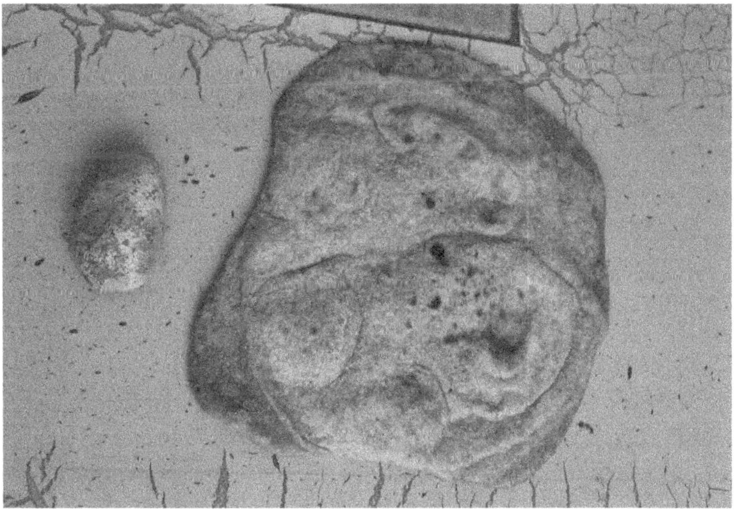

Picture 1.13 Fungus (left) and carbon (dark spots on the right) can be beneficial or harmful depending on the origin (organic lemon, or organic whole wheat bread) and process (atmosphere, or clay stove) involved. New science doesn't have a criterion to distinguish between the two pathways.

Picture 1.14 Volume 2 shows genetically altering is rendering real to artificial with a loss of value that can never be recovered.

edge of the flat earth to heat death à la Lord Kelvin. They are the smartest investors, politicians, industrialists, economists, scientists sand and mathematicians and they all made the same prediction. Anyone countering arguments of one corner would face the wrath of the other corner and they would call each other 'flat earth theorists' and if someone criticizes all of them, he certainly would become 'conspiracy theorist', 'pure lunatic', or worse 'terrorists'. This 600+ book sets out to establish scientific discourse that discredits and dismisses each and every theory of modern era and replaces all of them a truly scientific dogma free process that answers all questions involving human and environmental health. Chapter 10 presents the summary and conclusions of this book and explains how it is possible to deconstruct hundreds of years of work in all fields of New science in a 700+ book.

If this book teaches any lesson, it is: solutions to all ailments is in nature, in natural thinking and in natural mass and energy. This conclusion is complemented with future volumes that add appendix that lists many abstracts from contemporary work on cancer and highlights the cause and possible curative measures that would avert and indeed reverse cancer.

1.11 INTRODUCING CHAPTER EIGHT: About the References and Bibliography, or: Now Previous Knowledge Can Help

This establishes the fact that new science has little to offer in terms of explaining phenomena. Why then we have a list of references and bibliography (Chapter 11) that is over 60 long?

No one disputes, real knowledge is in history. However, how good is history if all we have is a collection of prejudice[19]? Furthermore, what do you make out of statements made by demonstrated liars? For instance, what one can gain from Former US President Nixon's comment, "In the long term we can hope that religion will change the nature of man and reduce conflict. But history is not encouraging in this respect. The bloodiest wars in history have been religious wars."? How do we deal with statements like the one made by Baroness Thatcher, who said, "A world without nuclear weapons would be less stable and more dangerous for all of us"?

Appendix 1.1

Part One - The Political Economy of U.S. Government-led Struggles to Control or Suppress Alcohol, Tobacco and Drug Addiction

The government of the United States is involved deeply and extensively today — sometimes as director, other times as financier and guarantor — in all areas and levels of theoretical and applied scientific research. Looking into the recent past, how such a state of affairs emerged can be readily demonstrated. Indeed: it was one of the most important results of a congressional-military-industrial complex of seemingly infinite plasticity over the last seven decades following the Second World War.

The way for its acceptance was paved by the ideological atmosphere of Prohibition and its aftermath in the depths of the Great Depression, when unemployment would hover near 30 per cent for almost a decade until the US declared war on Japan on 7 December 1941 following the Imperial Japanese Armed Forces' attack on Pearl Harbor. The inner meaning and unexpected consequences of that ideological atmosphere and its fruits — in which the road to intense corporatization was chosen as the road to irreversible progress —are discussed in the first part of this Appendix.

The second part of this Appendix discusses the transformations in the longest-embedded structures of the American state that enabled and accompanied the corporatization of contemporary scientific and technological development. This was and remains to this day a process whose depth and intensity have been peculiar to the United States alone among all the capitalist countries, and remain unique to this day.

A1.1.1 Introducing the Prohibition Conundrum

Today, alongside vast quantities and the widest available selection of alcohol and tobacco, all kinds of prescription and non-prescription substances are produced and marketed by Big Pharma to control or suppress various passing states of mind in addition to alcohol and-or tobacco as well as to alter the individual's psychological state or "mood". Reflecting the sheer scale and range of these products, it hardly comes as any kind of surprise to learn that, according to 2014 data[20], the profits of the pharmaceutical industry are the third-biggest of any major industrial grouping in the United States.

Much less obvious, meanwhile, is the huge and largely-unremarked role that these US-government-led struggles have played in transforming the medical sciences into the highest-profit branch of applied industrial biochemistry rolled out on industrial scale. A vast portion of the entire scientific research enterprise in the U.S. economy has become invested in perfecting the production, distribution and sales output associated with this effort.

The key moment that brought matters to this stage, paradoxically, became known to history as the Prohibition era. Most important — as far as our investigation is concerned — is this essential fundamental fact: regardless of all claims to the contrary, the aims of the actual powers that mattered when it came to dealing with resistance to, and open subversion of, Prohibition had absolutely nothing at all to do with any moralistic impulses to "improve" one's fellow human beings. The key to how this entire development eventually turned out lay in the ups and downs of the U.S. government's struggles to control or suppress alcohol production and consumption, and later tobacco and drugs addiction, in the schizophrenic-seeming relationship between the private marketers and public regulators. That is the subject of this Appendix.

The Americas were 'discovered' under a falsified pretext. A fairy tale was widely promoted to supplant the truth of how Christianity was actually spread in the New World, namely, by Spanish 'explorers' genociding more than 100 million indigenous people over a 150-year period beginning in the early 16th century. According to the more soothing fairy tale, a planted story of Columbus 'discovering' America at the end of the 15th century was hatched and spread. By the 19th century, however, the relentless optimism of the European enlightenment transplanted over the previous two centuries to Boston, New York and Philadelphia had been shattered as Americans were plunged into ideological chaos in the aftermath of the Civil War. "Saving the Union" and

[20]These data were reported on "Big Pharma: Getting high on big profits", the title of Episode 21 in the 'Economic Divide' series, written, produced, directed and narrated by Kaveh Taghvai. First broadcast on Press-TV [Iran] Friday 22 May 2015 0837 GMT (web link: http: //www. presstv.com/ Video/ 2015/ 05/ 22/412291/big-pharmaceutical-companies-profits)

"freeing the slaves" were each effected on the ruins of much of what had constituted the Southern states of that Union before the war. Meanwhile, large swaths of civil society in the most prosperous cities of the Northern states had supported elimination of chattel slavery. Now they were caught emotionally and ideologically utterly unprepared for the horrific scale of unrestrained slaughter that had been unleashed in the name of realizing such a lofty aim. An uplifting fresh mission was urgently called for, and the temperance movement, which had become a hobby-horse of various Protestant, mostly rurally-based, religious minorities, happened to be positioned at that moment in the foremost position to heed that call.

The temperance movement's response found its breeding ground as the chaos that had gripped a population displaced by four years of the most violent spasm to seize any country in Europe or the Americas to that time destroyed and displaced millions of families. In the same moment, the sources of the consequent *social* decay were then arbitrarily situated within the wrecking of *individual* families. Baptist and Methodist clergy sought to abolish the recreational use of alcohol as 'demon drink' which now became the newest the bogeyman of American Christianity. In less than two generations between the end of the Civil War and US entry into the First World War, a massive exodus to the large, increasingly industrializing cities took place from a countryside that had become an economic basket case in the aftermath of the elimination of outright slave labor. Within that physical displacement of millions of people, a rapid expansion of a new massive social movement against the modernized slavery of the wages system would soon be generated. In these unprecedented conditions, white American men found themselves challenged no longer by African Americans' presence but by the intensified competition for employment amid growing numbers of women and European immigrants. to the urban centres. Such were the changing social conditions out of which religious coalitions such as the Women's Christian Temperance Union emerged with their message of saving the humanity of men and women caught up in this maelstrom of profound social change. Such were the organizations that that would lead in creating the political pressure of nationwide support for banning alcohol consumption of any kind entirely.

Inspired by their success, Christian denominations charged ahead as they helped form the Anti-Saloon League to lobby the government to create prohibition legislation. Success came quickly and in 1920, Prohibition passed in the form of the Volstead Act which Congress was enabled to add the Eighteenth Amendment to the US Constitution.

Paradoxically in the same moment, however, Prohibition also tremendously damaged the American economy in the immediate short term. Thousands of jobs were lost, restaurants and theaters closed, and breweries were dismantled.

One of the profoundest immediate effects of Prohibition was felt on government revenues. Before Prohibition, many states relied heavily on excise taxes in liquor sales to fund their budgets[21]. In New York, almost 75% of the state's revenue was derived from liquor taxes. The legal basis for this state ruin being an act of a *federal* legislature amending the *constitution governing the entire country.* With Prohibition in effect, that revenue was permanently lost. At the national level, meanwhile, the immediate effects were even more dramatic: Prohibition cost the federal government a total of $11 billion in lost tax revenue, while costing over $300 million to enforce. The most lasting consequence was that many states and the federal government all came to rely on income tax — a federal measure intended to pay for the war —to replace the revenue to fund their budgets going forward.

The total production of distilled spirits rose from 187 million gallons in 1912 to 203 million gallons in 1926. Most of this increase, of course, can be attributed to the greatly expanded use of industrial alcohol. As a basic chemical, it was required in the manufacture of cosmetics, leather goods, dyes, and synthetic textiles (one rayon plant used 2 million gallons of denatured alcohol annually). The single most significant factor in the expanded production of denatured alcohol was the spectacular burgeoning of the auto industry. In 1919, there were 7.5 million passenger cars and trucks registered in the United States; in 1926, there were 22 million. Whereas before, 90 percent were open and therefore not suitable for cold-weather driving, by 1926, 70 percent were closed, suggesting the possibility of operation in cold weather. Antifreeze was thus a sudden necessity, and three-quarters of the total output of completely denatured alcohol went for that purpose.

The least anticipated consequence of Prohibition lacking any genuine moral consensus was its indirect fostering of criminal syndicates that extended nationwide and joined various cities in tightly-controlled networks dominated by ruthless gangsters like Al Capone and others. It was to these bootlegging operators that a considerable portion of the enormous increase in production noted in the previous paragraph

---

[21]The United States imposed income taxes briefly during the Civil War and the 1890s, and on a permanent basis from 1913. However, the rate was very small.

undoubtedly was diverted to these bootlegging syndicates. Emory Buckner, the United States Attorney for the southern district of New York, estimated that in 1925, 60 million gallons were sold as beverage alcohol (Dobyns 1940). Since it is likely that this congressional testimony was intended to secure additional appropriations for enforcement, the claims of such mammoth diversions are suspect. However, even more modest figures suggest systematic fraud.

During the Prohibition era, a wide range of entirely illegal economies flourished. The earliest bootleggers began smuggling foreign-made commercial liquor into the United States from across the Canadian and Mexican borders and along the seacoasts from ships under foreign registry. Their favorite sources of supply were the Bahamas, Cuba, and the French islands of Saint-Pierre and Miquelon, off the southern coast of Newfoundland. That type of smuggling became riskier and more expensive when the U.S. Coast Guard began halting and searching ships at greater distances from the coast and using fast motor launches of its own.

Figure A1.1 documents evidence for our claim that bootlegging helped lead to the establishment of American organized crime. Crime rates increased in all sectors, especially the ones dealing with violence and murder. Gradually, the gangs in different cities began to cooperate with each other, and they extended their methods of organizing beyond bootlegging to the narcotics traffic, gambling rackets, prostitution, labor racketeering, loan-sharking, and extortion. The American Mafia crime syndicate arose out of the coordinated activities of Italian bootleggers and other gangsters in New York City in the late 1920s and early' 30s. One would get the flavor of a corrupt entity that later morphed into being called: The Big Brother.

The conventional consensus is that Prohibition was lifted because the authorities "saw the light" and gave up trying to fight "human nature." The truth is far more interesting: thanks to the proliferation of these national bootlegging gangs, alcohol could now be linked to anything including a patriotic cause. And so it turned out: just as Lincoln had turned to the liquor trade in 1862 to finance the war effort, so too did Roosevelt campaign in 1932 on the promise to repeal the 18th Amendment and revive an industry that could provide both jobs and tax revenue! Nine days after his inauguration, Roosevelt sent before Congress a piece of legislation modifying the Volstead Act and legalizing the sale of beer. During the summer of 1933, the administration stumped for repeal; with James Farley in charge of the effort, and the AAPA and

WONPR providing organizational support, repeal was promoted as a key element of a recovery program.

The 18th Amendment was by this time so unpopular and so financially destructive that Franklin Delano Roosevelt incorporated its repeal into his 1932 campaign for President. Just over a year later it was formally repealed by the Twenty-First Amendment. People that didn't support the Prohibition were elated by the statistics that banning alcohol increased violence and murder (Figure A1.1). Just as glibly as the temperance advocates assured the public that crime and social decay would be reversed by abstinence and therefore by Prohibiion as well, no one looked at any of the science behind such data in that Figure which would suggest that Prohibition created a culture of underground crime.

So many other facts of the time reinforce our conclusion. Speakeasies, for example, replaced saloons. Black markets and gangsters took over the business. Nothing happened to the drinking habits of people. All "changes" were about money changing hands. As a byproduct of the cat-and-mouse nature of law enforcement, Prohibition led to a large number and wide range of unintended consequences. One of the most consequential of those unintended consequences was the embroilment of federal law enforcement in the pharmaceuticals business, both prescription and non-prescription.

This embroilment would prove to be of far greater moment over the long term. While the Eighteenth Amendment prohibited the manufacture, sale and transportation of intoxicating beverages, it did not outlaw the possession or consumption of alcohol in the United States. What this did, in fact, is make exceptions for two entities that benefited greatly from the seeming calamity of US economy. One of the legal exceptions to the Prohibition law was that pharmacists were allowed to dispense whiskey by prescription for any number of ailments, ranging from anxiety to influenza. Bootleggers quickly discovered that running a pharmacy was a perfect front for their trade. As a result, the number of registered pharmacists in New York State tripled during the Prohibition

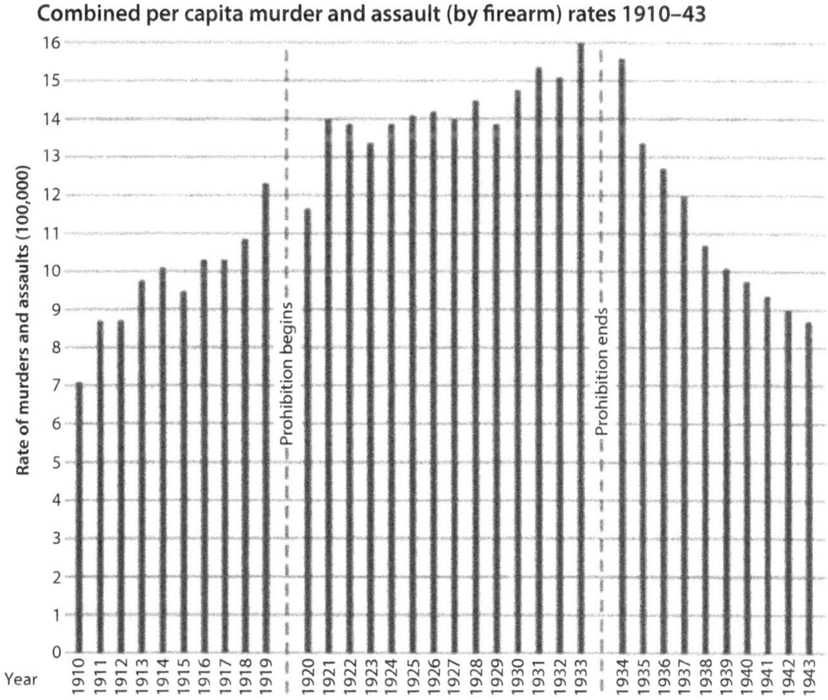

Figure A1.1 *Prohibition became synonymous with murder and mayhem.*

era. Benefits also came to physicians that were authorized to make exceptions. It was lucrative business for the 'health care' industry.

Willy-nilly, meanwhile, all manner of spontaneous extensions of federal law enforcement increasingly emerged as bootleggers also got involved in the pharmaceutical industry in various ways.[22] From this point forward, at an accelerating pace following the end of the Second World War, a profound shift unnoticed at the time would soon transform the entire involvement of the national government of the United States. As a byproduct of the unplanned and spontaneous focus of a

[22]They were also dealing with millions of bottles of "medicinal" whiskey that were sold across drugstore counters on real or forged prescriptions. In addition, various American industries were permitted to use denatured alcohol, which had been mixed with noxious chemicals to render it unfit for drinking. Millions of gallons of that were illegally diverted, "washed" of noxious chemicals, mixed with tap water and perhaps a dash of real liquor for flavor, and sold to speakeasies or individual customers. Finally, bootleggers took to bottling their own concoctions of spurious liquor, and by the late 1920s stills making liquor from corn had become major suppliers.

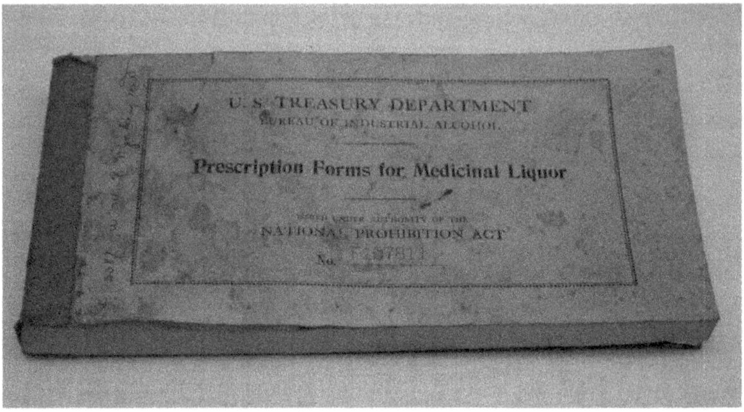

Figure A1.2 *"Medical" liquor was perfectly legal during the Prohibition period. All that was needed is enough money to convince a physician.*

federally-led law-enforcement effort against nationally-organized criminal bootlegging cartels, the food-drug axis that informed federal regulatory processes and interventions during the Progressive era just before the First World War would become increasingly displaced as the federal government became ever more involved supporting advanced research across all manner of medical fields.[23]

[23]For many of the same reasons that the 18th Amendment was repealed and drinking returned to respectability, smoking also underwent a rehabilitation in public opinion. By 1927, the bans had all been overturned. By the 1930s, cigarettes were grouped with alcohol as aids to economic revival through their provision of important tax revenues. In addition, smoking along with drinking became raised to the status of a normative, while non-smokers on the other hand — just like teetotalers — became suspect as antisocial eccentrics.

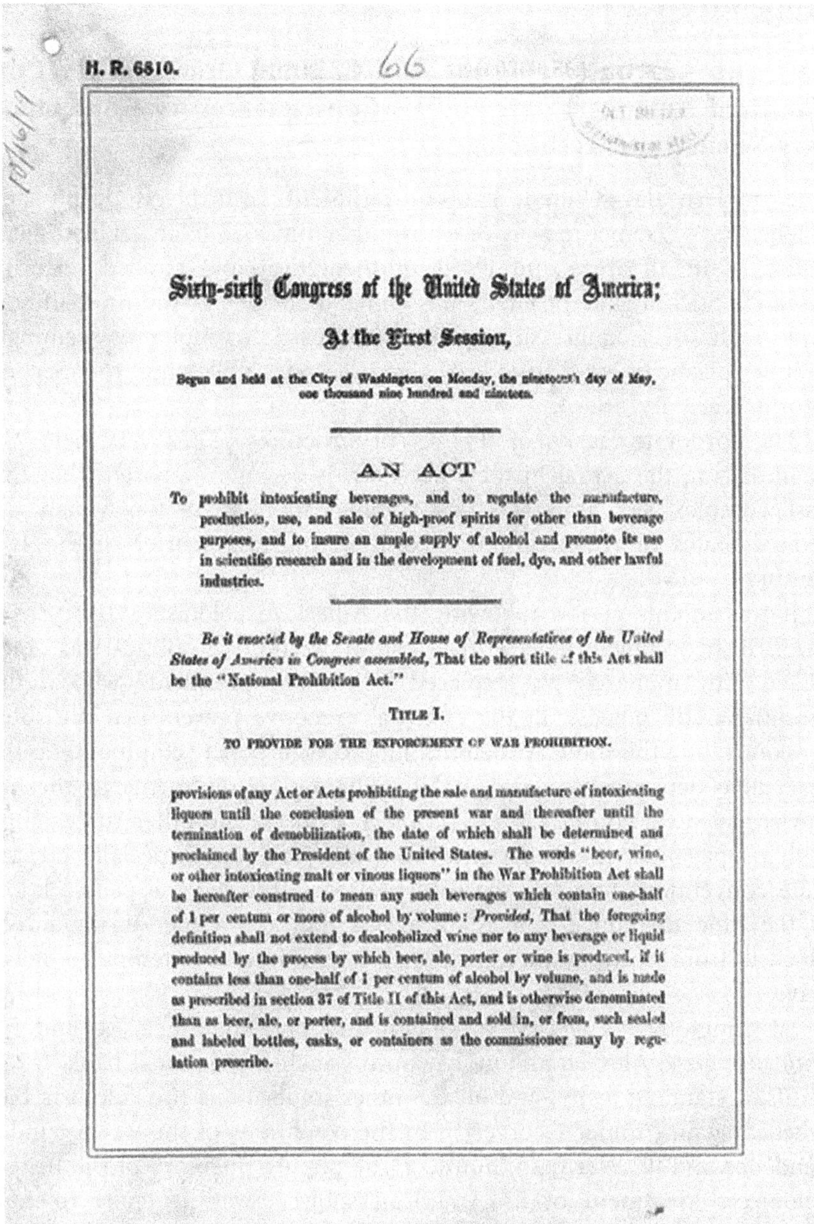

H. R. 6810.

66

Sixty-sixth Congress of the United States of America;

At the First Session,

Begun and held at the City of Washington on Monday, the nineteenth day of May, one thousand nine hundred and nineteen.

AN ACT

To prohibit intoxicating beverages, and to regulate the manufacture, production, use, and sale of high-proof spirits for other than beverage purposes, and to insure an ample supply of alcohol and promote its use in scientific research and in the development of fuel, dye, and other lawful industries.

Be it enacted by the Senate and House of Representatives of the United States of America in Congress assembled, That the short title of this Act shall be the "National Prohibition Act."

TITLE I.

TO PROVIDE FOR THE ENFORCEMENT OF WAR PROHIBITION.

provisions of any Act or Acts prohibiting the sale and manufacture of intoxicating liquors until the conclusion of the present war and thereafter until the termination of demobilization, the date of which shall be determined and proclaimed by the President of the United States. The words "beer, wine, or other intoxicating malt or vinous liquors" in the War Prohibition Act shall be hereafter construed to mean any such beverages which contain one-half of 1 per centum or more of alcohol by volume: *Provided*, That the foregoing definition shall not extend to dealcoholized wine nor to any beverage or liquid produced by the process by which beer, ale, porter or wine is produced, if it contains less than one-half of 1 per centum of alcohol by volume, and is made as prescribed in section 37 of Title II of this Act, and is otherwise denominated than as beer, ale, or porter, and is contained and sold in, or from, such sealed and labeled bottles, casks, or containers as the commissioner may by regulation prescribe.

Figure A1.3 *The Volstead Act implemented Prohibition for the next 13 years.*

Appendix 1.2

Part Two - Delinearizing the Anti-Colonial Origins of the Corporatized American State and Its Consequences for Contemporary Science and Technology

The modern day United States government is involved deeply and extensively — sometimes as director, other times as financier and guarantor — in all areas and levels of theoretical and applied scientific research. Such a state of affairs has come about one of the most important results of a congressional-military-industrial complex of seemingly infinite plasticity over the last seven decades following the Second World War.

The corporate façade of this set of structures seems very new. On delinearizing the actual history, however, it becomes apparent that this vast complex sits atop the post-colonial foundations on which the United States of America was erected in the last quarter of the 18th century.

It was in this period following the American colonists' victory over the British occupier that the true corporatization of politics was completed. The monarch was replaced by a "The president" who would essentially still possess all the essential executive powers run the show — controlling the military, determining foreign policy, commanding the "executive departments," and making lifetime appointments to the all-powerful federal courts. Using chicanery, bribery, and other ruses, along with propaganda, the Federalists were able to induce specially created state conventions to ratify the Constitution. Little did the people know at the time that once their state joined the New Order it wcould be blocked from a thousand directions should it ever attempt to never leave.

Washington was elected president, and once in office, he and his *éminence grise*, Alex Hamilton, began to establish a national bank, taxes, tariffs, a standing army, and all the other impositions the colonists had rebelled against under George III. In the conditions of the war for independence and its aftermath, however, the popular memory of the British colonizers' treatment of its American subjects was generally tolerant whereas punishments meted out by the Washington junta against those who fell afoul of its laws were widely resisted.

Thus for example, when a revolt broke out in Pennsylvania against the national tax on whiskey — the "Whiskey Rebellion" of 1794 — George Washington responded by leading 13,000 troops into the state to crush it. The army of the Second Republic enabled the new regime to put down dissent far more effectively than could the British in 1775 or Taxachusetts in 1786. Moreover, the American people had been propagandized into believing that, whatever happened, they were simply governing themselves. This was the Big Lie that took root sufficiently that the populace allowed the imposition of a far worse tyranny almost anything imposed by the British outsider. Of course, the likelihood of being hanged as a traitor to the state has served as a means of dampening the spark of Liberty down the ages. American history is rife with HSSA degradation.

A delinearized history makes it clear that George Washington was a rich man — not unlike Silvio Berlusconi today — whose wealth enabled him to construct an ongoing international role as an important statesman to reckon with, not unlike Silvio Berlusconi today. Gardner (2013) described the financial status of George Washington whose real worth at time of death was $25.9 billion, taking fourth place in the Forbes list of seriously wealthy Americans. Bill Gates is in first place at $59 billion, Warren Buffett in second at $39 billion, and Larry Ellison of Oracle fame gets the bronze medal with a $36 billion stash. Washington's $25.9 billion sneaks him in just ahead of Christy Walton of the Wal-Mart chain. The first president is in rich company. There, unlike what he claimed to be the case[24], George Washington's success came through his half-brother, 14 year older Army officer, membership in the Masonic lodge, engineered marriage with a rich widow, and numerous shady deals that have all the hallmarks of a mafia-like entity.

The next figure to approach Washington's iconic standingicon, Abraham Lincoln, was the person that said, "I, as much as any other man, stand in favor of having the superior position assigned to the white race... I have no purpose to introduce political and social equality between the white and the black races (still a backwoods Illinois politician and lawyer but married to Mary Todd, daughter of a wealthy & politically highly-connected family of slaveowners, 1858). These were not the saviors of human rights and civic dignity, they were simply the new Crusaders, obsessed with Money, Sex, and Control. So, why did they ever do anything that could touted as 'positive'? Abraham Lincoln gave the answer by himself. He wrote:

[24]Washington wrote: "My mother was the most beautiful woman I ever saw. All I am I owe to my mother. I attribute all my success in life to the moral, intellectual and physical education I received from her."

> My paramount object in this struggle is to save the Union, and is not either to save or to destroy slavery. If I could save the Union without freeing any slave I would do it, and if I could save it by freeing all the slaves I would do it; and if I could save it by freeing some and leaving others alone I would also do that. What I do about slavery, and the colored race, I do because I believe it helps to save the Union; and what I forbear, I forbear because I do not believe it would help to save the Union. (Letter to Horace Greeley, August 22, 1862)

As Thomas DiLorenzo and a number of non-court historians have conclusively established, Lincoln did not invade the Confederacy in order to free the slaves. The Emancipation Proclamation did not occur until 1863, by which time opposition in the North to the war was rising despite Lincoln's police state measures to silence opponents and newspapers. The Emancipation Proclamation was a war measure issued under Lincoln's war powers. The proclamation provided for the emancipated slaves to be enrolled in the Union army replenishing its losses. It was also hoped that the proclamation would spread slave revolts in the South while southern white men were away at war and draw soldiers away from the fronts in order to protect their women and children. The intent was to hasten the defeat of the South before political opposition to Lincoln in the North grew stronger.

DiLorenzo (2002) points out: "Lincoln spent his entire political career attempting to use the powers of the state for the benefit of the moneyed corporate elite (the 'one-percenters' of his day), first in Illinois, and then in the North in general, through protectionist tariffs, corporate welfare for road, canal, and railroad corporations, and a national bank controlled by politicians like himself to fund it all."

Lincoln was a man of empire. As soon as the South was conquered, ravaged, and looted, his collection of war criminal generals, such as Sherman and Sheridan, set about exterminating the Plains Indians in one of the worst acts of genocide in human history. Even today Israeli Zionists point to Washington's extermination of the Plains Indians as the model for Israel's theft of Palestine.

The War of Northern Aggression was about tariffs and northern economic imperialism. The North was protectionist. The South favoured free trade. The North wanted to finance its economic development by forcing the South to pay higher prices for manufactured goods. The North passed the Morrill Tariff which more than doubled the tariff rate to 32.6% and provided for a further hike to 47%. The tariff diverted the

South's profits on its agricultural exports to the coffers of Northern industrialists and manufacturers. The tariff was designed to redirect the South's expenditures on manufactured goods from England to higher-cost goods produced in the North.

On the personal side, Lincoln was depressed. Lincoln did suffer from what we now call depression, as modern clinicians, using the standard diagnostic criteria, uniformly agree. It is profoundly symbolic that he led a nation that would eventually become the icon of a depressed psyche in social sense. Schenk (2005) summarized Lincoln's condition thus:

> Was Lincoln's melancholy a "clinical depression"? Yes—as far as that concept goes. Certainly his condition in the summer of 1835 matches what the Diagnostic and Statistical Manual of Mental Disorders labels a major depressive episode. Such an episode is characterized by depressed mood, a marked decrease in pleasure, or both, for at least two weeks, and symptoms such as agitation, fatigue, feelings of worthlessness, and thoughts of death or suicide. Five and a half years later, in the winter of 1840–1841, Lincoln broke down again, and together these episodes suffice for modern clinicians to make an assessment of recurrent major depression.

Peele (2010) presents how Abraham Lincoln and his mentally unstable wife represented the two extremes of the psychological spectrum, extreme fear and extreme strength - typical of today's American psyche. It is also interesting to note that Abraham Lincoln was no't a Christian. nor did hHe does not invoke Jesus in any of his speeches, not even in the 163-word Bettysburg that US schoolchildren are routinely expected to memorize in grade school.

The official slogan for the Union was *E pluribus unum* (Latin for "out of many, one", symbolizing the union of many states) that was used on official seals, along with *Annuit cœptis* (Latin for "He/she/it approves (has approved) the undertakings") and *Novus ordo saeculorum* (Latin for "New Order of the Ages"), and adopted by an Act of Congress in 1782. Never codified in law, *E Pluribus Unum* was considered a *de facto* motto of the United States until 1956 when the United States Congress passed an act (H. R. Resolution 396), adopting *In God we trust* as the official motto.

None of these are Christian invocations, so what you hear and see in this case is exactly what you get. If the trinitarian God was an invention of the Roman Catholic Church, the Money god was formally canonized by an act of the US Congress.

Similarly, the original Pledge of Allegiance, composed in August 1892 by Francis Bellamy (1855–1931), who was a Baptist minister, a Christian socialist, and the cousin of socialist utopian novelist Edward Bellamy (1850–1898), didn't have the word 'God' in it. It read: "I pledge allegiance to my Flag and the Republic for which it stands, one nation, indivisible, with liberty and justice for all." The Pledge was supposed to be quick and to the point. Bellamy designed it to be recited in 15 seconds. As a socialist, he had initially also considered using the words equality and fraternity but decided against it – knowing that the state superintendents of education on his committee were against equality for women and African Americans.

It was formally adopted by Congress as the pledge in 1942. The official name of The Pledge of Allegiance was adopted in 1945. The last change in language came on Flag Day 1954 when the words "under God" were added. Ever since, it reads as: I pledge allegiance to the Flag of the United States of America, and to the Republic for which it stands, one Nation under God, indivisible, with liberty and justice for all. Ever since, the phrase, "founded on Christian or Judeo-Christian value" would become synonymous with US politics, in contradiction to the First Amendment to the United States Constitution that states: "Congress shall make no law respecting an establishment of religion, or prohibiting the free exercise thereof...." and Article VI specifies that "no religious Test shall ever be required as a Qualification to any Office or public Trust under the United States." In no uncertain terms, it was stated in Tripoli Treaty (Picture 1.10) in which Article 11 states, "As the Government of the United States of America is not, in any sense, founded on the Christian religion; as it has in itself no character of enmity against the laws, religion, or tranquility, of Mussulmen; and as the said States never entered into any war, or act of hostility against any Mahometan nation, it is declared by the parties, that no pretext, arising from religions opinions, shall ever produce an interruption of the harmony existing between the two countries."

2

Current State of the World of Big Pharma

2.1 Summary

Our civilization is defined by the way our human needs for matter and energy are planned with human thought material (HTM). If we start off with the premise that 'nature is good', we can measure the worth of any civilization by its proximity to natural thinking and usage of natural resources. This chapter describes the current state of the world on this question so critical to human survival, and illuminates the most important features that set this civilization apart from previos epochs. The cause of the current malfunction, termed 'technological disaster' by Nobel chemistry laureate Robert Curl, is also identified. What we have today is a display of profiteering in every aspect of the civilization. Of particular relevance is the fact that pharmaceutical and chemical companies are for-profit entities. To the extent that considerations of profitability overrule every other consideration in the Information age, "profit" has become an obscenity. In an era of maximizing profit and minimizing value, Big Pharma is widely alleged to push its wares by deceiving the public — even offering to pay doctors for prescribing their products — as well as bribing politicians. All this creates a vicious cycle of the health industry and government getting richer, feeding the oldest scheme of Trinity formed through Government, Church, and

Corporations. The scheme extends even onto the black market, where drugs are sold illegally, people become addicted and need to spend money for their fix. In countries, where health care is free of charge, this frequently implicates the judicial order by dragging the prison system drags the government, along with other government-sponsored corporations (e.g. prison system) in the most sinister form.

2.2 Introduction

We live in an interesting époque. Everything surrounding us seems to point to contradictions. We made unprecedented progress in technology development, only to hear from Nobel Laureate Chemists (e.g. Robert Curl) that ours is a 'technological disaster'. We claim to have progressed from dark ages of savagery and lack of empathy to modern enlightenment only to hear from some of the most ardent supporters of modern European mindset of Capitalism (e.g. Pope Francis) that unfair economic structures that creates huge inequalities are actually an act of terrorism. We claim to have reached the pinnacle of human equality and democratic values only to hear from Nobel Peace laureates (e.g. President Obama, Egyptian leader, Bardei) that such rights only belong to a small group as the enlightened world is fed the news of a military coup in the form of the news headline "Kerry Lauds Egypt Military for 'Restoring Democracy'. "We continuously hear about the threats to human existence that loom on us because of the process that we have followed as a human race. The most astounding contradictions are in the drivers of the modern époque, namely, Education and Economics. Education (the original meaning being 'to bring forth') has been reduced to training, prompting Albert Einstein to say, "It's a miracle curiosity survives formal education". In the meantime, Economics (the original meaning being 'to economize' domestic household) has been turned into an avenue for wasteful schemes by banks and industrial conglomerates, prompting World Bank employee and Nobel laureate economist Stiglitz to castigate International Monetary Fund (IMF) for offering 'remedies' that make things worse—turning slowdowns into recessions and recessions into depressions. This institutionalization of preposterous schemes is the hallmark of modern age, the word Physics (that means 'science of nature' and not the science of artificial) being just one of them. Each and every subject has been turned upside down, each seeing a reversal of original meaning of the word. Overall, this has been a departure from truth to falsehood, from intangible to tangible, from long-term to myopia, from love to lust, from empathy to antipathy. Dr. Moorehead summarized this paradox of modern society.

> "The paradox of our time in history is that we have taller buildings but shorter tempers, wider freeways, but narrower viewpoints. We spend more, but have less; we buy more, but enjoy less. We have bigger houses and smaller families, more conveniences, but less time. We have more degrees but less sense, more knowledge, but less judgment, more experts, yet more problems, more medicine, but less wellness. We have multiplied our possessions, but reduced our values. We talk too much, love too seldom, and hate too often.
>
> We've learned how to make a living, but not a life. We've added years to life not life to years. We've been all the way to the moon and back, but have trouble crossing the street to meet a new neighbour. We conquered outer space but not inner space.
>
> We've done larger things, but not better things. We've cleaned up the air, but polluted the soul. We've conquered the atom, but not our prejudice.
>
> We write more, but learn less. We plan more, but accomplish less.
>
> We've learned to rush, but not to wait. We build more computers to hold more information, to produce more copies than ever, but we communicate less and less.
>
> These are the times of fast foods and slow digestion, big men and small character, steep profits and shallow relationships. These are the days of two incomes but more divorce, fancier houses, but broken homes.
>
> These are days of quick trips, disposable diapers, throwaway morality, one night stands, overweight bodies, and pills that do everything from cheer, to quiet, to kill. It is a time when there is much in the showroom window and nothing in the stockroom."

Criticizing is easy. In fact, the current époque is infamously known for its tolerance to criticism while insulating status quo. It also ends with agreeing to disagree and finding consensus on matters that are truly irrelevant. In search of such 'consensus', no one goes back to the first assumptions that defined our civilization and created a culture of paradox.

Modern age is synonymous with paradoxes. Everything surrounding us points to contradiction upon contradiction. If the philosophy of dogma introduced the contradiction of mortal and immortal entity in one body, we are experiencing the worst of dogmas in today's society. We made unprecedented progress in technology development, only to hear from Nobel Laureate Chemists (e.g. Robert Curl) that ours is a 'technological disaster'. We claim to have progressed from dark ages of savagery and lack of empathy to modern enlightenment... only to hear from some of the most ardent supporters of modern European mindset of Capitalism (e.g., Pope Francis I) that unfair economic structures that creates huge inequalities are actually an act of terrorism. We claim to have reached the pinnacle of human equality and democratic values only to hear from Nobel Peace laureates (e.g., U.S. president Obama and the Egyptian politician Mohamed el-Baradei) that such rights only belong to a small group as the enlightened world is fed the news of a military coup in the form of the news headline "Kerry Lauds Egypt Military for 'Restoring Democracy'." We continuously hear about the threats to human existence that loom on us because of the process that we have followed as a human race. In this process, modern society has become an expression of contradictions, as depicted in Figure 2.1.

In the current époque, the biggest loser has been the wellbeing of humans that got caught in the paradoxes of modern life. The biggest industry is the 'defense' sector that promotes war to 'restore peace', the second biggest industry is the pharmaceutical sector that promotes medicines. There are trillions of dollars of medicines available in the market, yet there is not a single medicine that cures a single disease.

The most astounding contradictions are rooted in the very driver of the modern epoch, viz., Economics. Economics (the original meaning being 'to economize' in the sense of arranging discharge of responsibilities of domestic household management to a level of maximum

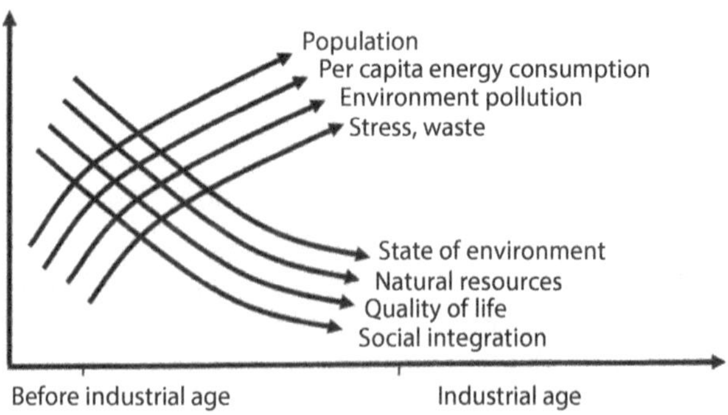

Figure 2.1 Our current epoch is an epic failure of intangible values.

efficiency) has been turned into an avenue for wasteful schemes by banks and industrial conglomerates, prompting World Bank employee and Nobel laureate economist Joseph Stiglitz to castigate the International Monetary Fund (IMF) for offering 'remedies' that make things worse, turning slowdowns into recessions and recessions into depressions. This institutionalization of preposterous schemes is the hallmark of modern age. It involves turning real into artificial while increasing the profit margin (Figure 2.2). Pharmaceutical industry has become the front for this institutionalization.

Every product of the pharmaceutical or chemical industry is an aberration from natural state. Further the departure higher is the profit. As Figure 2.2 shows, turning artificial value by counterfeiting real value (or naturally processed status) is the key to profiteering in the information age. For instance, raw food material can be processed through clay over wood fire to add real value to the final product. A counterfeit processing can be through microwave cooking and serving the final product with an artificial value addition. The latter process destroys the natural state of the food material adds the illusion that the food is cooked despite the fact that the essential nutrients are non-existent or worse turned into carcinogenic products.

Figure 2.3 elaborates this aspect for technology development. Ever since the plastic revolution began over a century ago, every chemical engineering product has been motivated by denaturing raw materials

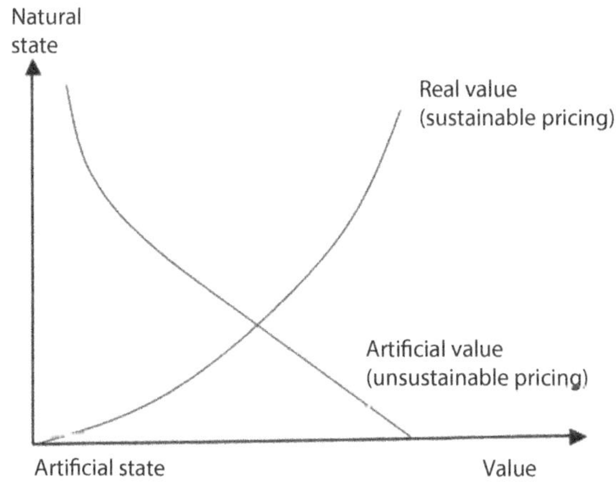

Figure 2.2 Modern science and technology development schemes focus on turning natural into artificial and assigning artificial values, proportional to the aphenomenality of a product.

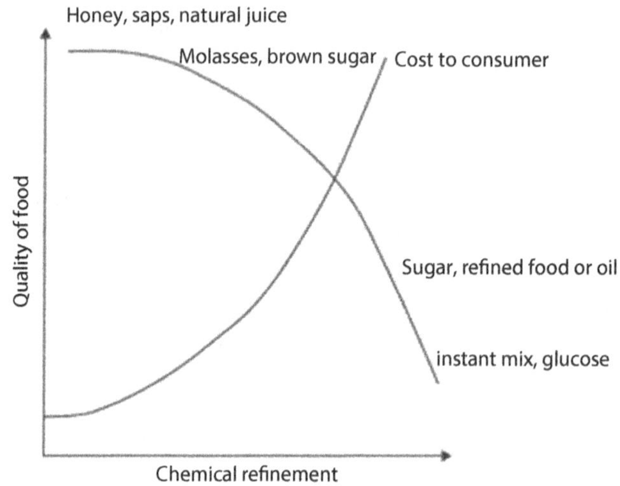

Figure 2.3 Chemical refining is synonymous with high cost to consumers and degradation of food quality.

and adding value through degeneration of natural state. The same tactic has been used in every discipline, ranging from politics to education, from journalism to medical science. As will be seen in latter chapters, all theories of modern hard science and social science, including Nobel prize winning ones, are based on this fundamental principle of turning natural to artificial. The result of such mode of social development has been catastrophic.

2.3 How We Got Here: A Delinearized History Of The Information Age

George Orwell wrote, "He who controls the past controls the future. He who controls the present controls the past." Our current époque is full of obsessive compulsion to distort the past, hence recreating a desirable future based on maximizing short-term gains. Early civilizations considered themselves the guardians and caretakers of all living things on the lands they inhabited, and held themselves responsible for future generations. Indigenous nations of the Americas considered themselves one with all around them. There was no special word for Nature, no separation: plants, animals, and humans were considered interdependent. In this world it was the coming of the European invader, funded by their own rulers at home, that led to the eventual corporatizing of the earth which all living things share in common, into a commodity to be broken up at will, through wars and land appropriation.

In the contemporary world, capital-centredness is ultimate source of the problems unleashed by an economic order based on oligopolies, monopolies and cartels all loyal to the same single-pillar shehadeh that declares that there is no god but Monopoly and Maximum is its profit. The main features of a properly delinearized history of how we have arrived at this point goes something like this:

The advent of property laws made "legal" after the fact what had actually been acts of misappropriation. In nineteenth-century America, following the Civil War, specifically in order to protect and encourage corporate property, this "right" to retain control or ownership of any form of property - especially property already accounted as a business asset (whether it originated as a natural resource or as a claim on someone else's labour) but acquired without "colour of right" (i.e., before there existed any law specifically defining or dealing with its legal existence as property (Latta 1961)- was consciously elaborated as an exception to the Rule of Law. As the result of wars and other struggles waged to protect this corporatized form of property and the technological development that stemming from it - including associated long-term toxic effects - the world now finds itself in an environmental crisis (Rich 1994).

The deepening of this crisis is marked by a simultaneous extension of corporate abuse of Humanity's rights of access to fresh air, clean water and other absolute necessities, alongside a growing rebellion by the human productive forces sustain- ing these corporations as their market, against government accommodation of the abusers and their abuses. Today, water and air have become commodities. Governments and corporations now own access to water. Overuse by industry and agriculture have made it into a scarce commodity over which future wars will be waged.(Barlow & Clarke 2002; Holden & Thubani 1996). Contaminated by industrial and agricultural run- off, the sale of bottled water, or home filters to the public who can afford it, is promoted as 'uncontaminated'.

Pollution itself, created by chemical poisons released into the air by industry, agriculture, and the automobile, has become a money-making commodity - with the sale of pollution 'credits' from one polluter to another. Home filters to 'clean' the air in homes and public buildings, promote clean air, again for those who can afford it, autos in many places are required to have catalytic converters to filter out poisons emitted from the burning of fuel, to keep down air pollution.

At the same time corporate activity, with its virtual immunity from prosecution, legal sanctions or legal responsibility in its home bases and main markets, is purchased and maintained by trying to dump unwanted wastes in various parts of Africa and Asia. This is arousing more and more people in the rest of the world against corporate fiat and dictate, energizing in its wake a rapidly widening discussion of alternative arrangements for Humanity's continued existence on this planet. Accordingly, the intention to control nature has become the last remaining pathway by which corporations hope to ensure a constant, never-ending stream of profit – and a battleground on which the fate of Humanity for generations to come may be decided.

The two most efficacious routes by which humans can begin to restore the earth to some kind of balance are a) by reining in the effects of capital-centredness and- or b) by eliminating the possibility of capital-centredness taking over from human- centredness. However, capital-centredness itself can conceal itself within a variety of private and public property regimes.

The drive to corporatization started in Britain in the 16th century. Systematically, most of the world's land was converted into property owned by individuals, and later by corporate entities (which enjoy the rights and personality of individuals under the law). The first act of the Bolshevik Revolution of 1917 was to eliminate such private property in land. This was a reaction to centuries of oppression by the rich. The entire social system elaborated from and on the basis of that act, on one-sixth of the earth's surface, threw into a chronic crisis the economy in the rest of the world. In the United States and Europe in particular the regime of private property – not only in land, but in means-of-production in general, including factories, distribution networks, etc. – still held sway.

However, increasingly in these countries, the State was interposed:

1. as guarantor of all parts of socially necessary services that were of minimal profit to private capital;
2. as guarantor of all high-risk energy development projects; and
3. as financier of parasitic spending on, and the endless and guaran- teed market for, weapons systems of every description. In these and other ways, the self-interest of private property and its unquenchable appetite for gain could still carry on – only disguised as government-approved and, therefore, somehow pro-social.

With the disappearance of the regime of socialist social property in the Soviet Union and eastern Europe after 1989–1991, however, the chronic crisis of the established capitalist world order did not come to an end. On the contrary, the crisis itself deepened by becoming more chronic, with serious decay in social indices from alcoholism to infant mortality and premature death throughout the former socialist countries as well as a change in some of the forms of this crisis.

Thanks to the emergence of United States as the only superpower in the world, the whole world is tightly focused on the race that sees profiteering as the only means for economic growth. In this, the traditional detractors of Capitalism has joined in in fact are leading the race. The Chinese Hu Jun magazine (Feb. 3, 2015) published its rating of dollar billionaires counting 2089 richest people of the world. Dominant positions are occupied by the US and China having taken 537 and 430 places correspondingly. For the first time over recent years India occupies the third place, leaving Russia behind. The magazine concluded that China is still strengthening its positions in the global economy. In the Chinese part of the list there have emerged 72 new names. The majority of them became rich owing to a significant stock market growth in 2014. The high-tech sphere remains the leading one for capital increase.

Goldberg (2014) reported China having second highest number of millionaires. China's staggering growth accompanies a nearly 50% increase in the country's private financial wealth, which includes bank deposits, securities and pension funds but excludes luxury goods and real estate. This growth is credited to China's growth in its "shadow banking" sector, wherein credit is exchanged through alternative financiers.

Figure 2.4. summarizes the extreme nature of world economy. Oxfam (2015) reported that 388 billionaires had the same amount of money as the bottom 50% of the Earth's population in 2010. The charity's report also said that the richest 1 percent of the population will own more than half the world's wealth by 2016. Oxfam said the world's richest people saw their share of global wealth jump to 48 percent last year from 44 percent in 2009.

The biggest impact of such extreme financial policies has been on the environment. Greenfact (2012) lists theme areas that are most affected. They are

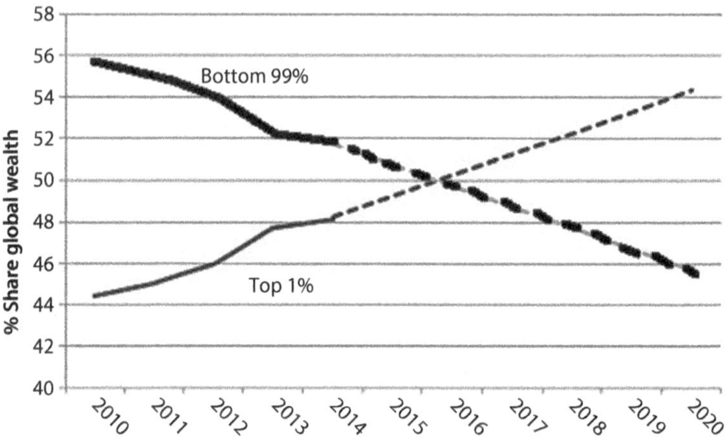

Figure 2.4 Past and future projection of shares of global wealth of the top 1% and bottom 99% (from Oxfam, 2015).

2.3.1 Agriculture and Development

Most affected aspects: biodiversity loss, global warming and water availability. The cause of such disastrous results are often overlooked. For instance, the role of chemical fertilizer and pesticide is rarely linked with loss in biodiversity, instead resorting to blaming over cultivation and deforestation (Chhetri and Islam, 2008). Similarly, global warming is not deemed connected to refining and other processes that add numerous artificial chemicals to the otherwise sustainable product (Islam et al., 2012). Water availability is rarely linked to chlorination, pollution from factory farms, industrial plants, and activities like fracking that uses artificial chemicals (Islam, 2014).

2.3.2 Desertification

Desertification is a huge problem. Drought is considered to be one of the biggest causes of famine and starvation all over Africa and South America. Because of climate changes, urbanization, deforestation and pollution, everyday thousands of acres of arable land are disappearing. In the mean time, less than 20% of many countries "arable" land is being used. While this topic is well talked about, most evade any tangible solution to the problem (Islam, 2015).

2.3.3 Ecosystem change

Any interference with the ecosystem with unsustainable technology will render irreversible change in the ecosystem (Khan and Islam, 2012). The only true model of sustainability to be found is in Nature. A truly sustainable process conforms to connected and/or underlying natural phenomena, in source and pathway. Scientifically, this means that true long-term considerations of humans should include the entire ecosystem. Some have called this inclusion 'humanization of the environment' and put this phenomenon as a precondition of true sustainability (Zatzman and Islam, 2007). The inclusion of the entire ecosystem is meaningful when the natural pathway for every component of the technology is followed. Only such design can assure both short-term (tangible) and long-term (intangible) benefits.

2.3.4 Fisheries

Environmental degradation due to unsustainable practices in fisheries serves as a great reminder as to how a seemingly infinite resource can be turned into a calamity. It is not an understatement to say that, for the first 470 years, the harvesting of these resources posed little or no threat either to the marine environment nor to the present or future prospects of the coastal communities most involved in this activity. However, in the last 30 years of that half-millennium, what remained was literally raped from stem to stern at unprecedented speed (Zatzman, 2013). The historical exegesis brings out in striking manner how far off-base both the promoters of this fishery and its critics actually were with regard to the conduct of this fishery in modern economic conditions of vertically-integrated resource extraction. None of them manifested the slightest awareness of how this fishery could have averted the dramatic collapse that eventually destroyed the livelihood of the families of more than 40,000 commercial fishermen from the Canadian provinces of Newfoundland and Labrador, Quebec and Nova Scotia after 1992. This dialogue of the deaf was manifest not only in the late 1970s — as the struggle over the northwest Atlantic fisheries' future heated up to become one of the sideshows of the global confrontation between the U.S. and Soviet superpowers over control of the world's oceanic spaces. The same thinking that failed to address the problems of that time was being repeated 30 years later by some of the most vociferous critics of the antics of the trawling fleets, Canadian and foreign, back in the 1970s. In 1992 the Canadian government declared a moratorium on the

Northern Cod fishery, which for the past 500 years had largely shaped the lives and communities of Canada's eastern coast. Today, there isn't an alternate model that can restore the fisheries industry.

2.3.5 Deforestation

The world's forestry plays a crucial role in maintaining balance. They provide renewable and sustainable raw materials (including medicine) and energy, maintain biological diversity, mitigate climate change, protect land and water resources, provide recreation facilities, improve air quality and help alleviate poverty. Each of the manmade activities in the post-renaissance period affect the forest in an irreversible manner.

In view of competing interests in the benefits of forest resources and forest land, the Food and Agriculture Organization of the United Nations has carried out global forest resources assessments at five to ten year intervals since 1946. The most recent and most extensive assessment was completed in 2005 and aimed at measuring progress towards sustainable forest management. The assessment focused on six themes representing important elements of forest management:

- Extent of forest resources
- Biological diversity
- Forest health and vitality
- Productive functions of forest resources
- Protective functions of forest resources
- Socio-economic functions

Information was collected from 229 countries and territories for three points in time: 1990, 2000, and 2005. None of these reports yielded any scientifically sound solution to the problem of deforestation (Khan and Islam, 2007). Claims have been made that by adopting the concept of sustainable forest management as a reporting framework, it is possible to provide a holistic perspective on global forest resources, their management and uses. However, in absence of a truly sustainable technology development schemes, these are but hollow claims. The current scenario is dismal. Despite their immense value, nearly half of the world's forests have been lost. What's worse, we're cutting them down at greater rates each year to plant crops, graze cattle and generate income from timber and other forest products. This deforestation affects immediately the climate change scenario as 11% of unnatural climate change is evoked by deforestation. This amount is similar to the amount of CO_2 emission from all cars and trucks on Earth combined. Then there is an estimated 50% tropical protected areas that are 'empty'. These "empty forests" contain trees but few animals as a result of overexploitation and uncontrolled hunting. As a result, animal species are in danger of extinction, tree species lose important seed dispersal, and local people lose an important supply of protein. Overall, this translates into extraordinary imbalance for the ecosystem.

2.3.6 Marine Litter

Marine littering is synonymous with the plastic culture that is just over 100 year old. Just like miniscule amount of Freon discharge in the environment can trigger gapping hole in the ozone layer, disposal of plastic can invoke marine disasters. Global production of plastics is increasing every year ever since the plastic revolution began (Figure 2.5). This plastic finds its way into the environment and into the oceans. Because plastic doesn't degrade or assimilate into the environment, the plastic content in the ocean increases proportionately. This increase can trigger many instances of imbalance, including CO_2 not being absorbed by the ocean.

About 30 percent of the carbon dioxide that people have put into the atmosphere has diffused into the ocean through the direct chemical exchange. In absence of absorption by the plants, dissolving carbon dioxide in the ocean creates carbonic acid, which increases the acidity of the water. Since 1750, the pH of the ocean's surface has dropped by

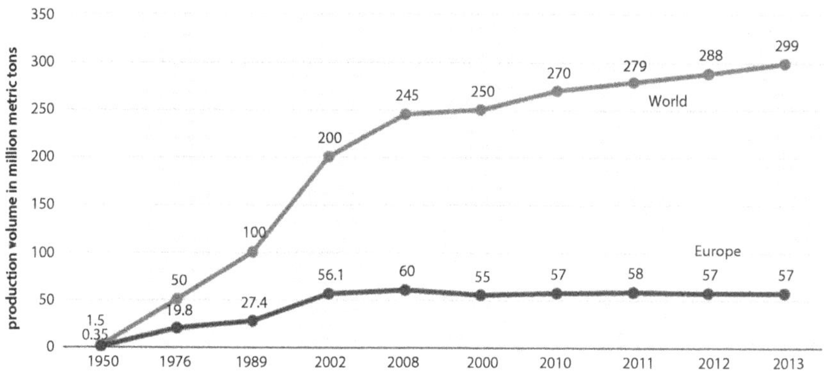

Figure 2.5 Global plastic production (data from http://www.statista.com/).

0.1, a 30 percent change in acidity. Such a change is drastic with numerous consequences.

The widespread occurrence of large plastic fragments in the sea and the direct impact this can have both on marine fauna and on legitimate uses of the environment has been well documented. In recent years the existence of smaller plastic particles referred to as micro-plastics and their potential impact has received increasing attention. This concerns particles smaller than 5 mm, and there is increasing evidence that such particles can be ingested by marine organisms, and could harm them.

2.3.7 Water Resources

According to Bertrand Russell, "Western philosophy begins with Thales. Thales' most famous philosophical position was his cosmological thesis, which comes down to us through a passage from Aristotle's Metaphysics. In the work Aristotle unequivocally reported Thales' hypothesis about the nature of matter – that the originating principle of nature was a single material substance: water. Islam et al. (2014) cited Quranic verses (11:7) that confirm water to be the original creation.

And it is He who created the heavens and the earth in six days - and His Throne had been upon water - that He might test you as to which of you is best in deed. But if you say, "Indeed, you are resurrected after death," those who disbelieve will surely say, "This is not but obvious magic." (Qur'an 11:7)

"Allah (Exalted is He) created seven heavens, one above another, and seven earths, one below another. [The time it takes to travel] from the highest earth to the lowest heaven is five hundred years, and the distance between each heaven and the next is also a journey of five hundred years. There is water above the seventh heaven, and the Throne of the All-Merciful One ['Arsh ar-Rahman] is above the water. Allah (Exalted is He) is upon the Throne, and beneath Him there are seventy thousand veils of light, of darkness, and of what else He knows best." (Hadith from Abu-Dawood)

Just like universe began with water, life also emerged from life. Water makes life as we know it possible. Every drop cycles continuously through air, land, and sea, to be used by someone (or something) else "downstream." Water covers 70% of Earth's surface, but only 3% is fresh, and only a fraction of one percent supports all life on land. Climate change and growing populations are increasing the pressures on that reserve.

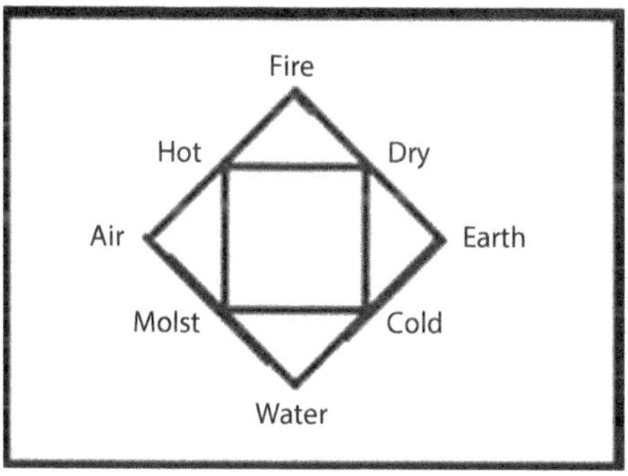

Figure 2.6 Aristotle's four-element phase diagram (steady-state).

Figure 2.6 characterizes matter and energy in four elements, but make them function of only composition, meaning one can move from water (cold and moist) to fire (hot and dry) by merely changing the composition of various elements. Similarly, by changing the properties one can introduce change in compositions. Consider water being the source as well as end of every mass transfer. If this cycle is broken, we create imbalance.

In every culture water is synonymous with life and liveliness. Opposite to water is fire. Water is the source while fire is the essence of dynamics, the process that sustains life. The role of fire is opposite to water, yet it is essential to life. Without fire, there is no carbon dioxide, the essence of plant, and therefore, life. Fire represents transition from cold to hot, from life to death, from tangible (water or liquid) to intangible (vapor or gas). This phase change is typical of every living object. In fact, the very fact that everything is moving (a function of time) makes it essential to go through this phase of tangible and intangible. Overall, this continues in an eternal circle. Figure 2.7 shows how it is natural to have such dual characteristic in any object.

The past few decades have witnessed a fundamental shift in public awareness of and concern about the threats to water resources and surrounding ecosystems. But when it comes to policy, little has changed. Most decisions about the management of water resources remain the product of economic criteria and politically charged reasoning – regardless of whether they concern a town, a region, a country or even several countries. Despite repeated calls from world experts, we continue to move in a wrong direction. Water pollution today is justified through 'water management',. This has an overwhelming impact on the

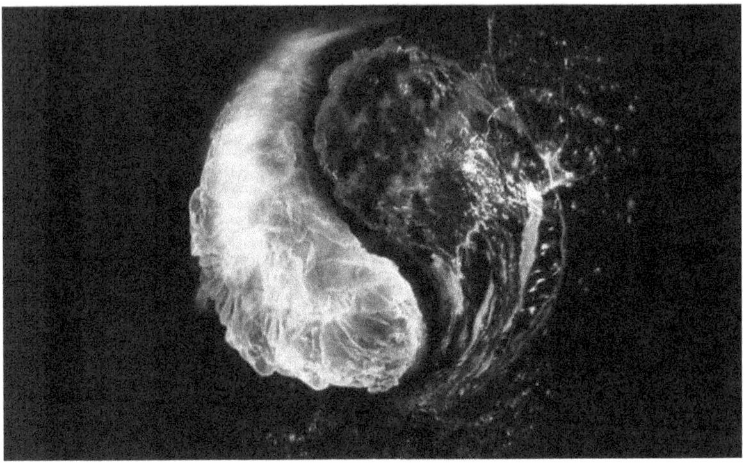

Figure 2.7 Water and fire are depicted through *taegeuk* (*yin yang*).

ecosystem, even more acute than what DDT and GMO has on the environment

The factors are considered to be the most important ones affecting water resources:

- population growth, particularly in water-short regions;
- movement of large numbers of people from the countryside to towns and cities;
- demands for greater food security and higher living standards;
- increased competition between different uses of water resources, and
- pollution from factories, cities, and farmlands.

Climate change and natural variability in the distribution and occurrence of water further complicate the sustainable development of our water resources. Once again, environmental insult began with commercialization of the insult. In a paper published in 1894, it was formally proposed to add chloride to water to render it "germ-free." Two other investigators confirmed this proposal and published it in many other papers in 1895. Early attempts at implementing water chlorination at a water treatment plant were made in 1893 in Hamburg, Germany, and in 1897 the town of Maidstone, England was the first to have entire water supply treated with chlorine. Permanent water chlorination began in 1905, when a faulty slow sand filter and a contaminated water supply led to a serious typhoid fever epidemic in Lincoln, England. Ever since chlorination has become a symbol of civilization. This event is synonymous with the 'plastic culture'.

Ever since, ozonation is used by many European countries and also in a few municipalities in the United States and Canada. Disinfection with chloramine is also becoming increasingly common. The issue of undesirable byproducts is less serious with chloramine than chlorine (gas or hypochlorite). Unlike chlorine, chloramine has a longer half life in the distribution system and still maintains effective protection against pathogens. The reason chloramines persist in the distribution is due to the relatively lower redox potential in comparison to free chlorine. Chloramine is formed by the addition of ammonia into drinking water to form monochloramine and/or dichloramine. Whereas Helicobacter pylori can be many times more resistant to chlorine than Escherichia coli, both organisms are about equally susceptible to the disinfecting effect of chloramine

UV radiation to disinfect offers a new genre of contamination that essentially moves from chemical pollution to energy pollution. Because New science has already disconnected energy from mass, it becomes impossible to compare the environmental impact caused by chlorine and UV (Khan and Islam, 2012).

The next series of planned water pollution is through water fluoridation. The pretext here is that fluoride will prevent dental cavities, leading to savings in dental care (Chhetri and Islam, 2008). Typically a fluoridated compound is added to drinking water, a process that in the U.S. costs an average of about $1.02 per person-year. Defluoridation is needed when the naturally occurring fluoride level exceeds recommended limits. A 1994 World Health Organization expert committee suggested a level of fluoride from 0.5 to 1.0 mg/L (milligrams per litre), depending on climate. Bottled water typically has unknown fluoride levels, and some domestic water filters remove some or all fluoride.

In Europe, only Ireland (73%), Poland (1%), Serbia (3%), Spain (11%), and the U.K. (11%) fluoridate any of their water. Most developed countries, including Japan and 97% of the western European population, do not consume fluoridated water. In the U.S., about 70% of public water supplies are fluoridated. This equates to approximately 185 million people, which is over half the number of people drinking artificially fluoridated water worldwide.

In terms of planned contamination of water, aspirin would come next. The popularity of aspirin declined after the market releases of paracetamol (acetaminophen) in 1956 and ibuprofen in 1969. In the 1960s and 1970s, John Vane and others discovered the basic mechanism of aspirin's effects, while clinical trials and other studies from the 1960s to the 1980s established aspirin's efficacy as an anticlotting agent that reduces the risk of clotting diseases. This followed revival of Aspirin sales well into the 21st century. The scheme has become so successful that today, the following question is a serious one:

"Should governments add low doses of aspirin to water supplies to save billions in healthcare costs? Or give free pills away?"

Then there are numerous sources of pollution that are unplanned. One such source, namely, birth control pills and other medications have been making headlines over the last decade. It is well known that trace amounts of birth control and other medications—as well as household and industrial chemicals of every stripe—are present in many urban and suburban water supplies around the world as long as there is a water treatment plant or the ground water is in communication with the surface water. While this would cause alarm because no medication is naturally degradable, the debate revolved around whether their levels are high enough to warrant concern. For instance, in 2008 the U.S. Geological Survey (USGS) tested water in nine states across the country and found that 85 man-made chemicals, including some medications, were commonly slipping through municipal treatment systems and ending up in our tap water. Another report by the Associated Press found trace amounts of dozens of pharmaceuticals in the drinking water supplies of some 46 million Americans. But according to USGS, such chemicals and medications are so diluted—at levels equal to a thimble full of water in an Olympic-sized swimming pool—that they do not pose a health threat. Scientific investigation, however, suggests that even extremely diluted concentrations of drug residues harm fish, frogs and other aquatic species, and have been shown to labs to impair human cell function. This effect cannot be tracked with conventional engineering analysis that relies heavily on fundamental premises that are based on misconceptions (Islam et al., 2010; Khan and Islam, 2012).

One of the common culprits is estrogen, much of which is inadvertently released into sewers through the urine of women taking birth control. Studies have shown that estrogen can wreak reproductive havoc on some fish, which spawn infertile offspring sporting a mixture of male and female parts. Researchers at the University of Pittsburgh found that human breast cancer cells grew twice as fast when exposed to estrogen taken from catfish caught near untreated sewage overflows. Today's science doesn't allow us to identify, let alone quantify risks associated to such artificial chemicals. What we know for fact is the number of breast cancer has skyrocketed in countries that use estrogen the most. Even though USA showed some decline in cancer rate, it is thought to be because of reduced usage of Hormone Replacement Therapy (HRT).

2.3.8 Climate Change

Climate change is considered to be the equivalent of global obesity (Islam et al., 2012) that captures the overall environmental insult caused by the modern day practices. In the late decades of the 20th Century, the term global warming became in vogue and substantial government funding became available for research in this area. As an interesting note, substantial government funding was available for global cooling in the preceding decades to the global warming phenomenon. At the time, living in Northern Alberta, Canada (which is still subject to cold winters) there were cries of beware, danger in the streets from those scientists who continually advocated the coming of the glaciers – another Ice Age was about to arrive. Within the recent decades, the cause of global cooling has been diminished and the term global warming took over. However, as late as 2010, Chilingar et al. (2010) claimed that we are well within a cycle of global cooling.

Global warming has been a subject of discussion since the late 1970s. It is thought that the accumulation of carbon dioxide in the atmosphere causes global warming, resulting in irreversible climate change. Even though carbon dioxide has been blamed as the sole cause for global warming, there is no scientific evidence that all carbon dioxides are responsible for global warming. Precisely to address this critical gap, this chapter includes a detailed analysis of greenhouse gas emissions from the Pre-industrial Age to the (for some) "Golden Era" of petroleum. A new theory has been developed, which shows that not all carbon dioxides contribute to global warming.

The climate of the Earth can be affected by natural factors that are external to the climate system, such as changes in volcanic activity, solar output, and the variance sof the orbit of the Earth around the Sun. Of these, the two factors relevant on timescales of contemporary climate change are changes in volcanic activity and changes in solar radiation. In terms of the energy balance of the Earth, these factors primarily influence the amount of incoming energy. Volcanic eruptions are episodic and have relatively short-term effects on climate. Changes in solar irradiance have contributed to climate trends over the past century but since the Industrial Revolution, the effect of additions of greenhouse gases to the atmosphere has also been recognized.

However, as Islam et al. (2010, 2012) recently pointed out, one must understand what is the most damning trait of modern-day technology development for one to discern major causes of climate change. The following events are extracted from history, much of which is discussed by Black (Black, 2013).

1712 - British ironmonger/inventor Thomas Newcomen invents the first widely used steam engine, paving the way for the Industrial Revolution and industrial scale use of coal. At that time, coal was not processed with any artificial chemical.

1800 - World population reaches one billion.

1824 - French physicist Joseph Fourier describes the Earth's natural "greenhouse effect". He writes: "The temperature [of the Earth] can be augmented by the interposition of the atmosphere, because heat in the state of light finds less resistance in penetrating the air, than in re-passing into the air when converted into non-luminous heat."

1846 – Canadian geologist Abraham Gesner invented kerosene making coal and petroleum practical raw materials for lighting fuel. This was the first case of natural oil processing.

1859 - Edwin Drake invented the first modern drilling process for deep oil wells. John Davison Rockefeller founded the Standard Oil Company, which dominated the oil industry and was the first great U.S. business trust. Rockefeller monopolized the petroleum industry.

1861 - Irish physicist John Tyndall shows that water vapour and certain other gases create the greenhouse effect. "This aqueous vapour is a blanket more necessary to the vegetable life of England than clothing is to man," he concludes.

1870 - Rockefeller founded Standard Oil Company and aggressively ran it until he officially retired in 1897.

1878 - Karl Friedrich Benz developed a petrol-powered automobiles.

1879 – Benz obtained a patent for the practical automobile. The invention of the internal combustion engine was the major influence in the rise in the importance of petroleum.

1886 - Karl Benz unveils the Motorwagen, often regarded as the first true automobile.

1892 - German scientist and inventor Rudolf Diesel invented his compression-ignition engine. Diesel originally designed his engine to use coal dust as fuel, and experimented with other fuels including vegetable oils such as peanut oil, which was used to power the engines which he exhibited at the 1900 Paris Exposition and the 1911 World's Fair in Paris.

1896 - Swedish chemist Svante Arrhenius concludes that industrial-age coal burning will enhance the natural greenhouse effect. He suggests this might be beneficial for future generations. His conclusions on the likely size of the "man-made greenhouse" are in the same ballpark - a few degrees Celsius for a doubling of CO_2 - as modern-day climate models.

1900 - Another Swede, Knut Angstrom, discovers that even at the tiny concentrations found in the atmosphere, CO_2 strongly absorbs parts of the infrared spectrum. Although he does not realise the significance, Angstrom has shown that a trace gas can produce greenhouse warming.

1913 - Thermal cracking was invented by William Meriam Burton, a process that employed heat and high pressures.

1927 - Carbon emissions from fossil fuel burning and industry reach one billion tonnes per year. It was mainly from coal.

1930 - Human population reaches two billion.

1937 - The catalytic cracking process was invented by Eugene Houdry. This was the beginning of chemical pollution of the atmosphere at a massive scale.

1938 - Using records from 147 weather stations around the world, British engineer Guy Callendar shows that temperatures had risen over the previous century. He also shows that CO_2 concentrations had increased over the same period, and suggests this caused the warming. The "Callendar effect" is widely dismissed by meteorologists.

By the 1950s - cars were becoming bigger and faster. Gasoline octane increased, and lead was added to improve engine performance.

1955 - Using a new generation of equipment including early computers, US researcher Gilbert Plass analyses in detail the infrared absorption of various gases. He concludes that doubling CO_2 concentrations would increase temperatures by 3-4C.

1957 - US oceanographer Roger Revelle and chemist Hans Suess show that seawater will not absorb all the additional CO_2 entering the atmosphere, as many had assumed. Revelle writes: "Human beings are now carrying out a large scale geophysical experiment..."

1958 - Using equipment he had developed himself, Charles David (Dave) Keeling begins systematic measurements of atmospheric CO2 at Mauna Loa in Hawaii and in Antarctica. Within four years, the project - which continues today - provides the first unequivocal proof that CO_2 concentrations are rising.

1960 - Human population reaches three billion.

1965 - A US President's Advisory Committee panel warns that the greenhouse effect is a matter of "real concern".

1972 - First UN environment conference, in Stockholm. Climate change hardly registers on the agenda, which centres on issues such as chemical pollution, atomic bomb testing and whaling. The United Nations Environment Programme (Unep) is formed as a result.

1975 - Human population reaches four billion.

1975 - US scientist Wallace Broecker puts the term "global warming" into the public domain in the title of a scientific paper.

1976 – Methylcyclopentadienyl manganese tricarbonyl (MMT) is introduced in Canadian gasoline at a concentration up to 18 mg Mn/l and in Australia in 2000.

1987 - Human population reaches five billion

1987 - Montreal Protocol agreed, restricting chemicals that damage the ozone layer. Although not established with climate change in mind, it has had a greater impact on greenhouse gas emissions than the Kyoto Protocol.

1988 - Intergovernmental Panel on Climate Change (IPCC) formed to collate and assess evidence on climate change.

1989 - UK Prime Minister Margaret Thatcher - possessor of a chemistry degree - warns in a speech to the UN that "We are seeing a vast increase in the amount of carbon dioxide reaching the atmosphere... The result is that change in future is likely to be more fundamental and more widespread than anything we have known hitherto." She calls for a global treaty on climate change.

1989 - Carbon emissions from fossil fuel burning and industry reach six billion tonnes per year.

1990 - IPCC produces First Assessment Report. It concludes that temperatures have risen by 0.3-0.6C over the last century, that humanity's emissions are adding to the atmosphere's natural complement of greenhouse gases, and that the addition would be expected to result in warming.

1992 - At the Earth Summit in Rio de Janeiro, governments agree the United Framework Convention on Climate Change. Its key objective is "stabilization of greenhouse gas concentrations in the atmosphere at a level that would prevent dangerous anthropogenic interference with the climate system". Developed countries agree to return their emissions to 1990 levels.

1995 - IPCC Second Assessment Report concludes that the balance of evidence suggests "a discernible human influence" on the Earth's climate. This has been called the first definitive statement that humans are responsible for climate change.

1997 - Kyoto Protocol agreed. Developed nations pledge to reduce emissions by an average of 5% by the period 2008-12, with wide variations on targets for individual countries. US Senate immediately declares it will not ratify the treaty.

1998 - Strong El Nino conditions combine with global warming to produce the warmest year on record. The average global temperature reached 0.52C above the mean for the period 1961-90 (a commonly used baseline).

1998 - Publication of the controversial "hockey stick" graph indicating that modern-day temperature rise in the northern hemisphere is unusual compared with the last 1,000 years. The work would later be the subject of two enquiries instigated by the US Congress.

1999 - Human population reaches six billion.

2001 - President George W Bush removes the US from the Kyoto process.

2001 - IPCC Third Assessment Report finds "new and stronger evidence" that humanity's emissions of greenhouse gases are the main cause of the warming seen in the second half of the 20th Century.

2005 - The Kyoto Protocol becomes international law for those countries still inside it.

2005 - UK Prime Minister Tony Blair selects climate change as a priority for his terms as chair of the G8 and president of the EU.

2006 - The Stern Review concludes that climate change could damage global GDP by up to 20% if left unchecked - but curbing it would cost about 1% of global GDP.

2006 - Carbon emissions from fossil fuel burning and industry reach eight billion tonnes per year.

2007 - The IPCC's Fourth Assessment Report concludes it is more than 90% likely that humanity's emissions of greenhouse gases are responsible for modern-day climate change.

2007 - The IPCC and former US vice-president Al Gore receive the Nobel Peace Prize "for their efforts to build up and disseminate greater knowledge about man-made climate change, and to lay the foundations for the measures that are needed to counteract such change". Later years, Al Gore goes on touring the Middleast in an attempt to soar up nuclear reactor sales on behalf of his state of Tennessee.

2007 - At UN negotiations in Bali, governments agree the two-year "Bali roadmap" aimed at hammering out a new global treaty by the end of 2009.

2008 - Half a century after beginning observations at Mauna Loa, the Keeling project shows that CO_2 concentrations have risen from 315 parts per million (ppm) in 1958 to 380ppm in 2008.

2008 - Two months before taking office, incoming US president Barack Obama pledges to "engage vigorously" with the rest of the world on climate change.

2009 - China overtakes the US as the world's biggest greenhouse gas emitter - although the US remains well ahead on a per-capita basis.

2009 - Computer hackers download a huge tranche of emails from a server at the University of East Anglia's Climatic Research Unit and release some on the internet, leading to the "ClimateGate" affair.

2009 - 192 governments convene for the UN climate summit in Copenhagen with expectations of a new global agreement high; but they leave only with a controversial political declaration, the Copenhagen Accord.

2010 - Developed countries begin contributing to a $30bn, three-year deal on "Fast Start Finance" to help them "green" their economies and adapt to climate impacts.

2010 - A series of reviews into "ClimateGate" and the IPCC ask for more openness, but clear scientists of malpractice.

2010 - The UN summit in Mexico does not collapse, as had been feared, but ends with agreements on a number of issues.

2011 - A new analysis of the Earth's temperature record by scientists concerned over the "ClimateGate" allegations proves the planet's land surface really has warmed over the last century.

2011 - Human population reaches seven billion.

2011 - Data shows concentrations of greenhouse gases are rising faster than in previous years.

2012 - Arctic sea ice reaches a minimum extent of 3.41 million sq km (1.32 million sq mi), a record for the lowest summer cover since satellite measurements began in 1979.

2013 - The Mauna Loa Observatory on Hawaii reports that the daily mean concentration of CO_2 in the atmosphere has surpassed 400 parts per million (ppm) for the first time since measurements began in 1958.

2013 - The first part of the IPCC's fifth assessment report says scientists are 95% certain that humans are the "dominant cause" of global warming since the 1950s.

2014 - Six countries in the world were still continue to use leaded gasoline: Afghanistan, Myanmar, North Korea, Algeria, Iraq and Yemen.

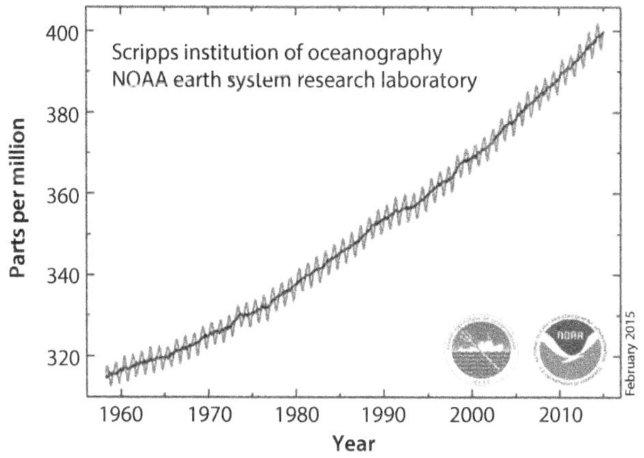

Figure 2.8 CO_2 concentration rise over the years (from NOAA site http://www.esrl.noaa.gov/gmd/ccgg/trends/).

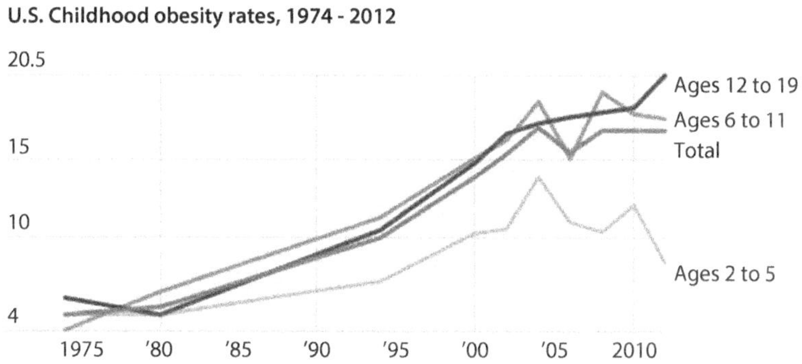

Figure 2.9 Obesity has gone up in USA even though there is little consensus on what causes it or how to remedy
it. Source: CDC National Health and Nutrition Examination Survey.

Overall, there is no consensus as to what causes sharp CO_2 increase in the atmosphere (Figure 2.8) even though recent work of Islam et al. (2010 and 2012) has made it clear that refining technologies are to blame for rendering organic CO_2 into non-organic 'stale' CO_2. The problem is similar to the rise of obesity in USA. There is consensus as to the fact that the obesity has gone up but few agree as to what is causing it (Figure 2.9).

2.4 Sociological Degeneration

Social degeneration is best manifested in the following list of social and environmental insults, all of which are topped by United States – the country that has led the modern-day civilization. Figure 2.10 highlights this point. The problem is broken down to incarceration rate in Figure 2.11 that shows the incarceration rate from around the world.

Figure 2.12 shows the incarceration rate over the years in USA. Figure 2.13 shows different types of crimes that took place over the years. This treacherous statistical condition is further supported by policies that have practically corporatized the justice system.

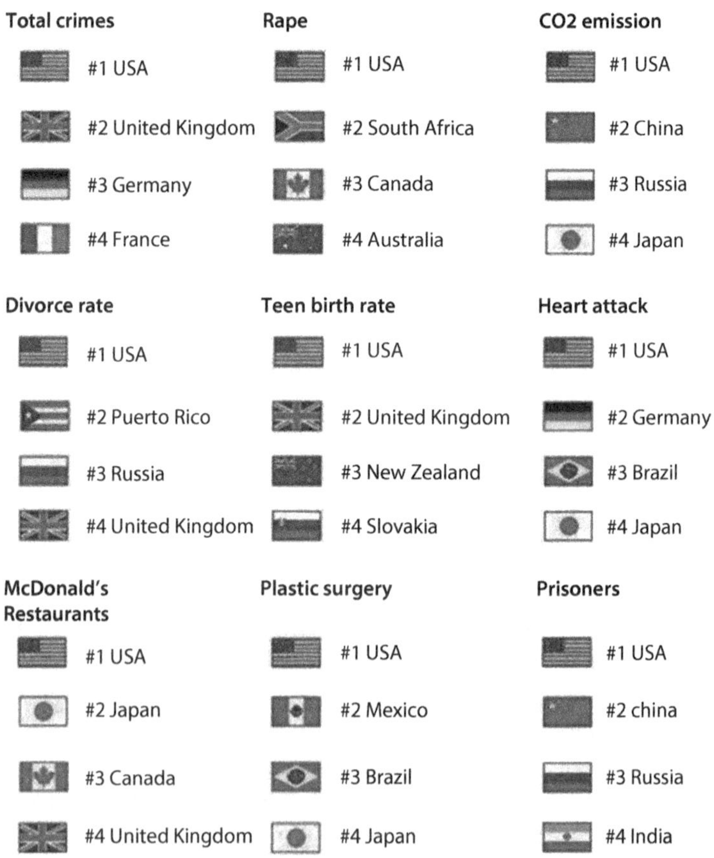

Figure 2.10 USA that leads the world in technology development also leads in practically all degenerations dreaded by mankind.

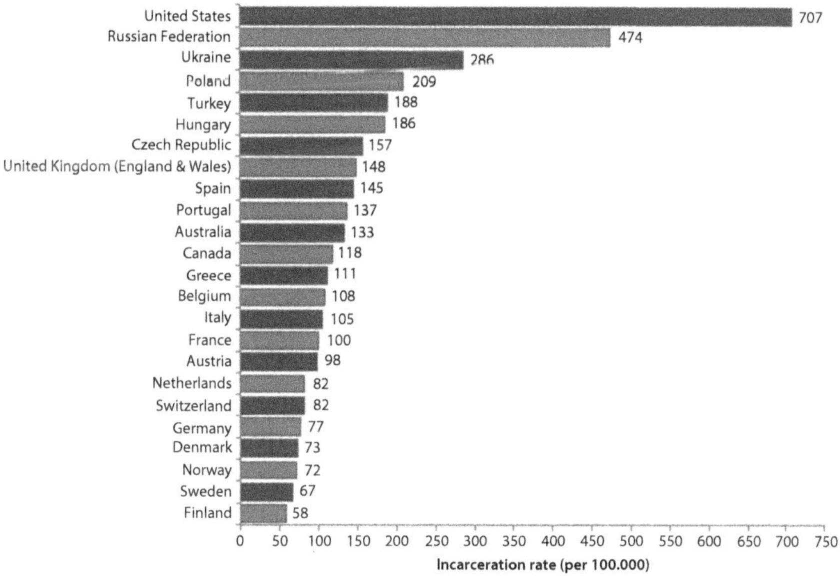

Figure 2.11 incarceration rate around the world (from Badger, 2014).

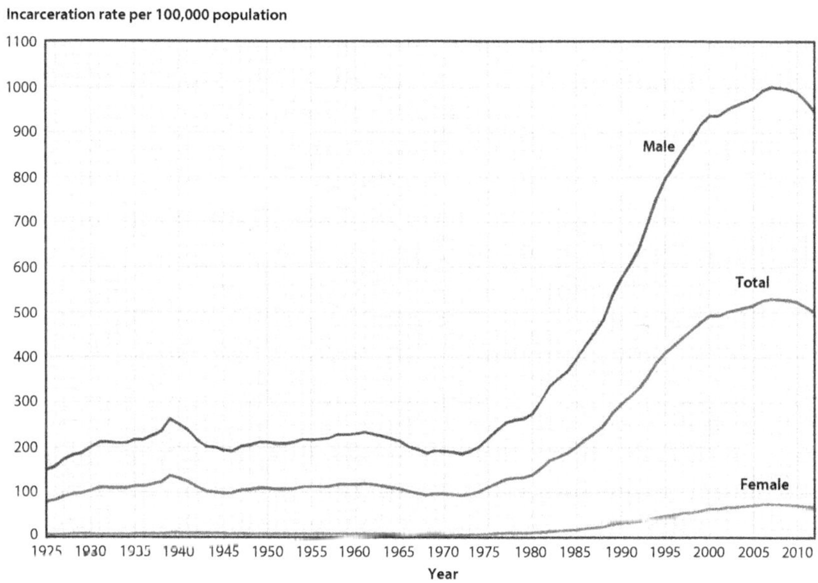

Figure 2.12 Incarceration rate over the years in USA.

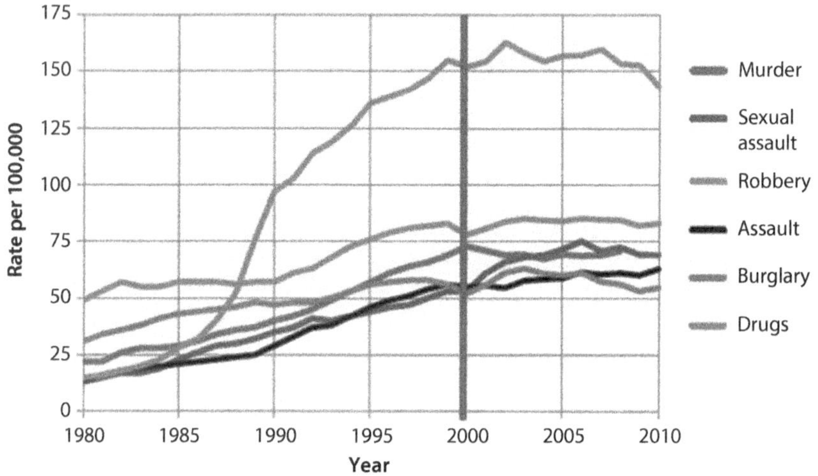

Figure 2.13 Incarceration rate by crime type in USA (from Badger, 2014).

Since the 1970s, Congress and state legislatures have enacted a number of changes to prison and sentencing laws that have mandated prison time for lesser offenses and ensured longer sentences for violent crimes and repeat offenders. The "war on drugs" — a corollary to Lyndon Johnson's "war on crime" — also ensured that drug crimes received more attention from police and harsher punishment in courtrooms.

During the 1980s, Congress began to enact "mandatory minimum" laws for drug crimes and violent offenses. The 1990s brought "three strikes" laws in Congress and more than half of states (giving a third offense mandatory sentences of 25 years or more). "Truth-in-sentencing" laws also required offenders to serve at least 85 percent of their sentences. Effectively, more people were going to jail, carrying with them longer sentences. And increasingly, they were serving almost all of that time.

As a result, between 1980 and 2010, the incarceration rate for drug crimes increased tenfold (Badger, 2014). This has resulted in direct boon for the prison corporations. New York department of correctional Services (DOCS) have led away the US version of this corporatization with \$2.7 billion of prison budget and a total per inmate cost of \$60,076. No other program than Medicaid is so draining on USA tax payers. Still what 'war on drugs' has done to the fiscal system is miniscule compared to what 'war on terrorism' has done in the post 9-11 era. In Guantanamo Bay, the average cost per inmate is around \$800,000 per year. This number is symbolic to what corporatization can do to a civilization. Scott Stringer, the comptroller who put together a report on the prison system, summarized the current status and the fate of corporatization when he said to an interview to the Time magazine: "We're spending more money on inmates and we're getting worse results."

The country that stands at the helm of modern civilization has yet another burden to carry. The United States has a higher percent of imprisoned minorities than any other country in the world. Comparing other countries with similar percentages of immigrants, Germany has a rate of 87 per 100,000 (as of 2012), Italy has a rate of 113 per 100,000 (as of 2010), and Saudi Arabia has a rate of 178 per 100,000 (as of 2009). Comparing other countries with a zero tolerance policy for illegal drugs, the rate of Russia is 577 per 100,000, the rate of Kazakhstan is 400 per 100,000, the rate of Singapore is 273 per 100,000 and the rate of Sweden is 78 per 100,000.

In the United States, women make up more than one tenth of the whole prison population. In most countries, the proportion of female inmates to the larger population is closer to one in twenty. Australia is the exception where the rate of female imprisonment increased from 9.2 percent in 1991 to 15.3 percent in 1999.

In addition, the United States has very abnormal statistics when observing the racial dimension of mass incarceration. According to Michelle Alexander, the United States "imprisons a larger percentage of its black population than South Africa did at the height of apartheid". According to the 2010 census of the US Census Bureau, Black Americans are the largest racial minority, comprising 13.6% of the population. Hispanic and Latino Americans comprise 16.3% of the population according to the 2010 US census, making up the largest ethnic minority. White, non-Hispanic or Latino population comprises 66% of the nation's total.

According to the US Bureau of Justice Statistics (BJS) non-Hispanic blacks accounted for 39.4% of the total prison and jail population in 2009. Hispanics (of all races) were 20.6% of the total jail and prison population in 2009. The Northeast has the highest incarceration rates of Hispanics in the nation. Connecticut has the highest Hispanic-to-White ratio with 6.6 Hispanic males for every white male. The national average Hispanic-to-White ratio is 1.8. Other states with high Hispanic-to-White ratios include Massachusetts, Pennsylvania, and New York.

In 2010 black non-Hispanic males were incarcerated at the rate of 4,347 inmates per 100,000 U.S. residents of the same race and gender. White males were incarcerated at the rate of 678 inmates per 100,000

U.S. residents. Hispanic males were incarcerated at the rate of 1,755 inmates per 100,000 U.S. residents. For female rates see the table below.

However, black majority cities have similar crime statistics for blacks as do cities where majority of population is white. For example, white majority San Diego has a slightly lower crime rate for blacks than does Atlanta, a city which has black majority in population and city government.

Census data for 2000, which included a count of the number and race of all individuals incarcerated in the United States, showed for each state that the proportion of blacks in prison populations exceeded the proportion of whites among state residents in every state. In twenty states, the percent of blacks incarcerated was at least five times greater than their share of resident population.

Islam et al. (2015) argued that such statistics on crime is a sign of social degeneration and overall indicative of the fact that the mental health of the most civilized nation on Earth is in fact the biggest liability. For anyone to propose a remedy, one must address the problem with a holistic approach.

2.5 The Deadliest 10 Diseases

Figure 2.14 shows the history of leading causes of death in United States. Even though the death rate due to cancer is increasing, heart disease remains the leading cause of death, followed by cancer. Share of death by heart diseases peaked during 1950 through 1980, synchronizing with the culture of sugar. After 1980s, cancer deaths started increase to become almost equal to heart disease deaths. This would coincide with the culture of artificial sugars, i.e., saccharine, aspartame, etc.

Figure 2.14 also indicates that heart disease, cancer, and stroke were among the five leading causes every year between 1935 and 2010. This is the time period that has seen the surge in artificial products, including medicines to combat the very same diseases.

Figure 2.15 shows the trend in developing new medicines dealing with cancer. It is no exaggeration to state that these medicines themselves contributed to worsening the conditions of the disease.

2.5.1 Corporatization and Healthcare

The golden era of plastic and artificial chemicals is synonymous with spending money on medicines before even discovering the cause of the

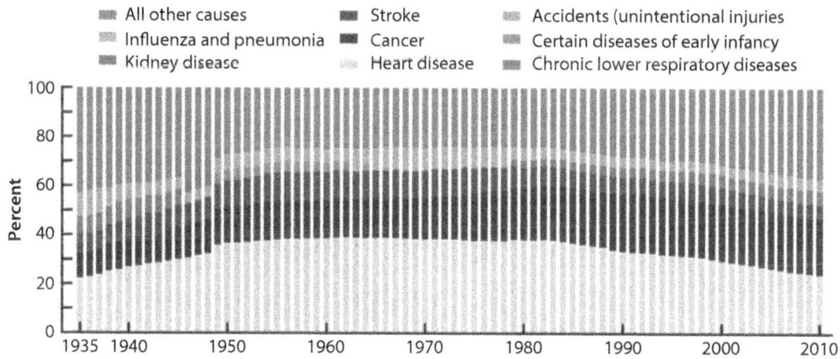

Figure 2.14 History of leading causes of death in United States. NOTE: 2010& data are preliminary. SOURCE: CDC/NCHS, National Vital Statistics System, Mortality.

Figure 2.15 New medicines in cancer treatment.

disease. In 1902 governments in the United States spent 0.25 percent of GDP on health care programs. In the early 21st century, governments spend over 7 percent of GDP on health care programs. Health care spending increased rapidly during the second half of the 20th century, most of which related to medicine and diagnostic tools.

Figure 2.16 shows percentage of GDP spent on healthcare throughout 1900s all the way up to 2020 (predicted for 2011-2020). Health care spending started out at the beginning of the 20th century at 0.25 percent of Gross Domestic Product (GDP). It increased slowly during the first half of the century, peaking at one percent of GDP in 1933 and then declining to 0.38 percent of GDP in World War II. It took until 1961 for health care spending to return to 1 percent of GDP. The federal government spent about 0.24 percent of GDP, state governments

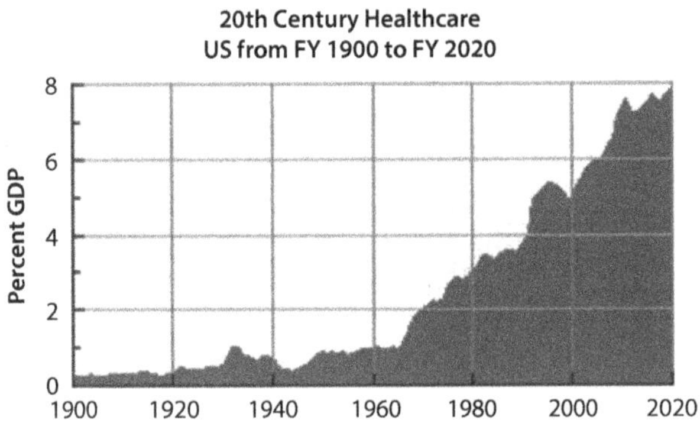

Figure 2.16 Percentage of GDP spent on health care in USA.

spend about 0.4 percent of GDP, and local government about 0.36 percent of GDP. In 1965 Congress passed Medicare, the federal health care program for Americans over 65 years old, and Medicaid, the joint federal-state health care program for the poor, and ever since health care spending has consistently grown much faster than GDP. The year that Medicare and Medicaid were passed, government health care spending at the three levels of government was: federal 0.24 percent GDP; state 0.4 percent GDP; local 0.36 percent GDP; about one percent of GDP overall. By 1970 the total health care spending had doubled to 2 percent of GDP, with federal 1.13 percent GDP including 0.25 percent GDP transferred to the states, state 0.70 percent GDP, local 0.45 percent GDP.

This increased health care spending drastically, reaching 2 percent of GDP in 1970 and 3 percent in 1980.

In 1980 the total spending on health care reached 3 percent of GDP, with federal at 1.93 percent GDP (with 0.5 percent GDP transferred to states); states at 1.02 percent GDP, and local at 0.59 percent GDP. During Reagan era in USA, the increase in health care spending slowed down somehow, still reached 4 percent of GDP in 1990. By 1990 total health care spending reached 4 percent of GDP, with federal at 2.61 percent GDP (with 0.8 percent GDP transferred to states); states at 1.28 percent GDP, and local at 0.65 percent GDP. During the Bush-Clinton era, healthcare cost increased rapidly in the early 1990s, reaching 5 percent of GDP in 1993 and peaking at 5.3 percent GDP in 1995. This was the Clinton era that was supposed to benefit from the end of cold war and significant funds were being allocated away from the most costly scheme, viz., Defence. However, the savings were hardly tangible in terms of health care cost, even though spending decreased slightly in the late 1990s, down to 5 percent of GDP in 2000. Rapid growth in health care spending resumed in the 2000s, reaching 6 percent of GDP in 2005 and 7 percent of GDP in 2009. In 2015 health care spending is estimated to be 7.54 percent GDP, reaching almost 7.9 percent of GDP by 2020.

In 2000 the total spending on health care reached 5 percent of GDP, with federal at 3.42 percent GDP (with 1.22 percent GDP transferred to states); states at 2.05 percent GDP, and local at 0.68 percent GDP. By 2010 total spending on health care reached 7 percent of GDP, with federal at 5.5 percent GDP (with 1.95 percent GDP transferred to states); states at 2.9 percent GDP, and local at 0.87 percent GDP.

In the 2010s the increase in health care spending continued, with federal spending for 2015 estimated at 5.7 percent of GDP (and 1.98 transferred to states); states at 3.1 percent GDP, and local declining to 0.77 percent GDP. By 2020, federal health care spending is expected at 5.95 percent of GDP, state at 2.93 percent and local at 0.77 percent of GDP.

It is worth noting that healthcare cost is the most draining one in US economy. In terms of defense spending, it stood at 6.8 percent of GDP at the height of the Reagan defense buildup. However, beginning even before the breakup of the Soviet Union it began a decline, reaching below 6 percent in 1990, below 4 percent in 1996 and bottoming out at 3.5 percent of GDP in 2001, about half the level of 1985. The 9/11 terrorist attack on US buildings in 2001 changed all that, and defense spending began a substantial increase in two stages. First, it increased to 4.6 percent by 2005 for the invasion of Iraq, and then to 5.0 percent in 2008 for the "surge" in Iraq. This trend continued for the Obama era and increased further to 5.7 percent in 2011 with the stepped up effort in Afghanistan.

The correlation between defence spending and healthcare cost is not trivial. However, repeatedly it has been asserted that these two spendings compete with each other and that doesn't hold true. There is another all pervasive misconception exists, that is, USA doesn't spend enough on healthcare, thereby abandoning its care for poorer community. The reality speaks otherwise. Figure 2.17 shows percentage GDP spent by various countries. In terms of percentage GDP, USA is by far the leading country. Health spending accounted for 16.9% of GDP in the United States in 2012 - the highest share among OECD countries and more than 7½ percentage points above the OECD average of 9.3%. In contrast to most OECD countries, health spending in the United

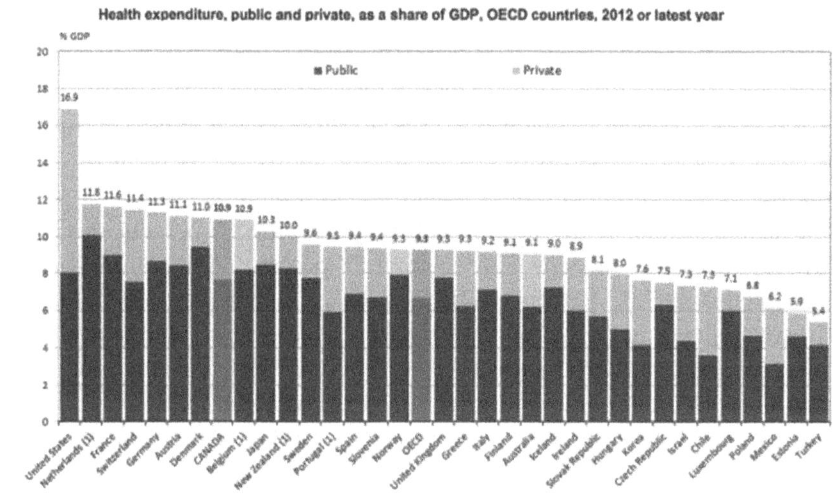

Figure 2.17 Spending in healthcare by various countries. Source: *OECD Health Statistics 2014.*

States is split evenly between public and private sources. In 2012, 48% of health spending in the United States was publicly financed, well below the average of 72% in OECD countries.

The USA is not lacking in spending. What it is lacking is efficiency. Figure 2.18 shows one aspect of this efficiency, i.e., availability of a medical doctor for the community. While the U.S. has 2.4 practicing physicians per 1,000 people, the OECD average is 3.1. Most OECD countries have an average of 3.4 beds, but the U.S. has only 2.6 for every 1,000 people. Similar inefficiency exists within pharmaceutical and hospital sectors. Health spending in the United States has been on a significant slowing trend, primarily due to price effects in the pharmaceutical and hospital sectors. The slowdown predates the economic crisis, but decelerated further to around 2% in 2011 and 2012. This lower rate growth remains higher than the OECD average which includes some European countries that made significant reductions in health spending.

Case in point – the country's most expensive hospital is located in Bayonne, NJ, where it costs nearly $100,000 to treat a case of chronic lung disease – five times as much as what most others hospitals charge for the same procedure.

TIME journalist Steven Brill's investigative piece (Brill, 2013), "Bitter Pill: Why Medical Bills Are Killing Us", explored the inefficiency of America's healthcare system. He evaluated hospital bills to determine why hospitals charge what they charge. What Brill found was that

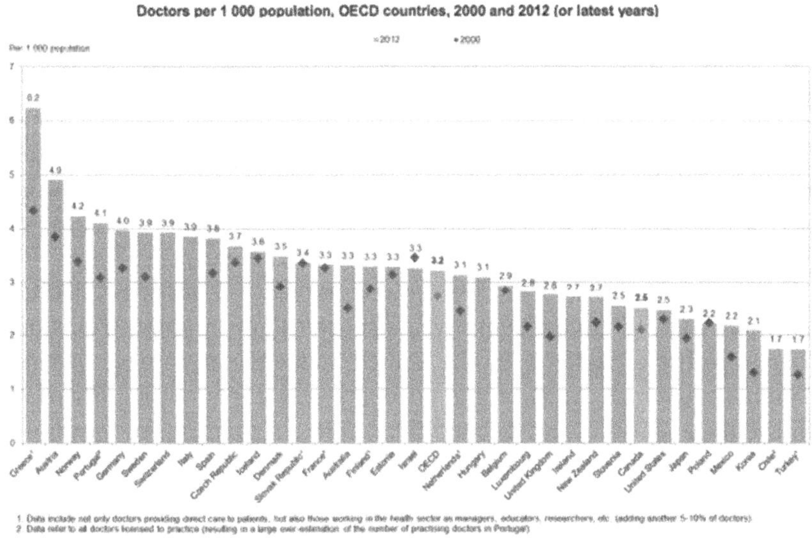

Figure 2.18 Availability of doctors for various countries.

hospitals were charging outrageous prices for everyday items – $1.50 for a generic version of a Tylenol pill, which one can buy 100 of on Amazon for $1.49 as well as overpriced gauze pads, hospital gowns and more. Even with the more complicated medicines, hospitals were still charging well beyond what they would need to make even a reasonable profit from the treatment. What was to blame? The chargemaster – a massive computer file that lists every product and service the hospital offers and how much it costs. These are the frightening "list" prices that uninsured patients see on their bills. When Brill asked hospital administrators why chargemasters listed such preposterous prices, they could only tell him, "It's complicated." What is not complicated is the profit these "nonprofit" hospitals are making. Billions. Hospital CEOs, take home salaries on par with any Wall Street executive – well into the six-figure range and often beyond.

In line with many other OECD countries, pharmaceutical spending in the United States has been on a downward trend, after years of stronger growth. The reduction of 1.1% in 2012 was mainly due to price effects caused in part by the loss of patent protection for some blockbuster drugs, and the resulting increase in the share of cheaper generics in the market. All these point to a complicated scenario resulting from corporatization of healthcare.

As further evidence of systemic inefficiency, one can cite life expectancy as well as overall health status. In 2011 (latest year available), life expectancy in the United States stood at 78.7 years, 1 ½ year less than the OECD average of 80.2 years. While life expectancy in the United States has increased by two years since 2000, this was less than the three year gain registered across OECD countries. This has widened the gap. The gap between the United States and leading countries has also widened. For example, the life expectancy for U.S. men in 2011 was more than 4 years shorter than in Switzerland (up from 3 years in 2000); for U.S. women, it was more than 5 years shorter than in Japan. The slower progress in life expectancy in the United States is due to gaps in health insurance coverage and proper primary care, poorer health-related behaviours and poor living conditions for a significant proportion of the U.S. population. Monetary inequality has created a symptom of social divide that spilled into lifestyle and health. Such inequality cannot be addressed with taxing the rich, by increasing overall expenditure of healthcare or social welfare.

Little has been researched as to the cause of deteriorating overall public health. For instance, United States has achieved remarkable progress in reducing the proportion of adults who smoke tobacco, with the rate of daily smokers coming down from 19% in 2000 to 14% in 2012. This is the lowest rate among OECD countries after Mexico, Sweden and Iceland. However, at the same time, obesity rates among adults in the United States have increased greatly to reach 35.3% in 2012, up from 30.9% in 2000 (based on actual measures of height and weight). This is the highest rate among OECD countries. As is the case in several other countries, the obesity rate in the United States tends to be higher among disadvantaged socio-economic groups, especially in women. Mortality from diseases including cardiovascular diseases and many cancers increases progressively once people become obese. Cigarettes used to be easy target for determining cause of mortality. However, little can now be said about cigarettes as the main cause of death with all indications pointing toward overall lifestyle, including food, drink, type of energy as the combined culprit.

2.5.2 Death, Health, and Lifestyle

Figure 2.19 shows number of deaths, along with crude and age-adjusted death rates for the time period of 1935-2010. The risk of dying decreased for all age groups but was greater for younger age groups

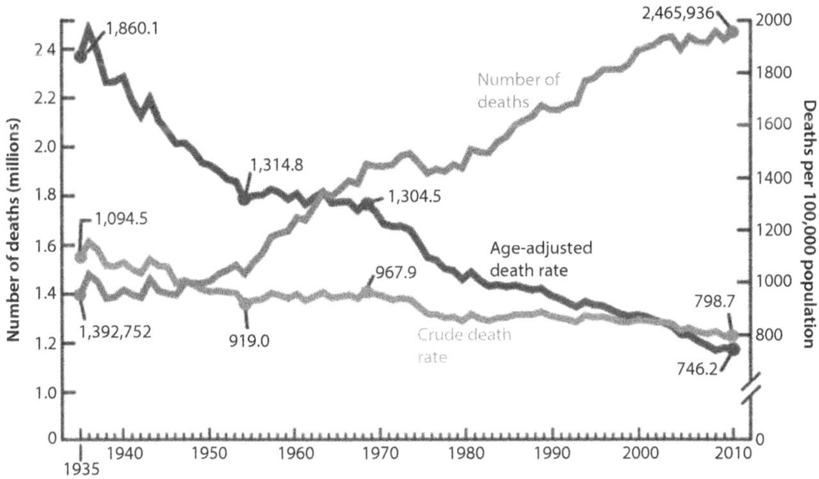

Figure 2.19 Number of deaths, crude and age-adjusted death rates: United States, 1935–2010 (NOTES: Crude death rates on an annual basis are per 100,000 population; age-adjusted rates are per 100,000 U.S. standard population. Rates for 2001–2009 are revised and may differ from rates previously published. (Source: CDC/NCHS, National Vital Statistics System, Mortality.)

with a 94 percent reduction in death rates at 1–4 years compared with a 38 percent decline at 85 years or more.

Interestingly, the reduction in death doesn't say much about quality of life as there is a different trend in terms of suicide rate. This is important because suicide itself is an indicative of (or lack of) mental health. The Centers for Disease Control and Prevention (CDC) collects data about mortality in the U.S., including deaths by suicide. In 2013, 41,149 suicides were reported, making suicide the 10th leading cause of death for Americans (Figure 2.20). After cancer and heart disease, suicide accounts for more years of life lost than any other cause of death.

From 1986 to 2000, suicide rates in the U.S. dropped from 12.5 to 10.4 suicide deaths per 100,000 people in the population. Over the next 12 years, however, the rate generally increased and by 2013 stood at 12.6 deaths per 100,000. No data is collected on the cause of suicide but this must be correlated with the number of mental patients.

Suicide death rates vary considerably among different groups of people. The CDC reports suicide rates by four key demographic variables: age, sex, race/ethnicity, and geographic region/state. Figure 2.21 shows suicide rate by age. In 2013, the highest suicide rate (19.1) was among people 45 to 64 years old. The second highest rate (18.6) occurred in

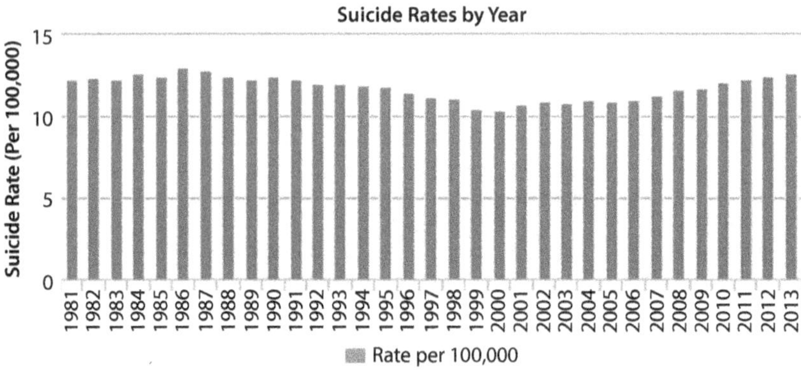

Figure 2.20 Suicide rate history of United States.

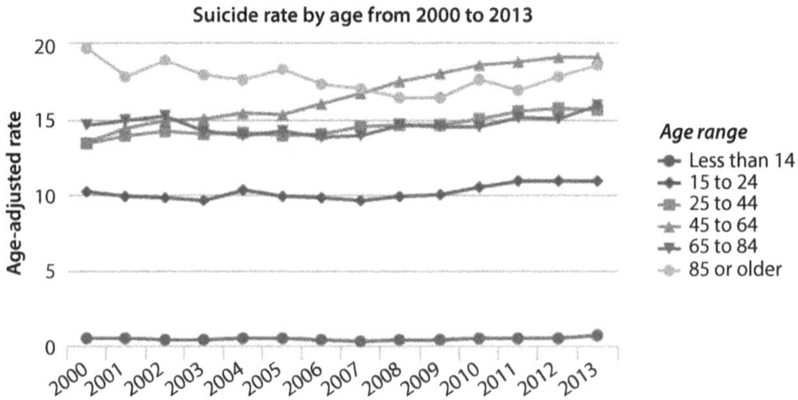

Figure 2.21 Suicide rate by age.

those 85 years and older. Younger groups have had consistently lower suicide rates than middle-aged and older adults. In 2013, adolescents and young adults aged 15 to 24 had a suicide rate of 10.9 (Figure 2.21). A society that takes pride in decreasing mortality rate and increasing longevity must take responsibility for the elderly that don't care to live longer.

What is even more alarming is, there is an overall trend of increasing suicide rate in all age groups since 2000. More recent data show that the rate is the highest among aged 45-64. This is the age group that is expected to be at its prime in a functional society. It is also true that this is the age group that has been indoctrinated to chemical medicines that are mainly taken for 'management' of health without any sign of ailment itself.

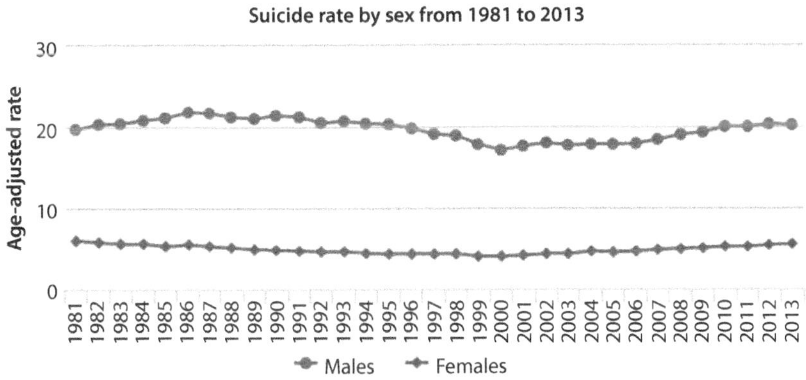

Figure 2.22 Suicide rate among males and females in USA.

For many years, the suicide rate has been about 4 times higher among men than among women (Figure 2.22). In 2013, men had a suicide rate of 20.2, and women had a rate of 5.5. Of those who died by suicide in 2013, 77.9% were male and 22.1% were female. This trend doesn't correlate with the number of mental disorder patients. What it shows in effect is men are less capable than women to handle the stressors of current civilization.

Similar conclusions can be made about ethnic disparity in suicide rates (Figure 2.23). White males accounted for 70% of all suicides in 2013. The highest rate of suicide during 1999-2013 is among whites, somewhat closely followed by American Indians. All other ethnic groups fall far behind these two groups, as shown in Figure 2.23. In 2013, the highest U.S. suicide rate (14.2) was among Whites and the second highest rate (11.7) was among American Indians and Alaska Natives. Much lower and roughly similar rates were found among Asians and Pacific Islanders (5.8), Blacks (5.4) and Hispanics (5.7).

During the same period, the use of anti-depressants has skyrocketed. Ever since, the 'miracle drug', Prozac was introduced to the market, Americans have been taking different forms of the medicine (Celexa, Effexor, Paxil, Zoloft, to name just a few) in astounding numbers. In 2011, National Center for Health Statistics (NCHS), the rate of antidepressant use in this country among teens and adults (people ages 12 and older) increased by almost 400% between 1988–1994 and 2005–2008.

The federal government's health statisticians figure that about one in every 10 Americans takes an antidepressant. And by their reckoning, antidepressants were the third most common prescription medication

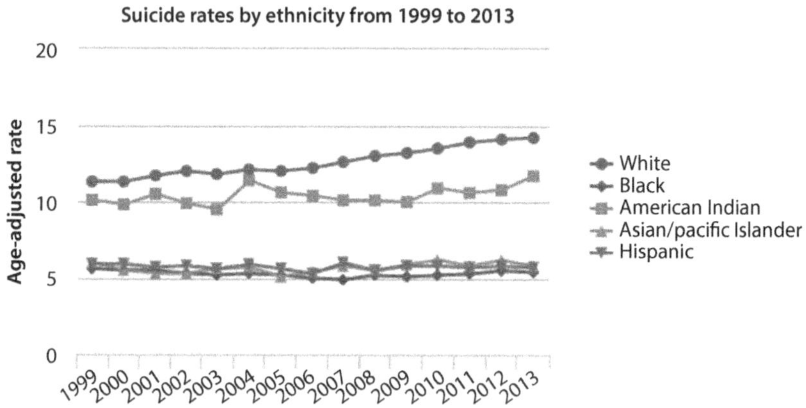

Figure 2.23 Suicide rates for different ethnic groups.

taken by Americans in 2005–2008, the latest period during which the National Health and Nutrition Examination Survey (NHANES) collected data on prescription drug use.

Among others, the following statistics stand out:

- 23% of women in their 40s and 50s take antidepressants, a higher percentage than any other group (by age or sex)
- Women are 2½ times more likely to be taking an antidepressant than men
- 14% of non-Hispanic white people take antidepressants compared with just 4% of non-Hispanic blacks and 3% of Mexican Americans
- Less than a third of Americans who are taking a single antidepressants (as opposed to two or more) have seen a mental health professional in the past year

Antidepressant use does not vary by income status. No matter what a person's income is or if he is covered with health care, the rate of anti-depressant usage remains unaffected. Of more significance is the correlation between antidepressant usage and suicidal behaviour. Healy and Aldred (2005) discussed such correlation. They stated that the idea that antidepressants might contribute to suicide in depressed patients was first raised in 1958. For 30 years antidepressants were primarily used in severely depressed and often hospitalized patients. The issue of suicidality on selective serotonin reuptake inhibitors (SSRIs) became one of public concern with reports in 1990 that Prozac could lead to suicidality in patients. Fourteen years later, warning labels were put on antidepressants suggesting particular difficulties "during the early phase of treatment, during treatment discontinuation, and when the dose of treatment is being changed, and that treatment related risks may be present in patients being treated for syndromes other than depression, such as anxiety or smoking cessation".

Gobbons et al. (2007) discussed how people under the age of 24 who suffer from depression can harbour suicidal thoughts if they take anti-depressants. Federal health officials unveiled proposed changes to the labels on antidepressant drugs in December 2006 to warn people of this danger (Lenzer, 2006).

Selective serotonin reuptake inhibitors (SSRI) prescriptions for children and adolescents decreased after U.S. and European regulatory agencies issued warnings about a possible suicide risk with antidepressant use in paediatric patients, and these decreases were associated with increases in suicide rates in children and adolescents in both the United States with a 14% increase, and 50% increase in the Netherlands (Lenzer, 2006).

Many seem to be puzzled as to why anti-depressants would cause more depression. Others are convinced that anti-depressants are only damaging to young people. There is no obscurity if one knows how SSRIs function. SSRIs ease depression by affecting naturally occurring chemical messengers (neurotransmitters), which are used to communicate between brain cells. SSRIs block the reabsorption (reuptake) of the neurotransmitter serotonin in the brain. The entire process depends on natural mechanism of dealing with grief and perception of fear. Chemical interference changes the balance of serotonin, forcing brain cells send and receive chemical messages, which in turn boosts mood. However, this can be done only for a short time as the natural system continues to fight these chemicals, eventually yielding to natural system that 'reads' the chemicals as an invasion.

Figure 2.24 shows top 10 leading causes of death in USA. In 2012, the 10 leading causes of death (heart disease, cancer, chronic lower respiratory diseases, stroke, unintentional injuries, Alzheimer's disease, diabetes, influenza and pneumonia, kidney disease, and suicide) remained the same as in 2011 and accounted for 73.8% of all deaths in the United States.

From 2011 to 2012, age-adjusted death rates declined significantly for 8 of 10 leading causes of death. The rate decreased 1.8% for heart disease, 1.5% for cancer, 2.4% for chronic lower respiratory diseases, 2.6% for stroke, 3.6% for Alzheimer's disease, 1.9% for diabetes, 8.3% for

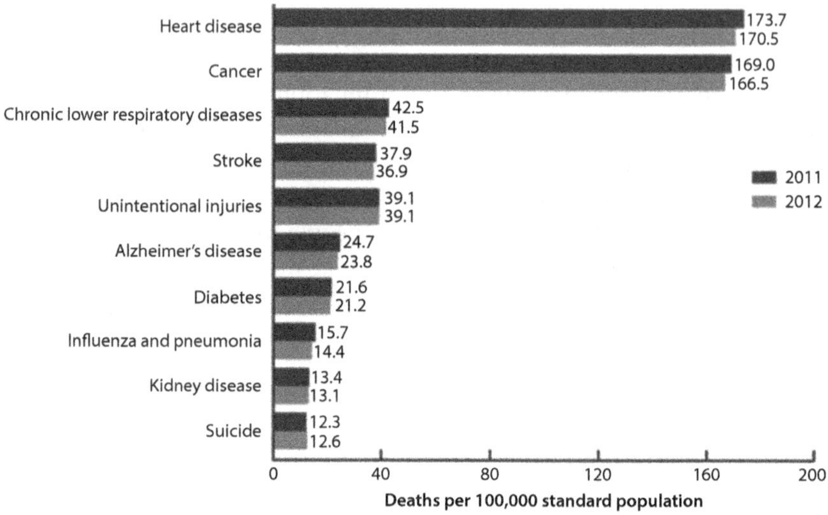

Figure 2.24 10 leading causes of death in USA in 2012.

influenza and pneumonia, and 2.2% for kidney disease. The rate for suicide increased 2.4%. The rate for unintentional injuries remained the same.

As previously shown in Figure 2.14, heart disease and cancer remain the 1st and 2nd leading causes of death, respectively, over the 75-year period. A 75-year perspective points to both stability and change in the leading causes of death: stability because three causes (heart disease, cancer, and stroke) remained among the five leading causes each year between 1935 and 2010; and change because other causes moved into or dropped out of the five leading causes at different points over the past 75 years. There were also changes in the proportion of all deaths that were due to each of the leading causes. For example, heart disease and cancer were the leading causes of death from 1935 to 2010, but it was in 1983 that these two conditions accounted for the highest percent (60 percent) of all deaths. In 2010, they constituted 47 percent of all deaths.

Other conditions have all occupied spots within the five leading causes for some but not all years between 1935 and 2010. Specifically, chronic lower respiratory diseases entered the five leading causes of death in 1979; certain diseases of early infancy was among the five leading causes only from 1949 to 1962 and 1964; accidents entered the five leading causes of death in 1946; kidney disease was among the five leading causes only from 1935 to 1948; and influenza and pneumonia was

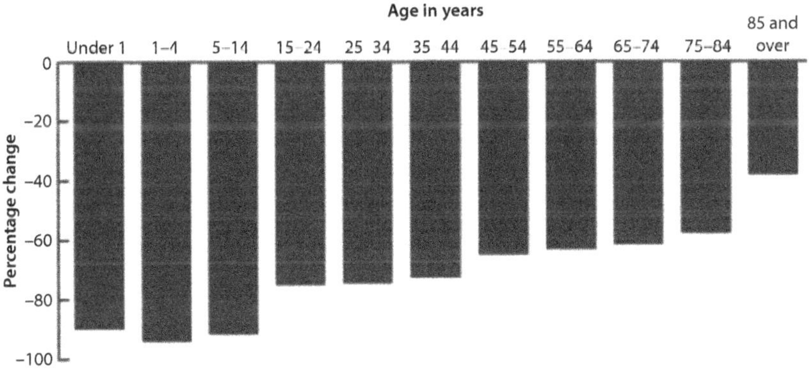

Figure 2.25 Percentage change in death rates by age: United States, 1935–2010. NOTE: 2010 data are preliminary. SOURCE: CDC/NCHS, National Vital Statistics System, Mortality.

among the five leading causes between 1935 and 1945, again in 1963, and between 1965 and 1978.

For all but the oldest age group (85 years and over), mortality risk fell more than 50 percent between 1935 and 2010 (Figure 2.25). The greatest reduction was among persons 1–4 years of age where the death rate declined from 440.9 to 26.6 deaths per 100,000 population, a decrease of 94 percent. Declining death rates were seen among the over 65 population, although not of the same magnitude as among the younger ages. For example, for persons 65–74 years of age, death rates declined by 62 percent, while death rates decreased by 58 percent for those 75–84 years of age, and declined 38 percent for persons 85 years or more.

Figure 2.26 shows age-adjusted death rates and ratio of rates by sex: United States, 1935–2010.

Between 1935 and 2010, age-adjusted death rates decreased by 56 percent for males and 62 percent for females. Female age-adjusted death rates declined from 1,690.6 in 1935 to 634.3 per 100,000 in 2010 while the age-adjusted death rate for males fell from 2,031.2 to 886.2 deaths per 100,000 population. Although males had consistently higher age-adjusted death rates than females throughout the 75 years, the ratio of male to female rates has fluctuated. For example, in 1935, the male to female death ratio was 1.2, rising to 1.7 in 1975–1981, and falling to 1.4 in 2000–2010.

Figure 2.27 shows age-adjusted death rates and ratio of rates by race: United States during 1935–2010. Between 1935 and 2010, age-adjusted

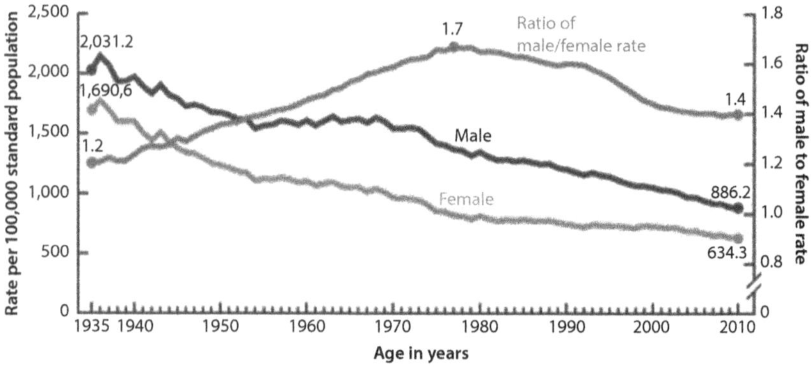

Figure 2.26 Gender dependence of death rate. NOTES: 2010 data are preliminary. Age-adjusted rates are per 100,000 U.S. standard population. Rates for 2001–2009 are revised and may differ from rates previously published. SOURCE: CDC/NCHS, National Vital Statistics System, Mortality.

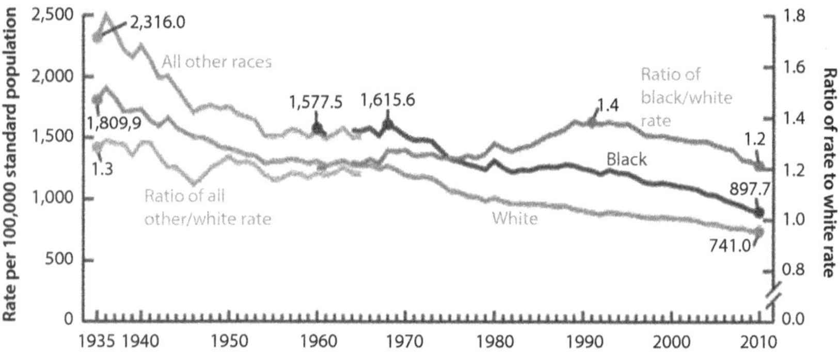

Figure 2.27 Death rate for various races. NOTE: 2010 data are preliminary. Age-adjusted rates are per 100,000 U.S. standard population. Rates for 2001–2009 are revised and may differ from rates previously published. Source: CDC/NCHS, National Vital Statistics System, Mortality.

death rates decreased by 59 percent for the white population. Death rates for the black population were first reported in 1960. Prior to that, rates were reported for "all other races." Since the black population made up 96 percent of those of all other races prior to 1960, it is possible to approximate the trend for the black population from 1935. Using this approximation, age-adjusted death rates decreased by 61 percent for the black population (35 percent for the all other races group between 1935 and 1959 and 43 percent for the black population between 1960 and 2010). Although the white population had consistently lower age-adjusted death rates than the all other races and the black populations throughout the 75 years, the ratio of the age-adjusted death rates fluctuated somewhat. For example, the all other races to white population death ratio was about 1.3 between 1935 and 1942 and 1.2 between 1943 and 1959. The ratio of black to white population death ratio was 1.2 in 1960, rose to 1.4 between 1988 and 1996, and then declined to 1.2 where it has remained since 2008.

2.5.2.1 Ischemic Heart Disease or Coronary Artery Disease (CAD)

Any disease starts off when certain part of the body or organ is 'forgotten' by the central control system. How that is triggered will be discussed in latter chapters of the book. It suffices to state here that ischemic heart disease is triggered by 'forgetting' the heart. As a symptom, heart doesn't receive enough blood and oxygen. Ischemic Heart Disease affects 1 out of 100 people, often middle-aged to elderly males. However, females are also affected. Risk factors include diabetes, high blood pressure and obesity.

The World Health Organization (WHO) estimates that about 7.4 million people died of ischemic heart disease in 2012. That was about 13.2 percent of all deaths. In the United States, about 600,000 people die of heart disease every year, according to the Centers for Disease Control and Prevention (CDC). That makes it the deadliest disease in the U.S., as well as the world. In the U.S. the most common type of heart disease is CAD, which takes about 380,000 lives each year.

Among the risk factors are high blood pressure, high cholesterol, and smoking. Each of these related to a lifestyle that is removed from nature. Ironically, the most devastating stimuli to this disease is chemicals that often are introduced in name of medicine.

2.5.2.2 Cerebrovascular Disease

Cerebrovascular disease is popularly known as stroke. A stroke happens when blood flow to a part of the brain is interrupted because a blood vessel in the brain is blocked (ischemic stroke) or bursts open (hemorrhagic stroke). If blood flow is stopped for longer than a few seconds, the brain can't get blood and oxygen. Brain cells then die, causing permanent damage. When brain cells die, the body "panics". High blood pressure is the No. 1 risk factor for a stroke.

A stroke is when an artery in the brain is blocked or leaks. Oxygen-deprived brain cells begin to die within minutes.

Stroke was responsible for 6.7 million deaths around the world in 2012, according to WHO. That figure represents about 11.9 percent of all deaths. CDC figures show that nearly 130,000 people in the United States die of stroke each year — that's one person every four minutes. About one in four strokes occur in people who have had a prior stroke. Stroke is also a leading cause of disability.

Risk factors for stroke are similar to those for CAD. As usual, good health habits and lifestyle can lower risks.

2.5.2.3 Chronic Obstructive Pulmonary Disease (COPD)

The lung is the equivalent of thermal engine, where main combustion reactions occur. This is essential to living. COPD is a chronic, progressive lung disease that makes it hard to breathe. Chronic bronchitis and emphysema are types of COPD.

About 3.1 million deaths were attributed to COPD in 2012, according to WHO. That represents about 5.6 percent of deaths, a rate that has held steady since 2000. In 2004, about 64 million people around the world were living with COPD.

It is commonly believed that the main cause of COPD is tobacco — and that means second hand smoke, too. Another factor is air pollution, both indoors and out. Note that pollution refers to artificial chemicals. COPD affects men and women at about the same rate. There's no cure for COPD, but its progression can be slowed down with medication that often contains chemicals that prevent natural functions of the lung.

The American Lung Association estimates that in 2011, 12.7 million adults in the United States had COPD, but even more showed some sign of lung problems. There's a great variation in the number of cases from state to state. In 2011, about 4 percent of people in Minnesota and Washington had COPD. In Alabama and Kentucky, it was more than 9 percent.

2.5.2.4 Lower Respiratory Infections

WHO estimates that lower respiratory infections caused about 3.1 million, or 5.5 percent of deaths in 2012. This group of diseases includes pneumonia, bronchitis, and influenza.

According to the CDC, about 20 percent of travelers returning to the United States seek medical attention for respiratory infection following a trip. Packed cruise ships, hotels, and other close quarters increase risk of transmission and outbreaks of disease.

2.5.2.5 Trachea, Bronchus, and Lung Cancers

Trachea, bronchus, and lung cancer are all respiratory cancers. The main causes of this type of cancer are smoking, second-hand smoke, and environmental toxins. Environmental toxins include wall paints, computer gadgets and numerous other objects that are made out of plastic and other toxic materials.

WHO estimates that in 2012, 1.6 million people died from trachea, bronchus, and lung cancers. These cancers represent about 2.9 percent of all deaths globally.

2.5.2.6 HIV/AIDS

HIV/AIDS was first reported in the 1980s, and throughout the decades this chronic life-threatening medical condition has affected more and more humans. HIV stands for human immunodeficiency virus, and it is one "bug" that we as a species can't seem to kick. HIV infection weakens the immune system to the point that it can't fight off anything anymore, and something as simple as the common cold can turn fatal. The virus attacks T cells and CD4 cells, both of which we need to fight off infection. HIV eventually fights off so many CD4 cells that the body cannot battle any infection or illness, at which point the person develops full-blown AIDS. AIDS, or acquired immune deficiency, is when your immune system is not working as it should. HIV/AIDs can be acquired through blood transfusions or the sharing of needles or bodily fluids.

HIV is short for human immunodeficiency virus. It's a virus that attacks the immune system. HIV can cause AIDS, or acquired immunodeficiency syndrome. AIDS is a chronic, life-threatening condition.

According to the Foundation for AIDS Research (amfAR), since the start of the pandemic, almost 39 million people have died due to HIV/AIDS. In 2013, about 1.5 million people lost their lives to AIDS. That's about 2.7 percent of deaths worldwide.

By the end of 2012, 35.3 million people around the world were infected with HIV. Every day, about 5,700 more become infected.

Rates vary dramatically by geographical location. HIV is rampant in sub-Saharan Africa, where almost one in 20 adults has it. The region is home to 70 percent of all people who have HIV. Sadly, it's also home to 91 percent of the HIV-positive children in the world.

2.5.2.7 Diarrheal Diseases

This common disease can lead to death due to dehydration. Diarrhea is usually caused by an intestinal infection transmitted through viruses, bacteria, or even parasites. This type of infection can easily spread through contaminated water or food. It's particularly widespread in developing nations that have poor sanitary conditions.

WHO estimates that 1.5 million people died from diarrheal diseases in 2012, which comprises about 2.7 percent of deaths. Fortunately, that's down from 2.2 million in 2000. Diarrheal disease is the second top killer of children under age five. Of concern is the fact that about 760,000 children die from diarrheal diseases each year.

According to a 2009 Unicef report, every year there are about 2.5 billion cases of diarrhea involving children under five years old. More than 50 percent occur in Africa and South Asia. More than 80 percent of child deaths due to diarrhea occur in those regions.

According to Unicef, healthy behaviors such as good handwashing technique can reduce the incidence of diarrheal diseases by 40 percent. Death due to diarrhea is possibly the most easily preventable with simple dietary supplements.

2.5.2.8 Diabetes Mellitus

Diabetes is a group of diseases that affect insulin production and use. In type 1 diabetes, the pancreas can no longer produce insulin. The cause is not known. In type 2 diabetes, the pancreas doesn't produce enough insulin, or it can't be used effectively. Type 2 diabetes can be caused by a number of factors, including poor diet, lack of exercise, and carrying too much weight.

In 2012, about 1.5 million people died from diabetes-related causes, according to WHO. People in low to middle income countries are more likely to die from complications of diabetes.

Type 2 diabetes is the product of modern-day lifestyle. The solution to this problem is simple as will be discussed in the following chapters.

2.5.2.9 Preterm Birth Complications

According to WHO, in 2012, as many as 1.1 million deaths were due to prematurity and complications due to low birth weight. Three-quarters of these deaths happen within the first week of life. Lack of skilled medical care makes this a huge problem in developing countries. Many newborn deaths could be avoided with good prenatal and postnatal care.

Every year, about half a million women worldwide die from complications related to pregnancy and childbirth, including severe bleeding/hemorrhaging, infections, unsafe abortions, obstructed labor and eclampsia, and more than 90 percent of maternal deaths occur in Asia and sub-Saharan Africa. These perinatal complications cannot only be fatal for mothers but for their young children as well, with medical conditions such as low birth weight contributing to more than one in five deaths in children. Of those deaths, more than 3 million infants die during the first week of life. Death caused by birth complications is preventable through healthy food habits and lifestyle.

2.5.2.10 Tuberculosis (TB)

TB is a lung condition caused by bacteria called Mycobacterium tuberculosis. It's an airborne disease that is often successfully treated. Some strains of TB are resistant to conventional treatments. Second-line drugs used to treat these patients are in limited supply. Some strains fail to respond to second-line treatment as well.

TB is a contagious bacterial infection that involves the lungs, but it may spread to other organs. The symptoms of this disease can remain stagnant for years or affect the person right away. People at higher risk for contracting TB include the elderly, infants and those with weakened immune systems due to other diseases, such as AIDS or diabetes, or even individuals who have undergone chemotherapy. Being around others who may have TB, maintaining a poor diet or living in unsanitary conditions are all risk factors for contracting TB. In the United States, there are approximately 10 cases of TB per 100,000 people.

In 2012, about 900,000 people lost their lives to TB, according to WHO estimates. The majority of TB-related deaths happen in poorer countries. It is one of the top causes of death for people who have HIV.

2.5.2.11 Malaria

Malaria is worth an honourable mention. This is the disease that drew a Nobel prize for Dr. Muller for his invention of DDT.

Malaria is caused by a parasite that is transmitted from one human to another through the bite of an infected mosquito. In humans, the parasites travel to the liver, where they mature. Once matured, they release another parasite that then enters the bloodstream and infects the red blood cells. The parasites multiply at a rapid speed, but symptoms could take up to one year to show. The disease is a major health problem in much of the tropics and subtropics. It is estimated that there are 300-500 million cases of malaria each year, and more than 1 million people die from the disease. It presents a major health hazard for travelers to warm climates — something definitely to be weary of when planning that once-in-a-lifetime trip to the rain forest.

2.5.2.12 Measles

Measles is a contagious viral disease that can be prevented through immunization. However, controversies surround the immunization shots. In 2010, it killed about 139,300 people around the world, according to the CDC.

2.6 Paradox and New Science

Caulfield (2013) wrote in an article, titled: "The paradoxes of pop science",

> "Science and popular culture have long mixed in unique and productive ways. Cutting-edge science has inspired great novels, movies, music and art. But as science becomes a larger part of our cultural landscape, we are seeing a concomitant increase in the profoundly and paradoxically unscientific use of scientific language and images. The advocates of dubious products, philosophies and therapies frequently lean heavily on scientific-sounding terminology as a way of capitalizing on the excitement surrounding cutting-edge areas of science, including genetics, stem cells and nanotechnology."

In truly scientific sense, New science suffers from the same problem. Every study is marred with paradoxical thinking that arises from spurious first premises. Removal of these paradoxes is the essence of true knowledge and happiness (Caulfield, 2012). Following is a discussion of some of the most-talked about paradoxes in medical science.

2.6.1 Obesity Paradox

Being fat is bad except when it's good. It's called "the obesity paradox." The adverse health effects of obesity are well established, but there are exceptions. Obesity appears to confer an advantage in certain subgroups with conditions like heart disease and diabetes. This paradox can be removed if the word 'fat' is defined with scientific basis.

An article in the *New York Times* asks:

> Obesity is the primary risk factor for Type 2 diabetes, yet sizable numbers of normal-weight people also develop the disease. Why?
>
> The answer to the question is a one-liner. Type 2 diabetes is a multifactorial disease involving interactions between genetic, environmental, and lifestyle factors. These three factors are also just one factor, i.e., delinearized history of the individual. This applies to every disease.
>
> Obesity involves gallstones, infertility, congenital defects, gastroesophageal reflux disease, stroke, carpal tunnel syndrome, multiple sclerosis, obstructive sleep apnea, erectile dysfunction, and many others.

2.6.2 Obesity/Mortality Paradox

Obesity predisposes to diabetes, heart diseases, sleep apnea, cancer and other diseases. Although several studies have shown an increase in mortality in obese people, other studies have suggested that obesity protects against death from all causes as well as death due to chronic diseases such as diabetes, heart failure, and stroke. This is called "obesity-mortality" paradox that suggests a beneficial influence of obesity. Similar to the previous paradox, this paradox can be erased by including all salient features of obesity.

2.6.3 Simpson's Paradox

Simpson's paradox for continuous data: a positive trend appears for two separate groups (blue and red), a negative trend (black, dashed) appears when the data are combined.

In probability and statistics, Simpson's paradox, or the Yule–Simpson effect, is a paradox in which a trend that appears in different groups of data disappears when these groups are combined, and the reverse trend appears for the aggregate data. This result is often encountered in social-science and medical-science statistics. Islam et al. (2010a) discussed this phenomenon as something that is embedded in Newtonian calculus that allows taking the infinitely small differential and turning that into any desired integrated value, while giving the impression that a scientific process has been followed. Furthermore, Khan and Islam (2012) showed that true trendline should contain all known parameters. The Simpson's paradox can be avoided by including full historical data, followed by scientifically true processing (Islam et al., 2014a; Pearl, 2009).

The above difficulty with statistical processing of data was brought to light by way of a comically- intended Web posting of the following correlation — between number of pirates vs. global temperature with the slogan: Join piracy, save the planet.

Similar "paradoxes" arise among so-called ecological fallacies. It is best described in Wikipedia with the following examples:

> Assume that at the individual level, being Protestant impacts negatively one's tendency to commit suicide but the probability that one's neighbor commits suicide increases one's tendency to become

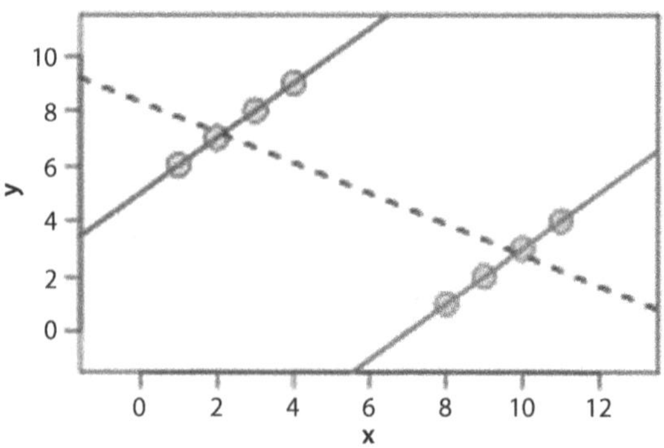

Figure 2.28 Simpson paradox highlights the problem of targeted statistics.

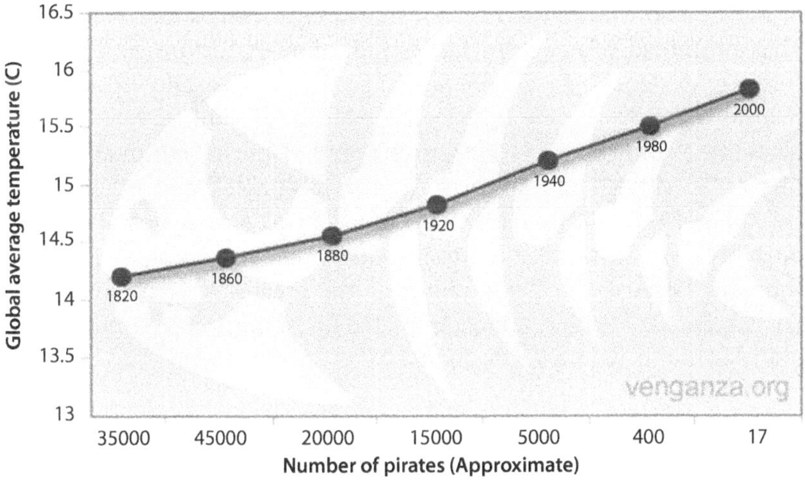

Figure 2.29 Using statistical data to develop a theoretical correlation can make an aphenomenal model appealing, depending on which conclusion would appeal to the audience.

Protestant. Then, even if at the individual level there is negative correlation between suicidal tendencies and Protestantism, there can be a positive correlation at the aggregate level.

Similarly, even if at the individual level, wealth is positively correlated to tendency to vote Republican, we observe that wealthier states tend to vote Democrat. For example, in 2004, the Republican candidate, George W. Bush, won the fifteen poorest states, and the Democratic candidate, John Kerry, won 9 of the 11 wealthiest states. Yet 62% of voters with annual incomes over $200,000 voted for Bush, but only 36% of voters with annual incomes of $15,000 or less voted for Bush.

Finally, The prosecutor's fallacy is a fallacy of statistical reasoning, typically used by the prosecution to argue for the guilt of a defendant during a criminal trial. In its crudest form, it involves the assertion that the probability of defendant to be guilty is 90% because the perpetrator and the defendant are known to be sharing the blood type that has a probability of 10% in the general population. It is purported that a DNA match is not a fallacy because the probability of match is much greater. However, this is also an example of how New science has refined techniques in favor of opacity instead of transparency. The use of DNA as the only match has the gravest risk of 'creating' evidence in case there is no other evidence. This is rarely talked about. Such mindset has promoted prosecution tactics that allowed torture as a means of extracting 'evidence'. In a broader sense, it has also allowed notorious conclusions, such as, the non-consideration of junk DNA, asserting probability of Bing Bang as 97%, probability of life as 'reasonable', probability of 'intelligent life' as even 'more reasonable', and others. All of them suffer from the fundamental basis that a 'fact is not a matter of probability'. For instance, the proclamation that the orangutan is linked to humans because the DNA match is the greatest (Derbyshire, 2011), what can be said when a greater match of some other component of genome is found with certain plants?

2.6.4 Low Birth Weight Paradox

The low birth weight paradox is an apparently paradoxical observation relating to the birth weights and mortality of children born to tobacco smoking mothers. As a usual practice, babies weighing less than a certain amount (which varies between different countries) have been classified as having low birth weight. In a given population, babies with low birth weights have had a significantly higher infant mortality rate than others.

However, it has also been observed that babies of low birth weights born to smoking mothers have a lower mortality rate than the babies of low birth weights of non-smokers. This paradox is often manifested in other cases. For instance, often a smoker lives a long life while a non-smoking partner dies or when someone with good health habits becomes more vulnerable to disease than his counterpart with bad health habits.

This paradox is removed if the entire individual history is considered. For instance, a person who is infested with non-organic food develops immunity to bad food. Another person who is used to organic food will have significant problems when he is suddenly exposed to non-organic food. Such response is entirely natural but cannot be standardized in absence of a theory that includes all salient factors.

A relevant anecdote exists from Ecuador. The village of Vilcabamba was identified as the "valley of longevity" by a Harvard Medicine professor (Leaf, 1973). French studies had shown that the diet and lifestyle of the inhabitants may have been a factor. Nobel Laureate Chemist, Richard Laurence Millington Synge, the man who discovered amino acids, claimed that there are remarkable medicinal qualities to be found in the plant life in certain places near the Ecuador in the proximity of the valley of Vilcabamba.

Thanks to scientific chemical assay techniques, analysis has now shown that the fruit, roots and herbs of this particular equatorial sub-area offer some of the strongest anti-oxidant protection in the world. In 1981, the Ecuadorian government hired medical journalist Dr. Morton Walker to study these people in depth. In his book, "The Secret to a Youthful Long Life", Walker reported that his research showed the mineral rich water that the Vilcabambans drank was key to their long lives and health. Laboratory analysis of the Vilcabamba water determined that the unique balance of enriched colloidal minerals in the local drinking water was ideal for promoting optimum human health.

In 1991, businessman Craig Keeland traveled to Vilcabamba to study the anti-oxidants. He developed and marketed a whole fruit puree made from Vilcabamba fruits and vegetables and sold it through his former company, Youngevity, which he sold in 2005. In July 2003, Keeland formed a new company that developed a whole food puree product called ViaViente which is now sold globally in over 26 countries and territories. Keeland founded the Andes Children's Foundation in 2003 to support the education of children in Vilcabamba. Furthermore, medical researchers had confirmed that the retinas of 100 year-old residents are often comparable with those of 45 year-old city-dwellers. The same village now forms the core of another epidemiological study. It has the highest rate of stomach cancer in the world (over 50 for 100,000 inhabitants). How is such drastic switch possible? The answer, of course, lies within consideration of each health practices that has changed in that community. Such practices may include, water purification, usage of European toilet (see picture below), processed food, smoking cigarettes, and numerous other artificial changes that are made in name of civilization. Natural processes have been replaced with new processes that increased the profitability and created a paradigm shift. This explains the emergence of paradoxes.

2.6.5 Prevention Paradox

The prevention paradox states that a large number of people at small risk may contribute more cases of a particular condition than a smaller number of people who are individually at greater risk (Rose, 1981). The prevention paradox describes the seemingly contradictory situation where the majority of cases of a disease come from a population at low

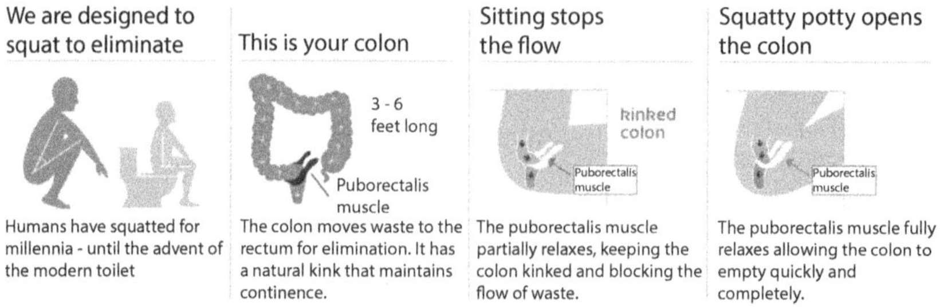

Figure 2.30 A recent Stanford University study reports squatting is more essential than dietary fibers.

or moderate risk of that disease, and only a minority of cases come from the high risk population (of the same disease). This is because the number of people at high risk is small. For instance, the case of Down's Syndrome where maternal age is a risk factor. Yet, most cases of Down's Syndrome will be born to younger, low risk mothers (this is true at least in populations where most women have children at a younger age). This situation is paradoxical because it is common and logical to equate high risk populations with making up the majority of the burden of disease.

Another example could be seen in terms of reducing overall alcohol problems in a population. Although less serious, most alcohol problems are not found amongst dependent drinkers. By achieving a small reduction in alcohol misuse amongst a far larger group of 'risky' drinkers with less serious problems, this will result in a greater societal gain than reducing problems amongst a smaller number of dependent drinkers. This paradox was identified by Spurling and Vinson (2003) in the context of alcohol consumption and injury.

The prevention paradox is a problem encountered when governments or organizations attempt to introduce a large-scale intervention to improve health. Many interventions that aim to improve health have relatively small influences and perceptible benefits on the health of most people. Therefore, for one person to benefit, many people have to change their behaviour—even though they receive no benefit, or even suffer, from the change.

Such paradoxes do not appear if it proper analysis of data is introduced (Islam et al., 2014a). The scientific community continues to use such paradoxical techniques in order to justify government interventions that help benefit the corporations.

2.6.6 The Novelty Paradox

This paradox arises from the demand that a person proposing to conduct research produce evidence from existing paradigm as a basis for the claim of novelty of an approach. Boudreau et al. (2013) showed that this paradox and the insistence by the funding agencies to adhere to this protocol. Their results confirm a systematic penalty for novel proposals. This works like a trap in which information effects rather than strategic effects account for the novelty penalty. Only a minority of the novelty penalty could be related to perceptions of lesser feasibility of novel proposals. Because there is little possibility of conducting research without grants from funding agencies, the novelty paradox acts as a gatekeeper to eliminate creative research. As a consequence, most funds end up supporting technology development projects that have premises that are the same as conventional ones, creating yet another paradox of costly health care technology (Skinner, 2013).

2.6.7 The Paradox of Worsening Conditions with Medications

Even though this paradox is highlighted for medicines of treatment of epilepsy (Guerrini, 1998), it can be applied to numerous medicines (Khan and Islam, 2012). Khan and Islam (2012) attributed this paradox to the first premise that artificial is the same as the real. Modern medicines are all artificial chemicals, thereby vulnerable to the outcome of aphenomenal first premise. The following transitions have taken place during the entire technology development phase, converting every natural to artificial.

Wood → plastic
Glass → PVC
Cotton → polyester
Natural fiber → synthetic fiber
Clay → cement
Molasses → Sugar
Sugar → Sugar-free sweetners
Fermented flower extract → perfume
Water filter (Hubble bubble) → cigarette filter
Graphite, clay → chalk
Chalk → marker
Vegetable paint → plastic paint
Natural marble → artificial marble

Clay tile → ceramic tile
Ceramic tile → vinyl and plastic
Wool → polyester
silk → synthetic
Bone → hard plastic
Organic fertilizer → chemical fertilizer
Adaptation → bioengineering

The above transition is the hallmark of modern technological developments that have been characterized by Nobel laureate chemist, Robert Curl as 'technological disaster'. The following table shows how every promise made in technology development has been reversed in the most spectacular way. Note how modern medicines are but one component of this disaster-bound technological train.

2.6.8 The Prostate Paradox

This paradox arises from the observation that victims of prostrate cancer die of other causes. This is further worsened by the lowering of life quality by the treatment of prostrate cancer (Website 1; Schröder et al., 2012). This paradox is also applicable to breast cancer. In addition, any cancer treatment degrades the quality of life, destroying the ability of immune system to fend from other diseases, thereby contributing to increased mortality.

2.6.9 The Health-Lifespan Paradox

Since the beginning of the last century, life expectancy at birth has increased tremendously, from 31 years to 67.2 today. During the same period, however, health (both physiological and mental) has been actually deteriorating way faster than we thought (Hulsegge et al., 2013). That study of 6,000 individuals published recently in the European Journal of Preventive Cardiology showed that men in their 30s were 20 percent more likely to be obese, while women in their 20s were 100 percent more likely to be so compared to 10 years ago. The researchers went so far as to conclude that today's 30-year-olds are as healthy as 45-year-olds a decade ago. Add to that is the fact the suicide rates among the nations with highest longevity record. Particularly, richest nations, such as Japan and USA have alarmingly high rate of suicide. The most vulnerable victims of suicide endemic are old and young

Table 2.1 Analysis of "breakthrough" technologies (revised from Khan and Islam, 2011).

Product	**Promise (knowledge at t= 'right now'**	**Current knowledge (closer to reality)**
Microwave oven	Instant cooking (bursting with nutrition)	97% of the nutrients destroyed; produces dioxin from baby bottles
Fluorescent light (white light)	Simulates the sunlight and can eliminate 'cabin fever'	Used for torturing people, causes severe depression
Prozac (the wonder drug)	80% effective in reducing depression	Increases suicidal behavior
Anti-oxidants	Reduces aging symptoms	Gives lung cancer
Vioxx	Best drug for arthritis pain, no side effect	Increases the chance of heart attack
Coke	Refreshing, revitalizing	Dehydrates; used as a pesticide in India
Transfat	Should replace saturated fats, incl. high-fiber diets	Primary source of obesity and asthma
Simulated wood, plastic gloss	Improve the appearance of wood	Contains formaldehyde that causes Alzheimer
Cell phone	Empowers, keep connected	Gives brain cancer, decreases sperm count among men.
Chemical hair colors	Keeps young, gives appeal	Gives skin cancer

(Continued)

Table 2.1 cont.

Product	**Promise (knowledge at t= 'right now'**	**Current knowledge (closer to reality)**
Chemical fertilizer	Increases crop yield, makes soil fertile	Harmful crop; soil damaged
Chocolate and 'refined' sweets	Increases human body volume, increasing appeal	Increases obesity epidemic and related diseases
Pesticides, MTBE	Improves performance	Damages the ecosystem
Desalination	Purifies water	Necessary minerals removed
Wood paint/varnish	Improves durability	Numerous toxic chemicals released
Leather technology	Won't wrinkle, more durable	Toxic chemicals
Freon, aerosol, etc.	Replaced ammonia that was 'corrosive'	Global harms immeasurable and should be discarded

adults. This health-lifespan paradox can be removed if true health is quantified the same way true lifespan is. For instance, lets consider the case of former Israeli prime minister, Ariel Sharon, who has been in a state of comma for many years. If artificiality is removed from his life-span, it becomes reasonable to correlate his natural life span with his natural state of health. Otherwise, a paradoxical relationship emerges. What New science has been lacking is a consistent standard of yardstick of anything, ranging from sustainability criterion (Khan and Islam, 2007) to what constitutes 'facts' (Islam et al., 2012; Islam et al., 2013).

2.6.10 Smoker's Paradox

Smokers have been shown to have lower mortality after acute coronary syndrome than non-smokers. This has been attributed to the younger age, lower co-morbidity, more aggressive treatment and lower risk profile of the smoker. Some studies, however, have used multivariate analyses to show a residual survival benefit for smokers; that is, the "smoker's paradox". Aune et al. (2011) demonstrated that such paradox doesn't appear at least for acute coronary syndrome if all known factors are included. However, their analysis falls short of the general statement that such paradox has an inherent false premise attached to it because it doesn't differentiate between natural and artificial.

2.6.11 Paradox of the Natural

Mauron (2003) outlines this paradox that entirely emerges because of false premises regarding nature and the nature of nature. His quotes are placed below, followed by correct statements that would erase the paradox.

- "Cornucopian" Nature
  "The abundance and resilience of nature never cease to surprise us. The individualist human ingenuity always finds a way to solve the problem." The correct statement should be, "Nature is perfect" (Khan and Islam, 2012). A natural human being is also perfect, something that occurs when a human is in 100% in conformance with universal order (through his intentions).
- Fragile Nature
  "The natural order is vulnerable. Even a small

transgression can lead to catastrophic effects. The egalitarian nature is exploited, oppressed." The correct statement is, "universal order is fixed, constant, and immutable" (Islam et al., 2013). Egalitarian nature is only possible in humans and it is oppressed, affecting only the long-term of the individual.

- Moderately tolerant Nature
  "Within margins, nature tolerates certain mistakes. Is causal properties are transparent. The "hierarchist" experts know best, we need them, to sort things out." The correct statement reads, "Only humans make mistakes. A mistake is a false intention. A false intention is the 'intention' that doesn't conform with universal order".
- Capricious Nature
  "Nature is essentially random. Its causal structure is impenetrable. The fatalist there is nothing we can do. Life is a lottery." The correct statement is, "Nature is universal order and is 100% pre-determined and immutable. Success is to accept whatever life has given us and make intentions to conform to the universal order".

2.6.12 The French Paradox

The November 1991 segment of CBS's 60 Minutes introduced millions of Americans to the cardio-protective effects of moderate alcohol consumption. However, the phenomenon has been known since 1926 when a study of tuberculosis patients in a sanatorium found that patients who drank alcohol moderately had approximately half the death rate as those who abstained. At that time, it was time of prohibition. This French paradox is similar to the smoker's paradox and is erased if the long-term criterion for selection of material is introduced (Khan and Islam, 2007; Islam et al., 2014a, 2014b).

2.6.13 Paradox of Aging

Traditionally, aging has been viewed as a period of progressive decline in physical, cognitive and psychosocial functioning, and aging is viewed by many as the "number one public health problem" facing Americans today. Yet, a scientific study shows after adjusting for age, a higher self-rating of successful aging was associated with higher education, better cognitive function, better perceived physical and mental health, less depression, and greater optimism and resilience. This paradox is removed by removing the false premise that time is an independent variable and disconnected from human conscience, which itself is perceived to be disconnected from human cognition (Islam et al., 2014a). A human being who sees experience with conscience and cognizes conscientiously is the one that benefits from experience. Another individual without similar conscientious participation continuously degrades with time, no matter what experience he/she gathers.

2.6.14 Paradox of Translational Medicine

Translational investigations typically rely on large research consortiums and population-based plasma banks that couple biomarker information with longitudinal observational data. Basic research, however, usually includes specific hypothesis-driven studies that have small sample sizes and are conducted by independent academic or industry researchers. Because of industry support of some basic research in academic institutions, some medical schools, especially the smaller ones, might need to make compromises for fear that companies would suspend funding. Some companies may take ownership of results although academic researchers have generated them. Biased reporting of the clinical benefits of various diagnostic or therapeutic methods may occur because some agreements signed between academic institutions and diagnostic or drug companies require confidentiality of data, often depending on the outcome. As a consequence, negative results may not be emphasized or submitted to medical journals for publication.

This paradox is removed only if first premises of the funding agencies are made transparent. Once it becomes transparent, researchers that continue to conduct research with an aphenomenal model will be exposed.

2.6.15 Peto's Paradox

The evolution of multicellularity required the suppression of cancer. If every cell has some chance of becoming cancerous, large, long-lived organisms should have an increased risk of developing cancer compared with small, short-lived organisms. The lack of correlation between body size and cancer risk is known as Peto's paradox. This paradox is the result of linearization of a complex process, infamously known as the Phillippe Rushton model, because of his assertion that human tangible features are amenable to correlation with intangible features.

2.6.16 TGF-β Paradox

TGF-beta is a potent anticancer agent that prohibits the uncontrolled proliferation of epithelial, endothelial and hematopoietic cells. Interestingly, tumorigenesis typically elicits aberrations in the TGF-beta signaling pathway that engenders resistance to the cytostatic activities of TGF-beta, thereby enhancing the development and progression of human malignancies. Moreover, these genetic and epigenetic events conspire to convert TGF-beta from a suppressor of tumor formation to a promoter of their growth, invasion and metastasis. The dichotomous nature of TGF-beta during tumorigenesis is known as the 'TGF-beta paradox', which remains the most critical and mysterious question concerning the physiopathological role of this multifunctional cytokine. By proper characterization of material as well as process involved during various stages of cancer formation and propagations, this paradox is removed.

2.6.17 Hispanic Paradox

While heart disease still edges out cancer among all Americans, cancer is the number one killer among Hispanics in Texas. Yet their prognosis remains superior to Caucasians, a Hispanic paradox that debunks the notion that income and education are key factors in health care. Hispanic Americans tend to survive illness and live longer than white Americans with the same diseases even though the Hispanics have less education, income and access to health care. This paradox is removed if individual history, along with immune system is included in the analysis.

2.7 The Cost Of Drugs

Healthcare that has been corporatized has fallen victim to capitalism' worst perpetrator. Drugs have become both costly and toxic. This list also tells us as to the continually degrading efficiency of the medications that don't cure anything. Forbes (Herper, 2010) presented a list of most expensive drugs. Forbes assembled its list by interviewing biotechnology industry experts and obtaining estimates of average price per patient from the companies themselves or Wall Street analysts who cover them.

Even though drugs, such as Lipitor for high cholesterol ($1,500 a year), Zyprexa for schizophrenia ($7,000 a year) or Avastin for cancer ($50,000 a year) don't make the most expensive drug list, their ubiquitous use makes them most lucrative. But none of these medicines come close to making Forbes' exclusive survey of the most expensive medicines on the planet. The nine drugs on their list all cost more than $200,000 a year for the average patient who takes them. Most of them treat rare genetic diseases that afflict fewer than 10,000 patients. For these diseases, there are few if any other treatments. So biotech companies can charge whatever they deem appropriate.

Alexion Pharmaceutical's Soliris, at $409,500 a year, is the world's single most expensive drug. This monoclonal antibody drug treats a rare disorder in which the immune system destroys red blood cells at night. The disorder, paroxysymal nocturnal hemoglobinuria (PNH), hits 8,000 Americans. In 2009, Soliris sales were $295 million. Since Alexion started selling Soliris in 2008, two years ago, its stock price is up 130% in first two years. In 2015, the rise stands at over 1200%.

In the inverted world of drug pricing, the fewer patients a drug helps, the more it costs. This creates a new paradigm on the supply and demand chart. Before testing Soliris for PNH, Alexion tested the drug for rheumatoid arthritis, which afflicts 1 million Americans. The trials failed. But if it had worked for arthritis, Alexion would likely have had to charge a much a lower price for this use, as would have to compete against drugs that cost a mere $20,000.

Three other drugs cost more than $350,000 per year. Shire Pharmaceuticals' Elaprase ($375,000 per year) treats an ultra-rare metabolic disorder called Hunter's syndrome. Just 500 Americans suffer from the disease, which causes infections, breathing problems and brain damage. Last year domestic sales of Elaprase were $353 million.

Naglazyme from BioMarin Pharmaceuticals treats another rare metabolic disorder and costs $365,000 a year, according to investment bank Robert W. Baird. Viropharma predicts that sales of its Cinryze, a treatment to prevent a dangerous swelling of the face, will increase from $95 million last year to $350 million several years from now. The drug costs an estimated $350,000 a year.

Amazingly, many brutally expensive cancer drugs don't make the cut. Targeted cancer drugs only help a small minority of patients for a few months. This reduces their average cost. Allos Therapeutics' Folotyn treats a rare type of lymphoma and costs $30,000 per month. But the average patient is only on the drug for just a few months, so it doesn't make the list.

Nonetheless, the price of each new rare-disease drug seems to get higher each year. It used to be that pricing a drug at $100,000 per year raised eyebrows. However, today such price rise is considered to be a norm.

Selling drugs for rare diseases has become immensely profitable. There are so few patients that companies don't have to invest as heavily in marketing. The medicines usually get paid for by insurers or governments. The problem starts with diagnosis to the treatment. Often, more tests mean more diagnosis and automatic treatment with expensive medicines before any established criterion for 'cure' has been established.

The success of specialty drugs for rare diseases comes at a time when the traditional drug business of selling medicines to the masses is in decline. Medicines touted as $1 billion sellers for Eli Lilly and Bristol-Myers Squibb have ended up with flat sales. Selling drugs for rare diseases is considered to be the future of the biotechnology industry.

Myozyme, which inspired the Harrison Ford movie Extraordinary Measures, costs up to $100,000 for a child. However, according to Genzyme, the average cost of adult treatment is $300,000 per year.

Specialty drugs have gotten more expensive than anyone imagined. For years drug companies ignored any disease that didn't afflict millions of patients. That started to change in 1983 when Congress, inspired by an episode of the television show Quincy, M.D., passed a law giving an extra monopoly for drugs for "orphan diseases" that hit fewer than 200,000 people in the country.

Initially biotechnology firms focused on more common orphan diseases like multiple sclerosis. In 1991 Genzyme launched Ceredase for Gaucher disease, a disorder in which a missing enzyme causes lumps of fat to build up in the spleen, heart and even the brain. Ceredase, made from human placentas, replaces that enzyme and initially cost $150,000 a year. Some observers naively expected the price to drop when a new version, made in genetically engineered hamster cells, hit the market in 1994. But the new version, Cerezyme, now costs $200,000 for the average patient and has annual sales north of $1 billion.

Some competition is finally arriving to the rare disease market. Shire is waiting for the Food and Drug Administration to approve its new Gaucher's drug and says it will price the medicine 15% less than Cerezyme. Pfizer and the Israeli biotech company Protalix, are testing a

Cerezyme competitor made in carrot cells that could cost even less. Novartis is also getting into the rare disease field. The entry of big drug companies, desperate for sales, could be what finally drives down prices of these drugs.

The above picture is complemented with the medicines that are most frequently used. These are the medications that make the guaranteed profit for the pharmaceutical industry. Table 2.2. shows the list of 300 most prescribed medicine in USA. For the first 30 medications, the usage is mentioned. Note that all these medications are so-called management pills with no specific function for curing a disease. They are all toxic chemicals that have no benefit other than immediate delay of symptoms.

2.8 "Non-Prescription" Drugs

It is important to note that all artificial chemicals are toxic, but only a few are distributed through prescriptions. Sceitnifically, they should be lumped as 'drugs', some of which are so-called "illicit drugs".

The 2014 Global Drug Survey (GDS) indicates that alcohol was the drug used most in 2013 (Huffinton Post, 2014). It was ranked ahead of tobacco and cannabis. Alcohol was also the most common drug to send people to the emergency room, and the vice that users' friends and relatives worried most about.

The GDS is the world's biggest annual drug survey, polling people's use of addictive substances. They surveyed nearly 80,000 participants in 18 countries to complete a questionnaire. Participating countries included the U.S., Australia, Germany, France, Mexico and Brazil, among others. The participants are not a random group of people, but self-selected. Thus they show a higher interest in substances than the general population and are generally also better-educated, since they need to be web-literate to complete the survey. The GDS found that alcohol, tobacco and cannabis were the most common drugs used by the survey group last year. In many countries, cocaine, amphetamines and MDMA often ranked just behind.

An interesting constant in the study were the misconceptions about the consequences of alcohol consumption among those that drink the most. The research showed that of all those respondents that can be categorized as heavily dependent on alcohol, according to WHO standards, less than 60 percent recognizes that their drinking habits put them at high risk of alcohol-related problems. The study indicates that

Table 2.2 The most prescribed (as listed by http://www.rxlist.com/).

1. Paxil	2. Lexapro	3. Hydrocodone (pain killer)
4. Xanax (anti-depressant)	5. Tramadol (pain killer)	6. Vicodin (pain killer)
7. Lyrica (anti-convulsion)	8. Oxycodone (pain killer)	9. Lisinopril (blood pressure pill)
10. Cymbalta (anti-depressant)	11. Lipitor (Cholesterol reducer)	12. Percocet (pain killer)
13. Zoloft (anti-depressant)	14. Metformin (anti-diabetes)	15. Effexor (anti-depressant)
16. Ambien (sleeping pill)	17. Asthma, COPD, CIDP, rheumatic disorders, allergic disorders, ulcerative colitis and Crohn's disease, adrenocortical insufficiency, hypercalcemia due to cancer, thyroiditis, laryngitis, severe tuberculosis, urticaria (hives), etc.	18. Atenolol (beta blocker for hypertension, treatment of heart attack)
19. Wellbutrin (anti-depressant)	20. Morphine (pain killer)	21. Naproxen (pain killer)
22. Phentermine (diet pill)	23. Tylenol (pain killer, reducing pain, reducing fever, and relieving the symptoms of allergies, cold, coughheadache, and influenza.)	24. Aspirin (pain killer, heart attack prevention, certain types of cancer)

(*Continued*)

Table 2.2 cont.

25. Norvasc (hypertension) or chest pain (angina) and other conditions caused by coronary artery disease)	26. Toprol (beta blocker, hyper tension, coronary artery disease)	27. Darvocet (mild pain killer, cough syrup)
28. Nexium (antacid)	29. Soma (muscle relaxant)	30. Diovan (high blood pressure, congestive heart failure, and to reduce death for people with left ventricular dysfunction after having had a heart attack)
31. Seroquel	32. Albuterol	33. Neurontin
34. Adderall	35. Protonix	36. Ultram
37. Zocor	38. Ativan	39. Metoprolol
40. Levaquin	41. Valium	42. Viagra
43. Prozac	44. Alprazolam	45. Cipro
46. Oxycontin	47. Gabapentin	48. Lortab
49. Clonazepam	50. Celexa	51. Synthroid
52. Plavix	53. Cyclobenzaprine	54. Prevacid
55. Celebrex	56. Vytorin	57. Lasix
58. Lorazepam	59. Amoxicillin	60. Trazodone
61. Zyrtec	62. Klonopin	63. Depakote
64. Clonidine	65. Codeine	66. Lamictal
67. Methadone	68. Flexeril	69. Topamax
70. Diclofenac	71. Promethazine	72. Mobic
73. Doxycycline	74. Augmentin	75. Cephalexin
76. Bactrim	77. Hydrochlorothiazide	78. Advair

(*Continued*)

Table 2.2 cont.

79. Ibuprofen	80. Omeprazole	81. Furosemide
82. Zetia	83. Diazepam	84. Fentanyl
85. Risperdal	86. Abilify	87. Skelaxin
88. Elavil	89. Amitriptyline	90. Coumadin
91. Flomax	92. Prilosec	93. Allegra
94. Clindamycin	95. Insulin	96. Lunesta
97. Fosamax	98. Zyprexa	99. Citalopram
100. Ranitidine	101. Fluoxetine	102. Hydroxyzine
103. Actos	104. Metronidazole	105. Altace
106. Crestor	107. Coreg	108. Digoxin
109. Phenergan	110. Benicar	111. Verapamil
112. Acetaminophen	113. Lithium	114. Zithromax
115. Dilantin	116. Tricor	117. Lotrel
118. Potassium	119. Norco	120. Keflex
121. Flagyl	122. Avandia	123. Ritalin
124. Ciprofloxacin	125. Inderal	126. Trazadone
127. Biaxin	128. Singulair	129. Aciphex
130. Enalapril	131. Zantac	132. Ultracet
133. Tamiflu	134. Diltiazem	135. Warfarin
136. Baclofen	137. Lidocaine	138. Lovastatin
139. Carisoprodol	140. Provigil	141. Ortho Evra
142. Medrol	143. Penicillin	144. Benadryl
145. Reglan	146. Remeron	147. Propoxyphene
148. Concerta	149. Cozaar	150. Hctz
151. Premarin	152. Levothyroxine	153. Methotrexate
154. Buspar	155. Glucophage	156. Glyburide

(*Continued*)

Table 2.2 cont.

157. Paroxetine	158. Avelox	159. Valtrex
160. Calcium	161. Allopurinol	162. Avapro
163. Erythromycin	164. Dilaudid	165. Aricept
166. Claritin	167. Imitrex	168. Simvastatin
169. Estradiol	170. Triamcinolone	171. Namenda
172. Meclizine	173. Lopressor	174. Glipizide
175. Cialis	176. Acyclovir	177. Azithromycin
178. Actonel	179. Etodolac	180. Robaxin
181. Lovenox	182. Trileptal	183. Atarax
184. Geodon	185. Propranolol	186. Temazepam
187. Zofran	188. Bupropion	189. Naprosyn
190. Detrol	191. Flonase	192. Voltaren
193. Cardizem	194. Amiodarone	195. Diflucan
196. Isosorbide	197. Byetta	198. Zanaflex
199. Heparin	200. Nabumetone	201. Nystatin
202. Toradol	203. Loratadine	204. Levoxyl
205. Magnesium	206. Demerol	207. Vancomycin
208. Motrin	209. Tegretol	210. Requip
211. Pravachol	212. Evista	213. Boniva
214. Haldol	215. Nifedipine	216. Keppra
217. Metoclopramide	218. Spironolactone	219. Lantus
220. Methocarbamol	221. Indomethacin	222. Quinine
223. Rozerem	224. Tizanidine	225. Lanoxin
226. Apap	227. Decadron	228. Omnicef
229. Triamterene	230. Strattera	231. Atropine
232. Fioricet	233. Relafen	234. Compazine

(Continued)

Table 2.2 cont.

235. Restoril	236. Endocet	237. Spiriva
238. Dexamethasone	239. Zelnorm	240. Vioxx
241. Hydrocortisone	242. Niaspan	243. Mirtazapine
244. Nasonex	245. Hyzaar	246. Tetracycline
247. Terazosin	248. Provera	249. Avalide
250. Lodine	251. Aldactone	252. Fluconazole
253. Lupron	254. Diphenhydramine	255. Septra
256. Thyroid	257. Methylprednisolone	258. Fexofenadine
259. Amlodipine	260. Ditropan	261. Famotidine
262. Accupril	263. Amaryl	264. Atrovent
265. Phenytoin	266. Colchicine	267. Vasotec
268. Atacand	269. Yasmin	270. Proscar
271. Flovent	272. Caduet	273. Lupron
274. Pepcid	275. Captopril	276. Doxepin
277. Macrobid	278. Entex	279. Lamisil
280. Prednisolone	281. Piroxicam	282. Sinemet
283. Bentyl	284. Dopamine	285. Lidoderm
286. Zestril	287. Imdur	288. Clarinex
289. Butalbital	290. Combivent	291. Levitra
292. Ampicillin	293. Versed	294. Duragesic
295. Bextra	296. Cardura	297. Hydralazine
298. Pseudoephedrine	299. Enebrel	300. Indocin

Americans and Australians in this heavy-drinking category are most aware of the harms that come with their habit, while the Portuguese are the least aware. Considering the glamorization of alcohol and incessant publication of 'science' articles that contradict each other, this result is not unexpected.

Equally disturbing is that awareness of national drinking guidelines was strikingly low in nearly every country. In Germany, for example, only one in three respondents knew their national guidelines. It brings into question the usefulness of such national guidelines.

The Republic of Ireland recorded the highest rates of employees turning up at work hung-over, with half of the respondents admitting to having gone to work after having a few too many. The U.K and Hungary followed the ranking, while the U.S. and Portugal recorded the lowest numbers. This is not necessarily a good news as other factors play a role. HR.com (2005) reported a survey of more than 2,000 people by Peninsula - cunningly timed to coincide with England's Euro 2004 match with Switzerland - found that 70 per cent of people in the UK admit to falling asleep on the job, while almost 80 per cent have taken a day off work to nurse a hangover. Peninsula also found that only a quarter owned up to taking hangover-related absence 'several times' during their working life, while one in ten said they had only done so once. Research also shows that it is light-to-moderate drinkers, not alcoholics as commonly perceived, that cause the most productivity problems. One study in the US found that more than half of all light drinkers, and 87 per cent of light-to-moderate drinkers, cause problems in the workplace. The primary illness that affects these amateur drinkers is the hangover, not the long-term consequences of alcohol abuse, such as inflammation of the liver and heart disease. Chronic alcoholism is responsible for only a small proportion of the total societal cost of alcohol use. They also found that people with hangovers posed a danger to themselves and others long after their blood alcohol levels had returned to normal, suggesting that hangovers could be more insidious than actual inebriation.

2.8.1 Illicit drug use

GDS surveyed some 1500 people to develop a comprehensive database on illicit drug consumption.

Figure 2.31 shows uses of different drugs, including illicit ones. Whilst alcohol, tobacco and cannabis remained the most common drugs

Table 2.3 Illicit drug use for different countries.

Countries	% of respondents that bought drug on internet
United Kingdom	22.1
Denmark	19.8
France	14.7
USA	14.3
Netherlands	10.4
Germany	10.3
Australia	9
Belgium	6.3
Hungary	5.4
Switzerland	5.3
New Zealand	4
Whole sample	11
Scotland*	20.5
Brazil*	11.1
Republic of Ireland*	10.7
Spain*	9.5
Mexico*	8
Portugal*	6.1

*Smaller(600-1500) survey samples

used within 2013, with cocaine, amphetamine in its various forms and MDMA frequently just behind them, countries showed marked variation in the use of other drugs. The increasing uptake of other preparations nicotine containing products namely shishas tobacco and electronic cigarettes demonstrate the pervasive presence of diverse nicotine based products in our culture. Considering the fact that e-cigarettes are worse than cigarettes, this poses a serious health concern for the future.

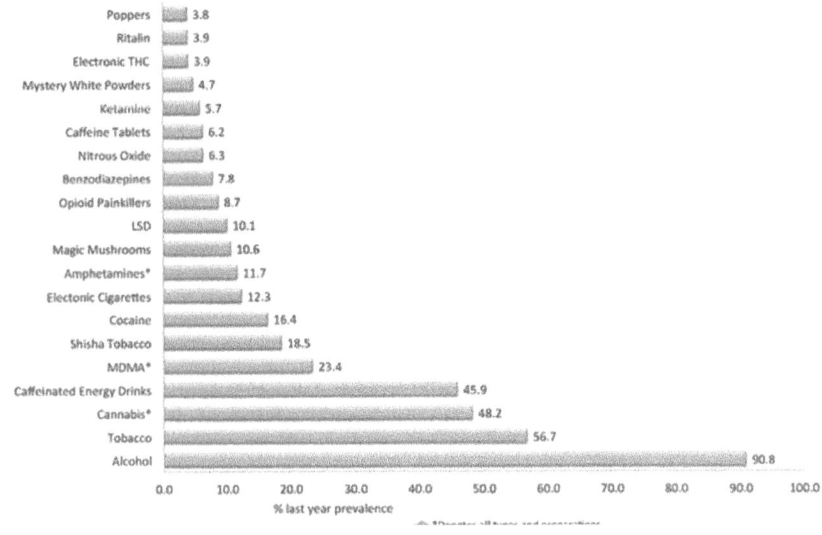

Figure 2.31 Usage of different types of drugs.

The high rates of caffeine energy drinks, caffeine tablets (and in some countries like Germany even intranasal caffeine) demonstrate the market for this legal stimulant is as strong as ever. Prescribed and non-prescribed psychoactive medication particular opioid painkillers and benzodiazepines were frequently in the top 10 drugs used by GDS populations in the last year, with their use, non-medical and problematic use being particularly high in the USA and New Zealand being dominant forces. Other medications that crossed over into the recreational drug scene such as GHB, methyphenidate (Ritalin) and ketamine were more sporadically distributed.

This damage to the overall public health comes with a price tag. Drugs prices varied widely – the average price of gram of high potency cannabis being 12 euros but varied almost fourfold from 6 euros in Spain to over 20 in Ireland. Cocaine remained the most expensive drug at mean global single gram purchase price of 100 euros / gram (ranging in price from 50 in parts of Europe to over 250 in NZ, which also had the most expensive MDMA as well).

Regardless of price, cocaine was voted the worst value for money drug in the world, with a mean score of 3.4/10. MDMA was voted the best value for money drug in the world. The Belgians were the most satisfied with their cocaine with a mean rating of 5.5/10 and the Australians the least with rating of 2.2.

2.8.2 Medical costs

Alcohol remained the biggest cause of concern among friends and the biggest culprit in sending people to Emergency Department. The percentage of last year drinkers who had sought emergency again varied widely from an average of just 1%, to 0.7% in France to over 2% in Ireland. Awareness of national drinking guidelines was universally poor with over 40% of drinkers being unsure of their countries drinking guidelines. The Germans were most clueless with 65% being unsure of them, the Danes the best informed with only 8% reporting they did not know them (that did not stop 1.5% of last year Danish drinker seeking emergency medical treatment following drinking last year).

The rates of seeking emergency medical treatment for other drugs other than alcohol varied widely. Further research is required to determine the factors that underlie the 3 fold difference in seeking emergency medical treatment (EMT) following the use of MDMA between Switzerland with the lowest rate of seeking EMT (03% of last year users) and the USA, The Republic of Ireland and France (09.1-1%).

As expected, cannabis may not be harmless in the short term for everyone who uses it, with an average of 0.5% of last year users seeking EMT. Again wide variation between countries was seen, the highest rates being in Germany where a similar proportion of last year user of cannabis sought EMT as drinkers (1%).

The danger of synthetic cannabis is clearly demonstrated in the fact that in the USA (for which it was 0.2%. Synthetic cannabis products, which remained the most commonly used novel psychoactive drugs across the sample) continued to be associated with high rates EMT visits. Compared to traditional cannabis research by GDS estimates their use is associated with a 30 times higher risk of seeking EMT.

In terms of severity of addiction, rates of tobacco use amongst the reported sample was high and although over half of all tobacco smokers expressed a desire to use less or stop and many wanted help to stop, only a tiny minority reported using forms of nicotine replacement therapies (NRT) to help them smoke less. This is a missed opportunity, since the use of NRT is not only associated with reduced tobacco use but can double the likelihood of a successful quit attempt. However, this book will discuss some of the misconceptions that are prevalent in this regard.

The migration of shisha tobacco use beyond its once tightly confined geographical boundaries accompanied the rise of electronic cigarettes though there was wide variations in the prevalence of their use between countries. Inefficacy of e-cigarettes was evident in the fact that only 10–20% of those who had tried them reported they had either stopped smoking altogether or had significantly reduced their consumption.

Our huge study of over 38,000 cannabis users showed that the USA was home to safest smokers - with only 7% choosing to smoke cannabis with tobacco followed by NZ (25%), compared to over 80% of smokers in most other countries. Although the most sensible cannabis smokers, the USA was the worst place to get caught with cannabis with over 17% reporting that it impacted on their education, employment, and travel.

3

HSS®A® Degradation[1] In New Science

3.1 Summary

There is a built-in prejudice in Eurocentric culture that we have evolved from the ancient 'savagery' to current 'enlightened' status through continuous evolution. Despite occasional realization that ancient cultures had life styles that are environmentally more sustainable than today's lifestyles, few give any thought that there may be a fundamental problem with the notion that our civilization is superior in terms of knowledge. Our education system is such that all such ingrained prejudices prevent us from invoking any paradigm shift, without which any deconstruction of existing theories is hopeless. In this chapter, major

[1]Since 2007 the authors have striven to popularize this Honey → Sugar → Saccharin® → Aspartame® pattern as a kind of shorthand reference to a particularly critical insight, often overlooked or only tangentially acknowledged, into the essence of the transformation of natural into artificial. This is a pattern so characteristic and widespread across every department of modern industrialized commodity production as to have become invisible. (The aha! character of that invisibility itself was best captured almost five centuries ago in Molière's *Le bourgeois gentilhomme*, where the hero M Jourdain famously remarks on his delight to discover that he had been "speaking prose all his life".) "HSS®A®" represents an entire class of other processes of degradation of a gift of Nature by its commodification as a byproduct of industrial-scale organic chemistry.

prevalent theories of the modern age are deconstructed. Deconstruction of such notions then can lead to the reconstitution of a correct system, based on a new paradigm.

The HSS®A® pathway is a metaphor representing many other phenomena and chinks of phenomena that originate from in some natural form but become subsequently engineered through many intermediate stages into "new products". These "new products" include materials, technology, and thought processes. This chapter identifies HSS®A® pathway in theories of physics as discussed by all major scientists and philosophers.

Most notably, the works of Newton, Maxwell, Einstein, Feynman, and Hawking are reviewed and their fundamental premises deconstructed. Once identified, it becomes clear how disinformation works in the current system in the context of laws and theories of physics. One can then figure out a way to reverse the process by avoiding aphenomenal schemes that lead to ignorance packaged with arrogance. As a consequence, fundamental change in theorization will be invoked, leading to the emergence of correct theories that can be then applied to develop sustainable technologies.

3.2 Introduction

Throughout history matter, energy and thought have been the primary focus of human civilization. In fact, we defined a civilization by the way it handles human needs for matter and energy and the thought process that predict these needs. This chapter reviews all major theories of matter and energy and presents a comprehensive and critical review, along with deconstruction that reveals the shortcomings of each of these theories. These shortcomings are listed as a fundamental premise that is aphenomenal. It is then demonstrated that such premises were allowed through dogmatic assertion as centuries old premises long became obsolete. In this process, every topic has reversed its meaning. Table 3.1 summarizes the original meanings and modern meanings of various keywords.

This chapter is aimed at shed light on the process that rendered scientific process into a propaganda tool for redemption of a hidden scheme. The metaphor of HSS®A® (Honey → Sugar → Sachharine®→ Aspartame® degradation is used.

Table 3.1 Modern Subjects and their original meaning as compared to modern implications (from Islam ***et al.***, 2013).

Subject	**Popular meaning in New Science**	**Most probable root meaning**
Algebra	Study rules of operations and relations and the construction of terms, polynomials (non-linear), and equations. Most commonly used algebra: Linear Algebra.	*Al-jabr* (Arabic) Reunion or restoration of broken parts (as in bone restoration)
Biology	Study of engineered bodies and artificial living.	Study of natural living bodies.
Calculus	Branch of mathematics focused on limits, derivative, integrals, and infinite series.	Small stone used for counting
Chemistry	Study of the properties and composition of artificial matter and their manufacturing (mass production)	Study of black (mineral rich) soil. Dates back to the knowledge of smelting in King David's time. *Al-Kimiya* (Arabic) means: the art of transmuting metals.
Economics	Analysis of production, distribution, and consumption of goods and services – all regulated by the government or the establishment. 2011 Nobel Prize in Economics on "two-way relationship between the economy and policy – how policy affects the economy and vice versa".	Natural laws of family estate (household goods). Distinct from politics or policy making. In Arabic, the equivalent root word (qsd) stands for optimizing (economizing) and dynamic intention.
Geometry	Study of straight lines, angles, circles, planes, and other idealized (yet artificial) shapes.	Measurement of Earth, its content, and other celestial bodies
Medicine	Study of artificial chemicals that delay the symptoms.	The art of healing

(Continued)

Table 3.1 cont.

Subject	**Popular meaning in New Science**	**Most probable root meaning**
Philosophy	Study of metaphysics, ethics, doctrinal logic, argument, etc.	Love of wisdom (that comes from logical thinking: homo sapiens means 'logical/thinking human'; in Arabic, it is: *hekma* (wisdom) or *aql* (logic)
Physics	Study of artificial matter and energy in such fields as acoustics, optics, mechanics, thermo-dynamics, electromagnetism, nuclear, and others.	Study of nature
Science	Study of theories as prescribed by New Scientists.	Logical deconstruction of phenomena or events.
Surgery	Study of radical procedures, specializing in controlling pain, bleeding, and infection, all through artificial means, such as laser surgery, antibiotic, anesthetic, etc. Preventive (breast, ovary, appendix) or cosmetic surgery is common.	Work of hand in order to heal

3.3 The HSS®A® (Honey → Sugar → Saccharin®→Aspartame®)$_{\text{Pathway}}$

3.3.1 Delinearized History of Honey

Honey dates back for 150 million years as the most natural food. Honey has been written about in hieroglyphics. When archaeologists opened an 18th Dynasty Egyptian tomb in the Valley of the Kings, they discovered vessels full of honey that remained edible and preserved its qualities even after 3000 years of storage. It is known that honey has been in use for medicinal as well as preservative in addition to its use as a sweetener. For instance, honey was used for in the mummification process and to-date it remains as the most effective preservation agent.

Honey was also used in all ancient cultures for as long as historical records have been preserved. The earliest known depiction of beekeeping appears on a carving from an Egyptian temple that dates to 4,500 years ago. Cave paintings that were discovered in Valencia, Spain showed the process of collecting honey. In the painting, two women are collecting both honey and honeycombs from the nests of wild bees. The painting has been dated to the Mesolithic period.

Ancient Chinese are known to have developed beekeeping techniques. In the book "Golden Rules of Business Success" written by Fan Li (or Tao Zhu Gong) during the Spring and Autumn Period there are some parts mentioning the art of beekeeping and the importance of the quality of the wooden box for bee keeping that can affect the quality of its honey. Similarly, every culture ranging from Mayan to ancient Greek, from ancient Indian to Roman, from ancient African to Roman had profound connection to bee keeping and honey. For instance, the research group of award winning Archaeologist, Amihai Mazar of the Hebrew University of Jerusalem, discovered evidence of bee keeping in the form of 30 intact bee hives from King Solomon's time. The hives were found in orderly rows, three high, in a manner that could have accommodated around 100 hives, held more than 1 million bees and had a potential annual yield of 500 kilograms of honey and 70 kilograms of beeswax. In ancient Greece, aspects of the lives of bees and beekeeping are discussed at length by Aristotle. Beekeeping was also documented by the Roman writers Virgil, Gaius Julius Hyginus, Varro, and Columella.

Honey has also appeared in many different ancient religious texts including the Hebrew Bible, the New Testament and the Qur'an. Greeks and Roman referred to honey as a food fit for the gods. Some are of the belief the honey is the ambrosia of the gods. Greek custom was to offer honey to the gods and deceased spirits, this tribute kept one out of harm's way and in a spirit or god's good graces. In Buddhism, honey plays a major role in some religious festivals. One in particular, Madhu Purnima, celebrates the retreat of Buddha into the wilderness as a way of making peace among his various disciples. In ancient India, yogurt and honey (*madhupak*) represented sacredness of a wedding. Such consistency is not unexpected. Qur'an, the only religious scripture that remains intact from the time of original revelation says (3:19):

> Indeed, Allah's prescribed lifestyle is Islam. And those who were given the Scripture did not differ except after knowledge had come to them - out of jealous animosity between themselves. And whoever disbelieves in the verses of Allah, then indeed, Allah is swift in [taking] account.

The English word 'honey' is known to have the Hebrew root, ע-נ (ayin-nun, read right-to-left) that means "joy". The root 'hani' means happy in Arabic. Interestingly, the Sanskrit word for honey is Madhu, which is a root word for honey wine or simply wine in a number of other languages. For instance in English, it is mead, Greek *μέθυ*, Avestan madu, Persian may, Latvian medus, German Met and Old Church Slavonic мєдъ (medŭ). It appears that this correlation of honey with alcohol or wine is more of a metaphoric one, e.g., "the wine of truth," and employed in that manner in Hindu religious literature. For example, the Brihadaranyaka Upanishad, believed to have been composed in the first millennium BCE, contains a chapter called the Madhu Brahmana and "the secret essence of the Vedas themselves, was called the madhu-vidya or honey doctrine." Similarity among various ancient cultures also exists in terms of honey being synonymous with something good as in honeymoon. The word 'moon' applies to the month, which is dictated by lunar calendar in most of the ancient calculations. In fact, both in Arabic and Turkish the word that stands for honeymoon literally translates into 'honey month'.

It is remarkable that the Qur'an (16:69) lists honey as a 'cure for mankind'. Only honey got that distinction by a book that is the oldest preserved book in history of mankind (Khan and Islam, 2012). The science of nature tells us honey is the most natural food we can find on earth. It also never rots and is the only food that prevents both bacterial and fungal growth. Today, it's recognized honey is the most wholesome food that has all nutritional components required for sustaining life by itself. Throughout history honey has been used as a medicine and immunity booster. Even in New Americas honey was used for all purposes included fighting allergy, infection (both bacterial and fungal) until the beginning of the era of artificial that saw the emergence of sugar as a substitute for honey and plastic as a substitute for natural material.

While scientifically honey is diametrically opposite to sugar, New Science characterized it as a "collection of sugar, fructose, and glucose", all of which are artificially manufactured (Khan and Islam, 2012). To-day, practically every scientist use this denomination of honey with no mention that these are not the same compounds that can be produced artificially (Mughal, 2013). Such omission is attributed to a number of myths that have plagued modern science (Islam *et al.*, 2012). Following is a list of most relevant (to human health) features of honey for both internal and external applications (Mughal, 2013):

1. Honey is rich in humectant compounds. This helps retain the moisture content in skin and restore its elasticity, making skin supple.
2. It helps remove dead skin cells and prevent the appearance of wrinkles.
3. The antibacterial and antimicrobial property of honey helps prevent the growth of certain bacteria and so it is used to treat wounds, cuts, burns and abrasions.
4. Honey helps clean wounds, reduces odour and pus, lessens pain and promotes speedy healing.
5. Honey is used to treat damaged skin and regenerate new skin cells. It also effectively cures eczema, dermatitis and other skin disorders.
6. Owing to the potent anti-fungal properties in honey, it can help cure infections like athlete's foot and jock itch.
7. Honey is loaded with natural antioxidants, which help protect the skin from ultraviolet damage.
8. Honey can be used as a sunscreen to protect the skin from the sun.
9. Honey deeply penetrates the skin's top layer, unclogs pores and sloughs off impurities. Thus, it helps fight infections and curb acne problems.
10. Honey is a good skin moisturiser and effectively tones and firms up the skin.

11. The application of honey on chapped and wrinkled lips will make them wonderfully smoother and softer.
12. Honey is a natural source of vitamins B1, B2, C, B6, B5 and B3, all of which vary according to the qualities of the nectar and pollen. Copper, iodine, and zinc also exist in honey in small quantities.
13. Honey is a natural source of organic glucose and fructose that can supply the body with immediate energy, which can boost endurance and reduce muscle fatigue.
14. Honey helps relieve morning sickness.
15. Regular consumption of honey will increase calcium absorption and haemoglobin count helping to fight anaemia.
16. Honey also helps lower total cholesterol while increasing HDL (good) cholesterol.
17. It can also be used to treat respiratory tract infections owing to its expectorant, soothing and demulcent properties.
18. As honey helps modulate the immune system, it prevents the recurrence of infections.
19. Honey plays a pivotal role in managing obesity. It helps accelerate metabolism, which in turn helps the body burn more fat.

A review of ancient documents reveals honey's usefulness in curing the following diseases, as listed in Table 3.2.

This exceptional quality of honey makes it a unique material that should be used as standard for material selection and determination of sustainability. One can argue that honey is the most sustainable nourishment nature has to offer. This conclusion is unavoidable if one takes the criterion of sustainability proposed by Khan and Islam (2007) and later elaborated by Zatzman and Islam (2007). In order to apply this criterion, we have to analyze the source and pathways of honey. Here are the components to consider:

- the honeybee;
- nectar; and
- process of honey production.

Table 3.2 Medicinal use of honey.

Acidity	**Anger**
Anorexia	Apnea and other nasal disorders
Asthma	Bed wetting
Birthing delay (delay in delivery)	Blister of the mouth
Bloor disorders	Brain disorder
Bug bite	Burning sensation (caused by any of the ailment listed in Table 2)
Cataract	Catarrh
Chicken pox	Conjunctivitis (pink eye)
Constipation	Convulsion
Cough	Cough with phlegm
Cracked lips	Deafness
Dementia	Dentition
Depression	Diarrhea
Dropsy (Edema)	Dysentery (with blood)
Earache	Ear diseases
Elongation of ovula	Epilepsy
Erysipelas	Eye inflammation
Eye pain	Fatigue
Fistula	Fungas (on nails, etc.)
Headache	Hemicrania
Hermatemesis	Hiccup
Hypertension	Hoarseness
Immune deficiency	Impotence
Indigestion	Influenza

(Continued)

Table 3.2 cont.

Acidity	**Anger**
insect bite	Insomnia
Itching and scabbies	Jaundice
Leprosy	Lung disease
Malaria	Menstrual disorder
Nervousness	Night blindness
Obesity	Otorrhoea
Paralysis	Piles
Pneumonia	Pyorrhea
Renal disorder	Ribs pan
Ringworms	Scorpion sting
Scrofula	Sinus of the eye
Somnabulism	Spasm of hands and feet
Sprain	Stomach ache
Stomach worms	Swelling or inflammation of tongue
Throat tumours	Throat inflammation
Tonsillitis	Toothache
Trachoma and corneal opacity	Tuberculosis
Urinary tract infection	Urticaria
Vomiting	Whooping cough

Table 3.3 Diseases that give burning sensation (information extracted from Wikipedia).

Beriberi	**Beriberi is a disease brought on by a Vitamin B-1 (thiamine) deficiency. It can lead to poor heart function and even heart failure. A symptom of dry beriberi is a tingling or numbness in the hands and feet.**
Calcium Deficiency Disease	A lack of calcium can lead to diseases such as osteoporosis and calcium deficiency disease. Severe deficiency can result in numbness and tingling in the mouth, lips, hands and feet, among other signs.
Canker Sore	A canker sore (aphthous ulcer) is a mouth ulcer or sore that is open and painful. The most common sign is a burning or tingling in an area of your mouth that the red sore appears.
Carpal Tunnel	Carpal tunnel syndrome is the compression of the median nerve, the nerve that passes through the wrist. A common symptom is numbness or tingling in the thumb and first three fingers.
Central Pain Syndrome	Central pain syndrome is a neurological disorder affecting the brain or spinal cord. It can be caused by many conditions, such as stroke or MS. Persistent pain is usually described as a burning or tingling.
Cervical Spondylosis	Cervical spondylosis is a condition that results from the wear and tear of the cartilage and bones of the cervical spine. One sign is tingling in the shoulders and arms, and sometimes the legs.
Frostbite	Frostbite occurs when the skin is exposed to extreme or prolonged cold. The skin and tissues beneath the surface of the skin freeze. This will cause a prickly sensation, discoloration, and pain.

(Continued)

Table 3.3 cont.

Beriberi	**Beriberi is a disease brought on by a Vitamin B-1 (thiamine) deficiency. It can lead to poor heart function and even heart failure. A symptom of dry beriberi is a tingling or numbness in the hands and feet.**
Gastroesophageal Reflux Disease (GERD)	Acid reflux occurs when stomach contents, including food, stomach acid, and digestive juices flow up from the stomach to the esophagus. It can cause heartburn, a burning sensation in the throat.
Herpes Simplex	The herpes simplex virus, also known as HSV, is an infection that causes herpes of the genitals or the mouth, depending on the type. It causes blistering sores that may burn or itch.
Hypoparathyroidism	Hypoparathyroidism is a rare condition that occurs when the parathyroid glands in the neck don't produce enough parathyroid hormone (PTH). It can cause burning or numbness in the fingertips, toes, or lips.
Leprosy	Leprosy is a chronic, progressive bacterial infection that primarily affects the nerves of the extremities, lining of the nose, and upper respiratory tract. It can cause numbness or dulled sensation.
Marine Animal Bites or Stings	Some see creatures deliver venom through their teeth, tentacles, spines, or skin. If a sea creature bites or stings you, the affected area may burn, swell, turn red, or bleed. Sometimes, it can be deadly.

(Continued)

Table 3.3 cont.

Beriberi	**Beriberi is a disease brought on by a Vitamin B-1 (thiamine) deficiency. It can lead to poor heart function and even heart failure. A symptom of dry beriberi is a tingling or numbness in the hands and feet.**
Megaloblastic Anemia	Megaloblastic anemia is a blood disorder marked by the appearance of very large red blood cells that crowd out healthy cells, causing anemia. Numbness or tingling in hands and feet is common sign.
Mononeuritis	Mononeuritis multiplex (MNM) is a disorder of the nervous system that affects two areas of the nervous system and can result in severe pain, loss of motor ability, and a tingling sensation in one or more areas.
Mononeuropathy	Mononeuropathy is a pathological effect of a single large nerve. This could be the result of injury, nutritional deficiency or autoimmune condition. A common symptom is numbness, tingling or burning that is in the same area of the malfuncting nerve.
Multiple Sclerosis	Multiple sclerosis is an autoimmune disease that affects the central nervous system (brain and spinal cord). Many people with MS experience stabbing or burning pain.
Neuralgia	Neuralgia is a stabbing, burning, and often quite severe pain that occurs along a damaged nerve. It has several causes, from shingles to diabetes or multiple sclerosis.
Peripheral Neuropathy	Peripheral neuropathy is a disorder that occurs when these nerves malfunction because they are damaged or destroyed. It causes tingling or numbness in the hands or feet.
Peripheral Vascular Diseases (PVDs)	Peripheral vascular diseases (PVDs) are circulation disorders that affect blood vessels outside of the heart and brain. Symptoms include cramping, achiness, fatigue, and burning.

(Continued)

Table 3.3 cont.

Beriberi	**Beriberi is a disease brought on by a Vitamin B-1 (thiamine) deficiency. It can lead to poor heart function and even heart failure. A symptom of dry beriberi is a tingling or numbness in the hands and feet.**
Pernicious Anemia	Pernicious anemia (PA) is an autoimmune disorder in which the body fails to make enough healthy red blood cells (RBCs), resulting in vitamin B-12 deficiency. Most people experience a burning or sore tongue.
Radiculopathy	Radiculopathy refers to disease of the spinal nerve roots. This causes pain and feelings of numbness or weakness to radiate to the extremities
Rosacea	Rosacea is a chronic skin disease marked by redness on the face. Signs of rosacea subtype 1 include flushing, redness in the center of the face, broken blood vessels, and stinging or burning skin.
Sciatica	Sciatica is the burning pain you feel in your back, buttocks, and legs when your sciatic nerve is irritated. The pain usually worsens with movement.
Shingles	Shingles is an infection caused by the virus varicella-zoster, which is the same virus that causes chickenpox. Burning pain on one side of the body usually occurs in small patches, followed by rash.
Slipped (Herniated) Disk	Injury or weakness can cause the inner portion of the cushions between vertebrae (disks) to protrude through the outer ring. This can result in pain and numbness, muscle weakness, and burning sensations.

(Continued)

Table 3.3 cont.

Beriberi	**Beriberi is a disease brought on by a Vitamin B-1 (thiamine) deficiency. It can lead to poor heart function and even heart failure. A symptom of dry beriberi is a tingling or numbness in the hands and feet.**
Snake Bites	A bite from a venomous snake can lead to an allergic reaction or infection and may be deadly. Common symptoms include pain, swelling, blurred vision, and numbness of the face and limbs.
Stasis Dermatitis And Ulcers	Stasis dermatitis is skin inflammation caused by blood pooling in the veins in your legs. It's a chronic condition that can cause discomfort such as burning and itching.
Ulnar Nerve Palsy (Dysfunction)	Damage to the nerve that runs from your shoulder to your little finger (ulnar nerve) is known as ulnar nerve palsy. It causes sensation loss and tingling or burning in your hand.
Whiplash	Whiplash results when the soft tissues—the muscles and ligaments—of your neck are extended beyond their typical range of motion. In addition to neck pain, a burning or prickling sensation may develop when the nerves are injured.

3.3.1.1 The Honey Bee

The Qur'an has a long chapter, titled "The Bee" (*Al-Nahl*, Chapter 16). This chapter is unique to creating sustainable material and healthy life-style. Indeed, no other insect serves human needs like honeybees. Bees pollinate an estimated one-third of all the food crops we consume. Here are some characteristic features of honey bees:

1. Honey bees can fly at speeds of up to 15 miles per hour. This speed is perfectly choreographed for short trips from flower to flower. Their tiny wings must flap about 12,000 times per minute just to keep their pollen-laden bodies aloft for the flight home.
2. A honeybee colony can contain up to 60,000 bees at its peak. Nurse bees care for the young, while the queen's attendant workers bathe and feed her. Guard bees stand watch at the door. Construction workers build the beeswax foundation in which the queen lays eggs and the workers store honey. Undertakers carry the dead from the hive. Foragers must bring back enough pollen and nectar to feed the entire community.
3. A single honeybee worker produces about 1/12th of a teaspoon of honey in her lifetime. From spring to fall, the worker bees must produce about 60 lbs. of honey to sustain the entire colony during the winter. It takes tens of thousands of workers to get the job done.
4. A queen honeybee stores a lifetime supply of sperm. The queen bee can live 3-4 years, but this time is not for finding a mate, instead the time is mostly spent for sustaining the beehive. Just a week after emerging from her queen cell, the new queen flies from the hive to mate. If she doesn't do so within 20 days, it's too late; she loses her ability to mate. If successful, however, she never needs to mate again. She holds the sperm in her spermatheca and uses it to fertilize eggs throughout her life.
5. The queen honey bee lays up to 1,500 eggs per day, and may lay up to 1 million in her lifetime. Just 48 hours after mating, the queen begins her lifelong task of laying eggs. She can produce her own body weight in eggs in a single day. In fact, she has no time for any other chores,

so attendant workers take care of all her grooming and feeding.

6. The honeybee uses the most complex symbolic language of any animal on earth. Their communication method as well as navigation style are subject of modern research and not yet fully understood.
7. Honey bees have a million neurons in a brain that measures a mere cubic millimeter, and they use every one of them. Worker bees must perform different roles throughout their lives. Foragers must find flowers, determine their value as a food source, navigate back home, and share detailed information about their finds with other foragers. Karl von Frisch received the Nobel Prize in Medicine in 1973 for cracking the language code of honey bees – the waggle dance.
8. Drones, the only male honey bees and they die immediately after mating. Male honeybees serve only one purpose: they provide sperm to the queen. About a week after emerging from their cells, the drones are ready to mate. Once they've fulfilled that purpose, they die.
9. Honey bees maintain a constant temperature of about 93° F within the hive year-round. A sophisticated temperature control system is place within a beehive. As temperatures fall, the bees form a tight group within their hive to stay warm. Honeybee workers cluster around the queen, insulating her from the outside cold. In summer, the workers fan the air within the hive with their wings, keeping the queen and brood from overheating. The sound of this hum of all the bees is audible from several feet away.
10. Honey bees produce beeswax from special glands on their abdomens. The youngest worker bees make the beeswax, from which workers construct the honeycomb. Eight paired glands on the underside of the abdomen produce wax droplets, which harden into flakes when exposed to air. The workers must work the wax flakes in their mouths to soften them into a workable construction material.
11. An industrious worker bee may visit 2,000 flowers per day. She can't carry pollen from that many flowers at

> once, so she'll visit 50-100 flowers before heading home. All day long, she repeats these round trip flights to forage, which puts a lot of wear and tear on her body. A hardworking forager may live just 3 weeks.

Even bee stings are beneficial to mankind (Telis, 2013). Bee venom is found to be useful for curing both parasite infection and immunity from allergies.

Honey bees also form an ideal of social behavior with optimum social structure and labour partition (Johnson, 2010). Group-level coordination of action and colony integration are effectively used by the bees to achieve a unity of purpose among the members of a colony (Seeley 1998). How societies of many thousands are able to achieve a unity of purpose has long intrigued biologists (Wilson 1965; Michener 1969, 1974; Calderone and Page 1988; Beshers and Traniello 1994; Seeley 1995; Jeanne 1999, 2003; Visscher 2007; Mertl and Traniello 2009; Johnson 2010; Johnson and Linksvayer 2010; Stabentheiner *et al.* 2010).

However, recent work of Islam *et al.* (2014) shows that there is no mystery in this behavior. Everything belonging to nature behaves in an optimal fashion. Only humans deviate from such optimum behavior. This distinct nature of humans has eluded scientists of modern age.

Biello (2006) summarized the unique features and 'superior' genetic traits of honey bees. Each aspect of honeybee's recognition of flower's color and shape as well as to waggle-dance that information back to its hive is a scientific miracle. It is recognized that honeybees possess the most complex symbolic language outside of our own family (primates) and they do it with a brain of only one million neurons–five orders of magnitude less than that of humans. This shows how hopeless the science of 'bigger the better' is. It is a puzzle that honey bees contain only four times as many neurons as possessed by fruit flies, flitting creatures that lack any form of society. Yet, the newly sequenced genome of the honeybee–Apis mellifera–reveals that some of the same genes that guide the development of a fruit fly's nervous system have been repurposed in the bee to control its genetic ability to switch social roles, among other hive secrets. It poses two problems for humans. First, it shows that our overwhelming them of counting neuron as a unit is flawed. Secondly, it shows that we are no where close to properly characterize function of a brain – any brain.

Biello reported the work of George Weinstock of the Baylor College of Medicine's Human Genome Sequencing Center and Gene Robinson of the University of Illinois at Urbana-Champaign. This work ultimately enlisted 112 scientists at 63 different institutions to help piece together the genetic puzzle of the honeybee, a process that spanned four years. "The honeybee was picked for sequencing because it is a popular system to study social behavior and because of its importance in agriculture through pollination as well as producing honey," Weinstock was quoted. "Honeybees are the premier pollinator on earth, contributing $10 billion to $20 billion worth of value to our economy," Robinson added. Using a large number of drones and one partially inbred queen, the researchers gathered more than three million DNA sequences. After deciphering the sequence composition, a computer helped piece them back together into a nearly complete genetic map–an approach called the whole genome shotgun strategy (also used to sequence the human genome quickly). Much like the human genome, some gaps remain, and the overall gene count is low at this point. "We described about 10,000 genes, about 30 percent less than the fruit fly or mosquito," Weinstock notes. "There are likely other genes to be found that were missed because at 300 million years distance from other [insect] genomes we cannot rely as much on sequence similarity to pick out genes."

Despite diverging from human ancestors more than 600 million years ago, the bee shares a number of genes with its vertebrate cousins that its insect brethren lack, such as those involving RNA interference, aging, DNA methylation and circadian rhythms. The honeybee was found to differ from its insect counterparts in a number of other ways as well, including more genes for smell as well as for making use of nectar and pollen, and fewer genes for innate immunity, protective outerwear, detoxification and taste. For honey bees and flowers, it's all a matter of signal as it appears that the flowers want them to come and pollinate. This is in sharp contrast to what most plants do for other insects, i.e., put up toxins for insects that are coming to feed on them. The honeybee didn't develop those defenses. This may help explain another mystery of honeybees: their precipitous decline in recent years. While, honey bees are inherently attracted to flowers, they are very vulnerable to artificial toxins. New science doesn't distinguish between artificial and real toxins (Islam *et al.*, 2010), leading to puzzlement of the scientists in explaining modern day decline of honeybees in the presence of pesticides, artificial energy source, and numerous other artificial components of modern day engineering.

It was also found that the honey bee exhibits very slow rate of evolution. At the same time, the bee has a relatively high rate of exchange of

DNA between chromosomes, or recombination, to preserve diversity with only one breeding female in a hive.

One can see amazing optimization of labour in a colony of honey bees. For instance, a colony of honey bees is able to make new comb at a rate sufficient to ensure that they do not run out of space, while at the same time collecting nectar and processing it into honey (Seeley, 1995). Furthermore, many of a social insect colony's tasks are such that tight cooperation between distinct groups is necessary (Anderson *et al.* 2001; Burd and Howard 2008). Nest building in Polybia wasps, for example, requires that a given ratio of workers be in each of three categories: collecting water, collecting pulp, and building the nest carton (Jeanne 1996, 1999). Workers in these categories must pass off materials to one another in an assembly line organization structure. To allocate the appropriate ratio of workers to each task group (a ratio that can change with changes in the environment) requires much information processing by the workers in each group (Seeley, 1998).

Of particular importance is the usage of signal and cues. They are important because of two reasons; 1. The signal and cue system may be altered by artificial signals, such as microwave and others, and 2. They serve as a great model for studying how nature optimizes itself, particularly in terms of dynamic optimization (Khan and Islam, 2012). The consensus view seems to be that cues are more important than signals because they are more numerous (reviewed in Seeley 1998; Beshers and Fewell 2001; Pankiw 2004). However, as Johnson (2010) pointed out, this creates something of a paradox. If signals are more sophisticated, because they have been the focus of selection, then colonies might be expected to rely on them in vitally important instances. An integrating hypothesis might therefore be that although cues are more numerous, signals tend to be disproportionately important because they play critical roles for which cues would be poorly suited. The entire contradiction and paradox disappear by realizing that both signals and cues are part of the universal order and are in perfect harmony.

Johnson (2010) presents an eloquent discussion on task partitioning of honey bees, focusing on colony integration. He discusses the type of information that needs to be transferred between groups to optimize task partitioning and explores possible ways of obtaining and transferring this information. He also presents general reasons why mechanical signals (physical signals dependent on body contact, sound, and substrate vibrations) are often superior to cues that could provide the same information. He explored the difference between 'active' versus 'passive' information transfer. In scientific term, this would be equivalent to

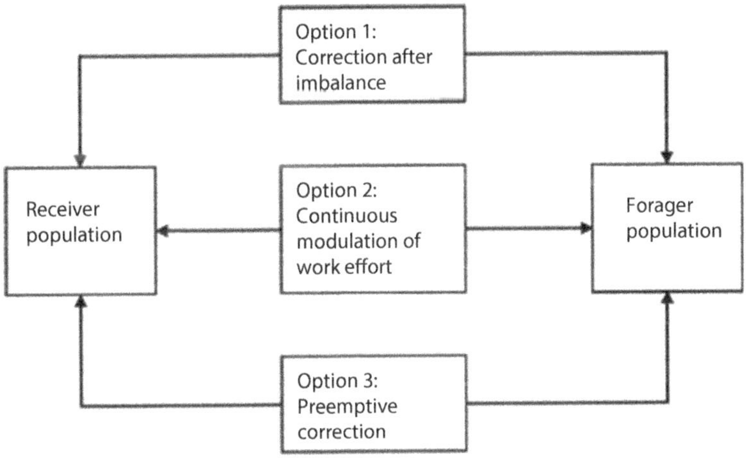

Figure 3.1 Strategies for dynamic optimization by bees (from Johnson, 2010)

different layers of dynamic optimization. He presented cues as passive information that workers may encounter, typically at random, whereas signals are actively sent (and often targeted) to their recipients. In truly scientific sense cue cannot detached from the 'purpose' unless it is assumed that nature is devoid of purpose.

Many cues are localized to one place and represent passive sources of information. However, some cues can travel rapidly over a distance and capture another worker's attention. Floral odors, for example, can reactivate unemployed foragers when they waft about the nest after the return of a successful forager (Reinhard *et al.* 2004). Similarly, pollen foragers that unload their pollen loads deep within the nest, transmit pollen cues throughout the nest (Calderone and Johnson 2002). The following section relies on the narration presented by Johnson (2010).

Honey production in the honey bee is a classic example of dynamically optimized food processing. Task partitioning is when one large task, such as food collection, is broken into 2 or more subtasks (Seeley 1992; Ratnieks and Anderson 1999; Anderson *et al.* 2001; Burd and Howard 2008). Foragers collect nectar in the field and transfer it within the nest to middle age bees that process the nectar into honey and store it in the honey comb. This is thought to increase the overall production of work because foragers can rapidly go back and forth to their flower patches, which have a limited period of productivity, without having to waste time in the nest looking for a cell in which to unload (Visscher and Seeley 1982; Seeley 1983). Although task partitioning can increase the overall production of work by allowing for such a "specialization" effect, the benefit is dependent on the smooth transfer of work from one group to the other (Jeanne 1986, 1996, 1999; Seeley 1992, 1995). For instance, let's assume that there are equal numbers of foragers and receivers and that individuals within both groups take 20 min to complete their task. In this case, the amount of nectar brought back to the nest will match the rate at which it is processed into honey and foragers and receivers should have short wait times for nectar transfer. However, if the foragers suddenly begin to take one half as long to collect nectar per trip (e.g., because a nearer flower patch comes into bloom), then there will be twice as much nectar coming in as before and the nectar receivers will be in short supply. Hence, the wait times of foragers needing to unload their nectar will drastically increase and time they could have spent foraging will instead be wasted within the nest (reviewed in Seeley 1995). Clearly, to increase honey production in this case, the colony would need some means of increasing the number of nectar receivers such that they once again can quickly process all the nectar coming into the nest (Seeley *et al.* 1996). Until today, how this feat is accomplished is a matter of conjecture in the scientific community. It suffices to state here that every step is optimized without which a bee colony cannot be sustained.

Honey bees are perceived to perform the task of matching work output between groups involved in task partitioning. This is shown in Figure 3.1. The first is to wait for the work output from one group to overwhelm the capabilities of the other and then call for help (Seeley 1992; Anderson and Ratnieks 1999). The benefit of such a system is that little information collection is necessary. All that is required of the insect brain is to determine if the level of some variable that correlates with the degree of mismatch between work outputs exceeds a given threshold. The costs are that work becomes out of balance before action is taken to correct it. Thus, although option one is an effective and common method for coordinating work, it not efficient at the colony level and requires further optimization.

Three options for coordinating activity in the context of task partitioning. The first option is to wait for the consequences of an imbalance between the work outputs of the two groups to build up such that it can act as a cue that action is necessary to realign work efforts. The second option is to modify work rates in time with changes in task demand. Honey bees could do this by noting their experience in the hive as they unload their nectar and changing their waggle dancing probability accordingly. Note that waggle dance is a term used in beekeeping and ethology for a particular figure-eight dance of the honey bee. By performing this dance, successful foragers can share, with other members of the colony, information about the direction and distance to patches of flowers yielding nectar and pollen, to water sources, or to new nest-site locations. This movement is complex and little understood by New scientists (Seeley *et al.*, 2006). A waggle dance with a very short waggle run used to be characterized as a distinct (round) recruitment dance. Austrian ethologist and Nobel laureate Karl von Frisch was one of the first who translated the meaning of the waggle dance (Frisch, 1967).

Option 3 is to preemptively signal to the other class of workers that a change in work-load is forthcoming. This is considered to be the optimal solution, from a work production perspective, as it prevents work stoppages. The honey bee shaking signal, which alerts within-nest workers, to future increases in work demand, is such a mechanism.

The second method is to modulate activity rates in time with changes in the demand for work. This would be the ideal solution, but it requires sophistication on the part of the workers. Basically, the workers would need accurate measures of current work flows and the ability to rapidly shift labor following an optimized distribution. Given that workers have a limited ability to sample information, they may not be able to obtain sufficiently accurate, and updated, measures of current task demand to do this. In general, although it is unlikely that social insects rely much on Method 2, it is still an open question that eludes most scientists. Work on how past experience within the hive affects current behavior in foragers may ultimately shed light on this question (Pirez and Farina 2004; Goyret and Farina 2005; Gruter and Farina 2009). However, this would require the study of the continuous time function – a task rarely undertaken in modern era.

The third option is to reallocate labor prior to a change in workload. Essentially, if it is known beforehand that workloads are going to change, then the workers at the receiving end should can be informed, or inform themselves, to get ready in preparation. The benefits of this are no work stoppages; while the potential costs relate to lost productivity should workers have to wait long for the arrival of work. Human organizations, for example, which have the capacity to make future predictions, often use this third method. However, as social insects are thought to rely on simple cues, this option has not been widely considered for them. However, contrary to the belief that predictive signals are too computationally demanding for social insects, a benefit of the third option is that predictive signals are often not computationally demanding. The reason will be explained below in conjunction with how the honey bee makes use of this option.

3.3.1.2 Nectar

Nectar is yet another marvel of nature. It is created by plants so insects and birds are attracted to them in order to help pollinate. Pollens are placed close to the nectar so they are carried to host plants. Nectars are created in such a way that they are attractive to birds and insects while they are also nutritious. This is a marvelous display of sustainability.

Flower nectar forms the source of all honey as well as bee pollen and part of beeswax. The attractiveness of nectar to pollinators is probably most affected by its taste (Adler, 2000), but may also be affected by odor (Raguso, 2004) and color (Thorp *et al.*, 1975). The taste of nectar is dominated by a high sugar concentration. Several studies have attempted to define whether honey bees prefer nectars that are rich in sucrose or hexose. Difficulties in this line of research lie within the fact that new science cannot distinguish between natural sugar and artificial sugar (Hossain *et al.*, 2009). Wykes (1952) showed that honey bees prefer a sugar ratio of 1:1:1 (sucrose/fructose/glucose) over a pure sucrose solution. Of course, this doesn't prove much on the actual attractive features of nectar as nectar doesn't have a single molecule of artificial sucrose. Without counting the history of nectar, it is possible to look at the 'sucrose' content. It is shown that avocado nectar contains almost exclusively sucrose and a low concentration of the unique seven-carbon sugar alcohol, perseitol (Ish-Am, 1994; Liu *et al.*, 1995; Dvash *et al.*, 2002; Dag *et al.*, 2003). Indeed, the high ratio of sucrose in avocado nectar has been suggested as the cause for the low attractiveness of avocado flowers to honey bees (Ish-Am, 1994). A preference for hexoses over sucrose is also suggested by physiological considerations, because sucrose has to be broken down before it can be utilized (Harborne, 1993). Other studies, however, have found that honey bees are indifferent to the ratio of sugars (Southwick *et al.*, 1981), or even prefer a high sucrose concentration (Bachman and Waller, 1977; Hagler *et al.*, 1990). The sensitivity of honey bees to perseitol has never been evaluated.

Other components of nectar, including minerals, phenolic compounds, and amino acids, may make a cardinal contribution to its attractiveness to honey bees. Minerals such as potassium (Waller *et al.*, 1972) and sodium chloride (von Frisch, 1950) deter honey bees. Phenolic compounds affect the taste of nectar even at very low concentrations (Baker, 1977), and in some cases they have been suspected of repelling honey bees (Hagler and Buchmann, 1993; Adler, 2000). It means contaminated nectar is naturally avoided by honey bees.

Several nectar amino acids have also been shown to affect preference (Kim and Smith, 2000; Gardener and Gillman, 2002; Carter *et al.*, 2006). Ish- Am (1994) found that glycine and histidine are the dominant amino acids in some avocado cultivars; at their naturally occurring concentrations, however, these amino acids do not repel honey bees (Inouye and Waller, 1984) and may even attract them (Kim and Smith, 2000). Avocado nectar is poor in proline, which is attractive to honey bees (Carter *et al.*, 2006), but so is citrus (cv. "Valencia") nectar, and yet it is highly attractive to bees (Ish-Am, 1994). Hence, it does not appear that amino acid composition can explain the low attractiveness of avocado nectar.

At the outset, odors guide bees toward flowers and may affect their attractiveness (von Frisch, 1967). The volatile components of nectar, including phenolic compounds, form particular odor bouquets (Anklam, 1998). Some of these compounds are more attractive to honey bees than others (Jay, 1986; Henning *et al.*, 1992; Winston and Slessor, 1993).

It is also believed that colors affect the attractiveness of flowers (von Frisch, 1967; Giurfa *et al.*, 1995), and nectar may contribute to their visual display (Thorp *et al.*, 1975). Color differences are especially salient when comparing honeys, which are derived from nectar, and can be used in choice experiments for testing the influence of nectar components on bees' preferences. However, it should be noted that the color factors come secondary to odor.

The dissolved solids in nectar consist mainly of carbohydrates (Luttge, 1977). However, a wide variety of minor components may define its nature (Adler, 2000). Because modern day material characterization depends on the main components that have 'significance' fraction present in the sample, the identification of identification of individual taste and odor compounds that may affect its attractiveness to pollinators is inherently difficult. Afik *et al.* (2006) separated various constituents of nectar and tested their effects on the preferences of honey bees. The response of honey bees to floral and honey odors evaluated the importance of volatile compounds. The odor of citrus flowers was more attractive to experienced bees than that of avocado flowers. These foragers were collected from a colony located among blooming citrus and avocado trees. Because honey bees tend to prefer citrus over avocado (Vithanage, 1990; Ish-Am and Eisikowitch, 1998; Gazit and

Degani, 2002), it is likely that they collected more bees that were foraging on citrus than on avocado. The higher response to the odor of citrus flowers probably reflected the bees' foraging experience in the field. The indifference of naïve bees to the floral odors supports the notion that odors act mainly as signals that bees learn to associate with their respective floral rewards, and do not themselves affect choice behavior greatly. This view was also supported by the indifference of bees to honey odors, while they clearly preferred nonavocado honey in taste assays. Taste, rather than odor, probably affected the choice of honey, because honey bees find it difficult to discriminate between honey odors (Bonod *et al.*, 2003).

The response of bees to different sugar solutions indicated that the sugar composition of avocado nectar cannot explain its low attractiveness. Their sensitivity to sucrose was found to be higher than their sensitivity to glucose or fructose, separately, thus supporting the findings of Wykes (1952), Waller (1972), and Bachman and Waller (1977). Sugar mixtures, however, are perceived differently relative to pure sucrose (Wykes, 1952; Waller, 1972; Bachman and Waller, 1977). The response to a hexose-rich mixture, "citrus nectar" in the present experiments, was similar to that of sucrose. Therefore, it appears that high sucrose content in avocado nectar does not diminish its attractiveness, in contrast to previous assumptions (Ish-Am, 1994). Bees did not respond to perseitol solution, and their response to sucrose solution containing perseitol was similar to their response to pure sucrose solution. Perseitol, which was suspected to deter honey bees from avocado (Ish-Am, 1994; Can-Alonzo *et al.*, 2005), seems to have no effect on preference. Thus, differences in sugar composition between avocado and citrus nectar cannot account for the bees'stronger preference for citrus flowers. Moreover, during the process of honey ripening, sucrose is inverted into glucose and fructose. As a result, the sugar compositions of avocado and non-avocado honeys are similar (Dag *et al.*, 2006). Nevertheless, higher consumption of non-avocado honey in the color and potassium experiments indicated that preference is determined by a component other than the dominant sugars. This finding also highlights the fact that our understanding of natural phenomena is very limited.

Honey color also cannnot explain the bees' preferences. Afik *et al.* (2006) discovered that bees were repelled by the avocado honey even when feeders were covered. Comparison of the consumption rate as the experiment progressed revealed a potential role of odor in establishing preference. The honey solutions would be easy to locate due to their aromas, whereas sucrose solution is odorless. Bees visiting the nonavocado honey feeder would further mark it with Nasanov pheromone (Winston, 1987), and bees visiting the avocado honey feeder would abandon it. Thus, a rapid preference for the nonavocado honey feeder would develop (Figure 3.2). Eventually, bees finding the sucrose solution would start marking it with pheromone, and by the end of the experiment, it did indeed attract the greatest number of bees, revealing that it is even more attractive than non-avocado honey.

Phenolic compounds have been suspected of being repellent to bees (Baker, 1977; Rhoades and Bergdahl, 1981; Adler, 2000), but actual repulsion has seldom been found. Hagler and Buchmann (1993) tested two phenolic compounds and found that they were repellent at high concentrations, but at low concentrations they did not deter bees and even increased attractiveness. They also found repulsion to phenolic-rich nectars and honeys from three different botanical sources. One of them was almond honey, which was later found to be repellent due to amygdalin (London-Shafir *et al.*, 2003). Another source was salt cedar, which contains high potassium concentrations (Waller *et al.*, 1972). Our results indicated that phenolics increase the attractiveness of sucrose solution to honey bees, probably by adding odor to the solution, making it easier to locate. A similar increase in visitation rate to phenolic-rich nectar was found for Apis cerana (Liu *et al.*, 2004). Hence, no support was found for a repellent effect of phenolics on honey bees.

Little is known about the effect of nectar minerals on honey bee foraging behavior (Nicolson and W.-Worswick, 1990). A repellent effect of sodium chloride in sucrose solution was demonstrated by von Frisch (1950), who showed that a 0.015 M salt solution deterred honey bees,

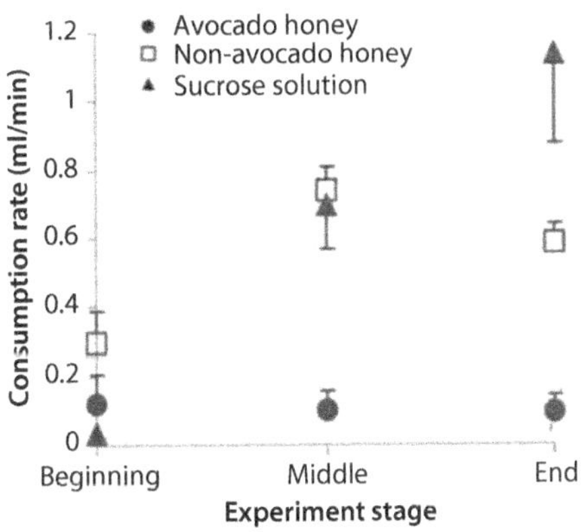

Figure 3.2 Attractiveness of various sweet components

although a 0.0075 M solution no longer deterred them. The sodium component in the latter solution was 173 ppm. Afik *et al.* (2006) found lower sodium concentrations than that in all honeys and nectars tested, and they were within a similar range in all samples. Hence, sodium does not seem to be responsible for deterring bees from avocado. Waller *et al.* (1972) found a repellent effect of potassium in onion nectar, at potassium concentrations similar to those found in the current study for avocado nectar. Their results indicate that, for three out of four potassium salts tested, dissolving 3500 ppm potassium in non-avocado honey to equalize its concentration with that of avocado honey decreases consumption by half. This indicates that potassium is a major cause for the low attractiveness of avocado honey, but not the only one. Cations such as potassium are usually accompanied by anions, although selective secretion to nectar is also possible (Luttge, 1977). In avocado nectar, potassium may be coupled with perseitol (Ishizu *et al.*, 2001). This is supported by the strong correlation between perseitol and potassium in avocado honey (Dag *et al.*, 2006). The most dominant anion found in avocado honey was phosphate, but its concentration was too low to equalize the potassium concentration. Adding K2-phosphate to non-avocado honey reduced consumption to a level similar to that of avocado honey. Thus, high mineral concentrations, mainly of potassium and phosphate, seem to be the main cause for the low attractiveness of avocado flowers. The effect of other minerals, such as magnesium, sulfur, and copper, whose concentrations differ between avocado and non-avocado honeys, still remains to be tested. Potassium concentration in avocado nectar may often be higher than that measured in this study and, therefore, high enough to repel honey bees from avocado flowers. Potassium concentration in avocado nectar is high, but not unique (Hiebert and Calder, 1983; Waller *et al.*, 1972). The effect of nectar minerals, potassium in particular, on the foraging behavior of pollinators may be widespread.

The reason for the presence of repellent components in nectar is not clear, but several possible roles have been suggested (Rhoades and Bergdahl, 1981; Adler, 2000). To answer this question for avocado, one can compare the honey bees' response with avocado nectar, these insects not being its natural pollinators, with the response of avocado's natural pollinators. These natural pollinators may not be as strongly repelled by the nectar. For example, various hummingbird species have been found to differ in their response to mineral-rich nectar (Bouchard *et al.*, 2000). Another approach would be to test whether potassium concentration in the nectar correlates with its abundance in the soil. It is possible that in avocado's natural habitat, the infertile soils of the neotropical rainforest (Wolstenholme, 2002), potassium concentration in the nectar is lower than in cultivated plots, and does not repel pollinators. The possible influence(s) of the agricultural practice of intensive fertilization of avocado (Lahav and Whiley, 2002) on the pollination effectiveness of bees is another area in need of further study.

Nectar and honey colors are produced by dissolved light-absorbing compounds. Although sugar solution is transparent, dark honeys such as avocado honey have relatively high concentrations of minerals (Petrov, 1970; Terrab and Heredia, 2004; Dag *et al.*, 2006). Other compounds, including flavinoids (Anklam, 1998) and organic acids (Mato *et al.*, 2003), may also affect honey color, in either the visible or ultraviolet spectrum; both are visible to bees (Hagler and Buchmann, 1993).

Previous work has revealed that avocado honey is relatively unattractive to honey bees when compared with honey from competing flowers (Afik *et al.*, 2006). These results suggest that some attributes of avocado

Table 3.4 Mineral contents in various types of honey

Mineral	Avocado honey	Nonavocado honey	Avocado nectar	Citus nectar
K	3768.3	324.8	3946.2	184.7
P	651.5	47.0	511.2	18.5
Mg	204.6	18.5	188.3	<5
S	188.3	27.7	170.4	<5
Ca	82.7	75.8	<150	<150
Na	58.9	79.1	53.8	18.5
Si	18.0	7.0	43.9	29.5
Zn	10.9	1.5	<30	<30
B	9.9	7.2	10.8	4.2
Fe	9.3	2.7	13.5	<5
Cu	3.2	0.1	3.1	<0.5
Pb	1.2	2.9	<1	<1

nectar are responsible for the relatively low attractiveness of avocado flowers. The following table was reported by Afik *et al.* (2006).

3.4 The Sugar Culture and Beyond

The invention of sugar correlates strongly with the most meaningful degradation in overall health and lifestyle in human society during recorded history. While molasses were being manufactured without introducing any health hazards, artificial sugar meant the introduction of numerous toxic chemicals that are simply not suitable for human consumption. However, this was introduced as a symbol of civilization. Sugar is white, and it is sweeter than molasses and ... far more profitable than honey.

Today, the world consumes some 110 million tones of sugar annually. Yet, sugar is prepared following a process that has never existed in nature. If one starts with the premise that unnatural is not sustainable, the production of sugar would mark a clear divergence from sustainable technology. If one has any confusion as to the existence of sugar as a natural process, one should be reminded of the process involved in manufacturing sugar. For instance, it involves crushing sugar cane, following by 'cleaning' with calcium hydroxide, a synthetic chemical that is not used by natural processes for anything of benefit, let alone for cleansing. At a later stage, brown sugar is 'refined' with chemical bleach, a potent toxin that oxidizes useful nutrients to render sugar 'white'. Then a series of other 'refining' material, such as, chalk, granulated carbon, etc. may be introduced.

While this color is appealing to the public that would consume sugar unconsciously, what does it say about the long-term sustainability of the food that just got reduced into a toxin? Indeed, anyone with conscious and minimum knowledge about how food is processed in a human body would have prevented engineers from employing such a technique. This intervention did not happen. Instead the entire engineering discipline focused on how quickly more sugar can be manufactured, while marketing agencies went out and found new markets to sell the product. Two questions arise here. First, why did an enlightened group of people resort to using toxins to process food? Were they that much intellectually bankrupt that they couldn't find a technique that nature has already in place? After all, what can be cleaner or whiter than milk? Why didn't the goodness of milk didn't have to be compromised in nature? Some say, it's about money. Natural processes can never be fully "cost-effectively engineered" in the sense of "subjected to command and control regimes consistent with maximizing profit in the shortest possible time". They are not amenable to the principle desideratum of mass production, which is profit based on minimum input costs. Minimizing input costs is not possible in a mass production context without sacrificing quality — usually through conversion of the real/natural into something artificial.

If a reality index were associated with pricing, the corrected profit would invariably always be negative (Khan and Islam, 2012). In reality, nothing is cheaper than natural products as long as sunshine and mother's milk are still available "for free", i.e., at no charge or at no cost to one's capital outlay.. How does this obvious logic elude modern-day academics? If the focus is so short-term that long-term benefits and short-falls are entirely disregarded in all economic calculations (Zatzman and Islam, 2007), it will indeed elude the powers of conventional observation.

This much about immediate practice is almost trivially obvious. In the absence of an economic theory that includes long-term elements, however, any engineered product can be marketed as anything else, covering the economic bottom line. This is far from obvious and the perfect cover for a system that is entirely artificial from root to surface. In this process, engineers have been playing a robotic role. They had no option to look into the natural order for solutions. This robotization starts early in the education system, and pervades all disciplines.

It didn't take humanity long to detect the effects of the sugar culture. For nearly a century, it has been known that sugar is responsible for non-genetic diabetes. Any reasonable consideration and rational reaction of this superflux of diabetes would lead to health warnings against sugar and to minimizing its consumption. Yet... the exact opposite happened. Sugar consumption skyrocketed as more and more processed food and fast food hit the marketplace. Sugar was introduced even as the first drink a newborn gets, displacing the age-old practice of giving honey to a newborn. Based on flawed analysis, honey was in fact banned from pediatric sections of the hospital and labels slapped on honey containers, warning people that honey can cause botulism – an utterly aphenomenal conclusion[2]. Today, sugar or similar sweeteners are ubiquitous, some food containing 75% sugar (Gillespie, 2010). Over

[2] Botulism is a rare paralytic illness caused by a toxin which is very poisonous to humans. As late as August 3, 2013 headline reads "New Zealand recalls dairy products over botulism fears".

time, more 'side effects' of this sweet poison have emerged. For example, addiction to refined sugar is more problematic than addiction to cocaine, and is associated with obesity, cancer, and diabetes (Goldwert, 2012).

Chemical engineering research has focused on several fronts, all maximizing short-term economic benefits. For instance, the notion that 'chemicals are chemicals' irrespective of their natural or artificial origin and components was used to sell the general public n the idea that natural sugar is the same as refined sugar, therefore, refined, i.e. artificial, sugar should be preferred because it's cheaper. After all, if honey has just as many calories as sugar but costs twice as much, the immediate practical reason to opt for honey disappears. Once this dogma of refinement trumps natural availability entrenched itelf, research could focus —and indeed has focused — on developing cheaper and more effective forms of sweeteners.

Biomedical engineering research revealed the addictive nature of sugar. As research began to reveal addictive nature of sugar, it was taken as a boon. It causes a euphoric effect that triggers dopamine, the chemical that controls pleasure in the brain. This should have triggered in a conscientious mind that natural sugar which produces natural glucose — the only food for the brain — cannot possibly be the same as artificial sugar that intoxicates the brain. The received message instead was: adding sugar to any product would make it addictive and sales will skyrocket!

The mindset of drug dealers is the moral equivalent of such a marketing marketing policy.

Medical research groups focused on 'fighting' symptoms of sugar. Because sugar consumption led to non-genetic diabetes, immediate replacement of natural insulin with artificial ones became the focus. Anyone with common sense and good conscience would be able to see this 'remedy' to diabetes as devastating as attaching an artificial limb because the limb had a cut that would otherwise heal naturally. Medical professionals, however, put diabetes patients on permanent insulin. Considering that insulin must be produced internally for it to have natural hold of the sugar burning process in an organic environment, how scientific is this?

A different campaign involved engineering the fabrication of poisons in order to fight bacteria that thrive under sugary conditions. It became common practice to use toxic chemicals, such as sodium nitrate, sorbic acid, sulfur dioxide, benzoate, and others to fight off bacterial growth. Magically, these toxins not only restored the original longevity of food (before adding sugar), they also increased shelf life! This was considered to be great technological progress in the eyes of modern corporations and corporatizers. The obsession to alter natural properties in order to 'fight' bacteria or natural decay was so intense that, by the 1960s, use of gamma-rays to kill bacteria became a common practice. It was assumed that the process of irradiation itself would not affect food.

A different marketing group began the campaign of vilifying natural fat. It was nothing but a publicity stunt as anyone with logical thinking should have known natural fat is necessary for sustenance of life. Even though Margarine, the first artificial fat was derived from animal fat in the Napoleon era, the biggest accomplishment of modern Margarine making was in rendering non-edible vegetable oil into artificial butter. Before, the fraud of trans fatty acid was detected, a huge campaign that started in 1970s culminated in 1990s when USA Health department actively campaigned in favor of artificial fat and calories against natural fat and natural sugar (as in fruits, etc.) The Food pyramid was replaced with a dart board that placed artificial food at the centre. Fat-free became the sign of good health. The sugar peddlers soon discovered "if you take fat out of food, it tastes like cardboard;" therefore, fat was replaced with sugar. The sugar consumption saw an unprecedented growth. See Figure 3.3. below. Figure 3.4. shows the chemical structure of a typical sugar molecule. Note how no information pertaining to catalysts and numerous chemicals is attached to the molecular structure.

Increased sugar consumption is likely to be concentrated in developing countries (Figure 3.4.) Asia and Africa will show the most growth, with growth in Asia attributable to population growth rates, economic development and changing tastes and preferences. In Africa the effect of population growth is expected to surpass the decline in per capita sugar consumption. Central America, South America and the Caribbean have shown a steady increase in consumption, mainly as a result ofpopulation increase as well as globalization. Sugar consumption in industrial countries will decline, although this fall should be more than offset by growth in Asia and Africa alone. In North America and the EU, consumption is stagnant: Population is only growing slowly and the effect of rising incomes on expenditure on sugar and sugar-containing products is minimal. In the USA, High Fructose Corn Sirup (HFCS/in Europe called HFS or High Fructose Sirup), is displacing ever more sugar, though at a slower growth rate than in the past. In Central Europe and the Former Soviet Union (FSU), consumption has decreased significantly as economic transformation takes place, however seems to be increased again.

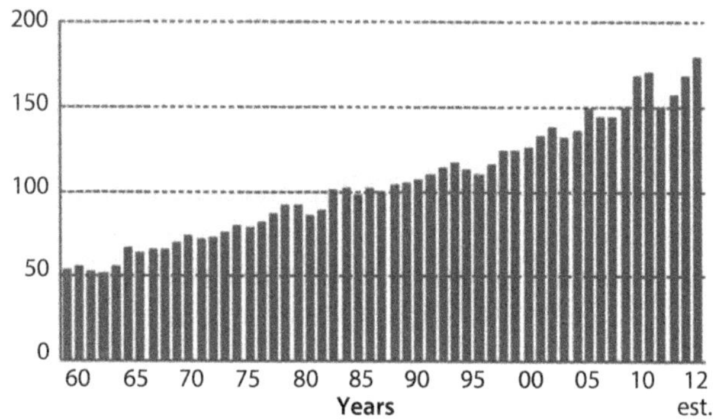

Figure 3.3 Millions of tons of sugar produced globally over the years (from **Website 2**)

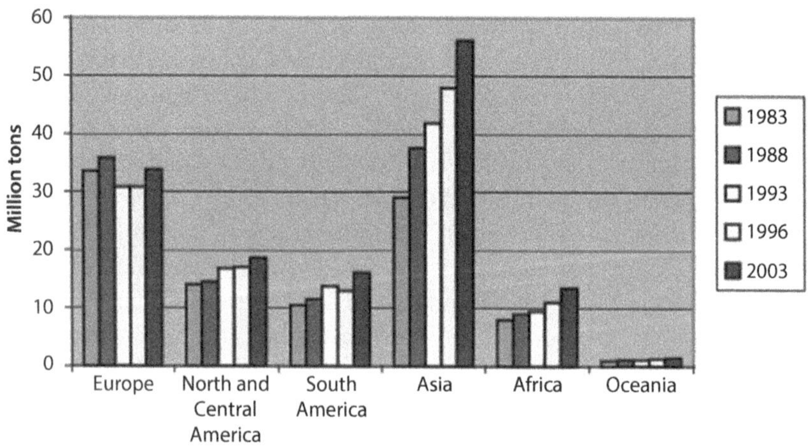

Figure 3.4 Sugar production history by region.

The three largest sugar consumers are India, the EU 15 and the Former Soviet Union (see Table 3.5). Consumption in the FSU and the USA has fallen sharply, but has risen significantly in India, China and Pakistan. The highest per capita consumption occurs in Brazil, with Mexico in second place. China has the lowest per capita consumption. However, China is the place that leads in saccharine production. These will all correlate with diabetes, cancer, and other ailments that are considered to be driven by genetics.

The consumption of sugar in Asian countries is increasing as a direct result of lower sugar prices and freer availability. In the last 20 years sugar consumption sugar consumption in Asia increase by 26 Million tons. 38% of world sugar consumption happens now in Asia.

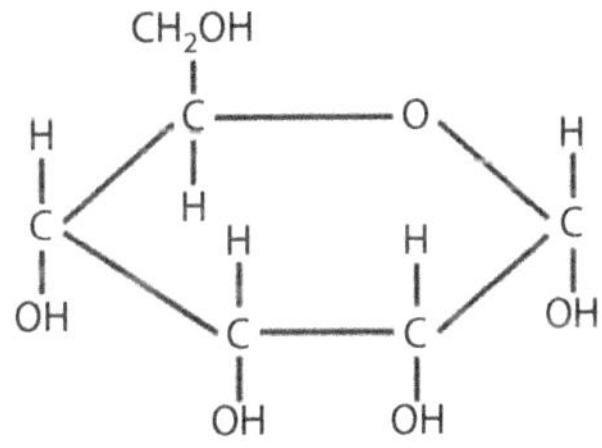

Figure 3.5 Sugar structure (note how all catalysts disappear).

As more and more people got addicted to sugar, an entire generation became afflicted with sugar-induced health problems. This crisis, along with the mantra that fat is evil, led to the development of alternative sweeteners that are "sugar-free". A new index started to surface. Everything is measured in terms of calories, calories being unhealthy. So, an entire line of artificial products was manufactured, all focusing on maximizing sweetness and minimizing cost. It meant the introduction of Saccharin®.

3.5 The Culture of the Artificial Sweetener

Even though sugar that is artificially processed was the beginning of the culture of artificial sweeteners, there is some consensus that truly artificial began with Saccharine.

Saccharin (C7H5NO3S) was discovered in 1878 in the Johns Hopkins University laboratory of Ira Remsen, a professor of chemistry. At age 21 Remsen had graduated with honors from the College of Physicians and Surgeons at Columbia University, earning an M.D. He abandoned his medical career to pursue organic chemistry[3], first at the University of Munich, then at the University of Göttingen, where he studied with Rudolph Fittig and began research on the oxidation of toluene isomers.

In Fittig's lab Remsen also studied sulfobenzoic acids, eventually publishing 75 papers on these and related compounds, laying the groundwork for the discovery of benzoic sulfinide—saccharin. Remsen returned to the United States in 1876—bringing with him influential German ideas about chemistry education—and accepted a professorship

3 At this time no university in the United States, and indeed very few outside Germany, supported advanced studies in organic chemistry. The field itself began with the accidental discovery in a German textile company's laboratory in the late 1850s of how to synthesize aniline blue.

Table 3.5 Sugar consumption for various regions/countries.

Country	Total sugar consumption (in million tons)				% of world consumption	Per capita consumption in kg		
	1980	1990	1996	2001		1980	1990	2001
India	5,60	11,07	14,75	20,0	14,5	8.3	13.4	15,7
EU	10,50	13,067	14,525	14,6	10,6	31.1	38.1	34,5
FSU*	12,40	13,40	10,27	10,5	7,6	46.7	46.2	37,0
USA	8,93	7,85	8,73	9,5	5,4	39.2	31.4	29,0
Brazil	6,55	6,62	8,30	9,8	7,1	54.	44.	53,1
China	4,30	7,13	8,50	10,2	7,4	4.3	6.2	6,3
Mexiko	3,23	4,43	4,25	5,0	3,6	46.5	54.5	46,7
Pakistan	0,89	2,29	2,93	3,6	2,8	10.7	20.4	24,5
Indonesia	1,73	2,65	3,25	3,2	2,3	11.8	14.8	15,8
Japan	2,70	2,83	2,60	2,4	1,7	23.1	29.9	19,0
Total	56,84	71,32	78,12	89,1	64,5	-	-	-

at Johns Hopkins. There he continued his research on the oxidation of methylated sulfobenzoic acids and their amides.

In 1877 a Russian chemist named Constantin Fahlberg was hired by the H.W. Perot Import Firm in Baltimore. Fahlberg studied sugar, while H.W. Perot imported sugar. The company enlisted him to analyze a sugar shipment impounded by the U.S. government, which questioned its purity. H.W. Perot also hired Remsen, asking him to provide a laboratory for Fahlberg's tests. After completing his analyses and while waiting to testify at trial, Fahlberg received Remsen's permission to use the lab for his own research. Working alongside Remsen's assistants, Fahlberg found the lab a friendly place. In early 1878 Remsen granted Fahlberg's request to take part in the institute's research.

One night that June, after a day of laboratory work, Fahlberg sat down to dinner. He picked up a roll with his hand and bit into a remarkably sweet crust. Fahlberg had literally brought his work home with him, having spilled an experimental compound over his hands earlier that day. He ran back to Remsen's laboratory, where he tasted everything on his worktable—all the vials, beakers, and dishes he used for his experiments. Finally he found the source: an overboiled beaker in which o-sulfobenzoic acid had reacted with phosphorus (V) chloride and ammonia, producing benzoic sulfinide. Though Falhberg had previously synthesized the compound by another method, he had no reason to taste the result. Serendipity had provided him with the first commercially viable alternative to cane sugar. Fahlberg and Remsen published a joint paper on their discovery in February, 1879. Remsen had a personal disdain for commercial ventures. However, Fahlberg aggressively pursued the commercial potential of the new compound. He named it "Fahlberg's saccharin" and patented it without informing Remsen, infuriating him.

Soon, the production and commercialization of saccharin became prominent. However, there were oppositions. The first one was a German company, Dye Trust, that brought down the price from $4.50 to $1.50 per kg. Locally, it was criticized for being 'false sugar' and having no food value. However, this burst of common sense was silenced by none other than the former US President, Theodore Roosevelt, who responded to the call of banning saccharin with his infamous line, "anyone who says saccharin is injurious to health is an idiot". 135 years later, the world would still try to figure out how anti-sugar this sugar substitute is (Suez *et al.*, 2014).

Saccharin® is produced from petroleum products. Scientifically speaking, sugar has a natural source that is food (e.g., sugar cane or beet) but

Saccharin® as a natural source that is not food (petroleum products are millions of years old vegetable). Logically, any movement toward aphenomenality would make the sweetener even more addictive, but that is considered to be a boon by a culture of greed. No in-depth health studies were necessary to ascertain that Saccharin®, being far more artificial than sugar, would be worse than sugar. However, no one commented on this aspect and Saccharin® enjoyed an unprecedented growth based on the following "qualities", each of which is based on dogmatic assumption; 1) it is not sugar; 2) it has little caloric content (because all calorie measuring techniques use burning ability as a measure of calorie), therefore "it must be good for weight loss"; 3) it doesn't cause tooth decay similar to sugar; and 4) it requires less preservatives (because it's actually toxic to bacteria) than sugar does. The sugar culture took a new turn, pointing to a new low in science and logical thinking. All signs show it was deliberate and sanctioned from the top of the Establishment. As stated earlier, even US president interfered. Theodore Roosevelt also wrote to Mr. Queeny on July 7, 1911:

> "I always completely disagree about saccharin both as to the label and as to its being deleterious...I have used it myself for many years as a substitute for sugar in tea and coffee without feeling the slightest bad effects. I am continuing to use it now: Faithfully your, T. Roosevelt"

Seven years later, Roosevelt would die in his sleep at Sagamore Hill as a result of a blood clot detaching from a vein and traveling to his lungs. He was 60 years old. The point here is the level of idiocy packaged in the form of power. Many talk about dogma and the role of religion, but many fewer see how this is about Money and control — the true 'axis of evil' today. Over 100 years later, another Nobel Peace laureate President, Barack Obama, how himself is smoker signed off on a bill to protect Monsanto and legalize uncontrolled use of genetically modified organisms (GMO) and genetic engineering (GE). Engdahl (2007) had outlined the evils of GMO, but technology development has not been about welfare of the society. In 2013, President Obama inked his name to H.R. 933, a continuing resolution spending bill approved in Congress days earlier. Buried 78 pages within the bill exists a provision that protects biotech corporations such as the Missouri-based Monsanto Company from litigation. In light of approval from the House and Senate, more than 250,000 people signed a petition asking the president to veto the spending bill over the biotech rider tacked on, an item that has since been widely referred to as the 'Monsanto Protection Act.' When taken in context,, GMOs are another in a long line of environmentally damaging practices for short term gain/profit. From the large-scale deforestation of the world's old-growth forests, to sustenance farming, to modern imported-fertilizer/pesticide/herbicide/fossil-fuel-dependent industrial agriculture, the trend has been consistent: GMOs are just another in that line of attempts to temporarily maintain/raise crop yields. Regardless of the type of agriculture or the location, there are limits to how long any land can remain productive. Applying imported fertilizers, or utilizing GMOs, only provides a temporary disruption of the land's transition to non-productive "wasteland", and ultimate desertification.

At this point we cannot resist sharing with readers Ronald Reagan's cigarette commercials glamourizing smoking (Picture 3.1). Ronald Reagan, the Star Wars president, was considered the most popular president ever, although the 'war' (named for a popular science fiction television program)' was actually about developing weapons systems to control 'conquer' outer space by the US against its perceived enemies. This scheme required such an enormous outlay of the collective wealth of the US that it was deemed, after long debate, unworkable. However, it has become a vastly successful commercial enterprise. Star Wars toys, stories, and movies are one of the to be most popular forms of entertainment even for adults. In the prevailing US culture, dominance is synonymous with weapons of mass destruction, which can be simultaneously morphed into consumer products engineered to becoming part of the human cultural space (Carey 1995). Few see the connection between mass destruction and onset or numerous chemical and pharmaceutical products. It has become a matter of continuous decline with increasing power to the group that pander artificial products and amass wealth at the expense of general welfare of the public.

Figures 3.6a shows the chemical structure of saccharin and related salts. The scientific name of this chemical is Benzoic sulfimide or Ortho sulphobenzamide. The popularity of this toxic chemical was entirely motivated by monetary gain.

A number of companies around the world manufacture saccharin. Most manufacturers use the basic synthetic route described by Remsen and Fahlberg in which toluene is treated with chlorosulfonic acid to produce ortho- and para-toluenesulfonyl chloride.

The original route by Remsen & Fahlberg starts with toluene; another route begins with o-chlorotoluene. Sulfonation by chlorosulfonic acid gives the ortho and para substituted sulfonyl chlorides. The ortho isomer is separated and converted to the sulfonamide with ammonia. Oxidation of the methyl substituent gives the carboxylic acid, which cyclicizes to give saccharin free acid (Figure 3.6b):

Subsequent treatment with ammonia forms the corresponding toluenesulfonamides. ortho-Toluenesulfonamide is separated from the para-isomer (this separation is alternatively performed on the sulfonyl chlorides), and ortho-toluenesulfonamide is then oxidized to ortho-sulfamoylbenzoic acid, which on heating is cyclized to saccharin (Mitchell and Pearson, 1991). The only producer in the United States currently uses the Maumee process, in which saccharin is produced from purified methyl anthranilate, a substance occurring naturally in grapes. In this process, methyl anthranilate is first diazotized to form 2-carbomethoxybenzenediazonium chloride. Sulfonation followed by oxidation yields 2-carbomethoxybenzenesulfonyl chloride. Amidation of this sulfonyl chloride, followed by acidification, forms insoluble acid saccharin. Subsequent addition of sodium hydroxide or calcium hydroxide produces the soluble sodium or calcium salt (Mitchell and Pearson, 1991). To be noted here that saccharin doesn't become any more acceptable because of its origin being grapes (e.g., US manufacturer). One has merely to cite the example of poppy seed that yields all sorts of toxins upon 'processing'.

Soon after Saccharin® was introduced in the market, it became popular as 'healthy' alternative to sugar, especially for the third world countries. All patients of diabetes and other sugar-related ailments were put on Saccharin® diet, often dubbed as 'diabetic sugar'. Instead of changing direction to move to natural substitutes to sugar, medical professionals were happy to make adjustments to the opposite side of the health spectrum. Diabetes diagnosis, 'treatment', 'sugar-free' diet, insulin, they all flourished in the west, soon followed by the third world countries. Sugar production started to see decline (per capita) as Saccharin® production climbed dramatically. Today, we consume some 40% less sugar globally than we did in 1970s. Chemical companies started to make huge profits. Also increased are all medical 'remedies' of diabetes, amassing profits for the pharmaceutical industries. Figure 3.7 shows the current market share of saccharin in 2001. This share has increased for saccharin, as well as other artificial sweeteners (Figure 3.8).

3.5.1 Delinearized History of Saccharin® and the Money Trail

In 1900, the annual production of saccharin in Germany was reported to be 190 tonnes. In 1902, partly at the insistence of beet sugar producers, saccharin production in Germany was brought under strict control, and saccharin was made available only through pharmacies. This was the first effort to link saccharin with some sort of medicinal value. As stated earlier, this linking worked particularly well in the third countries for whom the medical professionals are trained by the curriculum developed by the developed countries.

Saccharin use increased during the First World War and immediately thereafter as a result of sugar rationing, particularly in Europe. By 1917, saccharin was a common tabletop sweetener in America and Europe; it was introduced to the Far East in 1923. It was at that time that saccharin was being touted as a 'healthy alternative' to sugar, particularly for people that contracted non-genetic diabetes. The consumption of saccharin continued between the Wars, with an increase in the number of products in which it was used. The shortage of sugar during the Second World War again produced a significant increase in saccharin usage. In the early 1950s, calcium saccharin was introduced as an alternative soluble form (Mitchell and Pearson, 1991). Both forms of saccharin (Calcium and sodium) were banned in 1977 in the US, but reinstated subject to strict labelling stating: "Use of this product may be hazardous to your health, this product contains saccharin which has been determined to cause cancer in laboratory animals". It was known from early on that it interferes with normal blood coagulation, blood sugar levels and digestive function. They are banned in France, Germany, Hungary, Portugal, Spain. Banned as food additive in Malaysia and Zimbabwe. They are banned as a beverage additive in Fiji, Israel, Peru, and Taiwan. All calcium sachharin are banned from European Union.

In the aftermath of World War II, though, saccharin production remained high. Fundamental changes in the American diet meant fewer people prepared meals at home, relying instead on preprocessed food. Presweetened products, often containing inexpensive saccharin—the output of an increasingly large food-processing industry—alarmed nutritionists, regulators, and health officials. The same 1950's that brought a renewed hope for the country after two decades of Depression and War brought about the devastation of artificial food culture. From that period onward, food mainly consisted of processed foods, almost all containing artificial sweetener and other artificial chemicals, either as flavouring agent or preservative. In addition, the rise of the fast food industry, i.e. hamburger chains that sprouted up along side the newly build national highway system did not offer any better fare. Freeing Mom from the kitchen seemed to be the dominant theme as appliances and prepared foods became the 'norm". Empowered with the sex revolution of the 60's, the world started a journey of perpetual degradation, later dubbed as 'technological disaster'.

As America's economy boomed, women entered the workforce as never before and food lost its natural components. Housewives spent less time in the kitchen, women felt more empowered to 'not cook', so food companies came to the rescue with a buffet of processed foods. Foods were purchased in a can, package or pouch. Soups were available as liquids or in dry form. Tang landed on supermarket shelves and frozen dinners laid on trays in front of TV sets. TV dinners were introduced in 1953 by Swanson and with a flick of a wrist one could turn back the foil to display turkey in gravy, dressing, sweet potatoes and peas ready in about 30 minutes – all with no dishes to wash. Also, came "Better Living Through Chemistry" – a slogan that persists until today.

Just like any other technology development scheme in Europe, this change in processing came from the demand of the Army during WWII to provide needed ready-to-eat meals. The food industry responded by ramping up new technologies in canning and freeze-drying to feed the troops. The world hasn't seen such a great surge of food technology since the time of another wartime hero, Napoleon. The marketing of these foods presented a challenge, however. At first, many of them were less than palatable, so food companies hired home economists to develop fancy recipes and flooded magazines, newspapers and TV with ads to broadcast their virtues. TV itself became the greatest disinformation vehicle known to mankind, allowing everyone to believe in the most preposterous kind of fairytale stories. They also bought television sets in record numbers and watched shows that represented their new idealized lives like Ozzie and Harriet and Leave It to Beaver. Beaver's mother, June Cleaver was depicted as a housewife freed from household chores and often was serene and perfectly dressed with pearls and high heels pushing a vacuum cleaner and putting meals on the family table.

This followed with the Baby Boomer Generation. Fifty million babies were born from 1945 to 1960. Food marketing shifted to kids with Tony the Tiger and fish sticks leading the campaign. Fast food had its beginnings strengthened in 1955 when Ray Kroc bought a hamburger stand from the McDonald's brothers in San Bernadino, California.

Picture 3.1 Former President Ronald Reagan in a cigarette commercial. Such commercials are banned in USA.

Disneyland opened in 1955 and was so popular they ran out of food on the first day.

In 1958, the American scientist, Ancel Keys started a study called the Seven Countries Study, which attempted to establish the association

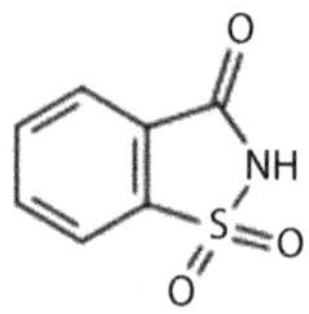

Saccharin

$C_7H_5NO_3S$ Relative molecular mass: 183.19

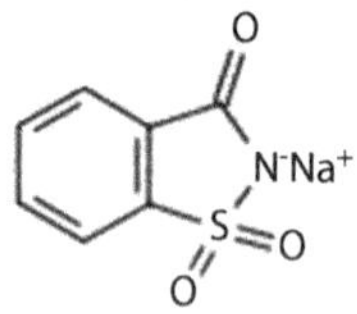

Sodium saccharin

$C_7H_4NO_3S.Na$ Relative molecular mass: 205.18

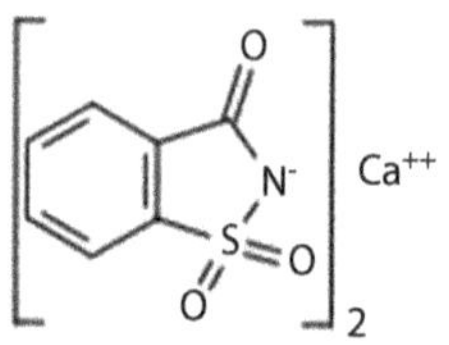

Calcium saccharin

$(C_7H_4NO_3S)_2.Ca$ Relative molecular mass: 404.46

Figure 3.6a Chemical structure of saccharin and related salts.

between diet and cardiovascular disease in different countries. The study results indicated that in the countries where fat consumption was the highest also had the most heart disease. This suggested the idea that dietary fat caused heart disease. He initially studied 22 countries, but only reported on seven: Finland, Greece, Italy, Japan, The Netherlands, United States, and Yugoslavia.

The problem was that he left out:

- Countries where people eat a lot of fat but have little heart disease, such as Holland and Norway and France.
- Countries where fat consumption is low but the rate of heart disease is high, such as Chile.

Basically, he only used data from the countries that supported his theory. Instead, if he understood the difference between real and artificial, this paradox could be solved easily. However, instead of cutting off artificial from the diet, the entire culture was based on cutting out

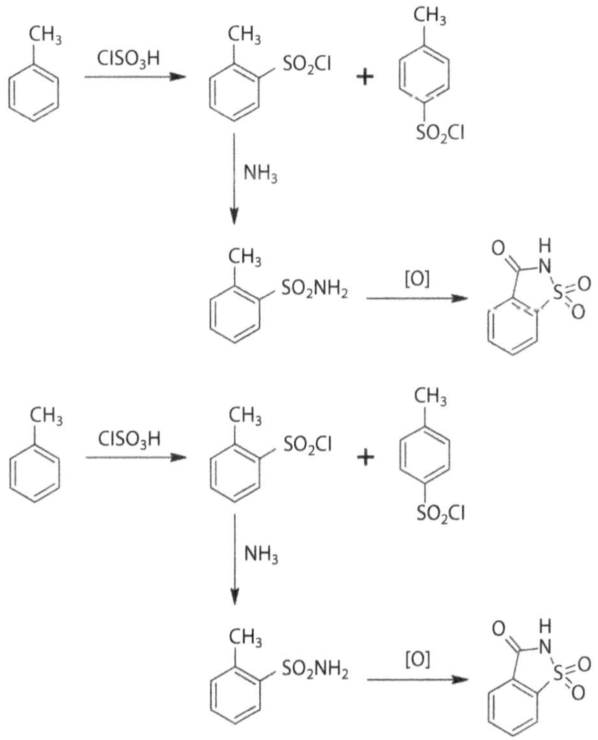

Figure 3.6b Chemical reactions used during saccharin manufacturing.

natural fat and sugar and replacing with artificial fat and artificial sweeteners.

Scientists began to discover the carcinogenic roles of saccharin. After all, it is based on toluene, a known carcinogen. While saccharin consumption increased, the debate over its safety was never truly settled. Science, to the public, had issued too many contradictory or inconclusive opinions, so when the decision about saccharin fell to individuals, most responded to their desire for a no-consequences sweetener. Partly in response to growing unease among regulators and the public, Congress passed the Food Additives Amendment in 1958. In preparing its legislation Congress heard testimony from members of the scientific community. For the first time in connection with food additives, scientists used the c-word: cancer. Representative James J. Delaney, a Democrat from New York, pushed hard for the addition of language specifically outlawing carcinogens. In its final form the "Delaney Clause" required the U.S. Food and Drug Administration (FDA) to prohibit the

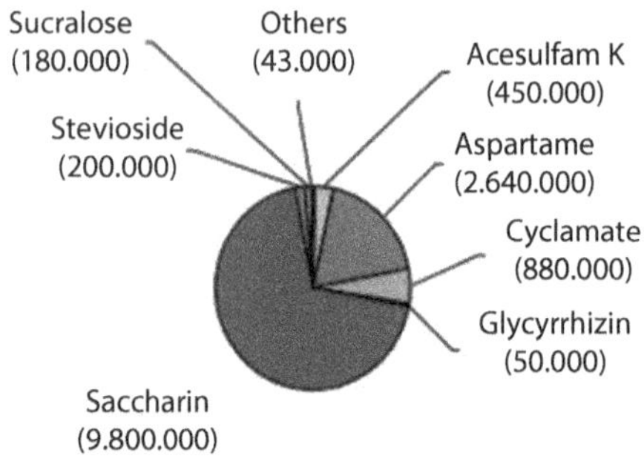

Figure 3.7 Saccharin consumption share in 2001.

use of carcinogenic substances in food. Seemingly uncontroversial at the time—who would support adding cancer-causing agents to food?—it later proved contentious.

In 1977, USA banned saccharin. The basis for the proposed ban was a study that documented an increase in cancer in rats being fed saccharin. The "Delaney clause" of the Food Additive Amendments to the Federal Food, Drug, and Cosmetic Act states that no substance can be deemed safe if it causes cancer in humans or animals.

Legislators had disastrously underestimated the data necessary to definitively declare a substance carcinogenic. The same year, the Cumberland Packing Corporation introduced Sweet'N Low, a mixture of saccharin and cyclamate, another artificial sweetener. The two chemicals balanced each other, with cyclamate blunting the bitter aftertaste of saccharin. Sweet'N Low arguably tasted more like real sugar, and those little pink packets brought artificial sweeteners into diners and coffee shops. Meanwhile, the use of artificial sweeteners continued to increase among weight-conscious consumers. Between 1963 and 1967 artificially sweetened soft drinks (Coca-Cola's Tab, for example) nearly tripled their market share, growing to over 10% of the soda market.

In 1882 Constantin Fahlberg had declared saccharin harmless because he suffered no adverse effects 24 hours after taking a single dose. Similarly, Harvey Washington Wiley's turn-of-the-century "poison squads" had declared a substance safe if the tester—a human guinea pig—remained healthy after ingestion. But post–World War II health science had begun investigating subtler, long-term effects. Research methodology had changed accordingly: studies observed a longer span of time, for example, and tried to control for a wider range of variables. Researchers shifted away from unstructured human testing toward animal testing that included control groups. Such research produced more and better data but increased complexity. No longer could a substance be labeled simply "poison" or "not poison." The results of these sophisticated tests demanded sophisticated interpretation.

In the late 1960s three trends converged: increasing government regulation in the food-processing industry, the rise of artificial sweeteners, and the growing complexity and sophistication of health science. One of the first results of this convergence was the ban on cyclamate. Two 1968 studies linked the chemical to bladder cancer. The FDA cited the Delaney Clause in recommending a ban, which was enacted the following year. That left only one artificial sweetener on the market: saccharin. In 1970 oncologists at the University of Wisconsin Medical School published the results of a clinical study showing a higher instance of bladder cancer among rats who consumed saccharin daily. Subsequent tests seemed to support the initial results, and in 1972 the FDA removed saccharin from the list of food additives "generally recognized as safe." Peter B. Hutt, chief legal counsel for the FDA, stated that, "If it causes cancer—whether it's 875 bottles a day or 11—it's going off the market."

Saccharin producers and commercial consumers recognized the FDA's move as a precursor to an outright ban. Large chemical companies—Monsanto, Sherwin-Williams, and Lakeway Chemicals—began assembling their own evidence to oppose prohibition. Soda companies expected a painful financial hit, as did makers of diet food. But they also knew the process could take years, as the FDA ordered new tests, analyzed the data, and—crucially—responded to public and political pressure.

By 1977 a saccharin ban looked likely. The Cumberland Packing Corporation, which had presciently reformulated Sweet'N Low in the shadow of the cyclamate ban, vowed to fight any regulation. Marvin Eisenstadt, the president of the company, appeared on television and radio to argue his case. He denied the scientific validity of animal testing and declared access to saccharin a consumer right. He helped draft a two-page ad from the Calorie Control Council, the industry group he headed. The ad appeared in the *New York Times* explaining "why the proposed ban on saccharin is leaving a bad taste in a lot of people's mouths." The ad described the ban as "another example of BIG GOVERNMENT" and recommended action. "Fortunately, we can all conduct our own experiment in this matter. It's called an experiment in democracy. . . . Write or call your congressman today and let him know how you feel about a ban on saccharin."

In the week after the saccharin ban went into effect in 1977, Congress received more than a million letters. Marvin Eisenstadt and other

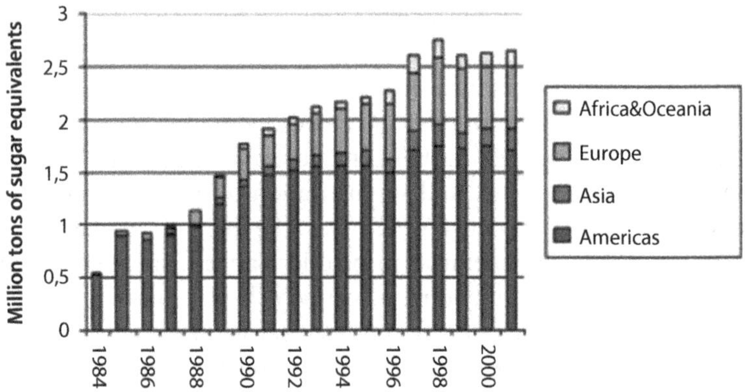

Figure 3.8 The dominance of saccharin has been continuing in last 3 decades.

public relations–savvy producers had turned the saccharin debate into a PR operation, and the public had responded. The Delaney Clause, as the FDA interpreted it, required a ban on any known carcinogen in the food supply. But the original legislation failed to account for the complexity of scientific data. The clause's premise of scientific consensus based on objective evidence and shared expertise no longer applied to the real world, if it ever had. Scientists couldn't agree on fundamental questions: What is a carcinogen? What daily dosage of a chemical might be reasonable for testing toxicity? Did the elevated risk of cancer in rats translate to an elevated risk in humans? Health science couldn't yet answer those questions definitively. But in the absence of incontrovertible scientific evidence, Marvin Eisenstadt could frame the debate as average citizens versus an encroaching big government.

The FDA understood the weakness of the existing laws and breathed a sigh of relief when, a week after the ban, Senator Ted Kennedy of the Senate Subcommittee on Health and Scientific Research moved to forestall the ban. The Saccharin Study and Labeling Act passed that year, declaring that all saccharin products would carry a warning label. It also imposed a two-year moratorium on any government action to remove saccharin from the market. More studies were needed, according to Congress.

In suspending the proposed saccharin ban, Congress ordered that products containing the popular sweetener must carry a warning about its potential to cause cancer. The FDA formally lifted its proposal to ban the sweetener in 1991 based on new studies, and the requirement for a label warning was eliminated by the Saccharin Notice Repeal Act in 1996.

In response, Sweet'N Low sales skyrocketed. Those sales included longtime buyers stocking up in case of a ban, but the free publicity also brought in new customers. By 1979, 44 million Americans used saccharin daily. Consumers voted with their dollars.

Congress renewed the moratorium every two years until 2000, when a National Institute of Environmental Health Sciences (NIEHS) study declared the earlier research invalid. The high dosages of saccharin given to the rats were a poor analog for human consumption, as rat digestion works differently from that of humans. The NIEHS recommended that Congress repeal the Labeling Act, officially declaring saccharin safe for human consumption.

Finally, though, it wasn't government regulation that toppled saccharin from its throne as king of the artificial sweeteners—at least not directly. The threat of a saccharin ban led producers to research alternatives. While saccharin—300 times sweeter than sugar—languished in the shadow of a potential ban, a new generation of artificial sweeteners flourished. In 1965 aspartame, which is 200 times sweeter than sugar, was discovered; in 1976 sucralose—600 times sweeter; and in 2002 neotame—7,000 to 13,000 times sweeter than sugar. Today, saccharin, once the undisputed king of artificial sweeteners, lags behind its newer counterparts, replaced by the next sweetest thing.

One of the most distinctive features of the world sweetener market in recent years has been the growing realization of the economic attractiveness of blending sweeteners; both intense-intense blends and intense-caloric blends. With the fall in the relative price of intense sweeteners noted in Table 3.3, alongside with the introduction of third generation sweeteners as sucralose, alitame and stevioside, and a relaxtion of regulations, for example, in adoption of the 1996 Sweetener Directive in the EU, the trend towards blending intense sweeteners has continued.

The same economic considerations and the desire to save money that have tempted EU food and beverage manufactures into using more intense sweeteners have also been observed. Regardless of prohibitive legislation, in other parts of the world such as Africa, Eastern Europe and the former Soviet states, blending is increasing in all parts of the world. Basically there are three main benefits that can be obtained by blending sweeteners: flavor-masking, enhanced potency and sweetener synergy. It turns out that these are also the most addictive chemicals known to date. This addiction factor became the reason for best selling feature of any chemical.

Table 3.6 Commodity price over last few decades.

Commodity	1998	1999	2000	2001
World Raw Sugar Price	127,4	90,0	117,7	123,7
Saccharin	91,9	81,8	76,7	72,7
Cyclamante (non US)	94,2	137,2	145,8	123,2
Aspartame (non US)	52,0	48,3	37,0	36,0

Table 3.7 Prices of various artificial sweeteners.

Sweetener	Price per kg
Acesulfame	80 Euro
Aspartame	40 Euro
Glycyrrhizin	50 Euro
Cyclamate	4.5 Euro
Saccharin	6.7 Euro
Sucralose	139 Euro
Stevioside	50 Euro

Information available in 1995 indicated that saccharin was produced in 20 countries, calcium saccharin was produced in five countries, and sodium saccharin was produced in 22 countries (Chemical Information Services, 1995). Table 3.7 shows how new alternatives to saccharin have been introduced at a higher price. They contain higher profit margin and greater chance of monopoly. This is reflected in the fact that new chemicals have lower cost of production attached to them. In comparison of sugar, intense sweetener are much more cheaper on sugar equivalent bases. The cost effectiveness in terms of sugar sweetness is one advantage of intense sweeteners. This do not reflect other properties, such as, bulking effects which are important for many food applications. Only sugar or starch sweeteners can provide those bulking effects. Intense sweeteners are unable to. The cost effectiveness of saccharine and cyclamate are unbeatable due to their cheap synthetic processes.

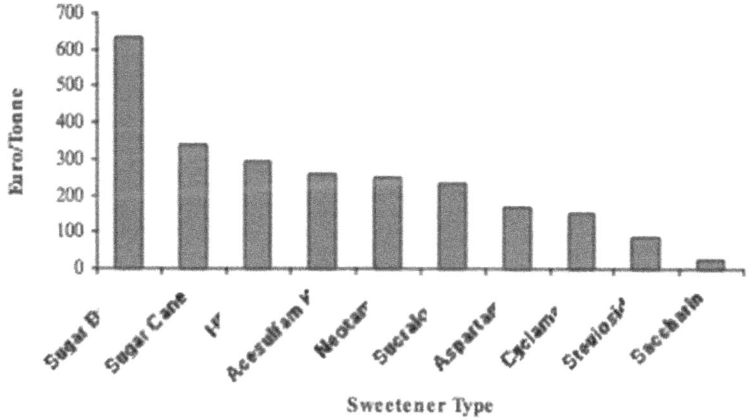

Figure 3.9 Cost per tonne for various sugar products (2003 value).

The comparison of the sweetener price related of intense sweetener to sugar is given in Figure 3.9.

Today, China is the world's largest producer of saccharin, accounting for 30–40% of world production, with an annual production of approximately 18 000 tonnes in recent years. Its exports amounted to approximately 8000 tonnes. In 1995, the United States produced approximately 3400 tonnes of saccharin and its salts, and Japan produced approximately 1900 tonnes. In the past, several western European companies produced sodium saccharin; however, by 1995, western European production had nearly ceased due to increasing imports of lower-priced saccharin from Asia (Bizzari *et al.*, 1996).

In 2001, Europe recorded the strongest growth, with demand up by 13% regarding to higher consumption in France, Spain and Italy, which more than outweighed declines in the UK and Germany. The growth in saccharin demand is driven by the growing popularity of blends within the EU, as well as rising demand from Eastern Europe. After a respite in 2000, Chinese exports of saccharin surged ahead in 2001, though local sales declined. This has squeezed other saccharin producing countries, some of which have gone out of business. During 1999 the Chinese media reported that the government had ordered the closure of nine of the 14 major saccharin plants, with the effect of reducing the overall production capacity from about 47.000 tons (14.1 million tons sugar equivalent) to around 20.000 tons (6.0 million tons sugar equivalents). The annual production capacity for each of the 14 major saccharin plants in operation in China ranges from 500 tons to over 10.000 tons. According to the press news the Chinese government intented to limit the saccharin production to about 24.000 tons (7,2 million tons sugar equivalent) together with a reduction in consumption to about 8.000 40 tons, which is about 60% of the current saccharin consumption level in China. Finally only smaller factories have been closed taking only 3.000 tons (0.9 million tons sugar equivalents) of capacity out of the market. Table 3.8 shows global exports of saccharin for various countries. They all show tremendous growth over last few decades.

3.6 The Culture of Aspartame

As soon it was discovered, Saccharin® doesn't give diabetes to its consumers, it gives cancer, it was deemed that the time for a new line of research to invent another alternative to sugar had arrived. None other than a neurotoxin invented through research in biological warfare became the best candidate to salvage the humanity suffering from sugar shock. Aspartame® was discovered and it was far 200 times sweeter than sugar. More importantly it needed no natural raw material – not even petroleum chemicals. sweeter than Saccharin® and logically it would also be more addictive. Most importantly, Aspartame® is not Saccharin® and it became the perfect candidate for 'Saccharine-free' diet. Aspartame® cannot be linked to cancer, but it is linked to freezing of the brain. After all, it is a neurotoxin. Scientifically, nothing is natural about Aspartame®, from the source to the process involved. Yet, the best even the nutritionists could come up with is, 'overconsumption' of such products would increase risk for heart attack and stroke, causes breast cancer, colon cancer, and increases harmful cholesterol, i.e., LDL that thickens arteries. The original invention document made Aspartame® an excellent neurotoxin fit for biological warfare. The following are well known contributions of Aspartame®:

- Marked changes in appetite and weight as reflected by paradoxical weight gain or severe loss of weight;
- Excessive insulin secretion and depletion of the insulin reserve;
- Possible alteration of cellular receptor sites for insulin, with ensuing insulin resistance;
- Neurotransmitter alterations within the brain and peripheral nerves;
- The toxicity of each of the three components of Aspartame® (phenylalanine; aspartic acid: the

Table 3.8 Global exports of saccharin (from USITC publication, http://www.usitc.gov/publications/701_731/pub4077.pdf).

Saccharin: Global exports, by country and by regions, 2003-07					
Source	**Calendar year**				
	2003	**2004**	**2005**	**2006**	**2007**
Quantity (1,000 pounds)					
United States	365	187	2,090	2,210	1,822
China	43,510	42,482	33,293	35,514	35,496
Nonsubject exporting countries:					
South Korea	3,922	4,417	4,256	3.933	5.117
Germany	1,931	2,564	2,156	2,410	3,062
Taiwan	1	150	612	732	878
Netherlands	584	606	778	952	875
Japan	549	372	557	746	802
Singapore	538	51	44	0	586
Belgium	324	564	542	324	478
Spain	527	589	505	542	434

(*Continued*)

Table 3.8 cont.

Saccharin: Global exports, by country and by regions, 2003-07					
Source	**Calendar year**				
	2003	**2004**	**2005**	**2006**	**2007**
India	214	663	262	305	388
United Kingdom	428	139	551	335	353
Switzerland	526	254	367	240	289
France	159	276	203	209	280
Austria	126	57	62	29	216
Turkey	6	22	8	10	198
South Africa	34	43	37	106	177
All other	403	545	481	366	761
Total nonsubject exporting countries	10,271	11,311	11,420	11,241	14,895
Total	54,146	53,980	46,803	48,965	52,213
Regions:					
EU15 (external trade)	1,797	2,174	1,975	2,125	2,440
EU27 (external trade)	1,360	1,812	1,587	1,237	1,709

methylester, which promptly becomes methyl alcohol or methanol), and their multiple breakdown products after exposure to heat or during prolonged storage.

3.6.1 Delinearized history of Aspartame

Aspartame, which is by far the most prominent artificial sweetener currently used in diet products, is also the most controversial of them all. Its origins are questionable, to say the least. Many claim it never should have been allowed on the market. The few who argue it is safe, have very little ground to stand on for those educated on how it became approved.

Many of the studies done to determine the safety of aspartame, in the process of its approval as a food additive, have had severe conflicts of interest mainly due to inappropriate by Searle, the very same company that produces NutraSweet (their "street name" for aspartame).

Dr. Robert Walton investigated the claims made that Searle essentially bought their way into the market. The results he found were quite shocking. In the 166 studies that he found to have relevance in regards to human safety, 74 of those studies had been funded by Searle. The 92 remaining studies were funded independently.

Unsurprisingly, of the 74 studies that were funded by Searle, 100% of them claimed that aspartame was safe for human use. As far as the independently funded studies, 92% of them found health concerns in regards to aspartame, and found it to be unsafe for human consumption.

The sugarlike taste of aspartame was discovered accidentally by James Schlatter, an American drug researcher at G.D. Searle and Co. in 1965. While working on an antiulcer drug, he inadvertently spilled some APM on his hand. Figuring that the material was not toxic, he went about his work without washing it off. He discovered APM's sweet taste when he licked his finger to pick up a piece of weighing paper. This initial breakthrough then led the company to screen hundreds of modified versions of APM. The company pursued and was granted United States patent 3,492,131 and various international patents, and the initial discovery was commercialized. The U.S. patent expired in 1992, and the technology is now available to any company who wants to use it.

Aspartame has been marketed since 1983 by Searle under the brand names NutraSweet' and Equal'. Currently, NutraSweet' is a very popular ingredient and is used in more than 4,000 products, including chewing gum, yogurt, diet soft drinks, fruit-juices, puddings, cereals, and powdered beverage mixes. In the U.S. alone, NutraSweet®'s sales topped $705 million in 1993, according to the company.

One of the earliest tests, done by the University of Wisconsin in 1967 by Dr. Harold Waisman, had been conducted on monkeys who drank milk, which contained aspartame. Of the seven monkeys being fed the mixture, one died and five others experienced grandular seizures. Despite these early warning signs, Searle pushed on.

In 1971, a neuroscientist by the name of Dr. John Olney, conducted several studies which showed that the aspartic acid found in aspartame, caused holes in the brains of baby mice. Later, one of Searle's own researchers conducted a similar study and concluded the same results as the ones demonstrated by Dr. Olney. Again, Searle pushed on.

In 1976, an FDA investigation of Searle was initiated, sparked by the many concerns that Searle's personal studies on aspartame were inconsistent with research from independent studies.

The investigation results found Searle's tests were not only full of inaccuracies, but also manipulated data. An investigator involved was quoted as stating they "had never seen anything as bad as Searle's testing."

Shortly after the investigation, the FDA sent a formal request to the U.S. Attorney's office to begin grand jury proceedings. Not surprisingly, one of the most significant events of this procession saw Samuel Skinner, the U.S. Attorney in charge of the investigation, resigning from the attorney's office and taking a position within Searle's law firm, allowing Searle to buy themselves out of a bad situation.

After many years of quibbling, the FDA initially approved aspartame's use as a sweetener in 1980. However, numerous objections surfaced and most scientific evidence showed the best application of Aspartame being as neurotoxin. However, the FDA and the Centers for Disease Control concluded in 1984 that the substance is safe and does not represent a widespread health risk. This conclusion was further supported by the American Medical Association in 1985, and aspartame has been gaining market share ever since. In addition to its use in the United States, aspartame has also been approved for use in over 93 foreign countries.

In 1985, "G.D. Searle and Co" (mentioned as Searle above) became the pharmaceutical unit of Monsanto – the company that received full protection in 2013 from US President Barack Obama. It was also the same company that developed the controversial drug, Celebrex.

Figure 3.10. show aspartame consumption over the years for various regions. This figure shows that Aspartame consumption slowed down in 1999 for the first time in history. The real growth has been in the areas of sports drink, fruit juices, and vitamin-added water. Ironically, these drinks have other components that are even more toxic than aspartame. In addition, there has been an increased trend of blending with other artificial sweeteners. Figure 3.11. shows the trend in major artificial sweeteners.

3.6.2 Timeline

DECEMBER 1965

While working on an ulcer drug, a chemist at pharmaceutical manufacturer GD Searle accidentally discovers aspartame, a substance that is 180 times sweeter than sugar, yet has no calories.

SPRING 1967

Searle begins safety tests, necessary for FDA approval.

AUTUMN 1967

GD Searle approaches eminent biochemist Dr Harry Waisman, director of the University of Wisconsin's Joseph P Kennedy Jr Memorial Laboratory of Mental Retardation Research and a respected expert in the toxicity of phenylalanine (which comprises 50 per cent of the aspartame formula), to conduct a study of the effects of aspartame on primates. Of seven monkeys fed aspartame mixed with milk, one dies and five others have grand mal epileptic seizures.

SPRING 1971

Dr John Olney, professor of neuropathology and psychiatry at Washington University in St Louis School of

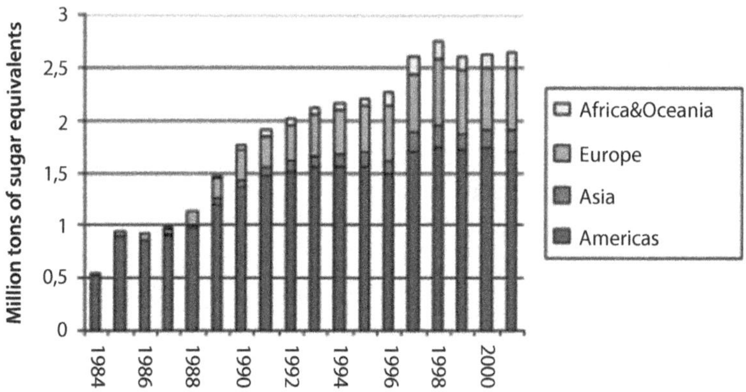

Figure 3.10 Aspartame market growth since 1984.

Medicine, whose research into the neurotoxic food additive monosodium glutamate (MSG, a chemical cousin of aspartame) was responsible for having it removed from baby foods, informs Searle that his studies show that aspartic acid, one of the main constituents of aspartame, causes holes in the brains of infant mice. One of Searle's researchers, Ann Reynolds, confirms Olney's findings in a similar study.

FEBRUARY 1973

Searle applies for FDA approval and submits over 100 studies it claims support aspartame's safety. Neither the dead monkeys nor the mice with holes in their brains are included in the submission.

12 SEPTEMBER 1973

In a memorandum, Dr Martha M Freeman of the FDA Division of Metabolic and Endocrine Drug Products criticises the inadequacy of the information submitted by Searle with particular regard to one of the compound's toxic breakdown products, diketopiperazine (DKP). She recommends that marketing of aspartame be contingent upon the sweetener's proven clinical safety.

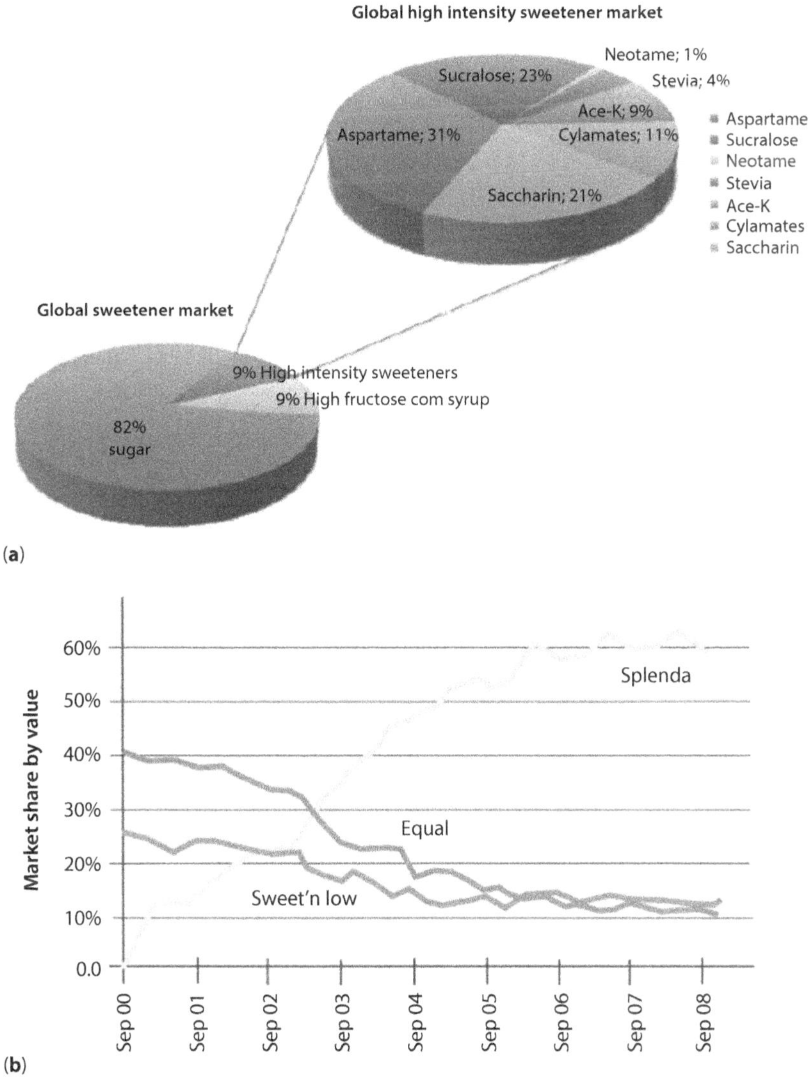

Figure 3.11 Market share of various artificial sweeteners.

26 JULY 1974

FDA commissioner Dr Alexander Schmidt grants aspartame its first approval as a 'food additive' for restricted use in dry foods. This approval comes despite the fact that his own scientists found serious deficiencies in the data submitted by Searle.

AUGUST 1974

Before aspartame can reach the marketplace, Dr John Olney, James Turner (attorney, consumer advocate and former 'Nader's Raider' who was instrumental in removing the artificial sweetener cyclamate from the US market), and the group Label Inc (Legal Action for Buyers' Education and Labeling) file a formal objection to aspartame's approval with the FDA, citing evidence that it could cause brain damage, particularly in children.

JULY 1975

Concerns about the accuracy of test data submitted to the FDA by Searle for a wide range of products prompt Schmidt to appoint a special task force to examine irregularities in 25 key studies for aspartame and Searle drugs Flagyl, Aldactone and Norpace.

5 DECEMBER 1975

Searle agrees to an inquiry into aspartame safety concerns. Searle withdraws aspartame from the market pending its results. The sweetener remains off the market for nearly 10 years while investigations into its safety and into Searle's alleged fraudulent testing procedures are ongoing. However, the inquiry board does not convene for another four years.

24 MARCH 1976

The FDA task force completes its 500 page report on Searle's testing procedures. The final report notes faulty and fraudulent product testing, knowingly misrepresented product testing, knowingly misrepresented and 'manipulated' test data, and instances of irrelevant animal research in all the products reviewed. Schmidt says: '[Searle's studies were] incredibly sloppy science. What we discovered was reprehensible.'

JULY 1976

The FDA forms a new task force, headed by veteran inspector Jerome Bressler, to further investigate irregularities in Searle's aspartame studies uncovered by the original task force. The findings of the new body will eventually be incorporated into a document known as the Bressler Report.

10 JANUARY 1977

FDA chief counsel Richard Merrill formally requests the US Attorney's office to begin grand jury proceedings to investigate whether indictments should be filed against Searle for knowingly misrepresenting findings and 'concealing material facts and making false statements' in aspartame safety tests. This is the first time in the FDA's history that it requests a criminal investigation of a manufacturer.

26 JANUARY 1977

While the grand jury investigation is underway, Sidley & Austin, the law firm representing Searle, begins recruitment negotiations with Samuel Skinner, the US attorney in charge of the investigation. Skinner removes himself form the investigation and the case is passed to William Conlon.

8 MARCH 1977

Searle hires prominent Washington insider Donald Rumsfeld as its new CEO to try to turn the beleaguered company around. A former member of Congress and defence secretary in the Ford administration, Rumsfeld brings several of his Washington colleagues in as top management.

1 JULY 1977

Samuel Skinner leaves the US Attorney's office and takes a job with Searle's law firm. Conlon takes over Skinner's old job.

1 AUGUST 1977

The Bressler Report is released. It focuses on three key aspartame studies conducted by Searle. The report finds that in one study 98 of the 196 animals died but weren't autopsied until later dates, making it impossible to ascertain the actual cause of death. Tumours were removed from live animals and the animals placed back in the study. Many other errors and inconsistencies are noted. For example, a rat was reported alive, then dead, then alive, then dead again. Bressler comments: 'The question you have got to ask yourself is: why wasn't greater care taken? Why didn't Searle, with their scientists, closely evaluate this, knowing full well that the whole society, from the youngest to the elderly, from the sick to the unsick. will have access to this product.' The FDA creates yet another task force to review the Bressler Report. The review is carried out by a team at the FDA's Center for Food Safety and Applied Nutrition and headed by senior scientist Jacqueline Verrett.

28 SEPTEMBER 1977

The FDA publishes a report exonerating Searle of any wrongdoing in its testing procedures. Jacqueline Verrett will later testify to the US Senate that her team was pressured into validating data from experiments that were clearly a 'disaster'.

8 DECEMBER 1977

Despite complaints from the Justice Department, Conlon stalls the grand jury prosecution for so long that the statute of limitations on the aspartame charges runs out and the investigation is dropped. Just over a year later Conlon joins Searle's law firm, Sidley & Austin.

1978

The journal Medical World News reports that the methanol content of aspartame is 1,000 times greater than most foods under FDA control. In high concentrations methanol, or wood alcohol, is a lethal poison.

1 JUNE 1979

The FDA finally establishes a public board of inquiry (PBOI), comprising three scientists whose job it is to review the objections of Olney and Turner to the approval of aspartame and rule on safety issues surrounding the sweetener.

1979

In spite of the uncertainties over aspartame's safety in the US, aspartame becomes available, primarily in pharmaceutical products, in France. It is sold under the brand name Canderel and manufactured by the food corporation Merisant.

30 SEPTEMBER 1980

The FDA's PBOI votes unanimously against aspartame's approval, pending further investigations of brain tumours in animals. The board says it 'has not been presented with proof of reasonable certainty that aspartame is safe for use as a food additive'.

1980

Canderel is now marketed throughout much of Europe (but not in the UK) as a low-calorie sweetener.

JANUARY 1981

Rumsfeld states in a Searle sales meeting that he is going to make a big push to get aspartame approved within the year. Rumsfeld vows to 'call in his markers' and use political rather than scientific means to get the FDA on side.

20 JANUARY 1981

Ronald Reagan is sworn in as president of the US. Reagan's transition team, which includes Rumsfeld, nominates Dr Arthur Hull Hayes Jr to be the new FDA commissioner.

21 JANUARY 1981

One day after Reagan's inauguration, Searle re-applies to the FDA for approval to use aspartame as a food sweetener.

MARCH 1981

An FDA commissioner's panel is established to review issues raised by the PBOI.

19 MAY 1981

Arthur Hull Hayes Jr, appoints a five-person commission to review the PBOI's decision. Three of the five FDA scientists on it advise against approval of aspartame, stating on the record that Searle's tests are unreliable and not adequate to determine the safety of aspartame. Hayes installs a sixth member on the commission, and the vote becomes deadlocked.

15 JULY 1981

Hayes ignores the recommendations of his own internal FDA team, overrules the PBOI findings and gives initial approval for aspartame to be used in dry products on the basis that it has been shown to be safe for its proposed uses.

22 OCTOBER 1981

The FDA approves aspartame as a tabletop sweetener and for use in tablets, breakfast cereals, chewing gum, dry bases for beverages, instant coffee and tea, gelatines, puddings, fillings, dairy-product toppings and as a flavour enhancer for chewing gum.

1982

The aspartame-based sweetener Equal, manufactured by Merisant, is launched in the US.

15 OCTOBER 1982

The FDA announces that Searle has filed a petition for aspartame to be approved as a sweetener in carbonated beverages, children's vitamins and other liquids.

1983

Searle attorney Robert Shapiro gives aspartame its commercial name, NutraSweet. The name is trademarked the following year. Shapiro later becomes president of Searle. He eventually becomes president and then chairman and CEO of Monsanto, which will buy Searle in 1985.

8 JULY 1983

Aspartame is approved for use in carbonated beverages and syrup bases in the US and, three months later, Britain. Before the end of the year Canderel tablets are launched in the UK. Granular Canderel follows in 1985.

8 AUGUST 1983

James Turner, on behalf of himself and the Community Nutrition Institute, and Dr Woodrow Monte, Arizona State University's director of food science and nutritional laboratories, file petitions with the FDA objecting to aspartame approval based on possible seriousadverse effects from the chronic intake of the sweetener. Monte also cites concern about the chronic intake of methanol associated with aspartame ingestion.

SEPTEMBER 1983

Hayes resigns as FDA commissioner under a cloud of controversy about his taking unauthorised rides aboard a General Foods jet (General Foods was and is a major purchaser of aspartame). He serves briefly as provost at New York Medical College, and then takes a position as senior scientific consultant with Burston-Marsteller, the chief public relations firm for both Searle and Monsanto.

AUTUMN 1983

The first carbonated beverages containing aspartame go on sale in the US.

17 FEBRUARY 1984

The FDA denies Turner and Monte's requests for a hearing, noting that aspartame's critics had not presented any unresolved safety questions. Regarding aspartame's breakdown components, the FDA says that it has reviewed animal, clinical and consumption studies submitted by the sweetener's manufacturer, as well as the existing body of scientific data, and concludes that 'the studies demonstrated the safety of these components'.

MARCH 1984

Public complaints about the adverse effects of aspartame begin to come in. The FDA requests that the US agency the Centers for Disease Control and Prevention (CDC) begins investigations of a select number of cases of adverse reactions to aspartame.

30 MAY 1984

The FDA approves aspartame for use in multivitamins.

JULY 1984

A study by the state of Arizona Department of Health into aspartame is published in the Journal of Applied

Nutrition. It determines that soft drinks stored at elevated temperatures promote more rapid deterioration of aspartame into poisonous methanol.

2 NOVEMBER 1984

The CDC review of public complaints relating to aspartame culminates in a report, Evaluation of Consumer Complaints Related to Aspartame Use, which reviews 213 of 592 cases and notes that re-challenge tests show that sensitive individuals consistently produce the same adverse symptoms each time they ingested aspartame. The reported symptoms include: aggressive behaviour, disorientation, hyperactivity, extreme numbness, excitability, memory loss, loss of depth perception, liver impairment, cardiac arrest, seizures, suicidal tendencies and severe mood swings. The CDC nevertheless concludes that aspartame is safe to ingest. On the same day that the CDC exonerates aspartame, Pepsi announces that it is dropping saccharin and adopting aspartame as the sweetener in all its diet drinks. Others quickly follow suit.

1 OCTOBER 1985

Monsanto, the producer of recombinant bovine growth hormone, genetically engineered soya beans, the pesticide Roundup and many other industrial and agricultural chemicals, purchases Searle for $2.7 billion.

21 APRIL 1986

The US Supreme Court, headed by Justice Clarence Thomas, a former Monsanto attorney, refuses to consider arguments from the Community Nutrition Institute and other consumer groups that the FDA has not followed proper procedures in approving aspartame, and that the liquid form of the artificial sweetener may cause brain damage in heavy users of low-calorie soft drinks.

16 OCTOBER 1986

Turner files another citizen's petition, this time concerning the risk of seizures and eye damage from aspartame. The petition argues that medical records of 140 aspartame users show them to have suffered from epileptic seizures and eye damage after consuming products containing the sweetener and that the FDA should ban aspartame as an 'imminent hazard to the public health'.

21 NOVEMBER 1986

The FDA denies Turner's new petition, saying: 'The data and information supporting the safety of aspartame are extensive. It is likely that no food product has ever been so closely examined for safety. Moreover, the decisions of the agency to approve aspartame for its uses have been given the fullest airing that the legal process requires.'

28 NOVEMBER 1986

The FDA approves aspartame for non-carbonated frozen or refrigerated concentrates and single-strength fruit juice, fruit drinks, fruit-flavoured drinks, imitation fruit-flavoured drinks, frozen stock-type confections and novelties, breath mints and tea beverages.

DECEMBER 1986

The FDA declares aspartame safe for use as an inactive ingredient, provided labelling meets certain specifications.

1987

An FDA report on adverse reactions associated with aspartame states the majority of the complaints about aspartame - now numbering 3,133 - refer to neurological effects.

2 JANUARY 1987

NutraSweet's aspartame patent runs out in Europe, Canada and Japan. More companies are now free to produce aspartame sweeteners in these countries.

12 OCTOBER 1987

United Press International, a leading global news-syndication organisation, reports that more than 10 federal officials involved in the decision to approve aspartame have now taken jobs in the private sector that are linked to the aspartame industry.

3 NOVEMBER 1987

A US Senate hearing is held to address the issue of aspartame safety and labelling. The hearing reviews the faulty testing procedures and the 'psychological strategy' used by Searle to help ensure aspartame's approval. Other information that comes to light includes the fact that aspartame was once on a Pentagon list of prospective biochemical-warfare weapons.

Numerous medical and scientific experts testify as to the toxicity of aspartame. Among them is Dr Verrett, who reveals that, while compiling its 1977 report, her team was instructed not to comment on or be concerned with the overall validity of the studies. She states that questions about birth defects have not been answered. She also states that increasing the temperature of the product leads to an increase in production of DKP, a substance shown to increase uterine polyps and change blood cholesterol levels. Verrett comments: 'It was pretty obvious that somewhere along the line, the bureau officials were working up to a whitewash.'

1989

The FDA has received more than 4,000 complaints from consumers about adverse reactions to the sweetener.

14 OCTOBER 1989

Dr HJ Roberts, director of the Palm Beach Institute for Medical Research, claims that several recent aircraft accidents involving confusion and aberrant pilot behaviour were caused by ingestion of products containing aspartame.

20 JULY 1990

The Guardian publishes a major investigation of aspartame and delivers to government officials 'a dossier of evidence' that draws heavily on the transcripts of the Bressler Report and demands that the government review the safety of aspartame. No review is undertaken. The Guardian is taken to court by Monsanto and forced to apologise for printing its story.

1991

The US National Institutes of Health publishes Adverse Effects of Aspartame: January '86 through December '90, a bibliography of 167 studies documenting adverse effects associated with aspartame.

1992

NutraSweet signs agreements with Coca-Cola and Pepsi stipulating that it is their preferred supplier of aspartame.

30 JANUARY 1992

The FDA approves aspartame for use in malt beverages, breakfast cereals, and refrigerated puddings and fillings and in bulk form (in large packages like sugar) for tabletop use. NutraSweet markets these bulk products under the name 'NutraSweet Spoonful'.

14 DECEMBER 1992

NutraSweet's US patent for aspartame expires, opening up the market for other companies to produce the substance.

19 APRIL 1993

The FDA approves aspartame for use in hard and soft candies, non-alcoholic flavoured beverages, tea beverages, fruit juices and concentrates, baked goods and baking mixes, and frostings, toppings and fillings for baked goods.

28 FEBRUARY 1994

Aspartame now accounts for the majority (75 per cent) of all the complaints in the US adverse-reaction monitoring system. The US Department of Health and Human Services compiles a report that brings together all current information on adverse reactions attributed to aspartame. It lists 6,888 complaints, including 649 reported by the CDC and 1,305 reported by the FDA.

APRIL 1995

Consumer activist, and founder of anti-aspartame group Mission Possible, Betty Martini uses the US's Freedom of Information Act to force the FDA to release an official list of adverse effects associated with aspartame ingestion. Culled from 10,000 consumer complaints, the list includes four deaths and more than 90 unique symptoms, a majority of which are connected to impaired neurological function. They include: headache; dizziness or problems with balance; mood change; vomiting and nausea; seizures and convulsions; memory loss; tremors; muscle weakness; abdominal pains and cramps; change in vision; diarrhoea; fatigue and weakness; skin rashes; deteriorating vision; joint and musculoskeletal pain. By the FDA's own admission, fewer then 1 per cent of those who have problems with something they consume ever report it to the FDA. This means that around 1 million people could have been experiencing adverse effects from ingesting aspartame.

12 JUNE 1995

The FDA announces it has no further plans to continue to collect adverse reaction reports or monitor research on aspartame.

27 JUNE 1996

The FDA removes all restrictions from aspartame use, and approves it as a general-purpose sweetener', meaning that aspartame can now be used in any food or beverage.

NOVEMBER 1996

Drawing on data compiled by the US National Cancer Institute's Surveillance, Epidemiology and End Results programme, which collects and distributes data on all types of cancer, Olney publishes peer-reviewed research in the Journal of Neuropathology and Experimental Neurology. It shows that brain-tumour rates have risen in line with aspartame consumption and that there has been a significant increase in the conversion of less deadly tumours into much more deadly ones.

DECEMBER 1996

The results of a remarkable study conducted by Dr Ralph G Walton, professor of clinical psychology at Northeastern Ohio Universities, are revealed. Commissioned by the hard-hitting US national news programme 60 Minutes, it sheds some light on the absurdity of aspartame-safety studies. Walton reviewed 165 separate studies published in the preceding 20 years in peer-reviewed medical journals. Seventy-four of the studies were industry-funded, all of which attested to aspartame's safety. Of the other 91 non-industry funded studies, 84 identified adverse health effects. Six of the seven non-industry funded studies that were favourable to aspartame were from the FDA, which has a public record of strong pro-industry bias. To this day, the industry-funded studies are the ones that are always quoted to the press and in official rebuttals to aspartame critics. They are also the studies given the greatest weight during the approval process and in official safety reviews.

10 FEBRUARY 1998

Monsanto petitions the FDA for approval of a new tabletop sweetener called Neotame. It is around 60 times sweeter than aspartame and up to 13,000 times sweeter than sugar. Neotame is less prone to breaking down in heat and in liquids than aspartame because of the addition of 3,3-dimethylbutyl, a poorly studied chemical with suspected neurotoxic effects. Strengthening the bond between aspartame's main constituents eliminates the need for a health warning directed at people suffering from PKU.

13 MAY 1998

Independent scientists from the University of Barcelona publish a landmark study clearly showing that aspartame is transformed into formaldehyde in the bodies of living specimens (in this case rats), and that this formaldehyde spreads throughout the specimens' vital organs, including the liver, kidneys, eyes and brain. The results fly in the face of manufacturers' claims that aspartame does not break down into formaldehyde in the body, and bolster the claims of aspartame critics that many of the symptoms associated with aspartame toxicity are caused by the poisonous and cumulative effects of formaldehyde.

OCTOBER 1998

The UK's Food Commission publishes two surveys on sweeteners. The first shows that several leading companies, including St Ivel, Müller and Sainsbury's, have ignored the legal requirement to state 'with sweeteners' next to the name of the product. The second reveals that aspartame not only appears in 'no-sugar added' and 'light' beverages but also in ordinary non-dietetic drinks because it's three times cheaper than ordinary sugar.

8 FEBRUARY 1999

Monsanto files a petition with the FDA for approval of the general use of Neotame.

20 JUNE 1999

An investigation by The Independent on Sunday reveals that aspartame is made using a genetic engineering process. Aspartame component phenylalanine is naturally produced by bacteria. The newspaper reveals that Monsanto has genetically engineered the bacteria to make them produce more phenylalanine. Monsanto claims that the process had not been revealed previously because no modified DNA remains in the finished product, and insists that the product is completely safe; though scientists counter that toxic effects cannot be ruled out in the absence of long-term studies. A Monsanto spokeswoman says that while aspartame for the US market is often made using genetic engineering, aspartame supplied to British food producers is not. The extent to which US brands of low-calorie products containing genetically engineered aspartame have been imported into Britain is unclear.

MAY 2000

Monsanto, under pressure - not least from the worldwide resistance to genetically manipulated food and ongoing lawsuits - sells NutraSweet to JW Childs Associates, a private-equity firm comprised of several former Monsanto managers, for $440m. Monsanto also sells its equity interest in two European sweetener joint ventures, NutraSweet AG and Euro-Aspartame SA.

10 DECEMBER 2001

The UK's Food Standards Agency requests that the European Commission Scientific Committee on Food conducts an updated review of aspartame. The committee is asked to look carefully at more than 500 scientific papers published between 1988 and 2000 and any other new scientific research not examined previously.

9 JULY 2002

The FDA approves the tabletop and general use of Neotame. The 'fast-track' approval raises eyebrows because, historically, the FDA takes at least 10 years to approve food additives. Neotame is also approved for use in Australia and New Zealand, but has yet to be approved in the UK.

10 DECEMBER 2002

The European Commission Scientific Committee on Food publishes its final report on aspartame. The 24-page report largely ignores independent research and consumer complaints, relying instead on frequently cited articles in books and reviews put together by employees or consultants of aspartame manufacturers. When independent research is cited, it is generally refuted with industry-sponsored data. An animal study showing aspartame's disruption of brain chemistry, a human study linking aspartame to neurophysiological changes that could increase seizure risk, another linking aspartame use with depression in individuals susceptible to mood disorder, and two others linking aspartame ingestion with headaches are all dismissed.

The report's conclusion amounts to a single sentence: 'The committee concluded that.there is no evidence to suggest that there is a need to revise the outcome of the earlier risk assessment or the [acceptance daily intake] previously established for aspartame.'

As with the FDA, there are concerns about the neutrality of some of the committee's members and their links with the International Life Sciences Institute (ILSI), an industry group that funds, among other things, research into aspartame. ILSI members include Monsanto, Coca-Cola and Pepsi.

19 FEBRUARY 2003

Members of the European Parliament's Environment, Public Health and Consumer Policy Committee approve the use of sucralose (see page 50) and an aspartame-acesulfame salt compound (manufactured in Europe by the aspartame-producing Holland Sweetener Company and sold under the name Twinsweet), agreeing to review of the use of both in three years' time. At the same time, a request by European greens that the committee re-evaluate the safety of aspartame and improve the labelling of aspartame-containing products is rejected.

MAY 2004

The feature-length documentary Sweet Misery is released on DVD (see http://www.soundandfuryproductions.com). Part-documentary, part-detective story, it includes interviews with people who have been harmed by aspartame, as well as credible testimony from advocates, doctors, lawyers and long-time campaigners, including James Turner, HJ Roberts and renowned neurosurgeon Dr Russell Blaylock. (UK orders: Namaste Publishing, info@namastepublishing.co.uk.)

SEPTEMBER 2004

US consumer group the National Justice League files a $350m class action lawsuit against the NutraSweet Corporation (the current owner of aspartame products), the American Diabetes Association and Monsanto. Some 50 other defendants have yet to be named, but mentioned throughout the lawsuit is the central role of Donald Rumsfeld in helping to get aspartame approved through the FDA. The plaintiffs maintain that this litigation will prove how deadly aspartame is when it is consumed by humans. Little progress has been made so far in bringing the action to court. The NutraSweet Company reopens its plant in Atlanta, Georgia, (dormant since 2003) in order to meet increased demand for its sweetener. Aspartame, sold commercially as NutraSweet, Equal, Equal-Measure, Spoonful, Canderel and Benevia, is currently available in more than 100 countries and used in more than 5,000 products by at least 250 million people every day. Worldwide, the aspartame industry's sales amount to more than $1 billion yearly. The US is the primary consumer.

JULY 2005

The Ramizzini Institute in Bologna, a non-profit, private institution set up to research the causes of cancer, releases the results of a very large, long-term animal study into

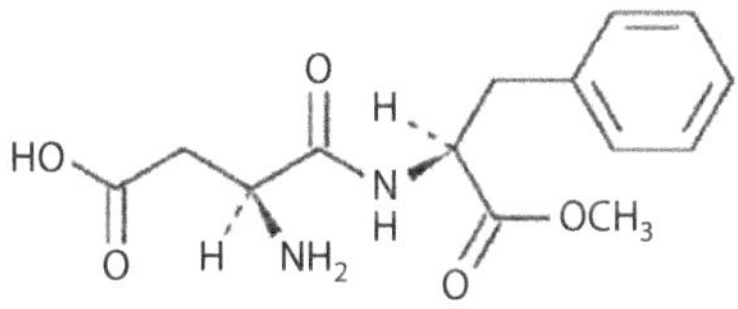

Figure 3.12 Chemical structure of Aspartame.

aspartame ingestion. Its study shows that aspartame causes lymphomas and leukaemia in female animals fed aspartame at doses around 20 milligrams per kilogram of body weight, or around half the accepted daily intake for humans.

3.6.3 The Hidden Epidemic of Aspartame

Figure 3.12 shows the chemical structure of aspartame. It is important to analyze aspartame compounds in order to appreciate its deadly consequences. Aspartame is comprised of 4 deadly compounds: phenylalanine, aspartic acid, methanol, and Diketopiperazine (DKP). The chemical bond that holds these constituents together is fairly weak. As a result, aspartame readily breaks down into its component parts in a variety of circumstances: in liquids; during prolonged storage; when exposed to heat in excess of 86° Fahrenheit (30° Centigrade); and when ingested. These constituents further break down into other toxic by-products, namely formaldehyde, formic acid and aspartylphenylalanine diketopiperazine (DKP).

3.6.3.1 Phenylalanine

The largest component of aspartame is called phenylalanine, making up 50% of the artificial sweetener.

Phenylalanine is an amino acid normally found in the brain. Persons with the genetic disorder phenylketonuria (PKU) cannot metabolize phenylalanine. This leads to dangerously high levels of phenylalanine in the brain (sometimes lethal). It has been shown that ingesting aspartame, especially along with carbohydrates, can lead to excess levels of phenylalanine in the brain even in persons who do not have PKU. A more subtle point is, phenylalanine is the artificial version of the amino acid that is present in brain. It's equivalent to replacing natural chemical with artificial one. When it comes to brain, it is equivalent to adding a neurotoxin. Similar effect has been demonstrated with natural vitamin C (that prevents cancer) and artificial vitamin C (ascorbic acid that gives cancer).

This presents a serious problems since high levels of phenylalanine in the brain can cause the levels of serotonin to decrease, leading to emotional disorders such as depression. It has been shown in human clinical trials that phenylalanine levels of the blood are increased significantly in those who chronically use aspartame.

Even a single use of aspartame raised the blood phenylalanine levels. In his testimony before the U.S. Congress, Dr. Louis J. Elsas showed that high blood phenylalanine can be concentrated in parts of the brain and is especially dangerous for infants and fetuses. He also showed that phenylalanine is metabolized much more efficiently by rodents than by humans. Thus, rodent studies showing minimal effects of phenylalanine should be considered with a grain of salt.

3.6.3.2 Aspartic Acid

The next deadly component with Asparatame is aspartic acid. Roughly 40% of aspartame is made up of aspartic acid. Dr. Russell L. Blaylock, a professor of neurosurgery at the Medical University of Mississippi, has cited 500 scientific references to show how excess free excitatory amino acids such as aspartic acid in our food supply are causing serious chronic neurological disorders and a myriad of other acute symptoms.

Aspartate (from aspartic acid) acts as a neurotransmitter in the brain but too much of it can kill certain neurons by allowing too much calcium into the cells, triggering an onslaught of free radical damage. This influx triggers excessive amounts of free radicals, which kill the cells.

Curiously, the blood brain barrier, which normally protects the brain from excess influx of aspartate (as well as many other toxins) does not fully protect against excess levels of this "excitotoxin" in the blood. Thus, aspartame ingestion has been associated with a number of neurological defects such as memory loss, multiple sclerosis, headaches, vision problems, dementia, brain lesions, and more.

One common complaint of persons suffering from the effect of aspartame is memory loss. Ironically, in 1987, G.D. Searle, the manufacturer of aspartame, undertook a search for a drug to combat memory loss caused by excitatory amino acid damage.

3.6.3.3 Methanol

Methanol (or wood alcohol) is a deadly poison and makes up 10% of aspartame. Methanol's toxicity is well documented. Methanol is gradually released in the small intestine when the methyl group of aspartame encounters the enzyme chymotrypsin. Methanolis then broken down into formic acid and formaldehyde in the body. Formaldehyde is a deadly neurotoxin. It's the same material that is used to preserve dead bodies.

An Environmental Protection Agency (EPA) of methanol states that methanol "is considered a cumulative poison due to the low rate of excretion once it is absorbed. In the body, methanol is oxidized to formaldehyde and formic acid; both of these metabolites are toxic."

The EPA recommends a limit of consumption of 7.8 mg/day.

To give one some perspective, a one-liter bottle of Diet Coke contains about 56 mg of methanol. Heavy users of aspartame-containing products consume as much as 250 mg of methanol daily or 32 times the EPA limit!

Symptoms from methanol poisoning include headaches, ear buzzing, dizziness, nausea, gastrointestinal disturbances, weakness, vertigo, chills, memory lapses, numbness and shooting pains in the extremities, behavioral disturbances, and neuritis. The most well known problems from methanol poisoning are vision problems including misty vision, progressive contraction of visual fields, blurring of vision, retinal damage, and blindness.

Formaldehyde, on its own, is a known carcinogen, causes retinal damage, interferes with DNA replication and causes birth defects.

As pointed out by Dr. Woodrow C. Monte, director of the food science and nutrition laboratory at Arizona State University, "There are no human or mammalian studies to evaluate the possible mutagenic, teratogenic or carcinogenic effects of chronic administration of methyl alcohol."

This is how sick and twisted our food supply has become...

Dr. Monte was so concerned about the safety issues of methanol (and aspartame) that he filed suit with the FDA requesting a hearing to address these issues. He asked the FDA to...

"Slow down on this soft drink issue long enough to answer some of the important questions. It's not fair that you are leaving the full burden of proof on the few of us who are concerned and have such limited resources. You must remember that you are the American public's last defense. Once you allow usage (of aspartame) there is literally nothing I or my colleagues can do to reverse the course. Aspartame will then join saccharin, the sulfiting agents, and God knows how many other questionable compounds enjoined to insult the human constitution with governmental approval."

Shortly thereafter, the Commissioner of the FDA, Arthur Hull Hayes, Jr., approved the use of aspartame in carbonated beverages, he then left for a position with G.D. Searle's public relations firm.

When aspartame was approved for use, Dr HJ Roberts, director of the Palm Beach Institute for Medical Research, had no reason to doubt the FDA's decision. 'But my attitude changed,' he says, 'after repeatedly encountering serious reactions in my patients that seemed justifiably linked to aspartame.' Twenty years on, Roberts has coined the phrase 'aspartame disease' to describe the wide range of adverse effects he has seen among aspartame-guzzling patients.

He estimates: 'Hundreds of thousands of consumers, more likely millions, currently suffer major reactions to products containing aspartame. Today, every physician probably encounters aspartame disease in everyday practice, especially among patients with illnesses that are undiagnosed or difficult to treat.'

As a guide for other doctors, Roberts, a recognised expert in difficult diagnoses, has published a lengthy series of case studies, Aspartame Disease: an ignored epidemic (Sunshine Sentinel Press), in which he meticulously details his treatment of 1,200 aspartame-sensitive individuals, or 'reactors', encountered in his own practice. Following accepted medical procedure for detecting sensitivities to foods, Roberts had his patients remove aspartame from their diets. With nearly two thirds of reactors, symptoms began to improve within days of removing aspartame, and improvements were maintained as long as aspartame was kept out of their diet.

Roberts' case studies parallel much of what was revealed in the FDA's report on adverse reactions to aspartame - that toxicity often reveals itself through central nervous system disorders and compromised immunity. His casework shows that aspartame toxicity can mimic the symptoms of and/or worsen several diseases that fall into these broad categories (see sidebar, below).

Case studies, especially a large series like this, address some of the issues surrounding real-world use in a way that laboratory studies never can; and the conclusions that can be drawn from such observations aren't just startling, they are also potentially highly significant. In fact, Roberts believes that one of the major problems with aspartame research has been the continued over-emphasis on laboratory studies.

This has meant that the input of concerned independent physicians and other interested persons, especially consumers, is 'reflexively discounted as "anecdotal"'.

Many of the diseases listed by Roberts fall into the category of medicine's 'mystery diseases' - conditions with no clear etiology and few effective cures. And while no one is suggesting that aspartame is the single cause of such diseases, Roberts' research suggests that some people diagnosed with, for example, multiple sclerosis, Parkinson's or chronic fatigue syndrome may end up on a regimen of potentially harmful drugs that could have been avoided if they simply stopped ingesting aspartame-laced products.

3.6.3.4 DKP

DKP is a breakdown product of phenylalanine that forms when aspartame-containing liquids are stored for prolonged periods. In animal experiments it has produced brain tumours, uterine polyps and changes in blood cholesterol. Before the FDA approved aspartame, the amount of DKP in our diets was essentially zero. So no claim of DKP's safety can be accepted as genuine until good-quality long-term studies have been performed. No such studies have been done.

The following diseases can be linked to aspartame toxicity:

Cancer

> Studies have found a dangerous connection between aspartame consumption and the development of brain tumors. When aspartame breaks down it produces a substance called DKP. As your stomach digests DKP, it produces a chemical that induces the growth of brain tumors.

Diabetes

> Aspartame consumption is extremely harmful to people with diabetes. It makes it more difficult to control sugar levels and aggravate diabetes-related conditions such as retinopathy, cataracts, neuropathy and gastroparesis. The sweetener also has been known to cause convulsions often mistaken for insulin reactions.

Psychological Disorders

Emotional and mood disorders have been linked to aspartame. Studies suggest that people with certain emotional problems are more sensitive to aspartame. High levels of aspartame cause changes in the serotonin levels which can lead to behavioral problems, depression and other emotional disorders. In some cases, the side effects were so dangerous that doctors were forced to put an end to the studies.

Hinders Weight Loss

Aspartame can be found in diet sodas and most other diet products. However, research indicates that the sweetener increases your hunger and can actually impede your weight loss. Phenylalanine and aspartic acid can cause spikes in insulin levels and force your body to remove the glucose from your blood stream and store it as fat. Aspartame also inhibits the production of serotonin and prevents your brain from signalling to your body that you are full. This can lead to food cravings and make it more difficult for you to lose weight.

Birth Defects

Aspartame is an excitotoxin or a substance that has the potential to damage or kill cells in the nervous system. The blood-brain barrier is a structure that stops harmful substances from penetrating the brain. The barrier doesn't completely form until a child is one year old, therefore, before a child is born, its nervous system is vulnerable to any dangerous excitotoxins that the mother may consume. Too much exposure to phenylalanine or aspartic acid may cause irreversible brain damage and other serious birth defects.

Vision Problems

Methanol, one of Aspartame's components, can damage the retinas and the optic nerves. Aspartame consumption has been connected to eye pain, blurred vision and, in some cases, blindness.

Brain Damage and Seizures

Aspartame can change the chemistry of the brain. Formaldehyde, a product of methanol, gathers in certain areas of the brain causing degenerative diseases such as Parkinson's, Alzhcimer's and ALS. Aspartame consumption can also trigger seizures in both epileptics and other individuals without a history of epilepsy.

3.7 The Honey-Sugar-Saccharin-Aspartame Degradation in Everything

HSS®A® is the most notorious accomplishment *par excellence* born of engineering a myriad of applications of the findings of so-called "New Science" — plastics to textiles to botox —from which intangibles like characteristic time are eternally banished. The HSS®A® label for this pathway generalizes the seemingly insignificant example of the degradation of natural honey to carcinogenic "sweeteners" like Aspartame® because, as Albert Einstein most famously pointed out, the environmental challenges posed by conventional suppression or general disregard of essential phenomena in the natural environment such as pollination actually threaten the continued existence of human civilization in general.

The HSS®A® pathway is a metaphor representing many other phenomena and chins of phenomena that originate from a natural form and become subsequently engineered through many intermediate stages into "new" products. In the following discussion, we lay out how it works. Once it is understood how disinformation works, one can figure out a way to reverse the process by avoiding aphenomenal schemes that lead to ignorance packaged with arrogance.

Ever since the introduction of the culture of plastic over a century ago, the public has been indoctrinated into associating an increase in the quality, and/or qualities, of a final product with the insertion of additional intermediate stages of 'refining' the product. If honey – taken more or less directly from a natural source, without further processing – was fine, surely the sweetness that can be attained by refining sugar must be better. If the individual wants to reduce their risk of diabetes, then surely further refining of the chemistry of "sweetness" into such products as Saccharin® must be better still. And why not even more sophisticated chemical engineering to further convert the chemical essence of this refined sweetness into forms that are stable in liquid phase, such as Aspartame®?

In this sequence, each additional stage is defended and promoted as having overcome some limitation of the immediately previous stage. But at the end of this chain, what is left in, say, Aspartame® of the 200-plus beneficial qualities of honey? Looking from the end of this chain back to its start, how many laboratory rats ever contracted cancer from any amount of honey intake? How many nerves become frozen by taking honey? Honey is known to be the only food that possesses all the nutrients, including water, needed to sustain life. How many true nutrients does Aspartame® have? From the narrowest engineering standpoint, the kinds and number of qualities in the final product at the end of this Honey → Sugar → Saccharin® → Aspartame® chain have been transformed, but from the human consumer's standpoint of the use-value of "sweet-tasting," has there been a net qualitative gain going from honey all the way to Aspartame®?

From the scientific standpoint, honey fulfils both conditions of phenomenality, namely: origin and process. That is, the source of honey (nectar) is real (even if it means flowers were grown with chemical fertilizers, pesticides, or even genetic alteration), and — even if the bees were subjected to air pollution or a sugary diet — the process is real (honeybees cannot make false intentions, therefore they are perfectly natural). The quality of honey can differ depending on other factors, e. g., chemical fertilizer, genetic alteration, etc., but honey remains real.

As we "progress" from honey to sugar, the origin remains real (sugar cane or beet), but the process is tainted with artificial inputs, starting from electrical heating, chemical additives, bleaching, etc. Further "progress" to Saccharin® marks the use of another real origin, but this time the original source (crude oil) is a very old food source compared to the source of sugar. With steady-state analysis, they both appear to be of the same quality.

As the chemical engineering continues, we resort to the final transition to Aspartame®. Indeed, nothing is phenomenal about Aspartame®, as both the origin and the process are artificial. So, the overall transition from honey to Aspartame® has been from 100% phenomenal to 100% aphenomenal. Considering this, what economic calculations are needed to justify this replacement? It becomes clear, without considering the phenomenality feature, that any talk of economics would only mean the "economics" of aphenomenality. Yet this remains the standard of neoclassical economics. Throughout the modern era, economics has remained the driver of the education system.

There is an entire economics of scale that is developed and applied to determine how far this is taken in each case. For example, honey is perceptibly "sugar" to taste. We want the sugar, but honey is also anti-bacterial and cannot rot. Therefore, the rate at which customers will have to return for the next supply is much lower and slower than the rate at which customers would have to return to resupply themselves with, say, refined sugar. Even worse: to extend the amount of honey available in the market (in many third world countries, for example), sugar is added. The content of this "economic" logic then takes over and drives what happens to honey and sugar as commodities. There are natural limits to how far honey as a natural product can actually be commodified, whereas, for example, refined sugar is refined to become addictive so that the consumer becomes hooked and the producer's profit is secured. The Education system has been commodified in the modern age and remains the most vulnerable under the new world order that is taking shape at the dawn of the information age.

The matter of intention is not considered in the economics of scale, leading to certain questions never being answered. No one asks whether any degree of external processing of what began as a natural sugar source can or will improve its quality as a sweetener. Exactly what that process, or those processes, would be is also unasked. No sugar refiner is worried about how the marketing of his product in excess is contributing to a diabetes epidemic. The advertising that is crucial to marketing this product certainly won't raise this question. Guided by the "logic" of the economies of scale, and the marketing effort that must accompany it, greater processing is assumed to be and accepted as being *ipso facto* good, or better. As a consequence of the selectivity inherent in such "logic," any other possibility within the overall picture – such as the possibility that as we go from honey to sugar to saccharin to aspartame, we go from something entirely safe for human consumption to something cancerously toxic – does not even enter the frame.

Such a consideration would prove to be very threatening to the health of a group's big business in the short term. All this is especially devastatingly clear when it comes to education and natural cognition. Over the last millennium, even after 'original sin' has been discredited as aphenomenal, it is widely and falsely believed that natural cognition is backward looking and humans are incapable of finding their own path of knowledge, they must be indoctrinated into being enlightened.

Edible natural products in their natural state are already good enough for humans to consume at some safe level and process further internally in ways useful to the organism. We are not likely to consume any unrefined natural food source in excess. However: **the refining that accompanies the transformation of natural food sources into processed-food commodities also introduces components that interfere with the normal ability we have to push a natural food source aside after some definite point**. Additionally, with externally processed "refinements" of natural sources, the chances increase that the form in which the product is eventually consumed must include compounds that are not characteristic anywhere in nature and that the human organism cannot usefully process without stressing the digestive system excessively. After a cancer epidemic, there is great scurrying to fix the problem. The cautionary tale within this tragedy is that, if the HSS®A® principle were considered *before* a new stage of external processing were added, much unnecessary tragedy could be avoided.

There are two especially crucial premises of the economics-of-scale that lie hidden within the notion of "upgrading by refining:" (a) unit costs of production can be lowered (and unit profit therefore expanded) by increasing output **Q** per unit time *t*, i.e., by driving $\partial \mathbf{Q}/\partial t$ (temporal rate of change of Q) unconditionally in a positive direction; and (b) only the desired portion of the **Q** end-product is considered to have tangible economics and, therefore, also intangible social "value," while any unwanted consequences – e.g., degradation of, or risks to, public health, damage(s) to the environment, etc. – are discounted and dismissed as false costs of production.

Note that, if relatively free competition still prevailed, premise (a) would not arise even as a passing consideration. In an economy lacking monopolies, oligopolies, and/or cartels dictating effective demand by manipulating supply, unit costs of production remain mainly a function of some given level of technology. Once a certain proportion of investment in fixed-capital (equipment and ground-rent for the production facility) becomes the norm generally among the various producers competing for customers in the same market, the unit costs of production cannot fall or be driven arbitrarily below a certain floor level without risking business loss. The unit cost thus becomes downwardly inelastic.

The unit cost of production can become downwardly elastic, i.e., capable of falling readily below any asserted floor price, under two conditions:

(a) during moments of technological transformation of the industry, in which producers who are first to lower their unit costs by using more advanced machinery will gain market shares, temporarily, at the expense of competitors; or
(b) in conditions where financially stronger producers absorb financially weakened competitors.

In neoclassical models, which all assume competitiveness in the economy, this second circumstance is associated with the temporary cyclical crisis. This is the crisis that breaks out from time to time in periods of extended oversupply or weakened demand. In reality, contrary to the assumptions of the neoclassical economic models, the impacts of monopolies, oligopolies, and cartels have entirely displaced those of free competition and have become normal rather than the exception. Under such conditions, lowering unit costs of production (and thereby expansion of unit profit) by increasing output **Q** per unit time *t*, i.e., by driving $\partial \mathbf{Q}/\partial t$ unconditionally in a positive direction, is no longer an occasional and exceptional tactical opportunity. It is a permanent policy option: monopolies, oligopolies, and cartels manipulate supply and demand because they can.

Note that premise (*b*) points to how, where, and why consciousness of the unsustainability of the present order can emerge. Continuing indefinitely to refine nature out by substituting ever more elaborate chemical "equivalents," hitherto unknown in the natural environment, has started to take its toll. The narrow concerns of the owners and managers of production are at odds with the needs of society. Irrespective of the private character of their appropriation of the fruits of production, based on concentrating so much power in so few hands, production has become far more social. The industrial-scale production of all goods and services as commodities has spread everywhere from the metropolises of Europe and North America to the remotest Asian countryside, the deserts of Africa, and the jungle regions of South America. This economy is not only global in scope but also social in its essential character. Regardless of the readiness of the owners and managers to dismiss and abdicate responsibility for the environmental and human health costs of their unsustainable approach, these costs have become an increasingly urgent concern to societies in general. In this regard, the HSS®A® principle becomes a key and most useful guideline for sorting what is truly sustainable for the long term from what is undoubtedly unsustainable.

The human being that is transformed further into a mere consumer of products is a being that has become marginalized from most of the possibilities and potentialities of the fact of his/her existence. This marginalization is an important feature of the HSS®A® principle. There are numerous things that individuals can do to modulate, or otherwise affect, the intake of honey and its impacts. However, there is little – indeed: nothing – that one can do about Aspartame® except drink it. With some minor modification, the HSS®A® principle helps illustrate how the marginalization of the individual's participation is happening in other areas.

What has been identified here as the HSS®A® principle, or syndrome, continues to unfold attacks against both the increasing global striving toward true sustainability on the one hand, and the humanization of the environment in all aspects, societal and natural, on the other. Its silent partner is the aphenomenal model, which invents justifications for the unjustifiable and for "phenomena" that have been picked out of thin air. As with the aphenomenal model, repeated and continual detection and exposure of the operation of the HSS®A® principle is crucial for future progress in developing nature-science, the science of intangibles and true sustainability.

Table 3.9 summarizes the outcome of the HSS®A® pathway:

The above model is instrumental in turning any economic process or chain of processes that are accountable, manageable and effective in matching real supply and real demand economic model into a model that is entirely perception-based. To the extent that these economic processes also drive the relevant engineering applications and management that come in their train, such a path ultimately "closes the loop" of a generally unsustainable mode of technology development overall.

3.7.1 Eurocentric Prejudice

What we see here is a breakdown of logical thinking. It turns out Europe and Eurocentric scientists and savants would not recover from this dogmatic pattern. At no time, they set the direction of an investigation/study in such a way that they would be open to the outcome and would embrace the outcome as the truth, thereby making decisions based on the new knowledge. Long before European scientists and savants were immersed in the discussion of how flat the earth is, or which date to celebrate Jesus' resurrection or Easter, or New year's day, or how to control people's life by frightening them with hell, another 'religion' was taking root in the deserts of Arabia. Muhammad, an unlettered man, with no exposure to previous scriptures, was born in 570 AD and passed away on 630 AD. While he is ranked as the most influential man of human history (Hart, 1992; 2000), little is discussed about the five centuries that followed the life of the Prophet Muhammad, of ancient learning by the Muslim scholars inspired by Islam. During the period of post-Thomas Aquinas (father of doctrinal philosophy) cognition in Europe, the work of Islamic scholars continued and today, if one just searches in Wikipedia, one will find: Ibn sina (Avicenna) is named the father of modern medicine and alchemy, Ibn Rushd (Averroes) the father of secular philosophy and Education, Ibn Haitham (Alhazen) the father of modern optics, Al-Kindi (Alkindus)-father of information technology, Ibn Khaldoun - father of modern social sciences, Al-Khwārizmī the founding father of algebra and mathematics, and Al-Farabi named the father of epistemology and metaphysics. Yet, all of them are listed as either polyscientist or polymath. In addition, all of them are considered to be inspired by prophet Muhammad. This is truly an unprecedented event in human history that has received little attention from the Eurocentric 'scientists' and 'scholars'. Prophet Muhammad was listed by Astrophysicist, Michael Hart (1992; 2000) as the most influential world leader. Who is second in that list? Sir Isaac Newton, the man who wrote more on Christian doctrine than on science. His Church wasn't based in Rome, it was the Church of England, headed by the Monarch. The transition from 'religion' to 'politics' wasn't even subtle. Some of our recent work only began to touch upon the original theories Islamic scholars that could serve humanity only if they were not altered with the intent of fitting a conclusion new scientists were trying to come up with in order to satisfy the Church, the Government, or the Corporation. Newton was unique, because he satisfied the Church and the Monarch simultaneously. It was possible because the Head of the Church of England was also the Monarch (true until today). While Newton had no reason to challenge the first premise of the Church that he belonged to, others (e.g., Russian scientists, Einstein) didn't dare question the first premise of anyone, but most notably that of Newton. If they did, they were quickly called 'the anarchist'.

As an example, the Nobel Prize winning work of Albert Einstein is noteworthy. Our recent book (Islam *et al.*, 2010) pointed out how Einstein's work simply took Maxwell's rigid sphere model as true and how Maxwell himself took that model from Newton. Consider how Ibn Haitham took the model of Aristotle and deconstructed it based on simple logic, discarded Aristotle's conclusion that light has infinite speed, and reconstructed a model that until today serves as the only model that can distinguish sunlight from artificial light. While it is well known that sunlight is the essence of life and artificial light is something that is used to torture people, Einstein's theory or any other optic theory cannot explain scientifically how this is possible. In addition, Ibn Haitham undid another one of old theories, that is, it is something that comes out of our eyes that make it possible for us to see. Instead of that theory, he introduced the notion of something entering your eye that make it possible to see. This 'something' was later proclaimed to be photon. This notion was correct, but the denomination as well as attribution of various properties made further research on the topic of light characterization impossible. For instance, because this theory postulates that all photons are alike and do not have mass, the source of the light cannot have an impact on the quality of light, leading to the same difficulty that made it impossible to discern between sunlight and artificial light. If, on the other hand, Ibn Haitham's theory was used correctly, one would be able to correlate the toxic nature of the light source (e.g., power-saving light) with long-term impact (e.g., breast cancer, brain dysfunction, numerous other reported correlations). This would also unravel the science behind skin-cancer causing chemicals that are often included in suntan or sun protection lotions. Another example is worth mentioning here. Only recently a new form of energy-saving light. This light was excellent in energy saving as well as in producing the 'white light' effect. However, it was also performing the so-called 'belly dance'. When the source was sent to the International Space Station for a probe, it was discovered that the 'belly dance' subsided or disappeared. Gravity was found to be the reason behind belly dance. Could this be explained with existing light theories? Of course not, because if photons have zero mass, how could gravity affect them? This paradoxical modus operandi continues when it comes to dark matters (infinite mass, but no energy) and dark energy (infinite energy but no mass) in the realm of discussion in Cosmic physics.

So, what's the catch? Ibn Haitham didn't read Aristotle's work to believe in it. He read the theory, used the criterion on truth and falsehood and was able to decipher true elements from the volumes of work. Einstein didn't dare use the same logic about Maxwell's 'laws' or he didn't have the criterion that Ibn Haitham was equipped with. Einstein was not a Christian, but he certainly was a believer of Eurocentric philosophy. This fundamental inability to discern truth from falsehood (called Furqan in Arabic) is missing from European new science or social science. As a result, what we see is constant confusion about

Table 3.9 The HSS®A® pathway and its outcome in various disciplines.

Natural state	First stage of intervention	Second stage of intervention	Third stage of intervention
Honey	Sugar	Saccharin®	Aspartame®
Education	Doctrinal teaching	Formal education	Computer-based learning
Science	Religion	Fundamentalism	Cult
Science and nature-based technology	New Science	Engineering	Computer-based design
Value-based(e.g. gold, silver) economy	Coins (non-gold or silver)	Paper money (disconnected from gold reserve)	Promissory note (electronic)

everything that governs our daily life. The article by Sardar (2009) talks about the philosophy that drove the mindset of Eurocentric scientists.

In Social science, take the work of Averroes, who is designated as the 'father of secular philosophy in Europe'. These are his quotes, review them and ask the question, Did any of the 'secular' scientists follow his logic?

> In his book titled: The Book of the Decisive Treatise, Determining the Connection Between the Law and Wisdom & Epistle Dedicatory (Averroes, 2001), Averroes wrote:
>
> "... [T]he Law makes it obligatory to reflect upon existing things by means of the intellect, and to consider them; and consideration is nothing more than inferring and drawing out the unknown from the known. . . "
>
> "You ought to know that what is intended by the Law is only to teach true science and true practice. True science is cognizance of God and of all the existing things as they are, especially the venerable ones among them; and cognizance of happiness in the hereafter and of misery in the hereafter. True practice is to follow the actions that promote happiness and to avoid the actions that

Table 3.10 HSS®A® Pathways of Economic Mismanagement

Natural state	1st stage of intervention	2nd stage of intervention	3rd stage of intervention
"Honey..."	"...Sugar..."	"... Saccharin®..."	"...Aspartame®"
Crude oil	Refined oil	High-octane refining	Chemical additives (for combating bacteria, thermal degradation, weather conditions, etc.)
Solar	Photovoltaics	Storage in batteries	Re-use in artificial light form
Organic vegetable oil	Chemical fertilizer, pesticide	Refining, thermal extraction	Genetically-modified crops
Organic saturated fat	Hormone, antibiotic	Artificial fat (transfat)	No-transfat artificial fat
Wind	Conversion into electricity	Storage in batteries	Re-usage in artificial energy forms
Water and hydro-energy	Conversion into electricity	Dissociation utilizing toxic processes	Recombination through fuel cells
Uranium ore	Enrichment	Conversion into electrical energy	Re-usage in artificial energy forms

promote misery; and cognizance of these actions is what is called 'practical science.'"

"... [T]he link between the physician and the health of bodies is [the same as] the link between the Lawgiver and the health of souls. . . "

"This health is what is called 'piety.'"

"... [I]njuries from a friend are graver than injuries from an enemy – I mean that wisdom is the companion of the Law and its milk sister. So injuries from those linked to it are the gravest injuries – apart from the enmity, hatred, and quarreling they bring about between both of them. These two are companions by nature and lovers by essence and instinct."

".. . [E]xistence is the cause and reason of our knowledge, while eternal knowledge is the cause and reason of existence."

We know Karl Marx's theory revolutionized political economy in the west. Yet read what is available on Wikipedia on Ibn Khaldoun:

Ibn Khaldun's outlines an early (possibly even the earliest) example of political economy. He describes the economy as being composed of value-adding processes; that is, labour and skill is added to techniques and crafts and the product is sold at a higher value. He also made the distinction between "profit" and "sustenance", in modern political economy terms, surplus and that required for the reproduction of classes respectively. He also calls for the creation of a science to explain society and goes on to outline these ideas in his major work the Muqaddimah.

Also see the comments of others on Ibn Khaldoun's work:

British historian Arnold J. Toynbee called the Muqaddimah "a philosophy of history which is undoubtedly the greatest work of its kind that has ever yet been created by any mind in any time or place."

The British philosopher Robert Flint wrote the following on Ibn Khaldun: "...as a theorist of history he had no equal in any age or country until Vico appeared, more than three hundred years later. Plato, Aristotle, and Augustine were not his peers, and all others were unworthy of being even mentioned along with him".

The British philosopher-anthropologist Ernest Gellner considered Ibn Khaldun's definition of government, "an institution which prevents injustice other than such as it commits itself", the best in the history of political theory.

Arthur Laffer, whom the Laffer curve is named after, noted that, among others, some of Ibn Khaldun's ideas precede his own.

So, what happened to the knowledge that was available for many centuries? Can we say we didn't know of them? After all, Arabic was the language of science for 1000 years and it was in Arabic that vast work of old Greek philosophers and polymath's work was translated into.

About dissemination of knowledge, prophet Muhammad ordered the following three (all three being an obligation): 1) seek knowledge; 2) act upon that knowledge; 3) disseminate that knowledge. It's exactly the same order for generating 'economic' activities. 1) Must work to sustain life; 2) must spend from the earnings for one's own sustenance; 3) must treat the earning as a trust and spend for others (Zatzman and Islam, 2007).

While ancient religions (such as Budhism, Hinduism) have talked about giving away in charity, disseminating knowledge is not there or not available in today's literature. While Islamic scholars (Ibn Haitham, Ibn Sina, Ibn Rushd, Al-Kindus and many others) have written dozens of books each, they really didn't relate dissemination of knowledge with being righteous.

Started thinking more deeply about this matter of "publication" — publication of results, sharing new discoveries (including sharing new findings about techniques etc. that lay long forgotten) and its relation to a thoroughly modern, contemporary knowledge-based culture.

There are a number of biases that operate continuously against the seeking, documentation, application and transmission of new knowledge/discoveries.

Modern society has this conceit that it is the most open in the entire history of Humanity to fostering the establishment and application of new knowledge and technologies. Of course, all scientific discovery depends on some kind of culturally-based sustaining framework for furthering and deepening such knowledge gathering, application and sharing/transmission.

What is the historical source of the contemporary globally-accepted European/Eurocentric framework? How does it stand up?

The Wikipedia article below sheds light on a deeply embarrassing feature endemic to that framework:

> "Techne" is a Greek term, etymologically derived from the Greek word τέχνη (ancient Greek: [tékʰnɛː], Modern Greek: ['texni]), that is often translated as craftsmanship, craft, or art.
>
> Techne is a term-of-art in philosophy which resembles epistēmē in its implication of knowledge of principles, although techne differs in that its intent is making or doing as opposed to disinterested understanding.
>
> As an activity, techne is concrete, variable, and context-dependent. As one observer has argued, techne "was not concerned with the necessity and eternal a priori truths of the cosmos, nor with the a posteriori contingencies and exigencies of ethics and politics. [...] Moreover, this was a kind of knowledge associated with people who were bound to necessity. That is, techne was chiefly operative

in the domestic sphere, in farming and slavery, and not in the free realm of the Greek polis."

Aristotle saw it as representative of the imperfection of human imitation of nature. For the ancient Greeks, it signified all the mechanic arts, including medicine and music. The English aphorism, "gentlemen don't work with their hands", is said to have originated in ancient Greece in relation to their cynical view on the arts. Due to this view, it was only fitted for the lower class while the upper class practiced the liberal arts of 'free' men. ...

Socrates also compliments techne only when it was used in the context of epistēmē. Epistēmē sometimes means knowing how to do something in a craft-like way. The craft-like knowledge is called a technê. It is most useful when the knowledge is practically applied, rather than theoretically or aesthetically applied. For the ancient Greeks, when techne appears as art, it is most often viewed negatively, whereas when used as a craft it is viewed positively because a craft is the practical application of an art, rather than art as an end in itself. In The Republic, written by Plato, the knowledge of forms "is the indispensable basis for the philosophers' craft of ruling in the city" ...

Techne is often used in philosophical discourse to distinguish from art (or poiesis). This use of the word also occurs in the digital humanities to differentiate between linear narrative presentation of knowledge and dynamic presentation of knowledge, wherein techne represents the former and poiesis represents the latter.

In other words, anything discovered by mere workers working away at solving practical problems was beneath any serious scholar's notice. This elite class bias regarding what constitutes knowledge and who is/are its progenitors and caretakers, is utterly breathtaking in the breadth and depth of its ignorance. It is excused as some folks behaving in character as ancient-Greek dudes. However, as between who or what societies in today's world possess or have a patented right to the real, i.e. infinitely money-making, "knowledge", has all that much changed? Why is Iran denied the right to figure out the nuclear power cycle for itself? etc. etc. Why is the identity of S. Bose — the brilliant Indian mathematician who was Einstein's co-discoverer — almost disappeared from the annals discussing the thought-processes that concluded in elaboration of the theory of relativity? Clearly, this elite class bias remains a serious obstacle to the widest possible dissemination, discussion and further development of new knowledge, even in its contemporary form of racist dismissal of the abilities of non-white peoples. The smell you may have begun to detect in this room is nothing but the rank hypocrisy of Western accounts about how ignorant and narrow-minded various ancient rulers in the East were in bottling up the productivity of their own scientific researchers and engineers so that no rival could copy and ultimately challenge their self-proclaimed exclusive right to rule.

3.7.2 Two Different Tracks of Cognition

It is commonly claimed that post-Renaissance Europe has got rid of its dogmatic past and developed a truly scientific cognition process. Recent work of Islam *et al.* (2010, 2012, 2014) demonstrated that this claim is false. Even though, Averröes is claimed to be the 'father of secular philosophy in Europe' and Thomas Aquinas the father of doctrinal philosophy, all evidence support the theory that European cognition adopted throughout modern history is anything but secular and addiction to dogmatic thinking has only become worse. The following reviews the cognition process presented by Averröes.

Nearly a millennium ago, long before the Renaissance reached Europe, Averröes (1126-1198 AD, known as Ibn Rushid outside of the western world) pointed out, that Aristotelian logic of the excluded middle cannot lead to increasing knowledge unless the first premise is true. In another word, the logic can be used only to differentiate between true and false, as long as there is a criterion that discerns the truth from falsehood. The difficulty, all the way to the present modern age, has been the inability to propose a criterion that is time-honored (Zatzman and Islam, 2007a). For Averröes, the first premise was the existence of the (only) creator. It was not a theological sermon or a philosophical discourse, it was purely scientific. Inspired by The *Qur'an* that cites the root word *ilm* (, meaning "science", the verb, *yalamu*, , standing for 'to understand', something that is totally subjective and is the first step toward acquiring knowledge), second most frequently (only second to the word, *Allah*), he considered The *Qur'an* as the only available, untainted communication with the creator and linked the first premise to the existence of such a communication. Averröes based his logic on post-Islamic Arabia (630 AD onward). As stated earlier, similar Ancient India and China, Islamic philosophy outlined in The *Qur'an* did include intangibles, such as intention and the time function. The modern-day view holds that knowledge and solutions developed from and within nature might be either good, or neutral [zero net impact] in their effects, or bad – all depending on how developed and correct our initial information and assumptions are. The view of science in the period of Islam's rise was rather different. It was that, since nature is an integrated whole in which humanity also has its roles, any knowledge and solutions developed according to how nature actually works will be *ipso facto* positive for humanity. Nature possesses an inbuilt positive intention of which people have to become conscious in order to develop knowledge and solutions that enhance nature. On the other hand, any knowledge or solutions developed by taking away from nature or going away from nature would be unsustainable. This unsustainability would mark such knowledge and solutions as inherently anti-nature.

Based on those intangibles, great strides in science, medicine, and all aspects of engineering was made in the Islamic era. Much of this was preserved, but by methods that precluded or did not include general or widespread publication. Thus, there could well have been almost as much total reliable knowledge 1400 years ago as today, but creative people's access and availability to that mass of reliable knowledge would have been far narrower. Only recently it is discovered that Islamic scholars were doing mathematics some 1000 years ago of the same order that are thought to be discovered in the 1970s (Lu and Steinhardt, 2007) – with the difference being that our mathematics can only track symmetry, something that does not exist in nature (Zatzman and Islam, 2007a). Knowledge definitely is not within the modern age. Recently, a three-dimensional PET-scan of a relic known as the 'Antikythera Mechanism' has demonstrated that it was actually a universal navigational computing device – with the difference being that our current-day versions rely on GPS, tracked and maintained by satellite (Freeth *et al.*, 2006). Only recently, Ketata *et al.* (2007a) recognized that computational techniques that are based on ancient, but nonlinear counting techniques, such as Abacus, are far superior to the linear computing. Even in the field of medicine, one would be shocked to find out what Ibn Sina ('Avicenna') said regarding nature being the source of all cure still holds true (Crugg and Newman, 2001) – with the proviso that not a single quality given by nature in the originating source material of, for example, some of the most advanced pharmaceuticals used to "treat" cancer remains intact after being subject to mass production and accordingly stripped of its powers actually to cure and not merely "treat", *i.e.*, delay, the onset or progress of symptoms. Therefore, there are examples from the history that show that knowledge is directly linked with intangibles and in fact, only when intangibles are included that science leads to knowledge (Islam *et al.*, 2010).

Thomas Aquinas (1225-1274 AD) took the logic of Averröes and introduced it to Europe with a simple yet highly consequential modification: he would color the (only) creator as God and define the collection of Catholic church documentation on what eventuated in the neighborhood of Jerusalem some millennium ago as the only communication of God to mankind (hence the title, *bible* – the (only) Book). If Aristotle was the one who introduced the notion of removing intention and time function from all philosophical discourse, Thomas Aquinas is the one who legitimized the concept and introduced this as the only Science (as in process to gaining knowledge). Even though, Thomas Aquinas is known to have adapted the logic of Averröes, his pathway as well as prescribed origin of acquiring knowledge was diametrically opposite to the science introduced by Averröes. This is because the intrinsic features of both God and *bible* were dissimilar to the (only) creator and *Qur'an*, respectively (Armstrong, 1994). For old Europe and the rest of the world that it would eventually dominate, this act of Thomas Aquinas indeed became the bifurcation point between two pathways, with origin, consequent logic, and the end being starkly opposite. With Aristotle's logic, something either is or is not: if one is 'true', the other must be false. Because, Averröes' 'the creator' and Thomas Aquinas's 'God' both are used to denominate monotheist faith, the concept of science and religion became a matter of conflicting paradox (Pickover, 2004). Averröes called the (only) creator as 'The Truth' (In Qu'ranic Arabic, the word 'the Truth' and 'the Creator' refer to the same entity). His first premise pertained to the book (*Qur'an*) that said, "Verily unto Us is the first and the last (of everything)"(89.13). Contrast this to a "modern" view of a creator. In Carl Sagan's words (Hawking, 1988), "This is also a book about God...or perhaps about the absence of God. The word God fills these pages. Hawking embarks on a quest to answer Einstein's famous question about whether God had any choice in creating the universe. Hawking is attempting, as he explicitly states, to understand the mind of God. And this makes all the more unexpected the conclusion of the effort, at least so far: a universe with no edge in space, no beginning or end in time, and nothing for a Creator to do."

This divergence in pathways was noted by Zatzman and Islam (2007a). Historically, challenging the first premise, where the divergence is set, has become such a taboo that there is no documented case of anyone challenging it and surviving the wrath of the Establishment (Church alone in the past, Church and Imperialism after the Renaissance). Even challenging some of the cursory premises have been hazardous, as demonstrated by Galileo. Today, we continue to avoid challenging the first premise and even in the information age it continues to be hazardous, if not fatal, to challenge the first premise or secondary premises. It has been possible to keep this modus operandi because new "laws" have been passed to protect 'freedom of religion' and, of late, 'freedom of speech'. For special-interest groups, this opens a Pandora's box for creating 'us *vs* them', 'clash of civilizations' and every aphenomenal model now in evidence (Zatzman and Islam, 2007b; Islam *et al.*, 2010, 2012, 2014).

Avoiding discussion of any theological nature, Zatzman and Islam (2007a) nevertheless managed to challenge the first premise. Rather than basing the first premise on the Truth à *la* Averröes, they mentioned the importance of individual acts. Each action would have three components: 1) origin (intention); 2) pathway; 3) consequence (end). Averröes talked about origin being the truth; they talked about intention that is real. How can an intention be real or false? They equate real with natural. Their work outlines fundamental features of nature and shows there can be only two options: natural (true) or artificial (false). The paper shows Aristotle's logic of anything being 'either A or not-A' is useful only to discern between true (real) and false (artificial). In order to ensure the end being real, the paper introduces the recently developed criterion of Khan (2006) and Khan and Islam (2007b). If something is convergent when time is extended to infinity, the end is assured to be real. In fact, if this criterion is used, one can be spared of questioning the 'intention' of an action. If any doubt, one should simply investigate where the activity will end up if time, *t* goes to infinity.

This absence of discussion of whatever happened to the tangible-intangible nexus involved at each stage of any of these developments is no merely accidental or random fact in the world. It flows directly from a Eurocentric bias that pervades, well beyond Europe and North America, the gathering and summation of scientific knowledge everywhere. Certainly, it is by no means a property inherent - either in technology as such, or in the norms and demands of the scientific method *per se*, or even within historical development - that time is considered so intangible as to merit being either ignored as a fourth dimension, or conflated with tangible space as something varying independently of any process underway within any or all dimensions of three-dimensional space. Recently, Mustafiz *et al.* (2007) identified the need of including a continuous time function as starting point of acquiring knowledge. According to them, the knowledge dimension does not get launched unless time as a continuous function is introduced. They further show that the knowledge dimension is not only possible, it is necessary. The knowledge is conditioned not only by the quantity of information gathered in the process of conducting research, but also by the depth of that research, *i.e.*, the intensity of one's participation in finding things out. In and of themselves, the facts of nature's existence and of our existence within it neither guarantee nor demonstrate our consciousness of either, or the extent of that consciousness. Our perceptual apparatus enables us to record a large number of discrete items of data about the surrounding environment. Much of this information we organize naturally and indeed unconsciously. The rest we organize according to the level to which we have trained, and-or come to use, our own brains. Hence, neither can it be affirmed that we arrive at knowledge directly or merely through perception, nor can we affirm being in possession at any point in time of a reliable proof or guarantee that our knowledge of anything in nature is complete. Figure 3.13 demonstrates this point.

Historically, what Thomas Aquinas model did to European philosophy is the same as what Newton's model did to the New Science. The next section examines Newton's models. Here it would suffice to say that Newton's approach was not any different from the approach of Thomas Aquinas or even Aristotle. One exception among scientists in Europe was Albert Einstein, who introduced the notion of time as the fourth dimension. However, no one followed up on this aspect of Einstein's work and it was considered that the addition of a time term in the Newton's so-called steady state models will suffice. Islam *et al.* (2010a) and Hossain and Islam (2009) recognized the need of including the time dimension as a continuous function and set the stage for modeling science of intangibles (Khan and Islam, 2012).

Table 3.11 summarizes the historical development in terms of scientific criterion, origin, pathway and consequences of the principal cultural approaches to reckoning, and reconciling, the tangible-intangible nexus.

Islam *et al.* (2014, 2014a) highlighted the contrast between these two starkly dissimilar pathways of cognition. From what we know from existing literature and history of our civilization, before the rise of Islam in the 7th century, no social order or ideology encouraged or supported the idea that the individual has a responsibility to increase his knowledge. Even though it has long been recognized that the thirst for knowledge is inherent to being a human (Marvin, 1917), the notion of seeking knowledge as an individual obligation is uniquely related to Islam and the teachings by Qur'an and the traditions of Prophet

Muhammad. In western society, this is an ideal that is paid much lip-service, but not taking up such a responsibility carries no downside, thus trivializing the aim.

Islam *et al.* (2014a) proposed a comprehensive model that shows cyclical nature of information and disinformation. Based on historical facts, separated from dogmatic assertions and 'beliefs', they showed how true knowledge oscillated throughout history and predicted a renewed surge soon after the current superflux of disinformation is overcome with the emergence of true knowledge. Figure 3.14 depicts their observation as well as future projection. Note how every sharp increase in corresponds to savant/visionary/prophet who consciously and conscientiously sought knowledge for the betterment of the society. Soon after their death, however, the society dipped back into a phase of disinformation and the situation didn't improve until another savant/visionary/revolutionary came to start afresh.

Today in countries of both the developed and developing world, "education" has become extremely professionalized. Accordingly, its focus has been narrowed down to the most pragmatic of concerns such as examination results and the checking-off of curriculum goals as they are achieved. At the same time, meanwhile, outside small circles of ivory-tower academics, hardly anyone ever mentions "curriculum development" in the same breath, much less the same room, as the notion of an individual's responsibility to increase his knowledge. Instead, accompanying the general takeover of everything by corporate power wielded in private hands through public institutions, an overwhelmingly rigid separation of educational theory from educational practice — including the displacement of human-centred aims by corporate-centred aims — has come into force. The hopelessness mentioned in the discussion of HSSA degradation can be averted only by returning to truly scientific research.

The ancient Greek schools (starting ca. 330 BCE) are generally acknowledged as the first organized approach to the education of the coming generation. These, however, had little or nothing to do with the notion — which modern societies all take for granted today — of an individual's responsibility to increase their own knowledge. Almost everyone involved in the discussion and efforts at educational reform today, meanwhile, unconsciously assumes that such a responsibility has always existed. Accordingly, they frame much of their discussion, theory and practice around the goal of educational reform around issues of form. At the same time, they subordinate issues surrounding the conscious participation of the learner in acquiring knowledge. As an unintended consequence, what they do not see is how the current arrangements actually operate to undermine the opportunities for each individual to realize this responsibility on his or her own.

At a time when the Eurocentric world remained mired in a crisis that was deepened by the consequences of the aftermath of the Crusades, how in the Muslim world did the matter of the responsibility of rulers to their subjects come to be consciously raised and discussed? To address this and illuminate the key surrounding issues, Islam *et al.* (2014) studied the impact and significance of certain Arab-world scholarship that proved most consequential in shaping the forms and content of subsequent approaches to matters of educational curricula. The main discussion focused on the life and work of two scholars in particular — ibn-Khaldun and his explanation of the origins and development of the origins of civilization, and the work of Averroës' regarding the relationship of logic to human reasoning. There is widespread acknowledgment throughout the Western and Western-influenced parts of the world of their accomplishments and contributions to human knowledge. There is much less discussion of the impact of their work on reforming and-or renewing the educational process. What the Eurocentric "appreciation" of this work misses entirely is the further development of both the forms of educational renewal and their content that the work of ibn-Khaldun and Averroës pushed forward. The relationship of the form to

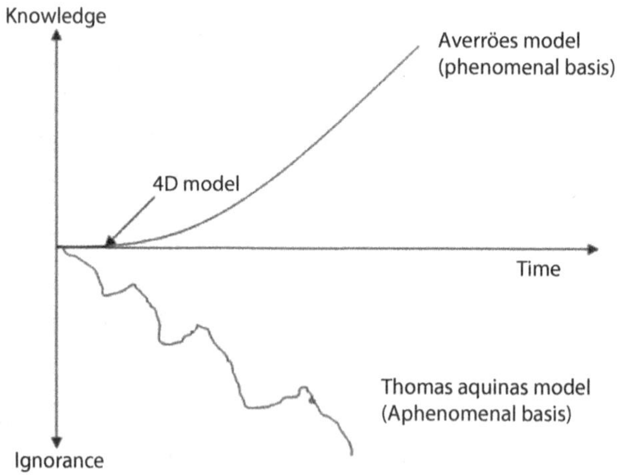

Figure 3.13 Logically, a phenomenal basis is required as the first condition to sustainable technology development. This foundation can be the Truth as the original of any inspiration or it can be 'true intention', which is the essence of intangibles (modified from Zatzman and Islam, 2007a and Mustafiz *et al.*, 2007).

Table 3.11 Criterion, origin, pathway and end of scientific methods in some of the leading civilizations of world history.

People	Criterion	Origin	Pathway	End
Zatzman and Islam (2007)	$\Delta t \rightarrow \infty$	Intention	*f*(t)	Consequences
Khan (2006)	$\Delta t \rightarrow \infty$	Intention	Natural	Sustainability
(Zatzman and Islam, 2007a)	$\Delta t \rightarrow \infty$	Intention	Natural	Natural (used $\Delta t \rightarrow \infty$ to validate intention)
Einstein	t as 4th-D	"God does not play dice…"	Natural	N/A
Newton	$\Delta t \rightarrow 0$	"external force" (1st Law)	No difference between natural & artificial	Universe will run down like a clock
Aquinas	Bible	Acceptance of Divine Order	All *knowledge* & *truth* reside in God; *choice* resides with Man	Heaven and Hell

(Continued)

Table 3.11 cont.

People	Criterion	Origin	Pathway	End
Averröes	Al- Furqan (meaning The Criterion, title of Chapter 25 of *The Qur'an*) stands for *Qur'an*	Intention (first *hadith*)	*Amal saliha* (good deed, de-pending on good intention)	Accomplished (as in *Muflehoon*, , 2:5), Good (+5) Losers (as in *Khasheroon*, , 58:19), Evil (-il
Aristotle	*A* or *not-A* (Δt=0)	Natural law	Natural or arti-ficial agency	*Eudaimonia* (*Eudaimonia*, tr. "happiness", actually more like "Man in harmony with universe")
Ancient India	Serving others; "world reveals itself"	Inspiration (*Chetna*)	*Karma* (deed with inspiration, *chetna*)	Karma, salvation through merger with Creator
Ancient Greek (pre-Socratics)	*t* begins when Chaos of the void ended	the Gods can interrupt human intention at any time or place	N/A	N/A
Ancient China (Confucius)	N/A	Kindness	Quiet (intangible?)	Balance

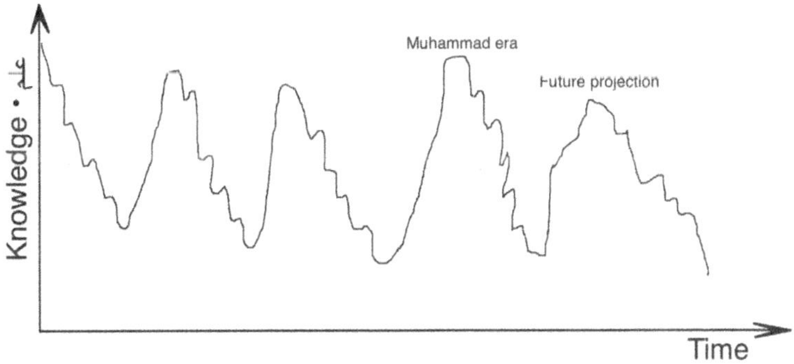

Figure 3.14 Oscillation of true knowledge throughout history.

the content of human thought-material and vice-versa, i.e., the destruction of the form and transformation of the content, are matters possessing the kind of philosophical depth that requires another series of papers. However, here we must add the following critique of pragmatism in order to alert the reader about where the rest of this paper is headed. Pragmatism and pragmatic criteria, whose practicality seduce many a researcher, are highly-developed reflexes of the Eurocentric outlook. Indeed: they form practically the unofficial universal religion among academics in the United States, for example. The bald fact remains that, although pragmatism and the pragmatic criterion demonstrate what is true on the seemingly irrefutable basis of "whatever works", this criterion itself refuses to grapple with and artfully dodges the need to clarify what is false, or what falsehood(s) had to be overcome, in order to arrive at the truth.

It is because of Pragmatism that the scientific meaning of every word has been transformed into something opposite to the original meaning. Here are some of the words related to Education.

A paradigm shift can be triggered by introducing the original meaning. Such a model was introduced by Averröes in Europe. Ironically, this knowledge model is the same one introduced by Prophet Muhammad and subsequently used in Islamic justice system, whereas Averröes is known as he 'father of secular philosophy' in Europe. In other part of the world, this knowledge model is known as the Islamic model. Prophet Muhammad's teaching style and the education system he instituted has been studied in great details.

3.7.3 The Scientific Cognition Model Used in Islam

Islam *et al.* (2014, 2014a) summarized the scientific cognition process adopted by Islamic scholars for some 1000 years, during which these scholars excelled in all aspect of scholarship (see section 3.64). It can be summarized in the following:

1. Start off with the Major premise: There is no Ilah (someone worthy of being obsessed with) except Allah and the Minor premise: Muhammad is the messenger of Allah.
2. Memorize the Qur'an (Qur'an being 100% from Allah and 100% in its original form). Start each deduction (Iqra, the first word revealed on Prophet Muhammad) from the Qur'an. This forms the axiom.
3. Use Hadith (Books of Hadith are preserved for some 1200 years) to form an axis in order to time-scale (qias) to any time or époque of interest.
4. Use stories of the past as recounted in the Qur'an or in the book of Hadith as case laws.
5. By inference, use 'rightly guided' Caliphs as the other case laws.

In the cognition process that can be called Islamic or endorsed by the Prophet [4] :

(a) ***cognition*** starts with a real question that has only Yes (1,+) or No (0,-) answer. Such question doesn't arise if it's already resolved in the Qur'an or Hadith;
(b) niyah (original intention) starts the cognition process solely in search of the truth (haq), so a right (haq) decision is made at the end of the cognition process;
(c) all available data is collected and questionable sources filtered out;

4 Note the all-encompassing reach of points (b) and (c). Even though not singled out, metadata (mentioned earlier in Chapter 1) are thus included.

(d) entering the "fuzzy logic" (manteq) phase, questions of a dialectical nature (manteq) arc posed that will have qualitative answer (by collecting a series of yes/no answers); each manteq question should be motivated by qsd (dynamic intention) that is in line with niyah;
(e) entering the logic (aql) phase, the final question is posed to determine the yes/no answer to the question asked at point (a)

This five-point process is essence to 'seeking knowledge', which is obligatory in Islam.

As shown in Figure 3.15, Prophet Muhammad formalized the cognition process that remains unparalleled today. It should be noted that conducting research is synonymous to seeking knowledge, which is obligatory in Islam.

Figure 3.16 shows how Islamic scholars used Qur'anic sources, then formed an axis with sayings of the prophet (Hadith) to be able to time scale any event. No other premise was allowed, thereby eliminating spurious sources of theories. Currently, such approach is known as approach of obliquity or the long-term approach. Figure 3.17 shows how long-term approach is crucial to increasing knowledge instead of increasing ignorance and arrogant adherence to dogma.

What Thomas Aquinas launched as doctrinal philosophy morphed into the following sequences in Europe:

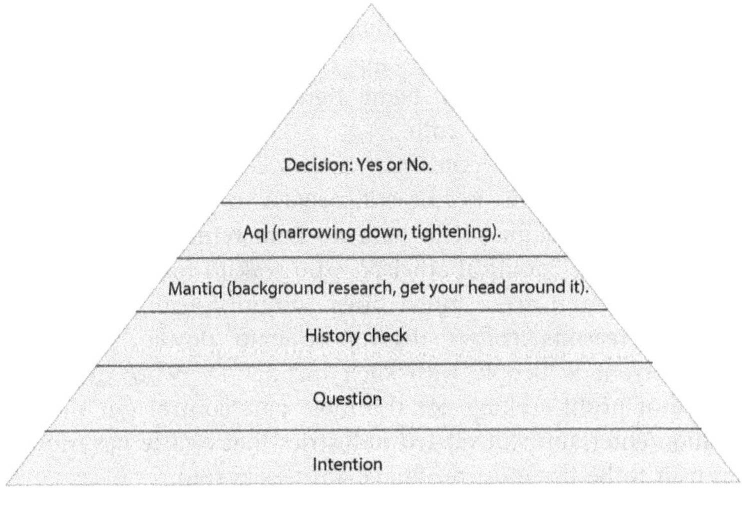

Figure 3.15 Decision making in Islamic jurisprudence.

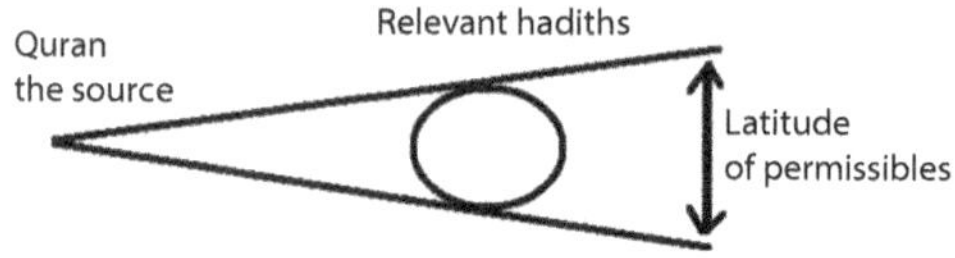

Figure 3.16 Use of major premise and minor premise to form the axis of time.

Dogma → Theism → Agnosticism → Secularism → Atheism

Following the HSSA degradation, Dogma (sugar) has been introduced, later changing to Theism (saccharine), followed by Agnosticism (Aspartame) and finally Atheism (Sucralose). During this process, a false perception exists that we are making progress, leading to increasing ignorance packaged within the most dogmatic arrogance. The same way, we have become obese (artificial fat) in physical sense, the world has become an institution for perpetrating extreme ignorance with arrogance. In reality, when considered at their root – that is, in the individual's psyche - one cannot see these phenomena as being genuine choices, mutually exclusive or a temporal progression as such, although when considering trends within our various forms of society or collective existence, the weight of numbers or, more to the point, power, clearly has set up a decadent digression which we can surmise involves a "cause and effect" based on the relative weight these mostly conflicting inclinations come to exert within the whole, not to mention on the extent to which other phenomena related to our spiritual condition emerge and take hold. Among these latter phenomena, one can include include individualism, neurosis, indulgence, hedonism, paganism, mysticism, consumerism, pleasure seeking etc. The bottom line is, they all start off with an intention of 'being right' and 'proving someone correct', instead of seeking the truth.

It is also important to note that surveys confirm more than 90% of the world's 7 billion-plus people still profess to embrace some form of faith so those who claim not to have no faith remain very much in the minority and those "political atheists" who seek to foster a world devoid of religion and faith are a much tinier minority again, albeit one that for obvious reasons enjoys disproportionate power and influence. Unfortunately, it is people with such inclinations who, driven by the imperative of profit seeking, for the most part control our mass media and fashion/entertainment related industries that clearly have found forbidden fruit to be the most profitable and easily sold.

All these concepts belong to the bottom half of the cognition graph (Figure 3.18). None of these concepts as describing a natural state or a

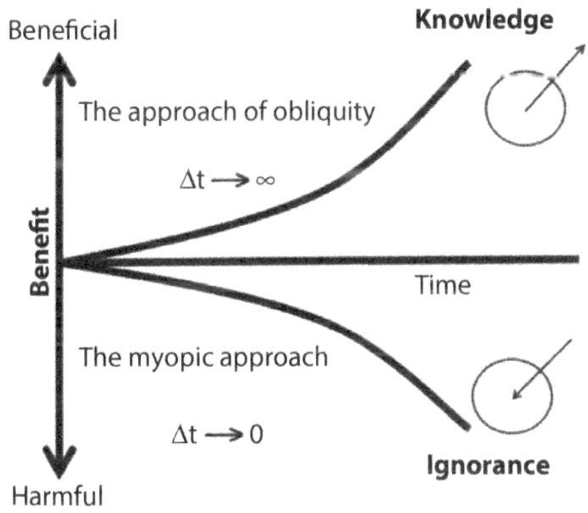

Figure 3.18 Approach of obliquity is the essence of truly scientific cognition.

choice we make, let alone one that is mutually exclusive with all others. Rather, these abstract concepts as being associated with inclinations that are, from time to time, experienced by all researchers of all creed. That said, the Sucralose (TM) concept of atheism is in many ways the odd one out since unlike the doubts embraced by agnosticism, what Atheism describes – an unqualified conviction that there is no God - does actually exist not in any reality that one knows of. After all, what it presumes is that people who fully embrace logic and the scientific method somehow arrive at the entirely illogical and unscientific premise, not of living in a state of extreme doubt about the origin of all things, but rather of illogically claiming to know that which they cannot possibly know with committing their own leap of faith. Secularism, in a world of people who overwhelmingly profess both faith and religious affiliation, is scarcely less unreal than the pretense of atheism and requires no fewer artificial and impractical distinctions, not to mention a historical disconnect.

It is also important to observe about these concepts which were formulated to describe and understand collective inclinations is that they have little meaning as devices to describe individuals, yet what we see is that these terms which describe individual strands of our human psyche have been politicized and turned into badges which people like to pin on their chests as a fashion choice as much as anything else. Then, these badges presume our spiritual state is static when in reality all people will go through periods of doubt about their core convictions, including both people of faith and so-called atheists. Of course, people of faith would not describe these doubts as agnosticism which projects a condition in which people who have never inculcate din dogma or for some reason abandoned their religious convictions, have conceded their spiritual ground to their preponderance of doubt. However, the agnostic condition and struggles by the faithful with their core beliefs are certain born of the same inclinations. The difference is that people of faith are encouraged by dogma to traverse these turbulent waters by continuing with their religious practices and by taking strength from the lifestyle they have constructed around their faith. Perhaps it is the fact that we presume safety in moments of doubt by tacitly understanding that our collective religious practices will see us through, that gives rise to sectarianism.

Finally, one should observe is that in terms of our societies as a whole, there are indeed progressions and decadent digressions or trends, which are set into play by the extent to which these various inclinations achieve disproportionate weight within our collective consciousness. Indeed we are born with a natural but unmanifested inclination to the truth. However, in an age dominated by all powerful, artificial entities created for the sole purpose of seeking profit and power; an age in which the cult of individualism, hedonism, personal liberty have seen a withering of conscientious thoughts while mass communication has at the same time facilitated the widespread promulgation of false dogma, corrupted ethics and general iconoclasm, there certainly seems to be a temporal progression in the manner we have described above, which, in turn, has created an environment that seems to make genuine long-term cognition something of an endangered species.

3.7.4 Islamic Scholars Who Were Founders of Their Fields

One of the least remarked yet probably the single most consequential innovation within the European/Eurocentric discourse developed around the concepts of learning and education is the appearance over the last couple of decades of an almost interchangeable usage of the terms "data", "information" and "knowledge". This is especially pronounced in the scholarly discourse surrounding learning and education coming out of North America.

This has turned up and-or influenced even the discourse surrounding discussions of Socratic methods and approaches to teaching and learning, centred on the role(s) served by collecting/formulating questions and focusing on the processes of answering these questions as the core content of the educational process. The latent difficulty here poses itself in the form of a failure to consciously distinguish between having questions in mind to eventually investigate versus or compared to the process(es) of the student actually investigating the answers to his/her own questions, and the highest and best use of a teacher-figure within such arrangements. This failure to distinguish is linked to certain proprietary assumptions with regard to the entire undercarriage of the educational vehicle: a great deal of data-information-knowledge is already someone's private property, and the accepted attitude to matters of property in most societies remains the single-pillar shehadeh that "there is no God but Monopoly and Maxxximum is its Profit." That is to say: there is a red line and nobody's knowledge, information or data is so free as to be permitted to transgress this red line. As a result it is no accident that the surrounding social order accepts having one's questions as what is called "curiosity", whereas anyone's actual investigation of specific answers to a specific chain of questions makes some people nervous about crossing the red line.

By applying an entirely different approach, as early as the mid-600s (immediately following the death of Propher Muhammad), various cultures, ranging from China to India, form Persia to Rome amalgamated, were brought together. Even as numerous local languages throve, Arabic remained the language of research and investigation (both in religion and hard science). Soon, some of the greatest Islamic scholars emerged from outside the Arabian peninsula, from India, Europe and North Africa. Cultural heritages of the region included Hellenic, Indic, Assyrian and Persian influences. Greek and Indian intellectual traditions were recognized, translated and studied broadly. As the population of the lands of Islam gained access to all the important works of all the cultures of the empire, a new common civilization formed in this area of the world based on the religion of Islam. A new era of high culture and innovation ensued, where these diverse influences were recognized and given their respective places in the social consciousness. Although the extent of Islamic scientific achievement is yet fully understood, it was certainly vast (Khan and Islam, 2012).

These achievements encompass a wide range of subject areas, most notably: Mathematics, Astronomy, Medicine, Physics, Alchemy and chemistry, Cosmology, Ophthalmology, Geography and cartography, Sociology and Psychology.

The sciences, which included philosophy, were viewed holistically. The individual scientific disciplines were approached in terms of their relationships to each other and the whole, as if they were branches of a tree. In this regard, the most important scientists of Islamic civilization have been the polymaths, known as *hakim* or doctors. Their role in the transmission of the sciences was central. The hakim was most often a poet and a writer, skilled in the practice of medicine as well as astronomy and mathematics. These multi-talented sages, the central figures in Islamic science, elaborated and personified the unity of the sciences. They orchestrated scientific development through their insights, and excelled in their explorations as well.

Following is a list of scientists from the Islamic era that are considered to be the 'father of their individual disciplines'.

3.7.4.1 Al-Kindi (801–873)

In all measures, he is the father of Information age (Khan and Islam, 2012). He was instrumental in the adoption of the Indian numbering system, later known as Arabic numerals. His developed algebra, which also had Indian antecedents, by introducing methods of simplifying the equations. He used Euclidian geometry in his proofs. He is one of the few 'pure polymaths' to collide with the ruling class for his adherence to pure logic, which is Islamic but wasn't popular with the rulers who only liked Islamic ruling when the ruling favored status quo.

Al-Kindi contributed a great deal in various subjects ranging from meta- physics, ethics, logic and psychology, to medicine, pharmacology, mathematics, astronomy, astrology and optics, and further afield to more practical topics like perfumes, swords, jewels, glass, dyes, zoology, tides, mirrors, meteorology and earthquakes. In the field of mathematics, al-Kindi played an important role in introducing Indian numerals to the rest of the world. He was a pioneer in cryptanalysis and devised several new methods of breaking ciphers. Using his mathematical and medical expertise, he was able to develop a scale that would allow doctors to quantify the potency of their medication. Al-Kindi was the first in known history to formalize theory of logic (Aql) with science (Ilm), the second most used word of the Qur'an. He saw no contradiction among Islamic theology, philosophy and science.

3.7.4.2 Abu al-Qasim al-Zahrawi (936-1013)

Al-Qasim Al-Zahrawi, also known as Abulcasis, has been called the "father of modern surgery" and the "father of operative surgery". He was an Arab/Muslim physician who lived in Al-Andalus. He has been described by many as the father of modern surgery. His greatest contribution to medicine is the Kitab al-Tasrif, a thirty- volume encyclopedia of medical practices. His pioneering contributions to the field of surgical procedures and instruments had an enormous impact in the East and West well into the modern period, where some of his discoveries are still applied in medicine to this day (Cosman and Jones, 2008). He was the first physician to describe an ectopic pregnancy, and the first physician to identify the hereditary nature of haemophilia. Donald Campbell, a historian of Arabic medicine, described Al-Zahrawi's influence on Europe as follows (Campbell, 2001):

The chief influence of Albucasis on the medical system of Europe was that his lucidity and method of presentation awakened a prepossession in favour of Arabic literature among the scholars of the West: the methods of Albucasis eclipsed those of Galen and maintained a dominant position in medical Europe for five hundred years, i.e long after it had passed its usefulness. He, however, helped to raise the status of surgery in Christian Europe; in his book on fractures and luxations, he states that 'this part of surgery has passed into the hands of vulgar and unculti- vated minds, for which reason it has fallen into contempt.' The surgery of Albucasis became firmly grafted on Europe after the time of Guy de Chauliac (d.1368).

In the 14th century, the French surgeon Guy de Chauliac quoted al-Tasrif over 200 times. Pietro Argallata (d. 1453) described Abū al-Qāsim as "without doubt the chief of all surgeons". Abū al-Qāsim's influence continued for at least five centu- ries, extending into the Renaissance, evidenced by al-Tasrif's frequent reference by French surgeon Jacques Delechamps (1513–1588).

3.7.4.3 'Ali ibn al-'Abbas al-Majusi (d. 982)

One scholar asserts al-Majusi "must be acknowledged as a founder of anatomic physiology". In addition, the section on dermatology in his Kamil as-sina'ah at- tibbiyah (Royal book-Liber Regius) has one scholar to regard him as the "father of Modern dermatology". He discussed Neuroscience and psychology in The Complete Art of Medicine. He described the neuroanatomy, neurobiology and neurophysiology of the brain and first discussed various mental disorders, including sleeping sickness, memory loss, hypochondriasis, coma, hot and cold meningitis, vertigo epilepsy, love sickness, and hemiplegia. He placed more emphasis on preserving health through diet and natural healing than he did on medication or drugs, which he considered a last resort (Haque, 2004).

3.7.4.4 *Ibn-Haitham (Alhazen, 965-1040)*

Alhazen is considered the "father of modern optics", the "father of physiological optics", and the "father of optics". His work has been summarized recently by Khan and Islam (2012).

3.7.4.5 *Al-Biruni (973-1043)*

According to Francis Robinson, Al-Biruni earned the "founder of Indology" and "first anthropologist" titles for his remarkable description of early 11th-century India. George Morgenstierne regarded him as "the founder of comparative studies in human culture". Al-Biruni is also known as the "father of pharmacy". Al-Biruni was well versed in physics, mathematics, astronomy, and natural sciences, and also distinguished himself as a historian, chronologist and linguist.

He was conversant in Khwarezmian, Persian, Arabic, Sanskrit, and also kne w Greek, Hebrew, Syriac and Berber. He spent a large part of his life in Ghazni in modern-day Afghanistan, capital of the Ghaznavid dynasty which ruled eastern Iranian lands and the northwestern Indian subcontinent. In 1017 he traveled to the Indian subcontinent and became the most important interpreter of Indian sci- ence to the Islamic world. He also made contributions to Earth sciences, and is regarded as the "father of geodesy" for his important contributions to that field, along with his significant contributions to geography.

3.7.4.6 *Al-Farabi (870-950)*

Al-Farabi formalized Islamic logic, Manteq, often translated as 'fuzzy logic'. Even though he is regarded by Eurocentric scientists as the "founder of Islamic/Arab Neoplatonism", his work was entirely built on the logic put forward by Prophet Muhammad and practiced by his companions, including the four 'rightly guided Caliphs'. As early as Imam Jafer As-Sadeq was an expert of Manteq. This logic is rooted in Islam and was not imported from Greek philosophy. Similar remark was also made by Peter Adamson, professor of Ancient and Medieval Philosophy at King's College London. He said Al-Farabi "was not a political philosopher in the sense of having given concrete political proposals for running a society: Rather, his aim was to describe the societal conditions that tend to produce virtue and vice".

3.7.4.7 *Al-Khawarizmi (780-850)*

He was most renowned as the "father of algebra"; the word "algorithm" immortalizes his name in one of algebra's keymost principles. Solomon Gandz states: "In a sense, Khwarizmi is more entitled to be called "the father of algebra" than Diophantus because Khwarizmi is the first to teach algebra in an elementary form and for its own sake, Diophantus is primarily concerned with the theory of numbers". Recently, Islam *et al.* (2011, 2012) discussed how algebra of his time was more accurate both in approach and in form than the linear (or non-linear) algebra of today's time.

3.7.4.8 *Averroes (1126-1198)*

He known in Arabic as Ibn Rushd and is considered to be the 'father of secular philosophy in Europe'. He was an Andalusian polymath born in Córdoba, Spain. Averroes was regarded by some Christian bishops as the "father of free thought and unbelief"and has been described by some as the "father of rationalism" and the "founding father of secular thought in Western Europe". Ernest Renan called Averroes the absolute rationalist, and regarded him as the father of free-thought and dissent. Khan and Islam (2012) argued that Averroes was the last dogma-free scientist of Europe as all modern scientists adopted some form of dogma, albeit used a secular title.

3.7.4.9 *Ibn Hazm (994-1064)*

He author of one of the earliest works on comparative religion and "honoured in the West as that of the founder of the science of comparative religion". Alfred Guillaume refers to him the composer of "the first systematic higher critical study of the Old and New testaments". However, William Montgomery Watt disputes the claim, stating that Ibn Hazm's work was preceded by earlier works in Arabic and that "the aim was polemical and not descriptive". With proper understanding of Islamic faith, it becomes clear that Ibn Hazm demonstrated that Islam is the only dogma-free cognition process and calling it a 'religion' is a misnomer.

3.7.4.10 Ibn Khaldun (1332-1406)

He is regarded by many as the father of sociology, historiography and modern economics. He is best known for his Muqaddimah. His description of social evolution is revolutionary and stands out as the most logical theory of political science and economics today.

3.7.4.11 Jabir ibn Hayyan (721-815)

He has often been referred to as "the father of chemistry" and is widely credited with the introduction of the experimental method into alchemy, as well as with the invention of numerous important processes that are still used in chemistry today.

3.7.4.12 Rhazes (865-925)

His treatise on Diseases in Children has led many to consider him the "father of pediatrics". He has also been praised as the "real founder of clinical medicine ".

3.7.4.13 Muhammad al-Shaybani (749-805)

He is considered to be the father of international law. He developed criteria for international laws based on the model demonstrated by prophet Muhammad, who implemented the model for maximizing natural justice. Such model is rarely used in today's legal system that is inherently biased toward the more powerful of two entities that are entering a legal contractual relationship.

3.7.4.14 Muhammad al-Shaybani (749-805)

He is considered to be the father of international law.

3.7.4.15 Ibn Sina (Avicenna) (908-946)

Ibn Sina (Avicenna) was a Persian physician, astronomer, physicist and mathematician from Bukhara, Uzbekistan. He is most known as the father of modern medicine. His work titled: The Canon of Medicine has been in use all over Europe for many centuries. Even though it is generally recognized that his legacy continued in modern Medicine, it is known that Ibn Sina advocated natural medicine, as inspired by prophet Muhammad's Hadith that every ailment has a cure in nature. This is in sharp contrast to modern medicine that has replaced all natural medicine with artificial ones. Ibn Sina also made important astronomical observations, and discussed a variety of topics including the different forms energy can take, and the properties of light. He contributed to the development of mathematical techniques such as Casting out nines. Such approach of combining Physics and Mathematics in order to better understand the cause of an ailment is non-existent in modern age.

3.7.4.15 Ibn al-Nafis (1213–1288)

Ibn al-Nafis was a physician who was born in Damascus and practiced medicine as head physician at the al-Mansuri hospital in Cairo. He wrote an influential book on medicine, believed to have replaced Ibn-Sina's Canon in the Islamic world - if not Europe. He wrote important commentaries on Galen and Ibn-Sina's works. One of these commentaries was discovered in 1924, and yielded a description of pulmonary transit, the circulation of blood from the right to left ventricles of the heart through the lungs. Until today, this important work remains elusive to the modern medicine experts.

3.7.4.16 Muhammad ibn Zakariyā Rāzī (854 CE - 925 CE)

Muhammad ibn Zakariya Razi was a Persian polymath, physician, alchemist and chemist, philosopher and important figure in the history of medicine and psychiatry.A comprehensive thinker, Razi made fundamental and enduring contributions to various fields of science, which he recorded in over 200 manuscripts, and is particularly remembered for numerous advances in medicine through his observations and discoveries. An early proponent of experimental medicine, he became a successful doctor; he was appointed a court physician, and served as chief physician of Baghdad and Rey hospitals. He was among the first to use humorism to distinguish one contagious disease from another and has been described as a doctor's doctor,[7] the father of pediatrics, and a pioneer of ophthalmology.

3.7.5 Purpose/Meaning of Life

For whatever one can gather from ancient cultures, there was no confusion about the purpose of human life. This purpose starts off with the designation of a formal status for humans. In ancient India, it was 'avatar' (from Sanskrit avatāra 'descent,' from ava 'down' + tar- 'to cross.') and Qur'an formalized that in verse 2:30 as 'khalifah', the word best translated as viceroy or vicegerent. In Islam, man's ultimate life objective is to worship (Arabic word 'abd' literally means 'obey with love') the creator Allah (English: God) by abiding by the Divine guidelines revealed in the Qur'an and the Tradition of the Prophet. It is stated: "And I (Allâh) created not the jinn and mankind except that they should be obedient (to Allah)." (Qur'an 51:56). Earthly life is merely a test, determining one's afterlife, either in Jannah (Paradise) or in Jahannam (Hell). The Qur'an describes the purpose of creation as follows: "Blessed be he in whose hand is the kingdom, he is powerful over all things, who created death and life that he might examine which of you is best in deeds, and he is the almighty, the forgiving" (Qur'an 67:1–2). So, a viceroy (human) is sent to the earth (Ardha in Arabic means 'habitat of humans') for a test period. If he carries himself well, he gets to return to heaven, if not he goes to hell. What does a human have to have a fair chance at going to heaven? He is created with best of features (Qur'an 95:4), even the word 'human' (insan in Arabic) meaning 'adorable', 'trusting', yet forgetful. This notion of forgetfulness makes it a test, a perfection would be impossible to test. However, this 'weakness' is anything but 'original sin' (Islam *et al.*, 2014). In fact, it is logical that the Creator equip humans with qualities that prepare them to be a viceroy. For instance, a good King would send a good person to represent him as a viceroy. What is a good viceroy? He keeps in contact with the King ('salat' in Arabic and yoga in Sankscrit both mean "communication") and manages the finances with utmost care (generosity and charity is fundamental trait that is mandated) and acts the same manner the King would have acted in case the viceroy were not there. So, how does one act like the King? He follow his traits. How is that trait embedded? The word 'Deen' in Arabic and Dharma in Sanskrit both mean 'natural traits' and not religion as commonly translated. In fact, the word religion and 'Deen' are opposite of the cognition spectrum. It is also logical that a good King prepare his viceroy well and give him enough provision. Finally, for him to be held accountable, a viceroy must have certain freedom. Islam gives the freedom of intention, everything else (including provision) being part of the universal order for which he has no accountability. The first Hadith of the book of Bokhari states that a man will be judged for his intention. In addition men are equipped with 99 good traits that re similar to Creator's, albeit at miniscule level. Let's review some of the traits:

1. Ar-Rahman (literally means 'a womb' that extends to infinity in space). Humans are, therefore, born with Rahma, the simplest translation is 'empathy' for everyone and everything;
2. Ar-Raheem (literally means 'a womb' that extends to infinity in time). Humans are, therefore, born with empathy at all times;
3. Al-Malik (literally means 'the owner of everything'). Humans are, therefore, born with one item that they have full control over. This happens to be intention. It is logical, because first hadith of Book of Bokhari confirms, we'll be judged by our intention. Now, does this ownership risk violating universal order? No. Intention has no continuity with anyone other than the individual whose intention is in question. So, what it does is gives people the freedom to intention, without disturbing the universal order, thereby holding him responsible for the intention that he had full control over;
4. Qur'an names a total of 99 such traits of the Creator and asserts that humans are created with those qualities so they are prepared to be the viceroy of the creator.

Chronologically, the first set of distortion was imparted by the Roman Catholic church. Metaphorically speaking that would be the sugar in human cognition. Although Islam accepted Jesus as a prophet/ messenger sent by Allah, Islam did not recognize Jesus as Son of God and therefore could not accept Christ as standing at the center of recognizing Allah. Yet, this was not the most critical difference.

The critical difference as far as religious doctrine was concerned lay elsewhere and went beyond the realm of purely religious doctrine. At the core of real Muslim-Christian differences lay the fact that Muslim outlook did not exclude the possibility that social and individual conditions of the believers were neither necessarily eternal nor static, and that the effort to harmonize and sort out internal differences under such conditions could strengthen the internal unity of a Muslim community.

For Christians, prior to the Reformation, all this was absolutely incompatible with their religion's conception of Original Sin. This doctrine, which in itself holds out no hope for reconciliation of any kind among individuals or reunification of community purpose, is a defining feature of Christian belief. Indeed, Christian observers of the apparent internal cohesion of most Muslim communities over the period of the Moorish empire preferred to interpret these phenomena as evidence of mass submission to rule under a variant of oriental despotism.

Throughout this period of the rise of the rise and spread of Islamic belief and the spread of Islamic rule, the Christian religious community itself was also divided or even crumbling. On the one hand, there were the followers of the Bishop of Rome, designating himself as Pope in most of Europe. On the other hand there were the followers of the Eastern or Greek Orthodox rite in Greece, Russia and parts of the Balkans influenced by Russia.

Life's purpose in Christianity is to seek divine salvation through the grace of God and intercession of Christ (cf. John 11:26). The New Testament speaks of God wanting to have a relationship with humans both in this life and the life to come, which can happen only if one's sins are forgiven (John 3:16–21; 2 Peter 3:9). This is where dogma comes into play and the nonsensical argument emerges as: "I (God) am going create man and woman with original sin. Then, I am going to impregnate a woman with myself as her child, so that I can be born. Once alive, I will kill myself as a sacrifice to myself. To save you (humans who accept me as the 'savior') from the sin I originally condemned you to". The only reference one gives is the bible, then one finds out that there is no such thing as bible (the book) that has 40 versions in English alone. How has that dogma changed when 'enlightenment' came in terms of western philosophy? In terms origin of universe, the big bang theory, the introduction of quantum nature, multiple (numerous) history of each point, Nature as the creator of 'numerous universes', and even the definition of what's true and false has introduced a logic that is more schizophrenic than dogma. What dogma did is introduced one aphenomenal standard (Zatzman and Islam, 2007), whereas the 'western philosophers' have introduced numerous such standards. The scientific equivalent of this transition is a quantum leap from bipolar to multipolar (schizophrenia) and is best described as 'deliberate schizophrenia' (Islam *et al.*, 2014). As an example, one can cite the word 'quantum'. In physics it means: a. The smallest amount of a physical quantity that can exist independently, especially a discrete quantity of electromagnetic radiation. b: This amount of energy regarded as a unit. This formulation that starts with the false premise that something 'independent' can exist in nature disconnects mass and energy matter and renders subsequent build up irrelevant. Apparently, such 'isolation' is necessary to observe phenomena. However, the invoking of false premise and subsequent truncation of history render the entire process aphenomenal. This has been going on for millennia. Ever since the concept of atom that was thought to be fundamental particle of even thought material or God, scientists have only made the scope of atom bigger or smaller, both ending up with an aphenomenal point. In the smaller scale, it has ended up with Higgs boson of zero mass and on the larger scale, it has become Big bang of a infinitely small object with infinitely large mass. Yet, scientists are celebrating the 'discovery' of these particles (Jha, 2013).

After the introduction of dogma, it has been all downhill. The purpose of life has been reduced to 'be happy', 'have fun', 'live to the fullest' (translation: maximize pleasure and minimize pain). Everything in history has been reconstructed to support this latest notion of the purpose of life. For instance one can cite the example of Antisthenes, a pupil of Socrates. He is known to have outlined the themes of Cynicism, stating that the purpose of life is living a life of Virtue which agrees with Nature. Happiness depends upon being self-sufficient and master of one's mental attitude; suffering is the consequence of false judgments of value, which cause negative emotions and a concomitant vicious character. This philosophy has no contradiction with the purpose of life outlined above. In fact, it is further stated that Cynical life rejects conventional desires for wealth, power, health, and fame, by being free of the possessions acquired in pursuing the conventional. Once again, there is no contradiction with the status of humans being 'viceroy'. However, how is this interpreted by New scientists? It is said: "as reasoning creatures, people could achieve happiness via rigorous training, by living in a way natural to human beings. The world equally belongs to everyone, so suffering is caused by false judgments of what is valuable and what is worthless per the customs and conventions of society." Then it is described whatever comes naturally is called 'natural' and whatever gives one instant pleasure and quick short-term victory is valuable, turning everything into a race for pleasure in this world. This is complete disconnection from the purpose that was known since the beginning of time.

Another case in point is: Epicurus, a pupil of Pamphilus of Samos. He allegedly taught that the greatest good is in seeking modest pleasures, to attain tranquility and freedom from fear (ataraxia) via knowledge, friendship, and virtuous, temperate living; bodily pain (aponia) is absent through one's knowledge of the workings of the world and of the limits of one's desires. This is entirely consistent with the original purpose of human life as stated earlier in this section and is supported by numerous verses of the Qur'an (e.g., 79:37-41). However, later interpretation took it to the other extreme making it in line with Roman Catholic church's stance on self emollition and avoidance of social responsibilities and such conclusions were made: "Combined, freedom from pain and freedom from fear are happiness in its highest form. Epicurus' lauded enjoyment of simple pleasures is quasi-ascetic "abstention" from sex and the appetites".

Another example is the evolution theory that characterizes humans as a continuity in the evolution of animals. The first person to posit the theory of natural selection and the co-discoverer of the theory of evolution by natural selection with Darwin, Alfred Russel Wallace, didn't think human evolution could be explained solely in terms of adaptation processes, at least as far as human cognition and behavior was concerned. This cost him his scientific reputation because science is run by elites who keep ancient knowledge to themselves. Darwin was less specific in theorizing why he also acknowledged man didn't fit the model like the other animals. We got upgraded, and when genius scientists like Wallace speak the truth they are rendered invisible by the media, and black balled rim academics, a wholly corporate subsidized enterprise. Today, we have Stanford scientists perpetrating the notion that homosexuality is natural because more than 400 species practice part-time homosexuality. Similar to dogma people, they cannot answer why incest or zoophilia is not natural then. In the mean time, the likes of Dawkins freely talk about pedophilia being natural while other talk about incest being natural (Kutner, 2015).

The 'enlightenment' phase of European history made the entire process of fulfilling purpose of life travel further down the degradation route. The original purpose was perverted by the Roman Catholic church to be 'salvation through Jesus' – a dogma the post 'enlightenment' replaced with replaced by notions of inalienable natural rights and the potentialities of reason, and universal ideals of love and compassion gave way to civic notions of freedom, equality, and citizenship. There the definition of 'natural' and 'universal' remained arbitrary, devoid of any reasoning of logical thought. That made these notions of 'freedom, equality, and citizenship' more dogmatic and original dogma itself. This has been the era of roller coaster ride of spiraling down of all values through a successive degradation through ever more lunatic dogmas and false premises:

1. Classical liberalism (humans as beings with inalienable natural rights (including the right to retain the wealth generated by one's own work), and sought out means to balance rights across society. Broadly speaking, it considers individual liberty to be the most important goal, because only through ensured liberty are the other inherent rights protected.)
2. Kantianism (all actions are performed in accordance with some underlying maxim or principle, and for actions to be ethical, they must adhere to the categorical imperative. Kant denied that the consequences of an act in any way contribute to the moral worth of that act, his reasoning being that the physical world is outside one's full control and thus one cannot be held accountable for the events that occur in it.
3. Utilitarianism ("Nature" has placed mankind under the governance of two sovereign masters, 'pain' and 'pleasure', then, from that moral insight, deriving the Rule of Utility: "that the good is whatever brings the greatest happiness to the greatest number of people".)
4. Nihilism (Life is without objective meaning. A natural result of the idea that God is dead, and insisting it was something to overcome. This is fighting the God that is now 'dead')
5. Pragmatism (Truth is whatever works, and "only in struggling with the environment" do data, and derived theories, have meaning, and that consequences, like utility and practicality, are also components of truth. Purpose of life is discoverable only via experience.)
6. Theism (God created the universe and that God and humans find their meaning and purpose for life in God's purpose in creating.)
7. Existentialism (Each man and each woman creates the essence (meaning and purpose) of his and her life; life is

not determined by a supernatural god or an earthly authority, one is free.)
8. Absurdism (the Absurd arises out of the fundamental disharmony between the individual's search for meaning and the apparent meaninglessness of the universe. As beings looking for meaning in a meaningless world, humans have three ways of resolving the dilemma: 1) Suicide; 2) "Religious" belief; and 3) Acceptance of the Absurd).
9. Secular humanism (the human species came to be by reproducing successive generations in a progression of unguided evolution as an integral expression of nature, which is self-existing. People determine human purpose without supernatural influence; it is the human personality (general sense) that is the purpose of a human being's life.)
10. Logical positivism (the question: what is the meaning of life? Is itself meaningless)
11. Postmodernism (seeks meaning by looking at the underlying structures that create or impose meaning, rather than the epiphenomenal appearances of the world.)
12. Naturalistic pantheism (the meaning of life is to care for and look after nature and the environment).

Benjamin Franklin famously claimed: "We are all born ignorant, but one must work hard to remain stupid." Margaret Thatcher famously stated, "there is no alternative". Overall, decline of human values can be summarized in the following transition.

Creator's viceroy → fallen sinner with original sin in need of salvation → inherently selfish and stupid (as part of the animal kingdom) → inherently selfish but intelligent → self sustained and capable of controlling his surrounding to maximize pleasure and minimize pain

3.7.6 What is True?

In 2006, the authors posed the question "What is true?" with the aim of uncovering and specifying the various ways of distinguishing truth from falsehood. After six months of research, the following criteria for truth were developed. This logic was used to define natural cognition or natural material or natural energy as follows:

(a) there must be a true basis or source;
(b) the truth itself must remain unrefuted continuously over time; and
(c) any break in continuity or similar exception must be supported by a true criterion or bifurcation point.

The third-mentioned item in the above list sets scientific cognition apart from doctrinal or dogmatic cognition. Notwithstanding the long-standing general acceptance of the distinction that Thomas Aquinas is the father of doctrinal philosophy and Averroes the father of secular philosophy, our research uncovers the fact that, regardless of the claim to be operating on an entirely secular basis utterly disconnected from 'religious bias' of any kind, all aspects of scientific developments in modern Europe have been based on doctrinal philosophy. If the assumption that modern New science is based on non-dogmatic logic is set aside, it becomes clear that, precisely because so many of its original premises are unreal/unprovable, unnatural or non-existent, modern science is full of paradoxes and contradictions.

Our research shows truth and falsehood had no illusion in ancient oriental culture (Islam *et al.*, 2010, 2012, 2014). Such was the case in ancient Greek civilization as well. From what is gathered from ancient Greek civilization, general consensus was that absolute truth cannot be fleeting. This is in line with the oriental concept that 'truth unravels itself' (Chhetri and Islam, 2008) 'truth unravels itself' (Chhetri and Islam, 2008). Every civilization recounted in history other than post-Roman Catholic church's Eurocentric era had a clear vision of what constitutes the truth. Plato understood it as synonymous with real that doesn't change with time (the physical world being fleeting or a function of time is not 'real'). Aristotle understood it as what really 'is'.

In ancient India, Truth was defined as "unchangeable", "that which has no distortion", "that which is beyond distinctions of time, space, and person", "that which pervades the universe in all its constancy". These, along with the national motto of India, "Satyameva jayate" (Truth alone wins), as well as "Satyam muktaye" (Truth liberates), are entirely consistent with what how Qur'an defined Truth as one of the 99 traits of the Creator. Other traits matching 'unchangeable', 'no distortion', beyond time, space, and person', etc. This was so well known in Islamic golden era that each scholar (e.g., Avicenna, Averroes,

Alhazen) started off re-stating what constitutes the truth. The first distortion was introduced by Augustine, who was the first one to break down the most important feature of truth – uniqueness. He wrote in his "Trinity", "God is truth for he, too, has a true [real] soul who has not a great soul; since the essence of body and soul is not the essence of the truth [reality] itself; as is the Trinity, one God, alone, great, true, truthful, the truth…"

Then came Thomas Aquinas. He had access to Avicenna, Averroes, as well as Aristotle (through Muslim scholars' translation). He wrote:

"A natural thing, being placed between two intellects, is called true insofar as it conforms to either. It is said to be true with respect to its conformity with the divine intellect insofar as it fulfills the end to which it was ordained by the divine intellect... With respect to its conformity with a human intellect, a thing is said to be true insofar as it is such as to cause a true estimate about itself. "

All of a sudden truth became subject to human intellect. This would be the beginning of today's insane definition of truth. Note the following degradation of the definition of truth.

1. Correspondence theory (stresses a relationship between thoughts or statements on one hand, and things or objects on the other).
2. Coherence theory (truth is primarily a property of whole systems of propositions, and can be ascribed to individual propositions only according to their coherence with the whole, i.e., Authority).
3. Constructivist theory (truth is constructed by social processes, is historically and culturally specific, and that it is in part shaped through the power struggles within a community.)
4. Consensus theory (truth is whatever is agreed upon, or in some versions, might come to be agreed upon, by some specified group. Such a group might include all human beings, or a subset thereof consisting of more than one person.)
5. Pragmatic theory (truth is verified and confirmed by the results of putting one's concepts into practice.)
6. Minimalist theories (truth does not denote a real property of sentences or propositions.)

7. Redundancy and related theories (truth is a redundant concept; that is, it is merely a word that is traditionally used in conversation or writing, generally for emphasis, but not a word that actually equates to anything in reality.)
8. Pluralist theories (there is a particular property the having of which makes a belief or proposition true. Pluralist theories of truth assert that there may be more than one property that makes propositions true: ethical propositions might be true by virtue of coherence. Propositions about the physical world might be true by corresponding to the objects and properties they are about.)

In summary, what we have is a comical assembly of research questions ranging from Harvard College's "How many angels can dance on a pin head?" to the 18th century classic "If a tree falls in a forest and no one is around to hear it, does it make a sound?" These have created a world in which systematic falsehoods have become the truth while the truth can be dismissed as 'conspiracy theory' whenever it happens to gore some powerful individuals' oxen. We have graduated from Dogma and trinity only to fall for multiple history of the same physical point — otherwise known as … quantum theory.

3.7.7 Origin of Universe

From all one can gather from ancient cultures is that the origin of universe was not a matter of speculation and theorization. In ancient India, before any creation it was 'Mahi dhara' (Holder of the earth), 'Maha Sunya' (Great Zero) and in ancient Greece it was **** (Chaos, the void). This status is later conflated with either 'nothingness' or state of disorder (hence the word 'chaos'). Then, Qur'an formalized in the following verses.

> "Allah is the Creator (*khalaqa*) of everything. He is the guardian over everything. Unto Him belong the keys of the skies and the earth." (39:62, 63)
>
> Originator (*bad'a*) of the skies and the earth. When He decrees a matter, He only says to it, "Be," and it is. (2:117)

> 'O my people, surely I quit that which you associate, I have turned my face to Him Who originated (*fatara*) the skies and the earth; a man of pure faith, I am not of the idolaters.'" (6:76-79)

Note that different root words are used for 'creation', namely, *Khalaqa, bad'a,* and *fatara.* The root word khalaq refers to creating something that never existed before and creating it in a perfect form and definite proportion. It also includes creating as in refashioning. The root word 'bad'a' refers to something that is originated from nothing. This would be the closest to the meaning of 'chaos' in ancient Greek. The root word '*fatara*' refers to creating out of nothing, cleave/split/crack/break, be broken into pieces, cleave asunder, begin to create. This is the onset of any grand phenomenon that didn't occur before. In addition, it was evident that the purpose of creation was to place humans as the viceroy. This special status is in line with the sequence of creation, i.e., earth first then, seven layers of skies (Qur'an 2:29), the first sky (*samaya dunya*) being for creating balance, decoration, and to protect the earth (Qur'an 67:5). This nearest sky is considered to be 'universe' in today's science.

The first level of fatal distortion was imposed by the following 'biblical words':

> Then God blessed the seventh day and made it holy, because on it he rested from all the work of creating that he had done. (Genesis 2:3)

For the first time in known history, the Creator was attributed traits of the creation (function of time) and the molestation continued.

The environment of the time was rife with flat earth theory and other obsessions with homogeneity, uniformity, steady state, and similar traits that enjoy no existence whatsoever anywhere in nature. From this point of departure, there could be no going back to the original premises. (All this matter was taken to an unprecedented low by Sir Isaac Newton, whose work will be deconstructed in latter chapters.) Newton, obsessed with the notion of steady state, also believed in the idea of an omnipresent God. For Newton, the divine presence operated as an immaterial "aether" that on the one hand offered no resistance to bodies, but could move them on the other hand through the sheer force of gravitation. This notion of 'zero resistance' but non-zero presence is in the core of all previous models based on 'atoms' and equally prevalent in later models, including quantum theories. Newton resorted to the notion of 'continual miracle' to explain why the Sun and the fixed stars are not pulled together. Newton envisioned an infinitely large universe, in which God had placed the stars at just the right distances so their attractions cancelled, as precisely as balancing needles on their points. Another possible solution was to place the fixed stars at such vast distances from one another that they could not attract each other perceptibly in the few thousand years since the Creation. Indeed, creation of the universe for these fellows were only a few thousand years old, in line with biblical narration of the time. [5]

The Qur'an revealed to Prophet Muhammad early in the 7th century CE states clearly that the sun and other celestial bodies move at a very fast speed (Qur'an 36:38). Europe remained mired almost an entire millennium longer in the 'steady-state', 'fixed stars', and other absurd models of the cosmos. Only in 1718, this absurd assumption was questioned when the English astronomer Edmond Halley reported a remarkable discovery. Three bright stars were no longer in the positions determined by ancient observations. The stars were freed to move like normal physical objects. It was with Halley's funding that Newton published The Principia in 1675. Together, these laws describe the relationship between any object, the forces acting upon it and the resulting motion, laying the foundation for classical mechanics.

RESUME HERE

Newton made clear his heliocentric view of the Solar System and recognized the "deviation of the Sun" from the center of gravity of the Solar System. For Newton, it was not precisely the center of the Sun or any other body that could be considered at rest, but rather "the common center of gravity of the Earth, the Sun and all the Planets is to be esteem'd the Centre of the World", and this centre of gravity "either is at rest or moves uniformly forward in a right line".

The French philosopher René Descartes, on the other hand, had proposed a non-mathematical model. He suggested that the universe consists of huge whirlpools ("vortices") of cosmic matter. Our solar system would be only one of many such whirlpools. Curiously, this notion would be more natural, but was rejected by subsequent new scientists.

Descartes' mechanical, mechanistic cosmology was highly acceptable within the general seventeenth-century conception of the world as a

5 By adding up the genealogies and reigns of the Old Testament kings, some scholars thought they could arrive at a fairly accurate date for creation. Using this method, Bishop James Ussher (1581-1656) dated Creation from 4004 BCE.

machine. His explanations, though, were but qualitative re-descriptions of phenomena in mechanistic terms. During the course of the eighteenth century, vortex theory proved unable to calculate the observed planetary motions. Meanwhile, the rival Newtonian theory advanced from one precise quantitative success to another.

Still unexplained were large anomalies in the motions of Jupiter and Saturn, and an acceleration of the Moon's orbital speed around the Earth. The French mathematical astronomer Pierre-Simon Laplace resolved these in 1785 and 1787. In his book Mécanique Céleste, published in five volumes between 1799 and 1805, Laplace summarized his studies of celestial mechanics. Here he proposed that all physical phenomena in the universe could be reduced to a system of particles, exerting attractive and repulsive forces on one another.

Laplace's writings were not just for scientists. His 1796 book Exposition du Système du Monde summarized for lay people the general state of knowledge about astronomy and cosmology at the close of the 18th century. In the book, Laplace advanced an idea that became known as the "nebular hypothesis." He suggested that our solar system, and indeed all stars, were created from the cooling and condensation of a massive hot rotating "nebula" (a gassy cloud of particles). Ever since, this has been the beginning of the so-called Big Bang logic.

In the late 18th century, observation at last entered stellar cosmology in a major way, in the person of the English amateur astronomer William Herschel. His discoveries were made possible by large telescopes of his own construction. Looking out from Earth (at center) toward us, a dense slab of stars becomes the band of the Milky Way. Looking away from us, a split in the Milky Way comes from a division in the distribution of stars. From his observations, William Herschel reported in 1784: "A very remarkable circumstance attending the nebulae and clusters of stars is that they are arranged into strata, which seem to run to a great length; and some of them I have already been able to pursue, so as to guess pretty well at their form and direction. It is probably enough, that they may surround the whole apparent sphere of the heavens, not unlike the milky way, which undoubtedly is nothing but a stratum of fixed stars."

Herschel's telescopes, culminating in 1789 with an awkward monster forty feet tall, were one of the wonders of the world. These powerful telescopes not only revealed more moons about planets and resolved some fuzzy-appearing nebulae into clusters of stars, but also enabled Herschel to reach farther into space than anyone had done before, and to begin to outline the structure of our galaxy. Herschel observed stars seemingly lying between two parallel planes and running on to great distances. He concluded that the Milky Way (a luminous band of light circling the heavens) is the appearance of the projection of the stars in the stratum.

Herschel proposed to determine the position of the solar system in the stratum of stars by " 'gaging' the heavens," that is, by counting the number of stars in different directions. This number, Herschel argued, assuming the stars to be equal in brightness and equally scattered, would be proportional to the distance to the edge of our galaxy in each particular direction. Although logically correct, his method failed in his day to get a true picture of the Milky Way, for his telescope could not detect ordinary stars at the farthest reaches of our galaxy. Further, as Herschel himself recognized, there was no good reason to assume that the stars are equally bright and equally scattered.

During the nineteenth century, Herschel's speculative cosmology failed to attract disciples. Professional astronomers could not accept his assumption that stars are equal in brightness, although it was needed as a working hypothesis to estimate the stars' distances.

Remarkable new observational techniques, photography and spectroscopy, did address cosmological questions, but indecisively. In 1835, the prominent French philosopher Auguste Comte, remarked that humans would never be able to understand the chemical composition of stars. Soon after that, spectroscopy and photography helped bring about a revolution in people's understanding of the cosmos and all of a sudden, we became convinced that we can measure chemical composition of matter that are outside of our galaxy, billions of lightyears away. This false perception 'revolutionized' cosmic studies with unprecedented arrogance and bizarre ignorance.

Spectroscopy held out a promise of differentiating between nebulae made of many stars and those made of glowing gases, and also of determining if nebulae were rotating. But here also the conclusions were questionable. As, indeed, was cosmology itself as a scientific endeavor. Advances in cosmology during the nineteenth century were considerable, but only in the twentieth century would cosmology be transformed from speculation, based on a minimum of observational evidence and a maximum of philosophical predilection, into a respectable observational science.

Approaching the beginning of the 20th century, the worldview pioneered by Herschel was vastly different from that of Aristotle or even Copernicus. No longer were human beings necessarily at or very near the center of the universe. The Milky Way was now understood to be an optical effect, with our solar system immersed in a much larger stratum of stars, a roughly disk-shaped stellar system. Possibly other island universes were scattered throughout a possibly infinite space. All these would be based on the premise that wasn't much different from the authors of the flat earth theory.

Changing cosmological understanding is manifested in changing social views. Now that we Earthlings were but one of possibly many intelligent inhabitants of a possibly infinite universe, there was less reason to believe that we had been created in the best of all possible worlds, and perhaps more sympathy for discontent with the established social hierarchy. The connection between social and scientific 'righteousness' has been completed.

In 1925, an ordained priest would develop the most important theory of modern time in matter of cosmology. Georges Lemaître published in 1927 in the Annales de la Société Scientifique de Bruxelles (Annals of the Scientific Society of Brussels), under the title "Un Univers homogène de masse constante et de rayon croissant rendant compte de la vitesse radiale des nébuleuses extragalactiques" ("A homogeneous Universe of constant mass and growing radius accounting for the radial velocity of extragalactic nebulae"). In theory, it is not very different from Democritus's atomic model – a model that has pervaded every theory, social or scientific in modern Europe. In this report, he presented his new idea of an expanding Universe (he also derived Hubble's law and provided the first observational estimation of the Hubble constant).

This single "primordial atom" is the beginning of another series of aphenomenal cognition. The idea subsequently received major boosts by Edwin Hubble's observations that galaxies are speeding away from us in all directions, and from the discovery of cosmic microwave radiation by Arno Penzias and Robert Wilson. The glow of cosmic microwave background radiation, which is found throughout the universe, is thought to be a tangible remnant of leftover light from the big bang. Yet, no other conclusion could be made with the premise of 'Big Bang' as the origin of universe. No one considered the premise that this detection which miniscule observation in time doesn't integrate over billions of years without making the process absurdly self-righteous. As contradictions will arise when 2011 Nobel prize winners Saul Perimutter and Brian Schmidt (2011 Nobel Laureates) observed that the universe is expanding with a diminishing rate, seemingly contradicting Big Bang theory, new line of dogmatic justification as well as new theories (e.g., Australian's 'big freeze' theory) would emerge.

The big bang theory leaves several major questions unanswered. One is the original cause of the big bang itself. Several answers have been proposed to address this fundamental question, but none has been proven—and there is no test procedure in the making that could prove them false. This is inherent problem of cognition in the era of New science.

In the following we deconstruct statements made by American theoretical physicist and cosmologist, Lawrence Krauss (2012) with premises and conclusions in parenthesis:

> "Conventional wisdom was the universe has been there for ever and will be for ever. Better than 1% accuracy that the universe is static [observation/interpretation]. It is flat like a sheet of paper [observation/interpretation]. It has a total energy of zero [Premise 1].
>
> Therefore, it could be begun from nothing [conclusion based on Premise 1].
>
> We don't need a deity [Premise 2].
>
> The actual theory is far more remarkable than the fairy tale that was made up in most religious situations [Conclusion based on Premise 2].
>
> The Universe has an end [Premise 3, guised as 'Conclusion']."

In summary, Krauss counters previous dogma (i.e. the universe has no beginning or end) with a new conclusion, which is actually another premise (Premise 3): "The Universe has an end". This conclusion has no premise attached to it and it doesn't come from any logical starting point. It would be interesting to find out how such conclusion is reached. Although not mentioned here, it is known that the premise Krauss is using involves the following:

For our own sun, after an initial bright youth it will burn up all of its fuel and then slowly begin to fade out. There will still a remnant left over, you just can't see it anymore.

Other types of stars, typically much heavier than our sun, after burning all their hydrogen start to collapse due to enormous gravitational forces. A catastrophic event called a supernova can occur. The resulting leftover may be a neutron star or black hole.

The above premise itself is premised on several others. For instance, Kelvin's premise was everything is degrading continuously and ultimately everything will become 'Heat dead" [6] . This concept is self-contradictory as it hypothesizes that the entropy increases as order decreases. Such postulation would imply there is no order when heat death had occurred. This premise is further premised on Roman Catholic church's as well as old Greek religion's assertion that 'desire degrades the soul' - a dogma that can lead to a tangible meaning of continuous material degradation.

One must also analyze the word 'end' in the proper context. Imagine a star 'burns up' and disappears. Conservation of mass dictates that material cannot be created or destroyed. If so, a star cannot be burnt up and disappear. The same principle applies to the 'end' of the Universe. There is no scientific theory that can be used to predict the universe will come to an end.

Similarly hopelessness persists in all levels of physics, matters involving subatomic to galactical scale. For instance, we still have no explanation as to what is in between the smallest perceived component of matter. This problem is inherent to Atomism and the dogma that was picked up by Roman Catholic church. Following is an excerpt from Islam *et al.* (2014) that described the problem related to definition of emptiness and zero.

> "A less elaborate 'atomic theory' as described by Democritus had the notion of atoms being in perpetual motion in a void. While being in constant motion (per- petual should not mean uniform or constant speed) is in conformance with natural traits, void is not something that is phenomenal. In Arabic, the closest word to describe void is 'cipher' (the origin of the word decipher, meaning removing the zero's or the fillers), which means empty (this word that has been in Arabic for over 1400 years was not used in the Qur'an). For instance, a hand or a bowl can be empty because it has no visible content in it, but it would never imply it has nothing it (for instance, it must have air). The association of 'cipher' with zero was done much later when Arabs came to know about the role of zero from Indian mathematicians. One very useful application of zero was in its role as a filler. That alone made the

6 This notion of entropy means that everything is moving toward a state of thermal (hence compositional) uniformity.

counting system take a giant leap forward. However, this zero (or cipher or 'sunya' in Sanskrit) never implies nothingness. In Sanskrit, Maha Sunya (Great Zero) refers to the outer-space, which is anything but void as in nothingness. Similarly, the equivalent word is As-sama'a, which stands for anything above the earth, including seven layers of stars in the entire universe (in conventional astronomical sense). In ancient Greek culture, however, void refers to the original status of the Universe which was thought to be filled with nothingness. This sta- tus is further confused with the state of chaos, ****, another Greek term that has void as its root. The word chaos does not exist in the Qur'an as it is asserted there is no chaos in universal order that would not allow any state of chaos, signaling the loss of control of the Supreme Authority. It is not clear what notion Liucippas had regarding the nature of atomic particles, but from the outset, if it meant a particle (undivided) that is in perpetual motion, it would not be in conflict with fundamental nature of natural objects. This notion would put everything in a state of flux. The mainstream Greek philosophy would view this negatively for its subversive implication that nature is essentially chaotic. Such an inference threatened the Greek mainstream view that Chaos was the Void that had preceded the coming into existence of the world, and that a natural order came into existence putting an end to chaos. As stated earlier, this confusion arises from misunderstanding the origin of the Universe."

Because the likes of Krauss didn't bother revisiting their first premises that are remarkably similar to dogmas, it is no surprise, he repeats the lines that are very old. He moved very little from flat earth theory to stumble on flat universe, from emptiness of 'chaos' to emptiness of 'dark matters', from dogmatic primordial beginning to 'scientific primordial beginning'. In order to justify such conclusions, 'experimental data' are routinely made available. For instance, the Wilkinson Microwave Anisotropy Probe (WMAP) satellite established the spatial curvature of the universe, Ω_k, to be between -0.0174 and +0.0051, where 0.0000 represents perfect flatness. Once again, this observation can also be used to develop a support base for any other first premise, such as spherical universe with a higher greater radius. This is eerily close to Newtonian gravitational energy of the universe being "zero" or "very close to zero". Then added the absurd conclusions that the the universe still contains a huge amount of heat left over from the cosmic creation event and enormous quantities of dark energy, exotic dark matter, ordinary dark matter, and visible galaxies, stars, planets, dust, and gas. By saying the universe came from "nothing" Krauss is reflecting, unwittingly, one of Christianity's foundational creeds. *Creation ex nihilo* (Latin for "creation from nothing") refers to the moment God created something (the universe) from nothing (that which lacks matter, energy, space, and time). Furthermore, nothing more is added to the Biblical assertion of *creation ex nihilo:* "The universe was framed by God's command, so that what is seen was not made out of what was visible." "Seen" and "visible" refer to the stuff investigators can and do detect, namely space, time, matter, and energy. Krauss' Alternative Explanations for the Universe's Origins *Virtual Creation* While Krauss ends up saying the same thing as Hebrews 11:3—that the stuff we humans detect was not made from detectable stuff—he does not start out that way. He first proposes that virtual particle production serves as an analogy for how the universe came to exist. Virtual particle production is a natural outcome of the uncertainty principle of quantum mechanics. This principle states, in part, that quantum fluctuations in the universe's space-time fabric will generate particles, provided those particles revert to quantum space-time fluctuations before any human observer can detect their appearance. Typically, the particles so produced must disappear in less than a quintillionth of a second. Since these particles cannot be detected directly, physicists refer to them as virtual particles. Krauss suggests that the entire universe may have popped into existence by the same means. These are just the same dogma recycled with a touch of arrogance and 'secular colour'. Then came the assertions that there are many of these 'baby universes' that are being formed instantly. However, no explanation is provided as to why such continuous phenomena should suddenly cease to exist and the universe should come to an end. It all boils down to matter forming from nothingness (the space-time fabric) and revert back to nothingness (the space-time fabric) in less than 10^{-102} seconds (this is yet another premise and unfounded, illogical time frame).

A second inadequacy in Krauss' suggestion comes from another principle of quantum mechanics. The probability of a quantum outcome occurring increases in proportion to the passage of time. That is, the larger the time interval, the greater the probability that a quantum outcome, like the production of a virtual particle, will take place. This principle implies that if the time interval is zero, the probability for any quantum event is zero. The space-time theorems prove that time has a beginning coincident with the beginning of the universe. Thus, the time interval at the beginning of the universe is zero. This eliminates quantum mechanics as a possible candidate for natural generator of the universe. *Hyper Quantum Mechanics* In *A Universe from Nothing*, Krauss never acknowledges the weaknesses of the virtual particle production analogy for cosmic creation. However, he does hypothesize a second way the universe could have arisen from nothing without divine agency. Krauss proposes that—in addition to the observable quantum mechanics constrained to space and time—there is an unobserved hyper quantum mechanics that exists beyond our universe. Here some dimension (or dimensions) of time, entirely distinct from cosmic time, would permit space-time bubbles, independent of the space or time dimensionality posited to exist beyond our universe, to pop into existence spontaneously. However, if the hyper quantum mechanics is anything like the quantum mechanics we observe, then the space-time bubbles must also disappear spontaneously within extremely brief time episodes. Krauss acknowledges that his appeal to some imagined hyper quantum mechanics to explain the origin of the universe leads to a time episode problem. He suggests that the problem might be solved if the universe experiences a very aggressive inflationary expansion event *before* the hyper quantum mechanics forced the newly generated space-time bubble (our universe) to disappear. Inflation is now an integral part of big bang cosmology. It refers to the brief but rapid exponential expansion of the early universe by a factor of at least 10^{78} in volume. For our universe, the inflation epoch lasted between 10^{-36} and 10^{-33} seconds. It occurred near the very beginning of the electroweak era, during which three forces of physics existed: gravity, the strong nuclear force, and the electroweak force. The electroweak force is actually a blending of electromagnetism and the weak nuclear force. This blending occurs only when the universe is very young and, hence, very hot. However, if the universe is too young, the electroweak force will blend with the strong nuclear force. When our universe was about 10^{-35} seconds old, the strong-electroweak force separated into the strong nuclear force and the electroweak force. Accordingly, an inflation episode cannot begin in our universe until the universe is 10^{-35} seconds old. A hundred billionth of a trillionth of a trillionth of a second might not seem like very much time, but it is far too long to make Krauss's hyper quantum mechanics a viable "creator" of our universe. This albeit extremely brief time interval is 10^{67} times longer than the time duration for a universe like ours to appear and then disappear via the quantum pathway that produces virtual particles.

Recently, Islam *et al.* (2014a) pointed out how all these 'intellectual' talks assume that the probability of having any phenomenon take place (including such absurd possibilities as multiple history, ubiquitous existence of same matter, having absolute vacuum or infinite mass, infinite temperature, infinite growth rate) is finite and non-zero but having a creator that is external to creation (Qu'ran 112:2) is zero.

The following anecdote perfectly illustrates the idiocy of Lawrence Krauss' approach to "scientific truth":

> The late Nobel physics laureate Richard Feynman, in his 1985 autobiographical memoir 'Surely You're Joking, Mr. Feynman': Memoirs of a Curious Character, set forth his vision of the meaningful content of the notion of "scientific integrity" in an arresting manner, with an anecdote from the annals of actual scientific research in the field of behavioral psychology.
>
> Although students in the field are carefully schooled in the concept of "conditioned stimulus – conditioned response"(CS-CR) including spending hours in the laboratory with mice running various experiments premised on the truth of this fundamental notion, what Feynman learned was that students are never steered towards actually testing the veracity of the CS-CR hypothesis itself, viz., that the response of the mammalian brain can be trained or conditioned by a consciously-planned series of stimuli.
>
> Looking further into the matter, he uncovered a research paper published before the Second World War whose author deliberately attempted to replicate the original experiments of John Watson purporting to demonstrate the CS-CR hypothesis. The investigator relaxed the condition that CS-CR be assumed true and sought alternative explanations for the lab mice's responses to various conditioned stimuli.
>
> He narrowed down one-by-one all other possible physical causes for the stampeding of the mice through the experiment's "maze" apparatus to get their "food" reward. He found the strongest and most positive correlation existed between the level of vibration of the boards of the raceways in the apparatus and when mice from various points inside the apparatus found their way through the correct series of tunnels in the maze to the reward-point.
>
> As he removed all possible sources of vibratory transmission from inside the apparatus, he reached a point where the response

of the mice could no longer be said to correspond to the original stimulus.

Feynman noted that he could find no references to this experiment anywhere in the literature: it disappeared, or perhaps was "disappeared" by those who sniffed its devastating implication that the CS-CR hypothesis itself was unproven. Feynman hailed the experiment for its profound pædagogical value as a basic primer in scientific integrity — and for teaching everything one would need to know about how to verify the CS-CR hypothesis experimentally.

3.7.7.1 The Qur'anic Narration

New Science has added conclusions such: the universe is flat like a sheet of paper: the universe has the total energy of zero.

Flat universe theory has the conclusion based on aphenomenal premises. It starts off with gravity being a function of mass. The upcoming chapters of the book will show that this is an aphenomenal premise that leads to numerous contradictions that can be only explained with more dogmatic justification.

Newton described gravity as a function of mass that controls motion of particles (with mass). For Newton, time was flat.

Gravity = f_1 (mass)

Acceleration = f_2 (gravity)

Time = external function independent of mass.

For Einstein, time was a function of perception as well as gravity. This triggered multiple solutions for each time function, making it non-unique. This is the first time in history, multiple solutions were attributed to the time function.

Time = f_3 (perception)

Perception = f_4 (orientation of an object)

Orientation of an object = f_5 (environment)

Qur'anic narration presents various facts with an entirely different starting point. Time is a creation that is controlled by the creator (Absolute Time). Time in turn controls all particles that form the environment and that environment controls the thought process as well as perception. Perception becomes reality only with good intention, something every human has freedom of. Everything (including time) is dynamic, meaning, a function of time or fleeting. Motion is a function of Absolute light (Creator). In brief:

Time = g_1 (Absolute Time, Creator) [7]
Matter = g_2 (Time)[4]
Gravity = g_3 (Absolute Light, Creator) [8, 9]
Perception = g_4 (Matter)
Intention = Flat (0,1) [10]

Therefore, an entirely different picture of 'everything' is created in the Qur'an. The same Qu'ran states:

يَوْمَ نَطْوِى ٱلسَّمَآءَ كَطَىِّ ٱلسِّجِلِّ لِلْكُتُبِ ۚ كَمَا بَدَأْنَآ أَوَّلَ
خَلْقٍ نُّعِيدُهُۥ ۚ وَعْدًا عَلَيْنَآ ۚ إِنَّا كُنَّا فَٰعِلِينَ ﴿١٠٤﴾

The Day when We will fold the sky like the folding of a [written] sheet for the records. As We began the first creation, We will repeat it. [That is] a promise binding upon Us. Indeed, We will do it. (Qur'an 21:104)

ٱلَّذِى خَلَقَ سَبْعَ سَمَٰوَٰتٍ طِبَاقًا ۖ مَّا تَرَىٰ فِى خَلْقِ ٱلرَّحْمَٰنِ مِن تَفَٰوُتٍ ۖ
فَٱرْجِعِ ٱلْبَصَرَ هَلْ تَرَىٰ مِن فُطُورٍ ﴿٣﴾

[And] who created seven skies in layers (plate-like). You do not see in the creation of the Most Merciful any inconsistency. So return [your] vision [to the sky]; do you see any discontinuity? (67:3)

This 'discontinuity' is invoked in New science

وَلَقَدْ زَيَّنَّا ٱلسَّمَآءَ ٱلدُّنْيَا بِمَصَٰبِيحَ وَجَعَلْنَٰهَا رُجُومًا لِّلشَّيَٰطِينِ ۖ وَأَعْتَدْنَا
لَهُمْ عَذَابَ ٱلسَّعِيرِ ﴿٥﴾

7 Hadith Kudsi,4: "Sons of Adam inveigh against (Absolute) Time, but I am the Time and in My hands are the night and the day".
8 Qur'an 2:74
9 Qur'an 24:35
10 Intention

And We have certainly beautified the nearest sky with stars and have made [from] them what is thrown at the devils (jinns that follow Satan) and have prepared for them the punishment of the Blaze. (67:5)

Then, there are numerous verses that talk about the end of the universe and beyond (the other six skies that don't have known galactic matter) – all in order to bring about the day of accountability. This is a consistent picture with everything that can be explained in line with the purpose of creation of everything. Curiously, these conclusions (e.g., flat universe, end of universes) are in line with quantum physics dogmas, however, quantum physics gives no reference to the purpose of creation as well as the role of humans within the universe.

3.7.7.2 *The Bottom Line*

Einstein's theory doesn't say matter and energy cannot be destroyed or created. That assertion comes from the Qur'an itself. For the longest time, every logic assumed that matter and energy cannot be created or destroyed. Also, every logic assumed that matter came into existence from nothing. Qur'an says it is by Allah's decree, "new science" says it came to existence without a God.

Now, coming to Einstein's theory itself, it is derived from Maxwell's equation, which is derived from Newton's. Newtonian model assumes matter to be composed to solid rigid balls. Maxwell model assumes energy composed of sold rigid balls, except these balls have no mass (e. g., photons). So, fundamentally nothing changed between matter and energy they simply applied the same absurd model to energy. We discovered that by studying Einstein's fundamental paper in 2005. It took us one day to discover, it was based on Maxwell's model. The following years we discovered Ibn Haitham (AlHazen) and his contemporaries didn't have such Aphenomenal model and they were quite accurate in predicting that sunlight is never the same as artificial light because they have real and artificial sources, respectively. Other aspects involved the notion that Aristotle used to believe humans see because there is something that comes out of the eye that make it possible to see. Ibn Haitham assumed the opposite and sure enough that was accurate because in reality we now know exposure to laser is harmful even if it is for a second. This simple observation was good enough for us to discover later that the media is of extreme importance, just like the source. This means if light travels through polluted media, it would alter for ever. This would be impossible for any other model to predict. Why? Because

Einstein invoked the speed of light is constant and cannot be superseded by anything else. So, in 2011 when neutrino particles were proven to be moving faster than light, everyone else was surprised but we were not. In 2005, Nobel prize was given to someone because they discovered speed of light can be changed depending on the media density (temperature in particular). Yet, that's what we discovered by using the logic of Ibn Haitham. This model can also explain most recent findings regarding light bending and other non-linear phenomena that would be considered absurd not too long ago (Firstenberg *et al.*, 2013).

It's actually a very easy exercise to look at the assumptions of various scientists and remove them with logical ones and sure enough new theories will emerge. So, let's now just look at quantum theory and what assumptions it has invoked.

1. It assumes that everything has multiple history. Logically there is only one version of truth, the true path. For Feynman and his followers, truth is a subjective matter and everything has many versions of the truth. In Arabic, the word *Noor* or *Haq* are all singular, *Dhulumat* (shades of falsehood or lack of Noor) are plural. What Feynman has done is replaced one truth with numerous (actually infinite number of) falsehood. If Trinity (changing one god to three) was a problem, imagine the problem here.
2. It also assumes, the observation itself can alter the history. This is not unlike Aristotle's assertion that we see because something from eye comes out and makes the matter visible. The extension of this model is, we actually influence the history by changing our observation pattern. George Orwell said, "He who controls the past controls the future. He who controls the present controls the past." What we have today in New science, is a rush to altering the history in order to control the future.

3.7.8 Dogma to Pragmatism

It is well recognized that dogmatic thinking set back natural cognition in Europe by centuries. It is less known, however, that the pragmatism that replaced dogma in the post-industrial revolution era is a veiled and more sinister form of dogma. Dogmatic cognition is deconstructed by laying it out this way: "God created humans with original sins, then got a woman pregnant with himself, so he can be born to sacrifice himself to himself in order to forgive humans of the original sin that he gave them to begin with." How has that dogma changed when 'enlightenment' came in terms of western philosophy? In terms origin of universe, the big bang theory, the introduction of quantum nature, multiple (numerous) history of each point, Nature as the creator of 'numerous universes', the 'effect of observation on the past' and even the definition of what's true and false has introduced a logic that's more schizophrenic than dogma. What dogma did is introduced one Aphenomenal standard, whereas the 'western philosophers' have introduced numerous such standards. To a neutral observer, it's a quantum leap from bipolar to multi-polar (schizophrenia) cognition. While it has become fashionable in the 'enlightened' population to criticize doctrinal philosophy and its offshoots (e.g., creationism), fundamental science has continued to introduce dogmatic definitions (e.g., 'black hole', 'dark matter', 'dark energy', 'selfish gene', 'gay gene') and supported them with aphenomenal logic, confounding the problem by diverting the general public from natural cognition.

When attacked as regards their blatant absence of any clear phenomenality, each of these additional notions (e.g., "origin of universe", "big bang theory", "quantum nature", multiple (numerous) history/histories of each point, Nature as the creator of 'numerous universes') is ultimately justified as an "explanatory framework". It is therefore important that one nails this fish-tailing and flipflopping for what it is. One cannot pick-and-choose the truth. The scholar-researcher has the responsibility to nail things down or honestly admit that they are not there yet. Most importantly, one must present one's first premise or basic assumptions clearly and be open to accepting alternate first premises in case contradictions arise in the process of subsequent reasoning. The illegitimate and unconscionable response to such contradictions is to rave about how some completely unpredicted or unanticipated "discovery" fits into some previously-advanced piece of theory (these "discoveries" being nothing but repeated assertions of the same assumptions that set off the cognition process). This is exactly what happened in the post-Newton era of human cognition.

Here is one example of the kind of question that needs to be answered honestly and not in a manner that fits/serves previously advanced dogma: I watch an oscilloscope trace every time a light pulses, and that trace is declared evidence, or confirmation, that light has a quantum character of a discrete albeit infinitely tiny piece of mass. Does it mean that the light I see after the sun rises is or is not the same light appearing as an oscilloscope trace, or does it mean that light can be both? And if you choose the latter, does this mean light is both a particle and a wave? Light is an actual phenomenon, but is the oscilloscope trace or its source the same phenomenon of not really germane for dealing with the natural phenomenon that is light? The pragmatic fix has been to say that these are different versions of the same thing, end of story. There's the slippery slope: material reality and my observation of material reality are "essentially" the same. But the moment we accept this, what remains definitive about material reality?!?

Pragmatism and pragmatic criteria, whose practicality seduce many a researcher, are highly-developed reflexes of the Eurocentric outlook. Indeed: they form practically the unofficial universal religion among academics in the United States, for example. The bald fact remains that, although pragmatism and the pragmatic criterion demonstrate what is true on the seemingly irrefutable basis of "whatever works", this criterion itself refuses to grapple with and artfully dodges the need to clarify what is false, or what falsehood(s) had to be overcome, in order to arrive at the truth.

Further evidence that dogmatic thinking has been reinforced with pragmatism, albeit being packaged as 'secular' is the inability of modern science in questioning the fundamental premise. The reluctance to question the logical validity of the first assumption may not be the fear of the Authority who might order an execution similar to that orchestrated by the Roman Catholic church for centuries, it is rather the marginalization by the academic community that didn't allow a single diversion from the new Authority that orders diktat based on self interest and short-term economic gains. The case in point is the evolution of electricity dependence. While the early 19th century had seen rapid progress in electrical science, the late 19th century would see the greatest progress in electrical engineering. Through such people as Nikola Tesla, Galileo Ferraris, Oliver Heaviside, Thomas Edison, Ottó Bláthy, Ányos Jedlik, Sir Charles Parsons, Joseph Swan, George Westinghouse, Ernst Werner von Siemens, Alexander Graham Bell and Lord Kelvin, electricity turned from a scientific curiosity into an essential tool for modern life, becoming a driving force of the Second Industrial Revolution. This entire evolution of the 'most important' branch of modern civilization is the assumption that lightening (natural process) is the same as electricity (artificial process). Unless realism is based on 'whatever exists is real' (a pragmatic premise), this assumption is false, as these two types have no source and process.

The dogmatic nature of cognition in Europe has only worsened due to repeated assertion that as a human race we have become increasingly enlightened. This assertion has enforced arrogance and sealed the possibility of ever questioning the validity of the fundamental premises no matter how illogical they have become based on new knowledge that has emerged.

3.7.9 Deification of Self

Deification of self and selfish desire as promoted by New science is in the core of quantum physics. This was in the works for sometime. Freud started the notion that everything we do is based on our desire. It's the desire that drives our actions, everything else is just crazy. The same principle applies to dreams. Only meaningful dreams are when they are inspired by sexual desire. In Islam, dreams are from God or from Satan. We call that the Cathleen Turner syndrome. She once said, "if someone walks in the room and doesn't stare at me, he must be gay". This, coupled with Margaret Thatcher's 'There is no alternative' mantra has defined our modern era. The entire notion of deification of self was aptly captured in a recent book and movie, The Secret (Byrne, 2006). The tenet of the film and book is that the universe is governed by a natural "law" called the law of attraction which is said to work by attracting into a person's life the experiences, situations, events, and people that "match the frequency" of the person's thoughts and feelings. From this, the book argues that thinking positively can create life-changing results such as increased wealth, health, and happiness. The sequence of natural cognition, as discussed in previous sections is completely reversed with a false promise of intellectual self gratification.

One of our co-authors wrote:

"Few months back, I met a gentleman, who was in his 50s, and he worked for a very reputable organization. When I talked to him - he seemed like a person who has a wealth of knowledge / information on various topics on life. He told me about the book "The Secret" (Byrne, 2006) and to prove his philosophy of promoting HSSA, he said that every time he smoke cigarette – he convinced his brain that he is eating an apple full of vitamins. His point was that he is completely immune from the side effects of the cigarettes since he tells his brain that he is NOT smoking cigarettes rather he is eating an apple. He suggested that a person should hypnotize himself into believing that he is doing the best things even if he is doing something bad for his health."

This translates into insult of material to insult of intangibles, for which the damage is limitless, literally spiraling down to negative infinity. In old Greek time, this was called 'desire leading to degradation of the soul".

This anecdote helps us answer the question: where does the HSSA-(Perpetual Decline) end? One can say that when someone no longer needs a physical 'cause and effect' and when someone resorts to willingly fool his brain into doing schizophrenic acts, it is then HSSA completes its vicious loop.

4

The Hopelessness of New Science

If learning the truth is the scientist's goal... then he must make himself the enemy of all that he reads

Ibn al-Haytham

4.1 Summary

The most important task of research is to differentiate truth from falsehood in order to answer any question that is crucial to solving a problem. New Science uses criteria that are recycled from the dogmas of 17th century Britain, offering no hope of distinguishing truth from falsehood or real/natural from artificial. As a consequence, whenever real data are collected, scientists line up to justify whatever first premise they wish to support. The process, therefore, becomes entirely aphenomenal and all conclusions bear no scientific merit.

This chapter discusses this crisis with three examples, namely, colony collapse disorder (CCD), cancer, and Autism spectrum disorder (ASD). These examples are selected in order to solve the mysteries around HSSD degradation, CCD affecting the bee and ASD affecting the purest form of humans. They both allegedly affect or emerges from malfunction of the brain. Scientifically, it means that both are related to intangible aspects of life. By using these examples, it is shown that the current state of hopelessness in determining cause of these disorders emerges from the fact that New science is deficient in form as well as content. Because the source of a disease also relates to onset and cause of a disease, this chapter offers an excellent avenue for seeking causes of disorders. Cancer epitomizes the worst of any 'incurable disease' that continues to elude new scientists. The roles of honey as well as mother's milk are discussed as they symbolize most natural cure and natural food, respectively. Finally, it emerges that all 'remedies' currently being offered are most likely to make matters worse and any real remedy should take a holistic approach that arrests the HSSA degradation.

4.2 Introduction

The phenomenon of colony collapse disorder (CCD), also known as vanishing bee syndrome (VBS), has been known to scientists for some time. However, it is still commonly claimed to be "poorly understood". This symptom is known to cause massive die-offs of bees for purportedly "mysterious" reasons. While not knowing the cause has become synonymous with modern civilization, the implications of bees disappearing are so great that everyone is paying attention, and the general public is focused on finding the causes rather than finding techniques that would delay the symptoms. This chapter discusses various hypotheses being proposed and advances the theory that currently used New Science, which is also the science of tangibles, is not equipped with any appropriate means for studying this phenomenon, let alone finding a cure for the underlying intellectual dis-ease. The science of tangibles looks for the sources of CCD in terms of accounting for one or more symptoms. This must necessarily lead to proposing "solutions" that address only those symptoms whose sources were investigated, rather than the entire natural surroundings in which those and all other CCD-related symptoms emerged. Truth lies within the holistic approach that considers the entire ecosystem. Such considerations are possible only when inherent features of the science of intangibles are used.

Autism Spectrum Disorder (ASD) covers a range of conditions classified as neurodevelopmental disorders in the fifth revision of the American Psychiatric Association's Diagnostic and Statistical Manual of Mental Disorders 5th edition (DSM-5). The DSM-5, published in 2013, redefined the autism spectrum to encompass the previous (DSM-IV-TR) diagnoses of autism, Asperger syndrome, pervasive developmental disorder not otherwise specified (PDD-NOS), and childhood disintegrative disorder. These disorders are characterized by social deficits and communication difficulties, stereotyped or repetitive behaviors and interests, sensory issues, and in some cases, cognitive delays. This new characterization has made it easier to identify ASD. However, it has made little contribution toward finding the cause of ASD. Asperger syndrome is closest to autism in signs and likely causes, even though people with Asperger syndrome have no significant delay in language development. PDD-NOS is diagnosed when the criteria are not met for a more specific disorder and effectively lumps any unexplained behavior within ASD. Some sources also include Rett syndrome and childhood disintegrative disorder, which share several signs with autism but may have unrelated causes; other sources differentiate them from ASD, but group all of the above conditions into the pervasive developmental disorders. This revised definition has made it more difficult to pinpoint the cause of ASD or even to analyze the date with any objectivity. For instance, 1970s and 1980s, about one out of every 2,000 children had autism. Today, the CDC estimates that one in 150 8-year-olds in the U. S. has an autism spectrum disorder, or ASD. However, immediately objections surface that this expanded definition refers not only to autism but also to a collection of brain development disorders such as Asperger's syndrome and a condition known as pervasive developmental disorder - not otherwise specified (PDD-NOS). Though all the disorders share some symptoms, they are different in other ways, including the timeline of symptoms and the severity, according to the CDC. At the same time, when news break about Amish children not having Autism at all, the objection multiplying because of the tacit suggestion that Autism might be related to vaccine - something the Amish society doesn't endorse. Overall, the key questions, such as the followings, remain unanswered:

1. Is autism truly on the rise, or do the new statistics simply reflect the growing awareness of the condition, the expanded definition, and other factors?
2. If autism is on the rise, as most experts believe, what is causing the increase?
3. Does it related to our lifestyle? Or, can lifestyle alteration evoke remedy?

This chapter then outlines the basis for the science of intangibles, which would be able to trace the causes of such a disaster. In addition, it is shown that if engineering practices are based on the science of intangibles, resulting technologies and their economics would become truly sustainable, implying that disasters can be averted. Finally, guidelines are provided as to how to extrapolate the findings of this chapter to detecting cause of diseases in general.

4.3 Colony Collapse Disorder (CDC)

Starting in 2006, there has been an acute alarm sounded due to the sudden surge in CCD in practically all major mass media (CNN 2007a, 2007b; BBC, 2007; Barrionuevo, 2007, NYT, 2007). Even though it has been portrayed in some media that the CCD crisis has been resolved (Johnson, 2010), most scientists argue that the fear of honey bee extinction is real (Laskaway, 2010). Fear of honey bee extinction has prompted all parties concerned to consider consequences that are beyond the science of tangibles, an approach to science that has been firmly in place since the Renaissance. In the United States, honey bees seem truly to be facing extinction as some of the regions reported bee disappearance percentages of 80-100% (Leidig, 2007). So far, the cause of this disappearance has proven elusive (Lovgren, 2007), despite efforts by numerous scientists (NYT, 2007). It is becoming clear that no single cause may be identifiable as the sole or main "suspect". On the contrary, the entire lifestyle of the modern age might have contributed to the sudden surge in CCD (Khan and Islam, 2012). Starting with complaints by beekeepers because of loss of financial revenue, this matter has reached the level of global crisis within a short period of time, as evidenced by recent government reports (Johnson, 2007b; Berenbaum, 2007).

The causes of CCD that have received worldwide attention are: 1) mystery illnesses; 2) pesticides; 3) mites; 4) microbes; 5) habitat declines; 6) genetically-modified crops; and 7) electromagnetic waves (including cell phones). The possible or probable causes are so many, and the consequences of bee loss are so profound, that for the first time in history there has come to the forefront some discussion as to whether modern lifestyle in its entirety can be sustained. Even though certain Nobel Laureates in New Science — e.g., Robert Curl, who called the current mode "technological disaster" — have alluded to the unsustainable nature of the modern lifestyle, few have proposed any alternatives. The paralyzing paradox is that any lifestyle needs design, but all of today's designs are based on New Science that is focused on tangibles. All of New Science has a first premise that is aphenomenal (Zatzman and Islam, 2007; Khan and Islam, 2007; Khan and Islam, 2012). Challenging the first premise has always been a thorny issue in the modern age (Islam *et al.*, 2010; 2012), and few have attempted it (Commoner, 2006). In the words of the Chair of the Sierra Club's Committee on Genetic Engineering (GE), in his letter to US Senator Thomas Harkin said: "GE and bee Colony Collapse Disorder - science needed!" This is in fact the science that has gone missing — far longer than the bees! — from the post-Renaissance modern world.

4.3.1 Facts About Honey and CCD

Honey is known to be the only food that has all the nutrients necessary, including water, to sustain life. Even though the overall goodness of honey has been known for a millennia to be the most important naturally available matter, its usefulness is only becoming known to post-Renaissance civilization in a slow, trickling-down manner. At least some 10,000 years ago, man started the hunt for honey (Wilson, 2004). As early as 6000 years ago, the Pharaohs of Egypt were known to use honey for various applications, ranging from preservation to medicine, in addition to being used as a sweetener. Yet, in the 1970's, Reader's Digest considered honey as comparable to sugar, but with a higher calorie content. In The Qur'an, there is an entire chapter (Chapter 16) entitled "The Bee" (Al-Nahl), in which it is stated (16:69) that honey is healing for all of mankind (...wherein is healing for mankind. Lo! herein is indeed a portent for people who reflect, فِيهِ شِفَآءٌ لِّلنَّاسِ ۗ إِنَّ فِى ذَٰلِكَ لَءَايَةً لِّقَوْمٍ يَتَفَكَّرُونَ ...). However, broad or general healing properties of honey have become known in our modern world only much more recently (AFP, 2007). Even these new findings are not devoid of prejudicial hangovers from the dominant discourse of the science of tangibles.

For instance, the news agency, Agence France-Presse (AFP, 2007), quoted a lead researcher who actually promoted the therapeutic value of honey, stating that "the honey kills bacteria because it is acidic and avoids the complication of bacterial resistance found with standard antibiotics". To take at face value, the assertion that "honey kills bacteria because it is acidic" is absurd. If acidity were even the predominant cause, bacteria would not be killed; honey would indeed harbor acidophilic bacteria. The science of tangibles defines "acidity" in terms of the presence of H+ ions (definition of pH). Wouldn't the time period over which honey is effective as a biocide far exceed the likely maximum amount of time that H+ ions would remain present in concentrations that would make a substance act as a biocide?

This is just one example of how the science of tangibles collapses at the first serious scrutiny. Another often cited reason is that honey has low water activity (Prescott *et al.*, 2004). This is another example of an incorrect conclusion out of aphenomenal reasoning. If low water activity were to be the reason behind honey's antibacterial nature, honey would destroy far fewer bacteria than even common motor oil, which has been engineered to maintain ultra-low water activity. A different outcome would emerge if both this pH and water effects were considered under the lights of 'natural' and 'artificial'. Honey is a natural antibiotic, which means its anti-bacterial effects are a result of an infinite number of factors. Thus, any analysis that would start and end with one factor at a time, would result in an inherently wrong conclusion. The most we know at this time, is that it kills more effectively than any other biocide.

Honey remains the only product that does not rot. Samples of honey were discovered after thousands of years in many places and in fact remained edible. In relatively pristine areas, even common explorers continue to discover honey samples that are hundreds of years old and are perfect for eating (Al-Mahrouqi, 2007).

The second component of the statement is similarly incorrect or misleading. "Standard antibiotics" are inherently harmful. Their only "advantage" — delaying the onset of symptoms — has not been proven (Chhetri *et al.*, 2008). Developing bacterial resistance is on one of the endless lists of harmful impacts invoked by antibiotics. These are not "complications". They are real reactions to antibiotics that are inherently anti-nature.

Based on these two premises, the researcher continued to make remarks that are inherently flawed (because they are based on an aphenomenal premise). For instance, the same researcher was quoted by AFP (2007) saying, "If we can prove that honey promotes healing in diabetic ulcers, we can offer new hope for many patients, not to mention the cost benefits, and the issue of bacterial resistance. The possibilities are tremendous." It is clear that with the conventional epidemiological study (one of the sciences of tangibles), the only positive outcome of honey therapy can be that a patented formula or technology will emerge with little regard to the overall benefit of honey, let alone its relationship to the sustainability of the entire humankind.

Chapter 2 demonstrated that honey is the ideal of natural cure. Even though it is known that honey is the only "complete" food, with all "necessary" nutrients, (including water), few modern studies have been reported on the comprehensive composition and structure of honey molecules. Even the most comprehensive study, simply points to the fact that honey has over 250 unknown ingredients. It is unknown because the science of tangibles is not equipped with a technique to identify, let alone quantify, various components. This is the paradox of the science of tangibles. If the criteria of "verifiable quantities" were to apply to honey, it would not pass regulatory requirements for the sale of food items. This exposes a serious deficiency of the science of tangibles. It has a great risk of rendering something perfect as "absurd".

Consider Table 4.1 (below). This table lists the inherent nature of natural and artificial products. It is important to note that the left hand side statements are true – not in the tangible sense of being "verifiable", but because there is no counter-example of those statements.

The left hand side of Table 4.1 lists characteristic features of Nature. These are true features, not based or dependent on perception. Each is true insofar as no example of the opposite has been sustained. It is important to note that the following table describes everything in existence as part of universal order and applies to everything internal, including time, and human thought material (HTM). However, the source of HTM, i.e., intention, is not part of these features.

At the same time, all the properties stated on the right-hand side, which assert the first premise of all "engineered products", are aphenomenal: they are only true for a time period approaching zero, resulting in being "verifiable" only when the standard itself is fabricated. In other words, every statement on the right-hand side only refers to something that does not exist. For instance, honey molecules are considered to be extremely complex. They are complex because they have components that are not present in other products, such as sugar, which is identified as made up of "simple" molecules. Why are sugar molecules simple? Because, by definition, they are made of the known structures of carbon and hydrogen. Here is an example from the website of Virginia Technical University (Honey 2007):

Honey is one of the oldest sweeteners known to man. Records of its use date back nearly 900 years. Many definitions and standards have been used to describe honey, although few, if any, are complete. One of the better definitions defines honey as a “”sweet, viscous fluid, extracted by bees from the nectar of plants and stored in their hives as food.", not completely sure of this is correct?, This definition will suffice under

Table 4.1 Typical features of natural processes, as compared to the claims of artificial processes (Modified from Khan and Islam, 2012).

Feature no.	Feature of natural	Feature of artificial
1	Complex	Simple
2	Chaotic	Ordered
3	Unpredictable	Predictable
4	Unique (every component is different), *i.e.*, forms may appear similar or even "self-similar", but their contents alter with passage of time	Normal
5	Productive	Reproductive
6	Non-symmetric, *i.e.*, forms may appear similar or even "self-similar", but their contents alter with passage of time	Symmetric
7	Non-uniform, *i.e.*, forms may appear similar or even "self-similar", but their contents alter with passage of time	Uniform
8	Heterogeneous, diverse, *i.e.*, forms may appear similar or even "self-similar", but their contents alter with passage of time	Homogeneous
9	Internal	External
1	Anisotropic	Isotropic

(*Continued*)

Table 4.1 cont.

Feature no.	Feature of natural	Feature of artificial
11	Bottom-up	Top-down
12	Multifunctional	Unifunctional
13	Dynamic	Static
14	Irreversible	Reversible
15	Open system	Closed system
16	True	Artificial
17	Self healing	Self destructive
18	Nonlinear	Linear
19	Multi-dimensional	Uni-dimentional
20	Zero degree of freedom*	Finite degree of freedom
21	Non-trainable	Trainable
22	Continuous function of space, without boundary	Discrete

(Continued)

Table 4.1 cont.

Feature no.	**Feature of natural**	**Feature of artificial**
23	Intangible	Tangible
24	Open	Closed
25	Flexible	Rigid
26	Continuous function of time	Discrete function of time
27	Balanced	Inherently unstable

*With the exception of humans that have freedom of intention (Islam *et al.*, 2014).

most circumstances but even this description fails to include honey made from honeydew or fruit and plant juices.

Composition: Honey is composed primarily of sugars and water. The average honey is 79.6% sugar and 17.2% water. The primary sugars are fructose (38.2%) and glucose (31.3%). These are "simple", 6-carbon sugars that are readily absorbed by the body. Other sugars include maltose (7.3%), a 12-carbon sugar composed of 2 glucose molecules, and sucrose (1.3%), a 12-carbon sugar composed of a glucose and a fructose molecule.

The first part of this website establishes the first premise, which is: "There was no man before Europe-centric civilization emerged". This was typical of the Thomas Aquinas model that replaced the Averröes model some 900 years ago (Islam *et al.*, 2010; Islam *et al.*, 2013; Islam *et al.*, 2014). Contrast this to the use of honey by the Pharaohs and others, some dating back some ten millennia, i.e., 10,000 years (Wilson, 2004).

In terms of written records, the books of hadith (some 1400 years old) outline clearly as something that had been recommended by the Prophet Muhammad with citations from The Qu'ran that gave high value to honey. The Qu'ran was complied in its tangible form some 1400 years ago and remains intact today.

The posting moves on to yet another aphenomenal premise. It talks about a standard to characterize something that is truly unique, then presents honey as first and foremost a "sweet" substance. It recognizes honey only as a food. Later on, the same website would actually end up mentioning that honey is not quite nutritious, laying the aphenomenal foundation of the slogan: honey is unreal and sugar is real.

This is the onset of the economics of tangibles, as shown in Figure 4.1. As processing is done, the quality of the product is decreased (along the HSSA syndrome). Yet, this process is called value addition in the economic sense. The price, which should be proportional to the value, in fact, goes up inversely proportional to the real value (opposite to the perceived value, as promoted through advertisements). Here, the value is fabricated, similar to what is done in the aphenomenal model that uses the "All Americans speak French," syllogism discussed in the previous section. The fabricated value is made synonymous with real value or quality (as proclaimed by advertisements), without any further discussion of what constitutes quality. This perverts the entire value addition concept and falsifies the true economics of commodity (Zatzman and Islam, 2007b). Only recently, the science behind this disinformation has begun to surface (Shapiro *et al.*, 2006).

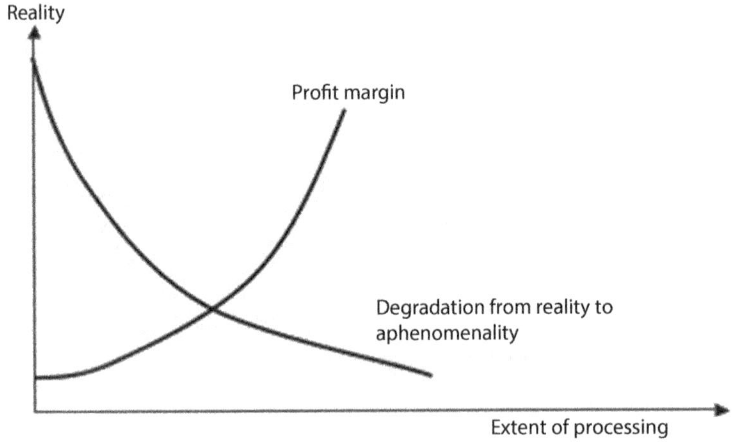

Figure 4.1 The profit margin increased radically with external processing.

The second paragraph deals with the composition of honey, as seen with the science of tangibles. Honey is reduced to "primarily sugars and water". The declaration that honey is to be made out of sugars, harbors within it the false premise that "chemicals are chemicals" (Khan and Islam, 2012). Based on this premise, mass production of practically all natural products has been affected only to discover, after many decades, that by mass-producing natural chemicals, the nature of the mass-produced products has been reversed.

Instead of conducting research on the true nature of honey, the website perpetrates the same misconception, while focusing on numbers (a tangible expression). Note the use of decimal points, signaling that this is The Composition of honey. Now, if the premise is true, why should honey behave any differently from any other syrupy substance that has the same composition, but is totally manufactured from different sources (e.g., synthetic fructose)?

Information has been lost by making the simple statement that "honey is primarily sugar and water". In terms of the discourse this chapter uses, this is equivalent to removing two components of a process, namely, 1) origin (or source); and 2) pathway (the continuous time function). Once these two are removed, the differences between true and false, natural and artificial, and good and bad have become obscured.

Several consequences follow this science of tangibles. The most important one is that the true nature of honey will not be investigated. The available data have been already filtered and tampered with

Table 4.2 Carbohydrate Composition of Honey NHB, (2007a).

Carbohydrates In Honey	Average	Range	Standard Deviation
Fructose (%)	38.38	30.91-44.26	1.770
Glucose (%)	30.31	22.89-40.75	3.040
Reducing Sugars (%)	76.65	61.39-83.72	2.760
Sucrose (%)	1.31	0.25-7.57	0.870
Fructose/Glucose Ratio	1.23	0.76-1.86	0.126

inherently prejudicial thought-processes that can only "verify" the false premise.

In keeping up with this obsession with number, the US National Honey Board NHB, (2007a) lists the composition, with standard deviations (see Table 4.2). While it is widely recognized that honey has numerous other chemicals, most of which are unknown (USDA, 1962), after this line of characterization of honey, the role of these 'unknowns' cannot be evaluated.

This process is further obscured by yet another post-Renaissance misconception, "whatever cannot be seen, does not exist" (Khan and Islam, 2007b), which is similar to the principle of "dilution is the solution to pollution" that has governed both regulatory agencies and other environmental industries in the post-Renaissance world.

So: if there are "other components" in honey that fall below the authorized level, they must not have any role in the entire pathway of honey. This will explain why honey from flowers that are grown with organic fertilizers and the one that are grown with synthetic fertilizers will, by these lights, be deemed equivalent. With the same reasoning, no impacts of any genetically modified crop can be assessed. Note that the same reasoning has allowed scientists to introduce numerous catalysts that are inherently toxic. The justification is in the premise that "catalysts only enhance a reaction, they do not participate".

A further review of Table 4.1 now will indicate how every item on the right-hand side is actually a matter of definition and a false premise. Sugar is not simple. Its composition is not static (why else would there be an expiration date on a product?). It is not predictable (who would have predicted 50 years ago that sugar would be the second most important cause of mortality in the USA). It is not unique. It is not

Table 4.3 True difference between sustainable and unsustainable processes (Reproduced from Khan and Islam, 2012).

Sustainable (Natural)	**Unsustainable (Artificial)**
Progressive/youth measured by the rate of change Non-progressive/resists change	Conservative/youth measured by departure from natural state
Unlimited adaptability and flexibility	Zero-adaptability and inflexible
Increasingly self evident with time	Increasingly difficult to cover up aphenomenal source
100% efficient	Efficiency approaches zero as processing is increased
Can never be proven to be unsustainable	Unsustainability unravels itself with time

symmetric, and the list continues. The only reason sugar is promoted to be following the right-hand side of Table 4.1 is that it can then be mass-produced and effectively used to replace whatever lies on the left-hand side, so that the profit margin is increased. Whoever became a billionaire selling honey?

If one considers the features of artificial products in Table 4.1 with those of Table 4.3, it becomes clear that any science that would "prove" the features (based on a false premise) in Table 4.1 is inherently spurious. However, the science of tangibles does exactly that and discards all natural processes as "pseudoscience", "conspiracy theory", etc. This also shows that the current engineering practices that rely on false premises are inherently unsustainable.

The case in point can be derived from any theories or "laws" advanced by Bernoulli, Newton (regarding gravity, calculus, motion, viscosity), Dalton, Boyle, Charles, Lavoisier, Kelvin, Poiseuille, Gibbs, Helmholz, Planck and a number of others who served as the pioneers of modern science. Each of their theories and laws had in common the first assumption that would not exist in nature, either in content (tangible) or in process (intangible).

4.3.2 CCD in Relation to Science of Tangibles

The race for finding the cause of CCD and the motivation behind it became evident soon after the alarm was raised (NHB, 2007b). Numerous funding opportunities were entertained and over following years many research findings were reported, ranging from creating super bee to robotic bees. In terms of finding the cause of CCD, the immediate suspect is some "mystery ailment" (Armas, 2007). This is quite characteristic of the science of tangibles that attempts to pinpoint a "suspect: and proposes counter measures to annihilate the suspect, without considering the long-term impact to the ecosystem. Not surprisingly, nothing substantial came out in terms of the cause of CCD. Ever since the alarm was raised in 2006, numerous publications came out but none provided one with a scientific model that would explain the CCD phenomena, let alone finding a remedy.. USDA reported the following (USDA, 2014):

> Scientists are looking in four general categories for the cause/ causes of CCD:A honey bee being inoculated with Nosema to determine bee infection rates and immune responses.
>
> Pathogens: Among others, scientists are considering Nosema (a pathogenic gut fungi), Israeli Acute Paralysis Virus, and possibly unknown pathogens as possible culprits in CCD. ARS research has indicated that no one pathogen of any class directly correlates with the majority of CCD incidents. Rather, a higher total pathogen load of viruses and bacteria correlates more directly with CCD than any one specific pathogen.
>
> Parasites: Varroa mites are often found in honey bee colonies that are affected by CCD. It is not known if the Varroa mites are directly involved or if the viruses that Varroa mites transmit (similar to the way mosquitoes transmit the malaria virus) are a factor in causing CCD.
>
> Management stressors: Among the management stressors that are possible contributors to CCD are poor nutrition due to apiary overcrowding and increased migratory stress brought on by the honey bees being transported to multiple locations across the country.
>
> Environmental stressors: Such stressors include the impact of pollen/nectar scarcity, lack of diversity in nectar/pollen, availability of only pollen/nectar with low nutritional value, and limited access to water or access only to contaminated water. Stressors also

include accidental or intentional exposure to pesticides at lethal or sub-lethal levels.

A survey of honey bee colonies revealed no consistent pattern in pesticide levels between healthy and CCD-affected colonies when pollen, bees, and beeswax were tested for the presence of 170 pesticides. The most commonly found pesticide in that study was coumaphos, which is used to treat honey bees for Varroa mites.

The pesticide class neonicotinoids (clothianidin, thiamethoxam, and imidacloprid) has been accused of being the cause of CCD. The neonicotinoids were developed in the mid-1990s in large part because they showed reduced toxicity to honey bees, compared with previously used organophosphate and carbamate insecticides.

In 2008, Germany revoked the registration of the neonicotinoid clothianidin for use on seed corn after an incident that resulted in the die-off of hundreds of nearby honey bees colonies. Investigation into the incident revealed that the die-off was caused by a combination of factors, including the failure to use a polymer seed coating known as a "sticker": weather conditions that resulted in late planting of corn while nearby canola crops were in bloom, attracting honey bees; use of a particular type of air-driven equipment used to sow the seeds, which blew clothianidin-laden dust off the seeds and into the air as the seeds were ejected from the machine into the ground; dry and windy conditions at the time of planting, which blew the dust into the nearby canola fields where honey bees were foraging; and a higher application rate than had been authorized was used to treat for a severe root worm infestation.

Several studies that reported a negative impact on honey bees by neonicotinoids relied on large, unrealistic doses and gave bees no other choice for pollen, and therefore did not reflect risk to honey bees under real world conditions. Nor have the studies demonstrated a direct connection or correlation to CCD.

ARS entomologist Jay Evans inspects a comb of honey bees. There have been scientific findings that imply that neonicotinoids have sublethal effects on honey bees at approved doses and exposures. ARS scientists and other researchers are looking into whether such sublethal effects may correlate with CCD or other bee health problems and whether they could be a contributing cause of CCD.

ARS held a workshop with the U.S. Environmental Protection Agency in early 2010 to discuss how potential sublethal effects could bc documented summarized in Pesticide Risk Assessment for Pollinators: Summary of a SETAC Pellston Workshop.

ARS researchers also have been analyzing samples from healthy and CCD-struck colonies and applying a variety of stressors from the four categories of possible causes to colonies in hopes of provoking a colony response that duplicates CCD.

While a number of potential causes have been championed by a variety of researchers and interest groups, none of them have stood up to detailed scrutiny. Every time a claim is made of finding a "smoking gun," further investigation has not been able to make the leap from a correlation to cause-and-effect. Other times, not even a scientific correlation has been demonstrated in the study claiming to have found "the cause" of CCD.

Researchers have concluded that no one factor is the cause of CCD. Most likely, CCD is caused by multiple factors. It is not possible to know at this time if all CCD incidents are due to the same set of factors or if the factors follow the same sequence in every case.

Two honey bees on a comb. One explanation for CCD being studied is that a perfect storm of environmental stresses may unexpectedly weaken colonies, leading to collapse when the colonies are exposed to the additional stress of a pathogen, parasite, and/or pesticide. Stress, in general, compromises the immune system of bees (and other social insects) and may disrupt their social system, making colonies more susceptible to disease.

Studies are being conducted by ARS scientists and collaborators to look at the combined impact of two or more factors on honey bees—most recently the impact of exposure to the neonicotinoid imidacloprid and Nosema. While the dual exposure indicated some sublethal effects on individual honey bees, the overall health of the colony did not show an adverse effect.

The focus of USDA is clearly away from taking a comprehensive approach.

This recent phenomenon of declining honey bees across the world, especially in North America has drawn a considerable attention to researchers, industrialists and ecologists, among others. This has been classified as a great mystery threatening the future of ecosystems crucial to farming (Hainey, 2007). As bees have the most important role for pollination, a decline in the number of bees would significantly affect the agriculture sector. Some reports indicated that thousands of honey bees have already vanished in Scotland. There are also reports showing the mysterious loss of entire hives in the United States, Italy, Spain, Greece, Poland, Portugal and France, among others. Some incidences have also been seen in the developing countries of Asia.

According to Hainey (2007), some bee keepers in the U.S. have reported the disappearance of up to 95% of their bees. Other commercial bee keepers in 22 states in the U.S., have reported the loss of tens of thousands. Each winter, U.S. bee keepers lose about 25% of their bees compared to 5-10% lossws some 35% years ago. Klein and Kremen (2006) carried out a study and reported that pollinators such as bees, birds and bats affect 35 percent of the world's crop production, increasing the output of 87 of the leading food crops worldwide. Out of 115 globally-produced crops, studied from some 200 countries, one-third depend on some degree of pollination. The report estimated that it takes about 1.4 million colonies of honey bees to pollinate 550,000 acres of California's almond trees alone. The report further indicated that the honey bees in North America are declining due to infestations of parasitic mites, and are battling antibiotic-resistant pathogens., the bees are battling antibiotoic-resistant pathogens?, The use of synthetic fertilizers during farming could be one of the many reasons that affect the populations of bees.

More than three quarters of 240,000 species of the world's flowering plants depend on pollinator insects, birds, bats and animals to various degrees for pollen to be transferred from the male to the female parts of flowers for reproduction (SPNA, 2007). Pollinators are not only vital to agriculture but to maintain the whole biodiversity globally. Pollinators are very important as most of the fruits, vegetables, and other crops providing fibers, drugs and fruits are pollinated by birds, insects and animals. Honey bees play a key role as pollinators in most parts of the world. Hence, a decline in honey bees eventually will pose an ecological disaster globally. The National Academics Report (2006), indicated that there is direct evidence of the decline of some pollinators, including honey bees in North America.

Sometime in the last century, Albert Einstein was famously quoted as saying: "If the bee disappeared off the surface of the globe then man would only have four years of life left. No more bees, no more pollination, no more plants, no more animals, no more man." (Häfeker, 2005). If the author's name were withheld (but not his physics background) or the crisis of CCD were not the burning topic, today's scientific community would have remarked, "This is pseudoscience, he should simply talk about probability and not some negative assertions", "He is not a biologist", "He is not an ecologist", "totally hypothetical, all bees will never disappear", and numerous other comments marginalizing the statement as "utter nonsense".

Because it is Einstein, and because the rate of bees disappearing is "real" (meaning measurable with tangible means), alarms are sounding everywhere. However, it is only a matter of time that the experts will begin to say, "Because we cannot find a 'significant' link between phenomenon and this effect, there is nothing that can be done." Of course, it won't be stated publicly and instead there will be a public outcry for funding so that "facts" can be collected, hypotheses can be "verified", tests can be "repeated", and the problem can be "solved" by proposing "counter measures". What would be absent in any future discourse is the questioning of what constitutes "facts", how a hypothesis can be "verified", what "repeating" a phenomenon means, and — most importantly—how one can measure the success of the proposed solutions (Zatzman and Islam, 2007a). Remarkably, the entire modern age is synonymous with a transition from honey to sugar to Saccharine® to Aspartame®. This transition also means more engineering leads to more revenues and more profits, even though the transition is actually a departure from real to artificial. Consider the following transition, as outlined by Zatzman (2007):

From the scientific standpoint, honey fulfils both conditions of phenomenality, namely, (1) origin; and (2) process. That is, the source of honey (nectar) is real (even if it means flowers were grown with chemical fertilizers, pesticides, or even genetic alteration). The process is real (honey bees cannot have a false intention, therefore, they are perfectly natural), even if the bees were subjected to air pollution or a sugary diet. The quality of honey can be different depending on other factors, e.g., chemical fertilizers, genetic alterations, etc., but honey remains real. None of these features are required to be recorded as per the guidelines provided by the regulatory agencies (e.g., EU-Council, 2002).

The science of tangibles is incapable of characterizing the quality of a product beyond very tangible features. For instance, only recently, the sale of "unpasteurized" honey became acceptable, at a higher price. Currently, there is no price structure to distinguish between honey that is produced by "organic" growers and the honey produced by chemical growers, who, for instance, give high sugar diets to their bees. As we "progress" from honey to sugar, the origin remains real (sugar cane or beet), but the process is tainted with artificiality, starting from electrical heating, chemical additives, bleaching, etc. Once again, the science of tangible does not offer any means of differentiating or controlling the quality degradation due to practices that are not sustainable (Chhetri and Islam, 2008). Further "progress" to Saccharin® marks the use of another real origin, but this time the original source (crude oil) is old, very old compared to the source of sugar. Even though crude oil is real, because it does come from natural processes, it is not comparable to sugar cane or beet, because "crude oil" was real as food millions of years ago as compared to sugar cane or beet that is food today. With steady-state analysis, they both will appear to be of the same quality. This steady-state analysis is the characteristic feature of the science of tangibles. This analysis has misconceptions embedded in it as outlined recently by Khan and Islam (2007b). As further processing continues, one witnesses the final transition to Aspartame®. Indeed, nothing is phenomenal about Aspartame®, as both the origin and the process are artificial. So, the overall transition from honey to Aspartame® has been from 100% phenomenal to 100% aphenomenal.

Considering this, what economic calculations are needed to justify this replacement? It becomes clear, without considering this phenomenality feature, that any talk of economics would only mean the "economics" of aphenomenality. Yet, this remains the standard of neo-classical economics (Zatzman and Islam, 2007b). Zatzman and Islam (2006) considered this aspect in the context of gas energy pricing and disclosed the science of tangibles behind the graph, depicted in 4.1.

Note that in this graph, it is impossible to quantify reality. For instance, one cannot say, the honey is 100% real (organic), as we have no way of determining, let alone guaranteeing, the complete composition of a product. Similarly, there is no way to determine what percentage of reality is lost by introducing an aphenomenal (un-natural) processing technique. Figure 4.2 shows how, during the external processing period, the profit margin is increased as the quality of the product declines. The right portion of the curves represents the bifurcation that represents continuous decline in quality (intangible) as profit margin (tangible) is increased. This bifurcation is reminiscent of the Knowledge vs. ignorance chart, presented by Khan and Islam (2012). The driving force in both these graphs is the fact that the short-term analysis (based on Δt approaching 0, or in other words: time being equal to 'right now') reverses the trends and makes the quality of the product appear to increase with increasing processing.

The case in point is a product that is marketed worldwide. It is called Tropicana Slim® (website www.tropicanaslim.com). (There is nothing particularly egregious about this product and/or its producer. It is quite typical of its kind. The same can be said for other products mentioned further *infra*. The problem addressed here is the distortion of basic scientific understanding that is used to push sales upwards.) This particular sweetener has a picture of a corn at the front of the 2.5g package. It is promoted as a health product, with the sign "nutrifood" clearly marked on the package. It also says, "low calorie sweetener for your coffee and tea". The back of the package outlines how low the calories are. It is only 10 calories per sachet of 2.5 g. Even though actual calorie content or the basis of this calculation means little to general consumers, this slogan of "nutrifood" along with "low calorie sweetener" gives the impression that the quality of the product is high. To reinforce that perception, the following statement is added: "No sugar, no cyclamate, no saccharine, no preservatives".

Even as it meticulously outlines what it does not contain, the package doesn't actually say what the contents are. One has to go to the website to find out what the package actually contains. Its ingredients are: Sorbitol (46.2 g/package) and Aspartame (40 mg/packet). To a non-researcher, this information means little and one would be hard pressed to rely on anything more than slogans that one easily comprehends, such as, "sugar substitute for weight control and diabetic diets. It has a low calorie sugar substitute to keep you healthy and slim. It is the real taste of corn sugar. There is also some "advice for using Tropicana Slim", such as, 1.The maximum daily of Aspartame is 40 mg/kg (does anyone keep count of aspartame consumed per kg?); 2. The Aspartame loses its taste in high temperature (so much for use with hot coffee and tea!); 3. It is not healthy for people who have Phenylketonurics because it is containing Phenylalanine (That would imply: This is healthy for those not ill with this disease. It also acknowledges substances other than the two ingredients mentioned). Then the website gives a long list of matters that the product does not have. They are: 1. sugar; 2. Cyclamate; 3. Saccharin; 4. Preservatives; 5. Sodium; 6. Fat; 7. Protein.

Now, it is known that on per kg basis, this product will sell at a price 10 times higher than sugar that is locally available. Contrast this product with another product for sale. It is called, Sugar Not®. The content of this product is shown in Figure 4.3. If a comparison of this product were to be made on the basis of sugar, this product would fare worse than the previous product. If the comparison basis is calorie (low calorie being better), this product would be seen as infinitely better than the other one (10 calorie/ 0 calorie = ∞). Any other basis for comparisons would become spurious, because they would have zero divided by zero.

275-0086 Dixie Sugar 2.82 oz			Diner Not
Nutrition Facts Serving Size: 1 Gm (1/5 tsp) Servings Per Container: 80			
	Amount Per Serving	% Daily Value*	
Total Calories	0		
Calories From Fat	0		
Total Fat	0	g	0%
Saturated Fat	0	g	0%
Cholesterol	0	mg	0%
Sodium	0	mg	0%
Total Carbohydrates	<1	g	0%
Dietary Fiber	0	g	0%
Sugars	<1	g	
Protein	0	g	

*Percent Daily Values are based on a 2,000 calorie diet.

Store in a cool, dry place.

Ingredients: Low glycemic fructose, all-natural Lo Han fruit concentrate, silica, No saccharin. No Aspartame. No chemicals.

Recommended Use: Sugar does not tastejust like sugar but is 15 times sweeter than table sugar. 1/5th measuring teaspoon is equal to 3 teaspoons of sugar.

Figure 4.2 The claim of "all-natural" is made, but there is no way to verify the claim.

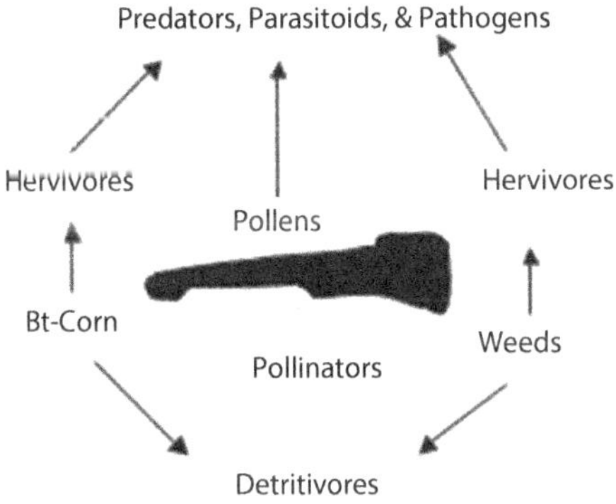

Figure 4.3 The role of genetically-engineered insecticidal Bt-corn in altering the ecosystem (redrawn from Losey *et al.*, 2004).

Other uncertainties arise from the fact that it is not explicit how the Sugar Not® product is actually extracted. This is also complicated by the misconception that "chemicals are chemicals", in which case fructose from natural origins is the same as that of artificial origin.

Overall, the confusion in deciding which product is better for consumers becomes quite arbitrary. Similar confusion exists in every product that is manufactured and marketed in the modern age. This confusion is deliberate as evidenced by numerous "scientific papers" that routinely promote disinformation (Zatzman and Islam, 2007a; Shapiro *et al.*, 2007). Examples of this disinformation are available in practically all product-oriented research.

As an example, the following quote is provided from Lähateenmäkia *et al.* (2002). Note that the paper is investigating how to overcome negative responses to genetically modified products. The title itself is quite telling: "Acceptability of genetically modified cheese presented as real product alternative". The focus here, as usual, is not to investigate the real effect of such products. Instead, the focus is on how to alter the taste and how to misinform so that "curious" consumers are persuaded to buy an artificial product that is packaged as real. The paper is funded by a consortium of industries:

"This study shows that, although consumers have an overall negative attitude towards gene technology, this negative attitude does not cause consumers to reject genetically modified options if they are presented as real product alternatives. Taste benefits seems to be a strong promoter for genetically modified products, and health benefits are also perceived positively. Consumers with the most positive attitudes are curious about gm-products and exposure to these products also influenced responses of those with the most negative attitudes towards gene technology."

In the context of the CCD, it is becoming clear that for the first time, such syndromes are being attributed to practically all gadgets that have glorified much of the modern age. In the Honey-Sugar–Saccharine-Aspartame degradation, the science of tangibles make it impossible to show the value of natural products, all the while obscuring any surrounding science that would make it possible to determine the cause of the disappearance of bees. In the words of Diana Cox-Foster, a member of the CCD Working Group, which was recently formed in order to investigate CCD, "It is particularly worrisome that the bees' death is accompanied by a set of symptoms which do not seem to match anything in the literature." This comment is in line with our assertion that the science of tangibles cannot account for the cause of CCD.

4.3.3 Possible Causes of CCD

The cause of decline of honey bees has not yet been determined. Several researchers have opined different possible causes for the disappearance of honey bees. Following is a discussion on various causes that are currently being discussed, albeit with the science of tangibles.

4.3.3.1 Genetically Engineered Crops

Genetic engineering (GE) represents the same type of degradation as outlined earlier in the context of the HSSA syndrome. While genetic engineering has increased tangible gains, in terms of crop yield, external appeal of the crop (symmetry, gloss, and other external features), it has also added potential fatal unavoidable side effects. In the context of honey bees, the most important impact of GE is through direct contact of genetically-altered crops (including pollen) and through plant-produced matters (including even organic pesticides and fertilizers). A series of scholarly publications have studied the effects of GE products on honey bees. Malone and Pham-Delègue (2001) studied the effects of transgenic products on both honey bees and bumble bees. Obrycki *et al.* (2001) studied genetically engineered insecticidal corn that might have severe impacts on the ecosystem. A comprehensive report was produced by Pham-Delègue *et al.* (2002), who attempted to quantify the impacts of genetically modified plants on honey bees. Similarly, Picard-Nioi *et al.* (1997), reported the impacts of proteins used in genetically engineered plants on honey bees. The need for including non-target living objects was highlighted by Losey *et al.* (2004). Figure 4.3 shows how various functional groups are linked to one another, accounting for an overall balance of the ecosystem. This figure shows clearly, that in nature, it is impossible to "engineer" nature and expect to keep track of its impacts. This is particularly true because of the special features of nature that do not allow any finite boundaries to be drawn. Even though mass balance, the most fundamental engineering equation, is based on a finite volume enclosed inside a boundary, this first premise of mass balance equation is clearly aphenomenal.

It is true that genetic engineering activities have been carried out at a pace unprecedented for any other technology. This subject has also been hailed to have made the most significant breakthroughs. Unfortunately, these "breakthroughs" only bear fruit in the very short term, within which period, the impacts of these technologies do not manifest in a measurable (tangible expression) fashion. Even though there is a general recognition that there are "unintended consequences", the science behind this engineering has never been challenged. Often these "unintended consequences" are incorrectly attributed to lack of precision, particularly in placing the location of the DNA in the new chromosome site. The correct recognition would be that it is impossible to engineer the new location of the gene and at the same time it is impossible to predict the consequences of the DNA transfer without knowing all possible sites that the DNA will travel throughout the time domain. This simple observation was made by Khan (2006) who contended that unless the consequences are known for the time duration of infinity, an engineering practice cannot be considered to be sustainable. Similar, but not as bold, statements were previously made by Schubert (2005) who questioned the validity of our understanding of genetic engineering technology and recognized the unpredictability of the artificial gene. Zatzman and Islam (2007a) recognized that an "artificial" object, as it comes to reality by its mere presence, behaves totally differently than the object it was supposedly emulating. This would explain why Vitamin C acts differently depending on its origin (e.g., organic or synthetic) as does every other artificial product, including antibiotics (Chhetri *et al.*, 2008; Chhetri and Islam, 2008). However, not uncharacteristically, a number of papers have appeared that deny any connection between GE and CCD (Dively, 2007). This is entirely expected. The nature of natural phenomena is such that there cannot be a single cause behind a phenomenon. As a consequence, each cause is a suspect, but with the science of tangibles, the suspicion can be buried within statistical uncertainties, depending on which agency is paying for the study. This is beyond the conventionally accepted conflict between corporate profit and public access to information (Makeig, 2002). It is rather about creating disinformation in order to increase profit margin, as noted earlier (Lähateenmäkia *et al.*, 2002).

4.3.3.2 "Foreign Elements"

The National Academics Report (2006) reported that one of the causes of the decline of bees in North America is the introduction of parasites carried out by bumblebees imported from Europe for greenhouse pollination. The other reason considered for bees' disappearance is caused by a fungal pathogen. Higes and Martin (2006) reported a new type of bacteria that was found in honey bees in Europe. A typical problem with this study is that the source of the parasite or the pathway that it followed is unknown. Also, this study does not explain such an elevated rate of CCD, particularly because the presence of this parasite is not widespread, even in the most affected areas. However, this report does indicate a very rapid collapse of a bee colony (within eight days).

Some evidence surfaced in recent days that a microbe that used to affect only Asian bees has evolved into a strain that can affect bees in Europe and in the United States (Ritter, 2001). The scientist who discovered the SARS virus in 2003 and later won a MacArthur Foundation "genius grant", asserted that parasitic elements are the culprit (Russel, 2007).

This is a simplistic conclusion of a serious problem. It is generally understood that despite media attention (Russel, 2007; Chong and Maugh, 2007), no link has been found between this strain of bacteria with CCD. Before Nosema is even linked with CCD, already the possibility of a remedy is being proposed, in the form of an antibiotic. It is called Fumagillin (Figure 4.4). At present, there is no means of determining the quality of honey, let alone the quality of honey bees, once they have been exposed to these antibiotics which is similar to what happens when bees are fed with high sucrose syrup, instead of honey.

Conventional science relies on "no distinction" between honey and sucrose, other than their caloric contents. Chhetri and Islam (2008) reported the inherent dangers of antibiotics, which, similar to all other artificial products, provide only a short-term remedy, while affecting the long-term more adversely than the disease that is being treated. The

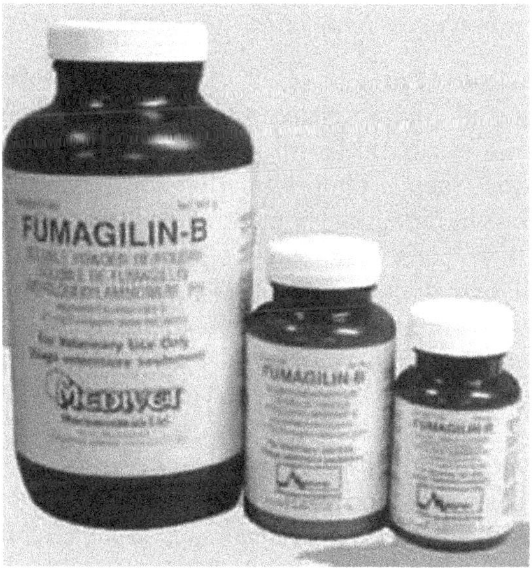

Figure 4.4 Antibiotic for bees (Blossomland, 2007).

same rule applies to antibiotics for honey bees. The long-term impact of these antibiotics is immeasurable using the science of tangibles.

Ingram *et al.* (1996) reported that a large number of insecticides used in agriculture are toxic to pollinating insects, including honey bees. Honey bees and other pollinating insects are highly vulnerable to organophosphate pesticides. Pesticides can affect the pollinator insects even at much further distances than their applications because of drift and other weather conditions. It is considered that when pesticides are applied in agricultural fields and forests by aircrafts, 50% to 75% of the chemicals sprayed can miss their targets, which leads to inadvertent exposure of non-target organismsm, such as pollinators Ingram *et al.* (1996).

According to Cheng and Cutkomp (1972), the adenosine triphosphatase (ATPase) system in worker honey bee brains carried out in vitro test showed an increased inhibition of activities of Mg^{2+}ATPase, Na^{+}ATPase and K^{+}ATPase by the use of 10 μm. DDT was between 40-50 and 20 % respcctively. A higher impact to DDT was reported in brains of older honey bees.

4.3.3.3 Electromagnetic Irradiation

Lean and Shawcross (2007) reported that the mobile phone could be one of the causes of colony collapse disorder syndrome in the mysterious disappearance of bees in North America. It has been suspected that the radiation from mobile phones interferes with bees' navigation systems, preventing the famously home-loving species from finding their way back to their hives. The same report referred to that German research that has long shown that bees' behavior changes near power lines. Similar to a once-popular research theme linking cell phones to brain cancer and numerous other ailments, the role of electromagnetic irradiation has been quickly downplayed (Johnson, 2007a; Sylvers, 2007).

The health effects of cell phone radiation exposure are one of the growing public concerns today. It has been accepted that Radio Frequency (RF)/ Microwave (MW) radiation is capable of inducing DNA damage in human cells (Aitken *et al.*, 2005). Weinberger, and Richter (2002) reported that frequent headaches and other neuropsychological symptoms usually occur in users of cellular telephones and relates to emerging controversy concerning risks for brain cancer. They further reported that brain cells and tissues demodulate the cell-phones' audio frequencies from the radio frequency carrier and that the low audio frequencies in the ranges of α and β waves affect these waves and hence influence brain function.

Some laboratory studies have reported alterations in permeability of the blood brain barrier, brain electrical activity, membrane permeability, and DNA breaks (Chia *et al.*, 2000). Von Klitzing (1995) reported that EEG brain α waves (8–14 cycles per second) and β waves (14–60 cycles per second) are altered after being exposed to cellular phone signals at field strengths as low as 0.1 uW/cm2. Zhao *et al.*, (2007) reported that even relatively short-term exposure to cell phone radiofrequency emissions can up-regulate elements of apoptotic pathways in cells derived from the brain, and that neurons appear to be more sensitive to this effect than astrocytes. If this is the effect on humans, the effects on honey bees that constantly travel around ubiquitous microwave signals would seem to cry out for investigation. The facts, however, are that certain already highly suggestive research findings, have been very capably misrepresented in a manner tending to demean — as the hobbyhorse of conspiracy theorists — the very suggestion of a need for further such research.

In April 2007, news of a University of Landau study (Stever *et al.*, 2006) appeared in major media, beginning with an article in *The Independent* [London, UK] that stated that the subject of the study was "mobile phones" (a term that mixes cellular with cordless phones). This media coverage somewhat sloppily related the study's results to CCD. However, one of the leaders of the study group, Dr Jochen Kühn, rescued the research from this slight discredit by repudiating the connection to CCD which was never an object of their work, and then going on to explain to the German news magazine of *Der Spiegel*, the actual connections suggested by the group's research findings between the decline in return of bees to their hives in the presence of electromagnetic radiation (EMR) from cordless phones.

Specifically, a 2006 University of Koblenz-Landau pilot study (Stever *et al.*, 2006), was looking for non-thermal effects of radio frequency ("RF") on honey bees (Apis mellifera carnica) and suggested that when bee hives have DECT cordless phone base stations embedded in them, the close-range electromagnetic field ("EMF") may reduce the ability of bees to return to their hives; they also noticed a slight reduction in honeycomb weight in treated colonies. In the course of their study, one half of their colonies broke down, including some of their controls which did not have DECT base stations embedded in them.

The group's actual focus is learning behavior, and electro-magnetic radiation was being investigated with regard to impacts in this broad area. Their earlier exploratory study (Stever *et al.*, 2005) on non-thermal effects on learning did not find any changes in behavior due to RF exposure from the DECT base station operating at 1880-1900 MHz.

Many possible biological effects of non-ionizing electromagnetic fields have been postulated, but it is generally accepted that the most significant effects are thermal. The amount of RF radiation routinely encountered by the general public is too low to produce significant heating or increased body temperatures.

Sharma and Kumar (2010) conducted a comprehensive study on the effect of cell phones on honey bees. They put two cellphones each in two hives for three months. The phones were turned on–thus the honeybees exposed to the electromagnetic radiation from the phones–two days out of each week, twice on each of those days. Each exposure period lasted 15 minutes, and was timed to coincide with the bees' peak period of activity, at 11 a.m. and again at 3 p.m. The experiment took place from February to April, and covered two brood cycles.

These honeybees were not exposed to daily radiation, as was misreported by CNN. The cell phones were turned on only twice a week, for

2 fifteen-minute periods per each of those days, for a total of one hour's exposure per week. A third hive had two dummy cellphones, thus no radiation from them, and a fourth had neither real nor dummy cellphones. These were the two control hives.

The study looked at both biological and behavioral differences between the bees which were exposed to radiation and those which were not exposed. The biological aspects measured were brood area (this refers to the space within the hive occupied by eggs, larva, and pupae) and the egg laying rate of the queen.

Behavioral aspects measured included colony growth (total quantity of bees, quantity of honey stores, and quantity of pollen stores in the hives) and foraging habits (efficiency and rate of activity, as well as rate of return to the hive).

Kumar (2010) wrote a comprehensive report on electromagnetic radiation and its effect on living organisms. In particular, the effect on honey bees was highlighted. Followings are the major findings:

- Bee strength: There were 9 comb frames in the unexposed colony, compared to only 5 comb frames in the colony exposed to cell phone radiation.
- Area under brood: Control colony, 1975.44 cm^2; exposed colony, 760.19 cm^2
- Egg laying rate of queen: Unexposed queen: 376.2 eggs per day; queen exposed to cell phone radiation: 144.8 eggs per day

There was a dramatic decrease in the number of bees returning to the hive exposed to cell phone radiation. By the end of the three months that the experiment lasted, "there was neither honey, nor pollen or brood and bees in the colony."

When honey bee colonies were exposed with radiation, the honeycomb weight and area were reduced and returning time of honey bees increased compared to similar non-exposed colonies.

Several other studies show that the high-frequency electromagnetic fields of mobile phones alter the resonant stimulus of living organisms and can cause modifications in certain areas of their brain. Changes in the brain structure of bees can be a cause of alterations of the returning capabilities of bees.

The honey bee's brain anatomy as well as the learning regions of the bee brain are well known and comparable to those of vertebrates and are well suited as a bio-indicator. If we can determine the damage caused by electromagnetic radiation on honey bees, it would explain to us how human brains are affected. Because human brains are like the motherboard for the human body, the damage to the brain can explain many ailments. However, we need a comprehensive theory that connects energy waves with mass. Such theory has been non-existent until recently (Islam *et al.*, 2014). It is also important to note, damage to the brain would invariably accompany cognitive disorders, including so-called 'mental diseases' that can be triggered with physical damage to the brain cells due to energy waves. Such connection has been discussed by Islam *et al.* (2014a, 2014b).

If indeed electromagnetic radiation is a major contributor to CCD, it would mean countries that do not have well developed networks would not suffer from CCD. This has been the case in most third world countries (Kumar, 2010).

Ever since the publication of papers by Sharma and Kumar (2010) on linking CCD and cell phone, there have been numerous arguments both in favor and against this theory. Favre (2010) published a comprehensive paper on the link of cell phones and CCD. He studied electromagnetic waves originating from mobile phones for potential effects on honeybee behavior. Mobile phone handsets were placed in the close vicinity of honeybees. The sound made by the bees was recorded and analyzed. The audiograms and spectrograms revealed that active mobile phone handsets have a dramatic impact on the behavior of the bees, namely by inducing the worker piping signal. In natural conditions, worker piping either announces the swarming process of the bee colony or is a signal of a disturbed bee colony.

Figure 4.5 shows some results. In this case reported by Favre (2010), a standby cell phone was activated 25 minutes after the onset of experiment (marked 1 in Figure 4.5a). The beginning of increased noise and frequency in the hive was observed 35 minutes later (marked 2 in Figure 4.5a). The cessation of the mobile phone communication is indicated as point 3 in Figure 4.5a. Figure 4.5b shows response to prolonged (20 hours) active cell phone.

This study found that in the presence of actively communicating cell-phones (those not in standby mode), bees produced the sounds known as "worker piping," which tends to indicate disturbance in a bee colony. Worker piping in a bee colony is not frequent, and when it occurs in a colony, that is not in a swarming process, no more than two bees are simultaneously active...The induction of honeybee worker piping by the electromagnetic fields of mobile phones might have dramatic consequences in terms of colony losses due to unexpected swarming.

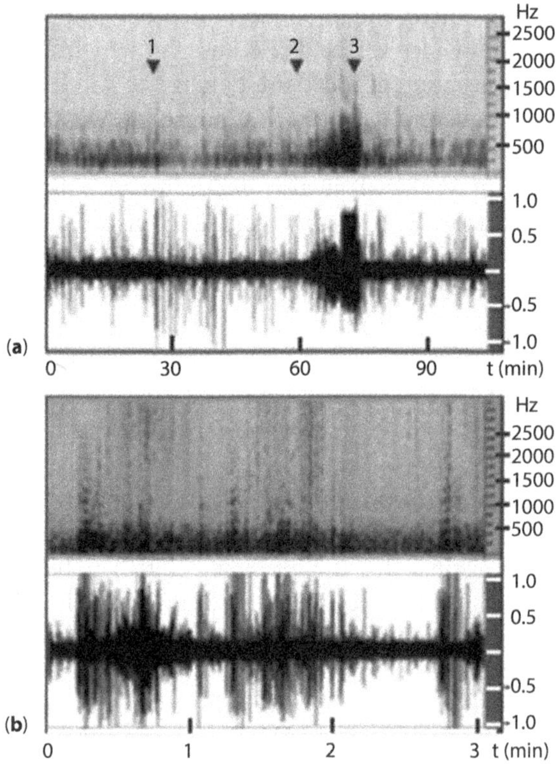

Figure 4.5 Induction of honebee worker piping by mobile phone handsets.

Soon after Favre's discovery, news items started to circulate with titles: "Cellphones Not Major Force Behind Worldwide Decline in Bees, Expert Says". This time, there was no need for research, there were only twists and disinformation. For instance, Dennis vanEngelsdorp of Penn State University's College of Agricultural Sciences said that while the Favre (2010) study may show that a cellphone placed artificially close to bees could harm the colony, it does not mean that cellphones are responsible for the unusually large losses of bees documented in recent years. "If cellphones were so responsible and so widespread, I think we'd see more consistent losses," he said.

Despite overwhelming evidence that cell phone and electromagnetic waves are harmful to the navigation system of honey bees, US government sites continue to assert there is no connection between the two. It is similar to the alarm raised in mid 2000 that microwave heating was responsible for destroying nutrients in vegetables and rendering protein into carcinogens. For instance, US government site says:

4.3.3.4 *Cell Phones and CCD*

Despite a great deal of attention having been paid to the idea, neither cell phones nor cell phone towers have been shown to have any connection to CCD or poor honey bee health. Originally, the idea was provoked by the media making a connection between CCD and a very small study done in Germany. But that study looked at whether a particular type of base station for cordless phones could affect honey bee homing systems. However, despite all the attention that this study has received, the base station has nothing to do with CCD. Stefan Kimmel, the researcher who conducted the study and wrote the paper, e-mailed The Associated Press to say that there is "no link between our tiny little study and the CCD-phenomenon... Anything else said or written is a lie."

In addition, apiaries are often located in rural areas, where cell phone coverage can be spotty. This makes cell phones or cell towers unlikely culprits. What were the recommendations of the USDA regarding CCD? Their official site states:

> "Honey bee flies to a flower. The best action the public can take to improve honey bee survival is not to use pesticides indiscriminately. In particular, the public should avoid applying pesticides during mid-day hours, when honey bees are most likely to be out foraging for nectar and pollen on flowering plants.
>
> "In addition, the public can plant pollinator-friendly plants—-plants that are good sources of nectar and pollen such as red clover, foxglove, bee balm, joe-pye weed, and other native plants. (For more information, visit www.nappc.org.)"

4.3.3.5 *Israeli Acute Paralysis Virus (IAPV)*

Cox-Foster *et al.* (2007) carried out a comprehensive study on colony collapse disorder phenomenon. The study identified Israeli acute paralysis virus (IAPV) as a potential cause for colony collapse disorder (CCD) and showed that there is a strong correlation between IAPV and CCD. The study further linked the CCD with the importation of packaged honey bees from Australia to USA as early as 2004. However, Anderson and East (2008) argued that honey bees importations from Australia to the United States only started after 2005. On the contrary, there are reports that isolation of IAPV from specimens of Apis mellifera collected within the United States existed in 2002. Hence, it is arguable that the IAPV from Australian bees may not be the reason for CCD. Anderson and East (2008) further argued that several CCD colonies were not affected by IAPV. Moreover, the study reported that the death of bees close to the hive similar to that of IAPV in Israel was not found during the CCD disappearance. This argument seems more valid than the other countries which have CCD symptoms, such as Greenland, Poland and Spain which never imported bees from Australia. However, since this seems to be a non-linear phenomenon, IAPV could be one of the several reasons that might have some correlation with CCD, but it cannot be concluded that IAPV is the sole reason for CCD phenomenon.

Interest in IAPV, however, has been maintained in order to justify research grants. According to the most recent surveys in North America and Europe, Varroa mites still appear to be public enemy #1 for bees, at least among the factors that were measured (Currie *et al.* 2010; Dahle 2010; Guzmán-Novoa *et al.* 2010; Peterson *et al.* 2010). Varroa mites transmit viruses to bees and also can suppress the bee immune system, causing latent viral infections to amplify which results in parasitic mite syndrome. Viruses can also be transmitted in the absence of mites and can often be detected in bees that appear perfectly healthy (Hunt *et al.*, 2012). Deformed Wing Virus (DWV) is a very common virus associated with mites and colony decline. Hunt *et al.* (2012) have set up DWV research facility at Purdue University, where they have the ability to detect viruses by molecular tests. As part of the Purdue University effort for the honey bee health CAP, Hunt *et al.* (2012) conducted experiments to try to select for bees that are resistant to IAPV. They inoculated bees with IAPV in cages through feeding. IAPV is one of the viruses transmitted by Varroa and has been associated with rapid colony declines. A sample of mostly IAPV obtained from the Purdue University Bee Lab was mixed with sugar syrup and fed to bees. IAPV has a dramatic effect on the life expectancy of bees in cages.

Hunt *et al.* report results of bees from many colonies. The sugar syrup contained MegaBee protein supplement for the first 3 days. When the bees were 3 days old, half of the cages got virus in their sugar water. After these tests, they selected hives that looked relatively susceptible and others that looked more resistant and used instrumental insemination to cross a susceptible hive to another susceptible hive, and also

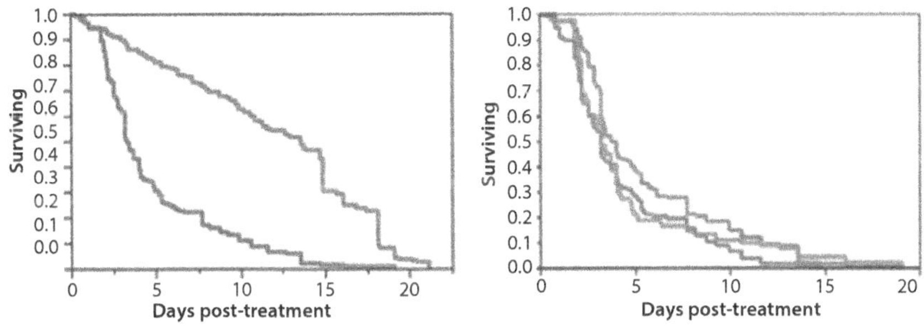

Figure 4.6 IAPV (left-hand panel) kills bees: bees kept in cages and fed plain sugar syrup (red line) survived longer than bees fed sugar syrup plus IAPV (blue). IAPV (right-hand panel): after inoculating bees with IAPV in sugar syrup, bees from two hives established from crosses between susceptible sources (green and blue) died faster than bees from the resistant x resistant cross (red). (from Hunt *et al.*, 2012)

resistants to resistants. After six months, the progeny of these crosses were tested and after averaging three different sets of tests, the resistant cross bees did live a little longer than the susceptible (Figure 4.6.)

4.4 Incurable Disease

It is often argued that there are only a few incurable diseases. However, a thorough survey indicates that no medication has real curative effect on any ailment. Modern medicines only offer delay of symptoms. This is mainly because modern medical science avoids the discussion of root causes of ailments, always focusing on externals or symptoms. Nevertheless, there is a long list incurable diseases that are easily available. Listverse (2007) makes such a list of top 10 incurable diseases and lists the following diseases in order of importance:

1. Common cold
2. Cancer
3. Asthma
4. HIV/AIDS
5. Diabetes
6. Creuzfeldt-Jakob Disease
7. Influenza
8. Lupus Erythematosus
9. Polio
10. Ebola

4.4.1 The Common Cold

The common cold is an acute viral infection that starts in the upper respiratory tract, sometimes spreads to the lower structures, and may cause secondary infections in the eyes or middle ears. Numerous agents are known to cause the common cold, including parainfluenza, influenza, respiratory syncytial viruses, and reoviruses. However, the most common one is Rhinoviruses. The popular term common cold reflects the feeling of chilliness on exposure to a cold environment that is part of the onset of symptoms. The feeling was originally believed to have a cause-and-effect relationship with the disease, but this is now known to be incorrect. The cold is caught from exposure to infected people, not from a cold environment, chilled wet feet, or drafts. People can carry the virus and communicate it without experiencing any of the symptoms themselves. Incubation is short — usually one to four days. The viruses start spreading from an infected person before the symptoms appear, and the spread reaches its peak during the symptomatic phase. Even though common cold killed millions of natives during the early period of colonization of the Americas, it is now considered to be non-life threatening. However, until today, many instances of death are reported every year and more importantly such occurrence of common cold has increased the burden on the health care system through the introduction of numerous vaccines. These vaccines themselves are stimulants of other ailments.

As influenza is caused by a variety of species and strains of viruses, in any given year some strains can die out while others create epidemics, while yet another strain can cause a pandemic. Typically, in a year's normal two flu seasons (one per hemisphere), there are between three and five million cases of severe illness and around 500,000 deaths worldwide, which can be considered as a yearly influenza epidemic. Although the incidence of influenza can vary widely between years, approximately 36,000 deaths and more than 200,000 hospitalizations are directly associated with influenza every year in the United States. On average 41,400 people died each year in the United States between 1979 and 2001 from influenza. In 2010, CDC in the United States changed the way it reports the 30 year estimates for deaths. Now they are reported as a range from a low of about 3,300 deaths to a high of 49,000 per year.

Roughly three times per century, a pandemic occurs, which infects a large proportion of the world's population and can kill tens of millions of people. One study estimated that if a strain with similar virulence to the 1918 influenza emerged today, it could kill between 50 and 80 million people (Murray *et al.*, 2006).

New influenza viruses are constantly evolving by mutation or by reassortment. Mutations can cause small changes in the hemagglutinin and neuraminidase antigens on the surface of the virus. This is called antigenic drift, which slowly creates an increasing variety of strains until one evolves that can infect people who are immune to the pre-existing strains. This new variant then replaces the older strains as it rapidly sweeps through the human population, often causing an epidemic. However, since the strains produced by drift will still be reasonably similar to the older strains, some people will still be immune to them. In contrast, when influenza viruses reassort, they acquire completely new antigens—for example by reassortment between avian strains and human strains; this is called antigenic shift. If a human influenza virus is produced that has entirely new antigens, everybody will be susceptible, and the novel influenza will spread uncontrollably, causing a pandemic (Parrish *et al.*, 2005). In contrast to this model of pandemics based on antigenic drift and shift, an alternative approach has been proposed where the periodic pandemics are produced by interactions of a fixed set of viral strains with a human population with a constantly changing set of immunities to different viral strains.

Numerous natural health research articles routinely show that the risk of pandemic increases with flu shots. However, these Epidemiological studies are usually dismissed by mainstream scientific community. Considering that the mainstream science continues to use the science of tangibles, it is no surprise that nothing gets properly defined and the issue becomes a matter of perpetual debate.

The biggest problem with the immunization industry is the fact that New science doesn't know the difference between natural and artificial and certainly doesn't acknowledge the difference between how natural vaccine (the type Dr. Jenner invented, see Riedel, 2005) works as opposed to an artificial one. Islam *et al.* (2010, 2012) and Khan and Islam (2012) demonstrated that there are fundamental differences between the two. It is clear that natural immunity is not appealing to the Establishment as it generates little profit and empowers the general public with less dependence on the pharmaceutical industry. In the United States, it is recommended that U.S. babies receive 26 doses of vaccines within the first 12 months of life (which, incidentally, is twice as many vaccinations as are given to babies in Sweden and Japan). These artificial agents interfere with natural immunity that is still developing among younger children. They build an inherent conflict with the natural "memory" system. This includes development of antibodies that become permanent agents working against natural systems. So, when a person comes in contact with a pathogene or another type of virus that is different from the immunization one, the body becomes more vulnerable and the degradation cycle continues. Here are just some of the ways vaccines can impair and alter your immune response:

- Some components in vaccines are neurotoxic and may depress one's immune response or cause brain and immune dysfunction, particularly heavy metals such as mercury preservatives (thimerosal) and aluminum adjuvants
- The lab altered vaccine viruses themselves may also affect the immune response in a negative way
- Vaccines may alter t-cell function and lead to becoming chronically ill
- Vaccines can trigger allergies or autoimmune disorders. Vaccines introduce large foreign protein molecules into a body. The body can respond to these foreign particles in a way that causes an allergic reaction or triggers autoimmunty, especially in persons genetically or biologically vulnerable to allergy and autoimmunity
- Getting a flu shot can affect one's cardiovascular system because vaccination stimulates an acute inflammatory response in one's body, which also could become chronic. One 2007 study published in the Annals of Medicine (Bodewes *et al.*, 2011) concluded that:

"Abnormalities in arterial function and LDL oxidation may persist for at least two weeks after a slight inflammatory reaction induced by influenza vaccination. These could explain in part the earlier reported increase in cardiovascular risk during the first weeks after an acute inflammatory disorder."

Picture 4.1 shows a seemingly 'over the top' picture that has some scientific merit. Today's vaccines are not like the ones Dr. Jenner once developed (against small pox). These vaccines are toxic heavy metal-infested artificial chemicals that fight with natural immune system throughout the life of person because they cannot be degraded, assimilated in the body, or purged from the body. So, why can't New science detect this simple fact? Because new science doesn't have a tool to differentiate between artificial and natural (Islam *et al.*, 2010). They believe

Picture 4.1 This anecdotal picture possesses scientific merit that New Science could never comprehend.

dogmatically that 'all chemicals are just chemicals' (once touted by Nobel Laureate Chemist Linus Pauling). This has been one of the greatest misconception of modern time. All the chemicals that are injected through a vaccine are present in nature, albeit in natural form and they represent Honey in the HSSA degradation process.

4.4.2 Cancer

Cancer refers to a group of more than 100 distinct diseases characterized by the uncontrolled growth of abnormal cells in the body. Cancer is a global epidemic. In 2008, it was estimated there were 12,332,300 cancer cases of which 5.4 million were in developed countries and 6.7 million were in developing countries (Othman, 2012). Over half of the incident cases occurred in residents of four WHO regions. The world population increased from 6.1 billion in 2000 to 6.7 billion in 2008. The increase in populations was much more in developing countries than in developed countries. Even if the age-specific rates of cancer remain constant, developing countries would have a higher cancer burden than developed countries. Cancer affects one in every three persons born in developed countries and is a major cause of sickness and death throughout the world. Cancer has been synonymous with modern age, its lifestyle, and the superflux of artificials, both in energy and mass. Figure 4.7 shows a general correlation between cancer rate and GDP. This is not to endorse an aphenomenal link between the two, but to demonstrate that lifestyle of modern time has induced overall increase in cancer.

Table 4.4 similar relationship between lifestyle of the rich with cancer death. For instance, in Africa the highest rate of cancer also goes with GDP and life expectancy in Africa. South Africa (202) has remarkably higher cancer rate compared to its neighbouring countries. For instance,

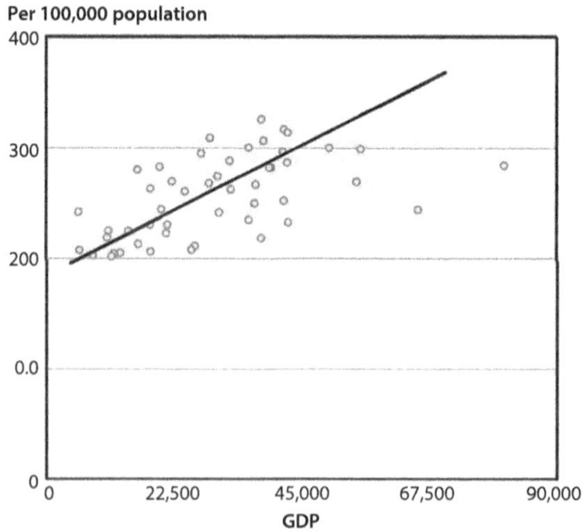

Figure 4.7 GDP and cancer rate correlation.

Botswana 85.7, Swaziland 103.1, Namibia 78.3, Lesotho 99.4, Zimbabwe 159.1, and Mozambique 123.5 all have much lower cancer rate than South Africa.

Metropolitan France has the highest rate of cancer followed by Spain. In Asia, Israel, the "capital of artificials" has the highest rate of cancer. Compare Israel: 288 with West bank & Gaza 54.9, Lebannon 169, Syria 72, Jordan 128.9, Egypt 108, Saudi Arabia 87.6, and Iraq 104.

Table 4.5 (below) shows rates of various types of cancer with various indices

Note that the definition of cancer death is such that if someone is 'cured' of cancer but dies 5 years later, the death is not recorded. Breast cancer occurrence is predominantly in richer countries.

Life expectancy correlates with rate of cancer. This paradoxical behavior can be explained with the fact that life expectancy is artificially boosted with the use of techniques and artificial chemicals that render the human body into a mechanical system on 'artificial life support', akin to what happened to late Israeli prime minister, Ariel Sharon. As discussed in Chapter 2, this life expectancy is full of desire to commit suicide and hopelessness.

It has been reported that the use of depleted uranium in Iraq has caused numerous problems, related to childbirth. Between October 1994 and October 1995, the number of birth defects per 1,000 live births in Al Basrah Maternity Hospital was 1.37. In 2003, the number of birth defects in Al Basrah Maternity Hospital was 23 per 1,000 livebirths. Within less than a decade, the occurrence of congenital birth defects increased by an astonishing 17-fold in the same hospital. While, these data are taken as facts, few researcher the use of artificial chemicals as being equivalent to exposure to depleted uranium.

Similarly, lung cancer has been unsuccessfully correlated with cigarette consumption, the top 10 countries not showing top 10 ranking in lung cancer. Following countries, for instance, have the highest rate of cigarette smoking per capita.

Armenia
Bulgaria
Greece
Russia
Maldova
Ukarane
Slovania
Bosnia and Hertzigovia

Table 4.4 Cancer and various lifestyle indices.

Countries	Cancer rate (100,000)	Life expectancy	Death rate for cancer	GDP per capita	Comment
South Africa	202	51	141	5600	Highest in Africa, high mortality
Botswana	85.7	53	63.6	6800	
Swaziland	103.1	47	81.7	2927	
Namibia	78.3	61	55.3	4017	
Lesotho	99.4	46	79.9	764	
Zimbabwe	159.1	47	130.6	355	High mortality
Zambia	140.8	47	113.8	1183	High in cervical cancer, 52.8/ 38.6
Mozambique	123.5	49	97.8	443	50.6/34.5 cervical
Guinea	111.6	53	92.7	395	56.3 /41.7cervical
Malawi	149	52	122.6	291	High mortality
Mongolia	242.2	67	185	2108	High incident and mortality; high stomach and liver cancer
Thailand	Total=150.5	74	93.6	3993	High on liver

(Continued)

Table 4.4 cont.

Countries	Cancer rate (100,000)	Life expectancy	Death rate for cancer	GDP per capita	Comment
Lao PDR	146.1	66	32.3	904	High mortality, high liver
Israel	288.3	81	102.1	27,592	Highest rate in Asia

Table 4.5 Types of cancer with lifestyle indices.

Countries	Cancer rate (100,000)	Life expectancy	Death rate for cancer	GDP per capita	Comment
Mongolia	Total=242.2 Liver=94.4 Stomach=34 Prostrate=2.4 Lung=20.8 Cervical=28.4 Breast=8	67	Total=185 liver=79.9 stomach=26.5 prostrate=1.5 lung=19.1 cervical=10.3 Breast=3.4	2108	High incident and mortality; high stomach and liver cancer
Thailand	Total=150.5 Liver=29.7	74	93.6 liver mortality=25.4	3993	High on liver
Lao PDR	Total=146.1 Liver=33.8	66	32.3 liver=32.3	904	
Vietnam	Total=138.7 liver=29.3	74	11.4 liver=27.3	1070	

(*Continued*)

Table 4.5 cont.

Countries	Cancer rate (100,000)	Life expectancy	Death rate for cancer	GDP per capita	Comment
China	Total=181 Liver=25.7 Lung=33.5 Stomach=29.9 Cervical=9.6 Breast=21.6 Prostrate=4.3	73	Total=124.6 Liver=23.7 Lung=28.7 Stomach=22.3 Cervical=4.2 Breast=5.7 Prostrate=1.8	3414	
Chinese Taipei	Total=244.1 Liver=35.7 Lung=27 Stomach=10.4 Prostrate=20.8 Cervical=10.2 Breast=52.8	Not assessed	Total=120.7 Liver=25.9 Lung=22.5 Stomach=6.7 Prostrate=5.2 Cervical=4.6 Breast=9.9		

(Continued)

Table 4.5 cont.

Countries	Cancer rate (100,000)	Life expectancy	Death rate for cancer	GDP per capita	Comment
Bangladesh	Total=124.8 Liver=3.8 Lung=19.4 Prostrate=1.9 Stomach=5.2 Breast=27.2 Cervical=29.8	68	Total=95.7 Liver=3.8 Lung=18.2 Prostrate=1.2 Stomach=5 Breast=13.6 Cervical=17.9		100% mortality in liver cancer
South Korea	Total=262 Liver=23.5 Lung=27.6 Prostrate=22.4 Stomach=41.4 Breast=38.9 Cervical=10.8	80	Total=100.5 Liver=17 Lung=21 Prostrate=4.1 Stomach=14.6 Breast=5.3 Cervical=2.7	19,028	High, very high stomach, high liver, high mortality

(Continued)

Table 4.5 cont.

Countries	Cancer rate (100,000)	Life expectancy	Death rate for cancer	GDP per capita	Comment
North Korea	Total=130.3 Liver=11.1 Lung=29.1 Prostrate=2.3 Stomach=12.1 Breast=30.5 Cervical=6.6	68	Total=92.2 Liver=10.4 Lung=26.1 Prostrate=1.4 Stomach=9.8 Breast=10.6 Cervical=3.3	Not assessed	Low cancer rate but high relative mortality
Japan	Total=201.1 Liver=11.2 Lung=24.6 Prostrate=22.7 Stomach=31.1 Breast=42.7 Cervical=9.8	83	Total=94.8 Liver=9.2 Lung=17.4 Prostrate=5 Stomach=13.5 Breast=9.2 Cervical=2.6	37,932	High rate, high mortality for liver
Israel	288.3	81	102.1	27,592	Highest rate in Asia

(Continued)

Table 4.5 cont.

Countries	Cancer rate (100,000)	Life expectancy	Death rate for cancer	GDP per capita	Comment
Iraq	Total=104.9 Liver=1 Lung=8.3 Prostrate=3.7 Stomach=3.6 Breast=31.1 Cervical=3.1	68	Total=84.2 Liver=1.1 Lung=4.9 Prostrate=5.1 Stomach=3.5 Breast=18.5 Cervical=2.3	2,867	High relative mortality
Afghanistan	Total=107.2 Liver=3.5 Lung=7 Prostrate=3.8 Stomach=15.8 Breast=21.3 Cervical=6.6	48	Total=90.7 Liver=3.5 Lung=6.6 Prostrate=3.1 Stomach=15 Breast=13.1 Cervical=4.9	367	Very high relative mortality rate

(Continued)

Belarus
Montenegro

Whereas

Table 4.6 shows the occurrence of lung cancer dominates different set of countries.

Lung cancer death (per 100,000)

1	Hungary	51.4
2	Serbia	42.4
3	Maldives	42.1
4	Poland	41.9
5	Armenia	39.3
6	Denmark	39.2
7	Netherlands	?
8	Croatia	35.7
9	United States	35.4
10	Cuba	33.8

Cancer trends are showing upward trends in many developing countries and a mixed pattern in developed countries. By 2050, the cancer burden could reach 24 million cases per year worldwide, with 17 million cases occurring in developing countries (Perkin *et al.*, 2000). Currently, there is a perception that cancers that are associated with diet and life style are seen more in developed countries while cancers that are due to infections are more in developing countries. However, if the real cause of cancer could be revealed, a different picture would emerge and the occurrence of cancer will correlate with the amount of artificial components (i.e. artificial chemicals) of our lifestyle.

According to the World Health Organization (WHO), death from cancer is expected to increase to 104% worldwide by 2020. While the number of total cancer is increasing, the trend of certain cancers is changing in developed and developing countries. In developed countries, the trend is declining (Belpomme *et al.*, 2007) since infections by microorganisms are declining and screening facilities are available. An older study (Seow *et al.*, 1996) showed that in Singapore, there was an

Table 4.6 Lung cancer occurrence (per 100,000).

Rank	Country	Age-Standardised Rate per 100,000 (World)
1	Hungary	51.6
2	Serbia	45.6
3	Korea, Democratic Republic of	44.2
4	FYR Macedonia	40.8
5	New Caledonia	40.1
6	Montenegro	39.6
7	Denmark	39.2
8	United States of America	38.4
9	Poland	38.0
10	Canada	37.9

average annual increase of 3.6% for breast cancers in women in the 1988–1992 period. Similar Bener *et al.* (2008) reported that in Qatar, there was a 57.1% rise of cancers 1991–2006, whereas and in Netherlands, there was an increase between 1.9% (females) and 3.4% (males) per year for oesophageal cancer 1989–2003 (Crane *et al.*, 2007). Overall, the data become very confusing if the big picture is not understood. For instance, there is no denying that the use of chemicals in name of modernization has been in rise in developing countries whereas the developed nations are experiencing some sort of 'going back to nature' movement with numerous people resorting to organic food and lifestyle that are practically extinct in the east.

4.4.2.1 Honey and Cancer

In Chapter 2 we showed how honey is inherently therapeutic and medicinal. In order to understand the usefulness of honey in cancer, we need to understand the various factors which could cause cancer. Carcinogenesis is a multi-step process and has multi- factorial causes. Development of cancers takes place long after initiation, promotion, and progression steps (Figure 4.8) have taken place. The cellular damage could be by one factor or multiplicity of these factors. The latter is more frequent. Cancer development could occur 10–15 years after exposure to the risk factors and closely related to continuous insult to the body.

Cancer is caused by damage in the genome of cells. While New Science assigns this damage in part to genetic defects, latter chapters will show that this conclusion's foundations includes false premises. Scientifically, all these damages are acquired through external stimuli. This is true even for infants that contract cancer, their cases being similar to children inflicted with HIV. It is the lifestyle that contributes primarily to the onset of cancer. After the onset, it is further aggravated by the 'intervention' with artificial chemicals as well as energy sources. Typically, it is said that one-third of cancer is due to tobacco use, one-third due to dietary and lifestyle factors, and one-fifth due to infections. However, it cannot explain why there is no direct correlation between cancer and cigarette smoking or why cancer (including lung) was not an issue before nicotine[1] was introduced. Also, one is unable to explain why organic wine is not carcinogen while non-organic ones are, why refined oils are carcinogen but unrefined organic ones are not. Other factors include chemical carcinogens, environmental pollutants, and alcohol (Figure 4.8). In the developing countries, cancers caused by infections by microorganisms such as cervical (by human papilloma virus) (Parkin *et al.*, 2008), liver (by hepatitis viruses) (Yuen *et al.*, 2009), nasopharynx (by Epstein-Barr virus) (Chou *et al.*, 2008), and stomach (by Helicobacter pylori) (Kuniyasu *et al.*, 2003) are more common than those in developed countries (DCP2, 2003). One should note here that bacterial or viral infection is not directly responsible for cancer's onset. Rather: such infections may aggravate the surrounding environment in a direction that renders the organism more vulnerable for the onset of cancer. Also to be noted that natural foods usually have all the necessary ingredients to combat both bacterial and viral infections. The developing countries have seen a shift toward artificial chemicals and lifestyle in an unprecedented pace and that must be the factor that contributes to the surge in cancer.

While cancers of the prostate, breasts, and colorectal are clearly more prevalent in developed than developing countries, the distinction is not very apparent as that for cancer of the lung which is as prevalent as

[1]Equivalent to artificial male hormone

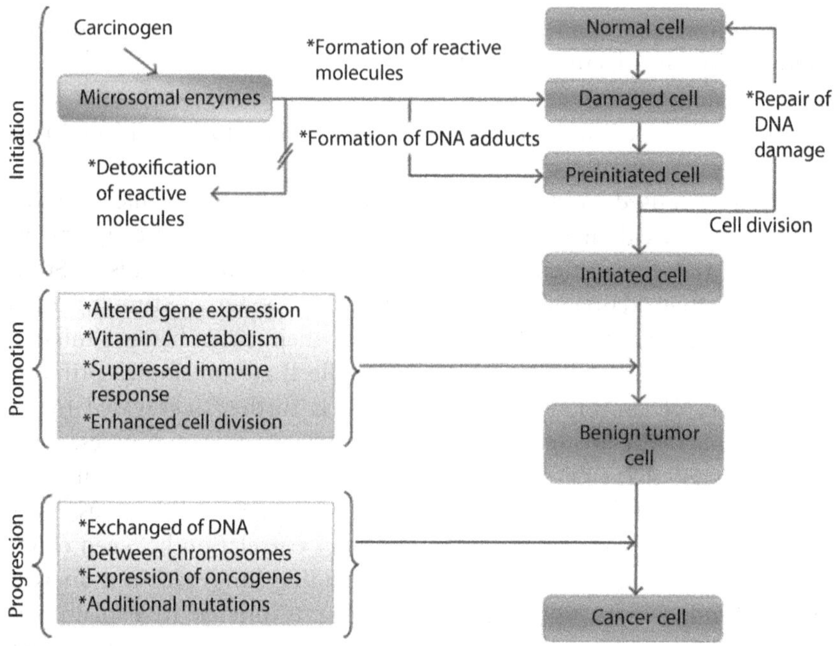

Figure 4.8 Steps in carcinogenesis. * Steps altered by alcohol consumption (Source: Garro *et al.* Alcohol Health & Research World 16(1):81–86, 1992 as reported by Othman, 2012).

that in more or less developed nations. Except for breast cancers, the top 5 cancers in males and females of developing nations are due to life-styles or infections (DCP2, 2003).

Followings are the factors that affect onset and propagation of cancer.

(i) ***Cigarette smoke***

Cigarette is associated with practically all forms of cancer. Cigarette, with its nicotine, epitomizes artificiality (both in mass and energy) and thereby can be characterized as the worst stimuli of cancer. Smoking is associated with a number of cancers such as larynx, bladder, breasts, oesophagus, and cervix. Smoking increases the risk of colorectal carcinomas by 43% (Huxley, 2007). Ever-smokers were associated with an 8.8-fold increased risk of colorectal cancers (95% confidence interval, 1.7–44.9) when fed on well-done red meat diet if they have NAT2 and CYP1A2 rapid phenotypes (Le Marchand *et al.*, 2001). No similar association was found in never-smokers. This finding gives out a valuable clue as to what triggers the onset of cancer.

(ii) ***Obesity and Physical Inactivity***

Obese subjects have an approximately 1.5–3.5-fold increased risk of developing cancers compared with normal-weight subjects (Pischon *et al.*, 2008). Obesity is associated in a number of cancers, particularly endometrium (Rapp *et al.*, 2008; Reeves *et al.*, 2007). Adipocytes have the ability to enhance the proliferation of colon cancer cells in vitro (Amemori *et al.*, 2007). As discussed in previous chapters, the trend of prevalence of overweight/obesity is in line with modern culture. In a study conducted in 2005 [34] in the Kota Bharu district in the state of Kelantan Malaysia, the overall prevalence of overweight/obesity was 49.1% (Nazri *et al.*, 2008), much higher than the figure reported earlier in 1996 (Jackson *et al.*, 1996). In this community, the rise of cancer is exponential in the period from year 2002 to 2007 (143.6% increment) compared to the previous 5-year period of 1996– 2001 (Othman *et al.*, 2008).

Obesity is not a disease. It is rather an expression of symptoms that arise from perpetual insult of the food that we consume. The greatest risk is for obese persons who are also diabetic, particular those whose body mass index is above 35 kg/m2. The increase in risk is by 93-fold in women and by 42-fold in men (Jung, 1997).

Worldwide the rate of obesity has nearly doubled since 1980, with just over 200 million adult men and just under 300 million adult women obese. The cause for women's increased vulnerability to obesity will be discussed in latter chapters. Obesity rates have been rising among children as well. In 2010, 43 million preschool children were overweight or obese, which would mark a 60% rise since 1990 (de Onis *et al.*, 2010). Children in affluent developing countries suffer the most in terms of obesity. A recent publication (Bell, 2013) reports a frightening picture of these societies. Researchers in UAE studied 1,440 children and teenagers aged 6 to 19 and found 14.2 per cent were overweight and a further 19.8 per cent were obese. There was also a correlation between obesity of parents and that of the children. Problems were especially acute in children aged 11 to 15, of whom 40 per cent were overweight or obese, and 16 to 19, of whom 39 per cent were overweight. This is also the age that children develop the most powerful mental state because this is the age that they gain mental maturity while retaining innocence of childhood. In the 6 to 10 age group, 22.8 per cent were overweight or obese. As the children grew older, the likelihood of obesity increased. The age that they could have used to prepare themselves as an adult, children are busy being obsessed with addictive food and other addictive stimulants. Not surprisingly, children who consumed more dairy products were less likely to become overweight or obese. The overall consumption of fruits and vegetables and amount of exercise in all children of all ages were "very sub-standard". While this rise in affluent developing countries is at par with North America, it is well over the global obesity rate. In the UK, about 30 per cent of children are overweight, the same as in the US. In 2010, more than 40 per cent of children in the North American and eastern Mediterranean regions were overweight or obese, 38 per cent in Europe, 27 per cent in the western Pacific and 22 per cent in southeast Asia (Ng *et al.*, 2014).

In USA, the obesity rate is expected to rise to around 50% by 2030 (Wang *et al.*, 2011). Of interest is the fact that the obesity rate is the highest among poorer and lower income families. While this is the case in the west, the east has a different trend. Richer families, just like more affluent third world nations, are more prone to being obese.

(iii) ***Diabetes***

Diabets, particularly Type 2 is a risk for cancer development. Obesity is closely related with diabetes. A community that has high prevalence of obesity also has high prevalence of diabetes. In Kelantan, Malaysia, the prevalence of diabetes in 1999 was 10.5%, and impaired glucose tolerance was 16.5% (Mafauzy *et al.*, 1999). For this place, high prevalence of diabetes correlates well with rapid rise in cancer rate. According to a review on diabetes, the WHO has estimated that, by 2030, there would be 2.48 million diabetics in Malaysia, a jump of 164% from 0.94 million in 2002 (Mafauzy, 2006). One of the most common cancers noted in community that has high diabetics and obesity is colorectal cancer (Ahmed *et al.*, 2007).

Epidemiological studies have demonstrated an association between type 2 diabetes and cancer. Type 2 diabetes is characterized by insulin resistance and hyperinsulinemia. Hyperinsulinemia may lead to cancer through insulin's effect on its cognate receptor and the insulin-like growth factor system. The effects of insulin and insulin-like growth factor I on cancer development and progression have been demonstrated in animal and human studies. Type 2 diabetes has been positively associated with cancers of the breast, colon, and pancreas. An inverse relationship has been observed between type 2 diabetes and prostate cancer, and this may be due to lower testosterone levels in men with type 2 diabetes. Medications used to treat type 2 diabetes may affect cancer cells directly or indirectly by affecting serum insulin levels. Hyperinsulinemia may be an important risk factor for cancer as well as a target for cancer therapy.

In a study of 138 colorectal cancers (CRC) seen in Hospital Universiti Sains Malaysia, 47.8% had metabolic diseases, of which 13.8% were diabetes type 2 (Othman *et al.*, 2008). Those diabetics with CRC often have distal cancers.

Mortality rates are higher for cancer patients who have diabetes than those without diabetes. It appears that insulin pathway crosses path natural pathway as well as pathway that leads to cancer, creating a snowball effect, spiraling down to collapse of the immune system. Several studies showed that the outcomes are the worst for diabetes patients with cancer who that are being treated with insulin, significantly increasing the mortality risk for both men and women compared with those with cancer who didn't have diabetes (Ranc *et al.*, 2014).

For their study, Ranc and colleagues assessed all patients diagnosed with cancer in Denmark between 1995 and 2009 - a total of 426,129 patients had incident cancer and 42,205 had diabetes before their cancer diagnosis. They found that cancer patients with diabetes had higher overall mortality than those without diabetes, and this was the case for all cancers combined as well as for most individual cancer sites.

They also found that the risk appeared highest for diabetic cancer patients treated with oral diabetes drugs or insulin, and when assessed further, insulin-treated patients were at highest risk. For instance, among all cancers combined and having a diabetes duration of 2 years at cancer diagnosis, insulin-treated patients had the highest mortality rates one year after cancer diagnosis:

Men: RR 3.7 (95% CI 2.7-5.1)

Women: RR 4.4 (95% CI 3.1-6.5)

Those rates were also higher for these patients 9 years after diagnosis:

Men: RR 5.0 (95% CI 3.5-7.0)

Women: RR 6.5 (95% CI 4.2-9.3)

Figure 4.9 shows the mortality rate ratios among patients in the three diabetic groups, with 2 years of diabetes duration at cancer diagnosis, relative to non-diabetic cancer patients, as function of time since cancer diagnosis in all diabetic groups or cancer overall. The rate ratios for cancer patients with preexisting diabetes were higher relative to non-diabetic patients and all three diabetic groups followed the same pattern. The mortality rate ratio curves revealed a small decline during the first

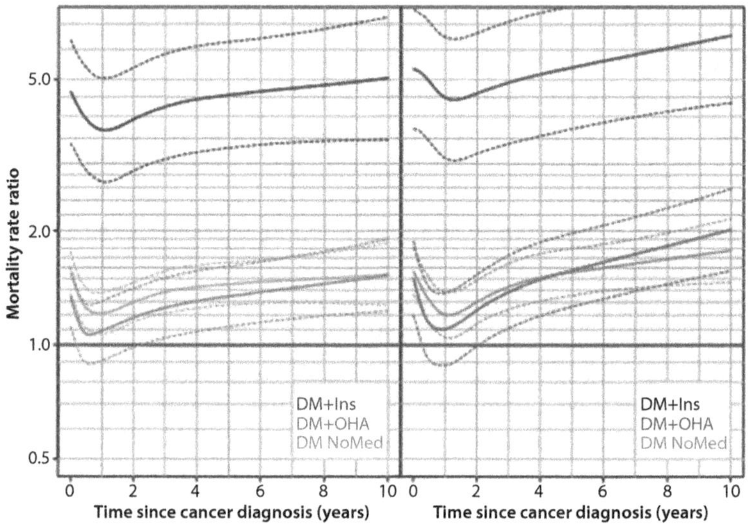

Figure 4.9 Cancer mortality rates compared among diabetic groups

year after the cancer diagnosis. However, after the first year there was a general increase in mortality rate ratios with increasing length of follow-up.

Figure 4.9 shows the mortality rate ratios among patients in the three diabetic groups, with 2 years of diabetes duration at cancer diagnosis, relative to non-diabetic cancer patients, as a function of time since cancer diagnosis in all diabetic groups for cancer overall. The rate ratios for cancer patients with preexisting diabetes were higher relative to non-diabetic patients and all three diabetic groups followed the same pattern. The mortality rate ratio curves revealed a small decline during the first year after the cancer diagnosis. However, after the first year there was a general increase in mortality rate ratios with increasing length of follow-up. (From Ranc *et al.*, 2014)

Ranc *et al.* (2014) explained the prevalence of the increased mortality risk among diabetic cancer patients through 1) an already increased burden of comorbidity; 2) delayed cancer diagnosis due to masking of cancer symptoms; 3) interaction with cancer therapy; and 4) direct effect of glucose-lowering regimens inducing more progressive cancer at diagnosis, or an impaired prognosis.

Cannata *et al.* (2010) reported epidemiological studies that demonstrated an association between type 2 diabetes and cancer. Type 2 diabetes is characterized by insulin resistance and hyperinsulinemia. Hyperinsulinemia may lead to cancer through insulin's effect on its cognate receptor and the insulin-like growth factor system. The effects of insulin and insulin-like growth factor I on cancer development and progression have been demonstrated in animal and human studies. Type 2 diabetes has been positively associated with cancers of the breast, colon, and pancreas. An inverse relationship has been observed between type 2 diabetes and prostate cancer, and this may be due to lower testosterone levels in men with type 2 diabetes. Medications used to treat type 2 diabetes may affect cancer cells directly or indirectly by affecting serum insulin levels.

(iv) ***Chronic Infections***

There are a number of microorganisms which could cause cancer. Common viruses causing cancers (Carrillo-Infante *et al.*, 2007) are Epstein-Barr virus (EBV) (Siddique *et al.*, 2010), human papilloma virus (cervical cancers and other squamous cancers) and Hepatitis B viruses (liver cancers). Viruses are oncogenic after long period of latency (McLaughlin-Drubin and K. Munger, 2008).

Bacteria which has been studied to have associations with cancer are Helicobacter pylori infections (stomach cancer), Ureaplasma urealyticum (prostate cancer), and chronic typhoid carrier (gall bladder cancer). Chronic fungi infections have also been studied to be associated with cancer. Parasites such as Schistosoma haematobium are associated with carcinoma of the urinary bladder; liver flukes Opisthorchis viverrini and Clonorchis sinensis associated with cholangiocarcinoma and hepatocellular carcinoma. There are three main mechanisms by which infections can cause cancer. They appear to involve initiation as well as promotion of carcinogenesis. Persistent infection within host induces chronic inflammation accompanied by formation of reactive oxygen and nitrogen species (ROS and RNOS). ROS and RNOS have the potential to damage DNA, proteins, and cell membranes. Chronic inflammation often results in repeated cycles of cell damage leading to abnormal cell proliferation. DNA damage promotes the growth of malignant cells. Secondly, infectious agents may directly transform cells, by inserting active oncogenes into the host genome, inhibiting tumour suppressors or stimulating mitosis. Thirdly, infectious agents, such as human immunodeficiency virus (HIV), may induce immunosup-pression.

As discussed in Chapter 2, honey has the ability to cure all the risk factors of cancer. Honey is known for centuries for its medicinal and health-promoting properties. It contains various kinds of phyto-chemicals with high phenolic and flavonoid content which contribute to its high antioxidant activity (Yao *et al.*, 2003; Pyrzynska and Biesaga, 2009;

Iurlina *et al.*, 2009). Agent that has strong antioxidant property may have the potential to prevent the development of cancer as free radicals and oxidative stress play a significant role in inducing the formation of cancers (Valko et al, 2007). Phytochemicals available in honey could be narrowed down into phenolic acids and polyphenols. Variants of polyphenols in honey were reported to have antiproliferative property against several types of cancer (Jaganathan and Mandal, 2009).

Honey stimulates inflammatory cytokine production from monocytes (Tonks *et al.*, 2003). Manuka, pasture, and jelly bush honey were found to significantly increase TNF-α, IL-1β, and IL-6 release from MM6 cells (and human monocytes) when compared with untreated and artificial honey-treated cells ($P < 0.001$) (Tonks *et al.*, 2003). A 5.8 kDa component of manuka honey was found to stimulate cytokine production from immune cells via TLR4 (Tonks *et al.*, 2007). Honey stimulates antibody production during primary and secondary immune responses against thymus-dependent and thymus-independent antigens in mice injected with sheep red blood cells and E. coli antigen (Al-Waili and Haq, 2004). Consumption of 80g daily of natural honey for 21 days showed that prostaglandin levels compared with normal subjects were elevated in patient with AIDS (Al-Waili *et al.*, 2006). Natural honey has been shown to decrease prostaglandin level, elevated no production in patients with a long history of AIDS. It was reported that oral intake of honey augments antibody productions in primary and secondary immune responses against thymus-dependent and thymus- independent antigens (Fukuda *et al.*, 2011).

Figure 4.10 shows how honey would suppress and/or remedy risks that lead to cancer. In addition to the above mentioned qualities, we have highlighted honey's therapeutic properties in Chapter 3. Othman *et al.* (2012) have identified such properties as listed below.

(i) ***Natural anti-inflammatory agent***

In routine everyday life, our cells may be injured by irritants from outside or within our bodies through microbial or viral actions. Cellular/molecular injuries result in inflammatory response, the body defense mechanisms in trying to rid of the irritants. In general inflammatory responses are beneficial and protective to us, but at times, inflammatory responses are detrimental to health if the condition persists. This is equivalent exposing oneself to allergens, especially the artificial kind. Honey is a potent anti-inflammatory agent. Infants suffering from diaper dermatitis improved significantly after topical application of a mixture containing honey, olive oil, and beeswax after 7 days (Al-Waili,

2005). Honey provides significant symptom relief of cough in children with an upper respiratory tract infection (URTI), comparable to natural speed wound healer (Heppermann, 2009). It has been shown to be effective in management of dermatitis and Psoriasis vulgaris (Al-Waili, 2003). Eight out of 10 patients with dermatitis and five of eight patients with psoriasis showed significant improvement after 2 weeks on honey-based ointment. Honey at dilutions of up to 1: 8 reduced bacterial adherence from 25.6 ± 6.5 (control) to 6.7 ± 3.3 bacteria per epithelial cell ($P < 0.001$) in vitro (Al-Naqdy *et al.*, 2005). Volunteers who chewed "honey leather" showed that there were statistically highly significant reductions in mean plaque scores (0.99 reduced to 0.65; $P = 0.001$) in the Manuka honey group compared to the control group suggesting a potential therapeutic role for honey for gingivitis, periodontal disease (English *et al.*, 2004), mouth ulcers, and other problems of oral health (Molan, 2001).

A case report of a patient who had chronic dystrophic epidermolysis bullosa (EB) for 20 years healed with honey impregnated dressing in 15 weeks (Hon, 2005) after conventional dressings and creams failed. This

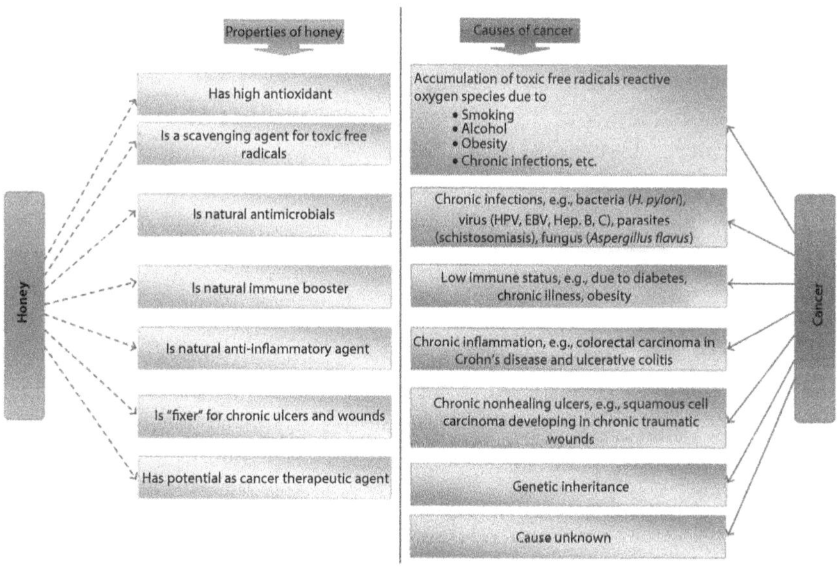

Figure 4.10 Properties of honey *vis-à-vis* causes of cancer

illustrates the usefulness of honey as an anti-inflammatory agent. Chronic inflammatory process has risk of cancer development.

Kirchheimer (2004) points out another fact, typically overlooked by others. In a small study, Rohrich and his colleagues at UT's Advanced Wound Healing and Tissue Regeneration Laboratory tested the effects of a new supplement called InflammEnz, available via the Internet but only with a doctor's prescription. InflammEnz, an herbal product, contains seven different nutrients and enzymes, including calcium and potassium. Twenty-six patients with open wounds from a biopsy received either the oral supplement daily or a placebo. Doctors monitored the healing process of their wounds. Those getting InflammEnz healed 17% faster, and experienced less redness and swelling at the site of the biopsy. Even though InflammEnz is a mixture of vitamin C, Bromelain (anti-inflammatory enzyme), Rutin (a blood vessel protector and accelerator for vitamin C action), and grape seed extract, honey contains all these ingredients in the most natural form and, therefore, can exhibit the same effect. The natural healer brings in the added advantage of not causing any undesirable side effect. For instance, vitamin C is not routinely given to boxers before a fight because it can "thin" the blood similarly to aspirin, and can be detrimental overall.

(ii) ***Honey as natural antimicrobial***

It is discussed in Chapter 3 that honey is the most potent natural antimicrobial agent. The most common infections humans get are from staphylococcal infection. Antibacterial effect of honey is extensively studied. The bactericidal mechanism is through disturbance in cell division machinery (Henriques *et al.*, 2010). Honey is also effective against coagulase-negative staphylococci (French *et al.*, 2005). Local application of raw honey on infected wounds reduced signs of acute inflammation (Al-Waili, 2004), thus alleviating symptoms. Antimicro-bial activity of honey is stronger in acidic media than in neutral or alkaline media. The potency of honey is comparable to some local antibiotics, with the exception that honey continues to have a positive impact on overall health as opposed to continuous degradation and insult with antibiotics. Honey application into infective conjunctivitis reduced redness, swelling, pus discharge, and time for eradication of bacterial infections. When honey is used together with antibiotics, gen-tamycin, it enhances anti-Staphylococcus aureus activity, by 22% (Al-Jabri, 2005). When honey is added to bacterial culture medium, the appearance of microbial growth on the culture plates is delayed [80]. Mycobacteria did not grow in culture media containing 10% and 20% honey while it grew in culture media containing 5%, 2.5%, and 1% honey, suggesting that honey could be an ideal antimycobacterial agent (Asadi-Pooya *et al.*, 2003) at certain concentrations.

Honey contains natural antioxidant properties that can destroy biologically destructive chemical agents which have been linked to many diseases such as cancer. Studies also found that dark-color honeys such as Buckwheat seem to possess more antioxidants than light-color varieties. Not only could honey's antioxidants help to eliminate free radicals in the body, they are also part of the nutrient supply for growth of new tissue.

(iii) ***Honey As possible agent for controlling obesity***

There is a close link among obesity, a state of chronic low-level inflammation, and oxidative stress (Codoner-Franch *et al.*, 2011). Obese subjects have an approximately 1.5–3.5-fold increased risk of developing cancers compared with normal-weight subjects, particularly endometrium, breasts, and colorectal cancers (Othman, 2012). Adipocytes have the ability to enhance the proliferation of colon cancer cells in vitro. The greatest risk is for obese persons that are also diabetic, particularly those whose body mass index is above 35 kg/m^2. The increase in risk is by 93-fold in women and by 42-fold in men (Jung, 1997).

In a clinical study on 55 overweight or obese patients, the control group (17 subjects) received 70 g of sucrose daily for a maximum of 30 days and patients in the experimental group (38 subjects) received 70g of natural honey for the same period. Results showed that honey caused a mild reduction in body weight (1.3%) and body fat (1.1%) (Yaghoobi *et al.*, 2008). Even though this result doesn't seem encouraging, there have been reports of honey stimulating metabolic activities that are conducive to weight loss and reduction in obesity. Honey contains 22 amino acids and a variety of minerals essential for its metabolism and hence is helpful in preventing obesity. It is believed that drinking lemon juice with a little honey the first thing in the morning is an effective anti cellulite treatment as it helps to increase body metabolism.

Honey and cinnamon mixture is another powerful natural cure. Cinnamon itself has benefits for diabetic patients. Studies show that consuming up to 6 grams of cinnamon per day "reduces serum glucose, triglyceride, LDL cholesterol, and total cholesterol in people with type 2 diabetes," and that "the inclusion of cinnamon in the diet of people with type 2 diabetes will reduce risk factors associated with diabetes and cardiovascular diseases." In addition, a certain cinnamon extract can reduce fasting blood sugar levels in patients (Mang *et al.*, 2006).

Cinnamon also has other benefits as well, including Alzheimer's disease (Frydman-Marom *et al.*, 2011). Similar benefits are reported even for HIV. Fink *et al.* (2009) reported that cinnamon may potentially be effective against HIV4. According to the study, the most effective extracts against HIV-1 and HIV-2 are respectively Cinnamomum cassia (bark) and Cardiospermum helicacabum (shoot + fruit). Mondal and Pahan (2015) reported that cinnamon may help stop the destructive process of multiple sclerosis (MS).

Researchers revealed that diets rich in honey can help reduce the body's negative responses to eating high-fat meals. For people who tend to overeat or feel discomfort in the stomach after meals, honey can be taken for better digestion. Honey pathway in human body is such that it produces usable final products immediately, affecting even brain functions.

Deibert *et al.* (2007) demonstrated that the intervention, consisting of meal replacement beverage, additional dietary counseling and guided physical activity sessions, was associated with considerable weight loss in both premenopausal and postmenopausal women. It contained honey drink. In addition, after the intervention, metabolic risk factors in women were decisively lower; indicating the usefulness of this lifestyle approach to reduce both metabolic syndrome and risk factors for cardiovascular disease in women.

McInnes (2014) reported a new line of weight loss regime that depends on introducing honey as the only stimulant for weight loss. He discovered that athletes who ate foods rich in fructose such as honey burnt a lot more fats and had increased stamina levels as well. Honey acts as a fuel to make the liver produce glucose. This glucose keeps the brain sugar levels high and forces it to release fat burning hormones. This conclusion is entirely logical. Wang *et al.* (2010) calculated activity levels and relating burning of calories for various organs. Table 4.7 shows vast majority of calories are burnt by the heart and the brain. This is where honey makes an impact.

(iv) ***Honey as the remedy for chronic ulcers and wounds***

Increasing numbers of antibiotic-resistant bacteria has made simple wounds become chronic and non-healing. However, all of them have one thing in common, that is, they all are destroyed with honey, making honey the most effective alternative treatment option (Sharp, 2009). Honey absorbs exudates released in wounds and devitalized tissue (Cutting, 2007). Honey is effective in recalcitrant surgical wounds (Cooper *et al.*, 2001). It increases the rate of healing by stimulation of angiogenesis, granulation, and epithelialization, making skin grafting unnecessary and giving excellent cosmetic results (Molan, 2001). In a randomized control trial, Manuka honey improved wound healing in patients with sloughy venous leg ulcers (Armstrong, 2009). Honey was shown to eradicate MRSA (Methylene resistant Staphylococcus aureus) infection in 70% of chronic venous ulcers (Gethin and Cowman, 2008). Honey is acidic and chronic non healing wounds have an elevated alkaline environment. The fact that honey thrives in acidic environment, which is same environment known to trigger cancer is an indication that honey can become the most effective medicine for cancer. Manuka honey dressings is associated with a statistically significant decrease in wound pH (Gethin et al, 2008). Available evidence in meta-analysis studies indicate markedly greater efficacy of honey compared with alternative dressings for superficial or partial thickness burns (Wijesinghe *et al.*, 2003). Honey is an inexpensive moist dressing with antibacterial and tissue-healing properties suitable for diabetic foot (Eddy *et al.*, 2008). The average cost of treatment per patient using honey dressing is much cheaper with conventional dressing (Ingle *et al.*, 2006).

(v) ***Honey as a sun blocker***

These precious honey properties help protect the skin under the sun and help the skin to rejuvenate and stay young-looking. As such, there have been an increasing number of manufacturers of honey skincare products such as sunscreens and facial cleansing products for treating damaged or dry skin. After all, the sunscreens commonly used today have so many potential carcinogens that it is no exaggeration that each sunscreen will eventually be implicated for the onset of cancer (Chhetri and Islam, 2008). In 2008, two studies by the CDC highlighted concerns about the sunscreen chemical oxybenzone (benzophenone-3). The first study detected the chemical in 97% of the 2,500 Americans tested, while the second found that mothers with high levels of oxybenzone in their bodies were more likely to give birth to underweight baby girls (CDC, 2008). While controversial, oxybenzone does provide broad-spectrum UV coverage, including UVB and short-wave UVA rays (Burnett and Wang, 2011). Similar investigation led Therapeutic Goods Administration of Australia carried out a review of sunscreen safety studies and concluded: "There is evidence from isolated cell experiments that zinc oxide and titanium dioxide can induce free radical formation in the presence of light and that this may damage these cells (photo-mutagenicity with zinc oxide). However, this would only be of concern in people using sunscreens if the zinc oxide and titanium dioxide penetrated

Table 4.7 Calories burnt by various organs at different age groups:

Organ/tissue and Elia's K_i value	**All subjects**	**21–30 y (young)**	**31–50 y (middle-age)**	**51–73 y**
Liver (200 kcal kg^{minus1} d^{minus1})				
95% CIs for corresponding K_i values	(182, 201)	(191, 227)	(181, 212)	(150, 181)
P value from *t* test of H0^2	0.087	0.31	0.67	<0.001
Marginal ^{2} value^3	0.92 (0.92)	0.93 (0.93)	0.93 (0.93)	0.93 (0.89)
Brain (240 kcal kg^{minus1} d^{minus1})				
95% CIs for corresponding K_i values	(220, 241)	(230, 267)	(219, 252)	(186, 219)
P value from *t* test of H0^2	0.073	0.34	0.57	<0.001
Marginal ^{2} value^3	0.94 (0.94)	0.95 (0.95)	0.94 (0.94)	0.94 (0.91)
Heart (440 kcal kg^{minus1} d^{minus1})				
95% CIs for corresponding K_i values	(348, 434)	(402, 554)	(338, 479)	(213, 338)
P value from *t* test of H0^2	0.024	0.32	0.37	<0.001
Marginal ^{2} value^3	0.72 (0.71)	0.79 (0.79)	0.73 (0.72)	0.68 (0.44)

(Continued)

Table 4.7 cont.

Organ/tissue and Elia's K_i value	**All subjects**	**21–30 y (young)**	**31–50 y (middle-age)**	**51–73 y**
Kidneys (440 kcal kg^{minus1} d^{minus1})				
95% CIs for corresponding K_i values	(336, 430)	(396, 568)	(328, 481)	(200, 332)
P value from *t* test of $H0^2$	0.016	0.34	0.35	<0.001
Marginal ^{2} value^3	0.68 (0.66)	0.75 (0.74)	0.69 (0.69)	0.65 (0.37)
SM (13 kcal kg^{minus1} d^{minus1})				
95% CIs for corresponding K_i values	(12.0, 13.0)	(12.5, 14.4)	(12.0, 13.6)	(10.3, 11.9)
P value from *t* test of $H0^2$	0.051	0.35	0.59	<0.001
Marginal ^{2} value^3	0.95 (0.95)	0.95 (0.95)	0.95 (0.95)	0.96 (0.93)
AT (4.5 kcal kg^{minus1} d^{minus1})				
95% CIs for corresponding K_i values	(3.21, 4.57)	(3.85, 6.37)	(3.14, 5.26)	(1.16, 3.34)
P value from *t* test of $H0^2$	0.077	0.34	0.58	<0.001
Marginal ^{2} value^3	0.50 (0.49)	0.61 (0.60)	0.56 (0.55)	0.32 (0.00)

(Continued)

Table 4.7 cont.

Organ/tissue and Elia's K_i value	**All subjects**	**21–30 y (young)**	**31–50 y (middle-age)**	**51–73 y**
Res (12 kcal kg^{minus1} d^{minus1})				
95% CIs for corresponding K_i values	(10.9, 12.0)	(11.5, 13.6)	(10.9, 12.6)	(9.3, 11.0)
P value from *t* test of H0[2]	0.051	0.32	0.59	<0.001
Marginal ^{2} value[3]	0.93 (0.93)	0.93 (0.93)	0.94 (0.94)	0.94 (0.91)

[1]AT, adipose tissue; Res, residual mass; SM, skeletal muscle. [2]H0: K_i = Elia's K_i value. [3]The value within parentheses is the corresponding reduction by taking K_{liver} = 200 (Elia's value). This is likewise defined for other organs and tissues.

into viable skin cells. The penetration is not significant enough to be detected, leading to the conclusion that it is limited to skin. However, if one studies the pathway of these chemicals, it becomes evident that they should be avoided.

Concerns have also been raised about potential vitamin D deficiency arising from prolonged use of sunscreen. Typical use of sunscreen does not usually result in vitamin D deficiency; however, extensive usage may (Norval and Wulf, 2009). In addition, it is not well understood how vitamin D gets affected when sunlight encounters artificial chemicals in the sunscreen. Sunscreen prevents ultraviolet light from reaching the skin, and even moderate protection can substantially reduce vitamin D synthesis.

In ancient time, Greeks used olive oil, while Egyptians used rice, jasmine, and lupine plants. On the other hand, the use of Zinc oxide (in its natural form) was common throughout known history. Also, honey was used for treating sunburn. These natural components are no longer used and should be revived.

(vi) ***Honey as natural cancer "vaccine"***

Decades after the invention of Dr. Jenner's natural vaccine (cow pox) of small pox, artificial vaccine was invented. Ever since, numerous vaccines have been invented for many diseases.

Synthetic vaccines like BCG or polio vaccine work by preventing vaccinated subjects from contracting tuberculosis and poliomyelitis. Honey has the element of a "natural cancer vaccine" as it can reduce chronic inflammatory processes, improve immune status, reduce infections by hardy organisms and so forth. Some simple and polyphenols found in honey, namely, caffeic acid (CA), caffeic acid phenyl esters (CAPE), chrysin (CR), galangin (GA), quercetin (QU), kaempferol (KP), acacetin (AC), pinocembrin (PC), pinobanksin (PB), and apigenin (AP), have evolved as promising pharmacological agents in prevention and treatment of cancer (Jaganathan and Manda, 2009). The antioxidant activity of Trigona carbonaria honey from Australia is high at 233.96± 50.95 microM Trolox equivalents (Oddo et al, 2008). The antioxidant activity of four honey samples from different floral sources showed high antioxidant properties tested by different essay methods (Hegazi and Abd El-Hady, 2009). Dark honey had higher phenolic compounds and antioxidant activity than clear honey (Estevinho *et al.*, 2009). The amino acid composition of honey is an indicator of the toxic radical scavenging capacity (Perez *et al.*, 2007).

Following is a list of various components added to the synthetic vaccine.

According to the U.S. Centers for Disease Control's vaccine additives page, all the following ingredients are routinely used as vaccine additives:

1. Aluminum - A light metal that causes dementia and Alzheimer's disease.
2. Antibiotics - Chemicals that promote superbugs, which are deadly antibiotic-resistant strains of bacteria that are killing tens of thousands of Americans every year.
3. Formaldehyde - A "pickling" chemical used to preserve cadavers. It's highly toxic to the nervous system, causing blindness, brain damage and seizures. It is also a carcinogen as opposed to natural formaldehyde that has no such implication.
4. Monosodium Glutamate (MSG) - A neurotoxic chemical called an "excitotoxin." It causes brain neurons to be overexcited to the point of death. MSG is toxic even when consumed in foods, where it causes migraine headaches and endocrine system damage.
5. Thimerosal - A methyl mercury compound that causes severe, permanent nervous system damage. Mercury is highly toxic to the brain. It can cross the blood-brain and placenta barriers.

(vii) ***Honey's potential in "cancer therapy"***

Honey may provide the basis for the development of novel therapeutics for patients with cancer and cancer- related tumors. Jungle honey fragments were shown to have chemotactic induction for neutrophils and reactive oxygen species (ROS), proving its antitumor activity (Fukuda *et al.*, 2011). Recent studies on human breast (Fauzi *et al.*, 2011), cervical, oral (Ghashm *et al.*, 2010), and osteosarcoma cancer cell lines using Malaysian jungle honey showed significant anticancer activity. Honey has been shown to have antineoplastic activity in an experimental bladder model in vivo and in vitro (Swellam *et al.*, 2003).

Honey is rich in flavonoids (Gomez-Caravaca *et al.*, 2003). Flavanoids have created a lot of interests among researchers because of its anticancer properties. The mechanisms suggested are rather diverse such as various signaling pathways (Woo *et al.*, 2004), including stimulation of TNF-alpha (tumor necrosis factor-alpha) release (Tonks *et al.*, 2001), inhibition of cell proliferation, induction of apoptosis (Jasanth and Mondal, 2006), and cell cycle arrest (Pichichero *et al.*, 2010) as well as inhibition of lipoprotein oxidation (Gheldoff and Engeseth). Honey is thought to mediate these beneficial effects due to its major com- ponents such as chrysin (Woo et al, 2004) and other flavonoids (Jasanth *et al.*, 2010). These differences are explainable as honeys are of various floral sources, and each floral source may exhibit different active compounds. Though honey has other substances of which the most predominant are a mixture of sugars (fructose, glucose, maltose, and sucrose) (Alijadi and Kamaruddin, 2004) which itself is carcinogenic (Heuson et al, 1972), it is understandable that its beneficial effect on cancer raises skeptics. The mechanism on how honey has anti-cancer effect is an area of great interest recently. The effects of honeys on hormone-dependent cancers such as breast, endometrial, and prostate cancer and tumors remain largely unknown. There is a lot we can learn from nature (Moutsatsou et al, 2007). For example, phytochemicals, such as genistein, lycopene, curcumin, epigallocatechin-gallate, and resveratrol have been studied to be used for treatment of prostate cancer (von Low et al, 2007). Phytoestrogens constitute a group of plant-derived isoflavones and flavonoids, and honey belongs to plant phytoestrogen (Moutsatsou, 2007;, Zaid *et al.*, 2007).

4.4.2.2 Mother's Milk and Cancer

Breast-feeding is well known to have a protective effect against infection in infants. Although the long-term effects of breast-feeding on childhood cancer have not been studied extensively, it is most likely because long-term health impact investigation is not part of New Science. Starting from 1960's, numerous studies appeared on protective effect against childhood cancers (e.g.. Hodgkin's disease and lymphoma). Back in the late 1980s, researchers from the U.S. Centers for Disease Control and Prevention (CDC) found that breast-fed children are far less likely than children fed formula and other unnatural foods to develop cancer. Shu *et al.* (1999) offered one of the earliest studies that tested the hypothesis that breast-feeding decreases the risk of childhood acute leukemia. They studied a total of 1744 children with acute lymphoblastic leukemia (ALL) and 1879 matched control subjects, aged 1-14 years, and 456 children with acute myeloid leukemia (AML) and 539 matched control subjects, aged 1-17 years. Ever having breast-fed was found to be associated with a 21% reduction in risk of childhood acute leukemias. A reduction in risk was seen separately for AML (OR = 0.77; 95% CI = 0.57-1.03) and ALL (OR = 0.80; 95% CI = 0.69-0.93). The inverse associations were stronger with longer duration of breast-feeding for total ALL and AML; for M0, M1, and M2 morphologic subtypes of AML; and for early pre-B-cell ALL. It was the beginning of numerous epidemiologic studies that showed the anti-infective and/or immune-stimulatory or immune-modulating effects of breast-feeding on leukemogenesis and other forms of cancer in children.

Leukemia is the most common childhood malignancy in Western countries and accounts for one third of all cancers occurring in children under the age of 15 years. Despite studies conducted over more than four decades, the etiology of childhood leukemia remains largely unknown. It is because New science offers no process that can establish the inefficacy of a process that yields the most money. Any time a risk factor is correlated with any aspect of lifestyle that is lucrative, numerous other publications surface that ridicule the finding based on cause and effect logic. The most that is allowed is to identify one component at a time and make comments about the prospect of mass production and commercialization of that component. Even though the benefits of human breast milk are known to be vast, researchers continue to 'discover' one item at a time. One such item was when Swedish scientists discovered one cancer fighting component. Mark *et al.* (2013) reported in a study published in the journal PLoS One, the substance, known as "Human Alpha-lactalbumin Made LEthal to Tumor cells" (HAMLET), effectively kills cancer cells, which in turn helps provide lasting protection against tumor development in young children. They demonstrated that HAMLET acts as an antimicrobial adjuvant that can increase the activity of a broad spectrum of antibiotics (methicillin, vancomycin, gentamicin and erythromycin) against multi-drug resistant Staphylococcus aureus, to a degree where they become sensitive to those same antibiotics, both in antimicrobial assays against planktonic and biofilm bacteria and in an in vivo model of nasopharyngeal colonization. They show that HAMLET exerts these effects specifically by dissipating the proton gradient and inducing a sodium-dependent calcium influx that partially depolarizes the plasma membrane, the same mechanism induced during pneumococcal death. These effects results in an increased cell associated binding and/or uptake of penicillin, gentamicin and vancomycin, especially in resistant stains. Finally, HAMLET inhibits the increased resistance of methicillin seen under antibiotic pressure and the bacteria do not become resistant to the adjuvant, which is a major advantageous feature of the molecule. These results highlight HAMLET as a novel antimicrobial adjuvant with the potential to increase the clinical usefulness of antibiotics against drug resistant strains of S. aureus.

Now it has become a little clearer why this is the case, as breast milk contains unique compounds like HAMLET that specifically target cancer cells while leaving healthy cells to grow and multiply as normal.

"Laboratory experiments have shown that HAMLET kills 40 different types of cancer, and the researchers are now going on to study its effects on skin cancer, tumors in the mucous membranes and brain tumors," explains Science Daily about the discovery. "HAMLET kills only cancer cells and does not affect healthy cells."

Scientists originally discovered HAMLET on accident while researching the antibacterial properties of human breast milk. Several years later, researchers from Lund University and the University of Gothenburg, both in Sweden, found that HAMLET contains a protein and a fatty acid that contribute to its cancer-fighting effects. And further followup research has revealed that the substance may actually form in the acidic environment of babies' stomachs, the natural byproduct of breast milk consumption.

"HAMLET is produced by combining alpha-lactalbumin in the milk and oleic acid which is found in babies' stomachs," explained Assistant Professor Roger Karlsson from the University of Gothenburg to the U. K.'s Telegraph about the findings. "So breast feeding has been linked to actually reducing the risk of cancer in babies. HAMLET also triggers some of the cell's apoptotic pathways - apoptosis is programmed cell death." The actual motivation of the study is revealed in its application. Human trials involving just the HAMLET component of breast milk have revealed that it can effectively treat male bladder cancer. Patients with bladder cancer who received injections of HAMLET experienced a reduction in cancer tumor size after just a few days of treatment. "A pilot study of bladder cancer patients were injected with a HAMLET solution through a catheter," Prof. Karlsson is quoted as saying to the Telegraph. "The solution killed cancer cells and the size of the tumors actually reduced within five days."

Breast milk may help treat cancer in adults as well. Stuebe (2009) reported that breastfeeding is a key modifiable risk factor for disease for both mothers and infants. Data suggest that variations in hospital practices account for disparities in breastfeeding duration. Improvements in the quality of antenatal and perinatal support could have a substantial impact on mother and infant health. There are specific and innate

Table 4.8 Effect of mother's milk (from Marks *et al.*, 2013)

Bacterial Strain	**Antibiotics**	**MIC (g/ ml)**	**MIC Fold reduction**	**Antibiotics, JV1BC lug/ml)**		**MBC Fold reduction**	**[HL]** to sensitize, uM**
	Meth	**Meth tHL»**		**Meth**	**Meth + HL***		
NRSl	>128	16	>8	>256	64	4	48
NRS70	32	1	32	128	4	32	3
NRS71	>12B	16	>8	>256	64	>4	54
NRSl00	>128	8	>16	>256	64	>4	18
NRS123	32	4i	8	>256	16	>16	12
	16	2	8	>256	8	>32	6
103075 70	16	2	8	>256	16	16	6
11090306	2	1	2	64	4	16	0

(Continued)

Table 4.8 cont.

Bacterial Strain	Antibiotics	MIC (g/ ml)	MIC Fold reduction	Antibiotics, JV1BC lug/ml)		MBC Fold reduction	[HL]** to sensitize, uM
	Meth	Meth tHL»		Meth	Meth + HL*		
	Erm	Em + HL*		Erm	Erm + HL*		
NRS3S4	16	4	4	>256	16	>16	6
NRS123	0.5	0.25	2	16	4	4	0
	Gent	Gent + HL*		Gent	Gent + HL*		
NR5384	1	0.00625	160	16	4	4	0
NRSl	>128	8	>16	>2S6	32	>8	48
	Vanc	Vanc + HL*		vanc	Vant + HL*		
NRS384	1	0.5	2	16	8	2	0
NRSl	8	4	2	32	8	4	18

immune factors present in human milk that provide specific protection against pathogens in the mother's environment. In addition, immune factors in milk provide protection against infections such as H influenzae, S pneumoniae, V cholerae, E coli, and rotavirus.

Not breastfeeding is associated with health risks for both mothers and infants. Epidemiologic data suggest that women who do not breastfeed face higher risk of breast and ovarian cancer, obesity, type 2 diabetes, metabolic syndrome, and cardiovascular disease.

Related studies conducted over the years have found that other natural compounds in breast milk help inhibit the growth and spread of cancer cells as well. A 1994 study published in the journal Cancer Research, for instance, found that the lactoferrin in breast milk inhibits the growth of tumors in mice. And another study published the same year in the journal Medical Hypothesis found that the fatty acids in breast milk also help deter cancer growth.

Additionally, reports have surfaced over the years of people actually using human breast milk to cure their own cancers. Several years ago, for instance, a man allegedly cured his own prostate cancer by drinking a little bit of breast milk every day. And in 2009, it was reported by ABC News that a U.K. man took daily doses of his own daughter's breast milk to help treat an allegedly terminal case of colon cancer that spread to his liver and lymph nodes.

The composition of human milk is the biologic norm for infant nutrition. With the purpose of human life, stated in Chapter 3, human milk should be the most beneficial food for children. This is in line with the premise: Nature is perfect (Khan and Islam, 2012). Human milk also contains many hundreds to thousands of distinct bioactive molecules that protect against infection and inflammation and contribute to immune maturation, organ development, and healthy microbial colonization. Some of these molecules, e.g., lactoferrin, are being investigated as novel therapeutic agents. In all these research activities, the holistic approach is not taken. Such approach would include specific individual needs of a child along with full knowledge of the purpose of human lives. Also to be considered is the dynamic nature of mother's milk. A dynamic, bioactive fluid, human milk changes in composition from colostrum to late lactation, and varies within feeds, diurnally, and between mothers. One must note that the quality of milk declines with expressed human milk. Further degradation occurs if milk is Pasteurized. Mother's milk is not conducive to commercialization or similar anti-nature activities. Many milk proteins are degraded by heat treatment and freeze-thaw cycles may not have the same bioactivity after undergoing these treatments. Ballard *et al.* (2012) reported a summary of human milk benefits.

The first fluid produced by mothers after delivery is colostrum, which is distinct in volume, appearance and composition. Colostrum, produced in low quantities in the first few days postpartum, is rich in immunologic components such as secretory IgA, lactoferrin, leukocytes, as well as developmental factors such as epidermal growth factor. Colostrum also contains relatively low concentrations of lactose, indicating its primary functions to be immunologic and trophic rather than nutritional. This is the essence of natural immunization. Contrast the composition of human milk (colostrum rich) with the composition of vaccines and it becomes clear human milk and vaccines are two opposites of the benefit spectrum. Georgiev (2008) highlighted the benefits of colostrum compared to cow's milk. He reviewed the differences in chemical composition between colostrum and milk in cows. The concentrations of many nutrients (proteins, vitamins, minerals etc.) and biologically active substances (immunoglobulins, enzymes, hormones, growth factors etc.) are many times higher in colostrum than in milk. A special attention is given to insulin-like growth factors (IGF-1 and IGF-2) – the predominant growth factors in the colostrum of cows (especially in the first portions). Table 4.9. shows some of the relevant concentrations in milk and colostrum.

Figure 4.11 shows major components of colostrum of human milk.

Colostrum contains:

- a high percentage of taurine, an amino-acid important for the development of the central nervous system and for the baby's retina;
- a high percentage of lactose which enables an excessive loss of weight in the first week of life;
  substances which stimulate the intestinal peristalsis;
- a high percentage of secreted immunoglobulin (IgA) which protects the baby from intestinal and respiratory infections, preventing the adhesion of bacteria to the mucosa cells. These antibodies have a local protective action against alimentary allergies. They stratify on the intestinal mucosa, making it not reactive to heterologous protein;
- immune cells like macrophagus, neutrophils, lymphocyte, and so on;

Table 4.9 Contents of some hormones in colostrum and milk

Hormone	Concentration
Insulin	Colostrum: 4.2–34.4 ng/mL milk: 0.042–0.34 ng/mL
Total cortisol	Colostrum: 4.4 ng/mL milk: 0.35 ng/mL
Free cortisol	Colostrum: 1.8 ng/mL milk: 0.3 ng/mL
Prolactin	Colostrum: 150 ng/mL milk: 50 ng/mL
Progesterone	Colostrum: 2.6 ng/mL milk: 0.8 ng/mL

- prostaglandins (PG) which have a very important protective action for the mucosa of the digestive apparatus. They release some enzymes during the transit of water, minerals and glucose.

Levels of sodium, chloride and magnesium are higher and levels of potassium and calcium are lower in colostrum than later milk. As tight junction closure occurs in the mammary epithelium, the sodium to potassium ratio declines and lactose concentration increases, indicating secretory activation and the production of transitional milk. The timing of secretory activation (lactogenesis stage II) varies among women, but typically occurs over the first few days postpartum. Delayed onset of lactogenesis is defined as onset >72 hours after delivery and appears to occur more often with preterm delivery and maternal obesity, and may be predicted by markers of metabolic health. Biochemical markers in early milk for onset of secretory activation include its sodium content, the sodium to potassium ratio, citrate, and lactose.

Transitional milk shares some of the characteristics of colostrum but represents a period of "ramped up" milk production to support the nutritional and developmental needs of the rapidly growing infant, and typically occurs from 5 days to two weeks postpartum, after which milk is considered largely mature. By four to six weeks postpartum, human milk is considered fully mature. In contrast to the dramatic shift in

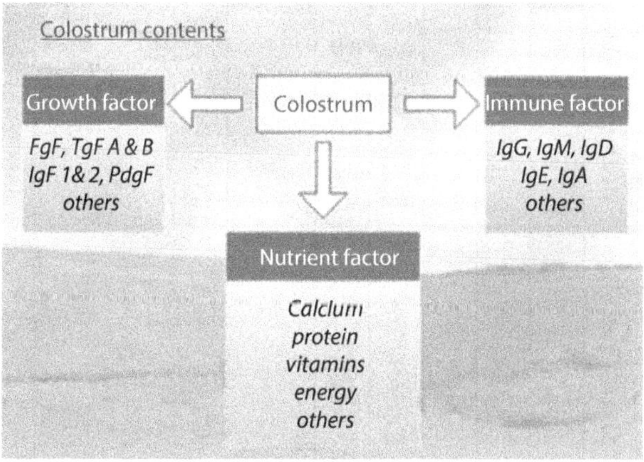

Figure 4.11 Colostrum content

composition observed in the first month of life, human milk remains relatively similar in composition, although subtle changes in milk composition do occur over the course of lactation. Table 4.10 shows general composition of mother's milk.

Hartmann (2012), Australian donor mothers of preterm infants (see article this issue by Landers and Hartmann)

Today, mother's milk has become a contentious issue and a matter of debate (Mead, 2008). Whenever, news breaks about finding toxic chemicals in mother's milk, the talk revolves around replacing mother's milk with artificial chemicals. Whenever, a mother produces milk that is deficient in nutrition, adding toxic chemicals to mother's diet is the first recommendation made, albeit in name of 'fortification'.

Similarly under attack is cow's milk. From a scientific perspective what we should make a case against is:

1. non-organic milk
2. Pasteurized milk
3. Homogenized milk
4. Addition of artificial chemicals (in name of 'supplements')
5. Storage in plastic container
6. ...

Yet, it is the sweeteners that are being talked about and people are talking about avoiding milk altogether (Ludwig and Willett, 2013).

Table 4.10 Mother's milk composition (from Ballard *et al.*, 2012) Macronutrient (g/dL) and energy (kcal/dL) composition of human milk from specified references

Author (year), n	**Protein**	**Fat**	**Lactose**	**Energy**
	Mean (± 2 SD)	**Mean (± 2 SD)**	**Mean (± 2 SD)**	**Mean (± 2 SD)**
Term infants, 24-hour collection, mature milk				
Nommsen et al (1991), n=58	1.2 (0.9, 1.5)	3.6 (2.2, 5.0)	7.4 (7.2, 7.7)	70 (57, 83)
Donor human milk samples				
Wojcik et al (2009), n=415	1.2 (0.7, 1.7)	3.2 (1.2, 5.2)	7.8 (6.0, 9.6)	65 (43, 87)
Michaelsen et al (1990), n=2553	[a] 0.9 (0.6, 1.4)	[a] 3.6 (1.8, 8.9)	[a] 7.2 (6.4, 7.6)	[a] 67 (50,115)
Representative values of mature milk, term infants				
Reference standard	0.9	3.5	6.7	65 to 70
Preterm, 24-hour collection, first 8 weeks of life				
Bauer & Gerss (2011)				
Born <29 weeks, n=52	2.2 (1.3, 3.3)	4.4 (2.6, 6.2)	7.6 (6.4, 8.8)	78 (61, 94)
Born 32-33 weeks, n=20	1.9 (1.3, 2.5)	4.8 (2.8, 6.8)	7.5 (6.5, 8.5)	77 (64, 89)

(*Continued*)

Table 4.10 cont.

Author (year), n	**Protein**	**Fat**	**Lactose**	**Energy**
	Mean (± 2 SD)	**Mean (± 2 SD)**	**Mean (± 2 SD)**	**Mean (± 2 SD)**
Preterm donor milk				
Hartmann (2012), n=47	1.4 (0.8, 1.9)	4.2 (2.4, 5.9)	6.7 (5.5, 7.9)	70 (53, 87)

[a]Median (lower bound)
Nommsen et al, 1991: Davis, CA mothers at 3 months postpartum
Wojcik et al, 2009, U.S. milk bank donors
Michaelsen et al, 1990, Danish milk bank donors
Reference standard based on expert review: Pediatric Nutrition Handbook, 6th edition, Table C-1, p. 1201 (Mature Milk)
Bauer & Gerss, 2011, preterm infants 23 to 33 weeks gestational age

4.4.2.3 Concluding Remarks on Cancer

Despite the claim of making progress over the last century in 'combating cancer', all claims of detection, prevention, and remediation have little scientific merit. Today, the most effective detection tools are the most powerful source of artificial energy that can trigger cancer, but New science doesn't know it yet. Also, the most 'successful' intervention tools, such as chemotherapy, radiotherapy, and gene therapy are also progressively more insulting to the body, triggering worse environment for the onset and propagation of cancer. New science has no knowledge of what triggers cancer and has no tool to investigate the role of artificial chemicals in causing cancer. The debate cannot even begin to start because New science falsely assumes that 'chemicals are chemicals' and 'energy is energy'. For instance, no one would accept natural radiation from mud can cure cancer whereas artificial radiation (as in radiotherapy) can trigger more cancer while killing certain cells. Similarly, there is no way to distinguish between honey and sugar in terms of its pathway in human body and medical practitioners routinely lump them in the same category. The same applies to mother's milk – the most natural food for a child and by extension cow's milk – the most natural food for humans.

4.4.3 Asthma

Asthma is a chronic disorder of the lungs in which inflamed airways are prone to constrict, causing episodes of breathlessness, wheezing, coughing, and chest tightness that range in severity from mild to life-threatening. Inflamed airways become hypersensitive to a variety of stimuli, including dust mites, animal dander, pollen, air pollution, cigarette smoke, medications, weather conditions, and exercise. It is theorized that the source of these stimulants doesn't matter, meaning the asthmatic reaction to pollen (organic or non-organic or GMO) or air pollution (artificial chemicals) all act the same way. It is known that stress can exacerbate symptoms but there is no means to characterize stress and link it to asthmatic episodes mainly due to the lack of science that would consider intangible aspects. Asthmatic episodes may begin suddenly or may take days to develop. Although an initial episode can occur at any age, about half of all cases occur in persons younger than 10 years of age, with boys being affected more often than girls. Among adults, however, the incidence of asthma is approximately equal in men and women. When asthma develops in childhood, it is often associated with an inherited susceptibility to allergens, substances such as pollen, dust mites, or animal dander that may induce an allergic reaction. In stead of taking this inheritance as a vulnerability and expose these children to less toxic chemicals, the opposite is done and children are put on chemical inhalers s early as infancy. In adults, asthma also may develop in response to allergens, but viral infections, aspirin, and exercise may cause the disease as well. Adults who develop asthma may have nasal polyps or sinusitis. Cure to asthma is in natural lifestyle, including use of honey, olive oil and natural antioxidants. Instead, totally artificial chemicals have been introduced. Following is a brief discussion of inhalers that are commonly prescribed, even to infants.

One of the most crucial components of a metered-dose inhaler (MDI) is its propellant. The propellant provides the force to generate the aerosol cloud and is also the medium in which the active component must be suspended or dissolved. Propellants in MDIs typically make up more than 99% of the delivered dose, so it is the properties of the propellant that dominate more than any other individual factor. As a career these are the only ones used and no research is available on organic dispenser. Suitable propellants must pass a stringent set of criteria, they must:

- have a boiling point in the range -100 to +30°C [10]
- have a density of approximately 1.2 to 1.5 g cm^{-3} (approximately that of the drug to be suspended or dissolved)
- have a vapour pressure of 40 to 80 psig
- have no toxicity to the patient
- be non-flammable
- be able to dissolve common additives. Active ingredients should be either fully soluble or fully insoluble.
- Chlorofluorocarbons (CFCs)

In the early days of MDIs the most commonly used propellants were the chlorofluorocarbons CFC-11, CFC-12 and CFC-114.

In 2008 the Food and Drug Administration announced that inhalers using chlorofluorocarbons as a propellant, such as Primatene Mist, could no longer be manufactured or sold as of 2012. This followed from the U.S. decision to agree to the 1987 Montreal Protocol on Substances that Deplete the Ozone Layer. It wasn't motivated by the fact that CFC is pure poison and applying that to lungs defies all logic. It was also unnoticed if CFC can create a hole in the ozone layer what type of harm it can cause to a lung. Instead, a new genre of toxic material was introduced, the era of HFA (hydrofluoroalkane) propellants was born. All subsequent research articles would focus on only safety and immediate suppression of symptoms (e.g. Ibiapina *et al.*, 2004), even though logically HFA is far worse than CFC.

4.4.4 HIV/AIDS

For over 30 years, acquired immunoodefficiency syndrome (AIDS) has captivated the headlines of medical profession. It is a transmissible disease of the immune system caused by the human immunodeficiency virus (HIV). HIV slowly attacks and destroys the immune system, the body's defense against infection, leaving an individual vulnerable to a variety of other infections and certain malignancies that eventually cause death. AIDS is the final stage of HIV infection, during which time fatal infections and cancers frequently arise. HIV/AIDS spread to epidemic proportions in the 1980s, particularly in Africa, where the disease may have originated. Since the beginning of the epidemic, almost 78 million people have been infected with the HIV virus and about 39 million people have died of HIV. According to UN report (WHO), globally, 35.0 million people were living with HIV at the end of 2013. An estimated 0.8% of adults aged 15–49 years worldwide are living with HIV, although the burden of the epidemic continues to vary considerably between countries and regions. Sub-Saharan Africa remains most severely affected, with nearly 1 in every 20 adults living with HIV and accounting for nearly 71% of the people living with HIV worldwide. As for USA, Figure 4.12 shows how exponential growth took place during the first two decades after the first occurrence was detected in 1981. The death rate is quite high as well. Figure 4.13 shows the number of deaths due to AIDS related complications during 1982 through 2004.

UN reports indicate, globally, the number of new HIV infections continues to fall. There were 2.3 million new HIV infections [1.9 million–2.7 million] in 2012. This is the lowest number of annual new infections since the mid-to-late 1990s, when approximately 3.5 million [3.3 million–4.1 million] people were acquiring HIV every year. The number of HIV infections declined by more than 50% in 26 countries between 2001 and 2012 and between 25% and 49% in an additional 17 countries. The drop in new HIV infections is most pronounced among

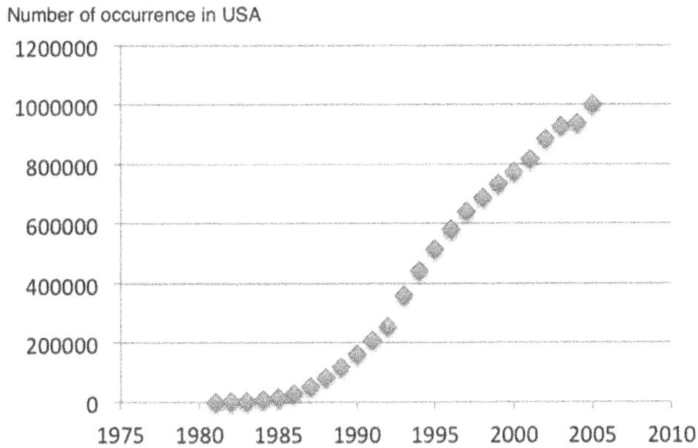

Figure 4.12 Occurrence of AIDS patients in USA

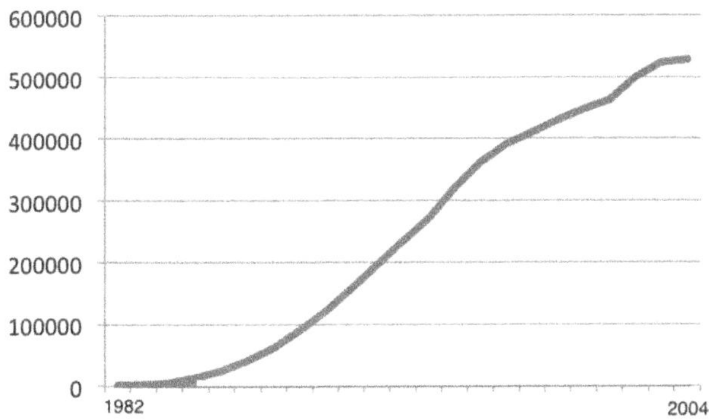

Figure 4.13 Number of deaths due to AIDS in USA

children. From 2001 to 2012 the number of children newly infected with HIV dropped by 52%--from 550 000 [500 000–620 000] in 2001 to 260 000 [230 000–320 000] in 2012. This decline is attributed to the following.

- Low cost of antiretroviral therapy: The cost of first line antiretroviral therapy in some low and middle-income countries has been reduced to around US$ 140 per person per year. In the mid 1990's the cost was around US$ 10 000 per person per year.
- Politics: Increased political commitment and smarter investments together with more strategic programming and massive reductions in the cost of treatment have led to a record 9.7 million people in low- and middle-income

> countries having access to antiretroviral therapy at the end of 2012. The rate of scale up has increased exponentially in recent years. In 2012 alone an additional 1.6 million people newly gained access to treatment.

Of particular importance is the decline in AIDS related death among children. Figure 4.14 captures this trend in HIV infection among children. The reason for this decline is attributed to access to better healthcare. What is absent from the analysis is the birth rate among HIV infected people. Also absent is if there has been a change in lifestyle because of increased awareness of HIV.

There is an widespread skepticism about government policies and intentions of pharmaceutical companies. Ionisio *et al.* (2008) presents the real reason behind these numbers. A detailed discussion of this topic will be made in Part II of the book.

The number of people dying from AIDS-related causes each year has declined from a high of 2.3 million 2.1–2.6 million] in 2005 to 1.6 million [1.4–1.9 million] in 2012. Since 2004, TB-related deaths among people living with HIV have declined by 36% world-wide. While this has been attributed to a number of causes, a comprehensive explanation eludes scientists.

In 1996, HAART - an effective combination therapy that delays the onset of AIDS - became available to those living with HIV in rich countries. 2 Within four years, death rates caused by AIDS-related illnesses in developed countries had dropped by 84 percent. 3

At a cost of US$10,000-15,000 per person per year, these antiretroviral drugs (ARVs) were far too expensive for the majority of people

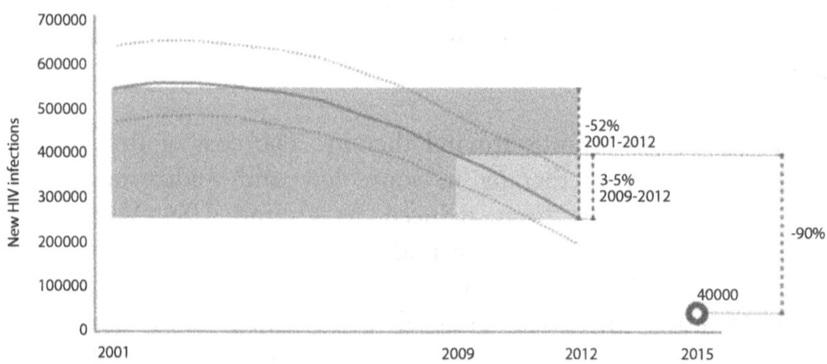

Figure 4.14 HIV infected children in recent period

infected with HIV in resource poor countries. Five years after HAART was introduced in the West, only 2 percent of people in developing countries were receiving the life-saving drugs. 4 In order for treatment to reach people living with HIV in the developing world, the price of the drugs clearly needed to come down to an affordable level.

At the beginning of the new millennium there was a breakthrough in treatment provision for resource poor areas when an Indian pharmaceutical company started to produce generic antiretrovirals that were exactly the same as those made by large pharmaceutical companies, but significantly cheaper. This sparked a price war between branded and generic drug makers, which forced the large pharmaceutical companies to lower the price of their HIV drugs. This competition, coupled with pressure from activists, organizations - such as the Clinton Foundation - and governments of poor countries with severe HIV epidemics, dramatically reduced the price of ARVs for developing countries. By the middle of 2001, triple combination therapy was available from Indian generic manufacturers for as little as $295 per person per year.

The price of ARVs for low and middle income countries (LMICs) has continued to fall. In 2013, the average cost of first-line antiretroviral treatment (ART) for LMICs was $115 per patient per year (PPY) and $330 PPY for second-line ART. The price of third-line ART has also decreased but LMICs still pay on average, more than $1500 PPY. Many middle-income countries in Asia, Latin America, Eastern Europe and Central Asia continue to pay higher prices due their inability to access cheaper generic ARVs.

In 2001, Indian generic drug manufacturer, Cipla, announced that it would sell a generic copy of a triple-therapy antiretroviral for US$350 per patient per year. This had an incredible impact as the competition this generated dramatically drove down the price of HIV drugs for developing countries, thereby increasing the range of affordable options for national treatment programs.

The graph below (Figure 4.15) illustrates the effect of generic competition on proprietary drug prices between 2000 and 2001. It shows the lowest world price per patient per year of triple combination therapy made up of stavudine (d4T) +lamivudine (3TC) +nevirapine (NVP)

The problem that rarely gets any attention is the fact that all HIV 'medications' are nothing but suppression of symptoms. The most effective procedure, the so-called Antiretroviral combination therapy, defends against resistance by suppressing HIV replication as much as possible, thus reducing the potential pool of spontaneous resistance mutations. Combinations of antiretrovirals create multiple obstacles to

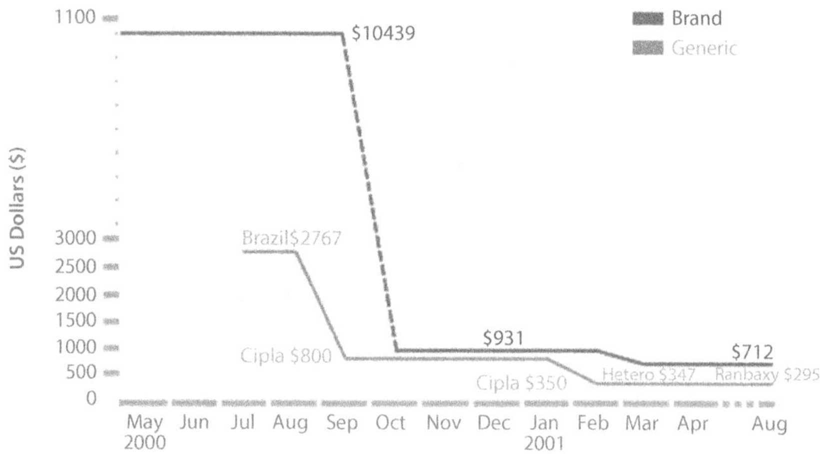

Figure 4.15 Drug prices during 2000-2001

HIV replication to keep the number of offspring low and reduce the possibility of a superior mutation. If a mutation that conveys resistance to one of the drugs being taken arises, the other drugs continue to suppress reproduction of that mutation. With rare exceptions, no individual antiretroviral drug has been demonstrated to suppress an HIV infection for long; these agents must be taken in combinations in order to have a lasting effect. As a result, the standard of care is to use combinations of antiretroviral drugs. Combinations usually comprise of three drugs from at least two different classes. This three-drug combination is commonly known as a triple cocktail. Combinations of antiretrovirals are subject to positive and negative synergies, which limits the number of useful combinations. All of them are very toxic to the human system and are prone to lack of effectiveness after a relatively short period.

A separate argument for starting antiretroviral therapy that has gained more prominence is its effect on HIV transmission. ART reduces the amount of virus in the blood and genital secretions. This has been shown to lead to dramatically reduced transmission of HIV when one partner with a suppressed viral load (<50 copies/ml) has sex with a partner who is HIV negative. In clinical trial HPTN 052, 1763 serodiscordant heterosexual couples in 9 countries were planned to be followed for at least 10 years, with both groups receiving education on preventing HIV transmission and condoms, but only one group getting ART. The study was stopped early for ethical reasons when it became clear that antiviral treatment provided significant protection. Of the 28 couples where cross-infection had occurred, all but one had taken place in the control group consistent with a 96% reduction in risk of transmission while on ART (Cohen et al, 2011). The term "treatment as prevention" has been used for the concept of treating patients with HIV to help prevent the spread of HIV. In 2011, the journal Science gave the Breakthrough of the Year award to treatment as prevention (Cohen, 2011).

In recent years, HIV and AIDS have been inflicted with skepticism, unparalleled in the information age. More and more scientists are entertaining the idea that no virus causes AIDS. In this, Kary Mullis, winner of the 1993 Nobel Prize in chemistry for discovering the polymerase chain reaction, has joined in, saying there is "no scientific evidence" proving that HIV causes AIDS. In June, 1994, the Pacific Division of the American Association for the Advancement of Science (publisher of Science) sponsored a daylong meeting at which the dissidents offered their points of view. Duesberg was the guest editor of an entire upcoming issue of the genetics journal *Genetica* that was devoted to alternative AIDS hypotheses. An article in Science, Dec. 1994, published an 8-page analysis of some dissent concerning HIV as a cause of AIDS (pp. 1642-1649). The article in several parts, authored by Jon Cohen, was headlined:

The Duesberg Phenomenon

A Berkeley virologist and his supporters continue to argue that HIV is not the cause of AIDS.

A 3-month investigation by Science evaluates their claims.

Science (and not just its reporter Jon Cohen) took responsibility for the investigation and its conclusions. Aside from the displayed responsibility of "a 3-month investigation by Science" in the heading, responsibility was further taken on the very first page, where Science was mentioned three more times as follows:

But because the Duesberg phenomenon has not gone away and may be growing, Science decided this was a good time to examine Duesberg's main claims. In a three-month investigation, Science interviewed more than 50 supporters and detractors... The main conclusions of Science's investigation are that:

- in hemophiliacs. .. there is abundant evidence that HIV causes disease and death (see p. 1645);
- according to some AIDS researchers, HIV now fulfills the classic postulates of disease causation established by Robert Koch (see p. 1647);

- the AIDS epidemic in Thailand, which Duesberg has cited as confirmation of his theories, seems instead to confirm the role of HIV (see p. 1647); and
- AZT and illicit drugs, which Duesberg argues can cause AIDS, don't cause the immune deficiency characteristic of that disease (see p. 1648).

One effect of the Science article of 9 December 1994 was to acknowledge officially, in the number-one magazine of the scientific establishment, the existence of an expanding challenge to the HIV/AIDS hypothesis and to the establishment's way of dealing with this challenge in the past.

On the other hand, the Science article of 9 December 1994 may be seen as tendentious and skewed. Although this is not the place to make a comprehensive detailed analysis, a a couple of examples may help illustrate and clarify.

First I [=Jon Cohen] object to personalizing dissent about the official line that "HIV causes AIDS" in the context of "The Duesberg Phenomenon." I object to lumping together different people such as Harry Haverkos (who sponsored the NIDA May 1994 meeting on nitrite inhalants), the co-authors of the article on AIDS in Africa14 referred to in footnote 4, or myself among many others, as part of "the Duesberg phenomenon." What has "not gone away" is that an increasing number of individual scientists, with different points of view, different backgrounds, and different responsibilities, have publicly documented reservations about the official position of the government or the scientific establishment concerning HIV and AIDS. Lumping together independent scientists under the single category of Duesberg "supporters" skewed the perspective on the dissenters and on their multiple reasons for dissent.

Second, the article completely omitted mention of dissenters such as Bialy and Haverkos, as well as many points raised by the dissenters. For example, the NIDA meeting of May, the position of Harry Haverkos on nitrite inhalants, the situation in Africa, the fact that malaria, tuberculosis, leprosy, and influenza, test false positive on the HIV antibodies test, were still not mentioned in the Science article. The AAAS June meeting was mentioned in only one sentence: "In June, the Pacific Division of the American

Association for the Advancement of Science (publisher of Science) sponsored a daylong meeting at which the dissidents offered their points of view." No indication was given what were these points of view.

Especially significantly, as we shall see below, the 9 December 1994 article also made no mention of Karposi's sarcoma.

A recent issue of Yale Scientific, which is published by Yale undergraduates in the sciences, carried a cover story by mathematician Serge Lang titled "HIV/AIDS: Have We Been Misled?"

All this controversy confounds AIDS researchers who think HIV has been decisively established as the cause of AIDS. Describing HIV as harmless is "irresponsible, with terribly serious consequences," says Warren Winkelstein Jr., a Berkeley AIDS epidemiologist who has known Duesberg for more than 20 years. Duesberg's message, Winkelstein and others say, undermines the value of campaigns for the use of condoms and clean needles. What is more, says Helene Gayle, associate director of the Centers for Disease Control and Prevention (CDC) office in Washington, D.C., that message is "very damaging" in the attempt to persuade other nations to stem the spread of AIDS, because "people already want to deny" the presence of HIV and AIDS in their countries. This Duesberg is no ordinary detractor of the Establishment. Peter H. Duesberg is a professor of Molecular and Cell Biology at the University of California, Berkeley and is well known for his work on cancer. He isolated the first cancer gene through his work on retroviruses in 1970, and mapped the genetic structure of these viruses. This, and his subsequent work in the same field, resulted in his election to the National Academy of Sciences in 1986. He is also the recipient of a seven-year Outstanding Investigator Grant from the National Institutes of Health. On the basis of his experience with retroviruses, Duesberg has challenged the virus-AIDS hypothesis in the pages of such journals as Cancer Research, Lancet, Proceedings of the National Academy of Sciences, Science, Nature, Journal of AIDS, AIDS Forschung, Biomedicine and Pharmacotherapeutics, New England Journal of Medicine and Research in Immunology. He has instead proposed the hypothesis that the various American/European AIDS diseases are brought on by the long-term consumption of recreational drugs and/or AZT itself, which is prescribed to prevent or treat AIDS. See The AIDS Dilemma: Drug diseases blamed on a passenger virus. He published his work in a book (Duesberg, 1996) and was supported by such notables as Nobel Laureate Chemist Kary B. Mullis.

Yet, Duesberg's findings have been a thorn in the side of the medical establishment and drug companies since 1987. Instead of engaging in scientific debate, however, the only response has been to cut-off funding to further test Professor's Duesberg's hypothesis. (Further discussion will be carried out in Part II of this book, including discussion of dogma-free scientific theories that explain the real cause behind AIDS. This line of discussion is rarely carried out in modern age [Ho, 2011]).

4.4.5 Diabetes

Type2 Diabetes is a disorder of carbohydrate metabolism characterized by impaired ability of the body to produce or respond to insulin and thereby maintain proper levels of sugar (glucose) in the blood. Diabetes epitomizes the onset of HSSA degradation in healthcare and public health. There are two major forms of the disease. Type I diabetes, formerly referred to as insulin-dependent diabetes mellitus (IDDM) and juvenile-onset diabetes, usually arises in childhood. It is an autoimmune disorder in which the diabetic person's immune system produces antibodies that destroy the insulin-producing beta cells. Because the body is no longer able to produce insulin, daily injections of the hormone are required to sustain vital functions. Type II diabetes, formerly called non-insulin-dependent diabetes mellitus (NIDDM) and adult-onset diabetes, usually occurs after 40 years of age and becomes more common with increasing age. It arises from either sluggish pancreatic secretion of insulin or reduced responsiveness in target cells of the body to secreted insulin. New science links genetics and obesity to diabetes. As unusual, such linkage arises from false premises, such as 'defective gene' and cannot explain any phenomena with reasonable logic. The solution to diabetes is given as introduction of more artificial doses of sugar, making the process inherently implosive.

Diabetes is synonymous with the sugar culture. International Diabetes Foundation's (IDF) most recent estimates indicate that 8.3% of adults – 382 million people – have diabetes, and the number of people with the disease is set to rise beyond 592 million in less than 25 years (Figure 4.16). Yet, with 175 million of cases currently undiagnosed, a vast amount of people with diabetes are progressing towards complications unawares. More importantly, New science is working hard toward labeling more people as diabetic and putting them in life-long 'management mode' with more artificial chemicals that make a diabetic patient

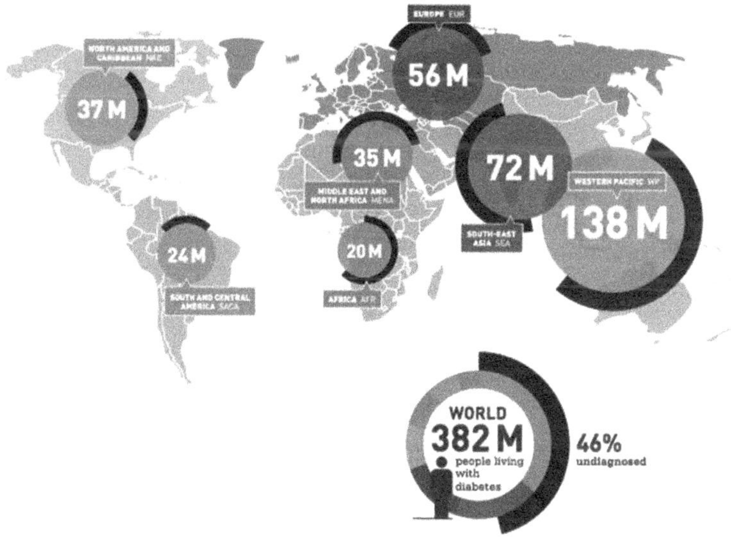

Figure 4.16 Diabetes around the world (from IDF.org)

more dependent on the very chemicals that created the mess to begin with. The spiraling down continues there on.

With diabetes comes other complications, none of which can be addressed in isolation with the root cause of diabetes. It just happens that the cause cannot be determined because new science has built-in prejudices in it. IDF lists the following complications (Figure 4.17):

4.4.5.1 Cardiovascular Disease

Cardiovascular disease is the most common cause of death and disability among peoplewith diabetes. The cardiovascular diseases that accompany diabetes include angina, myocardial infarction (heart attack), stroke, peripheral artery disease, and congestive heart failure. In people with diabetes, high blood pressure, high cholesterol, high blood glucose and other risk factors contribute to the increased risk of cardiovascular complications.

4.4.5.2 Kidney Disease

Kidney disease (nephropathy) is far more common in people with diabetes than in people without diabetes; and diabetes is one of the leading causes of chronic kidney disease. The disease is caused by damage to

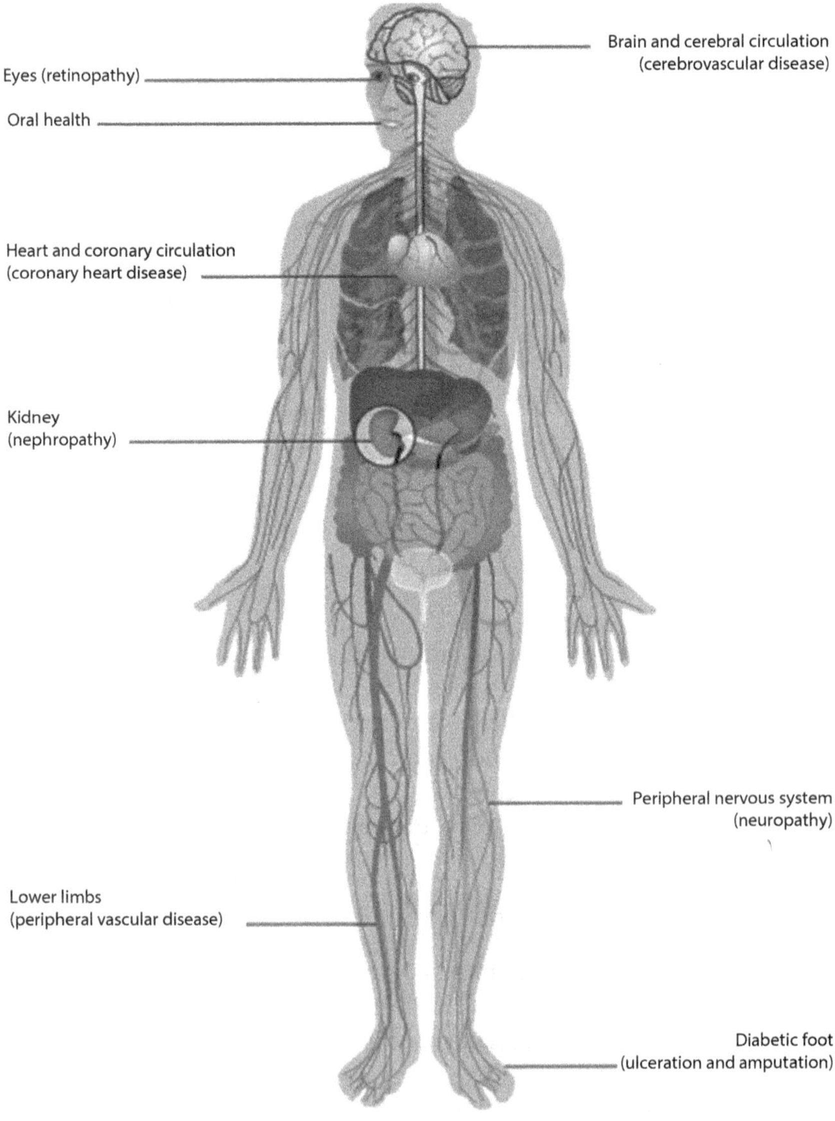

Figure 4.17 Complications arising from diabetes (from IDF.org)

small blood vessels, which can cause the kidneys to be less efficient, or to fail altogether. Maintaining near-normal levels of blood glucose and blood pressure can greatly reduce the risk of nephropathy.

4.4.5.3 *Diseases of the Eye*

Many people with diabetes develop some form of eye disease (retinopathy), which can damage vision or provoke blindness. Persistently high levels of blood glucose, together with high blood pressure and high cholesterol, are the main causes of retinopathy. The network of blood vessels that supply the retina can become blocked and damaged in retinopathy, leading to permanent loss of vision. Retinopathy can be managed through regular eye
checks and by keeping blood glucose levels close to normal.

4.4.5.4 *Nerve Damage*

When blood glucose and blood pressure are excessively high, diabetes can provoke damage to nerves throughout the body (neuropathy). This damage can lead to problems with digestion and urination, erectile dysfunction and a number of other functions.

4.4.5.5 *Diabetic Foot*

People with diabetes may develop a number of different foot problems as a result of damage to nerves and blood vessels. These problems can easily lead to infection and ulceration, which increase a person's risk of amputation. People with diabetes face a risk of amputation that may be more than 25 times greater than that in people without diabetes.

4.4.5.6 *Pregnancy Complications*

Women with any type of diabetes during pregnancy risk a number of complications. High blood glucose during pregnancy can lead to fetal abnormalities and cause it to gain excess size and weight, and overproduce insulin. These can lead to problems at delivery, injuries to the child and mother, and a sudden drop in blood glucose (hypoglycaemia) in the child after birth. Children who are exposed for a long time to high blood glucose in the womb are at higher risk of developing type 2 diabetes

4.4.5.7 *Oral Health*

Diabetes can pose a threat to oral health. For example, there is an increased risk of inflammation of the gums (gingivitis) in people with poor glucose control. Gingivitis in turn is a major cause of tooth loss and may also increase the risk of cardiovascular disease.

4.4.5.8 *Sleep Apnoea*

Recent research demonstrates the likelihood of a relationship between type 2 diabetes and obstructive sleep apnoea. Estimates suggest that up to 40% of people with sleep apnoea have diabetes, although the incidence of new diabetes in people with sleep apnoea is not known. In people with type 2 diabetes, sleep apnoea may have effects on their ability to control blood glucose.

4.5 Diseases with Implications for Fundamental Theoretical Concerns (Mass,Energy and HTM)

Modern science differentiates between physical ailment and mental disorders. Such characterization is based on the false premise that physical ailment is independent of human thought material (HTM) and mental ailment is independent of mass or chemical bodies. This is typical form of isolation that is commonly applied in Autism research (Aberar *et al.*, 2013). This in itself is violation of the law of continuity. Any subsequent cognition, therefore, is aphenomenal. As an example one can cite two diseases, namely, autism and bipolar disorder. For both, the examples of childhood disorders are implicated. Childhood is an important timeline in HTM because the premise: Nature is perfect implies that children are born perfect and most equipped with natural cognition. Also, the transition in fetus defines an environment that is the best nature has to offer, based on the same premise: Nature is perfect.

Epidemiological studies are often reported to demonstrate the extent of a disease. Figure 4.18 summarizes the occurrence of autism among children in USA. This figure shows a radical increase (Newschaffer *et al.*, 2007). Unfortunately these facts typically are not used to extract any useful information. Any such publication of facts quickly follows claims that such spike is because of the increase in diagnostic accuracy. When that is not satisfactory, some single cause is researched instead of diversifying the search for a spectrum of interrelated probable causes (Rice, 2011). Kim *et al.* (2011) combined data from various sources and

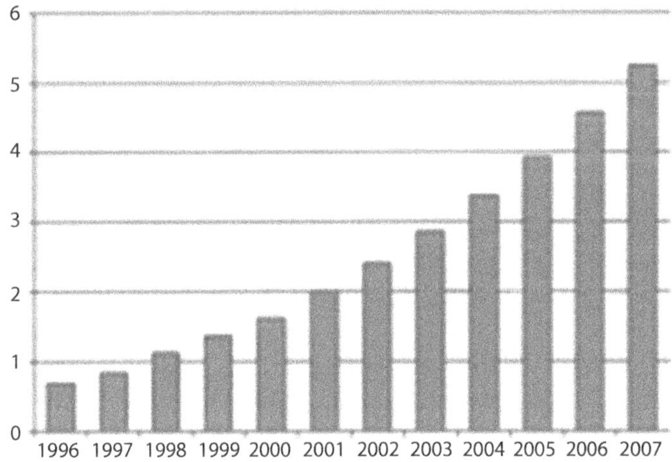

Figure 4.18 Occurrence of Autism among children in USA (per 1000 children).

reported that the number of such occurrences has exceeded 1 out of 88 in 2008. In any event, the increase in occurrence of autism is clearly exponential, a truth dealt with in a number of follow-up researches. Baio (2008) analyzed autism data based on ethnicity of different groups in USA. Missing from the analysis, however, is any discussion of how ethnicity is linked with type of immune system, upbringing, type of food, and social interactions. Not surprisingly, the report concludes:

> ASDs continue to be an important public health concern. The findings provided in this report confirm that prevalence estimates of ASD continue to increase in the majority of ADDM (The Autism and Developmental Disabilities Monitoring) Network communities, and ongoing public health surveillance is needed to quantify and understand these changes over time. Further work is needed to evaluate multiple factors affecting ASD prevalence over time. ADDM Network investigators continue to explore these factors in multiple ways, with a focus on understanding disparities in the identification of ASDs among certain subgroups and evaluating temporal changes in the prevalence of ASDs. CDC also is engaged with other federal, state, and private partners in a coordinated response to identify risk factors for ASDs and meet the needs of persons with ASDs and their families.

The lack of relevance and of anything other than self-serving statements are the most remarkable features of the above conclusion. Further scrutiny of the data reveals the following facts that are not used in the conclusion.

Fact 1. *On average, estimated ASD prevalence was significantly higher in ADDM sites that had access to education sources compared with sites that relied solely on health sources to identify cases, and ASD prevalence in the one Colorado county where project staff had access to both health and education sources was almost twice as high as in the six counties where only health records were accessed. Having access to education records continues to be associated with higher identified ASD prevalence when comparing across or within most ADDM Network sites.*

Even though the above fact is not used in the conclusion, it gives a clear message that the ASD prevalence is associated to lack of education and monitoring.

Fact 2. *The two sites with the highest ASD prevalence estimates in 2008 (Utah and New Jersey) identified higher proportions of children with IQ >70 compared with any site participating in 2008 or previous ADDM Network surveillance years. Among the seven sites reporting data on intellectual ability, ASD prevalence was typically higher in sites having a greater percentage of children with IQ >70 (Figure 4.19).*

The above statement of 'facts' is inherently flawed because IQ itself is based on certain assumptions that are skewed and come with fundamentally incorrect premises.

Fact 3. *The estimated prevalence of ASDs was significantly higher among boys (one in 54) than among girls (one in 252), with a male-to-female RR of 4.6 for all sites combined.*

Similar observations are recorded for other ailments. For instance, CDC 2008 reports that males had twice the prevalence of any Developmental disabilities (DD) than females and more specifically had higher prevalence of ADHD, learning disabilities, stuttering/stammering and other DDs.

This finding had the potential of trying to find a great clue as to what can trigger Autism, but this fact is not used in rest of the document and nothing in the conclusion indicated the possibility of any consideration of gender in dealing with autism. If one considered real difference in cognition pattern of boys and girls, one would be in a position to advance a theory that could unlock the mysteries of autism. It turns out that this difference is not obvious and little research is done on discovering the real difference (Eliot, 2010). In addition, the difference that New science has focused on such as size of the brain, verbal skill, physical skill, etc. are all part of the myopic approach and reveal nothing that would explain such discrepancy in occurrence of autism.

Fact 4. *ASD prevalence estimates also varied widely by race/ethnicity across and within most sites, with only one site (New Jersey) identifying approximately the same ASD prevalence among non-Hispanic white, non-Hispanic black, and Hispanic children. Most ADDM Network sites continue to identify higher ASD prevalence estimates among non-Hispanic white children compared with other racial/ethnic groups...*

This fact is another aspect that could give numerous clues as to what causes Autism. It would also show the conventional myopic approach that quantifies cognitive ability with measurable qualities will fail to explain such phenomena (Rushton and Jensen, 2005). This aspect is supported by most recent works on the topic of gender and intelligence (Naderi *et al.*, 2010).

Fact 5. *Studies in Asia, Europe, and North America have identified individuals with an ASD with an average prevalence of about 1%. A recent study in South Korea reported a prevalence of 2.6%* (Kim *et al.*, 2011).

South Korea would correlate with high GDP, which would correlate with 'high foot-print' lifestyle. This correlation is not cited by anyone.

Fact 6. *About 1 in 6 children in the U.S. had a developmental disability in 2006-2008, ranging from mild disabilities such as speech and language impairments to serious developmental disabilities, such as intellectual disabilities, cerebral palsy, and autism* (CDC, 2008).

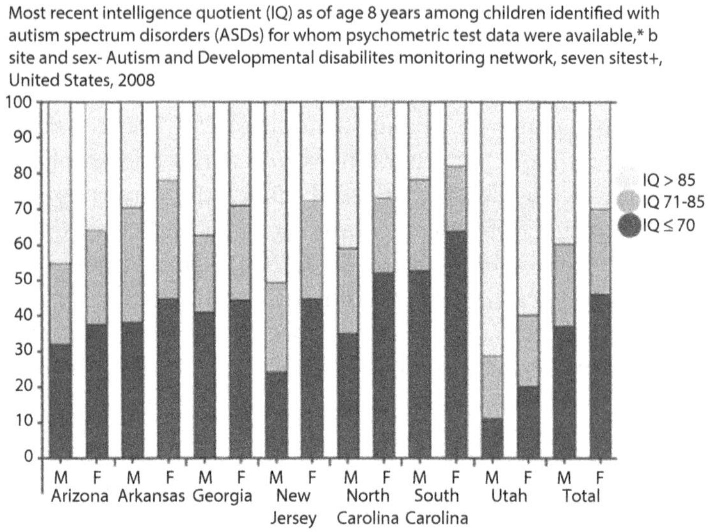

Figure 4.19 Correlation between Autism and IQ (taken from CDC report, Baio, 2008).

This is an observation that can indicate overall decline in cognitive abilities among children in USA. This is significant because, none of these occurrences can be isolated from each other despite the fact that New Science is incapable of considering them holistically. Table 4.11 lists several facts observed during 1997 through 2008.

The above facts can be used to determine vulnerable groups but, if the reason for vulnerability is not known, such characterization would have no merit.

In terms of physical facts, autism is known to be directly related to neurons of a human body. These neurons are affected by both mass and energy sources, family history, and the environment (Singh and Eroglu, 2010; Noor *et al.*, 2010, Rutter, 2005; Johnson and Myers, 2010). Even genetic modification is attributed to the environment and lifestyle (Kobor *et al.*, 2012). However, at the same time it is asserted that the environment has no impact on the occurrence of autism and little weight is given to the prospect of the connection between autism and the environment (Szpir, 2006). There are papers that connect various physical conditions with Autism, but once again the obvious link between energy and matter and lifestyle is missing (Strahana and Eldera, 2013; Matson and Goldin, 2013).

The situation is more complex and solutions more remote for other forms of mental illness. For instance, similar surge in bipolar disorder

Table 4.11 Observations during 1997-2008 (from CDC, 2008 and Boyles *et al.*, 2011)

The condition	Prevalence over 12 years period
Developmental disability (DD)	13.87%
Learning disabilities	7.66%
Attention deficit hyperactivity disorder (ADHD)	6.69%
Other developmental delay	3.65%
Autism	0.47%
	Increase over 12 years compared to decade before
DD	17.1%
Autism	289.5%
ADHD	33.0%
Hearing loss	-30.9%
	Group prevalence
	Greater prevalence among boys than girls
Autism, ADHD and learning disabilities;	Greater prevalence among white children than non-white children
Stuttering/stammering	Greater prevalence among black children than white children
General DDs	Children insured by Medicaid had a nearly two-fold higher prevalence compared to those with private insurance
General DDs	Children from families with income below the federal poverty level had a higher prevalence than others.
Autism	36-95% vulnerability among identical twins 0-31% vulnerability among non-identical twins 2-18% vulnerability among siblings

among children has also been reported (Judd and Akisal, 2003; Youngstrom, 2006; Blader and Carlson, 2007), however, the role of environment that would include both tangible material source as well as human thought material has been excluded. Instead Similar increase is observed among US children. At the end, the search for bipolar gene was intensified (Holden, 1993; Bernett and Smoller, 2009). Recently, Parry and Levin (2012) alluded to some of these frustrating aspects of bipolar research.

4.5.1 Violence and Mental Illness

Following facts are reported by American Psychiatric Association. (1994).

> **Fact 1**: The vast majority of people with mental illness are not violent.

Here is what researchers say about the link between mental illness and violence:

- "Although studies suggest a link between mental illnesses and violence, the contribution of people with mental illnesses to overall rates of violence is small, and further, the magnitude of the relationship is greatly exaggerated in the minds of the general population (Institute of Medicine, 2006)."
- "...the vast majority of people who are violent do not suffer from mental illnesses (American Psychiatric Association, 1994)."
- "The absolute risk of violence among the mentally ill as a group is very small.. . only a small proportion of the violence in our society can be attributed to persons who are mentally ill (Mulvey, 1994)."
- "People with psychiatric disabilities are far more likely to be victims than perpetrators of violent crime (Appleby, *et al.*, 2001). People with severe mental illnesses, schizophrenia, bipolar disorder or psychosis, are 2 ½ times more likely to be attacked, raped or mugged than the general population (Hiday, *et al.*,1999)."

Fact 2: The public is misinformed about the link between mental illness and violence.

A longitudinal study of American's attitudes on mental health between 1950 and 1996 found, "the proportion of Americans who describe mental illness in terms consistent with violent or dangerous behavior nearly doubled." Also, the vast majority of Americans believe that persons with mental illnesses pose a threat for violence towards others and themselves (Pescosolido, *et al.*, 1996, Pescosolido *et al.*, 1999).

Fact 3: Inaccurate beliefs about mental illness and violence lead to widespread stigma and discrimination

Discrimination and stigma associated with mental illnesses stem in part, from the link between mental illness and violence in the minds of the general public (DHHS, 1999, Corrigan, *et al.*, 2002).

The effects of stigma and discrimination are profound. The President's New Freedom Commission on Mental Health found that, "Stigma leads others to avoid living, socializing, or working with, renting to, or employing people with mental disorders - especially severe disorders, such as schizophrenia. It leads to low self-esteem, isolation, and hopelessness. It deters the public from seeking and wanting to pay for care. Responding to stigma, people with mental health problems internalize public attitudes and become so embarrassed or ashamed that they often conceal symptoms and fail to seek treatment (New Freedom Commission, 2003)."

Fact 4: The link between mental illness and violence is promoted by the entertainment and news media.

"Characters in prime time television portrayed as having a mental illness are depicted as the most dangerous of all demographic groups: 60 percent were shown to be involved in crime or violence" (Mental Health American, 1999).

"Most news accounts portray people with mental illness as dangerous" (Wahl, 1995).

"The vast majority of news stories on mental illness either focus on other negative characteristics related to people with the disorder (e.g., unpredictability and unsociability) or on medical treatments. Notably

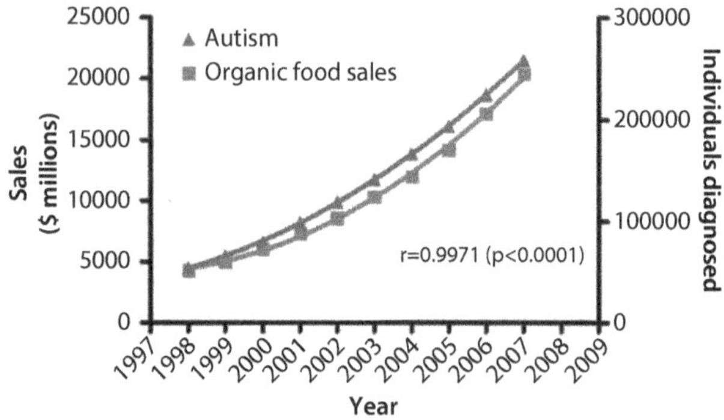

Figure 4.20 Any suggestion of correlation between Autism and vaccine is faced with ridicule.

absent are positive stories that highlight recovery of many persons with even the most serious of mental illnesses" (Wahl, *et al.*, 2002).

4.5.2 Role of Vaccine and Lifestyle

The American Academy of Pediatrics, the CDC, the World Health Organization, and the Institute of Medicine all agree that there is probably no relationship between autism and vaccines. It is said that all such correlation is worthy of mention only on talk shows or in other fora that produce nothing of probative value. Indeed, it took an actress, Jenny McCarthy, who went on Oprah to say she believes that a vaccination caused her son's autism and wrote a book about it, to Sen. John McCain, who, at a campaign event earlier this year, said he thought there was "pretty strong evidence" that some vaccines cause autism. In general this is ridiculed and in support of such ridicule, the likes of Figure 20 are presented.

However, parents continue to complain that infants never had symptoms of Autism until first shots of vaccination. Their beliefs may have been validated in March, 2008, when federal officials said that a Georgia girl was entitled to compensation because vaccines may have aggravated an underlying condition, causing autism-like symptoms. In addition, researchers at the National Institute of Mental Health (NIMH) said in March (2008) that they are still taking a careful look into parent concerns that vaccines are tied to the disorder.

"Parents are clamoring for a cause," said David Tayloe, MD, a pediatrician in Greensboro, N.C., and president-elect (2008) of the American Academy of Pediatrics (AAP). However, he continues: "It's a terrible condition. It upsets families, and it upsets me." But all the fear and anger about vaccines is misplaced, he says. "There's just nothing there." Few notice that such conclusions are nothing but re-statement of first premises.

When parents complain about symptoms of Autism surfacing only after vaccine, most medical researchers argue that this is probably a coincidence: Autism symptoms tend to become apparent around the same time that children are scheduled to get routine vaccines.

Although there are two separate issues concerning vaccines and autism, they're often lumped together. One has to do with the measles, mumps, and rubella (MMR) vaccine; the other involves vaccines containing the chemical preservative thimerosal, which contains a form of mercury that has been suspected of causing autism and has recently been removed from most vaccines.

The MMR scare started 10 years ago with a report published in *The Lancet* that described the cases of eight children that developed autistic symptoms and digestive ailments shortly after getting their first MMR dose. The researchers proposed that the vaccine might trigger a previously unknown form of regressive autism. They suggested that maybe the measles virus in the vaccine lodged in the intestine, causing some kind of reaction that then affected the brain.

After that, experts studied whether the MMR vaccine could cause autism. To do that, they looked for clues among kids who did and didn't get the vaccine. Since that initial finding, 14 studies including millions of children in several countries consistently show no significant difference in autism rates between children who got the MMR vaccine those who didn't. All researchers jumped to conclusion: It's very unlikely that the MMR causes autism.

Japan was the first country to ban MMR in favour of a single vaccine. Soon, there surfaced articles titled: Autism rises despite MMR ban in Japan (Coghlan, 2005). Once more, started with a premise, scientists marvel at the finding that their conclusions match their first premise. According to Coghlan, "A study of more than 30,000 children in Japan should put the final nail in the coffin of the claim that the MMR vaccine is responsible for the apparent rise in autism in recent years." He continued:

> In the UK, parents panicked and vaccination rates plummeted after gastroenterologist Andrew Wakefield claimed in a 1998 study that MMR might trigger autism, although the study was based on just 12 children and later retracted by most of its co authors.

> Soon the vaccine was being blamed for the apparent rise in autism, with Wakefield citing data from California, US (see Figure 4.21). In some parts of the UK, the proportion of children receiving both doses of the MMR vaccine has dropped to 60%. This has led to a rise in measles outbreaks and fears of an epidemic.
>
> Not one epidemiological study has revealed a link between the vaccine and autism. But until now they have all concentrated on what happened after MMR vaccination for children was introduced. Honda's is the first to look at the autism rate after the MMR vaccine has been withdrawn. Japan withdrew it in April 1993 following reports that the anti-mumps component was causing meningitis (it plans to introduce another version).

It would take over a decade to discover that Coghlan left out a few details from Japan. As pointed out by Hope in 2015, Japan continues to ban MMR and the increase in Autism that was attributed to not giving MMR was actually after introduction of a new variety of MMR. Hope (2015) pointed out that such decision was made only because a four-year experiment showed serious financial and human costs. Of the 3,969 medical compensation claims relating to vaccines in the last 30 years, a quarter had been made by those badly affected by the combined measles, mumps and rubella vaccine. Another fact left out by Coghlan was that prior to banning MMR in 1993, 1.8 million children had been given two types of MMR and a record number developed non-viral

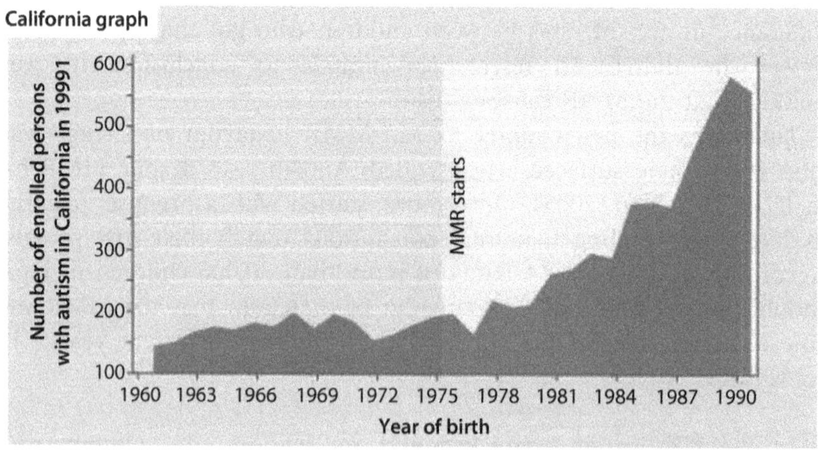

Figure 4.21 Correlation reported in California (data from Dept. of Developmental Services, Sacramento CA, 1999)

meningitis and other adverse reactions. Non-viral meningitis cases are counted by one group while aren't counted by the other. This cannot be how real science operates. When MMR was made mandatory in Japan, an analysis of over a three-month period showed one in every 900 children was experiencing problems. This was over 2,000 times higher than the expected rate of one child in every 100,000 to 200,000. The situation didn't improve when the ministry switched to another MMR vaccine in October 1991. The incidence was still high with one in 1,755 children affected. No separate record has been kept of claims involving autism. Tests on the spinal fluid of 125 children affected were carried out to see if the vaccine had got into the children's nervous systems. They found one confirmed case and two further suspected cases. Such desperation emerges from the absence of fundamentally sound theory of mass transfer, let alone energy transfer. In this book such theories are advanced in latter chapters.

When researchers started looking for a possible link between autism and MMR vaccine, all other childhood vaccines came under scrutiny. In 1998, 30 different vaccines with thimerosal in them were given to children. U.S. public health officials realized that the recommended schedule of vaccines could dose some children with mercury at a level exceeding the limit considered safe by government standards. This is no surprise. After all, the same companies once introduced birth control pills that exceeded toxin content 10 times the required amount to kill a human egg. In 1999, the U.S. Public Health Service and the AAP asked vaccine makers to reduce or remove thimerosal in vaccines.[3]?

The 2004 IOM review included five large-scale studies that compared autism rates in vaccinated and unvaccinated children. These and other recent studies, including one published in *The New England Journal of Medicine* in September 2007, have shown that children who received vaccines with thimerosal are not more likely to have been diagnosed with autism than those that weren't vaccinated or received less thimerosal from vaccines.

When the likes of Sallie Bernard, founder of the advocacy group Safe Minds, ask for scientific evidence or true scientific investigation, they are dismissed for focusing on "very superficial similarity." It is known that symptoms of mercury poisoning can include irritability, depression, anxiety, visual problems, speech problems, and sensory nerve problems.

---

[3]For what it's worth, by 2001, all routine childhood vaccines were reportedly available thimerosal-free. However, what replaced the 'thimerosal'? Could it be another episode of HSSA degradation

In autism, there can be findings of delayed speech, increase or decrease in response to sensory stimuli, and avoidance of human eye contact. But here's the thing: *no one is willing to consider the possibility of vaccine being inherently harmful because it is made out of artificial chemicals.*

Similar vilification arose when Kirby (2006) introduced in his book the controversies surrounding thimerosal-containing vaccines (TCVs), and whether TCVs have contributed to the apparent increase of autism, ADHD, speech delay and other childhood disorders in the United States. In the book, Kirby told the personal stories of parents of children who have autism, founders of the advocacy group SafeMinds, including Sally Bernard, Lyn Redwood, Mark Blaxill, Albert Enayati, Heidi Roger and Liz Birt. He wrote, "Many of the public health officials who discount the thimerosal theory were unwilling to be interviewed for this book (or [were] prohibited from speaking by superiors)...Readers are invited to reach their own conclusions on the evidence." Nevertheless, Kirby also acknowledges "(E)ach side accuses the other of being irrational, overzealous, blind to evidence they find inconvenient, and subject to professional, financial, or emotional conflicts of interest that cloud their judgment." Ultimately, Kirby finds the evidence leveled against thimerosal inconclusive but suggestive, and calls urgently for further research. Even then, he didn't escape the wrath of the Establishment. In May 2005, Evidence of Harm was reviewed negatively in the *British Medical Journal.* The reviewer described Kirby's book as "woefully one-sided", and wrote: "In his determination to provide an account that is sympathetic to the parents, Kirby enters into the grip of the same delusion and ends up in the same angry and paranoid universe into which campaigners have descended, alleging phone taps and other forms of surveillance as they struggle against sinister conspiracies between health authorities and drug companies." (Fitzpatrick, 2005)

At the core of all these 'controversies' lie numerous assumptions that are false but whose falsity cannot be demonstrated false with New Science today. For instance, Thimerosal contains ethyl mercury. The government's assessments of health risks for thimerosal are based on what is known about another chemical form of mercury, called methyl mercury. So, summarizing this portion of the deconstruction so far:

- there is no discussion about natural source and artificial source;
- there is no discussion about the fact that health assessments for methyl mercury are based on exposure

by eating or drinking it, which means stomach can process thcsc chemicals[4];

- there is no discussion of the fact that Thimerosal is not eaten (when experts called for reducing and removing thimerosal from vaccines in 1999, they assumed that methyl mercury and ethyl mercury were similar);
- Baker (2008) showed that ethyl mercury leaves children's blood faster than methyl mercury does, so the health risks may not be the same. Vaccine-lovers took this as a sign that Thimerosal is safe, imposing the false premise that: leaving blood quickly means 'leaving body' quickly. To enforce this premise, damage to kidneys was analyzed and declared 'safe' again, invoking yet another false premise, i.e., heavy metals travel to the kidneys by preference, even for transport through the blood stream.

Is it possible that Baker didn't know about fundamental facts about human anatomy and physiology, such as the following:

> "Intravenous injection of elemental mercury is comparatively less harmful as compared to that of its salts or the inhalation of mercury vapour. The heavy metal tends to gravitate to dependent areas and extravasates into tissues. **When carried by circulation, it tends to lodge in the lungs and mediastinum** and is also known to reach other organs over a period of time." (Gopalakrishna and Kumar, 2008)

Or is there is some other reason to focus on the kidneys?

4.6 The Need for the Science of Intangibles

Even though for millennia, the notion of intangibles was at the core of the Indian, Chinese, Egyptian, Babylonian, and other civilizations, this notion is largely neglected today (Khan and Islam, 2012). It can be argued that the lack of consideration of intangibles in the modern age is deliberate, due to a focus on the short-term. In the words of John Maynard Lord Keynes, who believed that historical time had nothing to

[4]Warnings about eating certain kinds of fish, for example, are all about fish contaminated with methyl mercury.

do with establishing the truth or falsehood of economic doctrine, "In the long run, we are all dead," (cited by Zatzman and Islam, 2007b).

"Economics", i.e. maximum profit in the minimum time, is the driver of modern engineering, and it follows that the shortest possible term has become the guiding principle behind all engineering calculations. This focus on the short term poses a serious problem in terms of scientific investigation. The science of tangibles says that, unless one can verify their presence and role with some experimental program – "experimental" meaning controlled conditions, probably in a laboratory, with experiments that are designed by means of the same science that one has set out to prove, there is no need or room otherwise for intangibles. By contrast, Khan and Islam (2007a; 2007b) argued that the science of tangibles, so far, has not been able to account for disastrous outcomes of numerous contemporary technologies. By the same token, scientists cannot determine the cause of global warming with a science that assumes all molecules are identical, thereby making it impossible to distinguish between organic CO_2 and industrial CO_2. Nor can scientists determine the cause of diabetes, unless there is a paradigm shift that distinguishes between sucrose in honey and sucrose in Aspartame® (Chhetri and Islam, 2008).

This same logic would indicate, unless the science that includes intangibles, the cause(s) of CCD cannot be determined. What remains uncharted are the roles of the pathways and the passages of time — something that cannot be followed meaningfully in lab-controlled conditions — in transforming the internal basis of changes in certain natural phenomena of interest. One example has been given by Khan and Islam (2012), regarding the use of catalysts. Tangible science says catalysts play no role in the chemical reaction equation because they do not appear in the result/outcome. No mass balance accounts for the mass of catalyst lost during a reaction, and no chemical equation accounts for what happens to the "lost" catalyst molecules when they combine with the products during extremely unnatural conditions. By using the science of tangibles, one can argue that the following patent is indeed a technological breakthrough (El-Shoubary *et al.*, 2003). This patented technology separates Hg from a contaminated gas stream using $CuCl_2$ as the main catalyst. At a high temperature, $CuCl_2$ would react with Hg to form Cu-Hg amalgam. This process is effective when combined with fire-resistant Teflon membranes.

According to the science of tangibles that describes the mass balance within the confines of the reactor, this is indeed an effective technology.

As a testimony to this statement, there has been a number of patents (all with El-Shoubary as an inventor) on the same topic. They are:

1. Patent # 6,841,513: "Adsorption powder containing cupric chloride", Jan 11, 2005.
2. Patent# 6,589,318: "Adsorption powder for removing mercury from high temperature, high moisture stream", July 8, 2003.
3. Patent# 6,5824,9: "Adsorption powder for removing mercury from high temperature high moisture gas stream", June 24, 2003.
4. Patent# 6,558,642: "Method of adsorbing metals and organic compounds from vaporous streams", May 6, 2003.
5. Patent# 6,533,842:"Adsorption powder for removing mercury from high temperature, high moisture gas stream", March 18, 2003.
6. Patent# 6,524,371: "Process for adsorption of mercury from gaseous streams", Feb 25 2003.

This high level of recognition of the technology is expected. After all, what happens to Teflon at high temperatures and what happens to Cu-Hg amalgam is a matter for the long term, or at least of a duration of time lying beyond the "time of interest": (Khan and Islam, 2007 describes this as "time t = right now").

If longer-term time is used for the analysis and a bigger area is considered for the mass balance, however, it would become clear that the same process has actually added more waste to the environment in the form of dioxins released from Teflon and Cu-Hg, both being in a more harmful state than their original states in the Teflon, $CuCl_2$, and gas streams, respectively. In the efficiency calculation, nearly 90% efficiency is reported within the reactor. This figure makes the process seem very attractive. However, if the efficiency calculation is conducted, *including the entire system in which the heater resides*, the efficiency — surprise!— drops drastically. In addition, by merely including more elements, the conversion of Hg in the natural gas stream and Cu in $CuCl_2$ solution into a Cu-Hg sludge, as well as the addition of chlorine in the effluent gas, pose difficult questions as to what has been accomplished overall.

Another example can be given from the chemical reaction involving honey, on the one hand, and Aspartame® on the other. With the science of tangibles, the following reactions take place:

Honey + $O_2 \rightarrow$ Energy + CO_2 + Water
Aspartame® + $O_2 \rightarrow$ Energy + CO_2 + Water

In fact, a calorie-conscious person would consider Aspartame® a better alternative to honey as the energy produced in Aspartame is much less than that of honey for the same weight burnt, should this be "the same number of calories burned?" An entirely different picture emerges if all components of the honey-system case and the Aspartame®-system case are included. Actual compositions of water as a product are very different for the two cases. However, this difference cannot be observed if the pathways are cut off from the analysis and if the analysis is performed within arbitrarily-set confines. Similar to confining the time domain to the "time of interest": or time t = "right now", this confinement in space perverts the process of scientific investigation. Every product emerging after oxidation of an artificial substance will come with long-term consequences for the environment. These consequences cannot be included with the science of tangibles. Zatzman and Islam (2007b) detailed the following transitions in commercial product development and argued that this transition amounts to an increased focus on the tangible in order to increase the profit margin in the short-term. The degradation in quality is obvious, but the reason behind such technology development is quite murky. At present, the science of tangibles is totally incapable of lifting the fog out of this mode of technology development.

The science of intangibles, on the other hand, includes all phenomena that occur naturally, irrespective of what might be detectable / perceived during some artificially-designated temporal or spatial confine. With regard to catalysis, for instance, it can be said that if a reaction cannot take place without the catalyst, clearly the catalyst indeed plays a role. The fact that at some given time (e.g., time t = "right now") the amount of the catalyst loss cannot be measured hardly proves that catalyst loss and/or some other role[s] for catalysts don't exist. The loss of the catalyst is real, even though one cannot measure it with current techniques. The science of intangibles does not wait for the time when one can "prove that catalysts are active". Because nature is continuous (without a boundary in time and in space), considerations are not focused on a confined "control' volume". For a science of tangibles, on the other hand, all that the absence of molecules of the catalyst in the reaction products means, is that one would not find that particular role at those particular space-time coordinates.

Tangible science says: If you can't find it in the reaction product, it doesn't count. The science of intangibles says: Obviously it counts, but -

just as obviously - not in the same way as what is measurable in the tangible mass-balance. This shows that the existing conventional science of tangibles is incomplete. Precisely to the extent that it remains incomplete, on this basis of disregarding or discounting qualitative contributions that cannot yet be quantified in ways that are currently meaningful, this kind of science is bound to become an accumulating source of errors.

The most serious, most important, most significant, most truly acid test of a proposed scientific characterization or analysis of any phenomenon, is that it accounts for everything necessary and sufficient to explain the phenomenon — its origin, its path and its end-point — thereby rendering it positively useful to human society. As mentioned earlier, the same criterion was used in previous civilizations to distinguish between real and artificial. Khan (2007) introduced a criterion that identifies the end-point by extending time to infinity. This criterion avoids scrutiny of the intangible source of individual action (namely, intention). However, Zatzman and Islam (2007a) pointed out that the end-point at time t = infinity can be a criterion, but it will not disclose the pathway unless a continuous time function is introduced. Mustafiz (2007) used this concept and introduced the notion of knowledge dimension — a dimension that arises from introducing time as a continuous function. In all these deductions, it is the science of intangibles that offers some hope. It is important to note that the insufficiency just mentioned is not overcome by doing "more" science of tangibles "better". It is already evident that what is not being addressed are intangible components that cannot be winkled, ferreted out, or otherwise measured by existing means available within the realm of the science of tangibles.

Any number of examples could be cited from the commercial world of product advertising to further illustrate the nub of the problem; this chapter will introduce some of the more egregious cases to illustrate the trends being noted here. Which discipline(s) from the science of tangibles, for example, could model the following?

"In every sense, a Whitestone Cheese is the embodiment of its environment. Pressed by hand, bathed by hand, turned by hand and packed by hand, it is a product of skill and mystery. Like original works of art, no two are alike. While their styles embrace a faint echo of Europe, Whitestone's cheeses are unto themselves as unique as the land that created them" (Delicious Organics, 2007).

We all know hand-made cheese is better tasting, but that mother's milk is the best. But do we have a criterion that should lead us to expect these assumptions to be true or to be best? How about hand-drawn milk as compared to machine-drawn? How about un-Pasteurized® milk as compared to Pasteurized®? Do we even have a choice? We truly don't, since commercialization is done after engineering calculations are made from the science of tangibles. Then, the economics of tangibles are applied to provide the justification with a guarantee.

Intangibles, which essentially include the root and pathway of any phenomenon, make the science suitable for increasing knowledge, as opposed to increasing confidence in a conclusion that is inherently false (Zatzman and Islam, 2007a). Zatzman and Islam (2007b) have introduced the following syllogism to make this point about the science of intangibles:

> All Americans speak French [major premise]
> François Hollande [current president of France] is an American [minor premise]
> Therefore, François Hollande speaks French [conclusion-deduction]

If, in either the major or minor premise, the information relayed above is derived from a scenario of what is merely probable (as distinct from what is actually known), the conclusion, which happens to be correct in this particular case, would be not only acceptable as something independently knowable, but reinforced as something also statistically likely. This, then finesses determining the truth or falsehood of any of the premises, and, eventually, someone is bound to "reason backwards" to deduce the statistical likelihood of the premises from the conclusion! Indeed, this latter version, in which eventually all the premises are falsified as a result of starting out with a false assumption asserted as a conclusion, is exactly what has been identified and labeled elsewhere as the aphenomenal model (Islam *et al.*, 2012). How can this aphenomenal model be replaced with a knowledge model? Islam *et al.* (2014a) emphasized the need of recognizing the first premise of every scientific discourse. They used the term "aphenomenality" (in contrast to truth or falsehood) to describe in general the non-existence of any purported phenomenon or of any collection of properties, characteristics or features ascribed to such a purported but otherwise unverified or unverifiable phenomenon. If the first premise contradicts what is true in nature, the entire scientific investigation will be false. Such investigation cannot lead to reliable or useful conclusions.

Consider the following syllogism (the concept of "virtue" intended here is "that which holds positive value for an entire collectivity of people:, not just for some individual or arbitrary subset of individual members of humanity):

All virtues are desirable.
Speaking the truth is a virtue.
Therefore, speaking the truth is desirable.

Even before it is uttered, a number of difficulties have already been built into this apparently non-controversial syllogism. When it is said that, "all virtues are desirable", there is no mention of a time factor (pathway) or intention (source of a virtue). For instance, speaking out against an act of aggression is a virtue, but is it desirable? A simple analysis would indicate that unless the time is increased to infinity (meaning something that is desirable in the long-run), practically all undesirable virtues, (even giving out in charity requires austerity in the short-term, defending a nation requires self sacrifice – an extremely undesirable phenomenon in the short-term). In the same way, if giving charity is a virtue, would that make giving away stolen goods a charity? Robin Hood may be an acceptable hero in the post-Renaissance culture, but is such a categorization scientifically grounded? Giving away stolen goods can be a virtue only if the history (time function) is obliterated. The third component is in the source of an act. For instance, is giving away with the intention of recovering something in the future a virtue? Is helping an oppressor a virtue? This logic shows the need for highlighting both the source (intention) and the pathway (time function going back to the origin) of an action in order to qualify it as a virtue.

The scientifically correct reworking of this syllogism should be:

All virtues (both intention and pathway being real) are desirable for time t approaching ∞.
Speaking the truth is a virtue at all times.
Therefore, speaking the truth is desirable at all times.

The outcome of this analysis is the complete disclosure of the source, pathway (time function), and final outcome (time t approaching ∞) of an action. This analysis can and does restore to its proper place the rational principle underlying the comparing of organic products to synthetic ones; free-range animals to confined animals; hand-drawn milk to machine-drawn; thermal pasteurization with wood fire compared to microwave and/or chemical Pasteurization®; solar heating compared to nuclear heating; use of olive oil compared to chemical preservatives; use of natural antibiotics compared to chemical antibiotics; and the rest of a long list of such artificial versus natural outcomes and pathways. When it comes to food or other matter ingested by the human body, natural components are to be preferred because we can expect that the source and pathway of such components, already existing in nature, will be beneficial (assuming non-toxic dosages of medicines and normal amounts of food are being ingested). Can we hold out such confidence when it comes to artificially-simulated substitutes for the natural versions? The pathway of the artificial substitutes' creation lies outside any process already given in nature—the most important feature of a food.

With the science of tangibles, the outcome being the same (e.g., an apparently similar chemical composition of both the nature-sourced and artificially-sourced food), provides the basis for asserting that there are no significant differences between the natural and the artificial. With the science of intangibles, source and pathway are taken into account, blocking the way to such premature and counterfeit declarations. This is not to say there may not be ways to improve upon the best uses humans can make of whatever has already been given in the natural realm. If, however, any such improvements are to be sustainable — the only test of truth in nature being that which eventuates over the passage of time — they cannot simply substitute something "chemically identical" from the laboratory or the synthesizing cycle in a chemical plant, completely discounting or ignoring the source(s) and pathway(s) of the natural original. In this, if the time consideration is not time t approaching ∞, then the validity of any conclusion would depend on the intention. If the intention is not phenomenal (real or consistent with environmental sustainability), every conclusion will be aphenomenal. Here, there is only one phenomenal intention, which coincides with the true first premise. The first premise of a logical train (the process of scientific investigation) is equivalent to the intention of an individual act. An intention may properly be considered "phenomenal" if or when it is driven by conscience (Zatzman and Islam, 2007a). It can also be characterized as "good", with similar applications as "for good", alluding to the infinitely long term (Khan and Islam, 2012).

4.7 The Need for Studying the Time Dimension in Implicit Form

Islam *et al.* (2010) indicated that the ultimate truth can be revealed only with an infinite number of dimensions. Abou-Kassem *et al.* (2008)

argued that by invoking Einstein's theory of relativity through expression of any event as a continuous function of time, one forces the solution to include infinite dimensions. This argument makes it possible to solve problems without extending to an infinite number of dimensions, which would be impractical at this point of human knowledge. The problem, then, is reduced to solving with only known factors, irrespective of how little the impact the variable may have on the outcome of scientific analysis. Kvitko (2007) discredited Einstein's relativity altogether. However, he did not elaborate on the first premise of the theory. More recently, Islam *et al.* (2014) and Speight and Islam (2015) argued that Einstein has introduced the wrong sequence of logic, misplacing functionality of time, event, and mass. Einstein's most revolutionary assumption was the connection of perception with time. Mathematically, it can be expressed as $t = f$ (perception). Perception is in itself function of the individual's past and the environment. Mathematically, it amounts to:

Perception = $g(t)$, $h(s)$,

Where $g(t)$ is individual history and conscious experience and h (s) is the environment which is a function of space at any given time period, Δt as long as Δt is very small.

This means that any time period becomes a function of infinite number of possibilities. This is the essence of quantum theory that asserts that there are numerous histories for every particle, ranging from the subatomic to the universe (Hawking, 2010). Apart from creating multiple histories in violation of the most important feature of creation — namely, its uniqueness — this manipulation of the time function distorts the definition of work, power as well as efficiency. As seen in earlier section, this is equivalent to creating different standard at the whim of some arbitrary 'authority'.

In addition, if the perception is not in conformance with reality, the same data will not produce phenomenal thought material. Similar to matter of energy and material, an aphenomenal basis will produce aphenomenal outcome and the entire cognition process will be reversed (Figure 4.22).

The matter of perception cannot have any influence on a reality that exists independent of the observer: a blind person doesn't see the difference between night and day but that perceptual state changes nothing about the phenomena of night and day. Recasting reality as a function of perception only serves to demonstrate the absurdity of Quantum theory. Reality cannot be a function of perception, unless there is no such thing as reality. According to Aristotle, one of the mistakes of

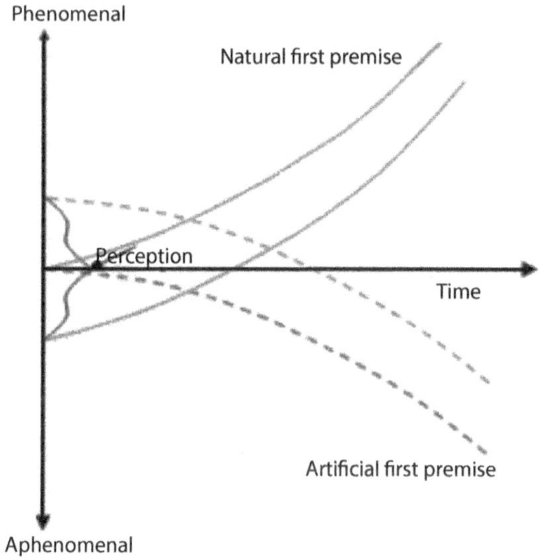

Figure 4.22 If the first premise is not real, perception becomes inherently false and leads to aphenomenal cognition (modified from Khan and Islam, 2012).

Zeno in his paradoxes of time and motion is that he did not distinguish between actual and potential infinities. Scientifically, it was the continuity of the time function that eluded Zeno in his famous paradox about Achilles and the turtle. Aristotle 'remedied' it by differentiating between actual (present) and potential (future). Then he asserted, "Everything is potential infinity and nothing is actual infinity." This in essence reversed the notion of reality as conceived in the works of Plato.

Invoking perception as the root is equivalent to introducing different sources to a non-linear system. It precipitates the onset of chaos. Such has been the case of the cognition processes at the core of New Science. As a chaotic system produces multiple solutions (depending on the origin), one of the outcomes is likely to be matched with the desired outcome, making it appealing to the pragmatic approach. This point is made in a cartoon in Picture 4.2.

Perception is a function of the environment, which might be "polluted" with disinformation and ignorance. Therefore, in the absence of an explicit criterion for truth, there is no guarantee (in fact, the opposite is guaranteed) there will be any beneficial outcome. In fact, if such cognition were to persist, any ability to align perception with knowledge over the long-term would be hampered to the extent that mental disorder could set in.

Picture 4.2 If reality is a function of perception, reality itself becomes a paradox.

Einstein's energy equation was deconstructed earlier. In his photoelectric theory, Einstein took Maxwell's rigid sphere model as true, whereas Maxwell himself took that model from Newton. By continuing to build on the same first premise, they have accumulated error and traveled the downward path.

Such was not the case in other parts of the world. Consider how Ibn Haitham took the model of Aristotle and deconstructed it based on simple logic (Khan and Islam, 2012). He discarded Aristotle's conclusion that light has infinite speed and reconstructed a model that, until today, serves as the only model that can distinguish sunlight from artificial light. While it is well known that sunlight is the essence of life and that artificial light is something that is used for torture, neither Einstein's nor any other optic theory can explain scientifically how this is possible. In addition, Ibn Haitham undid another one of the old theories, which is that something that comes out of our eyes makes it possible for us to see. Instead of that theory, he introduced the notion of something entering your eye that makes it possible to see. This "something" was later proclaimed to be a photon. This notion was correct, but the denomination as well as the attribution of various properties made further research on the topic of light characterization impossible. For instance, because this theory postulates that all photons are alike and do not have mass, the source of light cannot have an impact on the quality of light, leading to the same difficulty that made it impossible to discern between sunlight and artificial light.

If Ibn Haitham's theory was used correctly, one would be able to correlate the toxic nature of the light source (e.g., power-saving light) with long-term impacts, such as breast cancer, brain dysfunction, and numerous other reported correlations. This would also unravel the science behind skin-cancer causing chemicals that are often included in suntan or sun protection lotions. Another example is worth mentioning here, which is a recently touted new form of energy-saving light. This light was excellent for energy savings as well as producing the "white light" effect. However, it was also performing the so-called "belly dance". When the source was sent to the International Space Station for a probe, it was discovered that the "belly dance" subsided or disappeared. Gravity was found to be the reason behind the belly dance. Could this be explained with existing light theories? Of course not, because if photons have zero mass, how could gravity affect them? This paradoxical modus operandi continues when it comes to dark matters (infinite mass, but no energy) and dark energy (infinite energy but no mass) in the realm of the discussion in cosmic physics.

Where's the catch? Ibn Haitham did not read Aristotle's work to be persuaded to any particular conclusion. He read the theory, used the criterion on truth and falsehood and was able to decipher true elements from the volumes of work. Einstein did not dare use the same logic about Maxwell's "laws" or he didn't have the criterion that Ibn Haitham was equipped with. Einstein was not a Christian, but he certainly was a believer of Eurocentric philosophy. This fundamental inability to discern truth from falsehood (called Furqan in Arabic), is missing from European new science or social science. As a result, what we see is constant confusion about everything that governs our daily lives.

It is important to note that the observation of natural phenomena as a continuous function of time, including differential frames-of-reference for component processes, is a matter of documenting and reconstructing the actual pathways and steps of an overall process. Because of its implicit standpoint of the neutral external observer, conventional analysis is not capable of fully sorting out these pathways and their distinctive components. The form in which this standpoint expresses itself is embedded in the conventions that come with the "usual" linearizations, viz., viewing time as the independent variable that varies independently of the processes being observed. Both the approaches of Euler and Lagrange have the concept of external observer embedded in them. In Euler's scenario, the external observer is static – a physically impossible and hence absurd state anywhere within nature. For the scenario laid out by Lagrange, the external observer is in motion, but within the same pre-defined pathway (conditions for the independent variable). To an external observer, intermediate changes-of-states at the interface of successive sub-processes are "invisible", in much the same way that the third dimension is invisible at the interfaces of processes observed in two dimensions. (This is why analysis based on comparing output to input "works" so well — but only with the most linearized models). Within nature, there is no external observer, a state of affairs that renders the processes of tangible science "aphenomenal". Some researchers have indeed recognized the notion of the "external" as being aphenomenal. Rather than discarding this notion, however, they adapted the same principle, calling it "God's eye view" (He, 2005), while using Einstein's relativity (continuous time function) as the "human eye view". We consider this process of scientific investigation aphenomenal.

The following corollary is the core of the argument advanced in this section:

Just because an equation or set of equations, describing the transformation of an overall process from input to output, can or may be decomposed into a set of linear superpositions, does not mean that any or each of these superpositions describe or represent any actual pathway, or portion thereof, unfolding within Nature.

Consider the following logical train:

> Perfect is preferable;
> Nature is perfect;
> Therefore, anything natural is preferable.

Seeking perfection as something embedded already within humanity, the first premise sets the selection criterion for any conscience-driven human action. However, this one does not guarantee aphenomenality of the scientific process because the definition of "perfect" is linked to the notion of the ideal. If the "ideal" is aphenomenal, on the other hand, the meaning of "perfect" is reversed. The second premise, viz, "nature is perfect", is intricately linked with what nature is. The case in point is a Stanford professor's argument (Roughgarden, 2005). She argues that if more than 400 species are found to be practicing "part-time homosexuality", it must be natural for humans to engage in similar practices. In fact, this argument can be used to demonstrate "homosexuality is preferable". What is the problem with this logic? Only one dimension of the problem is being considered. If another dimension is used, it can also be deduced that incestuous relationships are natural, hence, preferable. When a generalization is made, one must not violate characteristic features of the individual or groups of individuals. Conscience, here, is not to be confused with moral or ethical values that are not inherent to humans, or at least that are subject to indoctrination, learning, or training. Humans are distinct from all other creatures that we know, because of the presence of conscience — the ability to see the intangibles (both past and future), analyze the consequences of one's action, and decide on a course of action.

Another example can be given as:

> Perfect is preferable;
> Nature is perfect;
> Earthquakes are natural;
> Therefore, earthquakes are preferable.

Reverse arguments can be made to curse nature. For example, on CNN, this was precisely the issue in Larry King's interview with two former US presidents following the December 2005 tsunami in the eastern Indian Ocean and Bay of Bengal. There are two problems with this argument:

First of all, it is not a matter of "preference". Anything that takes place without human intervention cannot be preferred or discarded. It is not a matter of intention; it is rather a matter of wish, which doesn't necessitate any follow-up human actions. Any natural phenomenon (including disasters and calamities) will take place as a grand scheme of natural order, as a necessary component of total balance. This total balance cannot be observed in finite time or finite space. All that can be observed of such a phenomenon in finite time and space are fragmentary aspects of that balance. The phenomenon may not appear to be balanced at all, or alternatively, there may occur some equilibrium state and because — the observation period is sufficiently finite —the equilibrium state is assumed to be "normal".

Secondly, if nature is perfect and dynamic at the same time, nature must be moving towards an increasingly better state with time. This logic then contradicts Lord Kelvin's assertion that nature is moving from an active to a passive state, reaching a state of useless "heat death". This is in sharp contrast to what has been found by the Nobel Prize winning work (2001) of Eric Cornell and others. As Eric Cornell outlined in his most popular invited lecture, titled. *Stone Cold Science: Things Get Weird Around Absolute Zero*, Kelvin's concept of nature and how nature functions is starkly opposite to the modern concept. At very cold temperatures, phase changes do occur but these have nothing to do with losing power or strength, as commonly understood by the term "death". This is further corroborated by later discoveries (Ginsberg *et al.*, 2007). Once again, unless the long-term is being considered over a large scale in space, this transition in universal order or in a laboratory cannot be observed. This is true for floods, lightning, and every natural phenomenon that we observe.

4.8 Assessing the Overall Performance of a Process

In order to break out of the conventional analysis introduced throughout the science of tangibles, we will proceed to discuss some salient features of the time domain and present how the overall performance of a process can be assessed by using time as the fourth dimension. Time *t* here is not orthogonal to the other three spatial dimensions. However, it is no less a dimension for not being mutually orthogonal. Socially available knowledge is also not orthogonal either with respect to time *t*, or with respect to the other three spatial dimensions. Hence, despite the training of engineers and scientists in higher mathematics that hints, suggests, or implies that dimensionality must be tied up "somehow" to the presence of orthogonality, orthogonality is not in itself a relationship built into dimensionality. It applies only to the arrangements we have invented to render three spatial dimensions simultaneously visible, i.e., tangible.

Between input and output, component phenomena can be treated as lumped parameters, just as, for example, in electric circuit theory, resistance/reactance is lumped in a single resistor, capacitance in a single capacitor, inductance in a single inductor, electromotive potential/force and current of the entire circuit lumped at a power supply, or at special gated junction-points (such as between the base and emitter of a transistor), etc.

Similarly, in the economic theory of commodity transactions, relations of exchange in the market lump all "supply" with the seller and all "demand" with the buyer — even though in reality, as everyone knows — there is also a serious question of a "demand" (need for money) on the part of the seller and there is a certain "supply" (of cash) in the hands of the buyer.

In Nature, or even within certain highly-engineered phenomena, such as an electric circuit, in which human engineering has supplied all the ambient conditions (source of electrical energy, circuit transmission lines, etc.), even after assuming certain simplifying conditions like a near-zero frequency and virtually direct current flow and very small potential differences, we still have no idea whether the current is continuous or how continuous, nor how stable or uniform the voltage difference is at any point in the circuit. The lumped-parameter approach enables us to characterize the overall result/difference/change at output compared to the input without worrying about the details of what actually happened between the input and the output. Clearly, when natural processes are being considered, such an approach leaves a great deal unexplained and unaccounted for. So long as the computed result match the difference measured between the input and the output, this approach opens the door to imposing any interpretation as a way to account for what happened.

Closely related to the technique of characterizing the operation of a process by means of lumped parameters, is the technique of assessing or describing overall performance of the process under study (or development), according to objective, external, uniform "standards" or norms. In the MKS system of SI units, for example, the meter is standardized as a unit of distance according to the length of some rod of some special element maintained in a vacuum bell at a certain temperature and pressure in some location in Paris, France. Similarly the NIST in Washington DC standardizes the duration of the "second" as the fundamental unit of time according to an atomic clock, etc.

The problem with all such standards is that the question of their applicability for measuring something about the process-of-interest is never asked beforehand. Consider the known and very considerable physical difference between the way extremely high-frequency [tiny-wavelength] EM waves, on the one hand, and much lower-frequency [much-greater wavelength] audible-sound waves on the other hand, each propagate. The meter may be quite reasonable for the latter case. Does it follow, however, that the nanometer — recall that it is based on subdividing the meter into one billion units — is equally reasonable for the former case? The physical reality is that the standard meter bar in Paris actually varies in length by a certain number of picometers or nanometers just within one Earth year. If the process-of-interest is EM radiation traversing light-years through space, however, variation of the standard metre by one nanometre or even 1000 picometers will make nonsense of whatever measure we assign to something happening in the physical universe at this scale.

What the objectivity, externality and uniformity of standards enable is a comparison based on what the human observer can directly see, hear, smell, touch or taste — or, more indirectly, measure — according to standards that can be tangibly grasped within ordinary human understanding. However, is science reducible to that which may be tangibly grasped within ordinary human understanding? If science were so reducible, we could, and should, have spent the last 350+ years since

Galileo fine-tuning our measurements of the speed of bodies falling freely towards the Earth. As a result, this feature might then be catalogued for different classes of objects according to Aristotle's principle, seemingly quite reasonable, perfectly tangible yet utterly erroneous, that the speed with which objects fall freely towards the Earth is a function of their mass.

This example hints at the solution to the conundrum. Once the principle of gravity as a force — something that cannot be directly seen, heard, smelt, touched or tasted — acting everywhere on the Earth was grasped, measuring and comparing the free fall of objects according to their mass had to be given up — because it was the attraction due to gravity that was the relevant common and decisive feature characteristic to all these freely-falling objects, not their individual masses. So, standards of measurements applied to phenomena and processes in Nature should cognize features that are characteristic to those phenomena and processes, not externally applied regardless of their appropriateness or inappropriateness.

Instead of measuring the overall performance of a process or phenomenon under study or development according to criteria that are characteristic, however, statistical norms are frequently applied. These compare and benchmark performance relative to some standard that is held to be both absolute and external. Public concern about such standards — such as what constitutes a "safe level of background radiation" — has grown in recent years to the point where the very basis of what constitutes a standard has come into question. Recently, Zatzman (2008) advanced the counter-notion of using units or standards that are "phenomenal" (as opposed to aphenomenal). For those who want a science of nature that can account for phenomena as they actually occur or appear in nature, standards whose constancy can only be assured outside the natural environment — under highly controlled laboratory conditions, for example, or "in a vacuum" — are in fact entirely arbitrary. Phenomenally-based standards, on the other hand, are natural in yet a deeper sense; they include the notion of a characteristic feature that may be cognized by the human observer. These are standards whose objectivity derives from the degree to which they are in conformity with nature. The objectivity of a natural standard cannot and must not be confounded with the vaunted neutrality of position of some external arbiter.

For all the work on intangibles (the mathematics of, the science of, etc.), one must establish:

actual, true source
actual, true science, pathway
actual, true end-point, or completion

Knowledge can be advanced even if the "true object" is not the entire truth. In fact, it is important to recognize the whole truth cannot be achieved. However, this should not be used as an excuse to eliminate any variable that might have some role but whose immediate impact is not "measurable". All of these potential variables that might have a role should be listed right at the beginning of the scientific investigation. During the solution phase, this list should be discussed in order to make room for possibilities which at some point, one of the variables will play a greater role. This process is equivalent to developing the model that has no aphenomenal assumption attached to it. There is a significant difference between that which tangibly exists for the five senses in some finite portion of time and space, and that which exists in Nature independently of our perceptual functioning in some finite portion of time and space. Our limitation is that we are not able to observe or measure beyond what is tangible. However, the model that we are comparing this with should not suffer from these shortcomings.

If we grasp the latter first, then the former can be located as a subset. However, errors will occur if we proceed from the opposite direction, according to the assumption that what is perceivable about a process or phenomenon in a given finite portion of time and space contains everything typical and/or characteristic of the natural environment surrounding and sustaining the process or phenomenon as observed in some given finite portion of time and space.

Proceeding according to this latter pattern, for example, mediaeval medical texts portrayed the human fetus as a "homunculus", a miniaturized version of the adult person.

Proceeding according to the former pattern, on the other hand, if we take phase [or "angle"] x as a complex variable, de Moivre's Theorem can be used to readily generate expressions for cos nx and sin nx, whereas (by comparison) if we struggle with constructions of right triangles in the two-dimensional plane, it is a computationally intensive task just to derive cos $2x$ and sin $2x$, and orders of magnitude more difficult to extend the procedure to derive cos nx and sin nx.

In technology developments, it is important to take a holistic approach. The only single criterion that one can be used is the reality criterion. A reality is something that doesn't change with time going to infinity. This is the criterion that Khan (2007) has employed to define sustainability. If the ranking of a number of options is performed based on this criterion that would be equivalent to the real (phenomenal) ranking. This ranking is absolute and must be the basis for the comparison of various options. This ranking is given in the left most column of Table 4.12 In technology developments, this natural (real) ranking is practically never used. Based on other ranking criteria, most of the rankings are reversed, meaning that the natural order is turned up-side down. However, there are some criteria that would give the same ranking as the natural one, but that does not mean that the criterion is legitimate. For instance, the heating value for honey is the highest. However, this does not mean the process is correct, or — putting it in terms of the syllogism that launched section 2 — it reaffirms that "all Americans do not speak French", i.e., something we already knew all along. This table is discussed in Section 8 *infra* as a starting-point for establishing a "reality index" that would allow a ranking according to how close the product is to being natural.

In engineering calculations, the most commonly used criterion is efficiency, which deals with output as a function of input. Ironically, an infinite efficiency would mean that someone has produced something out of nothing - an absurd concept as an engineering creation. However, if nature does that, it operates at 100% efficiency. For instance, every photon coming out of the sun gets used. So, for a plant the efficiency is limited (less than 100%) because it is incapable of absorbing every photon it is coming into contact with, but it would become 100% if every photon were to be accounted for. This is why maximizing efficiency as a man-made engineering practice is not a legitimate objective.

If the concept of efficiency is used in terms of overall performance, the definition of efficiency has to be changed. With this new definition (called "global efficiency" by Khan *et al.*, 2007c and Chhetri, 2007), the efficiency calculations will be significantly different from conventional efficiency that only considers small objects of practical interest. As an example, consider an air conditioner running outdoors. The air in front of the air conditioner is indeed chilled, while air behind the device is heated. For instance, if cooling efficiency calculations are performed on an air conditioner running outdoors, the conventional calculations would show finite efficiency, albeit not 100%, as determined by measuring temperatures in front of the air conditioner and dividing the work by the work done to operate the air conditioner.

Contrast this to the same efficiency calculation if temperatures all around are considered. The process will be proven to be utterly

Table 4.12 Synthesized and natural pathways of organic compounds as energy sources, ranked and compared according to selected criteria.

Natural (real) **ranking** ("top" rank means most acceptable)	**Aphenomenal ranking by the following criteria**			
	Bio-degradability	**Efficiency[1], *e.g.*,** $\eta = \frac{Outp-Inp}{Inp} \times 100$	**Profit margin**	**Heatingvalue (cal/g)**
1. Honey 2. Sugar 3. Saccharine 4. Aspartame	2 3 4 1	4 "sweetness lg" 3 2 1	4 3 2 1	1 2 3 4

(Continued)

Table 4.12 cont.

Natural (real) **ranking** ("top" rank means most acceptable)	**Aphenomenal ranking by the following criteria**			
	Bio-degradability	**Efficiency**[1], ***e.g.***, $\eta = \frac{Outp-Inp}{Inp} \times 100$	**Profit margin**	**Heatingvalue (cal/g)**
1. Organic wood	1 Reverses depending on applic'n, *e.g.*, durability	4 Reverses if toxicity is considered	4 Reverses if organic wood treated with organic chemicals	4
2. Chemically-treated wood	2	3	3	3
3. Chemically grown, Chemically treated wood	3	2	2	2
4. Genetically-altered wood	4	1	1	1

(*Continued*)

Table 4.12 cont.

Natural (real) **ranking** ("top" rank means most acceptable)	**Aphenomenal ranking by the following criteria**			
	Bio-degradability	**Efficiency**[1], *e.g.*, $\eta = \frac{Outp-Inp}{Inp} \times 100$	**Profit margin**	**Heatingvalue (cal/g)**
1. Solar 2. Gas 3. Electrical 4. Electromagn 5. Nuclear	Not applicable	5 # Efficiency can-not be calculated for direct solar 4 3 2 #	5 4 3 2 1	5 # - Heating value cannot be calculated for direct solar 4 3 2 #

(*Continued*)

Table 4.12 cont.

Natural (real) **ranking** ("top" rank means most acceptable)	**Aphenomenal ranking by the following criteria**			
	Bio-degradability	**Efficiency**[1], ***e.g.***, $\eta = \frac{Outp-Inp}{Inp} \times 100$	**Profit margin**	**Heatingvalue (cal/g)**
1. Clay or wood ash	1 Anti-bacterial soap won't use olive oil; volume needed for cleaning unit area	6 Reverses if global is considered	6	4 # 1 cannot be ranked
2. Olive oil + wood ash	3	5	5	6
3.Veg oil +NaOH	4	4	4	5
4. Mineral oil + NaOH	5	3	3	3
5. Synthetic oil + NaOH	6	2	2	2
6. 100% synthetic (soap-free soap)	2	1	1	#

(Continued)

Table 4.12 cont.

Natural (real) **ranking** ("top" rank means most acceptable)	**Aphenomenal ranking by the following criteria**			
	Bio-degradability	**Efficiency**[1], ***e.g.***, $\eta = \frac{Outp-Inp}{Inp} \times 100$	**Profit margin**	**Heatingvalue (cal/g)**
1. Ammonia 2. Freon 3. Non-Freon synthetic	1 2 3	Unknown	3 2 1	Not applicable
1. Methanol 2. Glycol 3.Synthetic polymers (low dose)	1 2 3	1 For hydrate control 2 3	3 2 1	Not applicable

(*Continued*)

Table 4.12 cont.

Natural (real) **ranking** ("top" rank means most acceptable)	**Aphenomenal ranking by the following criteria**			
	Bio-degradability	**Efficiency**[1], *e.g.*, $\eta = \frac{Outp-Inp}{Inp} \times 100$	**Profit margin**	**Heatingvalue (cal/g)**
1. Sunlight 2. Vegetable oil light 3. Candle light 4. Gas light 5. Incandescent light 6. Fluorescent light	Not applicable	6 5 4 3 2 1	6 5 4 3 2 1	Not applicable

inefficient and will become obvious that the operation is not a cooling process at all. Clearly, cooling efficiency of the process that is actually also creating heat is absurd. Consider now, with an air conditioner running on direct solar heating. An absorption cooling system means there is no moving parts and the solar heat is being converted into cool air. The solar heat is not the result of an engineered process. What would, then, be the efficiency of this system and how would this cooling efficiency compare with the previous one?

Three aspects emerge from this discussion. First, global efficiency is the only one that can measure true merit of a process. Second, the only efficiency that one can use to compare various technological options is the global efficiency. Third, if one process involves natural options, it cannot be compared with a process that is totally "engineered". For instance, efficiency in the latter example (as output/input) is infinity, considering no engineered energy has been imparted on the air conditioner.

[1]This efficiency is a local efficiency that deals with an arbitrarily set size of sample

*calorie/gms as a negative indicator—"weight watchers" (who are interested in minimizing calories) and is a positive indicator for energy drink makers (who are interested in maximizing calories).

No engineering design is complete until economic calculations are performed. Therein lies the need for maximizing profit margins. Indeed, the profit margin is the single-most important criterion used for developing a technology ever since the Renaissance that saw the emergence of short-term approach move at an unparalleled pace. As Table 4.12 indicates, natural rankings generally are reversed if the criterion of profit maximization is used. This affirms, once again, how modern economics have turned pro-nature techniques upside down (Zatzman and Islam, 2007b).

4.9 Aphenomenal Theories of Modern Era

Table 4.1 listed the properties of natural entities that are part of creation and form integral parts of the universal order. Table 4.12 lists the fundamental features of the external entity. The existence of an external entity is necessary condition in order to eliminate the notion of void that had been inherited from Atomism philosophy and was carried forward by first Thomas Aquinas and then by subsequent scientists, without exception (Islam, 2014).

This external entity was first recognized as God (ancient Greek philosophers all the way to Avicenna and Averroes of Islamic golden era), then conflated as plenum and aether (Islam *et al.*, 2013; 2014). While the existence of such entities has been denied and sometime 'proven' to be non-existent, the traits of this external entity have been included in all forms of 'fundamental' particles, ranging from photon to Higgs boson. In addition, such features have also been invoked in galactic models in the form of various entities, ranging from "dark matter", "black hole" to "absolute void". Newton introduced this as 'external' force and defined it as the originator of differential motion. The original Averroes concept, as supported by the Qur'an was that such originator of motion is the Creator, whose traits are all different from the traits of creation.

As will be discussed in latter chapters, all features of this external entity also has duality, which is a characteristic feature of that entity. For instance, Absolute time (Feature 17) and Absolute mass (Feature 18) are opposite to each other. Similar duality exists in other traits of the external entity.

Table 4.13 shows many currently used 'laws' and theories, all of which emerged from the New Science after the Renaissance. Note how the first premises of practically all of these theories violate fundamental features of Nature. Only conservation of mass — which in fact has its root in ancient times — does not have an aphenomenal first premise. New Science has given us only theories and 'laws' that have a spurious first premise, as evidenced by using Averröes' criterion for phenomenality (Zatzman and Islam, 2007a).

If all theories of New Science are based on premises that violate fundamental traits of nature, such laws and theories, if applied as universal laws and theories, should weaken considerably or worse, implode. They can be applied only to certain fixed conditions that pertain to 'idealized' situations existing nowhere in nature. For example, it can be said that the laws of motion developed by Newton cannot explain the chaotic motion of nature due to its assumptions that contradict the reality of Nature.

The experimental validity of Newton's laws of motion is limited to describing instantaneous macroscopic and tangible phenomena. However, microscopic and intangible phenomena are ignored.

Classical dynamics as represented by Newton's laws of motion, emphasize fixed and unique initial conditions, stability, and equilibrium of a body in motion (Islam *et al.*, 2010). However, as the list below in Table 4.14b serves to clarify, it is not possible with the 'laws' and

Table 4.13 Features of external entity (from Islam, 2014)

Feature no.	Feature
1	Absolutely external (to everything else)
2	All encompassing
3	No beginning
4	No end
5	Constant (independent of everything else)
6	Uniform
7	Alive
8	Infinity
9	Absolutely True
1	Continuous
11	All pervasive in space
12	All pervasive in time
13	Infinite degree of freedom
14	Unique
15	Open system
16	Dissimilar to everything else
17	Absolute Time that control time that controls mass
18	Absolute mass (pure light)

theories of Table 4.14a to make a distinction between the natural products and their corresponding artificial substitutes (see below). Consequently, the same theories that formed the basis of engineering the artificial products cannot be called upon to make the reversal.

The above transitions embody the main bulk of modern technological developments that have been characterized by Nobel laureate chemist, Robert Curl as 'technological disaster'.

4.10 The Law of Conservation of Mass and Energy

The law of conservation of mass was known to be true for thousands of years. In 450 B.C., Anaxagoras said, "Wrongly do the Greeks suppose that aught begins or ceases to be; for nothing comes into being or is destroyed; but all is an aggregation or secretion of pre-existing things; so that all becoming might more correctly be called becoming mixed, and all corruption, becoming separate."

However, Antoine Laurent Lavoisier (1743-94), is credited to have discovered the law of the conservation of mass. Lavoisier's first premise was "mass cannot be created or destroyed". This assumption does not violate any of the features of Nature. However, his famous experiment had some assumptions embedded in it. When he conducted his experiments, he assumed that the container was sealed perfectly. This would have violated the fundamental tenet of nature that an isolated chamber can be created. Rather than recognizing the aphenomenality of the assumption that a perfect seal can be created, he "verified" his first premise (law of conservation of mass) "within experimental error". The error is not in the experiment, which remains real (hence, true) at all times. No, the error is in fact embedded within the first premise—that a perfect seal had been created. By avoiding confronting this premise, and by introducing a different criterion (e.g., experimental error), which is aphenomenal and, hence, non-verifiable, Lavoisier invoked a European prejudice, linked to the pragmatic approach, that is "whatever works is true" (Islam et al, 2010). This leads to the linking of measurement errors to the outcomes, creating obstacle to the possibility of independent or objective validation of the theory (Islam et al., 2013). What could Lavoisier have done with the knowledge of his time to link this to intangibles? For instance, if he had left some room for a possible leak from the container, modern day air conditioner designs would have taken into account how much Freon is leaked into the atmosphere.

Lavoisier, nevertheless, faced extreme resistance from scientists who were still firm believers of the phlogiston theory (from the Greek word phlogios = 'fiery'). This theory was first promoted by a German physician, alchemist, adventurer, and a professor of Medicine—Johann Joachim Becher (1635 - 1682). This theory recognizes a form or state of matter, named phlogiston, existing within combustible bodies. When burnt (energy added), this matter was thought to have been released to achieve its "true" state. This theory enjoyed the support of the mainstream European scientists for nearly 100 years. One of the proponents of this theory was Robert Boyle, the scientist, who would later gain

Table 4.14a How natural features are violated in the first premise of various 'laws'and theories of the science of tangibles (*qv* Islam *et al.*, 2014)

Law or theory	First premise	Features violated (see Table 4.1)
Conservation of mass	Nothing can be created or destroyed	None, but applications used artificial boundaries
Quantum theories	Anything can be created from nothing; everything has multiple history	4, 6, 22, 23, 26
Conservation of energy	No energy can be created or destroyed in isolation with mass	22, 23, 26
Big bang theory	14 billion years ago, there was a super hot entity of infinite mass and zero volume that has been expanding after the big bang, super hot	1,3,6,9,14,24,26
Big chill theory	14 billion years ago, there was a super chill entity of infinite mass and zero volume that cracked and exploded into infinite pieces	1,3,6,9,14,24,26
Saul Perimutter and Brian Schmidt (2011 Nobel Prize)	Universe is expanding with acceleration	1,3,6,9,14,24,26
Higgs Boson (2013 Nobel Prize)	Uniform, discrete, symmetric, fundamental particles of zero mass, empty space in between	4, 7, 22, 23
Atomic theory	Uniform, symmetric, discrete, fundamental particles of finite mass	4, 7, 22, 23
Einstein's light theory	Photons of zero mass and constant speed	4, 7, 13, 22, 23, 26

(*Continued*)

Table 4.14a cont.

Law or theory	**First premise**	**Features violated (see Table 4.1)**
Genetic theories	Genes fundamental building blocks of living organism in isolation	4, 6, 13, 22, 23, 26
Defective genes	Inherent defects of genes in isolation	4, 6, 13, 17, 22, 23,
Probability theories	Steady state, repetitive and repeatable	4, 5, 6, 13, 16, 22, 23, 26
Relativity	Time function of perception, perception function of person	4, 7, 16, 22, 23, 26, 27
Gravitational	Force function of mass, steady state, time function of gravity	4, 7, 16, 22, 23, 26, 27
Cosmic theories	Empty space between celestial bodies that are expanding or contracting	4, 7, 16, 22, 23, 26, 27
Lavoisier's deduction	Perfect seal	15
Phlogiston theory	Phlogiston exists	16, 22, 23, 26
E = m c2	Mass of an object is constant	13, 22, 23, 26
	Speed of light is constant	13, 22, 23, 26
	Nothing else contributes to E	14, 19, 22, 20, 23, 24
Planck's law	If the medium is of homogeneous and isotropic constitution, then the radiation is homogeneous, isotropic, unpolarized, and incoherent.	5, 8, 10, 17, 22, 23, 26

(Continued)

Table 4.14a cont.

Law or theory	**First premise**	**Features violated (see Table 4.1)**
Aether theory	Zero mass, zero energy	16, 22, 23, 26
Quantum Cosmic theory	Infinite mass, zero energy	16, 22, 23, 26
Charles	Fixed mass (closed system), ideal gas, Constant pressure,	24, 3, 7
Boyles	A fixed mass (closed system) of ideal gas at fixed temperature	24, 3, 7
Kelvin's	Kelvin temperature scale is derived from Carnot cycle and based on the properties of ideal gas	3, 8, 14, 15
Thermodynamics 1st law	Energy conservation (The first law of the thermodynamics is no more valid when a relationship of mass and energy exists)	22, 23, 26
Thermodynamics 2nd law	Based on Carnot cycle which is operable under the assumptions of ideal gas (imaginary volume), reversible process, adiabatic process (closed system)	3, 8, 14, 15
Thermodynamics 0th law	Thermal equilibrium	10, 15
Poiseuille	Incompressible uniform viscous liquid (Newtonian fluid) in a rigid, non-capillary, straight pipe	7, 22, 23, 25, 26

(Continued)

Table 4.14a cont.

Law or theory	First premise	Features violated (see Table 4.1)
Bernouilli	No energy loss to the sounding, no transition between mass and energy	15, 22, 23, 26
Newton's 1st law	A body can be at rest and can have a constant velocity	13, 22, 23, 26
Newton's 2nd law	Mass of an object is constant Force is proportional to acceleration External force exists	7, 13, 14, 16, 18, 22, 23, 26, 27
Newton's 3rd law	The action and reaction are equal	3, 22, 23, 26
Newton's viscosity law	Uniform flow, constant viscosity	7, 13, 22, 23, 26
Maxwell's equation	Uniform, spherical, rigid balls form energy	4, 7, 22, 23, 26
Newton's calculus	Limit $\Delta t \to 0$	22, 23
Fractal theory	Single pattern that repeats itself exists	1-4, 6, 8, 1

Table 4.14b Transitions from natural to processed

wood → Plastic
Glass→ PVC
Cotton → polyester
Natural fiber → synthetic fiber
Clay → cement
Molasses → Sugar
Sugar → Sugar-free sweeteners
Fermented flower extract → perfume
Water filter (Hubble bubble) → cigarette filter
Graphite, clay → chalk
Chalk → marker
Vegetable paint → plastic paint
Natural marble → artificial marble
Clay tile → ceramic tile
Ceramic tile → vinyl and plastic
Wool → polyester
silk → synthetic
Bone → hard plastic
Organic fertilizer → chemical fertilizer
Adaptation → bioengineering

fame for relating pressure with volumes of gas. Mikhail Vasilyevich Lomonosov was a Russian scientist, writer and polymath who made important contributions to literature, education, and science. He wrote in his diary: "Today I made an experiment in hermetic glass vessels in order to determine whether the mass of metals increases from the action of pure heat. The experiment demonstrated that the famous Robert Boyle was deluded, for without access of air from outside, the mass of the burnt metal remains the same."

Albert Einstein came up with a number of theories, none of which is called "law". The most notable theory was the theory of relativity. Unlike any other European scientists of modern time, this theory recognized the reality of Nature as the proper standard of truth for the purposes of science. This was a refreshing approach considering that the 'steady-state model' had been in use since Aristotle's time. Ironically, the very first scientific article that mentioned relativity after Einstein, was by Walter Kaufmann, who "conclusively" refuted the theory of relativity. However, the point that Kaufmann didn't make is that Einstein's time function was in reverse order. Instead of making the mass, or environment or event a function of time, he made time a function of perception. As pointed out by Islam *et al.* (2014), perception as well as human thought material (HTM) is a function of the environment, which is a function of time.

Even though this "conclusive" refutation did not last very long, one point continues to obscure scientific studies, which is the expectation that something can be "proven". This is a fundamental misconception as outlined by Zatzman and Islam (2007a) and more recently by Islam *et al.* (2013, 2014b, 2014c). The correct statement in any scientific research should involve discussion of the premises a body of research is based on. The first premise represents the one fundamental intangible of the thought process. If the first premise is not true because it violates one or more fundamental feature(s) of Nature, the entire deduction process is corrupted and no new knowledge can emerge from this deduction.

Einstein's equally famous theory is more directly involved with mass conservation. Using the first premise of Planck (1901), he derived $E = mc^2$. Einstein's formulation was the first attempt by European scientists to connect energy with mass. However, in addition to the aphenomenal premises of Planck, this famous equation has its own premises that are aphenomenal (see Table 4.14a). However, this equation remains popular and is considered to be useful (in a pragmatic sense) for a range of applications, including nuclear energy. For instance, it is quickly deduced from this equation that 100 kJ is equal to approximately 1^{-9} gram. Because no attention is given to the source of the matter or the pathway, the information regarding these two important intangibles is wiped out from the science of tangibles. The fact that a great amount of energy is released from a nuclear bomb, is then taken as evidence that the theory is correct. By accepting this at face value (heat as the one-dimensional criterion), heat from nuclear energy, electrical energy, electromagnetic irradiation, fossil fuel burning, wood burning or solar energy, becomes identical. This has tremendous implications on economics, which is the driver of modern engineering.

4.11 Toward Uncovering Knowledge

Exposing aphenomenality not only helps researchers, engineers and other practical-minded people hew in more closely on the paths of nature-science. It also knocks at the door of an entire structure for reconstituting mathematics as a tool of furthering nature-science, rather than always encountering it as an obstacle. If the ways in which the scientific misadventures surrounding the handling of CCD have compelled us to reassess the nature of chaos, and if the chaos of Nature is true, then it cannot be happening only on the fringes or at the extremes. It must happen as one state of nature moves into a different state. Because it's a state of nature, if we're fortunate and clever enough, we may even develop, sooner than some think, a computational method that can capture it.

What does it mean to capture intangibles and make sense of them without throwing away tangibles? The CCD conundrum suggests that problems of this type require considering all energy sources and all masses, still using the mass balance equation, for example, but in this redefined form. Consider in particular, what is involved in the producing honey. Nectars are collected from flowers that may be grown with chemical fertilizers, pesticides, genetically modified seeds or herbicides, all of which will have a permanent foot print on the nectar. Then, there are bees; they collect the nectar and themselves can be raised on sugar, honey from organic sources and may be exposed to natural energy sources or artificial energy sources, and so on. The bee hive itself can be made totally natural or maybe out of plastic frames that will have their own features. During the honey collections, farmers would acknowledge, it is important to follow traditions, ie., not to stress the bees by going into the hive from the front. Many of these activities might sound ritualistic, but, just like hand-made cheese from hand-drawn milk of nature-grazing cows, their roles cannot be ignored. The final product (honey in this case) will have the signature of everything that came into contact with it. As stated in the previous section, just because modern science is not capable of quantifying their effects, does not mean these effects become irrelevant.

Quantifying quality—at first, it all seems counter-intuitive. Hasn't the aim of mathematicians for centuries been to reduce and simplify expressions and procedures? With the aid of electronic computing technologies, however, the drudgery which such clever workarounds were intended and developed to avoid, no longer poses practical problems of, for example, whether the solution(s) to a properly-formulated equation or set of equations can be obtained, say, during one's lifetime.

Modeling nature as it is, nevertheless, would still involve collecting and collating a large amount of data that takes at least initially the form of apparently discrete events. The temptation is to go with statistical methods. This, however, is also one of those points of bifurcation where the actual content of the data of nature has to be taken into account. The fact that events recorded from some processes in nature may be observed as discrete and distinct, does not mean or necessarily prove that these events are stochastically independent.

According to the prevailing theories of mathematical probability, it is legitimate to treat a sufficiently very large number of similar events, e.g., tossing dice, as though these discrete events approximated some continuous process. There is a "Strong Law of Large Numbers" [SLLN] and a more relaxed, less bounded version known as the "Weak Law of Large Numbers" [WLLN], which propose a mathematical justification for just such a procedure (Kolmogorov, 1930).

When we are examining moments in nature, however, which are defined to some extent by some actual passage of time, apart from continuous fluid flow or other motion that is similarly continuous in time, how legitimate or justifiable can it be to approximate discrete events using "nice", i.e., tractable, exponential functions that are continuous and defined everywhere between negative and positive infinity? If the event of interest, although in itself discrete, cycles in a continuum, it would seem that there should arise no particular problem (Of course, there is also no problem for any phenomenon that has been human-engineered and whose data output is to that extent based on human artifice rather than nature).

However, the fact that some recorded data of any large number of such discrete events, exists cannot be taken as sufficient. It is also necessary to be able to establish that the observations in question were recorded in the same time continuum, not in different continua attended by a different set or sets of external surrounding [boundary] conditions. To group and manipulate such data with the tools of mathematical statistics, however, as though the conditions in which the phenomena actually occurred are a matter of indifference, and cannot be justified on the basis of invoking the logic of either the SLLN or WLLN. The continuity of the number and of the characteristics of the abstract construct known as "the real numbers", which form the basis of the SLLN and WLLN, has nothing inherently to do with whether natural phenomena being studied or measured are themselves, actually continuous or occurring within a continuum possessing cyclical features. Some definite yet indeterminate number of such data measurements of the same event — recorded, however, in unrelated and distinct times and places — would likely be so truly "discrete" as not to form part of any actual time-continuum in nature.

Mathematically, working purely with numbers, it may not matter whether there was any physical continuum within which discrete data points were being recorded. In such cases, the strictures of the SLLN and WLLN are adequate, and the approximation of the discrete by the continuous generates no problem. But what we can "get away with" dealing in pure numbers is one thing. Interpreting the results in terms of physical realities is another matter. When it comes to interpreting the results in terms of physical realities in the natural environment in which the phenomena of interest were observed and recorded, the absence of a physical continuum means that any conclusions as to the physics or nature-science that may underlie or may also be taking place will, and indeed must necessarily, be aphenomenal. Correlations discovered in such data may very well be aphenomenal. Any inferences as to possible "cause-effect" relationships will also be aphenomenal.

Assuming abstract numerical continuity on the real-number line for an extremely large number of discrete data points generated for the same abstract event, lets us overlay another level of information atop the actual discrete data because the tendency of the numerical data purely as numbers is isomorphic to the envelope generated by joining the discrete data points. This isomorphism, however, is precisely what cannot be assumed in advance regarding the underlying phenomenon, or phenomena, generating whatever observations are being recorded from some actual process taking place in nature.

What does this mean? When it comes to the science of nature, the mere fact of some event's occurrence is necessary information, but in itself this information is also insufficient without other additional "meta"-data about the pathway(s) of the event's occurrence, etc. There are strong grounds here for treating with the greatest skepticism a wide range of quantitative projections generated by all the current models of global warming and climate changes.

4.12 Conclusions

This chapter studies the currently detected Colony Collapse Disaster (CCD) syndrome that has raised a lot of concern about the sustainability of the modern lifestyle. This chapter attempts to study this problem and highlights the need for addressing the causes rather than symptoms – a modus operandi that has become synonymous to post-renaissance world order. Based on the evidence presented in this chapter, the following conclusions can be reached:

- With the so-called hard sciences, it is not possible to determine the cause of CCD or any of the major diseases that we are faced today. Conducting more measurements with more science of tangibles will not be helpful.
- Any study should involve the recognition of the pathways of a product, including the origin. By tracking the origin and recognizing the role of matters that are conventionally not accounted for in the science of tangibles, one can begin to see the full picture. This analysis is typical of the science of intangibles.
- A new material characterization methodology, along with new theories of mass and energy interactions, is needed and will be discussed in latter chapters. The technique would include the role of intangibles automatically. With this approach, the importance of sustainability will be inherently featured in any product quality.

5

Mass, Energy and Time: A Delinearized History

5.1 Summary

This chapter serves up a delinearized history of mass, energy and time.[1] This approach uncovers the roots of many apparently contradictory

[1]Our path diverges from elaborating the apparent paradox of Professor Stephen Hawking's "brief history of time", where time is in the same moment both instantaneous and infinite. Our approach comprehends *all* phenomena in the natural order — including the consequences of human interventions and how phenomena come to possess a characteristic time. The regularity or repeatability of that time is subject to ambient conditions. Absent human intervention, such characteristic-time signatures possess one or more of the following characteristics: an onset period of some kind, a one-off character on the one hand and-or a rough periodicity of some kind on the other (these cannot be assumed to be mutually exclusive), and an approximate or on-average expected duration. The end-point is always here on earth, "terra firma", as opposed to, say, some colony sitting atop some distant piece of flotsam or jetsam in outer space (where Prof.

developments in the pharmaceutical and chemical industries (including Big Oil). At the core of these contradictions we unmask the current crisis within these industries. Several key theories and 'laws' are deconstructed in order to determine fundamental flaws. A historical discourse ranging from ancient civilization all the way up to the Information Age reveals numerous instances of conflation between traits of creation and creator at various stages, ranging from origin of universe to onset of life and death. This conflation permeated through every level of cognition ranging from Atomism of ancient Greek through dogma of Roman Catholic church all the way down to Newton's laws of motion and Einstein's quantum theory of the modern age. These theories all suffer from similarly spurious first premises. Once these premises are removed, a consistent pattern emerges that can be used to characterize mass and energy scientifically. This paves the way to developing technologies that can be considered truly 'green'. Considerable attention is paid to the theory of so-called "peak oil" and especially its foundation in the doctrines popularized by the Rev. Thomas Malthus and his followers.

5.2 Introduction

There was a time in the history of mankind that was blighted by the fear of hell or being pushed off the edge of the 'flat earth'. Today, those fears have been transmogrified into the fear of losing money and health. Ours is the post "9/11" era — a period in which the Iraq War, financial collapse, the Amber Alerts, the drone attacks, the cyber attack, and most recently ISIS have all combined — or been combined — to take our society to an unprecedented level of fear. Three out of four Americans say they feel more fearful today then they did twenty years ago (Glassner, 1999). The fear is all-pervasive (Picture 5.1). Corey (2004) defined this culture of fear as a "political tool, an instrument of elite rule or insurgent advance, created and sustained by political leaders or activists who stand to gain something from it, either because fear helps them pursue a specific political goal, or because it reflects or lends support to their moral and political beliefs — or both."

Sociologists call this a matter of perception (Glassner, 1999) and blame the news media for Americans' fear. Others blame politicians and

---

Hawking proposes to move and preserve the ultimate salvation of the human species...).

Picture 5.1 Culture of fear in contemporary civilization.

their lust for money. Glassner (1999) argues that the media bombards us with sensationalistic stories that are designed to increase ratings, a concept called the media-effects theory. Television news programs, for instance, survive on scares. Stories of crime, drugs, and disaster make up most of the newscasts because that is what gets people to tune in and watch. As stories of crime and disaster increase, so do ratings. Similarly, television newsmagazines such as Dateline, Primetime, or 20/20 commonly report on stories that are statistically nearly impossible to happen, yet they sensationalize one account of an event, letting emotional accounts trump objective information, and getting the general public to panic and fear over the event or subject.

News outlets, whether on television or in print, typically exaggerate events and inflate statistics, according to Glassner, and that leads to many of the fears that overcome Americans' lives. Other "peddlers of fear" that Glassner discusses are politicians and advocacy groups. In essence, they exaggerate statistics and stories to the American people so that their own personal causes and beliefs can benefit. For example, if a politician knows they can get more votes and win a re-election if they get funding for a program benefitting teen moms, he or she will inflate the problem of teenage pregnancy and instill fear in Americans over what would happen if such a program were refused funding. For instance, perhaps teenage pregnancy rates would skyrocket or teen moms on welfare would lose income and starve their children. Anyone hinting at the presence of a new Trinity comprising Government, Church, and Corporate personifications of the Money god that are determined to continue feeding on people's misfortunes regardless of the policy in place at the moment is dismissed as a crackpot conspiracy theorist. This level of dysfunction is the only visible signature of a secret real unelected part of government that runs the elected government and especially its foreign policy and global aims.[2] Meanwhile, the general public acceptance of scientists as the only true prophets of the truth has served to fortify such a view. Accompanying this thinking is the notion that we can make this entire culture of fear disappear once we sufficiently empower New Science and its peddlers.

This chapter focuses on the science of today's culture, particularly in relation to new science and social theories that have defined this civilization. It is shown the prophets of New Science are no better 'messengers of truth' and indeed are the pimps that have sold the truth for a meagerly price. In this process, every theory is tainted with bad intention and it is no accident that the same people touted the notion 'the road to hell is paved with good intention'. That 'good' was never intended to be real. Every technology be it for processing mass or producing energy is meant to maximize profit at the expense of its true value.

5.3 The Energy Crisis

Ever since the oil embargo of 1972, the world has been gripped with the fear of 'energy crisis'. U.S. President Jimmy Carter, in 1978, told the world in a televised speech that the world was in fact running out of oil

[2]The occasional emergence of an honest individual in the Presidency makes little or no difference to the functioning of these sources of U.S. governance hidden away among Washington's 17 intelligence agencies and the country's dozens of privately-endowed think-tanks. Thus for example Gen. Eisenhower on leaving office in 1960 famously warned his fellow citizens about the danger posed to the republic and its freedoms by a "***Congressional*** [emphasis added] -military-industrial complex" that sucks up all the resources of the US economy unto itself. However, the actual speech that Eisenhower was permitted to deliver ['permitted' only as a courtesy, because he was now out of office and therefore as powerless as any ordinary citizen] cut out the word "Congressional" that appears in Eisenhower's hand-written draft of the speech and was only discovered accidentally by historians 15 years after the speech. This crucial omission

at a rapid pace - a popular Peak Oil theory of the time - and that the US had to wean itself off of the commodity. Since the day of that speech, worldwide oil output has actually increased by more than 30%, and known available reserves are higher than they were at that time. This hysteria has survived the era of Reaganomics, President Clinton's cold war dividend, President G.W. Bush's post-9-11 era of 'fearing everything but petroleum' and today even the most ardent supporters of petroleum industry have been convinced that there is an energy crisis looming and it's only a matter of time, we will be forced to switch no-petroleum energy source. Then comes the peddler of 'renewable energy' and amass all resources from the public with the promise of offering 'clean energy'.

Peak oil is one of the concept that promotes the notion that global oil reserve is limited and at some point will start to run out, leading to sharp rise in oil price. This theory paints two possible pictures of the energy outlook, namely, 1) a worldwide depression will follow the peak in oil production as high prices drag down the whole world's economy; and 2) alternate energy sources have to be introduced in a hurry in order to prevent the energy crisis. Embedded in theory is the notion that per capita energy need is increasing globally and will continue to increase because of 1) modernization that inducts more people to the urban energy-intensive lifestyle and 2) increasing population. Because it is also assumed that oil reserves are limited, it follows that at some point oil production will peak, after which the amount of oil that is being produced will decline. It is believed that once the decline has started it will become a terminal decline and oil production will never again reach the levels they were at during the peak. When this happens there are going to be serious consequences to the world's economy. Since the demand for oil is unlikely to decline it inevitably means that the price will increase, probably quite dramatically. This crisis attributed to peak oil theory is proposed to be remedied with 1) austerity measures in order to decrease dependence on energy, possibly decreasing per capita energy consumption, and 2) alternatives to fossil fuel. None of

---

completely altered Eisenhower's intended message that there was a conscious underlying policy in place — the congressional part of the trinity. Instead, down to this very day, everyone discusses the misleading impression that their former President was upset over a machine called the 'military-industrial complex' running amok in need of the attention of honest human elements. Eisenhower's point was not just the danger posed by the physical scale of the apparatus by itself, but the need for the people to act to eliminate the human corruption feeding and sustaining the apparatus.

these measures seem appealing because any austerity measure can induce imbalance in the economic system that is dependent on the spending habit of the population and any alternative energy source may prove to be more expensive than fossil fuel. These concerns create panic, which is beneficial to certain energy industries, including biofuel, nuclear, wind, and others. Add to this problem is the recent hysteria created based on the premise that oil consumption is the reason behind global warming. This in itself has created opportunities with many sectors engaged in carbon sequestration.

The upcoming section of the chapter makes it clear that the underlying premises of the peak oil theory are entirely spurious.

5.3.1 Are Natural Resources Finite and Human Needs Infinite?

The notion of The Infinite has intrigued humanity for a long time. In ancient civilizations, infinity was not a 'large number'. It was something external to creation. In other words, only a Creator was considered to be infinite, along with many other traits that could not be part of Creation. However, this 'infinity' has nothing to do with the unboundedness of nature that has no boundary. Even though the ancient Greeks had a similar concept of infinitude, post-Aquinas Europe developed an entirely different take on infinitude, one highlighted recently by Islam *et al.* (2010; 2013).

In a study published more than a decade ago, Lawrence Lerner, Professor Emeritus in Physics and Astronomy at the University of Chicago, was asked to evaluate how Darwin's theory of evolution was being taught in each state of the United States (Lerner 2000). In addition to his attempt to find a standard in K-12 teaching, Prof. Lerner made some startling revelations. His recommendations created controversy, with many suggesting he was promoting "bad science" in name of "good science." However, no one singled out another aspect of his finding. He observed that "some Native American tribes consider that their ancestors have lived in the traditional tribal territories forever." He then equated "forever" with "infinity" and continued his comment stating, "Just as the fundamentalist creationists underestimate the age of the earth by a factor of a million or so, the Black Muslims overestimate by a thousand-fold and the Indians are off by a factor of infinity." (Lerner 2005). This confusion between "forever" and "infinity" is not new in modern European culture. In the words of Albert Einstein, "There are two things that are infinite, human stupidity and the Universe, and I am not so sure about the Universe." Even though the word "infinity" emerges from a Latin word, *infinitas*, meaning "unbounded-ness," for centuries this word has been applied in situations in which it promotes absurd concepts. In Arabic, the equivalent word (*la nahya*) means "never-ending." In Sanskrit, similar words exist (*Aseem*, meaning 'no end') and those words are never used in mathematical terms as a number. This use of infinity to enumerate something (e.g., infinite number of solutions) is considered to be absurd in other cultures.

Nature is infinite – in the sense of being all-encompassing – within a closed system that nevertheless lacks any boundary. Somewhat paradoxically, nature as a system is closed in the sense of being self-closing. This self-closure property has two aspects. First, everything in a natural environment is used. Absent anthropogenic interventions, conditions of net waste or net surplus would not persist for any meaningful period of time. Secondly, nature's closed system operates without benefit of, or dependence upon, any internal or external boundaries. Because of this infinite dimension, we may deem nature – considered in net terms as a system overall – to be perfectly balanced. Of course, within any arbitrarily selected finite time period, any part of a natural system may appear out of balance. However, to look at nature's system without acknowledging all the subtle dependencies that operate at any given moment introduces a bias that distorts any conclusion that is asserted on the basis of such a narrow approach.

From where do the imbalance and unsustainability that seem so ubiquitous in the atmosphere, the soil, and the oceans actually originate? As the "most intelligent creation of nature," men were expected to at least stay out of the natural ecosystem. Einstein might have had doubts about human intelligence or the infinite nature of the Universe, but human history tells us that human beings have always managed to rely on the infinite nature of nature. From Central American Mayans to Egyptian Pharaohs, from Chinese Hans to the Manichaeans of Persia, and from the Edomites of the Petra Valley to the Indus Valley civilization of the Asian subcontinent, all managed to remain in harmony with nature. They were not necessarily free from practices that we no longer consider (Pharaohs sacrificed humans to accompany the dead royal for the resurrection day), but they did not produce a single gram of an inherently anti-nature product, such as DDT. In modern times, we have managed to give a Nobel Prize (in medicine) for that invention.

Islam *et al.* (2010) and Khan and Islam (2012) have presented detailed accounts of how our ancestors dealt with energy needs and the knowledge they possessed that is absent in today's world. Regardless of the technology these ancient civilizations lacked that many might look for today, our ancestors were concerned with not developing technologies that might undo or otherwise threaten the perceived balance of nature that, today, seems desirable and worth emulating. Nature remains and will remain truly sustainable.

5.3.2 The Peak Oil Theory and its Connections to Population and Lifestyle

As is well known, the current version of the peak oil theory emerged from King Hubbert's 1956 presentation entitled "Nuclear Energy and the Fossil Fuels" (Hubbert, 1956). The peak oil theory says that for any given geographical area, from an individual oil-producing region to the planet as a whole, the rate of petroleum production follows a bell-shaped curve. This bell curve (Figure 5.1) that has been the basis of many theories popularized in almost all disciplines of modern era dictates that there would be a global peak in oil recovery, following which oil production would decline irreversibly and monotonically.

This graph shows typical increase in oil production, eventually reaching the peak. The increase occurs because of the global demand of energy that is dependent on world population as well as globalization

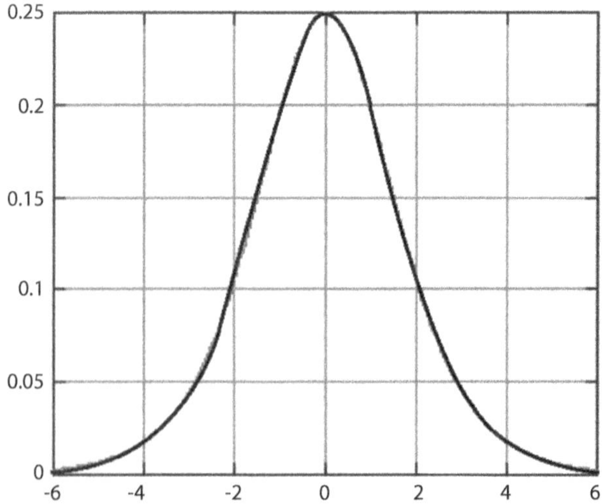

Figure 5.1 the bell curve has been the base curve of many theories in modern era (x-axis is replaced with time and y-axis with global oil production).

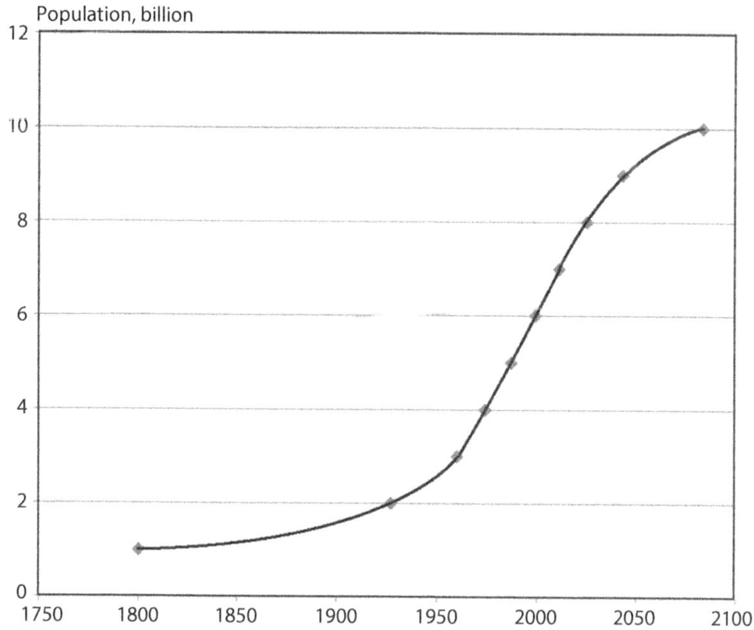

Figure 5.2 Population growth history and projection (data from CIA factbook, UN).

that is equivalent to urbanization that increases per capita energy consumption.

Figure 5.2 shows population growth since 1800. The actual data collection started in 1940's, therefore data from before is an assumption. The figure also shows prediction until nearly 2100. This prediction is based on a United Nations medium projection. In fact, previous models based their guesses about population growth on assumptions similar to those of the peak oil theory.[3]

[3]The authors first went after the connections between peak oil theory and the unfulfilled predictions of the Rev. Thomas Malthus back in (Zatzman & Islam, 2007). At that time we pointed out: "In 1798 Thomas Malthus published his *Essay on Population*. This asserted that population must always and everywhere expand to outstrip the capacity of societies to feed themselves. This has been repeatedly disproved everywhere – in developing countries as well as developed countries. Nevertheless in 1968, Paul R. Ehrlich published his work *The Population Bomb*, reiterating the same thesis with fancier computer projections. The first country that his model predicted would collapse calamitously was the People's Republic of China, the second was India. [Much of] the steady rise in the world oil price since 2004 [until the downturn in late 2014] is being "blamed" on China and India raising their level of consumption to the level of more

As early as 2006, the United Nations stated that the rate of population growth was visibly diminishing due to the ongoing global demographic transition. With that trend in the rate of population growth, it was foreseen that the rate of growth may diminish to zero by 2050, concurrent with a world population plateau of 9.2 billion (UN 2010). However, this is only one of many estimates published by the UN. Each report was based on different sets of assumptions, thereby leading to different projections. In 2009, UN population projections for 2050 ranged between around 8 billion and 10.5 billion (UN 2009). None of these projections include a negative growth even though some predict zero growth, leading to some sort of pseudo stable population. This is an exception because all other growth models dealing with natural systems have a bell-shaped curve as the foundation. Randers (2012) is the first reported one to predict negative growth after the global population reaches a plateau. This alternative scenario is a result of the argument that all existing projections insufficiently take into account the downward impact of global urbanization on fertility. Randers introduces population decline as the only outcome of urbanization the rate of which overshadows any other factor. Randers' "most likely scenario" reveals a peak in the world population in the early 2040s at about 8.1 billion people, followed by decline (Randers, 2012). Three different projections are plotted in Figure 5.3.

While researchers have attempted to characterize global population growth with little consensus, there appears to be definite trends in various geographical locations. For instance, Figure 5.4 shows how population has grown and declined in various geographical enclaves since 1950, the time when actual data became available. The *y*-axis shows population in millions. Clearly, Europe exhibited the lowest growth rate from 1950's, with a declining rate of growth as early as 1960's. This

---

developed countries of Europe and the Americas. There is however no longer any serious talk or threat of their population growth – which is still large in both absolute and relative (percentage) terms relative to any other part of the planet – overwhelming the ability of their economies to feed their population.

These doom-laden predictions lack any foundation anywhere in engineering practice or scientific discourse. As far as any notions about raw materials in general being in finite supply, technological breakthroughs have continually been finding new ways to make or do more per unit output of products or finished goods using less energy and-or less raw material per unit input. The reality of these technological revolutions has repeatedly refuted all previous claims in every other field that there are "limits to growth" beyond which human existence or social or progress cannot be sustained. In the last twenty years, the elaboration of cost-effective and profitable means for

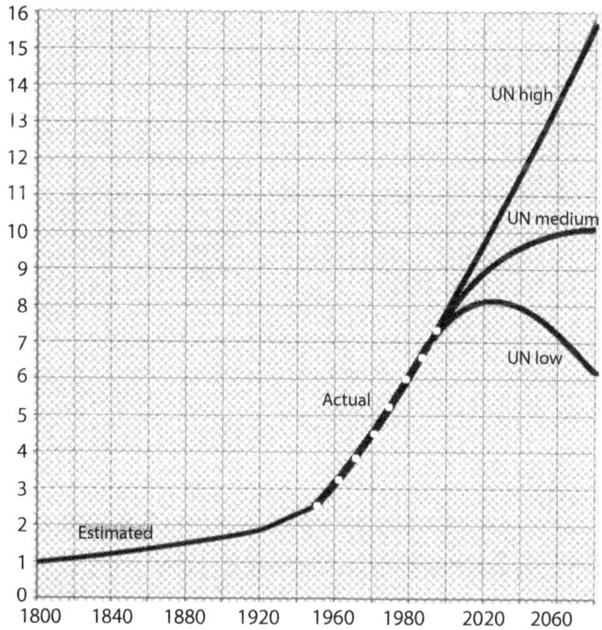

Figure 5.3 Estimated, actual, and projected population growth (decline).

resulted in reaching a plateau in 1990, following continuous decline in population, despite surge in immigrant population. To some extent, Europe represents the model that forms the basis of practically all resource models, used in modern time, including the peak oil theory.

Individually, Africa shows the highest rate of growth among all continents. Asia and Latin America started off with similar growth rate but Asia and then Latin America exhibited slowing growth rate, matching with that of North America. Of course, Northern America that includes

---

exploiting the extensive so-called "unconventional reserves" of oil — like the oil sands of western Canada — has completely turned upside down the notion that the world's lights must go out when the last barrel of oil has been pumped in Saudi Arabia, Libya, Iraq or Iran. Where Malthus imprudently asserted that population must grow exponentially while food production could at best be increased only arithmetically, the work of Lord Boyd-Orr's team at the UN Food and Agricultural Organisation in the decade following the end of the Second World War, carrying on from his own classic pre-war investigations, as a professional nutritionist, of Scottish (Boyd-Orr, 1937) and English (Boyd-Orr, 1943) diet among the working classes, decisively refuted all notions that there was anything like a finite capacity for food production relative to any actual rate of population increase recorded anywhere on the planet.

Hence, such repeated predicting followed by the failure of reality to meet the

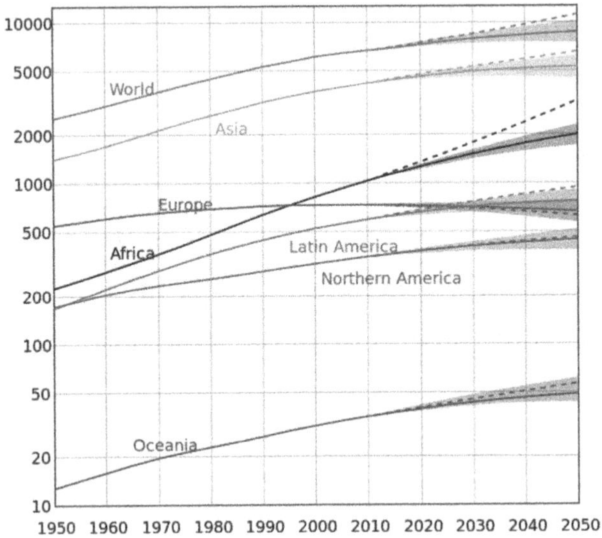

Figure 5.4 World population growth for different continents.

Canada would actually have a negative growth if it wasn't for the very high immigration rate in USA. The growth rate started to decline in Asia in 1980's. It was because of sustained campaign of birth control in some of the most populous countries in the world, namely, China, India, Pakistan, India, Bangladesh, etc. This campaign started in the 1960's but the results began to show up in 1980's. During the same period, the campaign of urbanization also began. Fueled with 'green revolution' that also took effect starting 1960's, urbanization has been in full gear, the most direct outcome of which has been the per capita energy consumption. This aspect will be discussed later.

---

prediction suggests that the activity of such prediction itself lacks any rational basis. It is a prejudice feeding the formation of yet another "devil theory of history": guess-who will be blamed as things fail to go so well for countries that presently think they are "on top" in world rankings...

"Peak oil" can thus only be understood as the latest attempt to prepare yet another devil theory of history. People will be blamed for consuming too much: governments were ignorant, corporations became excessively greedy, people became desperate... and the world went to hell. Apparently delivering another set of Cassandra-like warnings of impending doom, the proponents of the theory of peak oil and its purported consequences are also messing with people's ability to sort anything out rationally or scientifically. That is, they are turning prejudice into disinformation. Petroleum being the basis of plastics and much else, this dissemination of disinformation is exercising the most

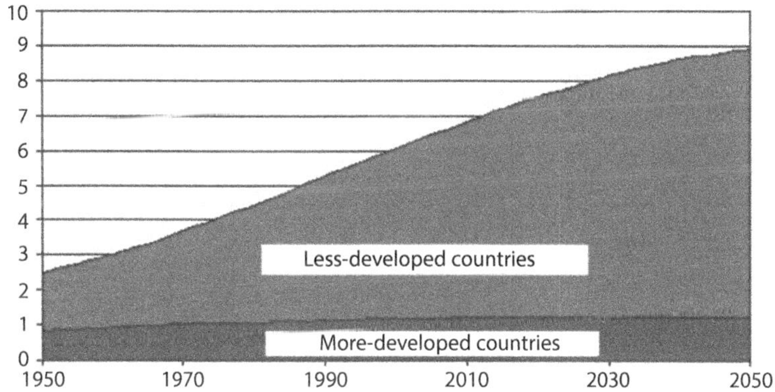

Figure 5.5 There are different trends in population growth depending on the state of the economy.

Figure 5.5 shows how population in more developed countries reached a plateau while that of less-developed countries continue to growth, albeit with a slowing down rate. In terms global energy need, this figure presents an interesting divide. In average, the energy consumption per capita of the 'less-developed countries' is an order of magnitude less than that of 'more-developed countries'. In mathematical terms, it means the world has a capacity of sustaining energy needs of the majority of the population even if the population is increased 10 fold. In practical term, it means, if we could contain the per capita energy consumption, we would have no worries about natural population growth. Indeed, the energy consumption of the 'more developed countries' has been contained. In last 20 years, the most populous 'developed country', USA has shown practically constant per capita energy consumption. USA is an important case as this country personifies global trend in terms of energy consumption. Historically, USA has set standards for all aspects of technology development and other tangible aspects of civilization for a duration that has been synonymous with petroleum golden era. Whatever USA does today is emulated by the rest of the world in years to come. Table 5.1 shows per capita energy consumption (in tones of oil equivalent per year) of USA in the last few decades, along with predictions for 2015. In this, Canada represents an

paralysing effect on developing and researching appropriate solutions to contemporary problems in all aspects of life."This leads to oscillatory behavior in the human population profile over a long-term scale. These assumptions will be discussed later in this section.

Table 5.1 Per capita energy consumption (in TOE) for certain countries.

Countries	1990	1995	2000	2005	2010	2015
USA	7.7	7.8	8.2	7.9	7.3	7.3
Canada	7.5	7.9	8.1	8.4	7.6	7.6
Japan	3.6	4.0	4.1	4.1	3.7	3.9
Germany	4.4	4.1	4.1	4.1	4.0	3.8
Russia	5.9	4.3	4.2	4.5	4.8	5.5
Saudi Arabia	3.9	4.8	5.1	6.0	6.6	7.7
China	0.8	0.9	0.9	1.3	1.8	2.2
India	0.4	0.4	0.5	0.5	0.6	0.7
Indonesia	0.6	0.7	0.7	0.8	0.9	1.2
Sri Lanka	0.3	0.3	0.4	0.5	0.5	0.6

interesting case. Canada follows USA trend closely in matters of per capita energy consumption but falls far behind in matters of population growth, expenditure in research and development (particularly in energy and environment), expenditure in defense and pharmaceutical industries, and other long-term economic stimuli. Japan, on the other hand represents other extremity of the energy consciousness spectrum. As can be seen in Table, Japan maintains steady per capita energy consumption at almost half the value of that of Canada. At the same time, Japan has maintained very high relative investment in education and research and development. However, Japan's population has been dropping keeping pace with Europe and unlike USA. Canada's population growth has been a mix of Europe/Japan (decline) and USA (mild growth). The difficulty involved in maintaining a balance between urbanization and per capita energy consumption is most sternly manifested in the case of Saudi Arabia. Both Germany and Russia show mild per capita energy consumption, signaling prudent usage of energy sources and high energy efficiency. Saudi Arabia is a 'developing country' in all measures except that it is projected to be the most energy-consuming country in the world by 2015. As early as 1995, it exceeded the per capita energy consumption of Russia and Germany and is slated to exceed that of USA by 2015. Saudi Arabia represents the global trend by

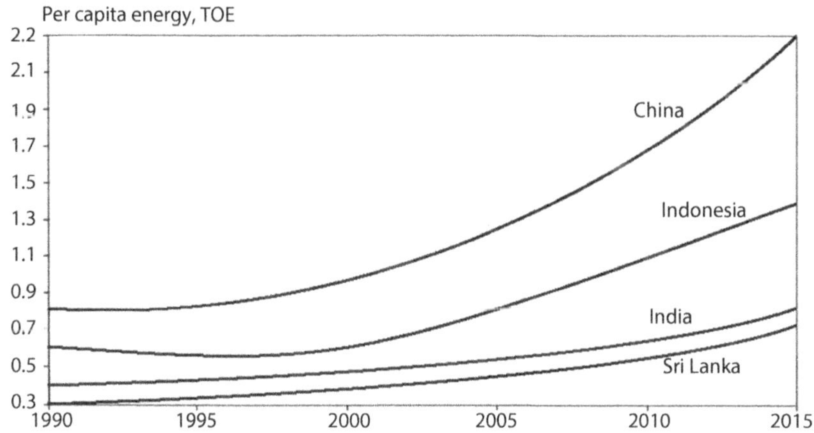

Figure 5.6 *Per capita* energy consumption growth for certain countries.

'developing countries' to emulate the wasteful habits of the USA while shunning positive aspects of USA in the areas of economic growth, education or research and development. This trend of Saudi Arabia is alarming and is a trademark of global obsession with wasteful energy habits. Saudi Arabia is just an example of this obsession that is all pervasive in the developing countries as can be seen in Figure 5.6.

Figure 5.6 shows the growth in per capita energy consumption for some key countries that are not characterized as 'more developed countries'. These countries all had very modest per capita energy needs in 1990. However, they all show exponential growth in energy needs in the last two decades. China leads the pack with the highest growth in energy needs. It nearly triples the energy need in 25 years. This trend shows that China could have dealt with its 'population crisis' by keeping the per capita energy consumption in check. This would have avoided many shortcomings of the one-child policy that China has imposed on its population for decades. Similar growth is shown by Indonesia – another country that attempted to decrease its population rather while increasing per capita energy needs. Over the two decades, Indonesia has doubled its per capita energy consumption. India has shown restraints in per capita energy consumption. However, still its per capita energy consumption has doubled during the decades of concern. Sri Lanka has been the lowest energy consuming country (from the list of countries) but still maintains growth very similar to India and Indonesia.

It has been recognized for some time that there is a strong correlation between per capita energy need and GNP (as well as GDP). Over

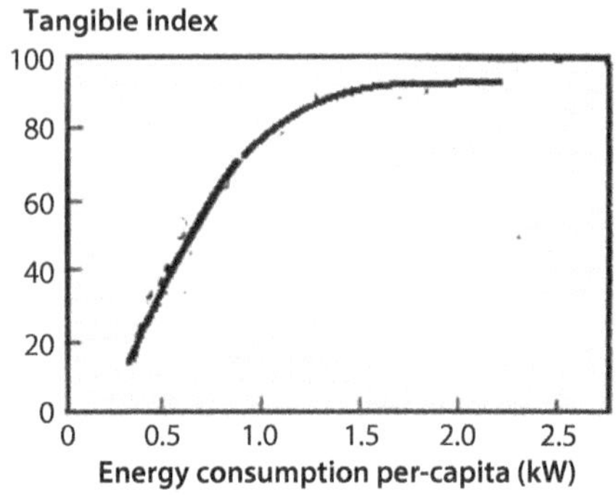

Figure 5.7 A strong correlation between a tangible index and per capita energy consumption has been at the core of economic development (from Goldenberg, 1985).

the last 30 years, the average consumption of the global 'South' has been nearly an order-of-magnitude less than that of the 'West' (Goldemberg *et al.*, 1985; Khan and Islam, 2012). As the West has been trying to boost its population and contain its per capita energy consumption, while increasing its GNP, the 'south' has been trying to contain its population while increasing the per capita energy consumption as well as GNP.

These contradictory measures have created confusions in both the west and the 'south'. This is most visible in the definition of GNP and GDP that reward an economy for increasing wasteful habits (e.g. per capita energy consumption). This contradiction has been discussed by Khan and Islam (2007), who introduced new techniques for measuring economic growth hat could take account of true sustainability. They showed that true sustainability would increase GNP by increasing efficiency (rather than increasing per capita energy consumption).

Figure 5.7 shows how energy consumption has become synonymous with welfare of the society, as expressed as tangible expression of the quality of life. Goldenberg *et al.* (1985) correlated per capita energy consumption with a Physical Quality of Life Index (PQLI), which is an attempt to measure the quality of life or well-being of a country. The value is the average of three statistical data sets: basic literacy rate, infant mortality, and life expectancy at age one, all equally weighted on a 0 to 100 scale. It was developed for the Overseas Development Council in the mid-1970s by Morris David Morris, as one of a number of measures created due to dissatisfaction with the use of GNP as an indicator of development. PQLI is best described as the measure of tangible features of the society, not unlike GDP (Khan and Islam, 2007). Ever since, numerous other indices have been proposed, including more recently developed Happiness index, but they all suffer from similar short-comings, i.e., focus on tangibles, as outlined by Khan and Islam (2012 and Zatzman, 2012, 2013). The following steps are used to calculate Physical Quality of Life:

1. Find percentage of the population that is literate (literacy rate).
2. Find the infant mortality rate. (out of 1000 births) INDEXED Infant Mortality Rate = (166 - infant mortality) × 0.625
3. Find the Life Expectancy. INDEXED Life Expectancy = (Life expectancy - 42) × 2.7
4. Physical Quality of Life = (Literacy Rate + INDEXED Infant Mortality Rate + INDEXED Life Expectancy)/3.

This trend goes back to the earliest times of the Industrial Revolution more than two-and-a-half centuries ago. Khan and Islam (2012) discussed the mindset that promoted such wasteful habits in all disciplines. Figure 5.8 summarizes the dilemma. At the dawn of industrial age, civilization began to be defined by consumption and wasteful habits. As the population grew, the energy consumption per capita should have been decreased in order compensate for the increasing energy demand. This would be in line with the claim that industrialization had increased human efficiency.

The opposite happened in the developed countries. For centuries, the per capita energy consumption increased, along with dependence of

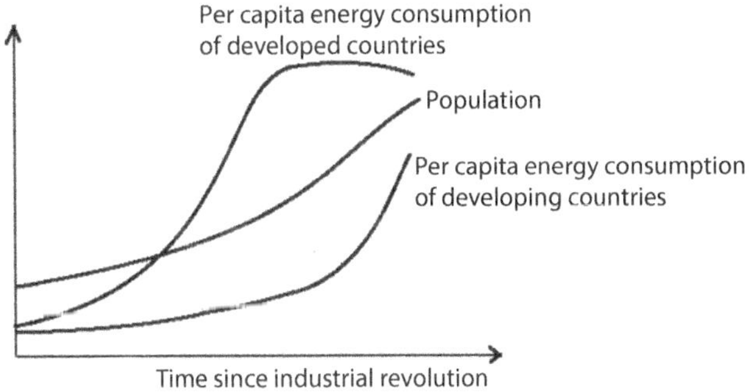

Figure 5.8 While population growth has been tagged as the source of economic crisis, wasteful habits have been promoted in name of emulating the west.

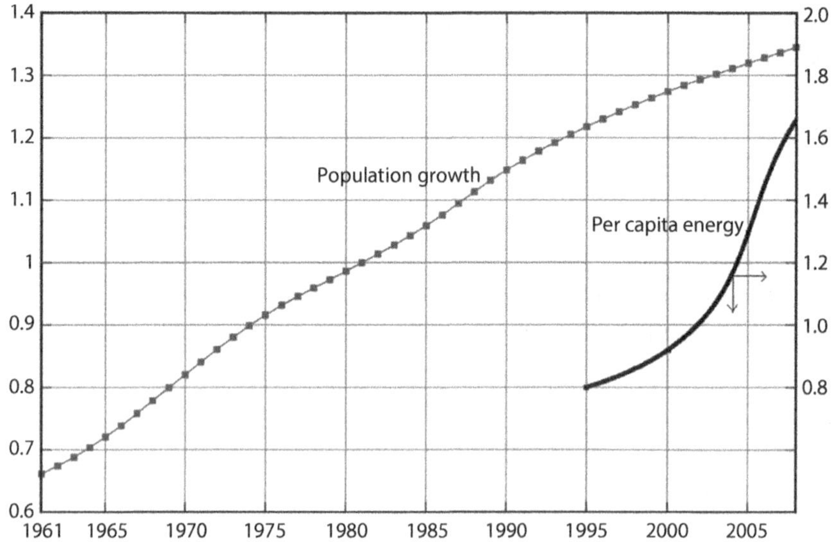

Figure 5.9 Population and energy paradox for China.

mechanization. It only stabilized in 1990s. By then, the population growth in the west has been arrested and have been declining in most part (the exception being USA). This population and energy paradox was further accentuated by encouraging the developing countries to emulate the west in wasteful habits. In every country, consumption per capita increased with time as a direct result of colonialism and imposed culture that is obsessed with externals and short-term gains. As a result, a very sharp increase in per capita energy consumption took place in the developing countries. As can be seen from Table 5.1, even with such increase, the "south" has not caught up with the "west", with the exception of some petroleum-rich countries.

A major case in point here is... China. For the last two decades, it attempted to curtail its population growth with a one-child per family law. The current Chinese government at the behest of the latest congress of the Communist Party of China has now repudiated this policy as practically unenforceable. Furthermore and even more interesting, however: Figure 5.9 shows that population growth has in fact been dwarfed by the increase in per capita energy consumption. A similar conclusion emerges from the comparable statistical profile for the Indian subcontinent, where infanticide and female-selective abortion is in order to boost male population in favor of female population that is considered to be a drain to the economy. This finding is meaningful considering India and China hold one third of the world population and can effectively change the global energy outlook either in favor or against sustainability.

In order to change the above trend, and address the population and energy paradox, several indices have been introduced. These indices measure happiness in holistic terms. Comparing one person's level of happiness to another's is problematic, given how, by its very nature, reported happiness is subjective. Comparing happiness across cultures is even more complicated. Researchers in the field of "happiness economics" have been exploring possible methods of measuring happiness both individually and across cultures and have found that cross-sections of large data samples across nations and time demonstrate "patterns" in the determinants of happiness. The New Economics Foundation was the first one to introduce the term "Happiness index" in mid 2000's (Khan and Islam, 2007; White, 2007). In first ever ranking, Bangladesh, one of the poorest nations of the time was found to be the happiest among some 150 countries surveyed. At that time, Bangladesh was among the lowest GDP countries along with very low per capita energy consumption. This study demonstrated that happiness is in fact inversely proportional to per capita energy consumption or GDP. Before, this study would set any trend globally in terms of energy policies, a number of similar happiness indices were introduced in succession, all showing a direct, albeit broad, correlation between GDP and happiness. One such index is the Happy Planet Index (HPI) that ranks 151 countries across the globe on the basis of how many long, happy and sustainable lives they provide for the people that live in them per unit of environmental output. It represents the efficiency with which countries convert the earth's finite resources into well being experienced by their citizens. The Global HPI incorporates three separate indicators:

(a) ecological footprint (the amount of land needed to provide for all their resource requirements plus the amount of vegetated land needed to absorb all their CO_2 emissions and the CO_2 emissions embodied in the products they consume);
(b) life satisfaction (health as well as "subjective well-being" components such as a sense of individual vitality, opportunities to undertake meaningful, engaging activities, inner resources that help one cope when things go wrong, close relationships with friends and family, belonging to a wider community);

(c) life expectancy (included is the child death, but not death at birth or abortions).

The first item couples CO_2 emission with footprint. This emission relates to only fossil fuel usage. This characterization doesn't take in account the fact that CO_2 that is emitted from refined oil is inherently tainted with catalysts that are added during the refining process. This creates bias against fossil fuel and obscures the possibility of finding any remedy to the energy crisis.

The Organisation for Economic Co-operation and Development (OECD) introduced the Better Life Index. It includes 11 topics that the OECD has identified as essential to wellbeing in terms of material living conditions (housing, income, jobs) and quality of life (community, education, environment, governance, health, life satisfaction, safety and work-life balance). It then allows users to interact with the findings and rate the topics against each other to construct different rankings of wellbeing depending on which topic is weighted more heavily. For the purpose of this analysis, what matters is the Life Satisfaction survey. Life satisfaction is a measure of how people evaluate the entirety of their life and not simply their feelings at the time of the survey. The OECD study asks people to rate their own life satisfaction on a scale of 0 to 10. The ranking covers the organization's 34 member countries plus Brazil and Russia.

The Happy Planet Index ranked Costa Rica as the happiest country in 2012. Particularly high score relates to high life expectancy and overall wellbeing. Vietnam and Colombia follow in second and third place. Of the top ten countries, nine are from Latin America and the Caribbean. Countries from Africa and the Middle East dominate the bottom of the ranking instead. Botswana is last after Bahrain, Mali, the Central African Republic, Qatar and Chad. Developed nations such as the United States and the European Union member countries tend to score high on life expectancy, medium-to-high in wellbeing, but rather low on their ecological footprint, which puts them in the ranking's second-tier.

The same year, Denmark placed first in the 2012 OECD Life Satisfaction survey, followed closely by Norway. Countries from Northern and Central Europe (such as Austria, the Netherlands Switzerland and Finland) dominate the top part of the ranking. Australia is the highest-ranking country outside of Europe, closely followed by Canada. At the bottom of the ranking are many countries from Eastern and Southern Europe. Hungary comes last, preceded by Portugal.

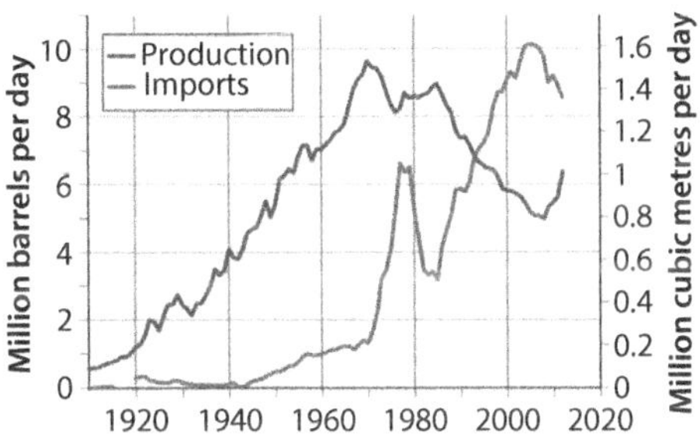

Figure 5.10 Oil production and import history of USA (data from EIA).

5.3.3 Evidence in Favor of the Peak Oil Theory

Hubbert's Peak was thought to have been achieved in the continental US in the early 1970s. Oil production peaked at 10,200,000 barrels per day (1,620,000 m^3/d) and then declined for several years since. This purported evidence for Hubbert's peak seemed to dissipate as Alaska's Prudhoe Bay came into production.[4] However, Alaskan production was not enough to sustain a longer-term production growth and overall oil production declined steadily until recent oil boom of unconventional oil and gas. Figure 5.10 shows US crude oil production, along with crude oil import.

As can be seen from Figure 5.10, until the early 1970s, oil production continued to rise monotonically even when oil import rose sharply. As oil production reached a peak, a peak in oil import followed. This was a time in which the United States was increasing its oil reserve. Oil imports dropped sharply after Iranian revolution and did not recover its pre-Iranian revolution import number until mid 1990s. During this period, the world oil price was exceptionally low and dividends from the end of the Cold War were affecting the US economy positively. Oil

[4]Ironically (and maybe more than a little absurdly?), as this is being written in mid-2015, "never is a heard a discouraging word" about present and future US oil supplies. On the contrary, an orgy of questionable claims make their appearance daily in both the general news and the more specialized energy industry media about the United States having allegedly reached the cusp of becoming a major world exporter of oil and gas based on what is being generated on its own territory by the widespread application of hydraulic fracturing technology (!) This drumbeat began in early 2014, while world oil prices still topped US$100 in Europe and the United States, and it persists to the present moment when the world oil price hovers between US$58-62 per barrel after collapsing below US$50 late in 2014. Unconventional oil and gas affected U.S. oil and gas production in an unprecedented manner. In 2012, U.S. oil production grew more than

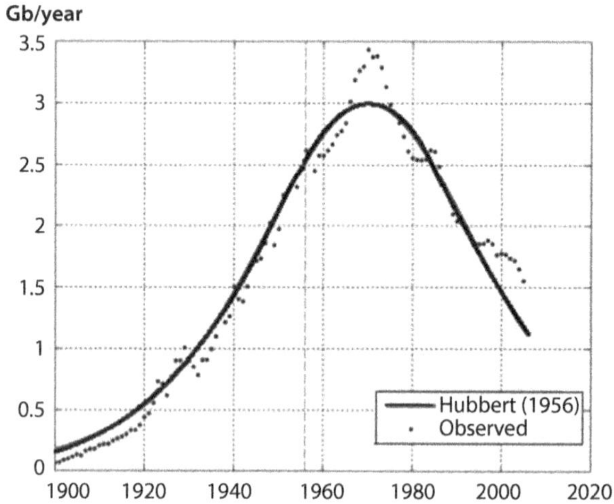

Figure 5.11 US data that appear to support Hubbert's "peak oil" hypothesis.

import continued to rise until the 2008 financial crisis and the sudden hike in the oil price at that same point.

Plotting Hubbert's prediction of US oil against observed production, there appears to be reasonable agreement. Figure 5.11 shows Hubbert's prediction against observed data in USA.

US EIA data show average oil production rose further to a level of 900,000 barrels a day. This surge was due to unconventional oil and gas, mainly enhanced through fracking technology as well as horizontal wells. Together, they have unlocked deposits of oil and gas trapped in formations previously thought to be unreachable. Kashi (2013) reports that USA is now the leading oil and gas producer of the world. Recent growth in USA domestic oil production is unprecedented in history (Fowler, 2013). A government report published in summer of 2013 revealed that U.S. domestic crude-oil production exceeded imports for the first time in 16 years (Smith, 2013).

The surge in US domestic oil production meant expansion of previously conventional oil reservoirs to include low-permeability formations, such as West Texas's Permian basin. In addition, huge expansions in

---

in any year in the history of the domestic industry, which began in 1859, and was set to surge even more in 2013. Daily crude output averaged 6.4 million barrels a day in 2012, up a record 779,000 barrels a day from 2011 and hitting a 15-year high (Fowler, 2013). It is the biggest annual jump in production since Edwin Drake drilled the first commercial oil well in Titusville, Pa., two years before the Civil War began.

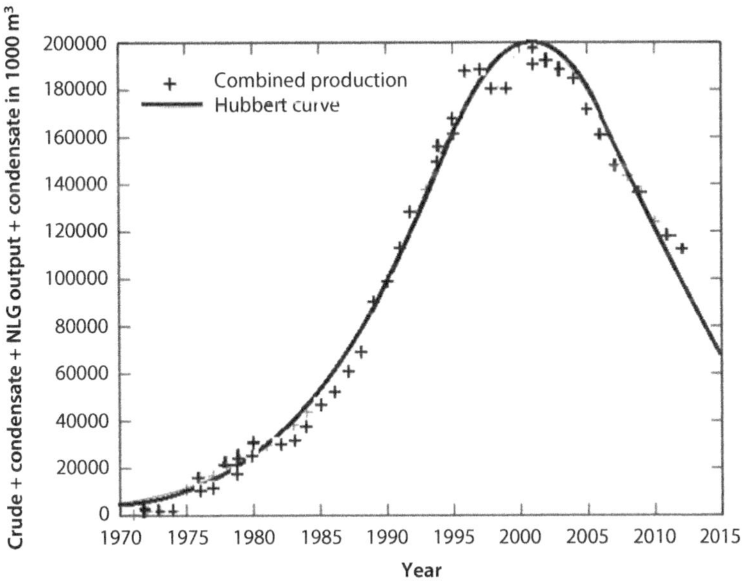

Figure 5.12 Comparison of Hubbert curve with Norwegian oil production.

areas that had been lightly tapped in the past also became available, such as North Dakota's Bakken shale region. The Bakken has gone from producing just 125,000 barrels of oil a day five years ago to nearly 750,000 barrels a day in 2013.[5]

There have been efforts to demonstrate that every country is undergoing peak oil behavior. Figure 5.12 shows such plot for Norway. There has been a society, called Association for the study of peak oil (ASPO), claiming world oil peak in various times, matching with Hubbert prediction. Figure 5.13 shows regional production history (until 2007) and prediction (2007+), all regions showing a Hubbert peak.

ASPO makes the following argument based on Figure 5.13:

[5]The production cost of hydraulically-fractured oil and gas was estimated in 2014 data from the Energy Information Administration of the United States Department of Energy at around US$85 per barrel-equivalent. Clearly this is economically unsustainable in the conditions of the current deep slump in the world oil price. Many medium-sized banks in the U.S. are worried about going bankrupt if those of its customers who invested in allegedly investment-grade bonds offered by many hydraulic fracturing start-ups through those banks (back when the oil price was well above US$100 per barrel-equivalent, of course) decide to dump their dubious investment.

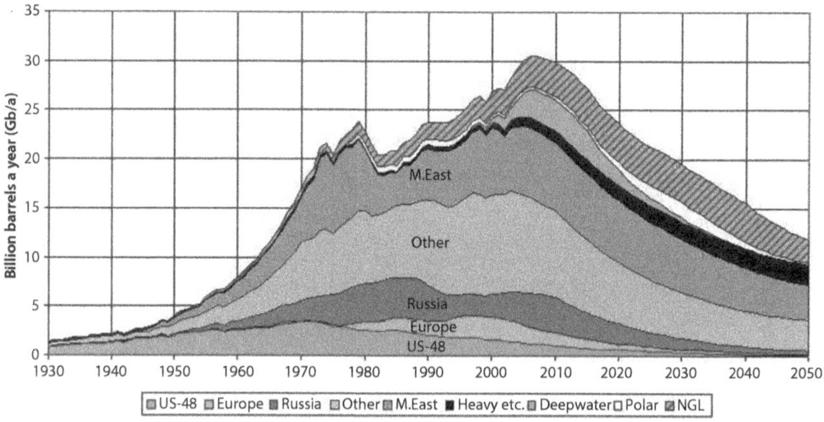

Figure 5.13 Association for the study of peak oil (ASPO) produced evidence of Hubber peak in all regions.

This graph is worth careful as a lot of world history is written into it. Note the steep rise in oil production after World War II. Note that 1971 was the peak in oil production in the United States lower 48. There is a sliver of white labled Arctic oil. That is mostly Alaskan Prudhoe Bay oil, which peaked in 1990. Prudhoe Bay was almost big enough to counteract the lower 48 peak of 1971. The sliver is very narrow now. The OPEC oil embargo of 1973 is very visible. The oil produced by non-OPEC countries stayed nearly constant while OPEC production nearly halved. The embargo caused the world economy to slow. But the high cost of energy spurred the development of energy efficient automobiles and refrigerators and a lot of other things. Note the effect of the collapse of the Russian economy in 1990 on Russian oil production. Note the rapid increase in oil production when the world economy boomed near the end of the twentieth century. Oil was $12 a barrel at that time. Note that European (North Sea) oil peaked in 2000. Note especially what would have happened if the 1973 embargo had not occurred. It is possible that the world would now be on the steep part of the right side of the Hubbert curve.

ASPO predicted something drastic looming in 2010. They wrote:

Beginning in 2010, the Middle East can no longer compensate for declining production elsewhere. Production and therefore consumption will decrease and demand will have little effect on production.

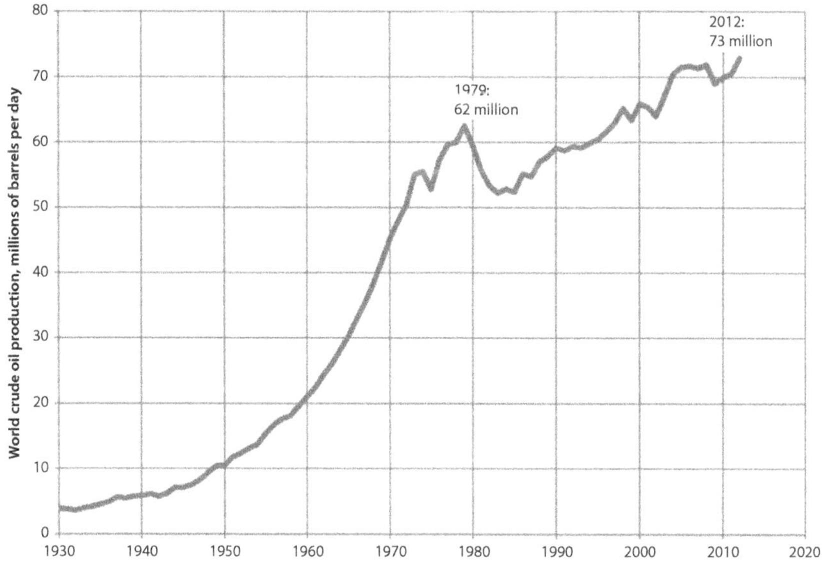

Figure 5.14 Actual global oil production (surface mined tar sand not included).

> *The decrease results from oil depletion and the realities of geology. Large increases in the price of oil will not greatly increase production but would reduce demand. Countries with no alternatives to oil will be forced to bid up the price of oil.*

As stated earlier, unconventional oil production created havoc for the Hubbert curve. Figure 5.14 shows the world oil production did not reach a peak in 2008 as predicted.

5.3.4 Historical Background: Foundations of Peak Oil Theory

A critical review of existing theories (scientific as well as social) demonstrates that the most important shortcomings of these theories are to be found in their fundamental premises.

To begin with, the core of the peak oil theory relies on elements from the unstated assumptions of a number of earlier theoretical expressions dealing with the most fundamental and most commonly-found features of conventional oil drilling. These include mathematically formulated theories that assume certain fundamental and more or less unchanging or features common to just about any enterprise undertaken in the context of generating an average profit from the most rational exploitation of some resource. A second unstated assumption is that the resource is privately-owned. A third unstated assumption is that the bed in which it has been found is itself also privately-owned. Tied to the second and third assumptions just mentioned is the assumption that the findings of any theory incorporating these two particular assumptions will be more or less universally or even generally applicable, and hence of broader theoretical value.

How can that be? In the countries of the Middle East that continue to dominate global oil production, as distinct form the United States, is it possible or even likely let alone probable that the land of the well-site is privately-owned and the resources to be captured in those lands can belong privately to a third entity (person or corporation)? The so-called "law of capture" in the United States makes it possible for the oil to be owned separately by some entity or individual different from the owner of the land on which the well is drilling, Such a distinction does not exist in Saudi Arabia or the Islamic Republic of Iran or the Kingdom of Qatar or Algeria or Libya or indeed any other country on the face of the earth.

Probably the single-most egregious unstated assumption, however, is that every oil producer is responding to its own view of market conditions within a framework of more-or-less free competition. Of course, such a thing would entail an absence of collusion among groups of producers to game the market sometimes in favor of this cartel member, at other times in favor of that cartel member. "Collusion", however, is the middle name of everyone participating in this game, and — in the conditions of monopolistic competition that characterize relationships among members of a cartel — what simply can no longer happen is the so-called "invisible hand" of Adam Smith, according to which profits become distributed in proportion each party's invested capital over some finite period of time.

Production itself, meanwhile, is assumed to be constrained on two principal fronts. It is assumed to be constrained first of all by the profit-generating potential of prevailing market incentives. It is also assumed to be constrained by the expected lifetime of the surrounding resource basin, and especially the median depth to be drilled for any individual well to reach the 'pay zone'.[6] Is it, however, a theory, i.e., something of a wider or more general application than the result of conventional statistical analysis of a particular dataset? "Hubbert's peak" is valid or potentially observable as a meaningful and correct generalization if and only if at least some of the aforementioned unstated assumptions happen to be operative. Considered from this standpoint, it seems misleading to dignify such an observation with the rather high-sounding label of "theory."

As for the theoretical foundations of the peak oil theory, some prefatory remarks are needed. One reaction comes along the lines of "what could theories about the nature of development of the broader social order have to do with a summary of patterns uncovered in oil-well drilling data from the petroleum sector of the economy of the United States collected and summarized in the middle third of the 20th century?"

This is a very good question. The starting-point of an answer is to point out that the economic order and how it works in any of ots parts — such as Big Oil —links and to a large extent organizes many seemingly diverse and unrelated parts of the social order. Here the *relevant* question, i.e., the question that compels examining previous and existing social theory in order to approach an answer, is: does the production output data examined and summarized in "Hubbert's peak" depend fundamentally on the capacities of nature, or does it depend rather on the cupidity of certain man-made human-guided economic organizations? To even hint at the notion of, much less assign any responsibility to, "the cupidity of certain man-made human-guided economic organizations" is instantly rejected in knee-jerk manner as the intellectual equivalent of defecating in church and a complete outrage: the one case that must be resolutely excluded from any consideration. Since class-interested "outrage" hardly constitutes a serious intellectual argument, however, the next response is to summon some part of social theory that supports the writer/researcher's personal bias: cue "social theory". The specific trend in social theory that is summoned to backstop the

[6]At the time King Hubbert was working on his Peak Oil theory, enhanced oil recovery (EOR) was not yet even a gleam in any petroleum engineer's eye, let alone a practical reality. The reality of EOR opens a possibility that a dormant well long believed to have been played out can be profitable revived at some time in the future. Some market analysts take this into account and some dismiss it as a marginal contribution to the overall profit picture.

peak-oil argument is the trend that says or assumes that humans and their behaviour operate at all times and places partly at the mercy of individuals' instincts and partly in relation to certain alleged limits of the natural order. Some selection of social theories based on or derived from the writings of the Rev. Thomas Malthus regarding food and population are then invoked, with Malthus famously arguing that it is part of man's condition that humans reproduce in numbers and at a rate that exceeds the possibilities of food production, leading to a cycle of famine followed by population decreases.[7]

The term "social science" first appeared in the 1824 book *An Inquiry into the Principles of the Distribution of Wealth Most Conducive to Human Happiness*. This 'happiness' applied to the *Newly Proposed System of Voluntary Equality of Wealth* by William Thompson (1775–1833). Economic theory as such meanwhile played little or no role in the development of what the world of the 19th century considered to be "social science". Only with the rise in Germany, France, Britain and the United States of a consciously-intended research effort to "expose" Karl Marx as a lying ill-intended fraud would the study of and research into economics broadened to fit existing social theory and especially theories of how societies develop.[8]

Some of this theorizing about society as some species of organic development (rather than some random accident) began a generation before Marx, in the works of Auguste Comte (1797–1857). Comte set the cat amongst the pigeons with his notion that ideas passed through three rising stages, theological, philosophical and scientific. He defined the difference as the first being rooted in assumption, the second in critical thinking, and the third in positive observation.

Comte's aim was to establish a framework in which to account scientifically for such phenomena as the republican upsurge that created the revolution in France in 1789. In the meantime, somewhat unexpectedly, this framework — although still rejected by many — also seemed to fit nicely with an entirely unrelated strand of thought. This was the strand that would push economic study away from the path of a descriptive discipline and onto the path of a mathematically-based discipline.

Comte and his German near-contemporary, Georg Wilhelm Friedrich Hegel, exercised little or no influence on the thinking of the other. Nevertheless, Karl Marx would deeply excavate Hegel's radical

[7]Malthus, Rev. Thomas R., *Essay on Population* (1798).
[8]This is discussed in depth in (Zatzman & Islam, 2007).

reconceptualization of the dialectic of ancient Greek philosophical school of Zeno to eventually conclude that the proper study of history — alongside the accepted traditional methods of documentary analysis — would and should be supplemented by some of the methods developed in the physical sciences to better specify the context of an historical question. Marx and his decades-long political and intellectual companion Frederick Engels jointly elaborated the basic principles of such a methodical approach under the name "dialectical and historical materialism."[9]

By the 1880s, based largely on the widely-embraced belief that the proper application of scientific method to any research should guarantee an answer, the "scientific" conception of social science taken up everywhere spurred an unprecedented research effort. Failure to reach a conclusion could only be the result of insufficient specificity of the research question. Similarly, any and every question, independent of any aim other than "uncovering the truth", now became potentially valid and worthy of further "scientific" research — complete with the appropriate graphs, data tables and equations. Certain researches became models of their kind. Newtonian mechanics (with its special position as the source of the differential calculus) was especially admired for its closed character as the complete description of any "conservative" system.

One of the key unstated and undiscussed assumptions of Newtonian mechanics is that differences internal to individual examples of phenomena of the same order are of little or no significance. In a conservative Newtonian system, phenomena of interest are to be distinguished only by their relative magnitude and "time signature" (meaning: at what point they occur in a process). By completely discounting any weight assignable to individual conscience or consciousness concerning the particular behavior under examination, the process of quantifying human behavior is radically simplified.

---

[9]As the political movement for the reform of the living and working conditions of the organized workers across Europe grew large and rapidly and entered the political mainstream, with many governments following the German and British examples of accommodating some of this movement's main demands piecemeal, a number of their comrades in the political struggles on the European continent back in the middle third of the 19th century, led by Karl Kautsky in Germany supported from Tsarist Russia by Georgii Plekhanov, would eventually scrap and even oppose the very idea of a dialectical and historical materialist *method* as "dangerous" and "destabilizing." The Russian political activist and leader of a politically-conscious section of the Russian workers' movement Vladimir Ilyich Ulyanov, also known as Lenin, however, stood in the way of this more or less deliberate defanging of the revolutionary implications of dialectical and

However, consider this implication: is there any longer anything identifiably human about such a dataset? This same conundrum plagues almost all European social theories down to this day: humans are 'just another species', hence individual conscience or consciousness is of virtually no concern. All that matters is some or any quantifiable piece of data, such as the presence or absence of some behavioral manifestation or other.[10]

A US writer, Professor Blumer (1954) has dared to question social theories advanced in the enlightened world, albeit limited to empirical science. He has written:

> *Now, it should be evident that concepts in social theory are distressingly vague. Representative terms like mores, social institutions, attitudes, social class, value, cultural norm, personality, reference group, social structure, primary group, social process, social system, urbanization, accommodation, differential discrimination and social control do not discriminate cleanly their empirical instances. At best they allow only rough identification, and in what is so roughly identified they do not permit a determination of what is covered by the concept and what is not.*[11]

Prof. Blumer's identification of the vagueness of many of the pet categories of social theory provides an important flag for the alert reader, a flag that the reader is about to be flim-flammed. Literally anything can fill such categories, and the end result is always the same: human beings are reduced by this process of decomposition to a collection of traits, and an overall status of just-another-species. In this process of reducing the human being to just-another-species, human conscience or consciousness become disconnected from human behavior. This

historical materialist method and rescued the revolutionary core of this method in a series of remarkable political pamphlets including *What Is To Be Done?* (1902), *Imperialism the Highest Stage of Capitalism* (1916) and *State and Revolution* (1917).

[10]In the 'natural' universe of Malthus, the absence of open-ended human agency is a defining feature whereas, by contrast, the idea that individuals or human collectives could act to consciously shape their own future is impossible. This unstated assumption sneaks — in partial form or full-blown —into the work of all those who accept any credence either for Malthus' original idea of allegedly natural checks on human population or for his more deeply-buried assumption that individual conscience or the consciousness of human collectives neither matter nor exist forms its underlying foundation. The list of such believers is, sadly, neither short nor undistinguished: Paul Ehrlich, Garrett Hardin, John Maynard Keynes, Alfred Russell Wallace, and William Thompson all

disconnection is precisely the bane of all modern European social theories (Islam *et al.*, 2013).

Here, the discussion moves onto one of the theories most widely accepted (consciously or otherwise) based on the premise that humans are liabilities. This notion forms part of the core of all other theories, including peak oil theory. Reverend Thomas Robert Malthus, a British scholar advanced the theory of rent. In his publications during 1798 through 1826, he identified various factors that would affect human population. For him, population was controlled by disease or famine. His predecessors believed that human civilization could be improved without limitations. Malthus, on the other hand thought that the "dangers of population growth is indefinitely greater than the power in the earth to produce subsistence for man". He added his religious fervor to this doctrine and considered it divine and wrote:

> *Must it not be acknowledged by an alternative examiner of the histories of mankind, that in every age and in every State in which man has existed, or does now exist that the increase of population is necessarily limited by the means of subsistence, that population does invariably increase when the means of subsistence increase, and, that the superior power of population is repressed, and the actual population kept equal to the means of subsistence, by misery and vice.*

Malthus flogged his counterfeited rediscovery of the law of population in support of the Corn Laws and in opposition to the Poor Law. The Poor Law had been in place since the 1500s and the rule of "Good Queen Bess", i.e. Elizabeth I, to deal with the 'nuisance' of beggars and 'impotent poor.' By the first generation of the industrial revolution (1750-1775), the Corn Laws, allegedly intended back a the time of their promulgation in the 1600s to protect local farmers from cheaper imports of wheat and other food grains, had entirely outlived their

---

subscribed consciously or otherwise to one or both these false premises. Even Charles Darwin's *Descent of Man*, although aware of the problem, insists nevertheless that the same kinds of evolutionary evidence uncovered for various species of reptiles and their airborne forbears is meaningful for the case of the so-called higher apes and men. Frederick Engels would later finesse that specific problem with a pamphlet significantly entitled *The Part **Played by Labor** in the Transition from Ape to Man* (emphasis added).

[11](Blumer, 1954).

usefulness. They became instead one of the main props for sustaining the pre-industrial regime of local nobility working hand-in-glove with government-financed Church of England clergy to keep individual farmers indebted to the hilt under the dictate of local landowning nobility, while trapping in the same moment the landless farm-labor population so they would be unable to leave their decaying rural districts to search for work in the rising industrial towns.

An entirely unexpected event would now intervened to upset everyone's best laid plans. In 1816, a volcanic eruption in the Indonesian archipelago brought tremendous short-term consequences to weather patterns literally around the globe. It spewed an enormous volume of dust into the atmosphere that travelled around the globe in the jet stream and led to the "year with no summer" in Europe and the northern half of North America. In 1817, grain crops on the continent of Europe failed. In industrial Great Britain, where the factory owners and their politicians boasted how the country's relatively (compared to the rest of the world) highly advanced industrial economy had overcome the "capriciousness of Nature," hunger and famine actually stalked the English countryside for the first time in more than a century and a half. The famine conditions were blamed on the difficulties attending the import of extra supplies of food from the European continent and led directly to a tremendous and unprecedented pressure to eliminate the Corn Laws – the system of high tariffs protecting English farmers and landlords from the competition of cheaper foodstuffs from Europe or the Americas.

In this moment, the industry lobby condemned the Corn Laws as the main obstacle to cheap food, winning broad public sympathy and support. Economically, the Corn Laws actually operated to keep hundreds of thousands employed in the countryside on thousands of small agricultural plots, at a time when the demands of expanding industry required uprooting and forced the rural population to work as factory laborers. Increasing the industrial reserve army would enable British industry to reduce wages. Capturing command of that new source of cheaper labor was, in fact, the industrialists' underlying aim. In the consciously-waged struggle that now rose to a fever pitch among these contending social classes, the supremacy of the industrial lobby and its supporters and the fate of the reactionary forces entrenching themselves in the countryside were mutually determined.

Without the famine of "the year with no summer," British industry would probably not have been so steadfast in its targeting the Corn Laws for elimination. thereafter. As matters stood back at that time, the final outcome was no slam-dunk. The anti-industrial lobby of aristocratic landlords dominated the House of Lords, delaying final elimination of the Corn Laws by British industry another nearly 30 years. Between 1846 and 1848 when Parliament finally eliminated the Corn Laws, industry captured access to a desperate workforce fleeing the ruin brought to the countryside. The sudden oversupply of laboring power, however temporary, was enough to drive overall industrial wages sharply downwards, greatly increasing the profitability of British industry. On this track of development, industry's appetite for new markets at home and abroad, including the export of important industrial infrastructure investments in "British North America," i.e., Canada, as well as in Latin America, and India, was greatly whetted..as England blasted its way to dominating world markets.

Major innovations pioneered within British industry at this time included extracting minerals and other valuable raw materials for processing into new commodities. This line of development just taken by itself soon brought an additional unpredictable level of further acceleration to the industrialization of the globe in regions where local industrial capital had not accumulated significantly, blocked either by traditional development blocked its role or lack of European immigration and settlement.

In sum: based on the political awareness of educated people living in the United Kingdom during the unfolding of the French Revolution and the later Parliamentary debates regarding the Corn Laws and free trade, the unexpected upshot and "icing on the cake" was that Malthus' theories were fiercely rejected. Indeed Malthus himself was shown to have reproduced the main core of his ideas from earlier writers without attribution, and to have done so with a very definite political axe to grind on behalf of his section of the ruling class.[12]

---

[12]The following deconstruction of Malthus' intellectual charlatanism is reproduced from a lengthy footnote published in Karl Marx's first volume of *Capital* as footnote 2 in Chapter 25 ("The General Law of Capitalist Accumulation"):

"If the reader reminds me of Malthus, whose "Essay on Population" appeared in 1798, I remind him that this work in its first form is nothing more than a schoolboyish, superficial plagiary of De Foe, Sir James Steuart, Townsend, Franklin, Wallace, &c., and does not contain a single sentence thought out by himself. The great sensation this pamphlet caused was due solely to party interest.

"The French Revolution had found passionate defenders in the United Kingdom; the "principle of population," slowly worked out in the eighteenth century, and then, in the midst of a great social crisis, proclaimed with drums and trumpets as the infallible

Science was now in a position to displace Prejudice as empirical observations confirmed that famine and natural disasters had no detectable long-term impact on population growth or decline, i.e., an impact beyond the immediate generation affected. Admittedly, this scientific acceptance was a grudging one. Nevertheless, many economists down to this day, including Nobel Laureate Amartya Sen (1998), have repeatedly confirmed the man-made factor playing a greater role in controlling human population than "natural"(so-called) increase or decrease.

The pre-industrial rural ruling classes were the heart and soul of political reaction for the first half of the 19th century. Unfortunately, as a result, the support of the entire English ruling class for this section ensured Malthus' theory lived on in every aspect of European social science and hard science throughout the 19th century and well into the 20th century. So, just to summarize the two key points about Malthus to carry forward from this portion of the discussion: ***i***) his doctrines were a counterfeiting of science and ***ii***) once their actual political target was defined, a political alliance formed by those with real-world interests opposed to the real-world interests of Malthus' supporters combined to defeat the immediate aim of the campaign for which Malthus' so-called "theory" was just a front. Unfortunately the reactionary classes and social forces lined up behind Malthus persisted and their ranks served to preserve Malthus' so-called theories down to this day, where they are even more irrelevant and anti-people than they were at their starting-point back in the early 19th century

One of the true scientific giants of 19th century Britain (and since) was Charles Darwin. Seeming to follow Malthus' lead in disconnecting human conscience from human being, at a certain level, Darwin seemed to share at some level with Malthus a sense of indifference as to whether humans are just another member of the animal kingdom.

---

antidote to the teachings of Condorcet, &c., was greeted with jubilance by the English oligarchy as the great destroyer of all hankerings after human development.

"Malthus, hugely astonished at his success, gave himself to stuffing into his book materials superficially compiled, and adding to it new matter, not discovered but annexed by him. Note further: Although Malthus was a parson of the English State Church, he had taken the monastic vow of celibacy-one of the conditions of holding a Fellowship in Protestant Cambridge University: "*Socios collegiorum maritos esse non permittimus, sed statim postquain quis uxorem duxerit socius collegii desinat esse.*" ("Reports Of Cambridge University Commission," p. 172).

"This circumstance favorably distinguishes Malthus from the other Protestant parsons, who have shuffled off the command enjoining celibacy of the priesthood and have taken, "Be fruitful and multiply," as their special Biblical mission in such a degree that

Similarly, both Malthus and Darwin seemed to focus entirely on nature-based causes of human population variations (as opposed to anything planned by government or other human agencies). But: after these two coincidences in their views, everything else diverges. Darwin put forward a testable series of hypotheses about the point at which either a new species emerges or an irreversible change from some earlier state of the human animal becomes visible. A certain number of characteristics that scientists today consider to be shared between humans and certain great apes may not have been visible earlier in the geological record: did some point in that earlier record mark the point in which the human emerges finally distinct from any of the apes?

It might take aeons to demonstrate but at least there is something to test, whereas this would be far past the point where Malthus drowns in his own dilettantism.

Darwin speculated on the basis of a lot of possibly supporting evidence that the emergence of a species distinct in definite ways from its immediate predecessor and new to the surrounding natural environment generally marked the final change in the sequence of steps in an evolutionary process. This led a number of otherwise quite sensible people into a deep swamp. Thus for example, the Anglo-American economist Alfred Marshall speculated from this that, for Darwin and others who hope he is correct, the preceding historical development matters not at all, only the last step — thus assuming precisely what we do not know!

The essence of Darwin's argument concerned the non-linearity of the final step, the leap from what was formerly one species to distinctly another species. Darwin was silent on the length of time that may have passed between the last observed change in a species-line, as well as about the point in time at which its immediate predecessor emerged – the characteristic time of the predecessor species – was the time period

they generally contribute to the increase of population to a really unbecoming extent, whilst they preach at the same time to the labourers the "principle of population." It is characteristic that the economic fall of man, the Adam's apple, the urgent appetite, "the checks which tend to blunt the shafts of Cupid," as Parson Townsend waggishly puts it, that this delicate question was and is monopolised by the Reverends of Protestant Theology, or rather of the Protestant Church. With the exception of the Venetian monk, Ortes, an original and clever writer, most of the population-theory teachers are Protestant parsons. For instance, Bruckner's "Theorie du Systeme animal," Leyden 1767, in which the whole subject of the modern population theory is exhausted, and to which the passing quarrel between Quesnay and his pupil, the elder Mirabeau, furnished ideas on the same topic; then Parson Wallace, Parson Townsend, Parson Malthus and his

in which all the changes so significant for later on were prepared. This latter could be eons, spanning perhaps several geological eras.

This part of Darwin, refusing to speculate beyond what his evidence can demonstrate, demonstrates his integrity as an investigator. It opens a window on some profound news that still eludes most European-American scientific theory in either the hard or social sciences. That news is the idea of "characteristic time" as an entirely natural feature internal to organisms and not attributable to extreme weather conditions or the like.

Should we be surprised? This idea of t_{NATURAL} as characteristic time is missing from every European theorist. This absence is not unexpected. Ever since the work of Thomas Aquinas, Europeans scientists simply repeated the dogmatic adherence to tangible timelines while distancing themselves from doctrinal philosophy. However, as Islam *et al.* (2012; 2013) have recently pointed out, they did not employ the scientific methodology of Averroes — even while accepting Averroes as the father of secular philosophy in Europe as well as claiming themselves secular!. This claim wasn't genuine.

A second, but equally telling source of pressure on social scientists has been to mathematize their research methodology was a sense that their work would not be taken seriously as scientific without some such mathematical rigor. As the models and mathematics from the so-called "exact" sciences would hardly be appropriate or seem credible in any field of study focusing on human beings and their incredible variety of needs, wants and impulses, another kind of mathematics would have to do. This fascination comes from another trail of cognition that was popularized when the term social science was introduced. Questions of history and historical phenomena were also a convenient target because of the lack of any means to describe them with any meaningful, non-trivial mathematical model (Zatzman and Islam, 2007).

The giant of economics and social theory in the English-speaking world during the 20th century has been John Maynard Lord Keynes, a fanatical adherent of Newtonian mechanism. Keynes believed that historical time had nothing to do with establishing the truth or falsehood of economic doctrine. "In the long run, we are all dead," he wrote. He tied this to a stance that attacked all easy acceptance without question of any of the underlying assumptions propping up all forms of orthodoxy. Accordingly, this retort was taken as the sign of a fresh and

---

pupil, the arch-Parson Thomas Chalmers, to say nothing of lesser reverend scribblers in this line."

rebellious spirit. However, in his own theoretical work he was frequently at pains to differentiate what happens to individuals who are driven by short-term considerations from what happens at the societal level at which he was theorising about broad historically sweeping movements of economic cause and effect (Keynes, 1936).

Keynes would emerge unscathed from the ideological wars among the Right and the Left in Britain, Europe and America. No one challenged his economic theories that were accepted at face value with doctrinal fervor. Down to the present, every Nobel Laureate in Economics ritually announces how much they derive their inspiration from Keynes.[13]

5.3.5 Petroleum in the Big Picture

Captain Drake is usually credited to have drilled the first-ever oil well at Titusville, PA in the United States in 1859. However, even if one discards the notion that petroleum was in use for thousands of years, there is in actual fact credible evidence that the first well in modern age was drilled in the present-day Canadian province of Ontario (abbreviated today as ON), specifically near Sarnia, in the still British colony of Canada West. Canadian engineer Charles Nelson Tripp was the first in North America to have recovered commercial petroleum products. The drilling was completed in 1851 at Enniskillen Township, near Sarnia, ON..

Soon after the "mysterious gum" bed was discovered, the first oil company was incorporated in British North America through a charter issued under the authority of thre British Parliament at Westminster. Tripp became the president of this company on December 18, 1854. The charter empowered the company to explore for asphalt beds and oil and salt springs, and to manufacture oils, naphtha paints, burning fluids. Even though this company (International Mining and Manufacturing) was not a financial success, the petroleum products received an honorable mention for excellence at the Paris Universal Exhibition in 1855. Failure of the company can be attributed to several factors. Lack of roads in the area rendered movement of machinery and equipment to the site extremely difficult. After every heavy rain the area turned

[13]One of them is Joseph Stiglitz, who was deconstructed by Zatzman and Islam (2007) as well as Zatzman (2012, 2013).

into a swamp and the gum beds made drainage extremely slow. This added to the difficulty of distributing finished products.

In subsequent years, James Miller Williams became interested and visited the site in 1856. Tripp unloaded his hopes, his dreams and the properties of his company, saving for himself a spot on the payroll as landman. The former carriage-builder formed J.M. Williams & Company in 1857 to develop the Tripp properties. Besides asphalt, he began producing kerosene. This 'refined' product, Kerosene is a combustible hydrocarbon liquid. The name is derived from the Greek κηρός (keros) meaning wax.

The word "Kerosene" was registered as a trademark by Abraham Gesner in 1854, and for several years, only the North American Gas Light Company and the Downer Company (to which Gesner had granted the right) were allowed to call their lamp oil "Kerosene" in the United States. In 1846 development of a process to refine a liquid fuel from coal, bitumen and oil shale. His new discovery burned more cleanly and was less expensive than competing products, such as whale oil. In 1850, Gesner created the Kerosene Gaslight Company and began installing lighting in the streets in Halifax and other cities. By 1854, he had expanded to the United States where he created the North American Kerosene Gas Light Company at Long Island, New York. Demand grew to where his company's capacity to produce became a problem, but the discovery of petroleum, from which kerosene could be more easily produced, solved the supply problem. This was the first time in recorded history artificial processing technique was introduced in refining petroleum products. Gesner didn't use the term 'refined' but made fortune out of the sale of this artificial processing. In 1861, he published a book titled: "A Practical Treatise on Coal, Petroleum and Other Distilled Oils," which became a standard reference in the field. As Gesner's company was absorbed into the petroleum monopoly, Standard Oil, he returned to Halifax, where he was appointed a Professor of Natural History at Dalhousie University. It is this university that was founded on pirated money while other pirates continued to be hanged by the Royal Navy at Point Pleasant Park's Black Rock Beach as late as 1844.[14]

[14]A cairn in front of its Administration building actually describes the university's origins two centuries earlier from a fund created to launder the ill-gotten gains of an early 19th century war crime committed by the Royal Navy against a customs house in the U.S. state of Maine several months after Anglo-American hostilities of the War of 1812 had officially concluded.

In the meantime, there was further parallel development in the present-day Canadian province of Ontario. In 1858, Williams dug a well in search of cleaner drinking water and came across oil at a depth of 15.5 meters. It became the first commercial oil well in North America, remembered as the Williams No. 1 well at Oil Springs, Canada West. The *Sarnia Observer and Lambton Advertiser*, quoting from the *Woodstock Sentinel*, published on page two on August 5, 1858:

> An important discovery has just been made in the Township of Enniskillen. A short time since, a party, in digging a well at the edge of the bed of Bitumen, struck upon a vein of oil, which combining with the earth forms the Bitumen.

Some historians challenge Canada's claim to North America's first oil field, arguing that Pennsylvania's famous Drake Well was the continent's first. But there is evidence to support Williams, not least of which is that the Drake well did not come into actual production until 28 August 1859. The controversial point might be that Williams found oil above bedrock while "Colonel" Edwin Drake's well had located oil within a bedrock reservoir.

History is not clear as to when Williams abandoned his Oil Springs refinery and transferred his operations to Hamilton. He was certainly operating there by 1860, however. Spectator advertisements offered coal oil for sale at 16 cents per gallon for quantities from 4,000 US gallons (15,000 L) to 100,000 US gallons (380,000 L). By 1859 Williams owned 800 acres of land in Oil Springs. Williams reincorporated in 1860 as the Canadian Oil Company. His company produced oil, refined it and marketed refined products. That mix of operations qualified Canadian Oil at the time as the world's first integrated oil company.

Exploration in the Lambton County backwoods quickened with the first flowing well in 1860. Previous wells had relied on hand pumps. The first gusher erupted on January 16, 1862, when struck oil at 158 feet (48 m). For a week the oil gushed unchecked at levels reported as high as 3,000 barrels per day, eventually coating the distant waters of Lake St. Clair with a black film. There is historical controversy concerning whether it was John Shaw or another oil driller named Hugh Nixon Shaw who drilled this oil gusher; the newspaper article cited below identifies John Shaw.

News of the gusher spread quickly and was reported in the Hamilton Times four days later:

> I have just time to mention that to-day at half past eleven o'clock, a.m., Mr. John Shaw, from Kingston, C. W., tapped a vein of oil in his well, at a depth of one hundred and fifty-eight feet in the rock, which filled the surface well, (forty-five feet to the rock) and the conductors [sic] in the course of fifteen minutes, and immediately commenced flowing. It will hardly be credited, but nevertheless such is the case, that the present enormous flow of oil cannot be estimated at less than two thousand barrels per day, (twenty-four hours), of pure oil, and the quantity increasing every hour. I saw three men in the course of one hour, fill fifty barrels from the flow of oil, which is running away in every direction; the flat presenting the appearance of a sea of oil. The excitement is intense, and hundreds are rushing from every quarter to see this extraordinary well.

Historically, the ability of oil to flow freely has fascinated developers and at the same time ability of gas to leak and go out of control has intimidated them. Such fascination and intimidation continues today while nuclear electricity is considered to be benign while natural gas considered to be the source of global warming, all because it contains carbon – the very component nature needs for creating an organic product. Scientifically, however, the need for refining stems from the necessity of producing clean flame. Historically, Arabs were reportedly the first ones to use refined olive oil. They used exclusively natural chemicals in order to refine oil (Islam *et al.*, 2010). However, such use of natural chemicals is non existent in modern day petroleum industry. When it comes to petroleum gas, it had been in use for millennia but only recent time 'processing' of such gas has been introduced.

Natural gas seepages in Ontario County, New York state were first reported in 1669 by the French explorer, M. de La Salle, and a French missionary, M. de Galinée, who were shown the springs by local native Americans. This marks the debut of natural gas industry in North America. Subsequently, William Hart, a local gunsmith, drilled the first commercial natural gas well in the United States in 1821 in Fredonia, Chautauqua County, NY. He drilled a 27-foot-deep well in an effort to get a larger flow of gas from a surface seepage of natural gas. This was the first well intentionally drilled to obtain natural gas.

Hart built a simple gas meter and piped the natural gas to an innkeeper on the stagecoach route from Buffalo to Cleveland. Because there was no pipeline network in place, this gas was almost invariably all used to light streets at night. However, in late 1800s, electric lamps were beginning to be used for lighting streets. This led to gas producers scrambling for alternate market. Shallow natural gas wells were soon drilled throughout the Chautauqua County shale belt. This natural gas was transported to businesses and street lights in Fredonia at the cost of $1.50 a year for each light (Website 3). In the mean time, in mid 1800's, Robert Bunsen invented the "Bunsen burner" that helped produce artificial flame by controlling air inflow in an open flame. This was significant because it helped producing intense heat and control the flame at the same time. This led ways to develop usage of natural gas for both domestic and commercial use.

The original Hart gas well produced until 1858 and supplied enough natural gas for a grist mill and for lighting in four shops. By the 1880s, natural gas was being piped to towns for lighting and heat, and to supply energy for the drilling of oil wells. Natural gas production from sandstone reservoirs in the Medina Formation was discovered in 1883 in Erie County. Medina production was discovered in Chautauqua County in 1886. By the early years of the twentieth century, Medina production was established in Cattaraugus, Genesee and Ontario Counties.

Gas in commercial quantities was first produced from the Trenton limestone in Oswego County in 1889 and in Onondaga County in 1896. By the close of the nineteenth century, natural gas companies were developing longer intrastate pipelines and municipal natural gas distribution systems. The first gas storage facility in the United States was developed in 1916 in the depleted Zoar gas field south of Buffalo.

By the late 1920s, declining production in New York's shallow gas wells prompted gas companies to drill for deeper gas reservoirs in Allegany, Schuyler, and Steuben Counties. The first commercial gas production from the Oriskany sandstone was established in 1930 in Schuyler County. By the 1940s, deeper gas discoveries could no longer keep pace with the decline in shallow gas supplies. Rapid depletion and over-drilling of deep gas pools prompted gas companies in western New York to sign long-term contracts to import gas from out of state.

It took the construction of pipelines to bring natural gas to new markets. Although one of the first lengthy pipelines was built in 1891 -it was 120 miles long and carried gas from fields in central Indiana to Chicago - there were very few pipelines built until after World War II in the 1940s. Similar to all other developments in modern Europe, World War II brought about changes that led to numerous inventions and technological breakthroughs in the area of petroleum production and processing. Improvements in metals, welding techniques and pipe

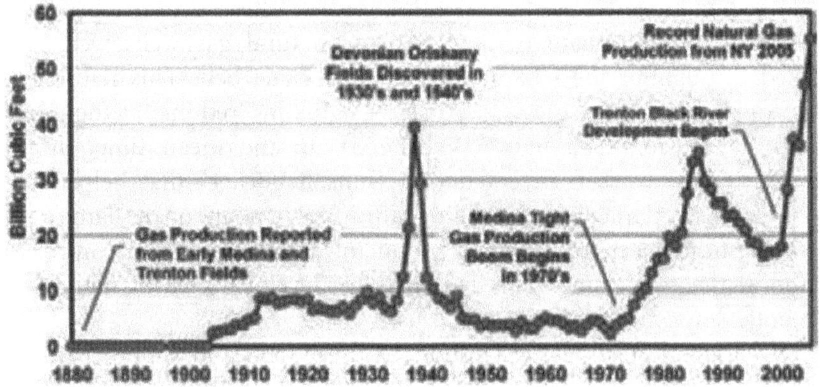

Figure 5.15 Natural gas production history in New York state (from Website 3).

making during the War made pipeline construction more economically attractive. After World War II, the nation began building its pipeline network. Throughout the 1950s and 1960s, thousands of miles of pipeline were constructed throughout the United States. Today, the U.S. pipeline network, laid end-to-end, would stretch to the moon and back twice. The phenomenon of pipelining is of significance. Because of this, there has been tremendous surge in the corrosion control industry.

Onondaga reef fields were discovered by seismic prospecting in the late 1960s. Seven reef fields have been discovered to date in southern New York. Today, the Onondaga reef fields and many Oriskany fields are largely depleted and are being converted to gas storage fields. This state of depletion was achieved after a long production period and extensive hydraulic fracturing throughout 1970s and 1980s. These were considered to be tight gas sands. Recently, the same technology has made a comeback. The rapid development of New York's current Trenton-Black River gas play is made possible by technological advances in three-dimensional (3-D) seismic imaging, horizontal drilling, and well completion. The surge in domestic oil and gas production through 'fracking' emerges from technologies popularized in the 1970s. However, 3D seismic or multilateral drilling technology was not in place at the time. Figure 5.15 shows how natural gas production evolved in the state of New York throughout history.

In this figure, the first spike relates to discovery of Devonian shale. That spike led to quick depletion. In early 1970s, production from 'tight gas' formations led to another more sustained spike in gas recovery. During that period, extensive hydraulic was introduced as a means for

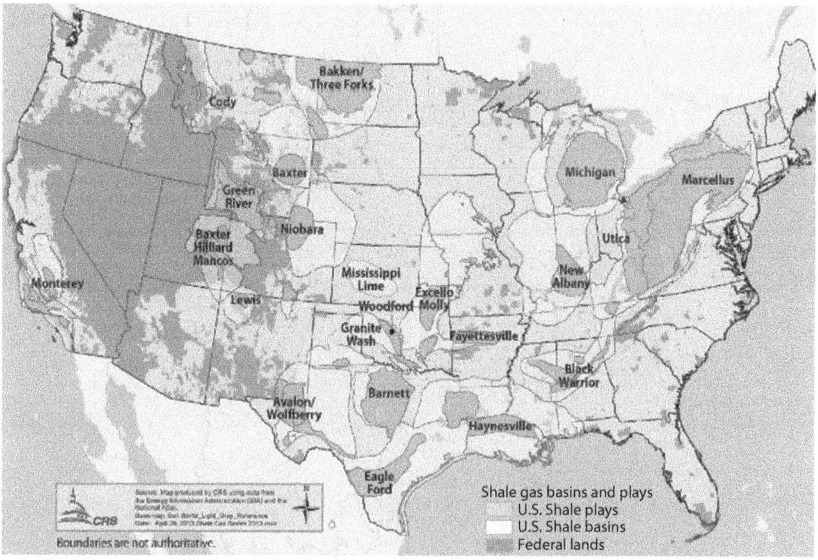

Figure 5.16 Locations of unconventional shale plays in lower 48 states (from Ratner and Tiemann, 2014).

increasing productivity. However, it was not considered to be a reservoir production enhancement scheme. In 2000, at the nadir of oil price, yet another spike took place in the state of New York. This related to the development of Trenton Black River field. This gas production scheme would lead to record gas production in that state in 2005. This spike continued and led the way to producing domestic gas and oil from unconventional reservoirs in USA. Today, production from unconventional gas reservoirs has taken an unprecedented turn. In 2013, production from shale gas, tight gas and coal bed methane accounted for domestic production surpassing imports for the first time in 30 years. Shale gas, tight oil, or other unconventional resources are found in many of the states that had already produced from conventional sources. Figure 5.16 shows the locations of these unconventional formations.

The primary recovery techniques from these shale plays involve multilaterals and intense hydraulic fracturing, now known as 'fracking'. Two significant differences between fracking and old-fashioned hydraulic fracturing are: 1) fracking uses multistage fractures with horizontal multilaterals; 2) fracking uses artificial sands as well fluid.

In 1997, based on earlier techniques used by Union Pacific Resources, now part of Anadarko Petroleum Corporation, Mitchell Energy, now part of Devon Energy, developed the hydraulic fracturing technique

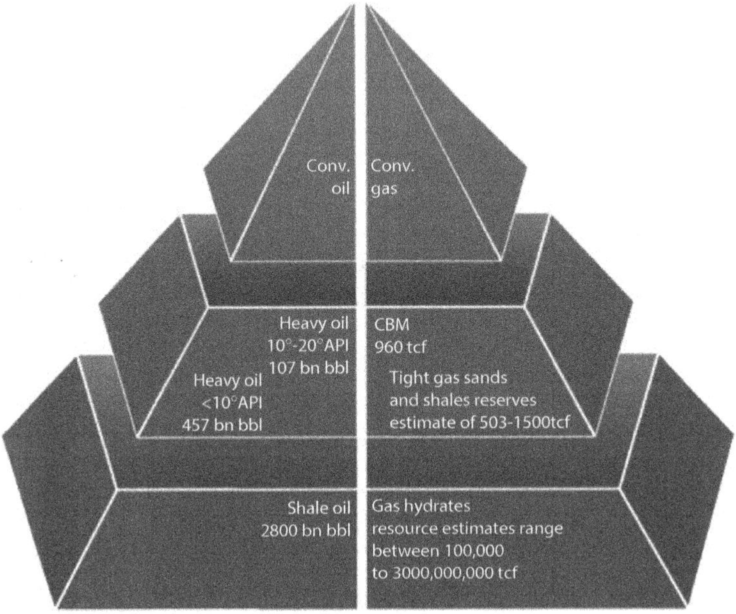

Figure 5.17 Moving from conventional to unconventional sources, the volume of the petroleum resource increases.

known as "slickwater fracturing" which involves adding chemicals to water allowing increase to the fluid flow, that made the shale gas extraction economical. These chemicals are both expensive and toxic to the environment.

5.3.6 Unconventional Oil and Gas Resources

The notion that conventional gas and oil is miniscule compared to unconventional gas reserve is decades old. With it comes the notion that it is more challenging to produce unconventional petroleum resource. In addition, at least for petroleum oil the notion that unconventional resource is more challenging to process is prevalent. This notion is false. With the renewed awareness of the environmental sustainability it is becoming clear unconventional resources offer more opportunities to produce environment-friendly products that conventional resources. Figure 5.17 shows the pyramid of both oil and gas resources.

On the oil side, the quality of oil is considered to be declining as the API gravity declines. This correlation is related to the processing

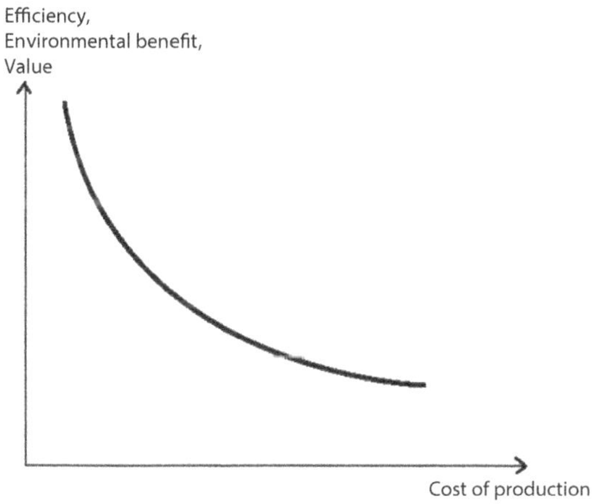

Figure 5.18 Cost of production increases as efficiency, environmental benefits and real value of crude oil declines (modified from Islam *et al.*, 2010).

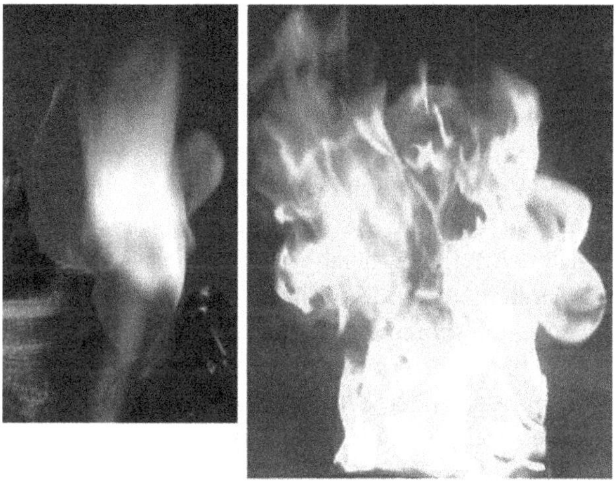

Picture 5.2 images of burning crude oil from shale oil (left) and refined oil (right).

required for crude oil to be ready for conversion into usable energy, which is related to heating value. Heating value is typically increased by refining crude oil upon addition of artificial chemicals are principally responsible for global warming (Chhetri and Islam, 2008; Islam *et al.*, 2010). In addition, the process is inefficient and resulting products harmful to the environment. Figure 5.18 shows the trend in efficiency, environmental benefit and real value with the production cost of refined crude. This figure shows clearly there is great advantage to using

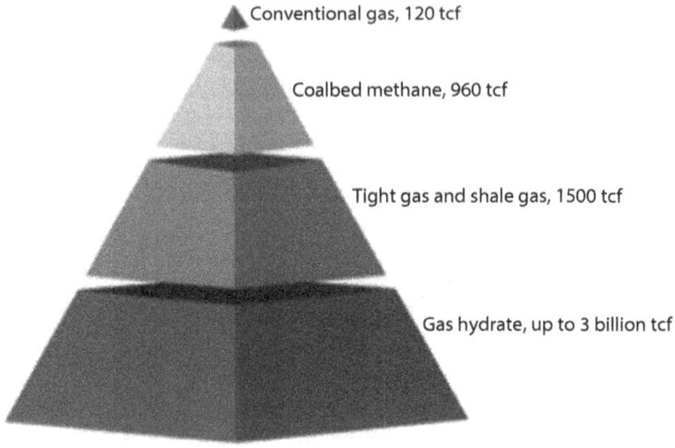

Figure 5.19 Current estimate of conventional and unconventional gas reserve.

petroleum products in their natural state. This is the case for unconventional oil. For instance, shale oil burns naturally. The color of flames (left image of Picture 5.2) indicates that crude oil produced from shale oil doesn't need further processing. The right image of Picture 5.2 emerges from burning gasoline and has similar colors to those of the left.

In addition, crude oil from shale oil is 'cleaner' than other forms of crude oil because of the fact that it is relatively low in tar content as well sand particles. Another crucial aspect is the fact that sulfur content or other toxic elements of crude oil have no correlation with unconventional or conventional sources. Also, heavier oils do not have more of these toxic elements and are not in need of refinement to be usable.

Lighter crudes are considered to be easier and less expensive to produce only because modern engineering uses refined version of the crude oil and all refining technologies are specially designed to handle light crude oil. If sustainable refining techniques are used, lighter or conventional oil offers no particular advantage over unconventional one and yet the volume and ease of production of unconventional are greater in unconventional resources.

For natural gas, the quality of gas actually improves with unconventional resources. For oil, the lighter the oil the more toxic it is considered. Whereas for gas, more readily available resources are less toxic. For instance, biogas is least toxic, whereas it is most plentiful. As can be seen in Figure 5.19, as one transits from conventional gas to coal bed methane (CBM) to tight gas and shale gas all the way to hydrates, one encounters more readily combustible natural resources. In fact, CBM

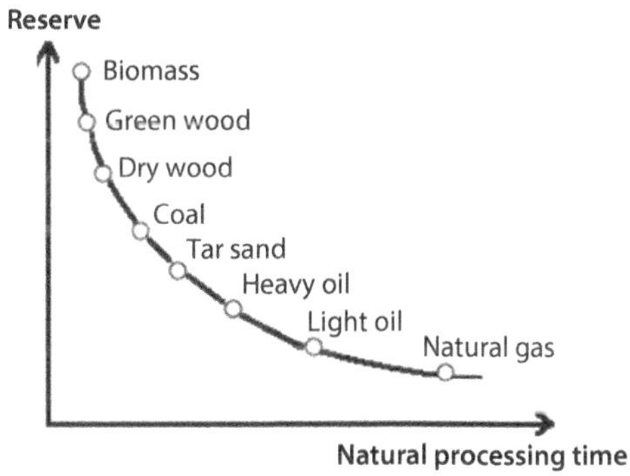

Figure 5.20 Abundance of natural resources as a function of time.

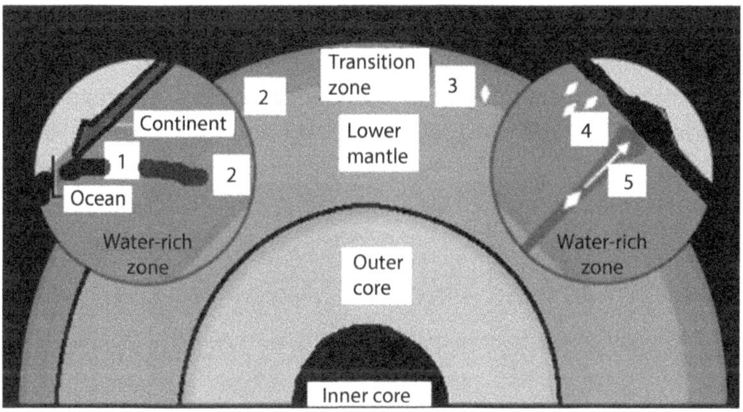

Figure 5.21 Water plays a more significant role in material production than previously anticipated (from the Guardian, March 12, 2014).

burns so readily that coal mine safety primarily revolves around combustion of methane gas. Processing of gas doesn't involve making it more combustible, it rather involves the removal of components that do not add to heating value or create safety concerns (e.g. water, CO_2, H_2S).

Figure 5.19 shows how the volume of resources goes up as one moves from conventional to unconventional resources.. In this process the quality of gas also increases. For instance, hydrate has the purest form of methane and can be burnt directly with little or no safety concern. At the same time, the volume of natural gas in hydrate is very large. The concentration of 'sour' gas components also decreases with abundance of the resources. Such trend can be explained by the processing time of a particular resource. There is a continuity in nature that dictates that the natural processing increases both the value and global efficiency of energy sources (Chhetri and Islam, 2008). Figure 5.20 depicts the nature of volume of natural resources as a function of processing time.

In this picture, 'natural gas' relates to petroleum products in conventional sense. This figure shows natural gas in general is most suitable for clean energy generation. Within unconventional gas sources, there exists another correlation between reserve volume and processing time. In general, the processing time for various energy sources is not a well understood science. Scientists are still grappling with the origin of earth or universe, some discovering only recently that water was and remains the matrix component for all matter (Pearson et al, 2014). Figure 5.21 shows how natural evolution on earth involved a distinctly different departure point not previously recognized. Pearson *et al.* (2014) observed a 'rough diamond' found along a shallow riverbed in Brazil that unlocked the evidence that a vast "wet zone" deep inside the Earth

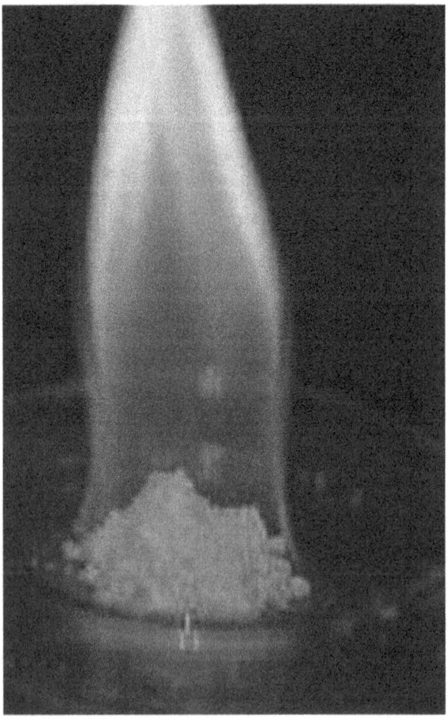

Picture 5.3 hydrate burns readily without any safety or environmental hazard.

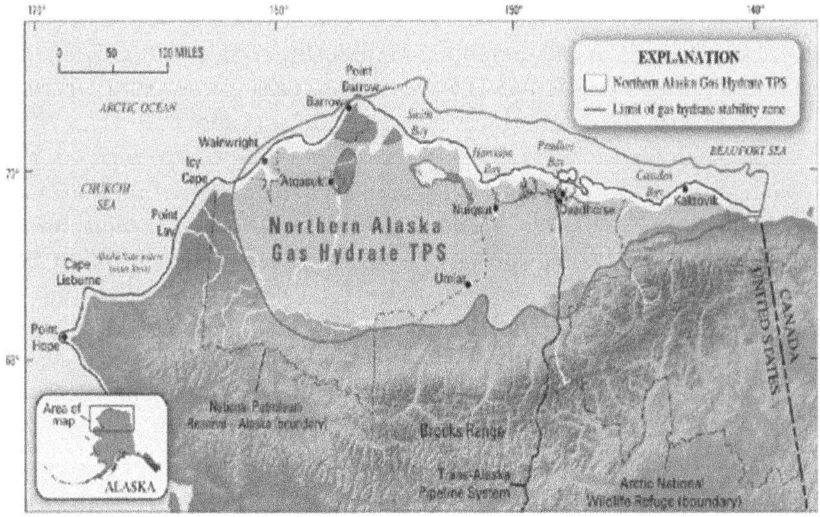

Figure 5.22 Gas hydrate deposits of Alaska.

that could hold as much water as all the world's oceans put together. This discovery is important for two reasons. Water and carbon are both essential for living organisms. They also mark the beginning and end of a life cycle. All natural energy sources have carbon or require carbon to transform energy in usable form (e.g. photosynthesis).

World petroleum reserve takes a different meaning if unconventional gas is added to the equation.

5.4 Gas Hydrates

It is well known that gas hydrates possess tremendous potentials as a source of natural gas. It is so vast that the estimates of global reserves are only sketchy, but range from 2,800 trillion to 8 billion trillion m^3 of natural gas. This is several times higher than global reserves of 440 trillion m^3 of conventional gas. While gas hydrate burns readily (Picture 5.3), exploitation of hydrate reserve is considered to be a difficult task.

EIA global estimates place the gas volume resident in oceanic natural gas hydrate deposits in the range of 30,000 to 49,100,000 trillion cubic feet (Tcf), and in continental natural gas hydrate deposits in the range of 5,000 to 12,000,000 Tcf. Comparatively, current worldwide natural gas resources are about 13,000 Tcf and natural gas reserves are about 5,000 Tcf. The current mean (expected value) estimate of domestic natural gas hydrates in-place is 320,222 Tcf. In comparison, as of 1997 the mean estimate of all untapped technically recoverable U.S. natural gas resources was 1,301 Tcf, U.S. proved natural gas reserves were 167 Tcf, and annual U.S. natural gas consumption was about 22 Tcf.

Large volumes of natural gas hydrates are known to exist in both onshore and offshore Alaska, offshore the States of Washington, Oregon, California, New Jersey, North Carolina, and South Carolina, and in the deep Gulf of Mexico. Most of the volume is expected to be in Federal jurisdiction offshore waters, although 519 Tcf of hydrated gas-in-place was assessed for onshore Alaska—more than three times the 1997 level of U.S. proved natural gas reserves. The USGS assessment indicates that the North Slope of Alaska may host about 85 TCF of undiscovered technically recoverable gas hydrate resources (Figure 5.22). According to the report, technically recoverable gas hydrate resources could range from a low of 25 Tcf to as much as 158 Tcf on the North Slope. Total U.S. consumption of natural gas in 2007 was slightly more than 23 TCF. Of the mean estimate of 85 TCF of technically recoverable gas hydrates on the North Slope, 56% is located on federally managed lands, 39% on lands and offshore waters managed by the state of Alaska, and the remainder on Native lands. The total area comprised by the USGS assessment is 55,894 square miles, and extends from the National Petroleum Reserve in the west to the Arctic National Wildlife Refuge (ANWR) in the east (Figure 5.22). The area extends north from the Brooks Range to the state-federal offshore boundary three miles north of the Alaska coastline. Gas hydrates might also be found outside the assessment area.

Global estimates by the committee for gas estimates reported methane in gas-hydrate deposits to be in the range of 3.1×10^{15} to 7600×10^{15} m^3 for oceanic sediments and from 0.014×10^{15} to 34×10^{15} m^3 for polar regions (Max, 2003). Boswell and Collette (2011) puts this reserve to a volume of $1\text{-}120 \times 10^{15}$ m^3 of methane trapped within global reserve. In near future, hydrates can alter the energy demography for the world. Such efforts are in progress in India (Sain, 2012), Japan (Pfeifer, 2014), and others. This has the potential of creating another energy revolution the likes of which didn't occur in last 100+ years. Following is an estimate of currently known hydrate reserve for some of the leading countries.

USA–318,000TCF
Alaska North Slope - 590 TCF
Japan - 1,765 TCF

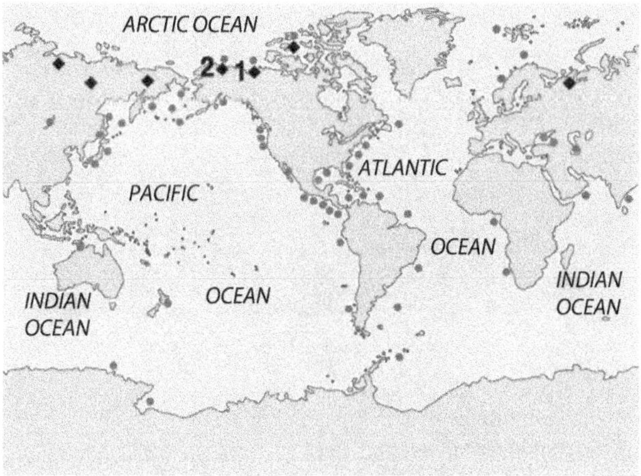

Figure 5.23 Known and inferred natural gas hydrate occurrences in marine (red circles) and permafrost (black diamonds) environments. (Figure from USGS http://pubs. usgs. gov/fs/fs021-01/fs021-01.pdf).

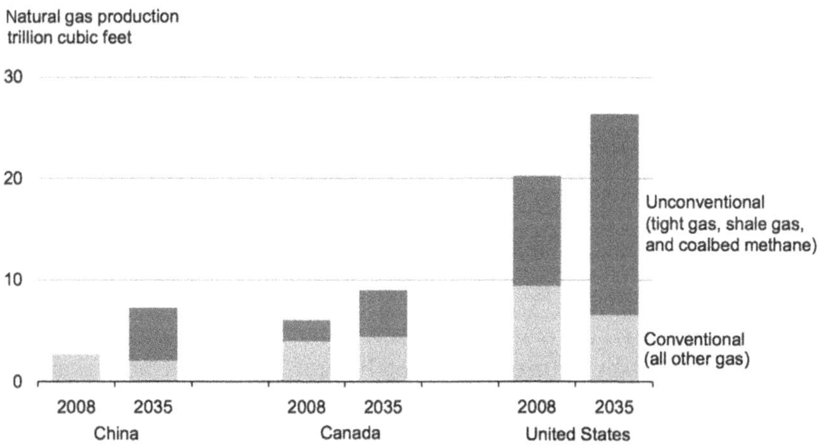

Figure 5.24 Future trends in some of the major future user of unconventional gas (from EIA report, 2013).

India–4,307 TCF
Canada - 1,550 – 28,600 TCF

A scientific approach to looking at natural gas resources of the world reveals that future of energy lies within exploiting unconventional gas. Unlike what has been promoted in recent decades, unconventional gas

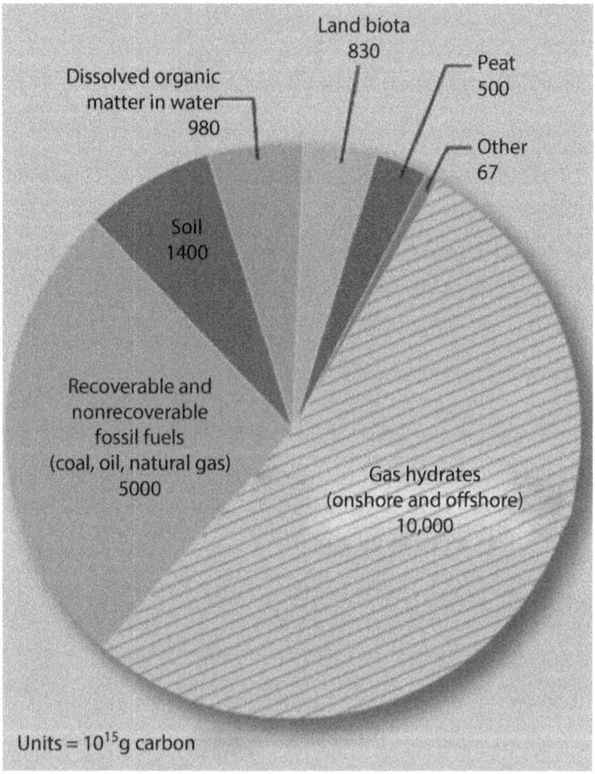

Figure 5.25 Gas hydrates that are the largest global sink for organic carbon offer the greatest prospect for the future of energy.

is more likely to generate truly 'clean' energy. Thankfully this fact is being recognized by future players of Figure 5.24 shows how there will be major shift in terms of usage of unconventional gas. This trend will shape the future of energy.

In even longer term future, gas hydrates offer the greatest promise of meeting energy needs. Gas hydrates are the most abundant of unconventional gas resources and they are the largest global sink for organic carbon (Figure 5.25).

At present, over 30 countries are actively involved at least in exploration of unconventional gas. For even the most active countries, this assessment of unconventional reserve is at its nascent state.

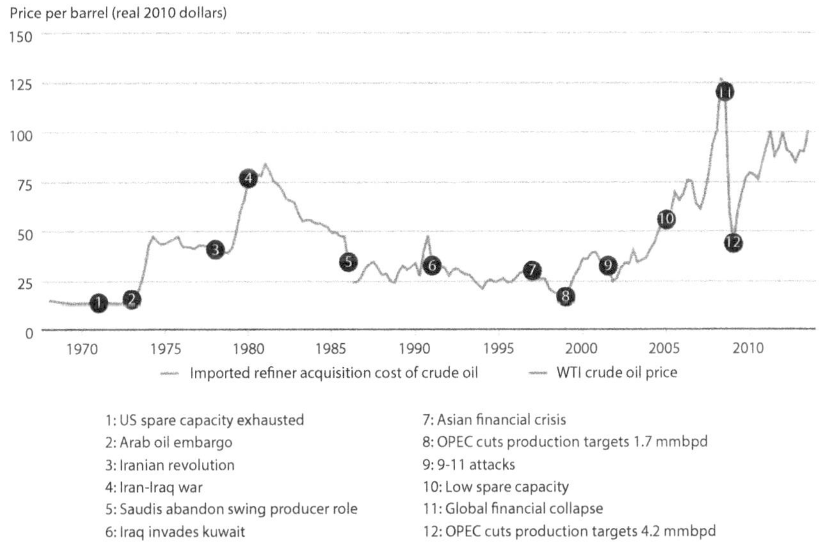

Figure 5.26 Petroleum is the driver of world economy and driven by political events (data from EIA).

5.4.1 Deconstruction of Peak Oil Theory

This theory advances the concept that after fossil fuel reserves (oil reserves, coal reserves, and natural gas reserves) are discovered, production at first increases approximately exponentially, as more extraction commences and more efficient facilities are installed. This is the first assumption. It is based on the following premises:

i. fossil fuel production technology is uniform and constant, with the same impact on the resource irrespective of the resource type and the nature of the technology; and
ii. demand for and need of energy is predetermined and independent of any intangible factors, including politics, social and environmental constraints.

Assumption i) is incorrect and demonstratively false (Islam, 2014). With the correct assumption, the fundamental tenet of the peak oil theory breaks down.

Assumption ii) needs some elaboration before deconstruction.

5.4.1.1 *Background of the Petroleum Industry*

The oil industry is past middle age in a business life cycle sense and is at a crossroads. Darcy's law that was proposed over 150 years ago still remains the only viable model used in all flow prediction schemes. As faster and bigger computers are developed, bigger and more robust reservoir simulators with all features of video games are deployed – without adding a single component that would describe the physics of the process any better than the one used a century ago. Two decades ago, using million blocks was the most sought after feature of a reservoir simulator. Now, oil companies brag about using a billion-block simulator, forgetting that this probably means increasing poorer reflection of the physics of the system. If science and engineering is this pitiful, how is it in management and energy pricing? Accounting principles from the eighteen hundreds continue to teach that the lowest cost producer will make the highest profit and therefore will be the most successful. With this lesson in mind, the oil companies are preoccupied with cost cutting, curtailing research and development, gaining economies of scale, and consolidating through mergers and buy-outs. Cadres of business consultants are hired to install governance and total quality

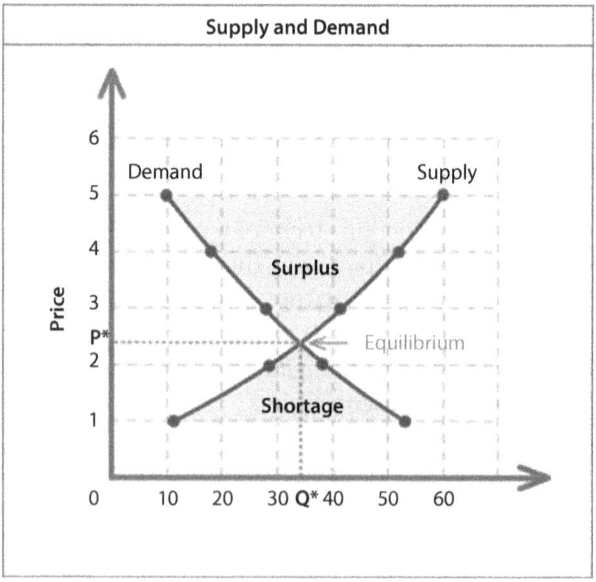

Figure 5.27 Price, quantity, supply, and demand according to conventional economic theory.

management programs. As these programs permeate the industry, returns sink to the same low level. In other words, there are numerous competing companies selling plentiful volumes of oil and gas while constantly bidding against each other for operating rights and product sales. No amount of cost cutting and further reduction in research can stop this spiral down movement.

It is by convention that economic theories are announced or explained as final finished products. However, in reality they are anything but. They express ideological and political priorities of the ruling forces of the establishment in the short term at various times in response to various pressures. Thus, the long-standing current defense of the conventional establishment economic theory takes the form of an argument to the effect that, so long as all economic players act according to their self-interest in the marketplace either as a buyer or seller of commodities, they will each maximize their own satisfaction.

This version of conventional theory replaced an earlier version that had declared that the marketplace was guided by an "invisible hand." This supposedly maximized the satisfactions of buyers and sellers, so long as neither buyers nor sellers combined to restrain the freedom of the other in the marketplace and so long as the government resisted all opportunities to interfere in the operations of the marketplace. If all these conditions were met, all markets would clear at equilibrium prices, as shown in Figure 5.27 below, and there would be no danger of overproduction or underconsumption. Subsequently in the Great Depression of the 1930s, the emergence of vast concentrations of ownership and production disastrously confirmed the validity of all the earlier warnings against sellers of finished goods combining in the marketplace. It also demonstrated conclusively that, once such monopolies emerged, overproduction had become endemic to the short term and long term of the economy. This, in turn, greatly strengthened arguments in favor of reorganizing production for the long term on a very different basis. The new basis proposed eliminating the capture of surpluses and profits as the main and sole driver of economic development and investment, either in the short term or the long term.

Almost mesmerizing in its simplicity, conventional theory tackles the production system as given for any commodity. The graph depicts the resulting situation provided that there are no interdependencies, all competing suppliers are in the market on an equal basis, and a current demand for any good is met entirely and only by its current supply.

Once a market is filled, whether by a quasi-monopoly supplier, a cartel, or all competitive suppliers, conventional economic theory asserts

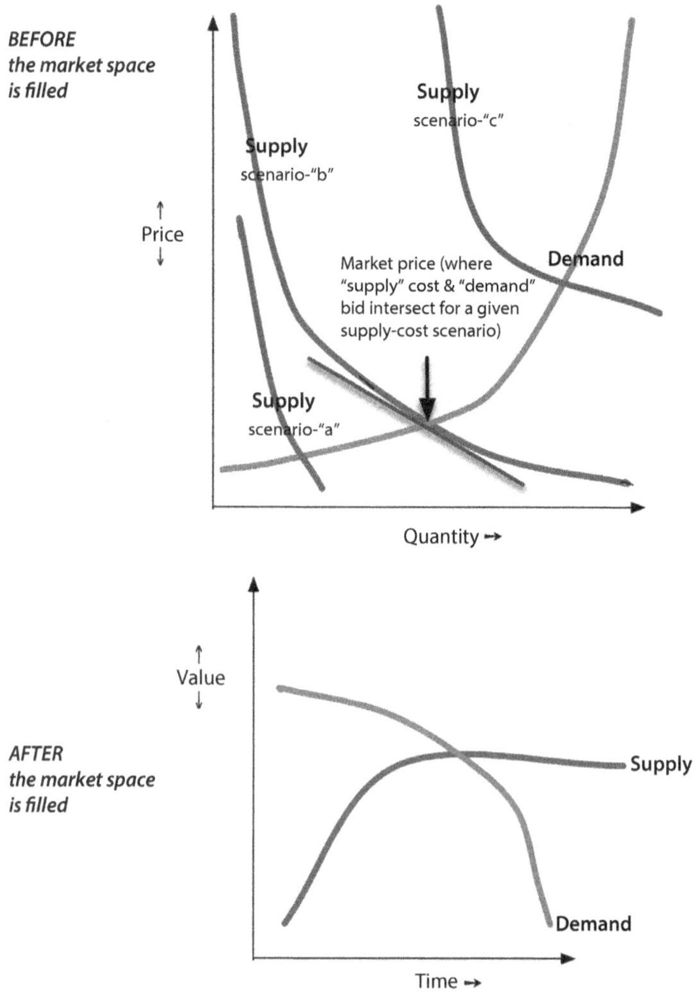

Figure 5.28 Production-Cost and Market-Price Realities "At The Margin."

that it also "clears." All goods that can be sold have been exchanged for money, and the production-consumption cycle is then renewed (Figure 5.28).

Reality demonstrates otherwise. Once actual total production has taken place, some proportion, which may increase over time, becomes stockpiled. As time passes, this surplus could be well in advance of current demand. Market demand, meanwhile, advances at rates far below this rate of increase in total production. In such a scenario, suppliers' costs are transformed from input costs framed by the exigencies of actual external competition into "transfer prices" between different levels of an increasingly vertically integrated structure of production and marketing. Consumers' costs then become predetermined in accordance with arrangements between owners of the forces of material production and owners or operators of wholesale and/or retail distribution networks.

5.4.1.2 Jevons' Paradox

"Jevons' Paradox" is the notion that lowering unit costs of production, on the basis of technological changes that lower the rate of energy consumption by using the same fuel source more efficiently than before, tends to bring about an increase in the production of commodities across all sectors that use or rely on that same energy source. The contradictory, paradoxical upshot is the tendency for the net consumption of energy to increase.

This "paradox" was first formulated in Jevons' *The Coal Question* published in 1865. A typical modern-day restatement from John H. Lienhard, an engineering professor at the University of Houston, follows below:

Herbert Inhaber and Harry Saunders take a disturbing look at energy conservation. They begin in 1865. An English mathematician, William Stanley Jevons, had just written a book titled *The Coal Question*. Watt's new engines were eating up English coal. Once it was gone, England was in trouble. And Jevons wrote:

> '... some day our coal seams [may] be found emptied to the bottom, and swept clean like a coal-cellar. Our fires and furnaces ... suddenly extinguished, and cold and darkness ... left to reign over a depopulated country.'

The answer seemed to lie in creating more efficient steam engines. Jevons may not have realized that steam engines were already closing in on thermodynamic limits of efficiency. But he did see that increased efficiency wouldn't save us in any case.

Look at the Watt engine, he said. It was invented because the older Newcomen engine was so inefficient. Did Watt cut coal consumption by quadrupling efficiency? Quite the contrary. By making steam power more efficient, he spread the use of steam throughout the land. Coal consumption was skyrocketing.

A few years later, Henry Bessemer invented a new highly energy-efficient scheme for smelting steel. Jevons's argument played out once more. Now that we could have cheap steel, we began making everything from it – plows, toys, even store fronts. Energy-efficiency had again driven coal consumption upward.

The existence of unoccupied economic space under free competition is a precondition for any further specialization within the industrial division of labor to increase the net addition of the stock of industrial productive forces (even as productive forces at the margin were destroyed or otherwise rendered economically redundant). This modus operandi enables technological changes that increase the efficiency of the utilization of an energy source to become linked with, and eventually bring about, the net increase in overall consumption of the said energy source observed by Jevons.

Fundamentally, the accounting trick that makes this "paradox" seem more real than it actually is involves an arithmetic sleight-of-hand. Lowering unit costs of production is associated mainly with technological change. (However, in Jevons' time, during a crash of the business cycle, the conditions in which the technological change took place are nowhere referenced.) The introduction of new technologies of this order is invariably undertaken at the upturn of the next cycle by the enterprise or enterprises that wiped out many rivals during the preceding slump. In other words, a mass of productive forces was destroyed, which new technology will render superfluous or displace. It follows that any associated lowering of unit costs of production (an increase in economic efficiency) will manifest only in specific industries or sectors where production became more concentrated as a result of the previous crisis clearing weaker economic players away, or otherwise severely marginalizing them in the market. Energy consumption per employed worker will therefore go up as production becomes more concentrated in fewer, more highly capitalized enterprises, probably employing fewer workers than the entire sector employed before the crisis. The reality is that the capitalization of certain sectors in any given crisis is strengthened through such processes while weakened in others. However, Jevons' method proceeds according to a fallacious assumption that what is true for any one sector will be true for all, just as whatever is true for the individual consumer will be true for consumption in general and in society as a whole.

If the losses of the whole of the economic system (occasioned by the bankruptcies and other epiphenomena of the crisis) were properly added back in, but in the post-crisis phase as deductions from overall

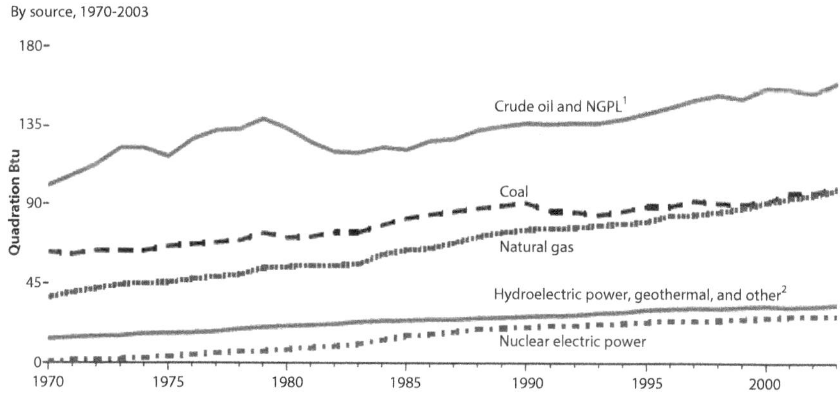

[1]Natural gas plant liquids
[2]Net electricity generation from wood, waste, solar, and wind. Data for the United States also include other renewable energy.

Notes: · Crude oil includes lease condensate. ·Because vertical scales differ, graphs should not be compared.

Figure 5.29 World energy supply (U.S. Department of Energy 2004).

energy consumption, what would the net energy consumption look like? It might have changed little, if at all. In effect, the increased production of commodities and the accompanying increase in energy consumption, referenced by this so-called "paradox," can also be seen as "over-compensating" the losses represented by sidelining or destroying other productive forces. In this respect, the "paradox" is only the outward appearance of an undamped oscillator.

Another feature, of which Jevons was not (and, indeed, could not be) conscious, was that economic space would become completely occupied as oligopolies and cartels consolidated their overall role throughout the economy. The space in which free competition once predominated would thus be eliminated. Already, in Jevons' day this space was being pushed to the margins, although this was not understood at the time for what it actually was. The English economist J.A. Hobson, who wrote about the British Empire as an economic proposition, wrestled with certain parts of the problem as early as 1902 (Hobson 1902). The Austrian economist Hilferding glimpsed some implications in 1910 of the rise and role of finance capital, which he defined as the merger of banking and industrial capital (Hilferding 1910).

The crucial fact about an economic space that has already been divided up and can only be re-divided is that an increase or decrease in energy consumption loses any clear-cut linkage to or dependence upon changes in "productivity." These cease to regulate each other in any predictable way. In effect, as oligopolies, cartels, and monopolies displace

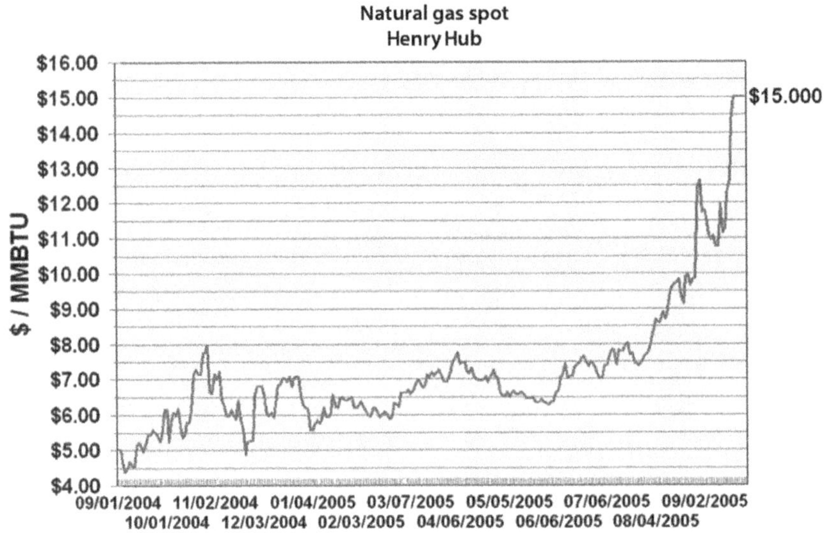

Figure 5.30 Henry Hub Natural gas spot market price,1 Sep '04-1 Sep '05 (WTRG Economics).

free competition while appropriation of the fruits and ownership of the means of production remain private, the impossibility of production and consumption mutually regulating one another spreads to encompass all other relationships engendered earlier when free competition reigned supreme. Thus, Jevons' paradox disappears. Why? Because the relationship it proposed (in order to account for phenomena that appeared sequentially related) no longer exists, even though the phenomena themselves persist.

Today, as the following chart (Figure 5.29) indicates, this energy production effort spans the entire globe and continues ceaselessly to grow.

Although the energy-market growth prospects shift away from North America and Europe towards China and India, as well as the increasingly global character of the continuous expansion of energy production, are masked by this presentation, it does disclose that the main change over time has been the rising production, distribution, and sale of natural gas on the global scale (US DOE, 2004). As discussed earlier, India and China are both energy efficient in terms of per capita energy consumption. Population growth of these countries is far less steep than the rise is per capita energy consumption that seems to be driven by the notion they must emulate the west. While this emulation is encouraged, the same countries are discouraged to pollute to the level of the

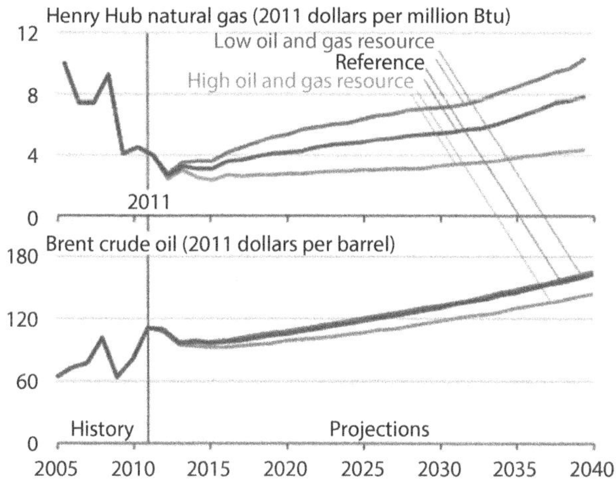

Figure 5.31 Henry Hub graphs for 2005-2012 and beyond.

west. This paradox is resolved by properly addressing petroleum refining and rendering petroleum production and utilization sustainable.

It is a curious fact that natural gas prices do not head towards a single world price, whereas the oil price continues to be a single price no matter the different conditions, local energy needs, and demands attending its production. Conventionally, the tendency towards a single world oil price has been explained as the outcome of an overwhelming concentration of the refining capacity in leading consuming countries. However, the tendency to maintain a single price in global export and import markets has been attenuated by the rise of the refining capacity in a number of the leading OPEC countries, including, in the cases of Venezuela and Iran, an increasing export trade of refined products and no longer just crude oil.

The temporary tripling of the Henry Hub spot price for natural gas, disclosed in Figures 5.30 and 5.31,, demonstrates the extreme potential effect of a short-term event. In this case, Hurricane Katrina, in the end, was found to have caused little consequential damage to the gas pipeline network along the Louisiana coast, whose throughput at Henry LA forms the basis for the so-called "hub price." This single event, plus the speculation before, during, and after of the potentially disastrous consequences of the 2005 hurricane season in the Gulf of Mexico, enabled sellers, basing themselves on the Henry Hub, into markets to double the price from the US$ 5.00-7.50/Mcf band in which gas had been traded from September 1, 2004 until the third week of August 2005. Such a jump would hardly seem possible if there were a world price for this commodity. By contrast, there was in the same time period an even shorter-lived "speculative premium" that bumped the price of oil from US$65 to US$70 bbl. It lasted less than 72 hours precisely because a world price exists that dampens the impact of such events.

After 2005, Henry Hub natural gas followed a trend opposite to that of crude oil for 3 years, up until the financial crisis of 2008. During and soon after that crisis, the natural gas price grew while the oil price declined. By the time the shock of the crisis was absorbed and recession contained, opposing trends took effect. Figure 5.31 shows that EIA predicts that such trend will be reversed and both natural gas and crude oil price will continue to grow all the way through 2040. In the 2013 Reference, Low Oil and Gas Resource, and High Oil and Gas Resource cases, industrial propane prices in 2040 range from \$22.13 per million Btu (2011 dollars) in the High Oil and Gas Resource case to \$27.48 per million Btu in the Low Oil and Gas Resource case, a difference of approximately 24 percent. The difference between the propane prices in the High and Low Oil and Gas Resource cases increases from \$3.49 per million Btu in 2015 to \$7.00 per million Btu in 2025 as natural gas prices and NGL production diverge in the two cases. Over time, however, as the divergence in NGL production narrows between the cases, the influence of oil prices on propane prices increases, and the difference in the propane prices narrows in the cases.

The secret of the divergence between the tendencies of oil and natural gas prices is buried in the historical foundations of modern fossil fuel exploration and production. Such examination discloses unexpected and surprising features of the true relations governing the exploration, production, and processing of fossil fuels, in general, as well as of natural gas, in particular.

For example, for some time, among those concerned about post-sanctions scenarios, anxiety surrounded the vexed question of re-introducing Iraqi crude in the world market. Adelman (2001) wrote, "When the sanctions regime finally erodes, Iraq will behave like an 800-pound gorilla: it will bring in foreign companies to invest and expand while leaving other members out." Instead, Iraq was marginalized by Russia's record oil output (1.5 MMBPD more than Saudi output, while oil prices remained higher than what was believed to be a "fair" \$25/bbl world price). Even with 140,000 troops on Iraq's territory since toppling the regime of Saddam Hussein in April 2003, and in the absence of any agreement with the Paris Club of long-term creditors of the former regime (Russia, France, etc.), the current regime remains unable to

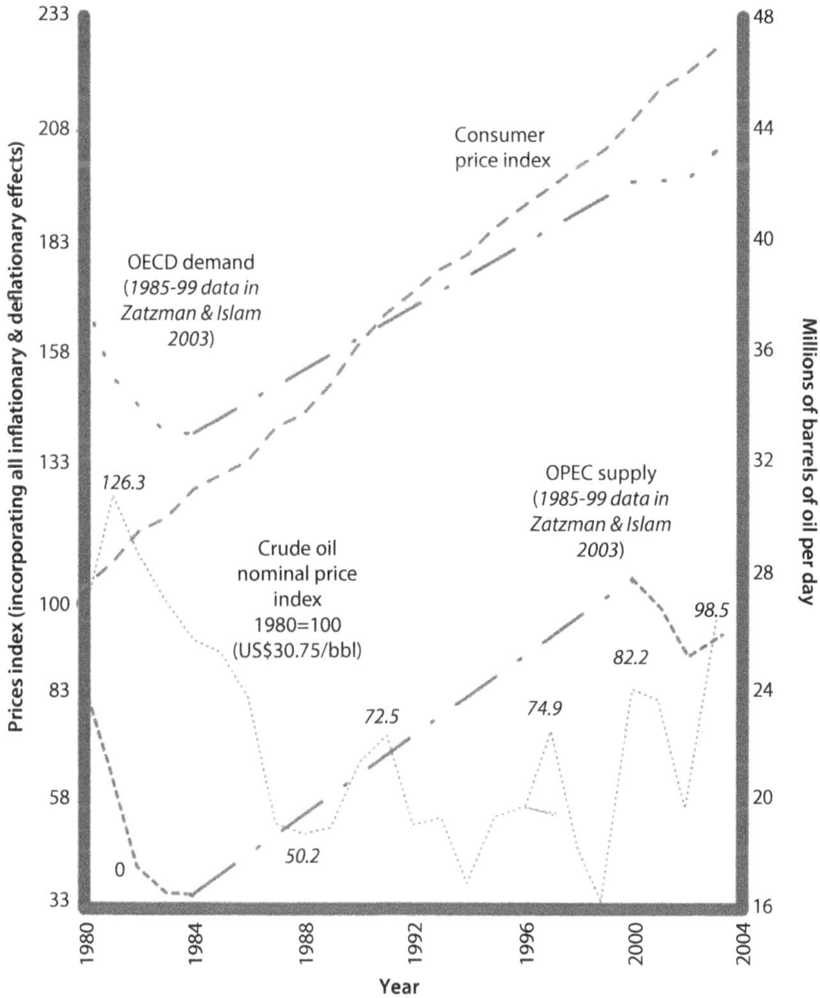

Figure 5.32 Crude Oil Nominal Price Index (CONPI) vs. CPI, and the permanent excess OECD demand compared to OPEC supply (EDO)(Sources: US BoLS, 2006, Tables 1.2, 2.2, and 7.1; EIA, 2003).

secure even pipeline transport of Iraqi oil out of the territory let alone bulk transportation contracts with international shippers. Meanwhile, the meter keeps ticking on the coalition's billion-dollar-a-day war and occupation, which was to be paid for from exports of Iraqi oil. Nevertheless, after dipping below US$24/bbl at the moment Saddam was toppled, the world oil price crept back up but only to the US$27-US $30/bbl range, which is typical for the period of the winter heating

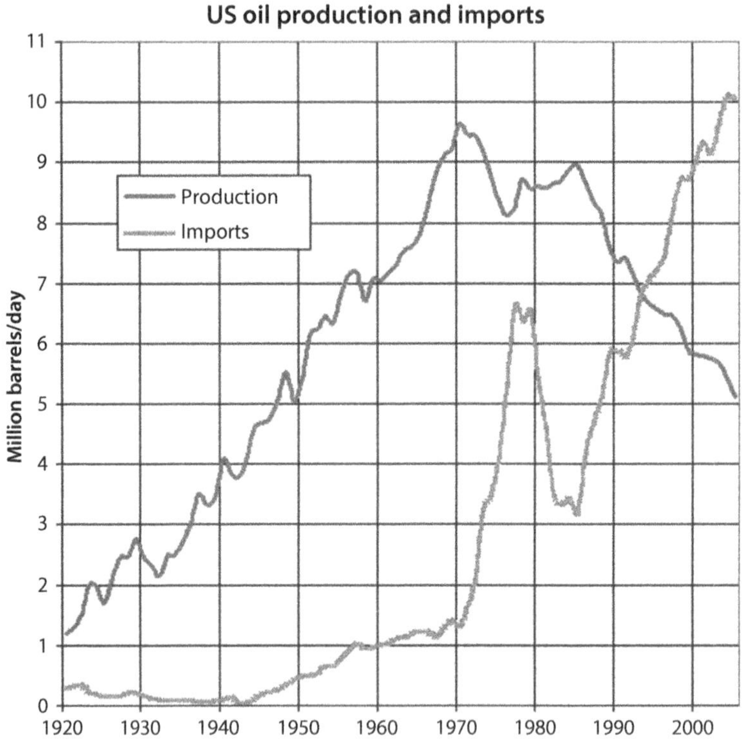

Figure 5.33 The diverging trends that started in Reagan era was reversed in 2013.

season in the northern hemisphere where the largest per-capita energy consumption markets are located. By the summer of 2005, the price had escalated to US$65-70/bbl.

The dashed (top) lines (Figure 5.32) show the price of oil, as indexed by its nominal price (Crude Oil Nominal Price index), fluctuates without reference to, and far below, the Consumer Price Index. The red lines show that the supply shortfall is permanent. The CONPI trend line relative to the demonstrated EDO shows neither oil supply nor demand regulate one another. This line of investigation based on CONPI is further elaborated by Islam and Zatzman (2004a) and Zatzman (2012).

Figure 5.33 shows diverging trends in export and import that was recently reversed. This was mainly due to a surge of 40% in domestic oil/gas production in USA. Such a surge was possible because of the exploitation of unconventional reserves of both oil and gas.

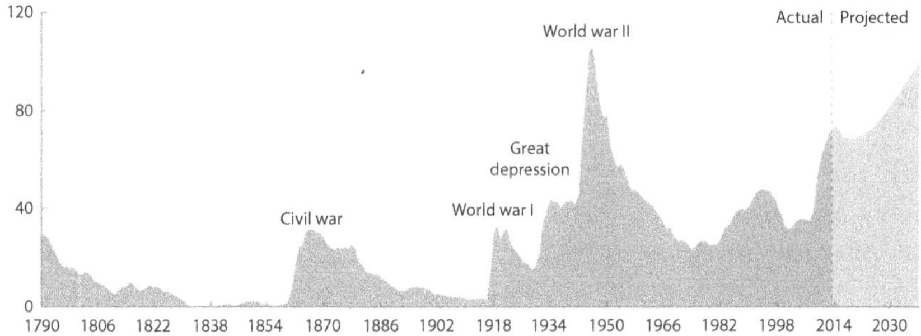

Figure 5.34 US public debt as percentage of GDP.

5.4.1.3 *Equity Shoulders Debt*

Another old accounting principle concludes that equity shoulders debt and therefore growth is limited by the ability of a company to raise capital. Such model has dictated the accumulated of public debt in the United States (Figure 5.33). However, such model is no longer valid in the information age. Cisco Systems gained a market capitalization of about one-half trillion dollars in 16 years, increased sales by 50-fold, has essentially no debt and hasn't sold any stock during the decade of 1990-2000. When an order is placed, it goes directly to the supplier who builds the unit and ships it directly to the buyer. Cisco invoices the customer, keeps the mark-up and pays the supplier. A few additional examples of decapitalized companies are Dell, Qualcomm, and Microsoft. The bottom line is that capital, like land, is losing its place as a governing metric for wealth creation.

Change is inevitable in this situation. The trend seems to be toward large oil and gas companies acting more like investment bankers. They have strong ties to host governments, the World Bank and other international credit and lending banks, to other oil companies and to oil service and construction companies. In this mode, the large oil companies bring together the various parties that are needed to develop new oil and gas fields, build pipelines, organize tanker transport, construct and operate refineries and chemical plants, distribute and market the products to consumers. Either before or soon after finding new reserves or establishing production from new fields, significant ownership portions are sold to competitors to recover initial investments and to limit risk. Retained partial ownership in the newly developed fields provides continuing cash flow to fund new investments. Partnering with competitors disseminates technology and limits risk. It also limits the opportunity to dominate industry segments via innovative technology application and therefore reduces the dęsire to fund research and development.

Oil companies becoming investment bankers have some similarities with the Cisco model. Although new oil and gas field development is highly capital intensive, the oil companies as investment bankers or virtual oil companies can spread the capital burden widely. As the model expands to allow suppliers, service and construction companies to partner in new fields, further risk and capital spreading is possible. In fact one scenario can be imagined whereby substantially less direct capital is required. Here, all development needs such as exploration, drilling, production and treating facilities, and transport are "capitalized" by produced volume sharing. Business to business (B2B) methods can be included in this model to further reduce capital and operating costs, and to generate revenue. However, the sad reality in all these is the import 'model' from Cisco to petroleum industry or any other application cannot be done and independent research must be done to develop its own solution that befits the information age. Cutting research money or selling consulting expenses in name of research to plunder public funds in name of tax benefit will not solve the problem. What oil companies are engaged in is similar to what a man in middle age does – he still likes glamour of the past but cannot think of a way to maintain that level of energy. Rather than being creative about it and finding a new solution, he engages in making cosmetic changes, coloring hairs, even resorting to plastic surgery. All these make him look more like a cadaver than a young energetic man. If someone points this out to him, he screams at him like a grumpy old man. I have observed many CEO's of oil companies – all behave this way when it comes to making company decisions. Of course, there are some exceptions, but they are to scared to make row with the establishment. Few, if any, have the courage to do things for the sake of overall good. Everyone is trying to look good today, lest there is no tomorrow for them.

5.4.1.4 The Finite/ Infinite Conundrum

The next assumption of peak oil theory is that the oil reserve is finite. The theory first assumes the ultimate recoverable reserve, then expresses cumulative oil production as a function of the ultimate recoverable reserve. Cavallo (2004) defines the Hubbert curve used to predict the U. S. peak as the derivative of:

$$Q(t) = \frac{Q_{max}}{1 + ae^{-bt}} \tag{5.1}$$

where Q(t) is the cumulative oil production and Q_{max} is the maximum producible reserve and a and b are constants. The year of maximum annual production (peak) then back calculated as:

$$t_{max} = \frac{1}{b} \ln(a) \tag{5.2}$$

The fixation of Q_{max} is in the core of the Hubbert curve. Theoretically, the recoverable reserve increases for two reasons: 1) the boundary of resource; 2) the technology. As discussed in earlier sections, the boundary of resource is continuously moving. The recent surge in unconventional oil and gas reserve makes an excellent point to this regard. In fact, the following section makes the argument that this boundary is fictitious and for a sustainable recovery scheme, this boundary should not exist. The second reason for the reserve to grow is the technology that becomes applicable to a broader resource base. The earlier section on EOR makes the argument that EOR schemes alone can continue to increase the reserve and has done so in the past.

There is a general misconception that Hubbert was concerned with "easy" oil, "easy" metals, and so forth that could be recovered without greatly advanced mining efforts and how to time the necessity of such resource acquisition advancements or substitutions by knowing an

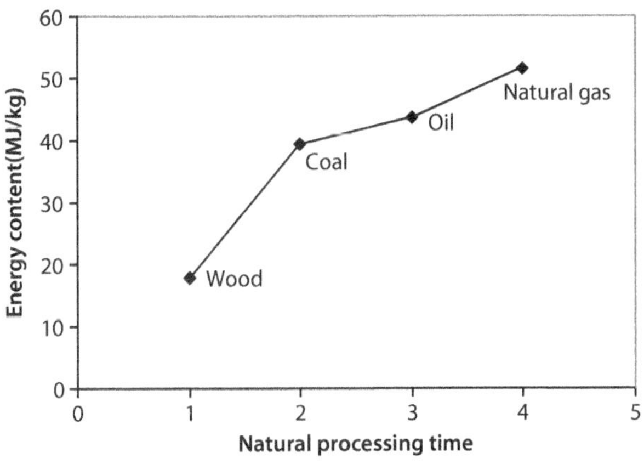

Figure 5.35 Energy content of different fuels (MJ/kg).

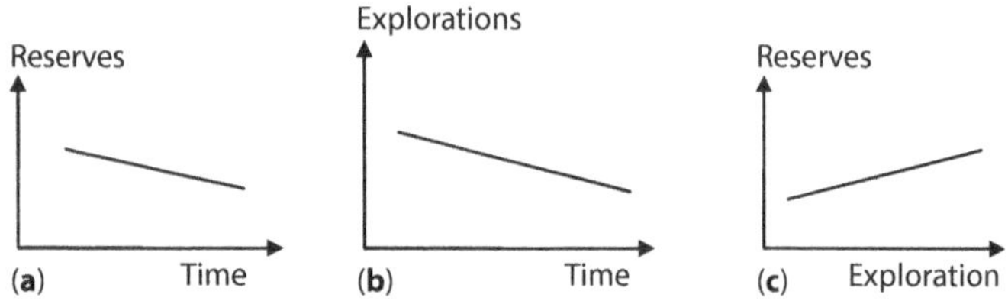

Figure 5.36 Fossil fuel reserves and exploration activities.

"easy" resource's probable peak. The difficulty of Hubbert curve is not its assumption that easy oil recovery is constant, it is rather the notion that resource that turns into reserve with time is finite. As shown in previous sections, accessing greater resource bases is not a matter of 'more difficult' technology, it is rather a matter of producing with sustainable techniques.

5.4.1.5 Renewable vs Non-renewable: No Boundary-as-such

Chhetri and Islam (2008) elaborated the notion that the 'finite resource' is not scientific. With sustainable recovery tools, resources are infinity and are part of the continuous cycle. Figure 5.35 shows that as the natural processing time increases, the energy content of the natural fuels increases from wood to natural gas. The average energy value of wood is 18 MJ/kg (Hall, and Overend, 1987) and energy content of coal, oil and natural gas are 39.3MJ/kg, 53.6MJ/kg and 51.6MJ.kg, respectively (Demirel, 2012). Moreover, this shows that the renewable and non-renewable energy sources have no boundary. It is true that solar, geothermal, hydro and wind sources are being renewed at every second based on the global natural cycle. The fossil fuel sources are solar energy stored by the trees in the form of carbon and due to the temperature and pressure, they emerge as coal, oil or natural gas after millions of years. Biomass is renewed from a few days to a few hundreds years (as a tree can live up top several hundred years). These processes continue for ever. There is not a single point where fossil fuel has started or stopped its formation. So, why these fuels are called non-renewable? The current technology development mode is based on extremely short term approach as our solution of the problems start with the basic assumption of 'Δt tends to =0'. Only technologies that fulfill the criteria of time approaching infinity are sustainable (Khan and Islam, 2007). The only problem with fossil fuel technology is that they are rendered

Table 5.2 US crude oil and natural gas reserve (Million barrels).

	Year	Reserve	% Increment
Crude Oil Reserve	1998	21,034	
	1999	217,65	3.5%
	2000	22,045	1.3%
	2001	22,446	1.8%
Natural Gas	1998	164,041	
	1999	167,406	2.1%
	2000	177,427	6.0%
	2001	183,460	3.4%

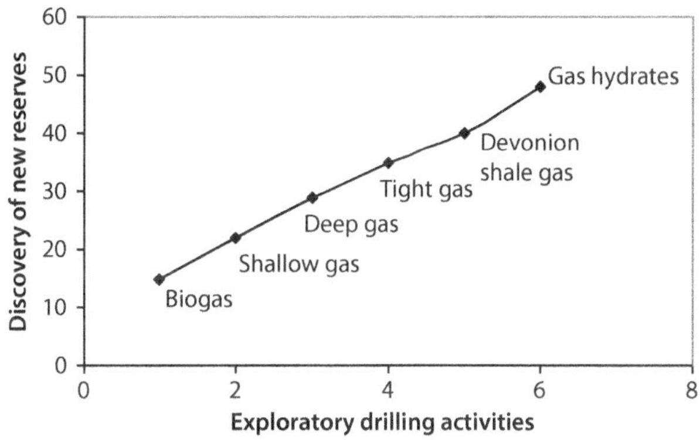

Figure 5.37 Discovery of natural gas reserves with exploration activities.

toxic after they are refined using high heat, toxic chemicals and catalysts.

From the above discussion, it is clear that fossil fuel can contribute significant amount of energy by 2050. It is widely considered that fossil fuels will be used up soon. However, there are still huge reserves of fossil fuel. The current estimation on the total reserves is based on the exploration to-date. If one assumes a priori that reserves are declining with time (Figure 5.36a), one fails to see the role of exploration and drilling activities (Figure 36b). As the number of drillings or exploration activities increases, more recoverable reserves can be found (Figure

5.36c). In fact, Figure 5.36 is equally valid if the abscissa is replaced by 'time' and ordinate is replaced by 'exploratory drillings' (Figure 5.36b). For every energy source, more exploration will lead to larger fuel reserve. This relationship makes the reserve of any fuel type truly infinity. This relationship alone can be used as a basis for developing technologies that exploit local energy sources.

The US oil and natural gas reserves reported by EIA consistently show that the reserves over the years have increased (Table 5.2 gives a sampler). This additional reserves were estimated after the analysis of geological and engineering data. Hence, as the number of exploration increases, the reserves will also increase.

Figure 5.37 shows that the discovery of natural gas reserves increases as exploration activities or drillings is increased. Biogas in naturally formed in swamps, paddy fields and other places due to the natural degradation of other organic materials. As shown in previous sections, there are huge gas reservoirs including deep gas, tight gas, Devonian shale gas and gas hydrates, which are not yet exploited. The current exploration level is limited to shallow gas, which is a small fraction of the total natural gas reserve. Hence, by increasing the number of exploration activities, more and more reserves can be found which indicates the availability of unlimited amount of fossil fuels. As the natural processes continue, formation of natural gas also continues for ever. This is applicable to other fossil fuel resources such as coal, light and heavy oil, bitumen and tar sands.

Figure 5.37 shows the variation of resource base with time starting from biomass to natural gas. Biomass is available in huge quantity on earth. Due to natural activities, the biomass undergoes various changes. With heat and pressure on the interior of the earth, formation of fossil fuel starts due to the degradation of organic matters from the microbial activities. The slope of the graph indicates that the volume of reserve decreases as it is further processed. Hence, there is more coal than oil and more oil than natural gas, meaning unlimited resources. Moreover, the energy content per unit mass of the fuel increases as the natural processing time increases. The biomass resource is renewable and the biological activities continue on the earth, the process of formation of fossil fuel also continues for ever. From this discussion, conventional boundary of renewable and non-renewable is dismantled and concluded that there is no boundary between the renewable and non-renewable as all natural processes are renewable. The only problem with fossil fuel arises from the use of toxic chemicals and catalysts during oil refining and gas processing. Provided the fossil fuels are processed using the natural and non-toxic catalysts and chemicals, or make use of crude oil or gas directly, fossil fuel will still remain as a good supplement in the global energy scenario in the days to come. These resources are totally recyclable.

5.5 Science of Healthy Energy and Mass

Scientific analysis of energy involves determination of energy source and content. Energy is known to be the causes of actions that are ubiquitous. Scientifically, every action and movement has a driver. Because every object is in motion, that driver is ubiquitous. New science has identified sun as the ultimate energy source for the earth. While this conclusion is true, the premise that defines energy in New science is spurious (Islam *et al.*, 2014). In this section, some of the scientific aspects of energy will be discussed.

Conventional notion of energy and the conservation of energy emerges from discrete description of mass and energy. It assumes that mass exists independent of energy. This disconnection between mass and energy is rooted in cognition paradox introduced during in Ancient Greece and later readily adopted by the likes of Thomas Aquinas and other Church-inspired scholars.

5.5.1 Role of Water, Air, Clay and Fire in Scientific Characterization

Around 450 B.C., a Greek philosopher, Empedocles, characterized all matter into – earth, air, fire, and water. Note that the word 'earth' here implies clayey material or dirt it is not the planet earth. The origin of the word 'earth' (as a human habitat) originates from the Arabic word *Ardha*, the root meaning of which is the habitat of the human race or "children of Adam"), lower status, etc. Earth in Arabic is not a planet as there are other words for planet. Similarly, the sun is not a star, it is precisely the one that sustains all energy needs of the earth. The word 'air' is *Hawa* in Arabic is air as in the atmosphere. Note that 'air' is not the same as oxygen (or even certain percentage of oxygen, nitrogen, and carbon dioxide, etc.) – it is the invisible component of the atmosphere that surrounds the earth. Air must contain all organic emission from earth for it to be 'full of life'. It cannot be reconstituted artificially. The term, 'fire' is '*naar*' in Arabic that refers to real fire, as when wood is burnt and both heat and light are produced. The word has the same

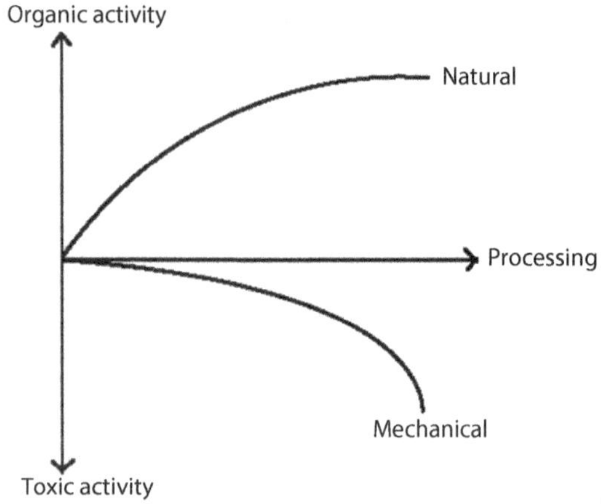

Figure 5.38 Water: a source of life when processed naturally but a potent toxin when processed mechanically.

root as light (*noor*), which however has a broader meaning. For instance, moonlight is called *noor*, whereas sunlight (direct light) is called *adha'a*. In Arabic, there is a different word for lightning (during a thunderstorm, for instance). The final material 'water' is recognized as the source of life in every ancient culture. This water is not H_2O. In fact, the H_2O of modern science that is best described as the combination of atomic hydrogen and oxygen is a toxic product that wouldn't allow the presence of any life. As the purity of H_2O increases, its toxicity goes up and it becomes poisonous to any life form. The word 'water' in ancient cultures is best defined as the source of life. *The Qur'an* recognizes water as the essence of life as well as the source of all mass. In that sense, water is recognized as the first mass created. Such beginning would contradictory to the Big Bang theory that assumes hydrogen to be the first mass form. However, Big Bang narration of nature is flawed and is not consistent with natural phenomena that doesn't show synthesis of elements to form new materials, instead showing transformation and irreversible merger of particles, much like merger of two galaxies. This has been called the Galaxy model by Islam *et al.* (2014). In summary, water represents the imbedding of all other forms of material. For water to be the source of life, it must have all ingredients of a life form. Figure 5.38 shows how depriving water from its natural ingredients can make it reactive to the environment and render it toxic. This graph needs further clarification.

Water is a great solvent. It has natural affinity to dissolve numerous salts and minerals that are necessary for life support. One can argue that every component necessary for life is in water. However, this can only occur in case of naturally occurring water. Water is routinely stripped off its solutes by the natural process of evaporation and subsequent precipitation through a series of highly complex and little understood processes. However, this processing prepares water for collecting organic matter that is necessary for life support. This rainwater is pure (in the sense that it has little solute) but it is not toxic or harmfully reactive to the environment. As rainwater comes in contact with soil, it immediately triggers organic transformation of matters and life flourishes. As rain water penetrates the outer crust it picks up minerals and the water becomes even more balanced for human consumption. Another component of human consumption is the water should be free of organic matters as well as bacteria. As water filters through the soil, it becomes free from these harmful components. So, whenever naturally processed water either becomes useful for human consumption or it becomes useful for other living organisms that are part of a life cycle that includes humans. At later stages of natural processing, a balance is struck, as reflected in Figure 5.38.

On the other hand, if the water is processed through artificial means (marked here as 'mechanical'), various life-supporting components are removed and then replaced with toxic artificial components, many of whom are not identified. It is commonly believed that artificially 'purified' water has great affinity to absorb back any component from external sources. That's why such 'pure water' is used to clean semiconductors. For the same reason, this water becomes harmful to humans. If ingested, this water starts to absorb all the valuable minerals present in the body. Tests have shown that even as little as a glass of this liquid can have a negative effect on the human body. This process produces water of particularly high toxicity when reverse osmosis and nanofiltration is used. The World Health Organization (WHO) determined that demineralized water increased diuresis and the elimination of electrolytes, with decreased serum potassium concentration. Magnesium, calcium and other nutrients in water can help to protect against nutritional deficiency. Recommendations for magnesium have been put at a minimum of 10 mg/L with 20–30 mg/L optimum; for calcium a 20 mg/L minimum and a 40–80 mg/L optimum, and a total water hardness (adding magnesium and calcium) of 2–4 mmol/L. At water hardness above 5 mmol/L, higher incidence of gallstones, kidney stones, urinary stones, arthrosis, and arthropathies have been observed. For fluoride the concentration recommended for dental health is 0.5–1.0 mg/L, with a maximum guideline value of 1.5 mg/L to avoid dental fluorosis (Kozisek, 2005).

A significant portion of essential minerals are derived from water. "Purified" water doesn't contain these essential minerals and thereby cause disruption to the metabolic process, thereby causing harm (Azoulay *et al.*, 2001). When the residual components in 'purified water' contains toxins, such as the ones released from membrane during the reverse osmosis process, the process becomes particularly toxic, as shown in the lower half of Figure 5.38.

Picture 5.4 shows the essence of natural processing of water. Formation of cloud through evaporation, rain, photosynthesis, filtration in the soil, and others form integral part of a life support system that is opposite to the mechanical system in every step of the way. It is also true that energy in an organic system emerges from water, just like life. As life cycle continues, mass transfer takes place simultaneous to energy exchange. By assigning zero mass to energy, this continuity is missed in the analysis adapted in New Science.

In all, the characterization credited to Empedocles and known to modern Europe conforms to the criterion of phenomena as outlined in the work of Islam *et al.* (2010) as well as Khan and Islam (2007). This fundamental criterion can be stated as not violating the properties of nature. In fact, this characterization has the following strengths: 1) definitions are real, meaning have phenomenal first premise; 2) it recognizes the continuity in nature (including that between matter and energy); 3) captures the essence of natural lifestyle. With this characterization, nuclear energy would not emerge as an energy source. Fluorescent light would not qualify for natural light. With this characterization, none of the unsustainable technologies of today would come to existence.

In the context of working out the systematic characterization of matter, the concept of fundamental substance was introduced by another Greek philosopher, named Leucippus who lived around 478 B.C. Even though his original work was not accessible even to Arabs who brought the annals of ancient Greek knowledge to the modern age, his student Democritus (420 B.C.) documented Leucippus' work which was later translated in Arabic, then to Latin, followed by modern Greek and other European contemporary languages.

That work contained the word 'atom' ($\alpha\tau o\mu\alpha\sigma$ in Greek), perpetrated as a fundamental unit of matter. This word created some discussion among Arab scientists 900 years ago. They understood the

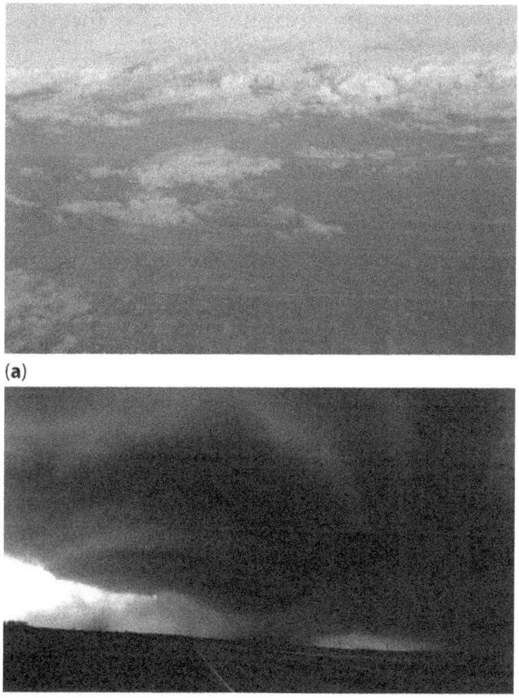

(a)

(b)

Picture 5.4 Role of natural processing in rejuvenating water is little understood by New Science in North America. This role was well understood elsewhere around the world as late as the Islamic era (7th to 17th century). a) Over Ontario [Canada] (23 May 2014) and b) Over Potter, NE (20 May 2014).

meaning to be 'undivided' (this is different from the conventional meaning 'indivisible' used in Europe in the post-Renaissance era. This would be consistent with Muslim scholars because they would not assign any property (such as indivisibility) that has the risk of being proven false (as in the case of the conventional meaning of atom). Their acceptance of the word atom was again in conformity with the criteria listed in Chapter 4, along with fundamental traits of nature.

Atom was not considered to be either indivisible, or identical, or uniform, or any other commonly asserted properties described in the contemporary Atomic theory. In fact, the fundamental notion of creating an aphenomenal basis or unit is a strictly post-Roman Catholic Church European one. Arab annals of knowledge in the Islamic era, starting from the 7th century, have no such tradition (Zatzman, 2008).

This is not to say they did not know how to measure. On the contrary, they had yardsticks that were available to everyone. Consider in this such a unit of time as the blink of an eye (*tarfa*) for small scale and bushel of grain from medium scale (useful for someone who does the milling of grains using manual stone grinders). The unit of matter was the dust particle (*dharra* means the dust particles that are visible when a window is opened to let the sunlight into a room – this word is erroneously translated as 'atom'). As cited by Khan and Islam (2012) and Zatzman *et al.* (2007), using this phenomenal basis, Islamic scholars were able to advance knowledge to a great extent. For example, some one thousand years before Europeans were debating the flatness of the Earth, researchers of Caliph Al-Mamoon already knew the earth is ovoid. When the Caliph wanted to know the 'circumference' of the earth, he sent out two highly competent scientific expeditions.

Working independently, they were to measure the circumference of the Earth. The first expedition went to Sinjar, a very flat desert in Iraq. At a certain point, on latitude 35 degrees north, they fixed a post into the ground and tied a rope to it. Then they started to walk carefully northwards, in order to make the North Pole appear one degree higher in the sky. Each time the end of the rope was reached, the expedition fixed another post and stretched another rope from it until their destination was reached: latitude 36 degrees north.

They recorded the total length of the ropes and returned to the original starting point at 35 degrees north. From there, they repeated the experiment heading south this time. They continued walking and stretching ropes between posts until the North pole dropped in the sky by one degree, when they reached the latitude of 34 degrees. The second of Almamon's expeditions did the same thing but in the Kufa desert. When they had finished the task, both expeditions returned to Al-Mamoon and told him the total length of the rope used for measuring the length of one degree of the Earth's circumference. Taking the average of all expeditions, the length of one degree amounted to 56.6 Arabic miles.

The Arabic mile is equal to 1973 metres. Therefore, according to the measurements made by the two expeditions, the Earth's circumference was equal to 40,252 kilometres. Nowadays, the figure is held to be 40,075 kilometres. How does this compare with the circumference of the earth as we know today? Today, It is known to be 40,075 km if measured through the equator, a difference of less than 200 km. This illustrates how powerful such a phenomenal basis was for conducting measurements and verifying theories.

Heraclitus (540 B.C.) argued that all matter was in flux and vulnerable to change regardless of its apparent solidity. This is obviously a more profound view, even though, like Democritus, he lacked any special lab-type facilities to investigate this insight further, or otherwise to look into what the actual structure of atomic matter would be. It would turn out, the theory of Heraclitus would be rejected by subsequent Greek philosophers of his time...

A less elaborate 'atomic theory' as described by Democritus had the notion of atoms being in perpetual motion in a *void*. While being in constant motion (perpetual should not mean uniform or constant speed) is in conformance with natural traits, void is not something that is phenomenal.

In Arabic, the closest word to describe void is '*cipher*' (the origin of the word decipher, meaning removing the zero's or the fillers), which means empty (this word that has been in Arabic for over 1400 years was not used in *The Qur'an*). For instance, a hand or a bowl can be empty because it has no visible content in it, but it would never imply it has nothing it (for instance, it must have air or dust specks, *dharra* that become visible under the sunlight). The association of 'cipher' with zero was done much later when Arabs came to know about the role of zero from Indian mathematicians. One very useful application of zero was in its role as a filler. That alone made the counting system take a giant leap forward.

However, this zero (or cipher or '*sunya*' in Sanskrit) never implies nothingness. In Sanskrit, *Maha Sunya* (Great Zero) refers to the outer-space, which is anything but void as in nothing-ness. Similarly, the equivalent word is *As-sama'a*, which stands for anything above the earth, including seven layers of skies, only the first one being 'decorated' with stars.

In ancient Greek culture, however, void refers to the original status of the Universe which was thought to be filled with nothingness. This status is further confused with the state of chaos, *Χαοσ*, another Greek term that has void as its root. The word chaos does not exist in *The Qur'an* as it is asserted there is no chaos in universal order that would not allow any state of chaos, signaling the loss of control of the Supreme Authority. However, 'nothingness' is used in terms of creation (*fatara*, in Arabic) from nothing.

It is not clear what notion Liucippas had regarding the nature of atomic particles, but from the outset, if it meant a particle (undivided) that is in perpetual motion, it would not be in conflict with fundamental nature of natural objects. This notion would put everything in a state

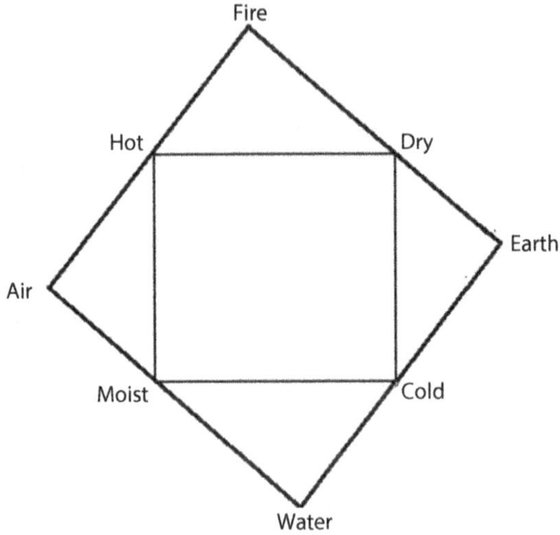

Figure 5.39 Aristotle's four-element phase diagram (steady-state).

of flux. The mainstream Greek philosophy would view this negatively for its subversive implication that nature is essentially chaotic. Such an inference threatened the Greek mainstream view that Chaos was the Void that had preceded the coming into existence of the world, and that a natural order came into existence putting an end to chaos. As stated earlier, this confusion arises from misunderstanding the origin of the Universe[15].

Even though this view was rejected by contemporary Greek scholars, this notion of nature being dynamic was accepted by Arab scholars who did not see this as a conflict with natural order. In fact, their vision of the Universe is, everything is in motion and there is no chaos. Often, they referred to a verse of *The Qur'an* (36:38) that actually talks about the sun as a continuously moving object – moving not just haphazardly but in a precisely predetermined direction, assuring universal order.

Another intriguing point that was made by Democritus is that the feel and taste of a substance is a function of "atom" (atomos) of the substance on the "atom" (atomos) of our sense organs. This theory advanced over thousand years before Alchemists' revolutionary work on modern chemistry was correct in the sense it supports the fundamental trait of nature.

[15]This confusion continues today when Quantum theory assumes that numerous universes are created from nothing (Hawking, 2010).

This suggestion that everything that comes into contact contributes to the exchange of "atoms" ($\alpha\tau o\mu o\sigma$) would have stopped us from making toxic chemicals, thinking that they are either inert (totally isolated from the system of interest) or their concentration is so low that the leaching can be neglected. This would prevent us from seeing the headlines that we see everyday. This theory that could have revolutionized Chemical engineering 1000 years before Alchemists (at least for Europe, as Egyptians already were much advanced in chemical engineering some 6000 years ago) was rejected by Aristotle (384–322B.C.) who became the most powerful and famous of the Greek scientific philosophers. Instead, Aristotle adopted and developed Empedocles's ideas of elemental substances, which was originally well founded. While Aristotle took the fundamental concept of fire, water, earth, and air being the fundamental ingredients of all matter, he added qualitative parameters, such as hot, moist, cold, and dry.

Figure 5.39 shows Aristotle's model for four fundamental elements of matter. This is the oldest form of phase diagram that can be found in Europe. This figure is in effect a steady-state model. The elimination of the time function made the diagram appear perfectly symmetrical, which is the essence of Atomism. Democritus is indeed most often cited as the source of the atomic theory of matter, but there's a strong argument/likelihood that what he had in mind was a highly idealized notion, not anything based on actual material structure. For the Greeks, symmetry was believed to be good in itself and was largely achieved by geometric rearrangement of [usually] two-dimensional space. There is an ambiguity as to whether Greek atomists thought of atoms as anything other than an infinite spatial subdivision of matter. Heraclitus' major achievement - which also marginalized him among the other thinkers of his time, unfortunately - was his incorporation of a notion of the effects of time as a duration of some kind, as some other kind of space in which everything played itself out.

On the matter of role of time sequence, and universal order, Heraclitus had a profound view that was considered to be a paradox and was rejected (Graham, 2006). Heraclitus wrote:

> This world-order [kosmos], the same of all, no god nor man did create, but it ever was and is and will be: everliving fire, kindling in measures and being quenched in measures.

This would be the first case of Agnostic assumption of 'self creation' and/or everlasting nature of universe, conflating 'infinity' as a trait of

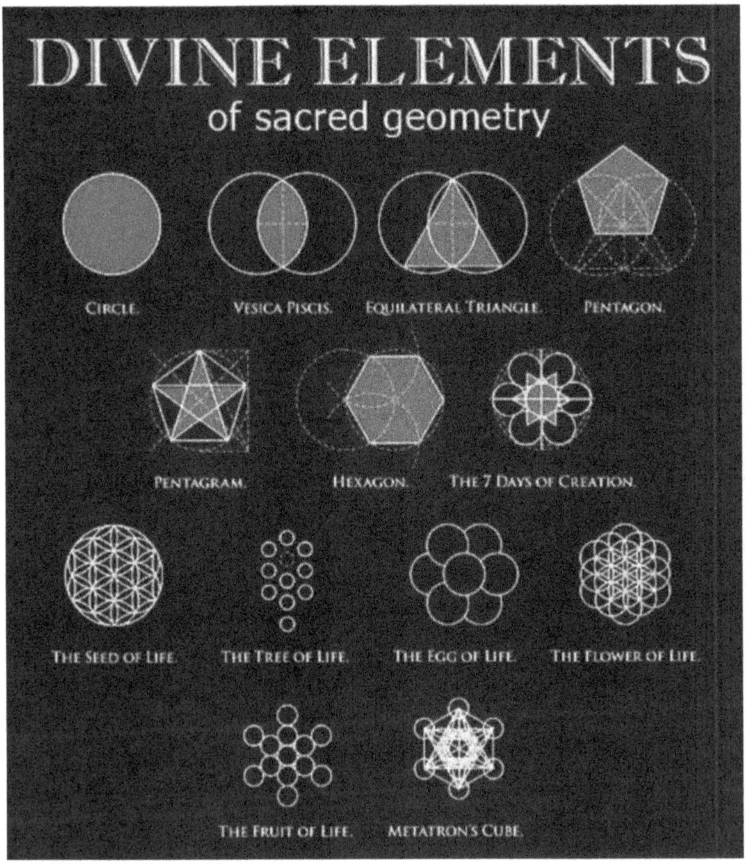

Figure 5.40 Divinity in Europe is synonymous with uniformity, symmetry, and homogeneity, none of which exists in nature.

creator with the trait of creation. In addition, he uses, for the first time in any extant Greek text, the word *kosmos* "order" to mean something perceived as "world." He identifies the world with fire, but goes on to specify portions of fire that are kindling and being quenched. Although ancient sources, including Aristotle as well as the Stoics, attributed to Heraclitus a world that was periodically destroyed by fire and then reborn, the present statement seems to contradict that view, as Hegel also noticed. If the world always was and is and will be, then it does not perish and come back into existence, though portions of it (measures of fire) are constantly being transformed. This contradiction and paradox are erased if "world-order" is replaced with "universal order"

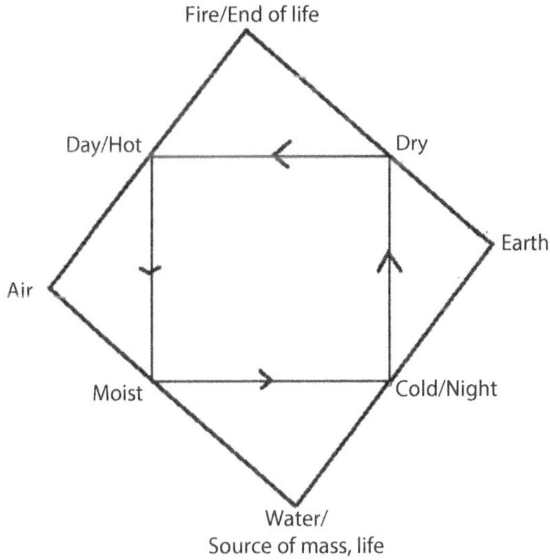

Figure 5.41 Recasting Figure 5.39 with the proper time function.

and the creation of time[16] is preceded to be before creation of everything else (known as 'matter' at later time). A consistent and non-paradoxical meaning emerges if the following sequence is used. Creator (Absolute time) created the Absolute plan (variable time) before creating everything as a function of time (Islam *et al.*, 2014). It also resolves the conflation of the term 'infinity' with 'never-ending'. While creator is infinity, creation is 'never-ending'.

Figure 5.40 is a depiction of what is sacred in European narration of philosophy. Even though this depiction is attributed to Roman Catholic Church, it is clearly related to features seen in 'Absolute being' in ancient Greek and later transformed into all aspects of scientific and mathematical cognition. Aristotle said, "The mathematical sciences particularly exhibit order, symmetry, and limitation; and these are the greatest forms of the beautiful". This can very well be the beginning of reversing science from long-term to short-term through mathematization. Linearization is the ultimate version of such mathematization.

This fascination with homogeneity, symmetry, and other 'godly' attributes and assigning them to creation is uniquely European (Livio, 2005). The orient is known to have a different description of nature. At no time, ancient India, Babylonia, or ancient China conflated nature of

[16]This time is a creation and is transient (changing).

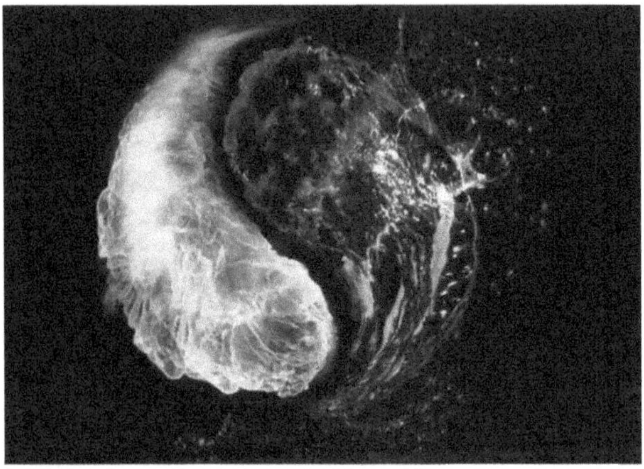

Figure 5.42 Water and fire are depicted through *taegeuk* (*yin yang*).

Figure 5.43 Korean national flag contains ancient symbol of creation and creator.

God with nature of humans, even for *Avatar*s. The most important "transformation" between creator and creation is through the notion of *Avatar* in India. An *avatar* is bound to change and is subject to all features of a mortal or natural objects.

Figure 5.41 is a recasting of Figure 5.39 that introduces the concept of water as the source of life and fire as the end of life, the whole process being connected through periodicity of day and night. This was the model used by the scholars and Alchemists of the Islamic golden era (7th to 17th century).

Note how water is considered to be the source of all mass that goes through natural processing during day and night, constantly giving rise to new life forms (e.g. photosynthesis, blossoming of flowers, sprouting of seeds) and ending with death, which then gives rise to ingredients necessary for new form of life. Therefore, both life and death were considered to be integral part of the universal order that follows a cyclic pattern, always conserving mass and energy simultaneously.

Of importance is the notion that Qur'an indicates that the word *Ardha* (Earth) is the habitat for mankind, who is made out of clay, the principal component of earth. This same earth also gives rise to plant lives that serve as the transformer of solar energy into biomass, The word 'air' in Arabic is *hawa*, which also means life. This description and characterization of mass and energy is also typical of the Yin Yang concept that is known in the Korean peninsula for the entire recorded period of history.

Figure 5.41 shows the depiction of fire and water in Yin Yang form. This figure shows the co-existence fire and water, the two main components of universe through asymmetrical pattern while keeping the complimentary nature of those components intact. The broader symmetry that is seen is more thematic than tangible.

The central message of this picture is stated through the word of famous mythology professor, Joseph Campbell (1904-1974), who said, "The goal of life is to make your heartbeat match the beat of the universe, to match your nature with Nature." This is a theme that is in the core of Islamic faith that describes religion as *Deen* (natural trait) and defines Muslim (root word *salama* that means surrender and peace) to be the one in tune with nature and natural order (*Qadr* in Arabic).

This natural order is unique and manifested through a unique function of time. With this characterization, there cannot be any multiple history of the same particle of event. Therefore, one of the biggest problems of quantum theory doesn't occur with scientific characterization of matter and energy. Note in Figure 5.42 how both fire and water parts contain global symmetry but have no local symmetry. Similarly, the flag of South Korea exhibits anti-symmetry, or symmetry with a property reversal (Figure 5.43). The white background is a traditional Korean color. It traditionally represents peace and purity. Note that this circle is not tangible and it signifies the never-ending (sustainable) nature of nature. It is anti-symmetrical because of the red/blue interchange, both in color and deeper meaning. The deeper meanings represent opposing functions of the universe. The most comprehensive meaning is the existence of tangible and intangible. These seemingly opposite or contrary forces are interconnected and interdependent in the natural universe that allows the existence of both in perfect harmony.

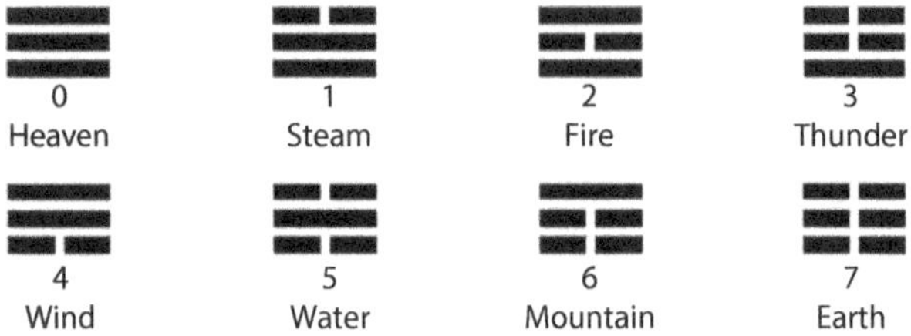

Figure 5.44 Combination of various fundamental elements make up the rest of the creation.

Such harmony is in sharp contrast with European notion that says the universe is constantly degrading or there is a continuous struggle between good and evil. The European notions emerge from a first premise that is similar to the dogmatic notion of 'fall' and 'original sin'. Nature being a union of opposites in harmony comes from the first premise that Nature is perfect and balanced (Khan and Islam, 2012). Many natural dualities (such as light and dark, woman and man, day and night, high and low, hot and cold, fire and water, life and death, and so on) are thought of as physical manifestations of the yin-yang concept.

This concept also reveals the fact that the duality is apparent and it is a matter of perception of the observer whereas it is absolute for the object. For instance, a blind person doesn't see the difference between night and day but that perception doesn't change anything about the night and day. This fact demonstrates the absurdity of Quantum theory that makes reality a function of perception. Reality cannot be a function of perception, unless there is no such thing as reality. According to Aristotle, one of the mistakes of Zeno in his paradoxes of time and motion is that he did not distinguish between actual and potential infinities. Scientifically, it is the continuity of the time function that eluded Zeno. Aristotle 'remedied' it by differentiating between actual (present) and potential (future). Then he asserted, "Everything is potential infinity and nothing is actual infinity." This, in essence, reversed the notion of reality concept of Plato. Such absurdity doesn't occur in oriental cognition that considers the time function properly. Furthermore, all objects (matter and energy) and events (time) in the world are expressed by the movement of "yin" and "yang." For instance, the moon is yin while the sun is yang; the earth is yin and the heaven (sky) is yang; a woman is yin and a man is yang; the night is yin and the day is yang; the winter is yin and the summer is yang; etc. Yin and yang are opposite, yet work in perfect harmony. This aspect is typical of Qur'anic philosophy as well as everything is reported to be created in pair (the Arabic word is *jaouz*, as in spouse or couple).

The duality in Yin Yang is further highlighted in the eight markings, called trigrams or Pa-kua symbols, are opposites of one another diagonally (Figure 5.43). Broken lines are unbroken and vice versa. The trigram together represents the principle of movement and harmony. Each trigram (hangul: kwae) represents one of the four classical elements, namely, heaven, earth, fire, and water. The word 'hangul' represents heaven, while *kwae* means movement and harmony. Each "Kwae" consists of three bars of divination signs that can be either broken or unbroken bars. A broken bar stands for yin while an unbroken bar stands for yang. Numerous combinations are possible, but four basic elements correspond to heaven, water, earth, and fire.

If one replaces each unbroken line in the trigrams by 0 and each broken line by 1, one can see that the symbols represent numbers in the binary (base two) number system. The symbols and their meanings appear below. The binary numbers are read from bottom to top (Figure 5.44).

The top left corner of the flag is three unbroken lines and represent heaven (0). This is significant because both 'heaven' and 0 signify the origin. Note that 'heaven' is not the one that 'righteous' people end up (in theological sense), this is the heaven that originated the entire creation. In English language, the use of the word 'heaven' is full of ambiguity that comes from Roman Catholic Church's interpretation of origin of universe and men as well as the myths that purport the notion of 'gods and angels' residing in heaven. However, in Arabic language, such ambiguity doesn't occur. The Creator doesn't reside in *Jannah* (literally meaning 'garden' and often translated as 'heaven') as He is not constrained by space or time. He is also the originator of everything. For instance, *Qur'an* (2:117) states: "Originator of the skies and the earth. When He decrees a matter, He only says to it, "Be," and it is." The Arabic word for skies, *Samawah*, is often incorrectly translated as 'heavens', highlighting confusion of the English language.

Equally important is the use of 'zero' to denominate the source of everything. This 'zero' can have two meanings, i.e., nothingness and origin. The nothingness would coincide with ancient Greek word, *Χαοσ* (Chaos). However, this zero ('*sunya*' in Sanskrit and *cipher* in Arabic) never implies nothingness as in void. It rather implies the originator, who originated everything from nothing. This is consistent with Islam as well as notable oriental religions. For instance, in Sanskrit, *Maha*

Sunya (Great Zero) refers to the outer space, which is anything but void as in nothing-ness and most often refers to the Creator. Interestingly, the Arabic word '*cipher*', while recognized as the origin of the word 'zero', does not represent void, it rather refers to 'emptiness'. In ancient Greek culture, however, void refers to the original status of the universe, which was thought to be filled with nothingness, in terms of tangibles, including time. Similarly, the equivalent word in Arabic is *As-sama'a*, which stands for anything above the earth, including seven layers of skies, is not 'heaven', which is *Jannah* (garden) in Arabic and literally means 'garden of paradise'. Qur'an often refers to the Creator as the one 'in the sky' (e.g. *Qur'an* 67:16 says: Do you feel secure that He who is in the sky (*sama'a*) would not cause the earth to swallow you and suddenly it would sway?)

Opposite to 'heaven' is the earth (designated by number Seven), placed at the lower right corner of the flag. This placement as well as the number 7 are both noteworthy and contain deeper meaning. While earth is known to be 'just' a planet in European science, it holds a much deeper meaning *The Qur'an* that defines humans as the Viceroy (*khalifa*) of the Creator (e.g. Chapter 2:30 of *The Qur'an* specifies man's role as the viceroy), charged with law and order on Earth. The Arabic word for 'earth' is *Ardha*, which means 'habitat for humans (the viceroy of the creator)'. This outlook is clearly different from the Eurocentric notions, ranging from the vastly discredited 'original sin' to widely accepted 'evolution' theories (McHenry 2009; Hall, 2008) that detach human conscience from its functioning in a society. Overall, they confused Creators's traits with traits of creation. This confusion is many

Picture 5.5 Natural occurrence of yin yang structure.

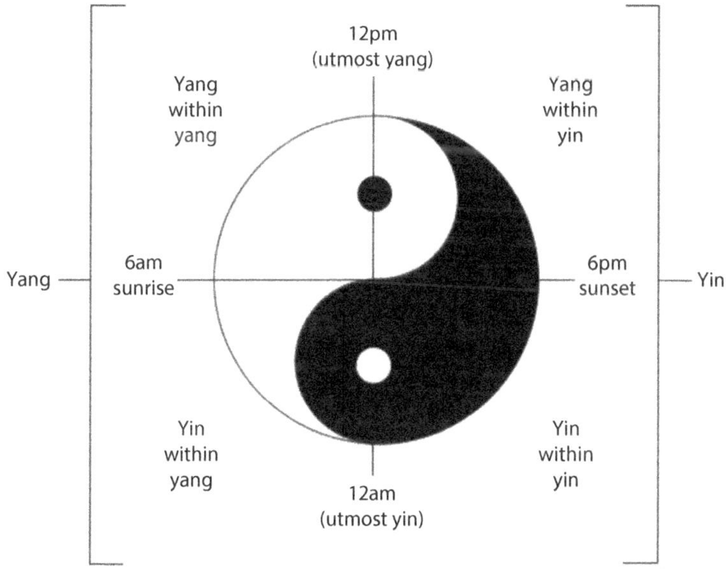

Figure 5.45 Evolution of Yin and Yang with time (from Website 4).

centuries old, as observed from the 'scientific' work of Thomas Aquinas. This immediately cut off the relationship between Creator and Creation. This involved confusion in understanding what is natural – a confusion that continues until today. Then they confused creations (other than humans) with humans due to lack of logical premise defining the purpose of humanity. For them, humans are just another set of animals and the earth is just another planet. This latter confusion cut off conscience (and ownership of intention) from humanity. The ground became fertile for onset of various forms of aphenomenal cognition, some of which called themselves Naturalism, Agnosticism, Secularism, Atheism, etc.

Water (number Five) is placed at the right top corner. The existence of water as a fundamental element is important. In every culture, water is synonymous with life and liveliness. Qur'an places the existence of water before anything else. Opposite to water is fire (number Two) at the lower left corner. The role of fire is opposite to water, yet it is essential to life. This life forming fire comes from carbon, another essential, but often overlooked, component of life. Without fire of carbon, there is no carbon dioxide, the essence of plant, and therefore, life. Fire represents transition from cold to hot, from life to death, from tangible (water or liquid) to intangible (vapor or gas). This phase change is

Table 5.3 The tangible and intangible nature of yin and yang (from Islam, 2014).

Yin	Yang
tangible	intangible
Produces form	Produces energy
Grows	Generates
Substantial	Non-Substantial
Matter	Energy
Contraction	Expansion
Descending	Ascending
Below	Above
Water	Fire

Picture 5.6 Depiction of Absolute Time, time and Yin Yang in nature.

typical of creation. In fact, the very fact that everything is moving (a function of time) makes it essential to go through this phase of tangible and intangible. Overall, this continues in an eternal circle. Picture 5.5 shows how it is natural to have such dual characteristic in any object. It is also important to note that these two components are often opposite but complementary. In most cases, one of them represents tangible aspect whereas the other represents the intangible aspect.

The next aspect of yin yang is the existence of such transition in everything at all scales. Figure 5.45 shows how a 24-hour clock of yin yang allows continuous transition as a cycle of life.

Table 5.3 shows the tangible and intangible nature of Yin Yang. Yin Yang show contrast as well as interdependence. For instance, no matter is produced without energy and no energy is produced without matter. Water is needed for plant, which is then needed for fire. This logic also shows nothing is real unless it is part of the positive negative cycle. For instance, fire without water isn't real. That would explain why diamond cannot be set on fire even though it is made out of carbon. Similarly, the presence of mass would indicate the presence of energy. This would make the existence of zero energy and infinite mass an absurd concept, even though new cosmic physicists routinely tout that notion (Krauss, 2012).

Figure 5.45 also shows how Yin and Yang encircle each other alternating as a continuous function of time. As time progresses, yin becomes yang and vice versa. This progression confirms the existence of characteristic time function for every object at every scale. Picture 5.6 shows the depiction of Yin Yang with relation to a mother. The mother here is represented by Time (as in time function) whereas TIME itself is surrounded by Absolute Time (*Dhahr* in Arabic), which is considered to be the trait of the creator in ancient Indian, Greek, as well as Qu'ranic traditions. This mother is significant as in ancient Hindu culture, the supreme God is symbolized by 'mother'. In Qur'anic narration, creator's first two traits literally mean 'womb that is infinitely continuous in space' and womb that is infinitely continuous in time. The kittens here represent yin and yang, while the mother forms a yang yin with the father of the kitten. The father here remains intangible whereas the mother cat is tangible. Absolute Time itself forms a yin yang within the same external object, i.e., the creator, whose other trait has been known to be Absolute Light (*Noor* in Arabic) since Ancient Greek time. Similarity within creation exists through matter (tangible) and energy (intangible). While the existence of these two components of nature is not controversial, New science has disconnected matter from energy by assigning zero mass to photons. The logic that without mass there cannot be any energy makes it clear that such disconnection is unwarranted. In addition, the notion of anisotropy should be understood in each of these relationships. For instance, Time is a function of Absolute

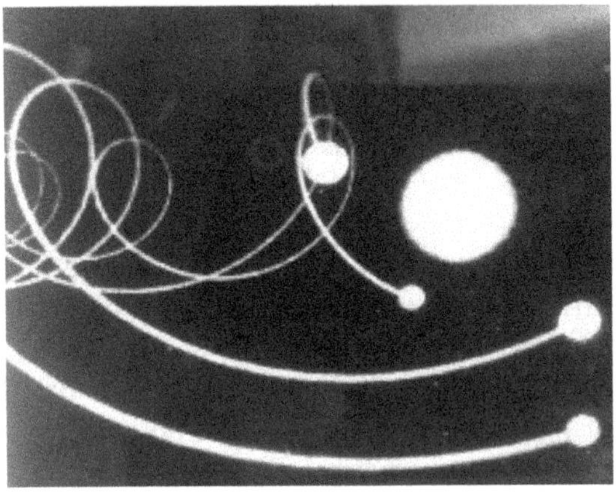

Figure 5.46 Sun, earth, and moon move at a characteristic speed in infinite directions.

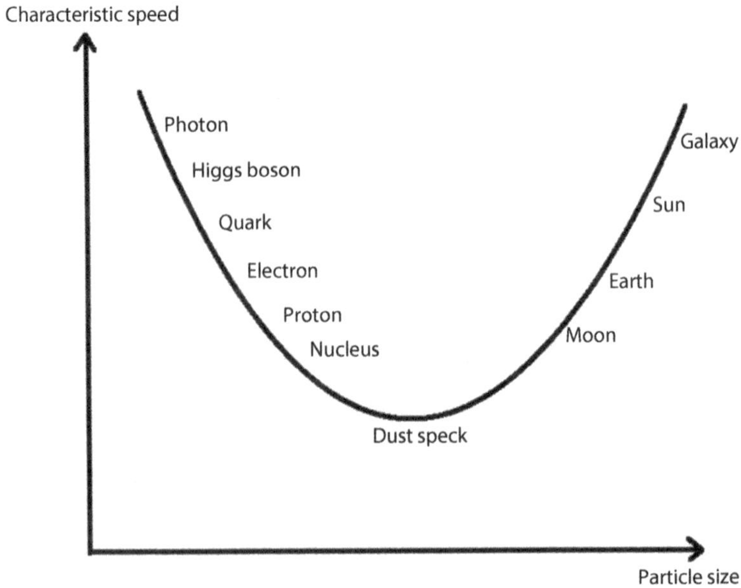

Figure 5.47 Orbital speed *vs* size (not to scale).

time even but Absolute time is free from any dependence on time. In Picture 5.6, this fact is symbolized by the mother cat, whose movement is confined by the wooden structure surrounding it but the mother cat has no influence on the wooden structure. Similarly, the mother cat controls the kittens and restricts their movement, whereas the kittens have no control over the mother cat. Finally, the role of intangible must be understood. How Absolute Time affects or is affected by Absolute Light is unknown to us. Thankfully, it is not necessary to have that knowledge in order to characterize matter and time on earth. However, the role of intangible within the realm of kittens is manifested through the mother cat (tangible) and father cat (intangible). The father cat doesn't affect mother cat's persona but affects the nature of kittens. There is no reversibility nor there is any real symmetry. This would explain the absence of real symmetry and the presence of uni-directionality in nature. It would also explain why everything in nature is transient and unique function of Absolute time, including Time itself. It is, therefore, expected that every phenomenal object would have a characteristic time function, often identified as frequency, which varies with time. In demonstrating this notion of characteristic frequency which is also a variable, our solar system offers an excellent example.

In the solar system, the moon is moving around its own axis, then around the earth, while keeping pace with the earth that is orbiting around the sun and keeping pace with the sun that is moving both around it's own axis as well as around an unknown object. Figure 5.46 shows a snap shot of the solar system. In this system, the Earth, the moon and the sun all are moving in many directions, but with variable/non-uniform motion. In a short span, the frequency may appear to be fixed or constant, but in reasonably larger time span, they vary. This is true for every object, including humans and human body parts (Islam *et al.*, 2014). It is reasonable to assume that such dependence of size on orbital speed reverses for invisible elements. Orbital speed of various known objects are plotted in Figure 5.47 as a function of size, along with a reverse relationship for invisible particles. If similar model is followed for invisible structure, smaller than dust speck, the following figure emerges. In this figure, dust speck (*Dharra* in Arabic) is identified as the smallest object This is in line with the Avalanche theory, recently advanced by Khan *et al.* (2008) and Khan and Islam (2012). From there, arises natural characteristic speed in the following scale.

In Figure 5.46, a dust speck represents reversal of speed vs. size trend. For so-called subatomic particles, speed increases as the size decreases. Higgs boson is assigned a smaller value than quark but larger value than photon. This is done deliberately in order to float the notion that fundamental particle and finality in determining such particle is a spurious concept. Note that the actual speed in absolute sense is infinity for smallest particle. It's because each element has a speed in every dimension. This dimensionality is not restricted to Cartesian coordinate. As the number of dimension goes up, so does the absolute speed,

Table 5.4 Characteristic frequency of "natural" objects.

Object	Nature of speed	Average speed sun	Comment
Sun	Orbital	240 km/s	Around unknown object, that's 2.55 10^{20}m away; estimated orbital time 200 million
	Drfit	19 km/s	Due to variation in galactic rotation
	Spinning		Unlear
Earth	Escape	240 km/s	to match with the orbital speed of the sun
	Orbital	30 km/s	Around the sun
	Spinning	0.44 km/s	At equator
Moon	Broad escape	240 km/s	to keep up with the sun
	Escape	30 km/s	To keep up with the earth
	Orbital	1 km/s	
	Spinning	12 km/s	To keep the same face exposed to one side
Atom, radius 10^{-9} m		Unknown	Rigid ball assumption
Electron, 10^{-15} m		2,200 km/s	under non-excited conditions (Bohr model, uniform speed assumption)
Proton, 3. 10^{-15} m		Unknown	rigid ball assumption
Quark		Unknown	Non-measurable dimension
Photon		300,000 km/s	rigid ball assumption
Higgs-Boson		300,000 km/s	rigid ball assumption

approaching infinity while projected in absolute scale. The characteristic speed also increases as the size of the entity goes down. For infinitely small entity, the speed would approach infinity. This analysis shows how both small and large scales are in harmony with infinitude, associated with 'void'. In pre-Thomas Aquinas period, such 'void' was synonymous with the creator within whom all the creation was believed to be embedded. Table 5.4 shows some of the characteristic speeds (and thus, frequencies) of various particles. Note that these characteristic speeds are all a function of time.

Furthermore, Figure 5.46 shows that there is no object in steady state. It is also true that there is no object at uniform motion. This arises from the original premise that time itself is a variable. As a consequence, characteristic speed of each object changes with time. Such characteristic time exists for every object. There is a quantum change characteristic features during phase transfer or when a life begins (from non-organic to organic) or ceases for an individual (from organic to non-organic). In this process, life and death are triggers or bifurcation points as associated time functions change drastically. It should be noted that such transition is subjective and death of one entity only means death for that prticular object. Here, it is the time function, $f(t)$ that defines the pathway of any entity, within the universal order. This time is not arbitrary and is tightly controlled by the external entity, the Absolute Time, as presented latter in this paper. In Figure 5.46, dust specks represent the most objects closest to stable and steady state. All ancient cultures, culminated in *The Qur'an*, consider that humans are created from clay or dust specks. As shown earlier, earth or clay is an integral part of organic systems that constitute the habitat for humans. Following is a list of some of the characteristic time as related to humans:

Earth: day and night, year, pace with the sun

Humans: blink of an eye, sunrise, mid-day, sunset, sleep and wake, week, month, menstrual cycle; 40 days, 40 years

Society: 40 years, centuries, millennia

Geology: millennia

Cosmos: billion years

The heart rate is typical of natural frequency of humans. Even though, heart rate is frequently talked about in the context of both physical and psychological conditions, brain waves are also characteristics of human activities (Figure 5.47). Change in brain waves is evident during sleep, alertness, meditation, etc. Little is available how such frequencies can affect overall human conditions, whereas most focus has been on how to alter natural frequencies. What makes it complicated is scientists have little knowledge of how they naturally vary with time as a person ages. Clearly, humans are not in control of their brain waves, thereby consolidating the theory that humans are integral part of the universal order and their degree of freedom lies only within their intention.

Also, heart beats themselves change over time. As a sign of every characteristic frequency itself being a function of time, the graph in Figure 5.48 is produced. In this, the data on puberty and older people are extrapolated from published papers (e.g. Larson *et al.*, 2013). Naturally, children are more dynamic and their body parts are renewed faster. This would necessitate faster replenishment of energy. The idea is to nurture natural frequencies, rather than fighting them. New science does the opposite and every 'treatment' is aimed at altering natural frequency, thereby countering forces of nature.

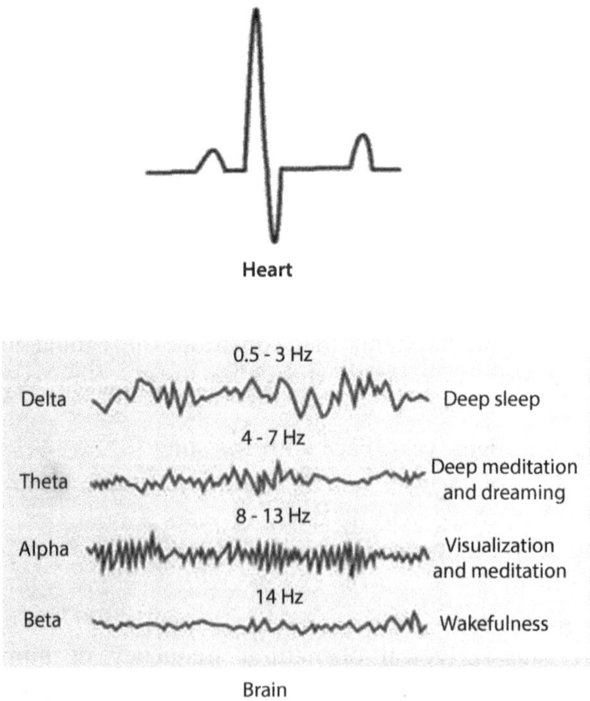

Figure 5.48 The heart beat (picture above) represents natural frequency of a human, whereas brain waves represent how a human is in harmony with the rest of the universe.

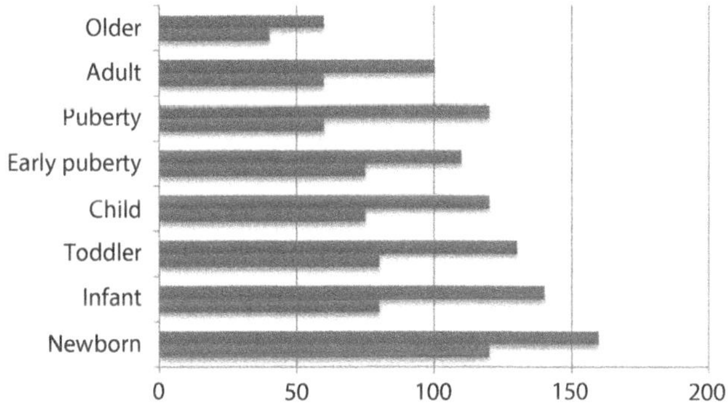

Figure 5.49 Maximum and minimum heart rate for different age groups.

Figure 5.50 Tangible/intangible duality continues infinitely for mega-scale to nanoscale, from infinitely large to infinitely small.

There is what is characteristic but there is also fundamental. Hard to believe/accept, for example, that other natural frequencies in one's body are unrelated to heartbeat frequency. Thus, it would be difficult to believe that an individual's brainwave frequency, for example, could be entirely accounted for from investigation of phenomena occurring within the cerebral cortex alone by itself. Our theory is: we have no control over these frequencies. What we have control over is our intention. Thankfully that doesn't affect the universal order but doubly thankfully does affect an individual's long-term future. This connection of human intention with long-term future as well as disconnection of the universal order from human intervention is a notion that has been absent in every scientific cognition of post-Roman Catholic church Europe.

The exchange between two dual and opposite objects continues and it is reasonable to assume that there is no distinction between energy particles and mass particles. The circle in the broader portion of the yin yang represents the same shape that would house another yin yang, which itself will house another set. Such trend continues until we reach a state that can be characterized as interface between Absolute light (or Absolute time), which is the external element. This is consistent with pre-Newtonian narration as well as Qur'anic narration of divine traits. Figure 5.49 shows the depiction of such infinitude. Figure 5.49 shows how 'pure light' that surrounds both tangible and intangible surrounds everything at all times. Such is the connection between time, the creation and the Absolute time, and 'radiative light' and pure light. Both absolute time and pure light represent infinity.

Furthermore, the notion of male contributing to the life of female and female in turn giving birth to male becomes an integral part of life cycle of humans. With this cyclic balance, there is no superiority of any particular entity over other as long as they belong to the group of creation in general. In addition, every object is in perfect harmony with nature, except humans that are non-ideal.

This 'non-ideal' feature has been a crucial matter of contention in European history, ever since the concept of original sin was introduced. If that premise is removed then the premise that everyone is born perfect is consistent with the premise that Nature is perfect. European dogma science defined Jesus Christ as the perfect man (a role model), but that is not consistent with his dual status as 'son of god'. In addition, very little is known about this 'role model'. In fact, modern scientists doubt he ever existed in the role of a messiah. European modern science doesn't define perfect human, abandoning the debate as 'religious mumbo jumbo'. In addition, it doesn't define good behavior or any purpose for humans, other than maximizing pleasure and minimizing pain. This is in contrast to Christian dogma but equally problematic as it gives rise to chaotic rules, akin to a roller coaster ride (Islam *et al.*, 2013).

In summary, both ancient oriental and Greek philosophers support the notion of separation of creator (external entity) from creation (internal entity), each of whom have completely different traits. The creation

Picture 5.7 Pure light contains within itself both tangible and intangible.

itself is further divided into tangible and intangible sets with opposite traits, each having continuously decreasing size and similar asymmetry (or duality). *The Qur'an* (that was compiled in mid 7th century) distinguishes Absolute light (or pure light, *Noor* in Arabic), which is the Creator's trait (in line with being omnipresent, Absolute Guide, Absolute womb, etc.) from radiative light (*adha'a*), which is associated to the movement of any entity. Only pure light (PL) is continuous in space and time. Everything else is embedded in PL.

This 'light' is not a collection of photons or any particle. It is pure energy and has no mass associated with it. The conventional presentation of light as a collection of photons each of zero mass fails to account for the fact that the sun is losing thousands of tons of mass every second. With *The Qur'an*ic denomination of pure light, as distinct from radiative or reflective light, there is no contradiction in transition from mass to energy. This distinction between pure light (PL) and radiating or reflective light (RRL) is necessary but no sufficient condition to truly scientific description of mass and energy. Sequential functionality must be described as a sufficient condition. This requires proper characterization of mass and energy. This depiction is supported by Greek philosophers as well as Augustine of Roman Catholic Church.

The Qur'an uses the similitude of a womb in describing how Creator sustains the entire creation. The relationship here is entirely non-symmetrical and non-reciprocal. A fetus is in need of the womb and cannot

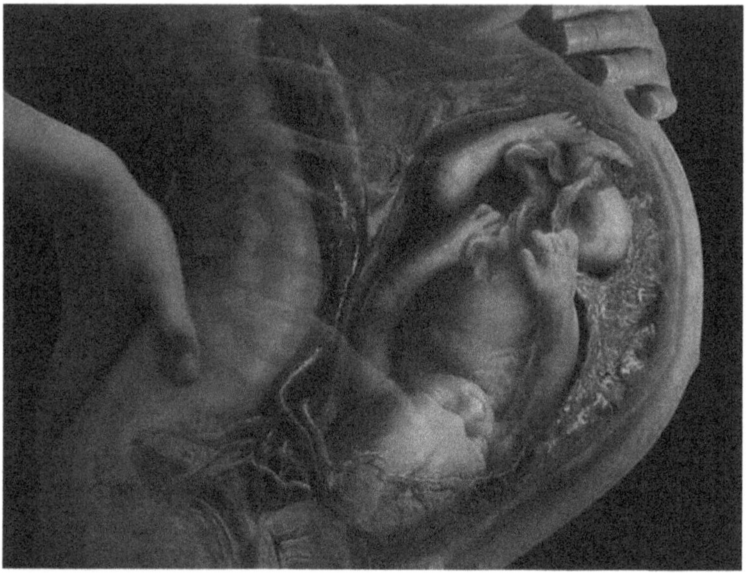

Picture 5.8 Closest analogy of pure light is the womb that surrounds the fetus and nourishes it.

exist without the protection and nourishment of the womb for the entire duration of the sojourn. In this case, the material existence of the fetus is the tangible and time is the intangible, whereas spatial existence of the womb is the tangible and temporal protection thereof is the intangible feature of the womb. This is reflected by the terms *Ar-Rahman* and *Ar-Raheem*, the first and second traits of the creator mentioned in *The Qur'an*.

The universe is the physical presence of the fetus, whereas universal order is the life span of the fetus. Just like the fetus goes through stages and moves from a drop, then to a blood clot, then to full grown fetus, ready to enter a new phase of life cycle.

5.5.2 Departure From Logical Thinking and the Onset of Dogma in New Science

Even though astronomers and alchemists and scholars of many other disciplines were active experimental scientists in other parts of the world for millennia, experimental science in continental Europe began only in the seventeenth century. Sir Francis Bacon (1561–1626) emphasized that experiments should be planned and the results carefully recorded so they could be repeated and verified. Again, there was no recognition of time as a dependent variable and a continuous function.

The work of Sir Isaac Newton (1643-1717) has exercised — and continues to exercise —the profoundest impact on European science and technology. Historically, what the Thomas Aquinas model did to European philosophy is similar to what Newton's model did to New Science (various aspects of Newton's laws of motion, gravity and light propagation are reviewed further in Chapter 6). Here, it suffices to indicate that Newton's laws suffered from the lack of a real — material, verifiable — first premise. With the exception of Einstein, every scientist took Newton's model as the ideal and developed new models based on the same, adding only factors thought to be relevant because experimental data were not matching with theoretical ones. Einstein, who was credited to have reinstated the time-concept of the ancient Greek era, made his own sets of assumptions that would be deconstructed in a latter section.

Boyle (1627-1691), an experimentalist, recognized the existence of constant motion in gas particles (*corpuscles*, in his word) - the same idea that Heraclitus proposed over 2000 years before Boyle (the idea that was rejected by Aristotle and subsequent followers). While this recognition was in conformance with natural traits of matter, his belief that the particles are: 1) in constant motion; and 2) uniform and rigid is in stark contradiction to real nature of matter. This fundamentally incorrect notion of matter continues to dominate kinetic molecular theory.

In the last half of the 18th century, John Dalton (1766-1844) reasserted the atomic view of matter, albeit now stripped of Heraclitus' metaphysical discussion and explanations. Newton's laws of motion dominated the scientific discourse of his day, so Dalton rationalized this modernized atomic view with Newton's object masses and, and we end up with matter composed of atoms rendered as spherical balls in three-dimensional space, continuously in motion throughout three-dimensional space - within time considered as an independent variable. This line of research seals any hope for incorporating time as a continuous function, which would effectively make the process infinite-dimensional.

The essential observations of Dalton are:

Elements are composed of *atoms* (themselves being unbreakable)

All atoms of a given element have identical properties, and those properties differ from those of other elements.

Compounds are formed when atoms of different elements combine with one another in small whole numbers (this one emerges from previous assumption that atoms are unbreakable)

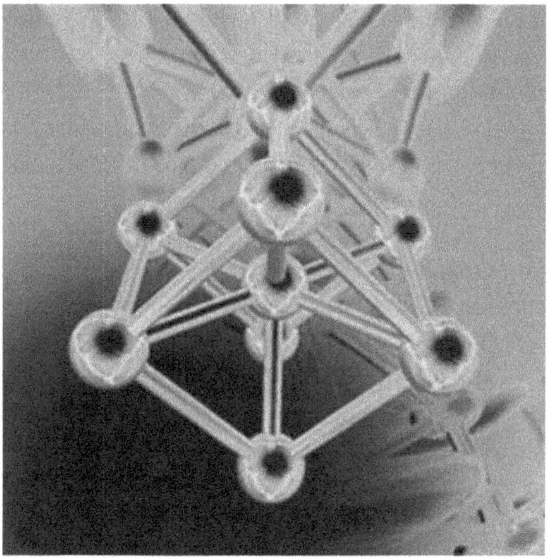

Figure 5.51 Dalton's atomic model.

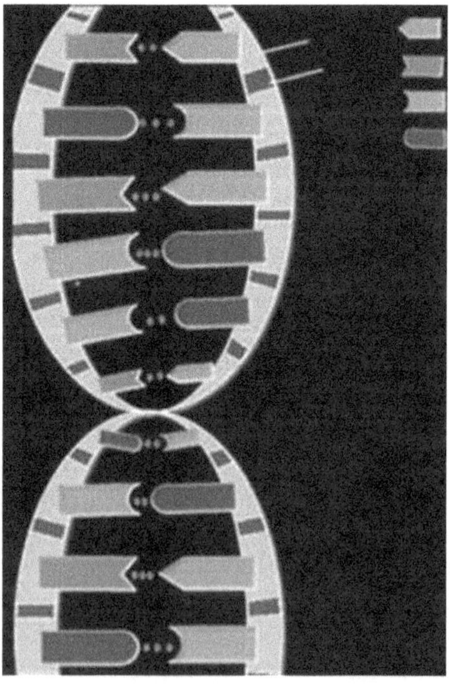

Figure 5.52 Symmetry and uniformity continue to be the main trait of today's scientific models (DNA model).

The relative numbers and kinds of atoms are constant in a given compound (this one asserts steady state, in contrast to notion of kinetic models).

Figure 5.51 shows Dalton's depiction of molecular structure. Note that in this figure, No. 28 denotes carbon dioxide. This depiction is fundamentally flawed because all four premises listed above are aphenomenal. This representation recapitulates Aristotle's depiction of matter and energy, as unique functions of composition and devoid of the time function.

Consider Figure 5.52 depicting organic molecules, such as DNA. The fundamentally flawed 'atomic' model of matter is used for every type of characterization of material. This became the basis for subsequent developments in chemical engineering. In later European-based studies, research into physical-chemical properties of matter was distinctly separate from research into energy and light. Even though Newton put forward theories for both, subsequent research went down different tracks focused uniquely on chemistry, or physics, or astronomy, or other branches of New Science.

Contrast the above figures with something that occurs naturally. Picture 4 shows naturally occurring patterns, ranging from tiny human fetus to constellation system. There is a broad symmetry but nowhere one can find local symmetry as envisioned in New Science.

The law of conservation of mass was known to be true for thousands of years. In 450 B.C., Anaxagoras said, "Wrongly do the Greeks suppose that aught begins or ceases to be; for nothing comes into being or is destroyed; but all is an aggregation or secretion of pre-existing things; so that all becoming might more correctly be called becoming mixed, and all corruption, becoming separate." When, Muslim scholars translated this work from old Greek to Arabic, they had no problem with this statement of Anaxagoras. In fact, they were inspired by verses from *The Qur'an* that clearly defined that the Universe was created out of nothing and ever since its creation all has been a matter of phase transition as no new matter or energy is created. In addition to this recognition, Islamic era scholars also knew about natural and artificial products. For creating new material or life form from another, the word used is '*khalaqa*', creator being the only good designer. Qur'an opines a true creator cannot be the same as the one that cannot create (humans). It is understood manmade products will become inherently flawed, thereby harming the environmental integrity. Picture 5.8 makes this point.

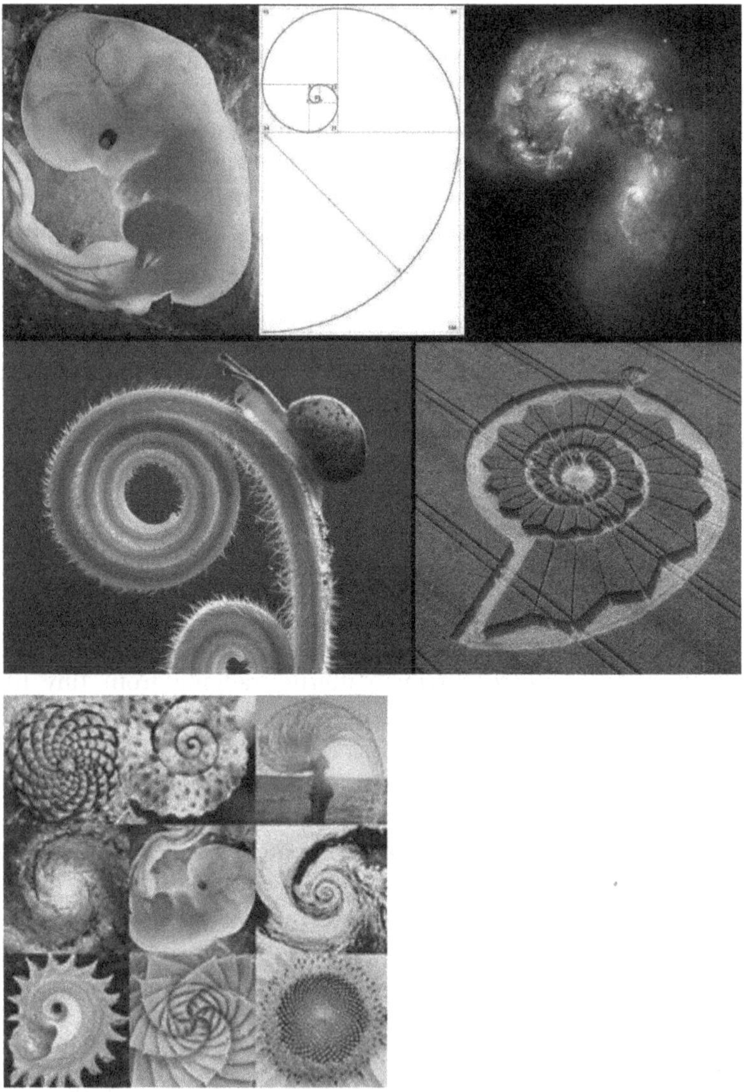

Picture 5.9 Naturally occurring patterns and symmetry.

History supports the notion that harmfulness of artificial was well known or previous civilizations didn't attempt to produce artificial products. Figure 5.53 shows how important it is to distinguish between artificial process and natural process.

The superflux of artificial started with Democritus' model that was first accepted by Aristotle and later glamourized by scholars affiliated

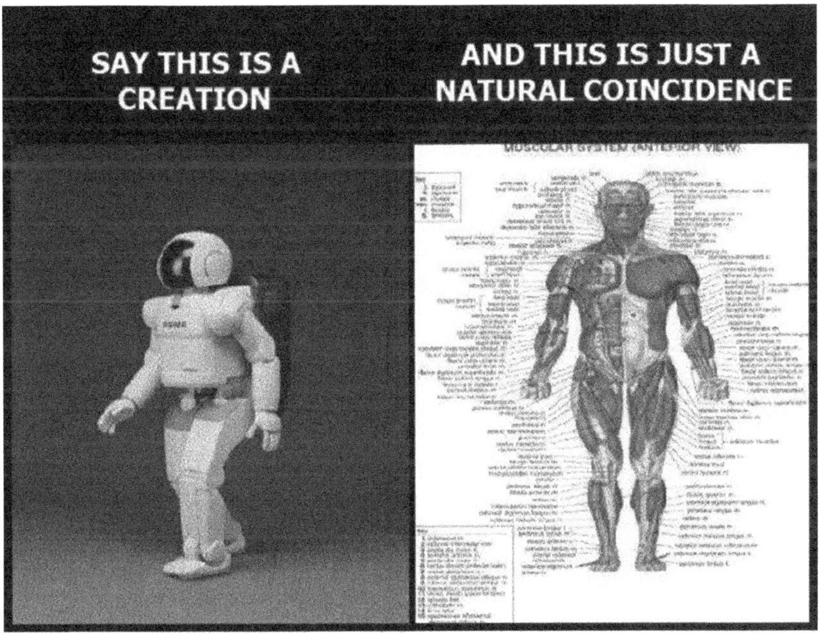

Picture 5.10 Throughout the ancient and medieval periods, there was no confusion about the opposite nature of organic and mechanical objects.

with the Roman Catholic church. Even though, New science claims that it has broken out of dogmatic cognition, in reality, every theory in New science emerges from aphenomenal premise, much like Atomism or dogma.

Even though for 1000 years of extensive research in all disciplines of science and engineering, all based on conservation of mass, was in front of the Eurocentric culture, modern scientific literature credits Antoine Laurent Lavoisier (1743-94) with discovering the law of conservation of mass. Lavoisier's first premise was "mass cannot be created or destroyed". This assumption does not violate any features of Nature. However, his famous experiment had some assumptions embedded in it. When he conducted his experiments, he assumed that the container is sealed perfectly. This would violate the fundamental tenet of nature that an isolated chamber can be created. Rather than recognizing the aphenomenality of the assumption that a perfect seal can be created, Lavoisier 'verified' his first premise (law of conservation of mass) '*within experimental error*'. The error is not in the experiment, which remains real (hence, true) at all times. Instead, within the first premise, a perfect seal has been created. By avoiding confronting this premise, and by

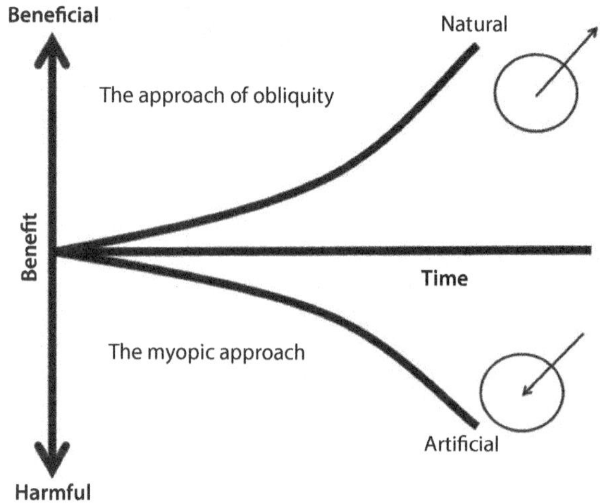

Figure 5.53 Historically, natural objects were synonymous with sustainability (from Khan and Islam, 2012).

introducing a different criterion (i.e.,' experimental error') — a criterion that is aphenomenal and hence unverifiable — Lavoisier invoked a European prejudice linked to the pragmatic approach, which is that 'whatever works is true'. This leads to the linking of measurement error to the outcome. What could Lavoisier have done with the knowledge of his time to link this to scientific reality? For instance, if he left some room for leak from the container, modern day air conditioner design would have room for how much Freon is leaked to the atmosphere.

Lavoisier nevertheless faced considerable resistance from scientists who were still firm believers in the phlogiston theory (In Greek, *phlogios* means 'fiery'). This theory was first promoted by a German physician, alchemist, adventurer, and a professor of Medicine Johann Joachim Becher (1635 - 1682). This theory recognizes matter named phlogiston existed within combustible bodies. When burnt (i.e., with addition of thermal energy), this matter was thought to have been released to achieve its 'true' state. This theory enjoyed support of the mainstream European scientists for nearly 100 years. One of the proponents of this theory was Robert Boyle, the scientist who would gain fame for relating pressure with volume of gas. Mikhail Vasilyevich Lomonosov (1711-1765) was a Russian scientist, writer and polymath who made important contributions to literature, education, and science. He wrote in his diary: "Today I made an experiment in hermetic glass vessels in order to determine whether the mass of metals increases from the action of pure heat. The experiment demonstrated that the famous Robert Boyle was deluded, for without access of air from outside, the mass of the burnt metal remains the same."

Ever since the work of Lavoisier, the steady state model of mass balance has been employed in all segments of chemistry and chemical engineering. These works focused on defining symbols and identifying new elements and classifying them. Current chemical symbols (formulas) are derived from the suggestions of Jöns Berzelius (1779–1848). He took oxygen to be the standard reference for atomic mass (O = 16.00 AMU). In contrast to Dalton's assertion that water had a formula of HO, Berzelius showed it to be H_2O. For Dalton, all atoms had a valence of one. This made the atomic mass of Oxygen to be 8. While this characterization yielded tractable result, it introduced inherent bias that haunts the scientists even today. It assumed that water is substance that is created by synthesizing oxygen and hydrogen. At no time does one find such a process anywhere in nature. At best this characterization made it impossible to discern between naturally processed water and artificially 'purified' water, even though they behave completely differently (see Figure 5.53).

The consideration of mass as independent of time forced all chemical models to be steady state or non-dynamic. More importantly, this model was embedded to the definition of time, coupling mass and energy in an intricate fashion that obscured the reality even from experts. See the following analysis in the context of time unit.

5.5.3 Standardization of an Aphenomenon

The need for standard is pivotal to any cognition. However, any flawed or non-existent standard would make the subsequent cognition 100% false. Atomism in ancient Greece was the beginning of aphenomenal standard of mass. The dogma culture introduced the notion of aphenomenal psychological standard, in name of 'Jesus the savior, who created humans with original sins, then impregnated a human so he could be born to sacrifice himself to himself in order to wipe the sins that he originally gave'. This became the beginning of cognitive degradation that was taken to a new low during Newton era, all the way down to Einstein era of quantum 'everything'. Newton connected mass with distance and time through his 'laws' of motion. This was in turn connected to power or energy by Lord Kelvin and into electromagnetism by Maxwell. Einstein took Maxwell model and introduced a series of assumptions to develop his theory of relativity. Before Einstein, however, massive efforts were made in standardizing Newtonian narration of distance, mass and time. This effort was to mathematize everything so anything can be measured in terms of tangibles and then in economic terms.

In an attempt to standardize distance as part of the universal measurement unit, France in the 1770's defined the meter as follows: one meter = 1/10,000,000 the distance from the North Pole to the Equator (going through Paris). They also discussed the spurious arrangement of introducing the unit of time as a second. In 1832, the concept of 'second' was attached to the SI arrangement. The original definition was 1 second = 1 mean solar day/864,000. As late as 1960 the ephemeris second, defined as a fraction of the tropical year, officially became part of the new SI system. It was soon recognized that both mean solar day and mean tropical year vary, albeit slightly, more 'precise' (apparent assertion being more precise means closer to the truth) unit was introduced in 1967. It was defined as 9,192,631,770 cycles of the vibration of the cesium 133 atom.

The assumption here is that vibration of cesium 133 atom is exact, this assumption being the basis of Atomic clock. Only recently, it has been revealed that this assumption is not correct, creating an added source of error in the entire evaluation of the speed of light. On the other hand, if purely scientific approach is taken, one would realize that the true speed of light is neither constant nor the highest achievable speed. Clayton and Moffat (1999) discussed the phenomenon of variable light speed. Also, Schewe and Stein (1999) discussed the possibility of very low speed of light. In 1998, the research group of Lene Hau showed that the speed of light can be brought down to as low as 61 km/hour (17 m/s) by manipulating the energy level of the medium (Hau *et al.*, 1999). Two years later, the same research group reported near halting of light (Liu *et al.*, 2001). The work of Bajcsy *et al.* (2003) falls under the same category except that they identified the tiny mirror-like behavior of the media, rather than simply low energy level.

More recent work on the subject deals with controlling light rather than observing its natural behavior (Ginsberg *et al.*, 2007). Khan *et al.* (2008) used the arguments provided by previous physicists and constructed the following graph. It is clear from the graph (Figure 5.53) that the assumption that 'speed of light', 'vacuum', and 'unit of time', 'unit of distance' are some arbitrarily set constants do not change the true nature of nature, which remains continuously dynamic. This was evident to Heraclitus over two and half millennia ago. Note that media density can be converted into media energy only when continuous transition between energy and mass is considered. This transition was even known to Democritus and accepted by Aristotle. Such transition, however, is rarely talked about in the context of engineering (Khan *et al.*, 2007). This graph also reveals that that once definitions and assertions have been accepted at face value and no longer subject to further scrutiny, the possibility of increasing knowledge (as in being closer to discovering the truth about nature) is diminished. Finally, this graph confirms that Aristotle's notion of infinite speed that was rejected by Arab scientists/philosophers was applicable only if the media density is zero – a scientifically absurd condition because it would mean the space is void, as in no matter present whatsoever. This is the state that Ancient Greeks accepted as the condition prior to the creation of the Universe. It is probable that the speed of light would be infinity in void, but the presence of light would simply mean the void is now filled with matter, unless the assumption is light carries no matter with it – yet another absurdity.

A more plausible explanation of the fact that infinite speed was accepted in void (or finite speed accepted for an arbitrary speed in 'vacuum') is in conflation between traits of creator and creation. To Ibn Haitham, this conflation would mean a mistake was made in terms of trait of creation and trait of creator (Khan and Islam, 2012). He indeed wrote a book deconstructing Aristotle's flawed premises. However, during dogma era this conflation was re-invoked and later picked up by New scientists that accepted it without contemplating the illogical consequences of the assumptions.

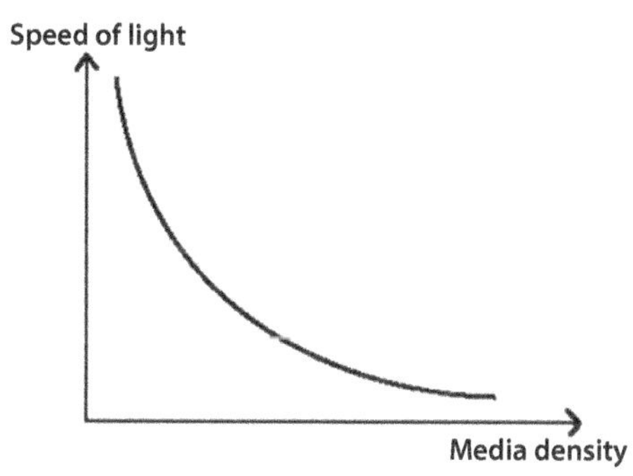

Figure 5.54 Speed of light as a function of media density (redrawn from Zatzman and Islam, 2007).

In 2001, Eric Cornell, Wolfgang Ketterle, and Carl Wieman were awarded Nobel Prize for "for the achievement of Bose-Einstein condensation in dilute gases of alkali atoms, and for early fundamental studies of the properties of the condensates". Their work, however, far from vindicating science of tangibles, has in fact made it more necessary to take an alternate approach. Eric Cornell's most popular invited lecture is titled: Stone Cold Science: Things Get Weird Around Absolute Zero. This 'weird'-ness cannot be predicted with the science of tangibles. One further citation of Eric Cornell is: "What was God thinking? Science can't tell" (Cornell, 2005). Far from being content that all discoveries have been made and, therefore, there is nothing more to discover, a true researcher in pursuit of knowledge readily picks up on these shortcomings of conventional thinking, and as history frequently shows, it is within the inquisitive mind that answers reside.

Ironically, this would also mean, discoveries are made because one is not afraid to make mistakes and, more importantly, theorize without fear of the Establishment. Satyendranath Bose, who did not have a PhD, himself did his part of the mistake. In fact, as the following quote from Wikipedia shows (Website 5), his findings were initially discarded by mainstream journals and he had to resort to sending Einstein his original work. It was only then that Einstein introduced the idea from light to mass (photon to atom) and the paper was finally published. If Bose did not have the tenacity and if Einstein did not have the decency to consider thinking "outside the box", one could not begin to think what would be the state of laser and all the gadgets that we take for granted today.

"Physics journals refused to publish Bose's paper. It was their contention that he had presented to them a simple mistake, and Bose's findings were ignored. Discouraged, he wrote to Albert Einstein, who immediately agreed with him. His theory finally achieved respect when Einstein sent his own paper in support of Bose's to *Zeitschrift für Physik*, asking that they be published together. This was done in 1924." (from Website 5)

In the core of any scientific research, the attitude must be such that there is no blind faith or automatic acceptance of an existing principle. In this particular case of Nobel-Prize- winning work of Eric Cornell, he would not be able to observe anomalies if he took Bose-Einstein theory as absolute truth (Tung *et al.*, 2006). In fact, all experimental observations indicate, there have to be major adjustments made to, if not a complete re-formulation of, the Bose-Einstein theory, as evidenced by follow up research that won Nobel Prize in 2005 (discussed below).

As recently as 2005, Nobel Prize in Physics was awarded jointly to Roy Glauber 'for his contribution to the quantum theory of optical coherence' and to John Hall and Theodor Hänsch 'for their contributions to the development of laser-based precision spectroscopy, including the optical frequency comb technique'. These discoveries have something to do with Bose-Einstein theory, but far from being a regurgitation or redemption of the theory, they open up opportunities for new discoveries.

Even after the above series of Nobel Prize quality research, several questions remain unanswered.

For example: Bose envisioned dissimilar characteristics of any two photons and Einstein extended that work to study with what degree of freedom two dissimilar atoms would behave within a balanced system. The work of the Nobel Physics laureate discovered the separation of certain atoms under very cold conditions, yet they concluded "all atoms are absolutely identical". The reason behind this conclusion? The difference between atoms could not be measured!

Fleischhauer (2007) began to question this assumption by comparing millimetre distance in a lab in quantum scale with the distance between the moon and the earth in mega scale and labels 'indistinguishable' as applicable to 'far apart'. However, soon (in the same article) he yields to the pragmatic approach and concludes: "But it also shows that we are entering a state of unprecedented experimental control of coherent light and matter waves. That could bring very real technological benefits: applications that spring to mind include quantum information interfaces that allow the transfer of a quantum bit encoded in a photon to a single atom, as well as ultra-sensitive rotation sensors and gravity detectors".

One must wonder: what would happen if this pragmatic reasoning was eliminated?

With the above conclusion, how can we observe asymmetry in experimental observations (as reported by Tung *et al.*, 2006)? It must be noted that perfect symmetry is aphenomenal: Nature does not have any example of perfect symmetry. Such asymmetry is rarely taken in account in conventional theory (Dorogovtsev *et al.*, 2000).

Later work of 2005, comes up with similar conclusion about photons, with tremendous implications on light source development and monitoring (e.g. GPS). How would this explain the difference in various light sources, particularly between natural light and artificial light?

For both sets of discoveries, the focus is on 'engineering' phenomena rather than observing. What are the possible implications of this 'engineering' on quality of energy source or matter, *e.g.*, nano materials that are being engineered?

What do these theories say about how the transition between mass and energy takes place? Satyendra Nath Bose wrote (Website 5) the real gas equation of state some 100 years ago, what can be said today about real fluids that are still modeled with aphenomenal (unreal) first premises?

What does "variable speed of light" mean in terms of the definition of the unit of time? What would a paradigm shift in standards and technology development entail if one were to consider truly natural unit of time (Zatzman, 2008). Should the shift be to more 'engineering' or more natural?

The notion of fixed speed of light and furthermore assigning that speed as the limit of 'achievable' speed has serious consequences fop rope conceptualizing of energy. Einstein's most famous theory is directly involved with conservation of mass. He derived $E = mc^2$ using the first premise of Planck (1901). Einstein's formulation was the first attempt by European scientists to connect energy with mass. However, in addition to the aphenomenal premises of Planck, this famous equation has its own premises that are aphenomenal. However, this equation remains popular and considered to be useful (in pragmatic sense) for a range of applications, including nuclear energy. For instance, it is quickly deduced from this equation that 100 kJ is equal to approximately 10^{-9} (i.e., 0.000000001) gram. Because no attention is given to the source of the matter nor the pathway, the information regarding these two important intangibles is wiped out from the science of tangibles. The fact that a great amount of energy is released from a nuclear bomb is then taken as evidence of the theory's correctness. By accepting this at face value (heat as the one-dimensional criterion), heat from nuclear energy, electrical energy, electromagnetic irradiation, fossil fuel burning, wood burning or solar energy become identical. This, of course, has tremendous implications for the underlying economics and future engineering and project planning.

Newton in a previous era and Einstein in the modern era are both credited with having the proper grasp of time and energy (e.g., light). Newton's work has been deconstructed by Islam *et al.* (2010). It has been demonstrated that post-Newton science recognized the inherent connection between mass and force, but disconnected mass from energy. Energy itself was connected rather to work — "useful work" — that submits to measurement of the spatial difference between two states in time.

The 'ability to work' is considered to be energy, while the term 'work' itself denotes displacement of an object. Thus, for example, if an object is moved around and brought back to the original place, no (net) work has been performed on it. By definition, the pathway or the time function mooted in the definition of the process, effectively 'disappears' the track of actual work performed.

In addition, and by way of a similar 'magic act', any 'work' is also related to 'heat'. This notion dates back to Lord Kelvin's notion of the universe as continually running down to a point projected in the future of becoming "dead" eventually. This approach incidentally also removes any dissimilarity between sunlight and solar heat from electric light and electrical heating. It also conflates energy produced from food with energy produced from, e.g., gasoline.

The core of this cognition has been in the definition of units of measurable energy. Thus, the BTU (British Thermal Unit) is defined as the amount of heat energy required to increase the temperature of one pound of water by one degree Fahrenheit, at sea level. This definition assumes and imposes a strictly linear property on water while concealing the actual chemical properties of water. The hyperbolic extension doesn't stop here. This BTU is then transformed into energy from food in a strictly organic setting.[17]

Einstein's concept of time, matter, and energy has also received critical review by recent work of Islam *et al.* (2014) and Islam (2014). This aspect is discussed in the following chapter.

[17]The designation itself is inherently absurd: one might just as well refer to a "Canadian Foot of Snow".The authors cannot help but bring readers' attention here to the remarkable similarity of the *modus operandi* of this "British" thermal unit to that of the British Empire in its heyday, on which it was said that "the Sun never sets". Since 1997 when London officially departed Hong Kong, there is no British Empire on which the Sun could any longer rise.

Appendix 5

Brief Historical Catalog, from Ancient China until Today, of Ages-Long Struggle to Clarify Mankind's Role in the Natural Order

A5.1 Convergence of Mass, Energy, and Human Thought Material (HTM)

Dogmatic cognition is deconstructed by laying it out this way: "I (God) am going to create man and woman with original sin. Then, I am going to impregnate a woman with myself as her child, so that I can be born. Once alive, I will kill myself as a sacrifice to myself to save you (humans who accept me as the 'savior') from the sin I originally condemned you to". How has that dogma changed when 'enlightenment' came in terms of western philosophy? In terms origin of universe, the big bang theory, the introduction of quantum nature, multiple (numerous) histories of each point, Nature as the creator of 'numerous universes', and even the definition of what is true or false has introduced a logic that is more schizophrenic than dogma.

As Stephen Hawking says, everything is true and probable. That is the new dogma. Quantum physics and its peddlers, the latest in the HSSA degradation mode, say, everything is possible and probable, except existence of one god. Why? Because all religions are false. When asked for proof, they offer the history of Christianity in Europe and they say, "Look at Christianity, it's all dogma!" For them, if you can visualize it, it is true except they don't agree on what is true, they say truth is a matter of perception, therefore, one man's truth can very well be another man's falsehood. They also say: if you can desire, it will come true. If it doesn't, it means you haven't desired enough. This literally transformed one deity to infinity while blaming Christianity for invoking Trinity in the place of one God. This is the classic definition of cognitive dissonance (Islam et al., 2015) and it turns out that the modern education system as well all Eurocentric models will allow only such cognition to dictate research activities.

A5.2.1 From Original Sin and Infallibility to the 'Murder Gene' and the 'God Gene'

Cognitive dissonance is described as follows:

Sometimes people hold a core belief that is very strong. When they are presented with evidence that works against that belief, the new evidence cannot be accepted. It would create a feeling that is extremely uncomfortable, called cognitive dissonance. Because it is so important to protect the core belief, they will rationalize, ignore and even deny anything that doesn't fit in with the core belief.

What dogma did is introduced one aphenomenal standard of trinity invoking cognitive dissonance (Zatzman and Islam, 2007), whereas the 'western philosophers' have introduced numerous such standards. The scientific equivalent of this transition is a quantum leap from bipolar to multi-polar (schizophrenia) and is best described as 'deliberate schizophrenia' (Islam et al., 2013).

As an example of what we are getting at, one can cite the word 'quantum'. In physics it means: a) The smallest amount of a physical quantity that can exist independently, especially a discrete quantity of electromagnetic radiation; and b) This amount of energy regarded as a unit. This formulation that starts with the false premise that something 'independent' can exist in nature disconnects mass and energy and renders subsequent build up irrelevant. Apparently, such 'isolation' is necessary to observe phenomena. However, the invoking of false premise and subsequent truncation of history render the entire process aphenomenal.

This has been going on for millennia. Ever since the concept of the atom as the fundamental particle of everything — presumably including even thought-material or God — scientists have only made the scope of atom bigger or smaller, ending up with an aphenomenal point.

At the small end of the mass scale, New Science has ended up with a Higgs boson of zero mass while, at the large end,, it has become the 'Big Bang' of an infinitely small object with infinitely large mass. Scientists celebrate the 'discovery' of these particles (Jha, 2013), utterly unconscious of the absurdity of such contradictions. Today's Nobel Prize-winning work involves such 'great discoveries' as the neutrino and the Higgs boson. Neutrino and photon are both zero mass, but whereas the neutrino travels through a solid, photons cannot. In a 'vacuum' (so-called), the neutrino travels slower than the photon mostly because Einstein said the speed of light cannot be exceeded and previous experiments showing the opposite trend were 'proven' to be faulty. As for the Higgs boson, exactly the same thing has been told before in the context of other 'fundamental' particles. Similarly, the same thing was told about ether in such a way that "ether = photon + neutrino". There is no mistake here because... they are all zero mass! As for Higgs' boson having mass, this is considered to be roughly constant averaged over a given time-interval essentially because... the heavier ones are not really heavy in mass, they are surrounded by more particles, because they attract them more, therefore they move slowly. It's a traffic jam! So, what is there in between traffic jam? Nothing? In that case, why is it a jam? If nothing, the speed should be infinity and they should just collide! We then have to ask: why are some more attractive than others?

None can be answered without resorting to dogma.

The "fundamental" character assigned to this particle derives from the circumstance that it fulfills a prediction posited by previous theory, i.e., dogma. The dogma is that there must be a fundamental particle, and after that there can be no other more fundamental particle(s). To say that it is not dogma because the equations that are satisfied by what has been detected so far about the boson are allegedly entirely "objective" won't do here. It's not enough; if the boson is truly fundamental then it must persist somehow as an actual phenomenon via some pathway. Is anyone even looking for such a thing? It sounds as though these scientists who swear they are doing nothing connected with religious belief actually have their own internal consensus that there can be no other 'deity' but the Boson, and Higgs is its messenger. One can accept their contention that they are not doing anything connected with religious belief. What they are doing is "closing the circle", i.e., perfecting dogma. That is the definition of cognitive dissonance.

The Big Bang theory, on the other hand, introduces an even more obtuse fairy tale„ complete with all sorts of numbers that make the theory seem 'scientific' and accurate. Here follows a brief description. Note how every statement is a premise, although they are packaged as a scientific deduction:

i. The universe was born with the Big Bang as an unimaginably hot, dense point. When the universe was just 10^{-34} of a second or so old, it experienced an incredible burst of expansion known as inflation, in which space itself expanded faster than the speed of light. During this period, the universe doubled in size at least 90 times, going from subatomic-sized to golf-ball-sized almost instantaneously. After inflation, the growth of the universe continued, but at a slower rate. As space expanded, the universe cooled and matter formed. One second after the Big Bang, the universe was filled with neutrons, protons, electrons, anti-electrons, photons and neutrinos:

ii. During the first three minutes of the universe, the light elements were born during a process known as Big Bang nucleosynthesis. Temperatures cooled from 10^{32} degrees K to 10^{9} degrees K, and protons and neutrons collided to make deuterium, an isotope of hydrogen. Most of the deuterium combined to make helium, and trace amounts of lithium were also generated;
iii. For the first 380,000 years or so, the universe was essentially too hot for light to shine. The heat of creation smashed atoms together with enough force to break them up into a dense plasma, an opaque soup of protons, neutrons and electrons that scattered light like fog;
iv. The globular cluster NGC 6397 contains around 400,000 stars and is located about 7,200 light years away in the southern constellation Ara. With an estimated age of 13.5 billion years, it is likely among the first objects of the Galaxy to form after the Big Bang;
v. Roughly 380,000 years after the Big Bang, matter cooled enough for atoms to form during the era of recombination, resulting in a transparent, electrically neutral gas. This set loose the initial flash of light created during the Big Bang, which is detectable today as cosmic microwave background radiation. However, after this point, the universe was plunged into darkness, since no stars or any other bright objects had formed yet;
vi. About 400 million years after the Big Bang, the universe began to emerge from the cosmic dark ages during the epoch of reionization. During this time, which lasted more than a half-billion years, clumps of gas collapsed enough to form the first stars and galaxies, whose energetic ultraviolet light ionized and destroyed most of the neutral hydrogen;
vii. Although the expansion of the universe gradually slowed down as the matter in the universe pulled on itself via gravity, about 5 or 6 billion years after the Big Bang, a **mysterious** force now called dark energy began speeding up the expansion of the universe again, a phenomenon that continues today;
viii. little after 9 billion years after the Big Bang, our solar system was born. 130 million years. In comparison, the solar system is only about 4.6 billion years old. This estimate came from measuring the composition of matter and energy density in the universe. This allowed researchers to compute how fast the universe expanded in the past. With that knowledge, they could turn the

clock back and extrapolate when the Big Bang happened. The time between then and now is the age of the universe;

ix. Scientists think that in the earliest moments of the universe, there was no structure to it to speak of, with matter and energy distributed nearly uniformly throughout. The gravitational pull of small fluctuations in the density of matter back then gave rise to the vast web-like structure of stars and emptiness seen today. Dense regions pulled in more and more matter through gravity, and the more massive they became, the more matter they could pull in through gravity, forming stars, galaxies and larger structures known as clusters, superclusters, filaments and walls, with "great walls" of thousands of galaxies reaching more than a billion light years in length. Less dense regions did not grow, evolving into area of seemingly empty space called voids;

x. Until about 30 years ago, astronomers thought that the universe was composed almost entirely of ordinary atoms, or "baryonic matter." However, recently there has been ever more evidence that suggests most of the ingredients making up the universe come in **forms that we cannot see**. It turns out that atoms only make up 4.6 percent of the universe. Of the remainder, 23 percent is made up of dark matter, which is likely composed of one or more species of subatomic particles that interact very weakly with ordinary matter, and 72 percent is made of dark energy, which apparently is driving the accelerating expansion of the universe;

xi. The shape of the universe and whether or not it is finite or infinite in extent depends on the struggle between the rate of its expansion and the pull of gravity. The strength of the pull in question depends in part on the density of the matter in the universe. If the density of the universe exceeds a specific critical value, then the universe is "closed" and "positive curved" like the surface of a sphere. This means light beams that are initially parallel will converge slowly, eventually cross and return back to their starting point, if the universe lasts long enough. If so, the universe is not infinite but has no end, just as the area on the surface of a sphere is not infinite but has no beginning nor end to speak of. The universe will eventually stop expanding and start collapsing in on itself, the so-called "Big Crunch";

xii. If the density of the universe is less than this critical density, then the geometry of space is "open" and "negatively curved" like the surface of a saddle. If so, the universe has no bounds, and will expand forever;

xiii. If the density of the universe exactly equals the critical density, then the geometry of the universe is "flat" with zero curvature like a sheet of paper. If so, the universe has no bounds and will expand forever, but the rate of expansion will gradually approach zero after an infinite amount of time. Recent measurements suggest that the universe is flat with only a 2 percent margin of error.

Similar assumptions are invoked regarding human cognition. In the 'dogma days' (of the European Middle Ages), 'original sin' was invoked. Subsequently, the notion of human beings emerging at birth with an allegedly 'blank slate' status was introduced. Finally, the notion of 'learning' has been introduced. This learning process is then assumed to be independent of the environment, and always an illogical function of something aphenomenal, akin to Eve's original sin being imparted on the entire humankind. An example of such manipulation of HTM is in medical diagnosis blaming a defective gene with such extreme extrapolation as to recognizing murder gene (Forzano, et al., 2010), gay gene (Hamer, 1994; Hamer et al., 1993), God gene (Hamer, 1999), whereas the logical consideration would be the analysis of personal history and the person's interaction with the environment (including fellow humans). Similar to invoking multiple history to every particle (ranging from subatomic to galactic), a mental 'illness' is made into a function of everything but the unique path traveled by the person. Thus HTM is isolated from its phenomenal source (i.e., conscience). In addition, the law of continuity is violated, thereby making it impossible to make phenomenal observation of HTM.

In terms of human thought material, the scientific logic has transited from dogma science to further obscurity of Agnostic narration to atheism. This is captured by the remarks of two contemporary scientists.

What I have done is to show that it is possible for the way the universe began to be determined by the laws of science. In that case, it would not be necessary to appeal to God to decide how the universe began. This doesn't prove that there is no God, only that God is not necessary. — Stephen Hawking (1942-)

The more you understand the significance of evolution, the more you are pushed away from the agnostic position and towards

atheism. Complex, statistically improbable things are by their nature more difficult to explain than simple, statistically probable things.
— Richard Dawkins (1941-)

Primary premises of Atheism are: We don't need 'deity'; we cannot prove 'god' doesn't exist, we cannot see 'him', 'life after death' we have to see it believe it (it's too much of an assumption that can never be verified), I can already explain creation and I am smart enough to figure out how to be a good person, therefore, absence of 'deity' is a pretty good first premise.

It is akin to yet another fatally flawed first premise that goes with Bush doctrine (Khan and Islam, 2012). The fuzzy logic (*Manteq* in Arabic, see Islam et al., 2013 for details) question to ask is: What if the person being tortured is actually innocent or the nation being bombed had

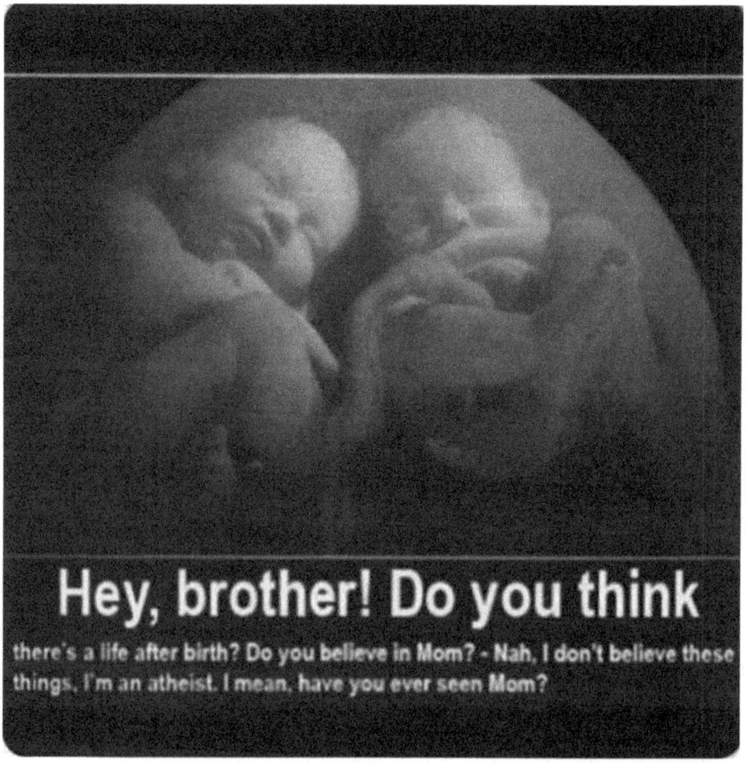

Picture A5.1 New Science has moved away from logic to 'faith' in all aspects of cognition.

no means to attack. How can you punish intention, something you have no idea of?

Each scientist bases his logic on an aphenomenal premise, resulting with conclusions that implode on us almost every day through scientific news items (e.g. Hogenboom, 2013). When attacked as regards their blatant absence of any clear phenomenality, each of these additional notions (e.g., "origin of universe", "big bang theory", "quantum nature", multiple (numerous) history/histories of each point, Nature as the creator of 'numerous universes') is ultimately justified as an "explanatory framework". This justification is not unlike 'you cannot handle the truth' of the dogma days. One cannot pick-and-choose the truth. The scholar-researcher has the responsibility to nail things down or honestly admit that they are not there yet. The illegitimate and unconscionable thing is to rave about how some completely unpredicted or unanticipated discovery fits into some previously-advanced piece of theory.

Here is one example of the kind of question that needs to be answered honestly and not in a manner that fits/serves previously advanced dogma: I watch an oscilloscope trace every time a light pulses, and that trace is declared evidence, or confirmation, that light has a quantum character of a discrete albeit infinitely tiny piece of mass. Does it mean that the light I see after the sun rises is or is not the same light appearing as an oscilloscope trace, or does it mean that light can be both? And if you choose the latter, does this mean light is both a particle and a wave? Light is an actual phenomenon, but is the oscilloscope trace or its source the same phenomenon of not really germane for dealing with the natural phenomenon that is light? The pragmatic fix has been to say that these are different versions of the same thing, end of story. There's the slippery slope: material reality and my observation of material reality are "essentially" the same. But the moment we accept this, what remains definitive about material reality?

In terms of life, humanity, and interaction of humans with the nature, the confusion runs very deep. Even the greatest of scientists of modern age didn't have a scientific understanding of purpose of humanity, the onset of life form, or the transition be society. We all know about Newton and he how he was 'religious' yet totally focused on self interest. We know about Einstein who was a Jew but wasn't certain about the role of Creator and Creation. He famously said,

> "Nature shows us only the tail of the lion. But there is no doubt in my mind that the lion belongs with it even if he cannot reveal himself to the eye all at once because of his huge dimension."[18]

This statement represents two levels of conflation and highlights the most important shortcoming of today's New Science..Overall the deficiency of New Science, dealing with both physical aspect ("hard science") and intangible aspect ("social science") can be summarized in the following:

i. aphenomenal first premise (source);
ii. aphenomenal function that connects events with the first premise. Most functionality involves the time function that is introduced as a fourth dimension (rather than an implicit continuous function) in order to render a 'steady state' into an 'unsteady steady state'. In addition to this spurious insertion of time function, it is expanded with a functionality that doesn't exist in nature.

A5.2.2 Convergence of Mass, Energy, and Human Thought Material (HTM)

Dogmatic cognition is deconstructed by laying it out this way: "I (God) am going to create man and woman with original sin. Then, I am going to impregnate a woman with myself as her child, so that I can be born. Once alive, I will kill myself as a sacrifice to myself to save you (humans who accept me as the 'savior') from the sin to which I originally condemned you." How has that dogma changed when 'enlightenment' came in terms of western philosophy? In terms origin of universe, the big bang theory, the introduction of quantum nature, multiple (numerous) histories of each point, Nature as the creator of 'numerous universes', and even the definition of what is true or false has introduced a logic that is more schizophrenic than dogma.

As Stephen Hawking says, everything is true and probable. That is the new dogma. Quantum physics and its peddlers, the latest in the HSSA degradation mode, say, everything is possible and probable,

[18]Letter to H. Zangger (10 March 1914), quoted in Jean Eisenstaedt, *The Curious History of Relativity* (2006), p. 126.

except existence of one god. Why? Because all religions are false. When asked for proof, they offer the history of Christianity in Europe and they say, "Look at Christianity, it's all dogma!" For them, if you can visualize it, it is true except they don't agree on what is true, they say truth is a matter of perception, therefore, one man's truth can very well be another man's falsehood. They also say: if you can desire, it will come true. If it doesn't, it means you haven't desired enough. This literally transformed one deity to infinity while blaming Christianity for invoking Trinity in the place of one God. This is the classic definition of cognitive dissonance (Islam et al., 2015) and it turns out that the modern education system as well all Eurocentric models will allow only such cognition to dictate research activities.

A5.2.3 From Original Sin and Infallibility to the 'Murder Gene' and the 'God Gene'

Cognitive dissonance is described as follows:

Sometimes people hold a core belief that is very strong. When they are presented with evidence that works against that belief, the new evidence cannot be accepted. It would create a feeling that is extremely uncomfortable, called cognitive dissonance. Because it is so important to protect the core belief, they will rationalize, ignore and even deny anything that doesn't fit in with the core belief.

What dogma did is introduced one aphenomenal standard of trinity invoking cognitive dissonance (Zatzman and Islam, 2007), whereas the 'western philosophers' have introduced numerous such standards. The scientific equivalent of this transition is a quantum leap from bipolar to multi-polar (schizophrenia) and is best described as 'deliberate schizophrenia' (Islam et al., 2013). As an example, one can cite the word 'quantum'. In physics it means: a) The smallest amount of a physical quantity that can exist independently, especially a discrete quantity of electromagnetic radiation; and b) This amount of energy regarded as a unit. This formulation that starts with the false premise that something 'independent' can exist in nature disconnects mass and energy and renders subsequent build up irrelevant. Apparently, such 'isolation' is necessary to observe phenomena. However, the invoking of false premise and subsequent truncation of history render the entire process aphenomenal. This has been going on for millennia. Ever since the concept of atom that was thought to be fundamental particle of everything, including even thought material or God, scientists have only made the scope of atom bigger or smaller, ending up with an aphenomenal point.

In the smaller scale, New Science has ended up with Higgs boson of zero mass and on the larger scale, it has become Big bang of a infinitely small object with infinitely large mass. Yet, scientists are celebrating the 'discovery' of these particles (Jha, 2013), as though the absurdity of such contradictions don't mean anything to them. Today's Nobel prize winning work involves such 'great discoveries' as Neutrino and Higgs Boson. Neutrino and photon are both zero mass, but neutrino travels through solid, but photons don't. In 'vacuum', Neutrino travels slower than photon mostly because 'Einstein said speed of light cannot be exceeded' and previous experiment that showed opposite trend is 'proven' to be faulty. As for Higgs Boson, exactly the same thing has been told before in the context of other 'fundamental' particles. Similarly, the same thing was told about ether in such a way that ether= photon+neutrino. There is no mistake here because they are all zero mass. As for Higgs boson having mass, this is considered to be equal just because the heavier ones are not really heavy in mass, they are surrounded by more particles, because they attract them more, therefore they move slowly. It's a traffic jam. So, what's there in between traffic jam? Nothing? In that case, why is it a jam? In nothing, the speed should be infinity and they should just collide! Then, we have to ask why some are more attractive than others? None can be answered without resorting to dogma.

The "fundamental" character assigned to this particle derives from the circumstance that it fulfills a prediction posited by previous theory, i.e., dogma. The dogma is that there must be a fundamental particle, and after that there can be no other more fundamental particle(s). To say that it is not dogma because the equations that are satisfied by what has been detected so far about the boson are allegedly entirely "objective" won't do here. It's not enough; if the boson is truly fundamental then it must persist somehow as an actual phenomenon via some pathway. Is anyone even looking for such a thing? It sounds as though these scientists who swear they are doing nothing connected with religious belief actually have their own internal consensus that there can be no other 'deity' but the Boson, and Higgs is its messenger. One can accept their contention that they are not doing anything connected with religious belief. What they are doing is "closing the circle", i.e., perfecting dogma. That is the definition of cognitive dissonance.

Big Bang theory, on the other hand, introduces an even more obtuse fairy tale story, complete with all sorts of numbers that make the theory sound 'scientific' and accurate. Here is a brief description. Note how every statement is a premise, although they are packaged as a scientific deduction:

i. The universe was born with the Big Bang as an unimaginably hot, dense point. When the universe was just 10^{-34} of a second or so old, it experienced an incredible burst of expansion known as inflation, in which space itself expanded faster than the speed of light. During this period, the universe doubled in sizc at least 90 times, going from subatomic-sized to golf-ball-sized almost instantaneously. After inflation, the growth of the universe continued, but at a slower rate. As space expanded, the universe cooled and matter formed. One second after the Big Bang, the universe was filled with neutrons, protons, electrons, anti-electrons, photons and neutrinos.
ii. During the first three minutes of the universe, the light elements were born during a process known as Big Bang nucleosynthesis. Temperatures cooled from 10^{32} degrees K to 10^9 degrees K, and protons and neutrons collided to make deuterium, an isotope of hydrogen. Most of the deuterium combined to make helium, and trace amounts of lithium were also generated.
iii. For the first 380,000 years or so, the universe was essentially too hot for light to shine. The heat of creation smashed atoms together with enough force to break them up into a dense plasma, an opaque soup of protons, neutrons and electrons that scattered light like fog.
iv. The globular cluster NGC 6397 contains around 400,000 stars and is located about 7,200 light years away in the southern constellation Ara. With an estimated age of 13.5 billion years, it is likely among the first objects of the Galaxy to form after the Big Bang.
v. Roughly 380,000 years after the Big Bang, matter cooled enough for atoms to form during the era of recombination, resulting in a transparent, electrically neutral gas. This set loose the initial flash of light created during the Big Bang, which is detectable today as cosmic microwave background radiation. However, after this point, the universe was plunged into darkness, since no stars or any other bright objects had formed yet.
vi. About 400 million years after the Big Bang, the universe began to emerge from the cosmic dark ages during the epoch of reionization. During this time, which lasted more than a half-

billion years, clumps of gas collapsed enough to form the first stars and galaxies, whose energetic ultraviolet light ionized and destroyed most of the neutral hydrogen.

vii. Although the expansion of the universe gradually slowed down as the matter in the universe pulled on itself via gravity, about 5 or 6 billion years after the Big Bang, a **mysterious** force now called dark energy began speeding up the expansion of the universe again, a phenomenon that continues today.

viii. little after 9 billion years after the Big Bang, our solar system was born. 130 million years. In comparison, the solar system is only about 4.6 billion years old. This estimate came from measuring the composition of matter and energy density in the universe. This allowed researchers to compute how fast the universe expanded in the past. With that knowledge, they could turn the clock back and extrapolate when the Big Bang happened. The time between then and now is the age of the universe.

ix. Scientists think that in the earliest moments of the universe, there was no structure to it to speak of, with matter and energy distributed nearly uniformly throughout. The gravitational pull of small fluctuations in the density of matter back then gave rise to the vast web-like structure of stars and emptiness seen today. Dense regions pulled in more and more matter through gravity, and the more massive they became, the more matter they could pull in through gravity, forming stars, galaxies and larger structures known as clusters, superclusters, filaments and walls, with "great walls" of thousands of galaxies reaching more than a billion light years in length. Less dense regions did not grow, evolving into area of seemingly empty space called voids.

x. Until about 30 years ago, astronomers thought that the universe was composed almost entirely of ordinary atoms, or "baryonic matter." However, recently there has been ever more evidence that suggests most of the ingredients making up the universe come in **forms that we cannot see**. It turns out that atoms only make up 4.6 percent of the universe. Of the remainder, 23 percent is made up of dark matter, which is likely composed of one or more species of subatomic particles that interact very weakly with ordinary matter, and 72 percent is made of dark energy, which apparently is driving the accelerating expansion of the universe.

xi. If the density of the universe is less than this critical density, then the geometry of space is "open" and "negatively curved" like the surface of a saddle. If so, the universe has no bounds, and will expand forever.

xii. The shape of the universe and whether or not it is finite or infinite in extent depends on the struggle between the rate of its expansion and the pull of gravity. The strength of the pull in question depends in part on the density of the matter in the universe. If the density of the universe exceeds a specific critical value, then the universe is "closed" and "positive curved" like the surface of a sphere. This means light beams that are initially parallel will converge slowly, eventually cross and return back to their starting point, if the universe lasts long enough. If so, the universe is not infinite but has no end, just as the area on the surface of a sphere is not infinite but has no beginning nor end to speak of. The universe will eventually stop expanding and start collapsing in on itself, the so-called "Big Crunch."

xiii. If the density of the universe exactly equals the critical density, then the geometry of the universe is "flat" with zero curvature like a sheet of paper. If so, the universe has no bounds and will expand forever, but the rate of expansion will gradually approach zero after an infinite amount of time. Recent measurements suggest that the universe is flat with only a 2 percent margin of error.

A5.2.4 Historic Disconnection

Averroes is known to be the "father of secular philosophy in western Europe". Even though the word secular is attached to his denomination in modern Europe, his teachings were Islamic and wholly inspired by *The Qur'an*. European scientists claim to be secular themselves but they claim to be following Averroes' philosophy. We studied this phenomenon and discovered this claim to be misleading. Europeans didn't follow Averroes' logic and even if they did it would be called secular unless Qur'an itself is secular (Khan and Islam, 2012). Let's review what Averroes had actually said. Followings are a few quotes from his work translated in English. The title of the book is: The Book of the Decisive Treatise, Determining the Connection Between the Law and Wisdom & Epistle Dedicatory (Averroes, 2001).

"... [T]he Law makes it obligatory to reflect upon existing things by means of the intellect, and to consider them; and consideration is nothing more than inferring and drawing out the unknown from the known... "

"You ought to know that what is intended by the Law is only to teach true science and true practice. True science is cognizance of God and of all the existing things as they are, especially the venerable ones among them; and cognizance of happiness in the hereafter and of misery in the hereafter. True practice is to follow the actions that promote happiness and to avoid the actions that promote misery; and cognizance of these actions is what is called 'practical science.'"

"... [T]he link between the physician and the health of bodies is [the same as] the link between the Lawgiver and the health of souls... "

This health is what is called 'piety.'"

"... [I]njuries from a friend are graver than injuries from an enemy – I mean that wisdom is the companion of the Law and its milk sister. So injuries from those linked to it are the gravest injuries – apart from the enmity, hatred, and quarreling they bring about between both of them. These two are companions by nature and lovers by essence and instinct."

"... [E]xistence is the cause and reason of our knowledge, while eternal knowledge is the cause and reason of existence."

In essence, Averroes promoted long-term over short-term, intangible over tangible, and natural over artificial. This notion wasn't Averroes'. It was actually inspired by *The Qur'an* and prophet Muhammad whose statements were preserved in the form of authenticated Books of Hadith. In modern history, *The Qur'an* is the oldest book preserved in its original form (over 1400 years) and the Books of Hadith (sayings of Prophet Muhammad) are the 2nd oldest preserved books (nearly 1200 years). Anyone familiar with *The Qur'an*ic approach knows that *Qur'an* establishes the long- term approach, along with focus on intangibles as the primary modus operandi for success in this life as well as hereafter, There are numerous *Qur'an*ic verses to this effect but I would suffice to read Chapter 2, verses 1 through 5 that define the criteria for success by taking the long-term approach. In addition, it is also important to recall, all Greek philosophers from ancient time whose work was translated by Islamic scholars/polyscientists (e.g. Averroes, Ibn Haitham, Avicenna) were also familiar with such approach of long-term and intangibles. For instance, Wikipedia lists the following principle of division of the entire universe between *Actus Purus* (absolutely perfect, denoted in *The Qur'an* as Creator) and the others (denoted in *The Qur'an* as Creation that can only be perfect in terms of all-perfect creational ability of the Creator).

In the metaphysical order, the highest determinations of Being are Actuality (entelecheia - Greek: a - Greek: rder, the highest ddynamis - Greek: δύναμις). The former is perfection, realization, fullness of Being; the latter imperfection, incompleteness, perfectibility. The former is the determining, the latter the determinable principle. Actuality and potentiality are above all the Categories. They are found in all beings, with the exception of the Supreme Cause, in whom there is no imperfection, and, therefore, no potentiality. God is all actuality, *Actus Purus*, following the principle of division of the entire universe between *Actus Purus* (absolutely perfect, denoted in *The Qur'an* as Creator) and the others (denoted in *The Qur'an* as Creation that can only be perfect in terms of all-perfect creational ability of the Creator).

Parmenides of Elea (early 5th century BCE) was an ancient Greek philosopher born in Elea, a Greek city on the southern coast of Magna Graecia. He can be called the first European philosopher who differentiated between correct and incorrect ways of cognition. The single known work of Parmenides is a poem, On Nature, which has survived only in fragmentary form. In this poem, Parmenides describes two views of reality. In "the way of truth", he explains how reality (coined as "what-is") is one, change is impossible, and existence is timeless, uniform, necessary, and unchanging. This notion is known attributes of the Creator in the orient as well as in Islamic post-*Qur'an* era. Consider Chapter 112 of *Qur'an* that says: "He is Allah, [who is] One, Allah, the Absolute Constant (immutable, immovable, unchanging,...). He neither begets nor is born, nor is there anything like Him." The other way that Parmenides poetry is known to describe is the "the way of opinion," the way that one uses one's sensory faculties to conceive is false and deceitful. This bifurcation of cognition is also typical of oriental thinking and encapsulated in *Qur'an*ic verses such as, "And have shown him the two ways?" (90:11). The chapter continues and describes one path involving long-term approach of selflessness and the other one the myopic approach of desire and self interest in the short-term. This bifurcating pathways are repeated many times in *The Qur'an* and form the basis for heaven and hell in the hereafter.

Zeno of Elea (490-430 B.C.), a disciple of Parmenides, is credited to have introduced dialectic technique of increasing knowledge of the truth. This technique is known around the world through practically all religions, most documented one being Islamic holly book, *Qur'an* (Islam et al., 2013). This questioning is called *Manteq* in Arabic and was introduced to the west in 1967 in the form of 'fuzzy logic'. However, Manteq was a science was well known throughout Islamic era and top scholars, judges, scientists were required to be proficient in *Manteq* (Islam et al., 2013).

Similar to Parmenides, Zeno gave little value to tangible and held the traits of the creator as the only truth. It is commonly understood that Zeno did introduce the concept of paradox of space, time, motion and change, allegedly calling them absurd. However, these are also the fundamental feature of all natural creation (Khan and Islam, 2007). Even artificial objects are subject to these traits (Khan and Islam, 2012). It seems reasonable that Parmenides or at least his interpreters have conflated creation with creator. If a distinction between the nature of creation (we know a lot about it) and creator (we know nothing about it) is made, no paradox appears. This particular aspect of philosophy appears repeatedly in European cognition, starting from Zeno then Plato and all the way to modern scientists. To Zeno's credit, however, he is not the author of the paradox and it is quite probable that such notion was invented by post-Roman Catholic church era that had vested interest in nurturing paradoxical thoughts. If the 'way of truth', as promoted by Parmenides is coupled with dialectic for which Zeno is credited, there would be no question anyone would be able to discover the truth that seems to have eluded all modern scientists, even though that technique was thoroughly used by Islamic scientists and scholars for over 1000 years (7th to 17th century).

Even Plato and Socrates held similar views about the logic of the separation of Creator and Creation. Famously known as Plato's Euthyphro dilemma. It involves the question: Is an act right because God says it's so, or does God say it's so because it's right? Rather than answering this question with logical answer, European Christianity resorted to dogma, prompting philosophers of the modern age to challenge Christianity at its dogmatic root, without realizing that there was profound change planted during the dogma era and logic was replaced with illogical assertions, turning the entire cognition process upside down. For instance,

Bertrand Russell argued in his book: *Why I Am Not a Christian* as follows (Russell, 1952):

> "If you are quite sure there is a difference between right and wrong, you are then in this situation: Is that difference due to God's fiat or is it not? If it is due to God's fiat, then for God Himself there is no difference between right and wrong, and it is no

longer a significant statement to say that God is good. If you are going to say, as theologians do, that God is good, you must then say that right and wrong have some meaning which is independent of God's fiat, because God's fiats are good and not good independently of the mere fact that he made them. If you are going to say that, you will then have to say that it is not only through God that right and wrong came into being, but that they are in their essence logically anterior to God."

If one gleans historical facts, one can gather all Greek savants and philosophers as well as oriental scholars from pre-Islamic renaissance are very similar in promoting long-term approach that gives priority to intangibles over tangibles. Here we will list a few with their most notable features as stated in Wikipedia.

- Euclid (435 BC-365 BC):

Euclid himself wrote six dialogues — the Lamprias, the Aeschines, the Phoenix, the Crito, the Alcibiades, and the Amatory dialogue — but none survive. The main extant source on his views is the brief summary by Diogenes Laërtius. Euclid's philosophy was a synthesis of Eleatic and Socratic ideas. Socrates claimed that the greatest knowledge was understanding the good. The Eleatics claimed the greatest knowledge is the one universal Being of the world. Mixing these two ideas, Euclid claimed that good is the knowledge of this being. Therefore this good is the only thing that exists and has many names but is really just one thing. He identified the Eleatic idea of "The One" with the Socratic "Form of the Good," which he called "Reason," "God," "Mind," "Wisdom," etc.

Compare the above statement with *The Qur'an*'s assertion that "God" (*Allah*) has 99 traits that are stated explicitly, some being "The Absolute Good" (*Al-Quddus*), "The Absolute Knower" (*Al-Aleem*), "the Absolute Wisdom" (*Al-Hakeem*), "the Absolute Visionary" (*Al-Basīr*), and "the Absolutely Powerful" (*Al-Qawi*), "The One" (*Al-Ahad*).

The same Euclid of Alexandria (300 BC): In response to a demand for faster solution to mathematical model, Euclid is credited with telling the ruler of Egypt that 'there is no royal road to Geometry'.and that 'the laws of nature are but the mathematical thoughts of God.' What European philosophers/scientists have done is to collate Euclid's theorems into something called 'Euclidean geometry' and mathematize these 'thoughts of God' through linearization. The result would be labeled a component part of 'natural law'.

This has all changed. As early as Newton, for instance, said, "God is an exact clock", then he proceeded to define time. How does he know about God so much to ascribe Godly accuracy to his calculations? A more contemporary genius, Albert Einstein, said: "I cannot imagine a God who rewards and punishes the objects of his creation, whose purposes are modeled after our own — a God, in short, who is but a reflection of human frailty." What is the first premise? God created humans whose purposes are the same as God's. Then, humans are frail, therefore, God must be frail. How does Einstein know this? What if it is not true? After all, it is the same Einstein that said, "Whoever undertakes to set himself up as a judge of Truth and Knowledge is shipwrecked by the laughter of the gods." He is actually making the pretentious claim of knowing God's intention. This has become the symbol of European cognition.

- Confucius (551BC-479 BC):

Confucianism discusses elements of the afterlife and views concerning Heaven, but it is relatively unconcerned with some spiritual matters often considered essential to religious thought, such as the nature of souls... Confucius presents himself as a "transmitter who invented nothing". He puts the greatest emphasis on the importance of study, and it is the Chinese character for study (學) that opens the text. Far from trying to build a systematic or formal theory, he wanted his disciples to master and internalize the ancient classics, so that their deep thought and thorough study would allow them to relate the moral problems of the present to past political events (as recorded in the Annals) or to past expressions of commoners' feelings and noblemen's reflections (as in the poems of the Book of Odes).

In matter of ethics) One of the deepest teachings of Confucius may have been the superiority of personal exemplification over explicit rules of behavior. His moral teachings emphasized self-cultivation, emulation of moral exemplars, and the attainment of skilled judgment rather than knowledge of rules.

Compare the above with Hadith that says "seeking knowledge is obligatory for each male and female believer" and later confirms the follow-up of knowledge in practice as well as in making knowledge known to be essential to the best of one's ability. Also, note how the word 'science' is the second most used word in *The Qur'an*. Top that with the final sermon captured in the following Hadith, "An Arab is no better than a non-Arab, and a non-Arab is no better than an Arab; a red man is no better than a black man and a black man is no better than a red man – except if it is in terms of piety."

- Damascius (458-538 CE):

His chief treatise is entitled Difficulties and Solutions of First Principles. It examines the nature and attributes of God and the human soul. This examination is, in two respects, in striking contrast to that of certain other Neoplatonist writers. It is conspicuously free from Oriental mysticism, and it contains no polemic against Christianity, to the doctrines of which, in fact, there is no allusion. Hence the charge of impiety which Photius brings against him. In this treatise Damascius inquires into the first principle of all things, which he finds to be an unfathomable and unspeakable divine depth, being all in one, but undivided. His main result is that God is infinite, and as such, incomprehensible; that his attributes of goodness, knowledge and power are credited to him only by inference from their effects; that this inference is logically valid and sufficient for human thought. He insists throughout on the unity and the indivisibility of God. This work is, moreover, of great importance for the history of philosophy, because of the great number of accounts which it contains concerning former philosophers.

Even Augustine (A.D. 354-430), who notoriously believed in flat earth theory as well as 'original sin' and 'fall' is known to have said,

God is timeless and is creator of time, and universe; God holds all knowledge and has no beginning or end; God is perfection, eternal, infinity, incomprehensible, simplicity, and unique; God is "the inexhaustible light" that enlightens all and "physical light" is derivative; God created "everything out of nothing") at the moment chosen by God; God created everything at its 'perfect' time; the universe was created simultaneously in "logical framework" and not in "six calendar days" because creavit omni simul ("he created all things at once"); Recognized that 'original sin' and 'fall' are difficult to explain;

Compare that with some of the 99 traits of God mentioned in *The Qur'an*, such as, "The Absolute Light" (*Al-noor*), "The creator who creates everything out of nothing" (*Al-Khaleq*). Or, compare that with the Hadith *Qudsi* (saying of Allah but not part of *The Qur'an*) that says: "Allah said: The son of Adam curses Absolute Time (*dahr*) but I am Time, for in my Hand are the night and day" (Sahih Muslim 2246).[19]

A5.2.5 Delinearized Brief History of Civilization

In the 14th century, Ibn Khaldoun (1332-1406) wrote extensively on the history of civilization. Ibn Khaldoun's work was later recognized as 'undoubtedly the greatest work of its kind that has ever yet been created by any mind in any time or place' (Encyclopaedia Britannica, 2010). The British philosopher-anthropologist Ernest Gellner characterized Ibn Khaldoun's definition of Government, "an institution which prevents injustice other than such as it commits itself", as the "best in the history of political theory" (Gellner, 1988). Here, we strive only to summarize Ibn Khaldoun's key observations relevant to our present objectives, supplemented with further observations of our own.[20]

Throughout history, civilizations[21] started to build alongside rivers, ranging from Huang Ho River Valley in China (4000 BCE) to Indus River valley in Pakistan (3300-1300 BCE), from Nile River in Egypt (5000 BCE) to Tigris and Euphrates rivers of Mesopotamia in Iraq (5000 BCE). How these rivers nourished fertile lands and then gave

---

[19]Descendants of the same Europeans, however, would not hold similar thoughts. The former US President, Abraham Lincoln (still a backwoods Illinois politician and lawyer but married to Mary Todd, daughter of a wealthy & politically highly-connected family of slaveowners), is known to have said in 1858:

I, as much as any other man, stand in favor of having the superior position assigned to the white race... I have no purpose to introduce political and social equality between the white and the black races.

These thoughts are not aberrations, nor are they isolated personal views. They stand at the core of Eurocentric cognition. While philosophical thought has flip-flopped between 'faith' and atheism, the nature of departure from the standard that was known millennia ago has not changed. Indeed: Weissert (2013) reported the following on

easy access to trade and commercial routes, eventually transforming business and trading hubs into major cities.

These cities gave rise to civilization for humans that are uniquely adept to social infrastructure. Ibn Khaldoun got his inspiration from the Quran that defines humans as the Viceroy (*khalifa*) of the Creator (e.g. Chapter 2:30 of the Quran specifies man's role as the viceroy), charged with law and order on Earth that seems to be abandoned not withstanding the grand plan of the Creator in the form of universal order. This outlook is clearly different to the Eurocentric notions, ranging from the vastly discredited 'original sin' to widely accepted 'evolution' theories (McHenry, 2009; National Geographic, 2008) that detach human conscience from its functioning in a society.

Of interest here is the fact that the only Eurocentric theory that goes deeper than superficial features of humans is the one that deals with 'aggressiveness' as the unique feature of human beings. This was popular during Nazi era in name of 'theory of ethology' and later promoted by the likes of Konrad Lorenz and Robert Ardrey (see specifically Leakey and Ardrey, 1975). Konrad Lorenz and Karl von Frisch jointly won the Nobel Prize in Medicine in 1973 for their theory of ethology. Of significance is the fact that the original notion of original sin was never replaced by 'original conscience', the closest European scientist went is the notion of 'original aggressiveness' as a distinctive feature of humans. Even when the concept of tool was synonymous with uniqueness of humans, the making of tools was linked to aggressive behavior, rather than a sign of ingenuity or technological marvels. Indeed, modern age is synonymous with technological leaps during wartime or due to fear of war. Such wasn't the case in previous era that was known to be more connected to conscience.[22]

---

federal judge (Edith H Jones, 5th circuit Court of Appeals) in modern-day America. Notice how the notion of God and logic of 'mentally retarded' is not different from the dogmatic assertions that are two millennia old:

Jones is accused of saying that certain "racial groups like African-Americans and Hispanics are predisposed to crime," and are "prone to commit acts of violence" and be involved in more violent and "heinous" crimes than people of other ethnicities.

The judge also allegedly said Mexicans would prefer to be on death row in the United States than serving prison terms in their native country, and that it's an insult for the U.S. to look to the laws of other countries such as Mexico.

The complaint also states that Jones said defendants' claims of racism, innocence, arbitrariness, and violations of international law and treaties are just "red herrings" used by opponents of the death penalty, and that claims of "mental retardation" by capital

Originally, there was no sign of violence in the civilizations that developed along the ancient rivers. With the passage of time, some worriers or external forces would invade/interact with these cities and towns and bring them under some uniform political entities. The definition of politics being subjective, it would imply that the system used by the invader would be applied to the vanquished.

Why did such takeovers occur? European scholars see these as invasions and are motivated by 'aggressive' behavior of humans. They advance greed, 'religion', feud, and other aggressive reasons as justification behind such invasion (Martel, 2012). This is true in Roman history, including Crusades but Ibn Khaldoun's take on war was different. If humans have to be considered to be distinct from other animals because of their conscience, the possibility that a war can be motivated by the sense of justice must be kept open. Indeed, many of these civilizations went through transformations without any war.

In ancient era, violent clashes were more common among nomads. Nomads often take over a city or town and sometime settle within the city. Ironically, city dwelling softened Nomads and eventually they assimilated with local population, soon becoming part of the indigenous culture. Eventually, they would become the local population only to be later become overcome by other nomadic invasion and the cycle would continue. The pattern identified here is: conquest à consolidation à expansion à degeneration à conquest. The degeneration aspect is not expanded but it puts the external invention in perspective.

The Qur'an mentions about civilization of Thamud, an extraordinarily powerful race, that used to build houses with curved out rocks as found in Madain Saleh (city of prophet Saleh) in Saudi Arabia. Ptolemy referred to this nation as 'Tamudaer'. Quranic description involves destruction of this city as a form of punishment for a nation that was

---

defendants disgust her. The fact that those defendants were convicted of a capital crime is sufficient to prove they are not "mentally retarded," the complaint alleges Jones to have said. The complaint further alleges she said a death sentence provides a service to capital-case defendants because they are likely to make peace with God only just before their execution.

[20]Readers are reminded that the level and extent of what any "civilization" might accomplish can be no higher nor go beyond the parameters and perimeter of existing property relations. Underpinning the capability to follow the pathways to such wide vistas of inquiry as reintroduced by Ibn Khaldoun was the fact that the ruling authority of the Muslim society in which he worked would ensure that his work lacked for nothing when it came to furnishing or ensuring the requisite time, resources and freedom from other demands that such enterprise required.

degenerating beyond repair. Similarly, the story of Ad, another race with tall features, is also mentioned. This city was also destroyed, according to Quranic recount and is believed to be near Hadramout in Yemen. Similarly, *The Qur'an* describes the destruction of the cities of Sodom and Gomorrah (close to Dead Sea, Jordan) with some details and relates their destruction to degenerative social behavior.

Ibn Khaldoun only makes passing comments about these degenerations, without mentioning them as a socio/political phenomena. In today's, terms, this destruction would be attributed to "natural disasters" and their link to human behavior unacknowledged.

First civilization known to European social scientists is that of Mesopotamia and Babylon, between Tigris and Euphrates rivers. Sumerians are credited to connecting related cities and dwellings on river banks in an apparent effort to build an empire, called Sumer. They are known to have scripts, had invented wheels, carts, etc. Akkadians, the race from the northern mountainous regions conquered Sumer. Akkadians were led by Sargin Akkad, whose empire included modern day Iran, Asia minor, and Syria. He is known to be the first person in recorded history to rule over a multi-ethnic, centrally ruled empire (Liverani, 1993). He is believed to have famously said, "Now any king who wants to call himself my equal, wherever I went, let him go". This challenge was soon met because in less than two centuries, the Akkadian dynasty was over and it was conquered by another nomadic tribe that themselves became part of the local tribes and the cycle continued. Gutians ransacked Mesopotamia and within a few generations, they became 'civilized' and reached their golden era. However, only about a century later, they were expelled by the rulers of Uruk – a dynasty led by Lugalzagesi, who was later defeated by Ur-Nammu of the Ur dynasty. After that would come Babylonian, Achaemenid, Seleucid, and Parthian Empires. Each of these empires excelled in specific themes of social science and

---

[21]What is called "civilization" here is some form of social order based ultimately on redistribution of a collective surplus. Such an arrangement presupposes some notion(s) about property, especially the right assumed by a ruler to enjoy others' property.

[22]Implicit in the Eurocentric view based on the Christian concept of "original sin" is that, on the one hand, there is no civilisation worthy of the name without hewing to the central Christian belief in Jesus Christ as one's personal Saviour. At the same time, on the other hand, however, within this Eurocentric outlook, a latent prospect of genocide stalks all human civilization. That is the same line — dressed up as the so-called "science" of ethology — expressed in the theories of built-in aggression as the defining feature of humanity. The key point about civilisation as elaborated by Ibn Khaldun and those who further developed theories along the same lines is that civilisation is possible

technological marvels. For instance, Babylonians were pioneers of astronomy and mathematics. The most famous European geometric theory (Pythagoras theorem) is known to be taken from the Babylonians that knew that relationship some 1000 years prior to Pythagoras (website 6).

The powerful empire was eclipsed by the Assyrians that brought civilization to yet another zenith. Their capital, Nineveh, was considered to be one the greatest cities ever built. Among their greatest achievements are the establishment of first library, paved roads, as well as Imperial administration, something that is comparable to civil service in today's world. History indicates they also became tyrants, soon after which they succumbed to a coalition of Babylonian descendants, Medes, Scythians, and others. More than 50 years after that, Assyria and Babylonia became provinces of Persia. Despite their demise, they continued to influence subsequent cultures and empires. For instance, Chaldeans were well known for their great achievements in astronomy, medicine, architecture, and mathematics. It was the Chaldeans that built the hanging garden of Babylon. Despite technological marvels, as a government, Chaldeans followed the same tradition of oppression as that practiced by the Assyrians. It was them that invaded Jerusalem and kept Hebrews as captives. Language-wise Aramaic was still spoken but Akkadian language was being revived.

The downfall of the Chaldeans came with the invasion of the Persians. The Persian emperor of the time, Cyrus the Great, would extend the empire from Indus River to the Nile. Interestingly, Persians didn't follow through Assyrian or Chaldean policies of divide and rule or ruthless tyranny. They made significant political reforms, including freeing the Hebrews in Jerusalem. In a way, they introduced the policy of multiculturalism. People were allowed to keep their cultural/religious traditions as long as they paid taxes. This was a new dimension to centralized governing. They also issued a common currency and built a vast network of roads. They also built libraries that would contain more books than European libraries that were built even in modern age. We are not certain about the motivation of the Persian emperor but such tactics were sustainable.

Later on, Greeks supported revolts against the Persian empire, prompting the Persian emperor to order revenge. Persian emperor, Darius the Great amassed a huge army but the war ended up with an

only as a function of the just and measured application of human conscience to the solving of social problems.

effective stalemate, despite early gains by Persian general Mardonius. Eventually, Darius the Great died leaving his son, Xerxes to follow up with a second campaign to invade Greece. Initially, the Persian army almost overran the entire Greece, but eventually lost the war and settle for a peace treaty, called the Peace of Callius. One hundred and fifty years later, Alexander of Macedonia conclusively defeated the Persian Emperor, Darius III, and established the largest empire of the time.

Alexander left a controversial legacy in the history of mankind. He left any city that surrendered to his invading army relatively unharmed while ransacking anyone that resisted the invasion. For instance, the capital of Persia, Persepolis, was burnt to the ground. This city was a technological masterpiece. It had major underground tunnels for sewage, had a gigantic water storage tank carved on a mountain, cisterns built, and had natural air conditioning systems. The city reportedly had three walls with ramparts. The first wall was 7 meters tall, the second 14 meters, and the third wall, which covered all four sides, was 27 meters in height. No trace of these walls exists today. The same Alexander built the lighthouse of Alexandria – one of the seven ancients wonders of the world. Alexander had planned to name Babylon his new capital and fuse Greek and Persian cultures into one. He also recommended to his generals that they should take Persian wives. He died before implementing these policies. At his death bed, Alexander's generals asked, "Who should succeed you?" He replied, "the strongest". Such question was legitimate considering Alexander had only an illegitimate child (with his Persian concubine) and his second child wasn't yet born. Both children were murdered before they reached adulthood.

The Greek empire would eventually degenerate, leading way to the rise of Parthians that revived old Persian traditions but didn't bring back sophistication in governance of the past. They were mainly excellent as warriors and despite ruling a large empire, their contribution to human civilization is limited to warfare tactics and war ammunitions. The Parthians were little concerned about Roman-Byzantine invasion and the rise of Roman form of Christianity. The Parthians were overthrown by the Sassanid dynasty, which would go down in history as the last Persian dynasty before Islam's rise in the world. Sassanids were more interested in erasing Greek influences than excelling in warfare. They also excelled in building monuments and cities. Meanwhile, the Roman empire was falling apart. The western part collapsed leading Europe to plunge into darkness. The eastern empire lived on with the Byzantine empire that had all but erased the original traditions of the Roman empire.

A5.2.6 Counterposing "Science" to "Religion": When Did the Divergence Occur?

This apparent divergence (of "science" from "religion") is seen everywhere today. For the last three centuries, its date of birth was conventionally asserted to be sometime during the French Encyclopaedists' movement of the middle decades of the 18th century, which proclaimed atheism as a viable alternative to a Church-backed Establishment that oppressed vast segments of the ordinary people while terrorizing into silence any conscious observers of these processes. On closer examination, however, our investigation has located the earliest impregnation of the social order with this peculiar protoplasm in the peculiar conditions attending the Roman occupation of Palestine during the first century CE. This was an occupation that came, on the one hand, to rely partly on the Jewish moneylender — who made the continuing Roman presence a source of mutual profit for both the money lenders and the occupier — and partly on the growth, especially among the ordinary humble *fellahin* that formed the bulk of the following of the sect organized around the life and figure of Jesus of Nazareth, of deep-going hostility to the yoke of the moneylender and his principal ally, namely, the Sanhedrin or Jewish rabbinical court.[23]

A5.3 The Phenomenon of Muhammad, the Last Prophet of Islam

Then came the most significant change in human history. A man, who never attended school, who was illiterate, and orphan (father died before birth and mother died when he was 5 year old, then grandfather died when he was six year old) claimed he had received divine revelations that would continue for 23 years and would make the only book that remains preserved in its original form for over 1400 years. The book is the Quran and the man is prophet Muhammad, who was ranked as the most influential world leader of human history (Hart,

[23]The authors are indebted to the late US citizen-of-the-world Gore Vidal for his brilliant political satire in general, and his insight in particular into the more than strange twists and turns in the history of Christianity's first century. In a brilliant flash of inspiration, in his *Live from Golgotha: The Gospel according to Gore Vidal* (New York: Penguin Books, 1992), Vidal projects onto centre stage an unwritten alliance between the Roman occupier on the one hand, the Jewish moneylenders and high religious court on the other hand and the struggle of the fellahin to break the chains that subjugated them in semi-bondage and endless indenture.

1992). His approach was fundamentally different from anything Europe has seen in modern age. Hart, an American Astrophysicist, wrote, "My choice of Muhammad to lead the list of the world's most influential persons may surprise some leaders and may be questioned by others, but he was the only man in history who was supremely successful on both the religious and secular level". No such leader with similar 'religious' and 'secular' standing emerged after his time. He is the first man in known history that made it mandatory on each of his followers to acquire knowledge. How did he propose that man acquire knowledge? He said, 'cure to ignorance is to question'. Today, Muhammad is the only man in history whose biography is preserved, through his own sayings and description of the context of those sayings, for nearly 1200 years. He claimed to bring only one miracle with him. It was the Quran, a book he claimed would remain unaltered and guide mankind until the day of Judgement. Today, *The Qur'an* is the only written document that is preserved for over 14 centuries in its original form.

His influence catapulted 1000 years of unparalleled boost in research and development, a period for which the language of science was Arabic. The book that he claimed was a divine revelation had the word 'science' (*ilm* in Arabic) over 700 times (second most used word only second to the word 'Allah') and has no dogma. It lays out the foundation of a society based on conscious and conscientious participation of every individual. It promotes memorization of the Quran that remains exact. With Quran as the starting point, it adds an axis that is the practice of Prophet Muhammad, then asks its followers to time scale (*qias* in Arabic) that scenario to the époque of interest and treat worldly belonging (including time) as a trust and defines human role as the 'viceroy of the creator'.

Muhammad challenged the fundamental premise of all creeds and stated the first premise should be: There is no *Ilah* (someone worthy of being obsessed with) but Allah, changing the center for any approach of cognition. Then, he claimed he was the last of numerous messengers that were sent by Allah, all carrying the same message of Islam (peace through submission to Allah). If this premise is true, he would be the most educated person ever and a role model for the rest of the humanity. It would also make his companions the most rightly guided in human history.

We considered the list provided by Michael Hart and assigned impact factor on human history based on the merit of the first premise of each of the individuals listed. If the premise advanced was false or illogical, the individual had a –ive sign assigned to him. The list follows.

Figure A5.1 shows the "top 14" from the above list, along with Adam, the first human. Of interest is the positioning of Jesus, son of Mary and well as Moses, the prophet of Islam. If Prophet Muhammad's first premise is true, then Jesus, the son of Mary would be a prophet and would have very positive influence on mankind. However, the Jesus that Michael Hart reported to be the 2nd most influential person in history is a creation of St. Paul and others, whose narration is tainted by their own dogmatic cognition that puts their credibility to the negative territory. Therefore, 'Jesus the son of God' fairs on the negative side. The same applies to Moses. According to Prophet Muhammad, he was a great prophet. New science has questioned the existence of this character and others have accepted him as the founder of the State of Israel. Consequently, their positioning falls on the negative territory. This figure, among other things shows how a false premise amounts to disinformation that continues to hurt the knowledge gathering process for years to come. In this, if the disinformation is insidious or if many people are bought into the falsehood, the impact on the society is disastrous.

Throughout history, important characters appeared and increased the knowledge base to a great level by planting the seed of knowledge that can be gathered by conscious and conscientious participation of the general public. In Europe, it has been told for many centuries that such knowledge deals strictly with the existence of God or the concept of heaven and hell. This notion is a scientific fraud planted with the doctrinal philosophy that stripped off conscious participation of general public. A logical conclusion that arises from Prophet Muhammad's first premise is knowledge is from God but it's principally for functioning a society that needs rule of law in order to establish peace and justice. Einstein famously said, "Great spirits have always encountered violent opposition from mediocre minds". Prophet Muhammad's starting point is, everyone is a born genius but some choose to violently oppose conscientious activities because they fail to focus on the long term. Myopic vision makes one blind. This focus on tangible and short-term is tantamount to the work of devil that always opposes knowledge and promotes ignorance. Therefore, shortly after the emergence of a great leader, one invariably finds gradual decay in the knowledge base, so much so that the society falls in a chaos until a new leader shows up.

This is shown in Figure A5.2 (above). Added to this figure future of knowledge that is predicted to gain a boost upon arrival of Jesus, who, according to Islamic faith, is to return to establish knowledge-based system for another thousand years. Table A5.2 (below) meanwhile summarizes major philosophical thoughts of ancient and medieval era. This table shows consistency in many respects, with the exception of the Roman Catholic church.

Table A5.1 Ranking of most influential people of history (footnotes follow this table).

Person	Time	Impact factor	Fundamental premise/discovery or invention
Muhammad	570-630AD	100	Allah (creator) is the only *îlah.* Muhammad, human) is the last messenger; Islam is the only *Deen* (*Dharma* in Sanskrit); You are judged by your intention and only intention (*niyah*), which is the root of all conscious actions (*Ama'al;* see also *Karma**). You are guided by Allah through Salat[#] (first thing to be questioned on the day of Judgement) and *ilm* of *Ayahs* of Allah (all tangible and visible objects, see also *Ayatana***) to reach salvation (see *Nirvana****).
Newton	1643-1727	-99	Dogma; wrote more on Dogma Christianity than on Dogma Science. Most notable first premise: There is external within creation.
Jesus Christ	30 BC-1 AD	98 (-/+)	Jesus: Son of God: Dogma Jesus: Son of Mary, prophet of Islam; Allah is the only *Ilah.* (see Muhammad)
Buddha	563BC-483BC	97	Chetna → Yoga[#]will give one *Nirvana****.
Confucius	551BC-579	96	What you don't want yourself, don't do to others. Benevolence, charity, humanity, love, kindness are fundamental virtues.
St. Paul	5-67AD	-95	"Resurrected Jesus" and Ananias of Damascus were being truthful; Jesus son of God..

(*Continued*)

Table A5.1 cont.

Person	Time	Impact factor	Fundamental premise/discovery or invention
Cai Lun	50AD-121AD	94	Unknown
Johannes Gutenberg	1398-1466	-93	Dogma: turned religion/secret knowledge into Money making enterprise.
Chris Columbus	1451-1506	-92	Dogma and European Christianity; Money is the *Ilah*.
Einstein	1879-1955	-91	Discontinuity in logic as well as matter and energy
Pasteur	1822-1895	-90	Dogma; Money is the *Ilah*.
Galileo	1564-1642	-89	Dogma; Money is the *Ilah*.
Aristotle	384BC-322BC	88	Nothing can be A and not A at the same time.
Euclid	300 BC	87	Unknown (some question if Euclid existed at all) May be a creation of Roman Catholic church.
Moses	unknown	86 (-/+)	Moses: Christian prophet and lawgiver: Father of State of Israel (zionism): Dogma Moses: prophet of Islam; Allah is the only *Ilah*. (see Muhammad)
Darwin	1809-1882	-85	Idealism, Nature is God.
Shih Huang Ti	247 BC-221BC	84	Unknown
Augustus Caesar	27BC-14AD	-83	Money is the *Ilah*.
Nicolaus Copernicus	1473-1543	-82	Dogma; Stationary sun at the center of solar system
Antoine Lavoisier	1743-1794	81	Knowledge in search of truth and serving others

(*Continued*)

Table A5.1 cont.

Person	Time	Impact factor	Fundamental premise/discovery or invention
Constantine the Great	272-337	-80	Dogma, European Christianity
James Watt	1736-1819	79	Unknown/steam engine
Michael Faraday	1791-1867	78	Unknown/magneto-electricity
James Clerk Maxwell	1831-1879-	-77	Speed of light independent of mass; energy independent of mass/ Electromagnetism theory
Martin Luther	1483-1546	-76	Dogma;, Original sin
George Washington	1789-1797	-75	Money is *ilah*; All humans are not equal.
Karl Marx	1818-1883	-74*	Nature is God; society is self correcting; people can be trusted to do good; anti-dogma
Orville and Wilbur Wright	1871-1948;1867-1912	73	Money is *Ilah*/Airplane
Gengis Khan	1162-1227	-72	Control is *Ilah*; Father of Othoman empire
Adam Smith	1723-1790	-71	Anti-dogma, western philosophy
Edward de Vere ()	1550-1604	-70	Monarch is God; Dogma, wrote about caselaws (in plays); wrote six volumes on Christianity and philosophy
John Dalton	1766-1844	69	Wasn't rich; experimental observations accurate but procedure crude.
Alexander the Great	336BC-323BC	-68	Imperialist; Control is *Ilah*.
Napoleon	1769-1821	67	Dogma, Money, Sex, and status

(*Continued*)

Table A5.1 cont.

Person	**Time**	**Impact factor**	**Fundamental premise/discovery or invention**
Thomas Edison	1847-1931	66	Money, Sex, and status; light bulb, phonograph
Anthony van Leeuwenhoek	1632-1723	65	Discovered microbes,
William T.G. Morton	1819-1868	-64	Pioneer in chemical Anesthesiology,
Guglielmo Marconi	1874-1937	-63	Dogma; radio
Hitler	1889-1945	-62	Homogeneity is a sign of civilization
Plato	428BC-347BC	61	Truth lies within intangibles
Oliver Cromwell	1599-1658	-60	Dogma; Money is *Ilah*; "intensely religious man"
Alexander Graham Bell	1847-1922	-59	Dogma, Money is Ilah; telephone
Alexander Fleming	1881-1955	58	Dogma, penicillin, chemotherapy
John Locke	1632-1704	-57	Western philosophy, political philosophy, "government with the consent of the governed"
Beethoven	1770-1827	56	Dogma; Money is *Ilah*; music
Werner Heisenberg	1901-1976	-55	Principle of uncertainty; quantum mechanics?
Louis Daguerre	1787-1851	-54	Dogma; photography
Simon Bolivar	1783-1830	53	Dogma; L

(Continued)

Table A5.1 cont.

Person	Time	Impact factor	Fundamental premise/discovery or invention
Rene Descartes	1596-1650	-52	Dogma; rationalization of philosophy and math; father of western philosophy
Michelangelo	1475-1564	-51	Dogma; painting and architecture
Pope Urban II	Unknown-230	-50	Dogma, Money is *Ilah*; crusade; Church-Monarch Imperialism
Umar ibn al-Khattab	579-644	49	Same as Muhammad
Asoka	304BC-232BC	48	unknown
St. Augustine	354-430	-47	Dogma; Money is *Il*ah; Flat earth; Eurocentric supremacy, original sin, just war
William Harvey	1578-1657	-46	Dogma; circulation of blood; modern embryology
Ernest Rutherford	1871-1937	-45	Money is *Ilah*; Subatomic physics
John Calvin	1509-1564	-44	Dogma; Money, Sex, Status is the trinity.
Gregor Mendel	1822-1884	-43	Dogma; genetics; similar to Newton's premises
Max Planck	1858-1947	-42	Dogma, similar to Newton's
Joseph Lister	1827-1912	-41	Dogma; chemical bacteria killers
Nikolaus August Otto	1832-1891	40	Unknown; internal combustion engine
Francisco Pizarro	1471-1541	-39	Dogma; Money is *Ilah*.(Spanish conqueror of South America)
Hernando Cortes	1485-1547	-48	Dogma, Money is *Ilah*.(Conqueror of Mexico)

(Continued)

Table A5.1 cont.

Person	Time	Impact factor	Fundamental premise/discovery or invention
Thomas Jefferson	1743-1826	-47	Eurocentricism;
Queen Isabella I	1451-1504	-46	Dogma; Spanish Queen who sent off Columbus, ordered conversion or exile of Muslims and Jews; Money is *Ilah*.
Joseph Stalin	1878-1953	-45	Control is *Ilah*. Marxism.
Julius Caesar	100 BC-44 BC	-34	Control is *Ilah*.
William The conqueror	1028-1087	-33	Dogma; Money and control are *Ila*h.
Sigmund Freud	1856-1939	-32	Sex is *Ilah*. Short-term is everything.
Edward Jenner	1749-1823	31	Dogma, small pox vaccine, immunology
Wilhelm Roentgen	1845-1923	-30	Manufacturing of X-ray
Johann Bach	1685-1750	-29	Dogma; music
Lao Tzu	600 BC	28	Anti-authoritarian, hermit-style
Voltaire	1694-1778	-27	Western philosophy
Johannes Kepler	1571-1630	-26	Plenatary motion
Enrico Fermi	1901-1954	-25	Dogma, Money is *Ilah*; father of atom bomb, theory of beta decay
Leonhard Euler	1707-1783	-24	Dogma; 'devout christian'; Newtonian calculus revisited
Jean-Jacques Rousseau	1712-1778	-23	Dogma to western philosophy
Nicoli Machiavelli	1469-1527	-22	Dogma, western philosophy, white supremacy, political realism

(*Continued*)

Table A5.1 cont.

Person	Time	Impact factor	Fundamental premise/discovery or invention
Thomas Malthus	1766-1834	-21	Dogma, humans are liability, population control by natural phenomena
John F. Kennedy	1917-1963	-20	Dogma; Sex, Money and Control
Gregory Pincus	1903-1967	-19	Males are superior;; Money is *Ilah*, birth control pill
Mani	216-276	18	Salvation possible through education, self-denial, fasting and chastity; proclaimed to be a messenger of God (see Muhammad)
Lenin	1870-1924	-17	Marxism; Sex, Money, Control is the Trinity.
Sui Wen Ti	541-604	-16	Control is *Ila*h. Unified china
Vasco da Gamma	1469-1524	15	Dogma, Money is *ilah*, discovered passage to India.
Cyrus the Great	600 BC-530 BC	-14	Control is *ilah*. Persian empire.
Peter the Great	1672-1725	-13	Dogma; Control is *Ila*h. Russian empire. Monarch-Church Imperialism.
Mao Zedong	1893-1976	-12	Control is *ilah*.
Francis Bacon	1561-1626	-11	Dogma; Inductive scientific method, empiricism
Henry Ford	1863-1947	-10	Money is *Ilah*; motor vehicle manufacturing.
Mencius	372BC-289BC	-9	Confuciusm
Zoroaster	600BC	8	unknown

(*Continued*)

Table A5.1 cont.

Person	Time	Impact factor	Fundamental premise/discovery or invention
Queen Elizabeth II	1926-	-6	Money is *Ilah*. See George Washington. Created Church-Monarch-Industry trinity.
Mikhail Gorbachev	1931-	-5	Dogma; end of USSR
Menes	unknown	4	Control is *Ilah*; Unified Egypt.
Charlemagne	742-814	-3	Dogma; creator of Church Imperialism.
Homer	800 BC	1	Poetry
Justinian I	482-565	-0	Dogma; Church-Monarch imperialism
Mahavira			
Thomas Aquinas	1225-1274	-	
Archimedes	287BC-212BC	+	
Charles Babbage			Computer forerunner
Cheops			pyramid
Marie Curie	1867-1934	-	radioactivity
Benjamin Franklin			
Mohandas Gandhi			
Abraham Lincoln			
Ferdinand Magelian			
Leonardo da Vinci			

* Karma in Indian religions is the concept of "action" or "deed", understood as that which causes the entire cycle of cause and effect (i.e., the cycle called saṃsāra, meaning 'continuity') originating in ancient India and treated in the Hindu, Jain, Buddhist, and Sikh religions (from Wikipedia)
** Āyatana (Pāli; Sanskrit) is a Buddhist term that has been translated as "sense base", "sense-media" or "sense sphere."[1] In Buddhism, there are six internal sense bases (Pali: ajjhattikāni āyatanāni; also known as, "organs", "gates", "doors", "powers" or "roots") and six external sense bases (bāhirāni āyatanāni or "sense objects"; also known as vishaya or "domains"). (From Wikipedia on Buddhist philosophy)
*** Nirvāṇa is an ancient Sanskrit term used in Indian religions to describe the profound peace of mind that is acquired with moksha (liberation). In shramanic thought, it is the state of being free from suffering. In Hindu philosophy, it is union with the Brahman (Supreme Being). (from Wikipedia on Nirvana)
**** San Kalpa (sahn KAL-pah) is an ancient Sanskrit sutra that means "My intentions have infinite organizing power." (from Chopra, 2012).
# Yoga means 'connection' in English and *Salat* in Arabic.

Table A5.2 Major philosophic moments identified over (historical) time.

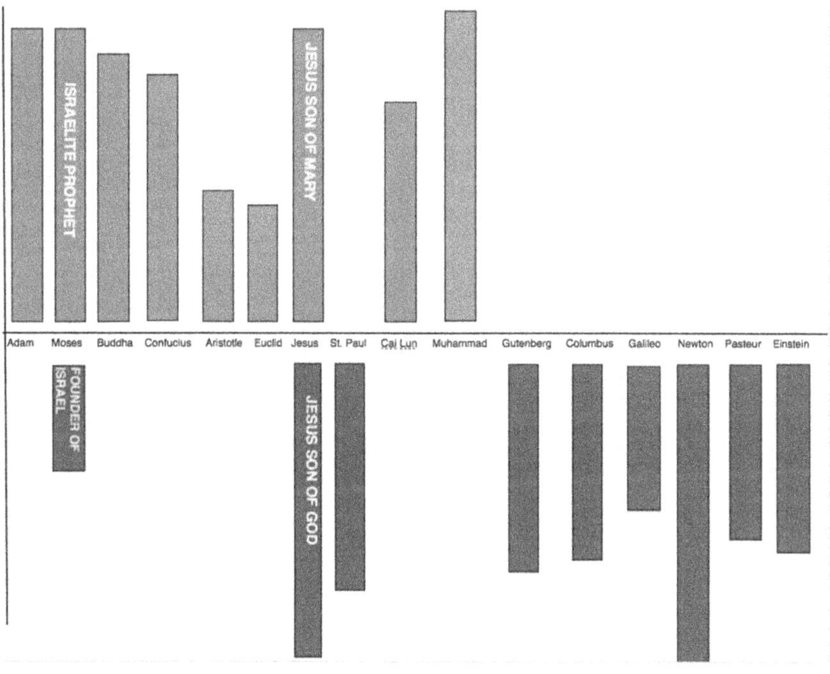

Figure A5.1 Impact of various historical figures on knowledge base of the human race.

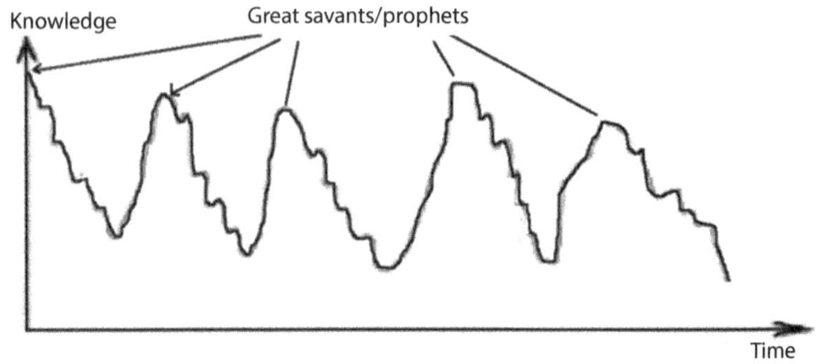

Figure A5.2 Cycles of knowledge if the first premise of Muhammad is true (from Islam et al., 2014a).

Epoch	Historical figure/ document	Philosophy
Ancient Indian	Mahabharat/Geeta	God is the truth and she unravels herself with time, *karma* (infinite cycle) is *chetna* (inspiration); God is the "Great Zero/void" (*Maha Sunya*) and is the holder of the universe (*mahidhara*)
Ancient Chinese	Budhha/Ancient Chinese philosophers	God is the source, everything else is His manifestation (*avatar*) and is in perfect harmony; peace is in harmony with nature, made out of fire, water, earth, and air. Everything in nature is in motion and transits through dual properties, hot and cold, summer and winter, day and night, male, female, etc.
Ancient Mesopotamia (2000 BC)	Abraham	Qur'anic recount (e.g. 6:76-78): Allah is the only creator, only truly constant, only truth, and intangible. Everything else is fleeting and tangible.

(*Continued*)

Table A5.2 cont.

Epoch	Historical figure/ document	Philosophy
Ancient Greek (early 5th century BCE)	Parmenides of Elea	Truth is absolute; two paths exist, "path of truth", "path of opinion", questioning will unravel the truth.
Ancient Greek (490-430 B. C.)	Zeno of Elea	Dialectic reasoning to unravel the truth; God is absolute knowledge.
Ancient Greek (469-399 BC)	Socrates	Questioning to eliminate contradiction and unravel the truth; the greatest knowledge is understanding the good.
Ancient Greek (5th BC)	Leucippus	Nothing is continuous and everything (earth, fire, water, and air) is constituted of fundamental spatial entities that cannot be divided any further; "Void" is empty (Caos, chaos in Greek)
Ancient Greek (460-370 BC)	Democritus	Same as Leucippus, named fundamental unit as atom ($\alpha\tau o\mu\alpha\Sigma$, uncuttable); all atoms are the same, uniform, symmetric; the feel and taste of a substance is a function of $\alpha\tau o\mu\alpha\Sigma$ of the substance on the $\alpha\tau o\mu\alpha\Sigma$ of our sense organs
Ancient Greek	Heraclitus (540 BC)	Everything (creation) is in flux (transient); Universal order is independent of time.
Ancient Greek	Plato (429-347 BC)*	Reality is intangible, everything in flux is unreal (God is the only reality)

(*Continued*)

Table A5.2 cont.

Epoch	Historical figure/ document	Philosophy
Ancient Greek	Aristotle (384-322 BC)*	Exclusion of the middle; Atoms are fundamental spatial unit; hot, moist, cold, and dry are the fundamental properties (steady state model); speed of light is infinity; human eyes see because 'something is emitted from their eyes'; God is infinity; time is infinity.
Ancient Greek	Euclid (435 BC-365 BC):	Goodness is the knowledge of God.
Ancient Egypt	Euclid of Alexandria (300 BC):	The laws of nature are but the mathematical thoughts of God.
Roman Catholic Church	Augustine (A.D. 354-430)	God is timeless and is creator of time, and universe; God holds all knowledge and has no beginning or end; God is perfect, eternal, infinity, incomprehensible, simple, and unique; God is "the inexhaustible light" that enlightens all; God created "everything out of nothing") at the moment chosen by God; God created everything at its 'perfect' time; the universe was created simultaneously in "logical framework" and not in "six calendar days" because *creavit omni simul* ("he created all things at once"); Trinity and flat earth.
Roman Catholic Church	Damascius (458 AD-538 AD)	Nature is of unfathomable and unspeakable divine depth, being all in one, but undivided; God is infinite; he is absolutely good, knowledgeable, and powerful.

(*Continued*)

Table A5.2 cont.

Epoch	Historical figure/ document	Philosophy
Arabian Medieval period	Muhammad (570-632 AD)	Major premise: There is no god (*ilah*, in Arabic means someone worthy of being obsessed with) but Allah; Minor premise: Muhammad is the last of many messengers (including Jesus, Moses, Abraham) of Allah; Documents left behind are: Qur'an as entirely God's words and Hadith (his sayings, actions, and actions of others that he didn't criticize). Allah is the only independent entity whose 99 traits are mentioned; rejects Trinity, calls Gabriel an angel (messenger between God and prophets), and states the purpose of human as the viceroy (*khalifah* on earth) and are created inherently good, opposite to the notion of 'original sin'. Most famous Hadith states intention (*niyah*) to be the source of all actions and accountability and ask followers to optimize dynamic intention (*qsd*), in line with *niyah* and *niyah* in line with his role as a *khalifah* (viceroy). He said, "Cure to ignorance is to question."

*Islamic scholars (Arab and non-Arab) were principally responsible for bringing back these philosophical thoughts from ancient Greek archive through translation into Arabic after 10th century.

6

Newton & Einstein: A Delinearized Deconstruction

6.1 Summary

In previous chapters, we discussed how New Science is incapable explaining natural phenomena. In this chapter, the most important theories that shaped New Science are deconstructed and the ground work for presenting a comprehensive energy and mass balance is made. Of importance is the finding that energy and mass alone can alter human thought process and thereby change their perception, rendering humans into robots. Unless a new theory is developed, one cannot begin to propose solutions to the crisis humanity is facing today.

6.2 Introduction

The modern age has been characterized as being both a time of "technological disaster" (as per Nobel Laureate Chemist, Robert Curl), and of "scientific miracles" (as the most predominant theme of modern education). Numerous debates break out every day, resulting in the formation of various schools of thoughts, often settling for "agreeing to disagree". At the end, little more than band-aid solutions are offered in order to "delay the symptoms" of any ill-effects of the current technology developments. This modus operandi is not conducive to knowledge and cannot be utilized to lead the current civilization out of the misery that it faces, as is evident in all sectors of life. In this regard, the information age offers us a unique opportunity in the form of 1) transparency (arising from monitoring space and time); 2) infinite productivity (due to inclusion of *intangibles, zero-waste*, and transparency); and 3) custom-designed solutions (due to transparency and infinite productivity). However, none of these traits has any meaning if we don't have a theory with correct hypothesis. This chapter addresses the most important theories advanced in modern age and deconstructs them in order to set stage for a comprehensive theory that can explain natural phenomena without resorting to dogma. These theories are widely deemed to be 'revolutionary' in the sense of having caused a 'paradigm shift' in their respective fields. Our contention on the other hand is that all these theories are rooted in fundamentally flawed theories and 'laws' from the time of Atomism. Here is the list of theories.

1. Information theory: Claude Shannon, 1948
   Information theory is a branch of applied mathematics, electrical engineering, and computer science involving the quantification of information. Information theory was developed by Claude E. Shannon to find fundamental limits on signal processing operations such as compressing data and on reliably storing and communicating data. This theory is important because it changes intangibles into tangibles. Since its inception it has broadened to find applications in many other areas, including statistical inference, natural language processing, cryptography, neurobiology, the evolution and function of molecular codes, model selection in ecology, thermal physics, quantum computing, linguistics, plagiarism detection, pattern recognition, anomaly detection and other forms of data analysis. This theory is based on fundamental frequencies and waveforms that are non-existent in nature.
2. Game theory: John von Neumann and Oskar Morgenstern, 1944 (with important embellishments from John Nash in the 1950s)
   Even though originally developed for Economics, Game theory is used in political science, psychology, logic, computer science, biology, and various other disciplines in relation to risk management. The fundamental premise of this theory is that the resources are finite and non-renewable. It also adds other spurious premises related to

probability, Atomism, human-non-human interactions, and static nature of the society.

3. Oxygen theory of combustion: Antoine Lavoisier, 1770s
Lavoisier did not discover oxygen, but he figured out that it was the gas that combined with substances as they burned. Lavoisier thereby did away with the prevailing phlogiston theory and paved the way for the development of modern chemistry. Underlying premises of Lavoisier have been deconstructed in Chapter 4.
4. Plate tectonics: Alfred Wegener, 1912; J. Tuzo Wilson, 1960s
Wegener realized that the continents drifted around as early as 1912. But it wasn't until the 1960s that scientists put the pieces together in a comprehensive theory of plate tectonics. Wilson, a Canadian geophysicist, was a key contributor of some of the major pieces, while many other researchers also played prominent roles. (Keep in mind that plate tectonics should not be confused with Plates Tectonic, a good name for a revolutionary science-theme restaurant.) This theory uses the same premise that other scientists have used to describe origin of universe. Such premises have been deconstructed in Chapter 3.
5. Statistical mechanics: James Clerk Maxwell, Ludwig Boltzmann, J. Willard Gibbs, late 19th century
By explaining heat in terms of the statistical behavior of atoms and molecules, statistical mechanics made sense of thermodynamics and also provided strong evidence for the reality of atoms. Besides that, statistical mechanics established the role of probabilistic math in the physical sciences. Modern extensions of statistical mechanics (sometimes now called statistical physics) have been applied to everything from materials science and magnets to traffic jams and voting behavior. This theory invokes spurious premises related to Newtonian mechanics, Atomism, and probability, which will be deconstructed in this chapter.
6. Special relativity: Albert Einstein, 1905
This theory is revolutionary for the fact that it includes time as an implicit function. However, it introduces spurious premises that will be discussed in this chapter. In itself, it is

based on Maxwell's theory, which in turn is based on Newtonian description of matter.

7. General relativity: Einstein, 1915
   It is conventionally perceived that General relativity was much more revolutionary than special relativity, because it replaced Newton's law of gravity in favor of curved spacetime. It gave rise to the emergence of a series of cosmic theories, ranging from Big bang to blackholes. This aspect has been deconstructed in Chapter 3 and will be deconstructed further in this chapter.
8. Quantum theory: Max Planck, Einstein, Niels Bohr, Werner Heisenberg, Erwin Schrödinger, Max Born, Paul Dirac, 1900–1926
   Quantum theory replaced the entire fabric of classical physics that was based on Newtonian mechanics. In the HSSA degradation mode, Quantum theories represent the worst form of cognition. It has been deconstructed in previous chapters will be deconstructed further in this chapter and beyond.
9. Evolution by natural selection: Charles Darwin, 1859
   Darwin showed that the intricate complexity of life and the intricate relationships among life-forms could emerge and survive from natural processes. This theory has been deconstructed by Islam *et al.*, (2010), especially in relation to its extension to human society. Fundamentally, this theory is similar to Quantum theory and applies similar spurious premises.
10. Heliocentrism: Copernicus, 1543
   Eurocentric prejudices dictate that such 'great insight' belonged to ancient Greek. While Copernicus was the first to challenge the Establishment in Europe in favour of natural cognition, we argued that Islamic scholars have been using far more powerful cognition tools for some 1000 years prior to Copernicus. This 1000 years of history is wiped out from New Science, triggering a cognition tool far worse than dogma itself. This aspect has been discussed in Chapter 3 and will be discussed in this chapter and beyond.

In previous chapters, we identified that almost all the theories and "laws" of the modern age have spurious assumptions behind them. It was also established that New Science is insufficient to account for natural phenomena, thereby making it impossible to design processes that are insightful in true sense of knowledge.

At present, numerous debates break out in favor and against any study that appears in the mainstream literature. Both sides use New Science to make their points, without questioning the validity of the "laws" and theories of New Science. In this book, the premises behind all of these laws and theories are challenged. Just like what happened in global warming for which each party calls the other party 'the flat-earth theorist' or 'conspiracy theorist', debates rage on as to every point of modern medical and chemical industries. Ironically, scientists all believe in the "chemicals are chemicals" or "energy is energy" mantra debate over why organic food and wood stove are better than their toxic alternatives but they all agree that it's carbon or heat that cause cancer. Just like the global warming debate, for which no one asks 'how could carbon dioxide be the enemy when we need carbon dioxide for producing life-sustaining plants, no one wonders how high temperature or carbon can give cancer when carbon is the essence of life and clay ovens produced healthy breads for thousands of years - all at high temperatures. No amount of doctrinal sermon can explain these contradictions, particularly as the same group, which promotes nuclear as "clean" energy, considers genetically modified, chemical fertilizers and pesticide infested crops derivatives processed through toxic means as "renewable". This same group also proclaims that electricity collected with toxic silicon photovoltaics and stored with even more toxic batteries -all to be utilized through the most toxic "white light"—as sustainable. In the past, the same logic has been used in the "I can't believe it's not butter" culture that saw the dominance of artificial fat (transfat) over real fat (saturated fat) as geared toward creating a similar crisis involving water (CBC, Dec. 19, 2008; Icenhower, 2006).

6.3 Historical Context

In medieval Europe, the resort to experimental methods did not arise on the basis of rejecting or breaking with Church authority. Rather it was justified instead by a Christian-theological argument, along the following lines:

(a) knowledge of God is what makes humans right-thinking and good and capable of having their souls saved in Eternity;

(b) this knowledge should be accessible wherever humans live and work; and
(c) the means should be at hand for any right-thinking individual to verify the Truth or eliminate the Error in their knowledge.

These "means" are then formulated as the starting point of what becomes the "scientific method". So, as a result (combining here the matter of the absence of any sovereign authority for the scientific investigator's conscience, and the Christian-theological justification for certain methods of investigation that might not appear to have been provided by any previously-existing authority), even with scientific methods, such as experiments, the conscience of an investigator who separated his/her responsibility for the Truth from the claims of Church authority — but without opposing or rebelling against that authority — could not ensure that his/her investigation could or would increase knowledge of the truth.

There is another feature that is crucial, regarding the consequences of vesting authority in a Central Knowledge-Certifier. For thousands of years, Indian mathematics were excelling in increasing knowledge, yet nobody knew about its findings for millennia outside of the villages or small surrounding territories — because there did not exist any notion of ***publication of results and findings*** for others. Contrast this with the enormous propaganda ascribing so many of the further advancements in the New Science of tangibles to the system that emerged of scholarly publication and dissemination of fellow researchers' findings and results. This development is largely ascribed to "learning the lessons" of the burning of the libraries of Constantinople in 1453 (by those barbaric Ottomans, remember), which deprived Western civilization of so much ancient learning (...)

The issue is publication, and yet at the same time, the issue is *not just* publication. Rather, it is, on what basis does publication of new findings and research take place? Our point is that publication will serve to advance knowledge in rapid and great strides ***if and only if*** authority is vested in the integrity and depth of probing by earlier investigators and investigations into all the various pathways and possibilities. Otherwise, this societal necessity and usefulness for publication becomes readily and easily subverted by the Culture of Patents, the exclusivity of "intellectual property", or what might be described today as "Monopoly Rights".

If & only if we put first the matter of the *actual conduct* of scientific investigations and the "politics" attached to that conduct (meaning: the *ways and means* by which new results are enabled to build humanity's store of knowledge), ***then & only then*** may we hope to reconstruct the actual line of development. With the actual knowledge of this line of development, for any given case, we can then proceed to critique, isolate and eliminate the thinking and underlying ideological outlooks that kept scientific work and its contents travelling down the wrong path on some given problem or question. The issue is not just to oppose the Establishment in theory or in words.[1]

6.3.1 Shedding Off Built-in Bias and Anti-Knowledge Motive

Human civilization is synonymous with working with nature. For thousands of years of known history, we know that man marveled at using mathematics to design technologies that created the basis of sustaining life on this planet. In this design, the natural system had been used as a model. For thousands of years, the sun was recognized as the source of energy that was needed to sustain life. For thousands of years, improvements were made over natural systems without violating natural principles of sustainability. The length of a shadow was used by ancient civilizations in the Middle East to regulate the flow of water for irrigation – a process still in existence in some parts of the world, known as the *fallaj* system. At nights, stars and other celestial bodies were used to ascertain water flow. This is an old, but by no means an obsolete, technology. In fact, this technology is far superior to the irrigation implanted in the modern age that relies on deep- water exploitation.

For thousands of years of known history, stars were used for navigation. It was no illusion, even for those who believed in myths and legends: stars and celestial bodies are dynamic. This dynamic nature nourished poetry and other imaginings about these natural illuminated bodies for thousands of years. The Babylonians started these stories, as far back as one can see from known history. Babylonian civilization is credited with dividing the heavenly bodies into 12 groups, known as the Zodiac. The Babylonians are also credited with the sexagesimal principle of dividing the circle into 360 degrees and each degree into 60 minutes. They are not, however, the ones responsible for creating confusion between units of time (seconds and minutes) and space (Zatzman, 2008). Their vision was more set on the time domain. The Babylonians had noticed that the sun returned to its original location among the stars once every 365 days. They named this

[1]The issue is rather to oppose the Establishment in practice, beginning with *vested* authority, regarding matters of science and the present state of the integrity and depth of our knowledge by probing earlier investigators and investigations to date, into all the various pathways and possibilities of a given subject-matter.

length of time a "year". They also noticed that the moon made almost 12 revolutions during that time period. Therefore, they divided the year into 12 parts and each of them was named a "month". Hence, the Babylonians were the first to conceive of the divisions of the astronomical clock.

Along came Egyptian civilization, which followed the path opened by the Babylonians. They understood even in those days, that the sun was not just a star, and that the earth was not just a planet. In a continuous advancement of knowledge, they added more constellations to those already identified by the Babylonians. They divided the sky into 36 groups, starting with the brightest star, Sirius. They believed (on the basis of their own calculations) that the sun took 10 days to cross over each of the 36 constellations. That was what they were proposing, ***thousands of years before the Gregorian calendar fixed the number of days to some 365***. Remarkably, this latter fixation would actually violate natural laws. In any event, it was something in which the Egyptians had no part. The Gregorian "solution" was larded with a Eurocentric bias, one that solved the problem of the days that failed to add up by simply wiping out 12 days.[2]

It was the Greeks — some of whom, *e.g.*, Ptolemy, travelled to Egypt to gather knowledge — who brought the total number of constellations to 48. This was a remarkable achievement. Even after thousands more years of civilization and the discovery of constellations in the southern sky, — something previously inaccessible to the peoples to whose history we have access — the total number of constellations was declared to be 88 in 1930. Of course, the Greek version of the same knowledge contained many myths and legends, but it always portrayed the eternal conflict between good and evil, between ugliness and beauty, and between right and wrong.

The emergence of Islam in the Arabian Peninsula catapulted the Arabs to gather knowledge on a scale and at a pace unprecedented in its time. Even before this, they were less concerned with constellations as groups of stars, and far more focused on individual stars and using them effectively to navigate.[3] Arabs, just like ancient Indians, also gave particular importance to the moon. Based on the movement of the moon among the stars, the Arabs divided the sky and its stars into 28 sections, naming them *manazil*, meaning the mansions of the moon. The moon is "hosted" in each mansion for a day and a night. Thus, the pre-Islamic Arabs based their

[2]Unix users can see this for themselves if they issue the command "cal 1752" in a terminal session.

[3]Not by accident, star constellations' names are of Greek origin, while the names of individual stars are mostly of Arabic in origin. Indeed: in the modern astronomical atlas, some 200 of the 400 brightest stars are given names of Arabic origin.

calendar on the moon, although they noted the accumulating differences between the solar and lunar calendars.[4]

Prehistoric Indians and Chinese assumed that the Earth had the shape of a shell borne by four huge elephants standing on a gigantic turtle. Similarly, some of the inhabitants of Asia Minor envisaged that the Earth was in the form of a huge disk, carried by three gigantic whales floating on the water. The ancient inhabitants of Africa believed that the sun sets into a "lower world" every evening and that huge elephants pushed it back all night in order for it to rise the next morning. Even the ancient Egyptians imagined the sky in the shape of a huge woman surrounding the Earth, decorated from the inside with the stars.[5]

The ancient Greeks, including Aristotle, assumed that the orbits of these celestial bodies were perfectly circular and that the bodies would keep revolving forever. For Aristotle, such perfection manifested symmetric arrangements. His followers continue to use this model. Scientifically speaking, the spherical model is nothing different from the huge elephant on a gigantic turtle model and so on. What precipitated over the centuries following Ptolemy, is a Eurocentric bias that any of the models that the Greeks proposed were inherently superior to the models proposed by ancient Indians, Africans, or the Chinese. In the bigger picture, however, we know now that the pathways of celestial bodies are non-symmetric and dynamic. Only with this non-symmetric model can one explain retrograde motion of the planets – a phenomenon that most ancient civilizations even noticed. Eurocentric views, however, would continue to promote a single theory that saw the Earth as the centre of the Universe. In Ptolemy's word: "During its rotation round the Earth, a planet also rotates in a small circle. On return to its orbit, it appears to us as if it is going back to the west." Of course, this assertion, albeit false, explained the observation of retrograde motion. Because it appears to explain or account for a phenomenon, it becomes true – the essence of a pragmatic approach which led to the belief that the Earth is indeed the centre of the Universe.[6]

---

[4]They also had many myths surrounding the sun, moon, and the stars. While Greek myths focused on kings and gods, however, Arab myths were more focused on individuals and families.

[5]This was in sharp contrast to the ancient Greek belief that the stars were part of a huge sphere. Ptolemy refined the ancient Greek knowledge of astronomy by imagining a large sphere with the stars located on the outer surface. He thought that all the planets known at the time - Mercury, Venus, Mars, Jupiter and Saturn - were revolving within this huge sphere, together with the sun and the moon.

[6]This belief would dominate the Eurocentric world for thousands of years.

The knowledge gathered about astronomy by the ancient Chinese and Indians was both extensive and profound. The Chinese were particularly proficient in recording astronomical incidents. The Indians excelled in calculations and had established important astronomical observatories. It was the Arabs of the post-Islamic renaissance that would lead the world for many centuries, setting an example of how to benefit from knowledge of the previous civilizations. Underlying this synthesizing capacity, was a strong motive to seek the truth about everything.

Among other reasons, a most important one was that every practicing Muslim is required to offer formal prayer five times a day, each relating to the position of the sun in the horizon. They are also required to fast one month of the year and offer pilgrimage to Mecca once in a lifetime, no matter how far away they resided (as long as they can afford the trip).

Most importantly, they were motivated by the *hadith* of The Prophet that clearly outlined, “It is obligatory for every Muslim man and woman to seek Knowledge through science (as in process)”. This was a most significant point of departure, diverging extremely sharply away from the Hellenized conception that would form the basis of what later became “Western civilization” at the end of the European Middle Ages. Greek thought from its earliest forms associated the passage of time, not with the unfolding of new further knowledge about a phenomenon, but rather with decay and the onset of increasing disorder. Its conceptions of the Ideal, of the Forms etc., are all entire and complete unto themselves, and — most significantly— standing *outside* Time, since truth was being identified with a point assumed to exist somewhere in time at which everything stands still.[7] Implicitly, on the basis of such a standpoint, consciousness and knowledge exist in the here-and-now — after the Past and before the Future unfurls.[8] All this has a significant, but rarely articulated consequence, for how Nature and its truths would be cognized. According to this arrangement, the individual’s knowledge of the truth at any given moment, frozen outside of Time, is *co-extensive* with whatever is being observed, noted, or studied.

[7]Even today, conventional models based on the “New Science” of tangibles unfolded since the 17th century disclosed its debt to these Greek models by virtue of its obsession with the steady state as what is considered the “reference-point” from which to discuss many physical phenomena, as though there were such a state anywhere in nature.

[8]Again, today, conventional scientific models treat time as the independent variable, in which one may go forward or backward, whereas time in nature cannot be made to go backward — even if a process is reversible.

The Islamic view diverged sharply by distinguishing its beliefs, knowledge (*i.e.*, some conscious awareness of the truth), and truths (or actuality). In this arrangement, the individual's knowledge of the truth or of nature is always fragmentary and also time-dependent. Furthermore, how, whether or even where knowledge is gathered cannot be subordinated to the individual's present state of belief(s), desires or prejudices. In the Islamic view, a person seeking knowledge of the truth cannot be biased against the source of knowledge, be it in the form of geographical location or the tangible status of a people. Muslims felt compelled to become what we term as "scientists" or independent thinkers—each person deriving their inspiration from the *Qur'an* and the *hadith* of Prophet Muhammad. Hence, they had no difficulty gaining knowledge from the experience of their predecessors in different fields of science and mathematics. They were solely responsible for bringing back the writings of Greek Aristotle and Ptolemy and the Indian Brahmagupta in the same breath. Neither were they role models: they were simply their ancestors whose knowledge Muslims did not wish to squander.

They started the greatest translation campaign in the history of mankind, to convert the written works of previous civilizations into Arabic. In due course, they had gained all prior knowledge of astronomy that enabled them to become world leaders in that field of science for five successive centuries.[9]

Some eight centuries later, we arrive at the heart of the Eurocentric attitude to science and knowledge. In the 16th century, Copernicus identified that "the Earth is not located in the center of the universe but the sun is. The earth and the planets rotate around the Sun." This simple observation of the truth could not be tolerated by the Catholic Church. For the Catholic Church, the Earth was the centre of the universe[10]

They could not realize there was no room for blind faith and prejudice while seeking knowledge or the truth. Galileo, yet another Church

[9]Even their political leaders were fond of science and knowledge. One remarkable pioneer of knowledge was Caliph Al-Mamoon, one of the Abbasid rulers. Some one thousand years before Europeans were debating how flat the Earth is, Al-Mamoon and his scholars already knew the earth was spherical (although — significantly — *not* in the European perfect-sphere sense), but he wanted them to find out the circumference of the Earth. Al-Mamoon sent out two highly competent scientific expeditions. Working independently, they were to measure the circumference of the Earth. The first expedition went to Sinjar, a very flat desert in Iraq. At a certain point, at a latitude of 35 degrees north, they fixed a post into the ground and tied a rope to it. Then, they started to walk carefully northwards, in order to make the North Pole appear one degree higher in the sky. Each time the end of the rope was reached, the expedition fixed another

member, offered another blow to the Eurocentric dogma. Galileo saw the earth moving. In his words: "O people! Beware that your Earth, which you think stationary, is in fact rotating. We are living on a great pendulum." Galileo wasn't just a "terrorist", interested in blasphemy. He discovered the four great moons of Jupiter. He was the inventor of the clock pendulum and the "Laws of Motion". The Church could not bear Galileo's boldness. He was put on trial. Confronted with such tyranny, Galileo, who was by then old and weak, yielded and temporarily changed his mind. But while he was going out of the court, he stamped his feet in anger saying: "But you are still rotating, Earth!" This was the beginning of a New Science that would dominate the world until today.

Galileo marks the *Eureka!* moment in Western "science". Science finally had broken out of the grip of the Church, and, therefore, was free from the bias that had a chokehold on clear thinking.[11] This has engendered unfortunately, yet another misconception. The earth science that was unleashed after Galileo, remains the science of tangibles. Within this science, the earth is not flat or at steady state, but it still is not part of the science of knowledge (Islam, 2008).

Take, for instance, the case of Earth itself. In Ibn Kordathyah, an Arab scientist mentioned the earth is not flat in early in his books, *Al-Masalik* and *Al-mamlik* in the 800's. So, what shape did he think the earth was? It was the word بَيضاوي أو بَيض (*baidh* or *baidha*). In modern the Europe-dominated word,, world or word?, it is translated as, elliptical. In reality, elliptical is an aphenomenal shape, meaning it doesn't exist anywhere in nature. The true meaning of this word is Ostrich's egg or its nest, which, obviously, is not elliptical. The inspiration of Ibn Kordathyah came from the Qur'an (Chapter 79, verse 30: في ت عالى وق ال). Ideal in Islamic culture is the Qur'an (Zatzman and Islam, 2007b). Contrast this with Western "science", for which the starting point would be the outline circumference

post and stretched another rope from it until their destination was reached—latitude 36 degrees north. They recorded the total length of the ropes and returned to the original starting point at 35 degrees north. From there, they repeated the experiment , heading south this time. They continued walking and stretching ropes between posts until the North pole dropped in the sky by one degree,when they reached the latitude of 34 degrees. The second of Almamon's expeditions did the same thing, but in the Kufa desert. When they had finished the task, both expeditions returned to Al-Mamoon and told him the total length of the rope used for measuring the length of one degree of the Earth's circumference. Taking the average of all expeditions, the length of one degree amounted to 56.6 Arabic miles. The Arabic mile is equal to 1973 meters. Therefore, according to the measurements made by the two expeditions, the Earth's circumference is equal to 40,252 kilometers. Nowadays, the figure is held to be 40,075 kilometers. How

of a circle rendered as an ellipse which has "degenerated" into some kind of ovoid. Then the egg is elaborated as an extrusion into 3-D of a particular case or class of a non-spherical, somewhat ellipsoidal circumference. Why not just start with the egg itself, instead of with circles and ellipses?[12]

Until the works of the Spanish Muslims, all translations were limited to science without particular attention to what would be considered philosophical. Averröes was the first to venture into even purely philosophical work, particularly that of Aristotle.[13] What emerges is a difference in attitude between standpoints maintained pre- and post-Thomas Aquinas, the father of Eurocentric philosophy. Before his time, truth was bound up with knowledge, and could be augmented by subsequent inquiries. After that point,, the correctness or quality of knowledge is rendered as a function of its conformity with the experience or theories of the elite (called "laws"). Before this time, personal experience was just that: personal. After this point, the experience of the elite had become a commodity that could be purchased as a source of knowledge. Before this point, the source of knowledge was an individual endeavor of research and critical thinking. After this point, it became dogma, blind faith, and the power of external (aphenomenal) forces.

After Thomas Aquinas, few Europeans emerge who can be said to be engaged in **increasing** knowledge *per se*. If they did so, they were severely persecuted. Copernicus (1473-1543) is just one example. What was his offence? The Earth moves around a stationary sun. It was not complete knowledge (it is important to note that "complete" knowledge is anti-knowledge), but it was knowledge in the right direction. His theory contradicted that of Ptolemy's and in general that of the Catholic church. Yet, Wikipedia wrote this about him: "While the heliocentric theory had been formulated by Greek, Indian and Muslim savants centuries before Copernicus, his reiteration that the sun — rather than the Earth — is at the

does this compare with the circumference of the earth as we know it today? Today, it is known to be 40,075 km if measured through the equator, a difference of less than 200 km. Contrast that with the debate that was taking place in Europe over the earth being flat many centuries later. Another important aspect of this was that this was the first time in known history that a state sponsored fundamental research. The fact that the motive of Caliph Mamoon was not to capture more land and history shows that these rulers were not the recipients of any taxes. In fact, at the height of the Islamic era all rulers in the Islamic world paid zakat, the obligatory charity, for the wealth they possessed, the entire amount going to the poor. Also, the judicial system was separate from the administration, with the judicial system being always in the hands of the "most righteous", rather than most "powerful". In fact, during the entire Ottoman period, even the state language was not Arabic. In the administration, it was Turkish for

center of the solar system is considered among the most important landmarks in the history of modern science" (Copernicus, 2007).

While there is some recognition that Copernicus's knowledge was not ***new*** knowledge, it did not prevent European scientists from making statements that would sanctify Copernicus. Goethe, for instance, wrote:

> Of all discoveries and opinions, none may have exerted a greater effect on the human spirit than the doctrine of Copernicus. The world had scarcely become known as round and complete in itself when it was asked to waive the tremendous privilege of being the center of the universe. Never, perhaps, was a greater demand made on mankind — for by this admission so many things vanished in mist and smoke! What became of our Eden, our world of innocence, piety and poetry; the testimony of the senses; the conviction of a poetic — religious faith? No wonder his contemporaries did not wish to let all this go and offered every possible resistance to a doctrine which in its converts authorized and demanded a freedom of view and greatness of thought so far unknown, indeed not even dreamed of (Copernicus, 2007).

In the above statement, there are three items to note: 1) there is no reference to Copernicus's knowledge being prior knowledge; 2) there is no comment on what the problem was with Copernicus's theory; 3) there is no explanation as to why religious fanatics continued to stifle knowledge and how to handle them in the future.

What would be the knowledge-based approach here? To begin with it might be to ask whether the theory contradicts the truth. European scholars did not ask this question. They compared, compared what?, with words in the *Holy Bible* — a standard whose authenticity, impossible to establish unambiguously, was itself subject to interpretation. When we ask the

---

communication with the headquarters while local languages were used for the local communication.

[10]...with possibly the Vatican at its epicenter!

[11]

[12]Eggs are real, actual. We can know all their properties directly, including everything important to know about the strength and resilience of their shapes as containers for their particular contents, without having to assume some so-called simple ideal and then extrapolate everything about what's inside eggs from these abstractions that exist solely in someone's imagination. , is this correct?, Going the other direction, on the other hand, is the much richer scientific path. Once we have explored real eggs and generalized everything we find out, we can anticipate meaningfully what will happen in

question "Is such-and-such true?", we cannot simply define the truth as what we wish. We have to state clearly the standard measure of this truth. For Muslim scientists prior to the European Renaissance, the *Qu'ran* formed the standard. Here is the relevant passage, Chapter 36 (36-40), from the *Qu'ran* addressing the matters of whether the sun is "stationary" or whether the earth stands at the centre of the solar system, or if the moon is a planet:

لا (كَالْعُرْجُونِ الْقَدِيمِ (٣٩ وَالْقَمَرَ قَدَّرْنَاهُ مَنَازِلَ حَتَّىٰ عَادَ (الْعَزِيزِ الْعَلِيمِ (٣٨ وَالشَّمْسُ تَجْرِي لِمُسْتَقَرٍّ لَهَا ذَٰلِكَ تَقْدِيرُ
(٤٠) الْقَمَرَ وَلَا اللَّيْلُ سَابِقُ النَّهَارِ وَكُلٌّ فِي فَلَكٍ يَسْبَحُونَ الشَّمْسُ يَنْبَغِي لَهَا أَنْ تُدْرِكَ

One possible translation: "And the sun runs on its fixed course for a term (appointed). That is the Decree (the word comes from 'qadr' as in "proportioned" or "balanced") of the All-Mighty (Al-Aziz) and the All-Knowing (Al-Aleem, the root word being *ilm* or science). And the moon, we have measured (or proportioned, again coming from the root word, "qadr") for it locations (literally meaning "mansion") till it returns like the old dried curved date stalk. It is not for the sun to overtake the moon, nor does the night outstrip the day. They all float, each in an orbit."

When did we find out that sun is not stationary? What is the speed and what does the solar orbit look like?

With 20/20 hindsight, many people write these days that the speed of the sun could be predicted using Newton's law. What is missing in this assertion is the assumption that Newton's law is absolute and all hypotheses behind Newton's gravitational laws are absolutely true. In addition, it also assumes that we know exactly how gravitational attractions are imparted from various celestial bodies — a proposition that stands (not to put too fine a point on it) is "over the moon"!

Along came Galileo (1564-1642). Today, he is considered to be the "father of modern astronomy," the "father of modern physics", and the "father of science". As usual, the Church found reasons to ask Galileo to stop promoting his ideas. However, Galileo really was not a "rebel". He

---

the relation between the forms of other exterior surfaces found in nature and their interior contents.

[13]It is widely acknowledged that without this translation, Aristotle's works would have disappeared from public access. This is indeed a bifurcation point in time. Muslims used older knowledge to increase their knowledge. Eurocentrics took that knowledge to increase their ignorance. As a consequence, one can easily see how all atrocities ranging from the mediaeval Crusades (at the Mediterranean edge of western Asia) to modern-day Crusades (in the Gulf region of western Asia) have focused only on promoting the aphenomenal model and how prior knowledge was destroyed or distorted in order to achieve "quick victories".

remained submissive to the Church and never challenged the original dogma of the Church that promotes the aphenomenal model. Consider the following quotations (Galile, 2007):, is the spelling of this citation correct?,

Psalm 93:1, Psalm 96:10, and Chronicles 16:30 state that "the world is firmly established, it cannot be moved." Psalm 104:5 says, "[the LORD] set the earth on its foundations; it can never be moved." Ecclesiastes 1:5 states that "the sun rises and the sun sets, and hurries back to where it rises."

Galileo defended heliocentrism and claimed it was not contrary to those Scripture passages. He took Augustine's position on Scripture—not to take every passage literally, particularly when the scripture in question is a book of poetry and songs, not a book of instructions or of history. The writers of the Scripture wrote from the perspective of the terrestrial world, and from that vantage point, the sun does rise and set. In fact, it is the earth's rotation which gives the impression of the sun in motion across the sky.

Galileo's trouble did not come from the Establishment because he contradicted Aristotle's principles. Galileo contradicted Aristotle's notion that the moon is a perfect sphere and that heavy objects would fall faster than lighter objects, directly proportional to weight, etc. Amazingly, both the Establishment and Galileo continued to be enamored with Aristotle while bitterly fighting with each other. Could the original premise that Aristotle worked on be the same premise as that of the Church as well as that of Galileo's? Why didn't he rebel against this first premise?

Galileo's contributions to technology, as the inventor of the geometric and military compasses, suitable to be used by gunners and surveyors, are notable. There, even Aristotle would agree, this was indeed *τεχνε* (techne) or "useful knowledge"—useful to the Establishment, of course.

What happens if we embark from another starting point, employing a different mode of discourse, apparently more familiar in its uses of conventional mathematical conceptions?

The starting point is important in this discourse. However, all starting points are arbitrary. However, according to a well-worn notion, "if you don't know where you're going, any path can take you there." (This idea has been recorded among peoples of different times and places — in the U.S., by the U.S. baseball celebrity, Yogi Berra, of the late 20th century to the Arabian Desert, attributed to the Quran). This section sets out to investigate the notions of delinearized historical rendering of scientific and technological developments. This process has become firmly established in the world's thinking as entirely Western, if not indeed overwhelmingly based upon the United States., is this a better way of saying this?, Our starting-point? It is the synthesis, over the five centuries that followed the life of the Prophet Muhammad, of ancient learning by the Muslim scholars inspired by Islam. During the period of post-Thomas Aquinas (father of doctrinal philosophy), cognition in Europe, the work of Islamic scholars continued and today, if one just searches in Wikipedia, one will find: Ibn sina (Avicenna) is named the father of modern medicine and alchemy, Ibn Rushd (Averroes), the father of secular philosophy and Education, Ibn Haitham (Alhazen), the father of modern optics. Al-Kindi (Alkindus) is the father of information technology, Ibn Khaldoun, the father of social sciences, Al-Khwārizmī , the founding father of algebra and mathematics, and Al-Farabi is named the father of epistemology and metaphysics. Yet, all of them are listed as either polyscientists or polymath, polymathematicians. In addition, all of them are considered to be inspired by the prophet Muhammad. This is truly an unprecedented event in human history and only goes with the latest work of Michael Hart that ranked Prophet Muhammad as the most influential world leader. Who is second in that list? That would be Sir Isaac Newton, the man who wrote more on Christian doctrine than on science. His Church wasn't based in Rome; it was the Church of England, headed by the Monarch. The transition from "religion" to "politics" wasn't even subtle. Some of our recent work only begins to touch upon the original theories of Islamic scholars who could have served humanity only if they were not altered with the intent of fitting a conclusion new scientists were trying to come up with in order to satisfy the Church, the Government, or the Corporation. Newton was unique because he satisfied the Church and the Monarch simultaneously. It was possible because the Head of the Church of England was also the Monarch (true until today). While Newton had no reason to challenge the first premise of the Church that he belonged to, others (e.g. Russian scientists, Einstein) didn't dare question the first premise of anyone, but most notably that of Newton's. If they did, they were quickly called "anarchists".

As an example, the Nobel Prize winning work of Albert Einstein, is noteworthy. Our recent book (Islam *et al.*, 2010), pointed out how Einstein's work simply took Maxwell's rigid sphere model as true and how Maxwell himself took that model from Newton. Consider how Ibn Haitham took the model of Aristotle and deconstructed it based on simple logic. He discarded Aristotle's conclusion that light has infinite speed and reconstructed a model that, until today, served as the only model that can distinguish sunlight from artificial light. While it is well known that sunlight is the essence of life and that artificial light is something that is used to torture people, Einstein's theory or any other optic theory cannot explain scientifically how this is possible. In addition, Ibn Haitham undid another one of the old theories, which is that something that comes out of our eyes makes it possible for us to see. Instead of that theory, he introduced the notion of something entering your eye that makes it possible to see. This "something" was later proclaimed to be a photon. This notion was correct, but the denomination as well as the attribution of various properties made further research on the topic of light characterization impossible. For instance, because this theory postulates that all photons are alike and do not have mass, the source of light cannot have an impact on the quality of light, leading to the same difficulty that made it impossible to discern between sunlight and artificial light. Whereas, if Ibn Haitham's theory was used correctly, one would be able to correlate the toxic nature of the light source (e.g. power-saving light) with long-term impacts, such as breast cancer, brain dysfunction, and numerous other reported correlations. This would also unravel the science behind skin-cancer causing chemicals that are often included in suntan or sun protection lotions. Another example is worth mentioning here, which is a recently touted new form of energy-saving light. This light was excellent in energy savings as well as producing the "white light" effect. However, it was also performing the so-called "belly dance". When the source was sent to the International Space Station for a probe, it was discovered that the "belly dance" subsided or disappeared. Gravity was found to be the reason behind the belly dance. Could this be explained with existing light theories? Of course not, because if photons have zero mass, how could gravity affect them? This paradoxical modus operandi continues when it comes to dark matters (infinite mass, but no energy) and dark energy (infinite energy but no mass) in the realm of the discussion in cosmic physics.

What's the catch? Ibn Haitham didn't read Aristotle's work to believe in it. He read the theory, used the criterion on truth and falsehood and was able to decipher true elements from the volumes of work. Einstein, in the other hand, either did not dare use the same logic about Maxwell's "laws" or lacked Ibn Haitham's criterion. Einstein was not a Christian, but he certainly was a believer of Eurocentric philosophy. This fundamental inability to discern truth from falsehood (called Furqan in Arabic), is missing from European New Science and social science. One result is continual confusion about everything governing daily life.[14]

These questions are not new. What is new is the realization that the answers to these questions could have been found hundred years earlier

[14]An article by Sardar (2009), talks about the philosophy that drove the mindset of Eurocentric scientists.

and today's environmental, technological, and moral disasters possibly averted by rejecting the path of doctrinal philosophy. Instead of looking at our previous scientists with a kind of veiled contempt as though they were somewhat inferior or defective human beings — we could be living in a different world today had we taken up natural cognition processes and continued with the theories that they had advanced,.

6.3.2 A Reflection on the Purposes of Science

There is a widespread notion that, ***with Science, many practical problems of daily existence can be solved that could not be solved otherwise***. This notion is especially popular among engineers. However, it must be admitted that this idea is also very much part of contemporary culture. Certainly in the countries of what is broadly known as the Global Troika — the Anglo-American bloc (*i.e.*, the United States, Canada, Britain and Australia), the other countries of Europe & Russia, plus Japan — it is one of the hallmarks of what is generally accepted as part of the modern outlook.

Here we have rendered the proposition in its most innocent-sounding form. If we zero in on the essence of this proposition, however, we stumble across something potentially far more contentious. People are inured to associating "***many practical problems of daily existence***" quite unconsciously with the immediate, the short term, that period in which time t stands at "right now". In order to define that period objectively, it must first be appreciated that the moment a particular point in time has been identified, actual time has already moved on. So, in reality, time t is always and everywhere actually "$t + \Delta t$". Hence, it follows that "right now" may be understood, and even "experienced", as the point at which $t + \Delta t \rightarrow t$, because at this point, it is in effect, $\Delta t \rightarrow 0$.

Here, however, we have to ask: has Science actually undertaken to address and deal with this rather singular point at which $t + \Delta t \rightarrow t$? Are these the highest and best purposes of Science, or are there indeed any other socially positive purposes, particularly well-served by addressing and dealing only or mainly or primarily with an immediate reality conditioned by the "$\Delta t \rightarrow 0$" criterion?

Elsewhere (Islam, 2003), this point has been identified and associated with what is called "the roller-coaster ride of the Information Age". For example, no sooner is a "daily life problem" like insomnia identified, than the pharmaceutical industry produces a pill to help one fall asleep. Then, however, the problem of meeting the demands of daily life, such as waking up on time to get to work, etc., emerges, and another pill — this one to help one "wake up" to full consciousness — is developed and marketed. Similarly, sugar is identified as a latent dietary time-bomb and in almost no time flat a "solution" is proffered, consisting of a choices among "sugar-free" substitutes. It hardly ends there, as the sugar-free substitutionalists then end up in deadly warfare, *e.g.*, the "Aspartamers" *versus* the "Saccharinisers" so to speak, over whose effects cause more lab rats to die more horrible deaths from dosages that—when scaled to anything in the range of adult human body-mass indices—bear no relation to any known pattern of human consumption of sugar, even over an entire lifetime (Zatzman, 2008).

The serious point masked throughout this dreadful comedy is precisely the temporal criterion that each of these "solutions" was supposed to be addressing in the first place. In every case, the litmus test applied to the proposed intervening substance is its performance over the period of time t = "right now". Everything else — including consequences for the human body or even initial acceptability to a functioning living organism — is deemed to fall in the category of "side effects" before being rudely sloughed off as someone else's problem.[15]

6.4 Time Conceptions, Tangible-Intangible Nexus, and Social Roles of Knowledge

As already mentioned above, the very moment a particular point in time has been identified, actual time has already moved on. Therefore, in reality, time t is always and everywhere actually "$t + \Delta t$". There is a subtlety introduced here: the observer cannot be detached from the process or from the phenomenon being observed. This is indeed an important truth that cognitive psychology, with reference to the human personality in general, explored in some depth during the 20th century. However, its application to the actual practice of scientific investigation has yet to be elaborated. Quite to the contrary, Newton, one of the bedrock premises of the so-called "hard sciences" (physics, chemistry, biology, geology and of all the engineering disciplines associated with these bodies of scientific knowledge), has been the observer who observes and speaks from *outside* the reference-frame of the phenomenon or process being studied.

[15]The assumption underlying this practice is that the Science that addresses problems at the point where $t + \Delta t$ à t is the only useful or profitable "science". It follows that any science that takes of care of anything *other than* the short-term is "blue sky" or otherwise "doesn't count", *i.e.*, isn't going to make anyone a billionaire.

Of course, the immediate consciousness of the scientific investigator as a human observer, at time t = "right now", is precisely that s/he is indeed independent of the phenomenon or process being studied. Far from being a merely static contradiction, however, this indicates that it is an inevitable byproduct of uncritically accepting the immediate consciousness as the definitive reality, which is that an actual loss of information has already taken place; i.e., information about the relationship between the observer and the phenomenon or process being studied. In this relationship, the "I" of the individual has become the "relate", meaning: the element that connects the observer as part of the reference frame to everything else in the reference frame.

What about this information loss? It should alert us to the risk we start to run if we conflate knowledge of the truth entirely and exclusively with what has been observed and recorded from within the moment in which $t + \Delta t \rightarrow t$ occurs. The chief error and limitation in this doctrine known to philosophy as empiricism — which says the only reliable knowledge is inductive and is generated by experiment, experience, and/or the human sensorium (sight, hearing, taste, touch or smell) — resides in its failure to grasp the objective fact of the continuance of time, independently of anyone. This state of affairs is rife with profound implications for assessing what constitutes reliable knowledge of the truth and distinguishing it from a catalogue of perceptions larded with pattern-matchings of varying degrees of elegance.

What is needed now is to advance the analysis further from this point. The difficulties involved are palpable. First, it becomes necessary to propose a new level of generality. This would be a level of generality of consideration in which the evident contributions to our knowledge of the truth, deriving from moment-to-moment observations based on the human sensorium and various data-collection methods and measuring systems on the one hand, and on the other hand, the positioning of where these phenomena fit into a bigger picture with time, considered over the long- term, can each be accommodated. For this task, the existing methods of what has been called, since the 17th century, "New Science" — *viz.*, to effect a "division of labor" between, on the one hand, the collection and cataloguing of observations of a process or phenomenon, and, on the other hand, the generalizing of larger meanings and patterns from the riot of data collected and catalogue, seem ever clumsier and more inadequate. In today's Information Age, we stand literally awash in collections of data about anything and everything on scales heretofore unimagined.

Here, a very old idea can be summoned into service with a highly modern twist. The old idea is to distinguish what is tangible from what is

	Physically incommensurable	Quantifiable
TANGIBLE?	to some degree: **NO**	to large degree: **YES**
INTANGIBLE?	to large degree: **YES**	to some degree: **NO**

Figure 6.1 Tangibility/Measurability matrix (Zatzman and Islam, 2007b).

intangible. The modern twist is to classify as "intangible"[16] all those features that exist within a process or phenomenon but have not yet acquired or developed a tangible expression. Then, the differences between the physical commensurability and measurability of tangible and intangible elements have to be arranged systematically. The following matrix is suggestive:

This matrix and its spectra suggest *pairing* tangible with intangible components, according to some relate, relationship?, that combines their relative commensurability and quantifiability, and to consider phenomena or processes as a **nexus** of tangible and intangible components. Note that this position stands 180 degrees opposite the general dispositions of conventional "New Science". According to the conventional standpoint, in the name of ensuring that the truth of the tangible doesn't get mixed up with superstition, religion, gobbledy-gook, or subjective biases, all notions of the intangible are to be banished to the margins (Figure 6.1).

It is the progress of (actual) ***time***, which provides the implicit connection between greater or lesser tangibility/intangibility. That "actual time" is what has been discussed extensively elsewhere (Zatzman and Islam, 2007b) as "characteristic time" in the case of natural processes, designated symbolically as $t_{NATURAL}$, or as "historical time" in the case of social processes, designated symbolically as $t_{HISTORICAL}$. Either of these is consistent with the idea of **time as a fourth dimension**. This stands in stark contrast, however, to the conventional notion of **time as the independent variable** developed throughout New Science since Newton — what is best characterized as t_{LINEAR} (Zatzman and Islam, *ibid*).

A more explicit index of this tangible-intangible nexus, also related to the passage of actual time, is the state of our information, i.e., our previous or existing relevant knowledge. This is less like data and more like understanding, taking "understanding" to refer to the fruit of conscious

[16]The term of art is "latent".

participation of individuals in definite acts of "finding out" (Zatzman and Islam, ibid).

Time and information here have their tangible expressions, e.g., as "duration", or as "data". However, it is their intangible roles that are decisive in mediating our ability as humans to make use of knowledge of the truth for socially positive ends and humanize the environment, natural or social. These are the activities that secure humanity's long term. No human social problem is without some human social solution. Whoever would increase knowledge is bound to disturb the status quo, but even so, a person must increase his/her knowledge of the truth. The essence of human social agency lies on the path of pursuing knowledge.

6.4.1 More About Time: Newton's "Laws of Motion" – Versus Nature's

A scientific approach to investigating phenomena requires examining both things in and of themselves and things in relation to other things. This entails mastering "laws of motion" relevant to the matter under investigation. Newton set out one particular model which was long assumed to be, and accepted as, the most profound summation of everything about matter in nature. His summary, in the form of the Three Laws of motion is one thing, but their underlying assumptions were quite another.

According to Newton's First Law, objects in motion remain in motion, and objects at rest remains at rest, unless acted upon by external forces. According to his Second Law, a force acting on a body is proportional to the acceleration of the body and acts in the same direction of that acceleration. The first assumptions are that:

(a) there is such a thing as a "steady state" in Nature; and
(b) motion is always and everywhere the result only of external force acting on matter, not the mode of the matter's very existence.

Newton's Third Law avers that for every action there is an equal and opposite reaction; algebraically, $\sum F=0$. Here the first assumptions are that:

(a) every physical system is closed and therefore conservative; and
(b) within any such closed system, $\Delta t = > 0$ and $\Delta s = > 0$

The remarkable feature is that each of these assumptions is aphenomenal, i.e., anti-Nature. None of the assumed conditions can exist, in real and changing time or in real and changing space, anywhere in four-dimensional Nature. The Third Law is often translated metaphorically into social contexts involving human political or economic intervention, and this translation poses embarrassing questions, however, of whether good actions are to be countered by bad actions. Even more problematic, according to the underlying logic of this law, how does one ensure that a good intention is fulfilled and an aphenomenal, anti-Nature intention is rejected?

From the standpoint of nature-science, **motion is the mode of existence all matter**. Whether it is energy, or matter that has become transformed into energy, or energy that became transformed into matter, there is no form of material existence that is not in motion. Thus, in nature in general, the tangible-intangible nexus is expressed as a matter-energy nexus, in which the one is transformed into the other. There is no such thing yet discovered anywhere in the universe as a vacuum devoid of matter or energy, and therefore the processes of the transformation of each into the other, are detected as well as modulated throughout the operation of these laws of motion. As already indicated, for grasping the connection between the tangible and intangible components of a phenomenon or process, **time** and **information** *taken in their intangible aspects* are crucial. When it comes to laws of motion, however, a difficulty leaps out: the long-established procedures of conventional "New Science" have insisted on rendering time and information only in their tangible aspects. As will now be shown, this approach has created far more problems than it has solved.

The first untidy knot arises with the notion of "continuity", as utilized in the mathematics that Sir Isaac Newton elaborated in order to formulate his famous Three Laws of Motion. On the front of the scientific work undertaken to investigate and determine laws of motion, his work stands at the watershed. His elaboration of the general laws of motion of all matter was a huge advance over the incoherent and conflicting notions that previously prevailed. Various limitations appeared since his day at certain physically measurable/detectable boundaries – at speeds approaching the speed of light, for example, or within space approaching the measurable minimum limit of (approximately) 10^{-32} m. This led researchers to make important corrections and amendments to Newton's formulae. Nevertheless, Newton's fundamental breakthrough in the very idea of summarizing the laws of motion itself, common to all discrete forms of matter understood and observed to that time (*i.e.*, not atomic, molecular or sub-atomic), is indisputable. Equally remarkably, in order to take account the temporal components attending all matter in motion, Newton invented an entirely new departure in mathematics. It was a new departure rendered necessary because existing mathematics were useless for describing any aspect of change of place while matter was undergoing such changes.

Newton's mathematics made it possible to treat time as though it were as infinitely divisible as space. This was something no one had ever previously accomplished. Newton's solution worked extremely well for the purposes of involving the relative motion of masses acting under the influence of the same external forces, especially force due to gravity and acceleration due to gravity. Extended to the discussion of the planets and other celestial bodies, it appeared that Time throughout nature – Time with a capital "T" – was indeed highly linear. For Newton and for all those applying the tools of his calculus to problems of time and space comprehensible to ordinary human perception, t_{LINEAR} and t_{NATURAL} were one and the same.

Newton's approach was based on a bold and unprecedented maneuver. It arrived as the fruit of an unpredicted turn in the revolution in human thought already under way since the start of the Renaissance during the century and a half predating Newton. Launched from the leading centers of the Bourbon and Hapsburg Empires, with the aim of reversing the correct verdicts of the "new science" of Copernicus, Kepler, Galileo and others who emerged during the European Renaissance in increasingly open revolt against the authority of Church dogma, the Catholic counter-reformation had failed, and failed utterly. Throughout the continent of Europe, Catholic monarchs and the authority of the Holy Roman Catholic Church were placed on the defensive. In England, the "Catholic forces" were entirely routed. Among that country's scientific and philosophical circles, Newton, along with many of his colleagues in the Royal Society, were standard-bearers of the newly-victorious forces.

In his *Principia Mathematica (1687),* Newton built on and refined the implications and tentative conclusions of a number of contemporaries and near-contemporaries. Although lacking an overarching theoretical framework, quite a number of them were already working with processes of infinite summation that converged to some finite value. He proposed differentiation as a method for deriving rates of change at any instant within a process.

However, his definition of the derivative as the limit of a difference quotient involving changes in space or in time, is, as small as anyone might like, but not zero, *viz.*:

Set the cat among the pigeons. For one thing, it became apparent soon enough that, without further conditions being defined as to when and where differentiation would produce a meaningful result, it was entirely

$$\frac{d}{dt}f(t)=\lim_{\Delta t\to 0}\frac{f(t+\Delta t)-f(t)}{\Delta t}$$

Figure 6.2 Formulation of Newton's breakthrough idea *(expressing Leibniz' derivative notation in Cauchy's "limits" notation).*

possible to arrive at "derivatives" that would generate values in the range of a function at points of the domain where the function was not defined or did not exist. It took another century following Newton's death before mathematicians would work out the conditions – especially the requirements for ***continuity*** of the function to be differentiated within the domain of values – in which its derivative (the name given to the ratio-quotient generated by the limit formula), could be applied and yield reliable results. Dominating the English universities of Newton's day, the Anglican clergy led by Bishop George Berkeley condemned Sir Isaac's *hubris* in assuming that there could be any notion or source of continuity without affirming that the universe was an act of Divine Creation; i.e., anything involving the Infinite was entirely the province of Almighty God and not men. The typical English compromise that emerged for nearly the next two centuries was that practical men interested in engineering new inventions, retained and mastered Newton's difficult notation and basic calculus principles. However, the theoretical work, which was needed to place the continuity concept on firm footing. could not develop in English universities and crossed over The Channel to France.

In the period 1740-1820, as the basic theory of differential equations also came to be elaborated, Newton's notation was almost universally replaced by that of calculus' cofounder Leibniz. This facilitated several breakthroughs in the theory of analysis for the Swiss mathematician, Euler, among others, most notably with regard to the techniques which were developed using the techniques of superposition (Kline, 1972). The notion of superposition was an ingenious solution to a very uncomfortable problem implicit in (and left over from) Newton's original schema. Under certain limiting conditions, his derivative would be useful for dealing with what today we call vectors – entities requiring at least two numerical quantities to fully describe them. All the important and fundamental real-world entities of motion – velocity, acceleration, momentum etc.– are vectorial insofar as, if they are to usefully be manipulated mathematically, not only in their magnitude but also in regards to their direction which must be specified., is this right?

Here, however, there also inhered a limiting condition for applying Newton's calculus. So long as magnitude and direction change *independently* of one another, no problems would arise in having separate derivatives for each component of the vector or in superimposing their effects separately and regardless of order. (This is what mathematicians mean when they describe or discuss Newton's derivative being used as a "linear operator".) The moment it is not possible to say whether these elements are changing independently, however, a linear operation will no longer hold., last sentence is confusing?

Modeling is always an approximation. For a long time, this provided many researchers a license to simplify and relax requirements, to some degree or other, as to just how precisely some part of natural reality had fit into the chosen or suggested model. Provided the assumptions – boundary conditions or initial conditions – could be retrofitted to exclude unwanted dependencies, one could generate some sort of model and results. The interior psychology of this act of choice seems to have been that, since the linearized option would reach a result, it could and should be used. Over the centuries, the implication of this choice has turned out to be rather more mischievous. Everything non-linear has been marginalized, either as exceptional, excessively intractable in its "native" non-linear state, or usable only insofar as it may be linearized.

In the actual evolution and development of what became of the field of real analysis, every step was taken incrementally. Newton's discoveries were taken up and re-used as tools. Meanwhile, however, the theoretical work needed to explain the conditions under which analytic methods in general, and the derivative in particular were applicable which had not yet reached the stage of explicit elaboration. The notion of the derivative as a linear operator, and even aspects of a more generalized theory of linear operators, began to develop and to be utilized ***before*** the continuity criteria underpinning the entire field of real analysis were made explicit. This led to associating **linearity,** principally with **superposition techniques** and **the possibility of superposition**. By the time Cauchy published his work elaborating the importance of continuity, no one would connect continuity with linearization. In real analysis, discontinuity became correlated mainly and even exclusively with undifferentiability (Kline, *ibid*).

With the rigorizing of real analysis by Cauchy and Gauss, applied mathematics in the middle third of the nineteenth century, developed a powerful impetus and greatly broadened its field of action throughout all the natural sciences. This went especially deep in all areas of mechanical engineering. There arose a preponderating interest in steady and/or equilibrium states, as well as in the interrelations between static and dynamic states. To penetrate the mathematical reasoning process best suited for modeling such conditions, linearizing assumptions were great time-savers.

Newton's method itself, long described as "Newton's method of tangents" because it could be illustrated geometrically by picturing the derivative as the slope of a straight-line segment tangent to the curve of any function's graph, relied implicitly on the notion of approximating instantaneous moments of curvature, or infinitely small segments, by means of *straight lines*.

As bold and utterly unprecedented as Newton's approaches were, the drive to linearize these various approaches all contain a trap for the unwary. Going backward or forward in space or in time becomes a matter of indifference. If, however, the door is not closed on the possibility of treating time as reversible, can natural reality be modeled as if it actually unfolds? What does it mean for describing anything happening in nature according to naturally-conditioned temporal factors, where time cannot be reversed? To engineer anything in Nature, applying Newton's calculus often requires suppressing or otherwise sidelining such considerations. For example, it has long been accepted, as a pragmatic matter, that fudge factors and ingenious work-arounds are needed to linearize the non-linear.

If time is treated as an independent variable, its irreversibility appears at first to present no problems. On the other hand, if time is to be emulated as a fourth dimension, possessing an all-important irreversibility property, and varying no longer independently but actually dependent on multiple processes that overlap within some portion of the same duration, then the problem becomes unavoidable. Merely backing up a few steps on the path that brought matters to this stage, back to the point where everything still looked more or less linear and the non-linearities had not yet taken over, is not going to help overcome this fundamental difficulty. The starting-point itself contains the core of the problem: Newton's calculus edifice, in its very foundation, is truly *anti*-Nature. That is to say: starting anywhere on this path, one will diverge ever further from Nature.

6.4.2 Science and the Problem of Linearized Time

The development of objective descriptions of relationships in social science was profoundly affected by the fact that t_{LINEAR} à *la* Newton had been monopolizing European scientific discourse from the early 18th century onwards. Even t_{NATURAL} was partially fitted by resorting to periodically predictable regularly-spaced cycles, while exponential time was readily fitted by means of Euler's famous discovery that $\boldsymbol{e}^{i\pi} = \mathbf{-1}$. Other timescales or models of time were adapted to fit these parameters. Those that did not or could not fit, like $t_{\text{HISTORICAL}}$, were by and large dismissed. Such a marginalization of reference-frame, scrapped a potentially huge source of information of a kind obtainable in no other form. This loss is not a purely passive one. Marginalizing the reference-frame is also a tremendous weapon to wield against the challenge that a new discovery might pose to established knowledge. The struggles waged in European intellectual circles throughout the 17th, 18th and 19th centuries may no longer have involved stakes as high as they had been during the Catholic inquisition of previous centuries. However, even if it now stood at a certain remove from life-and-death, the struggle to establish scientific method and differentiate scientific investigation from self-interested assertions by persons said to speak with "authority", was no less intense. Instead of the immortal soul of the individual, now the stakes became the freedom to research and establish the truth. Here was laid the foundation of all subsequent aphenomenal modeling in the sciences (Khan *et al.*, 2005b).

By the middle of the 19th century, the challenges posed to established notions in particular by the works of Karl Marx and Charles Darwin were not small. Darwin's explanation of speciation was particularly subversive. The emergence of new species only made sense as the non-linear outcome of a lengthy series of processes that must precede and prepare the way for the emergence of a new species. At the same time, knowledge about these earlier processes, no matter how complete, still would not enable a specific and absolutely reliable prediction of all the features expressed in the new species. The story of how upsetting this was to a few religious figures, concerned about the authority of the Biblical story of Creation, is an old and well-told one (Irvine, 1955). The upset actually went much further, however, and it represents one of the longest-lasting acts of intellectual terrorism still active in the world. One of Darwin's closest collaborators was the geologist, Sir Charles Lyell. For the first 10 years after Darwin published his landmark work, Lyell would not publicly defend the theory of evolution. Fear of unknown consequences outweighed any other consideration, including even the fact that Lyell's own work established the notions of the fossil record and geological time, as well as the fact that it was Lyell himself who had encouraged Darwin through the more than two decades that would elapse between the completion of the voyages of the *Beagle* to the Galapagos and the readying of his *Origin of Species* for publication. Until he openly defended his friend, he officially retained public doubts about Darwin's assertion of the mechanism of "natural selection", even as Darwin was corresponding with him about these ideas (Darwin, 1892).

In the natural sciences, during the 20th century, work continued in many fields using t_{LINEAR} à *la* Newton. Some theoretical work on the frontiers such as Einstein's theory of relativity seriously tackled, at the level of the universe, the need to correct, at least in part, Newton's assumptions and implications about temporal factors, and to render time's irreversibility explicit. Other theoretical and applied work such as quantum mechanics took the path of applying probability measures of uncertainty to the coordinates of elemental matter at the inter-atomic and sub-atomic levels. In general, the response in the natural and engineering sciences to this exposure of the inadequacy of existing temporal reference frames were incoherent.

6.4.3 Reproducibility and the Extinction of Time

The precision and especially the reproducibility of results achieved using a Newtonian calculus, albeit a calculus that had become cluttered with refinements and special recipes of all kinds by the 19th century, were indeed remarkable. The physical sciences were written about and spoken of as "exact sciences". There were not a few who understood,, nevertheless the price of such progress: *everything depended on establishing appropriate initial and/or boundary conditions in which a given differential equation could be applied.* An inappropriate selection could render meaningless any results from using the equation. There were probably rather fewer who also understood that preparatory research would be required. *Before* selecting and applying any existing linearising model equation to the task of extracting possible solutions, it would first have to be established just how invariant with time any actual initial and/or boundary conditions of a process taking place in nature might be. Against this overwhelming current, who was going to look back and question fundamentally the applicability in general to the reality of nature, of methods and models emerging from the linearising assumptions of real analysis? Unfortunately, this set the context for researchers, not only in the natural and engineering sciences, but also those in the social sciences, who became concerned with rigorizing their methods... is this sentence clear?

Dynamism is inherent in all social or natural development. The idea that equilibrium is normal and anything other than the steady state is a disturbance and disruption, is a notion that has served every Establishment in all times and places. How true a description has it ever been of social, economic or political reality anywhere? Such striving for the steady state, emerged clearly in Newton's time, invading and permeating his scientific work. The issue for this chapter is: if such a thing as steady-state equilibrium is possible, and actual, anywhere in Nature, how is it **also** possible that matter and energy can neither be created nor destroyed, but only change form, sometimes even changing one into the other? One or the other: ***either*** steady state, in which case neither matter nor energy can be changing form, *or else* motion is the mode of existence of matter.

This can be, and has been, fudged in various ways. For example, repetitive forms like reciprocal or cyclical motion are represented as a kind of pseudo-steady state within a clearly delimited range. However, the maintenance of real life reciprocal motion, like that of pistons in an internal combustion engine, requires a directed expenditure of energy in a bounded chamber that ceases once the supply of combustible fuel is cut — either by turning the engine off, or as a result of running out of fuel, both of which are engineered phenomena found nowhere in Nature. Cyclical repetition in Nature does not repeat the exact same path in each circuit, any more than the Earth repeats the identical path in its orbit around the sun. The repetitive cycles of "chaotic attractors" (like Julia or Mandelbrot sets) generate an infinite number of "self-similar", but unique, non-identical cycles.

Nevertheless, the reality – that, regardless of what can be engineered to happen for some finite period, there exists no such thing anywhere in Nature as a steady state – still remains masked by widespread acceptance of Newton's First Law of Motion as the first and last word on the inertial properties of *matter*. This law provides that "an *object* at rest tends to stay at rest and an *object* in motion tends to stay in motion with the same speed and in the same direction unless acted upon by an unbalanced force."

On its face, this law indeed does appear to provide definitive criteria for the analysis of inertia in all possible cases – at rest, or already in motion. However, in fact, it is at the very least a potential source of disinformation. First, ***resistance to*** motion is identified, at the empirical level of "objects", with ***absence of*** motion. Secondly, even at apparent equilibrium, *i.e.*, at a point between a previous completed state of motion and a pending resumption of motion at some subsequent stage, something at the microscopic level of matter remains in motion, *e.g.*, at the molecular level. If Newton's First Law is loosely applied to all forms of matter in general, however, motion ceases to be an inherent property of matter. Once such a separation is effected, all kinds of mischief is unleashed.

It is widely accepted without further thought that we cannot have motion without equilibrium moments. It is also widely accepted that all motion tends anyway, eventually, towards steady-state equilibrium in the presence of an appropriate balance of forces. According to Newton's First Law of Motion, any *external* force can disturb an equilibrium condition. However, **these mutually contradictory expectations rest on the evidence gathered by a stationary observer gathering observations from outside the reference frame.** Can such observations be accepted as the final word or as definitive descriptions of what is actually taking place in the unceasingly dynamic environment that continues to exist outside and around that same observer? Within any actual natural or social process, such an observer is, of course, a complete fiction, an aphenomenal conceit. There is only one possible time and place in which such an aphenomenal observer could exist: that would be where the process of motion being observed and described is analyzed while $\Delta t \rightarrow 0$, *i.e.*, in the context of $t + \Delta t \rightarrow t$. From this emerges a crucial insight: infinite reproducibility from a process becomes possible neither mainly nor only because its underlying science is entirely tangible. It becomes possible mainly as the product of an "applied science" designed to negate the particularity of nature and thereby suppress any concern for its resilience to the onslaught of industry and its multifarious consequences.

6.4.4 The Long Term as an Infinite Summation of "Short Terms", or $T = \sum_{i=1}^{\infty} f_i(t)$

To this point, the focus of this chapter has been on the departure from science based on what actually occurs in nature or society in favor of the idealized — in fact, aphenomenal — models that fit the data of reality according to smooth, linear, continuous pathways. Our main interest is the natural sciences, but the same patterns also emerged in the social sciences during the 19th century and additional refinements to this generally corrupting process have continued since the latter part of the 20th to have had an impact on the natural and engineering sciences. This is especially so when it comes to matters having to do with the relationship(s) between the long-term and the short-term.

Over the decades, the advertisement of the rapid advances of 19th century European society associated with the rigourisation of real analysis *and its applications throughout the natural and engineering sciences*, has become a well-burnished coin. The dark side of this coin, however, was a full-scale counterfeiting of genuine scientific understanding of actual natural processes with aphenomenal modeling of material commodity production for private profit. The ready reproducibility of mass-industrial commodity production, with its promise of massive gain for the rightly-positioned individual in the short-term, had an especially remarkable distorting effect in society-at-large. In that context, it served to mask the systematic enslavement of hundreds of millions based on stripping them completely of any power over their long-term. From the 19th century to date, meanwhile, these societal consequences of "progress" — engineered according to aphenomenal models of natural processes — could not help but influence, and indeed did influence, the development of the social sciences as well. In these fields, however, it was stochastic, probabilistic and other statistical systems of modeling, rather than models developed from analysis of real-valued functions, that would play and continue to play particularly crucial roles.

Developed for the purpose of creating uniformly reproducible outcomes in the sphere of commodity production and the profits system based on that production, the aphenomenal idealized approach to modeling natural processes became the greatest victim of its very success. Its benefits to the owners and managers of Capital having proved themselves, there was no going back to investigating natural processes in any of the other ways that societies outside Europe had pioneered millennia earlier. As part of consolidating its grip over entire societies and their interrelations, this Eurocentric version of modern industrialism went much further than subjugating the natural sciences. From the middle of the 19th century onwards, it proceeded to marginalize and eliminate all theories and approaches in the social sciences that could potentially challenge its economic and political hegemony. The equilibrium steady state so desired in the scientific annals, the church pews, the factories and the marketplace was now to be re-engineered throughout the social sciences as well. The problem at the commanding heights of scientific study and research in the universities lay with the paradigms initiated by defenders of the French Revolution and, especially, by Karl Marx and other socialists. Whether calmly implying or shouting from the rooftops that modern society was a dynamic organism in which the factory owners and bankers of today might be gone tomorrow, all views along this line were now to be expunged from formal academic discourse. The social sciences were to be renovated on the basis of an ideological identification of the Present, *i.e.*, the status quo, as Progress coupled to a mathematical modeling system based on statistical probability.

In the developing discourse of social science as an academic discipline, the positivist doctrines of Auguste Comte emerged to prominent positions. These lauded the societal model of Napoleon III's France and the industrial model of Great Britain as representing, in combination, the epitome of "Progress" in general with a capital "P", tending to the always desired steady state equilibrium. Many researchers in social sciences of this era already saw development as a struggle between "forces of progress" and forces opposing progress. By identifying the present moment in western European social development as "progress", all opposition was portrayed as potentially or actually opposed to progress. At the same time, all tendencies reinforcing the current line of development within the status quo, especially everything tending towards equilibrium, were presented as supports for this vaunted "progress". Effectively, an inherently linearized conception of time, *viz.*, time based on the steady state, was thereby confounded with the notion of time as a measure of "progress". This was effected, however, in a way that, the irreversibility of *actual* time — time now subsumed by the achievement of equilibrium — was made to disappear (Karl Marx, 1867, Comte,1848, and Butterfield, 1968).

Furthermore, according to this logic, any disturbance of such equilibrium appeared illegitimate, as something triggered by deranged, deviant, alien, *i.e.*, external, sources and forces. At the end of the 19th century, the French sociologist, Emil Durkheim, for example, acknowledged that society itself could be the seed of many of these disturbing phenomena. However, the individual was ultimately responsible in any particular case since deviance itself manifested itself individually (Durkheim, 1897).

Meanwhile, no one would believe or credit as applicable to human beings any kind of social science that followed the highly deterministic models of physics and chemistry. In the alternative, normative statistical models were quickly developed. This was mathematical modeling allegedly liberated from the deterministic "necessity" of the aphenomenal modeling of natural processes by means of differential equations. Here was a mathematics that would account for individual choices in all their varieties. And guess where the junctures occur at which choices are exercised and may be measured? Of course: at every point where $\Delta t \to 0$, *i.e.*, the points where $t + \Delta t \to t$! This was accomplished by treating a large number of individual discrete cases in actual human society as approximating the continuous. Random-variable analysis in one form or another — normal distribution, binomial distribution and many, many others — has ever since emerged as a major tool of social science statistical modeling.

Long-term thinking is one of the keys by which people can become empowered, indeed by which they can empower themselves. The moment one becomes hooked on long-term thinking as one's regular habit, it becomes blindingly clear that all true empowerment derives from oneself and one's relations with other people, not from any externally-imposed or externally-induced conditions. Long-term thinking means thinking about the consequences of one's own actions. It also means re-examining everything reported from the standpoint of extracting its longer-term significance, beyond what is being immediately reported. Within the process itself of living, it ought theoretically to be well within everyone's interest to apply long-term thinking in all times and places, yet obviously this has not happened. There are what are called the "pressures of daily life" – which actually means pressure to produce some outcome in the short- term – which are usually blamed for this. But the fact remains, and it cannot be made to disappear by being glossed over, that it is actually in the *interest* of the vast majority of individuals, viewed either in the short-term or for the long-term, to apply long-term thinking as a habit in all times and places. Human conscience exists in all times and places and it will always assert its claims in the field of human action. Just because individuals can and do frequently suspend listening to their conscience, does not make it go away or disappear. The key to maintaining long-term thinking is to suspend listening to, or being pressured by, anyone and anything that places the interests and needs of the short-term ahead of the long-term. If the long-term is not continually attended to, there will be not only no long-term, but the short-term will become far shorter. There is nothing at all mysterious about long-term thinking. Start with clarifying *where* whatever it is you are thinking about fits or exists with respect to the Past, the Present and/or the Future, and with *why* it is of any significance or importance to you. That already takes care of two profoundly significant intangibles: time and intention. Just as there cannot be such thing as matter without motion (*i.e.*, energy), there is no such thing as understanding, *i.e.*, meaningful knowledge, without the individual taking action(s) to "find out". A positive, *i.e.*, pro-social and long-term, intention ensures that the seeker will find something useful. Finding out something for oneself is the most empowering thing there is. "Learning" something on the say-so of some "authority", on the other hand, is the most disempowering and enslaving thing there is. What is so especially empowering about long-term thinking is not whether this-or-that piece of knowledge was already known or even previously thought about, but rather the journey on which it sets the seeker.

The solutions obtained by integrating the area under the curve of a random-variable distribution, $T = \sum_{i=1}^{\infty} f_i(t)$, expressed the long term essentially as a summation of an infinite number of short terms.

6.4.5 Erasing History in Order to "Disappear" the Long-Term and Enshrine the Steady State

At the end of the 19th century, a crucial part of this attack on long-term thinking took the form of discrediting any examination of, or giving of weight to, the history of a phenomenon or process from its present form back to its earliest onset. This was initially, most sharply expressed at that time in the social sciences. Alfred Marshall (1890), for example, the founder of neoclassical economics as an academic discipline, typically argued that such historical considerations and analysis were useless and irrelevant anywhere in the social sciences because, allegedly, Darwin's theory of evolution demonstrated that the only factors decisive in any process of change would be found among those most recently generated, not among those historically handed down, (Zatzman and Islam, *ibid.*). In this dismissal of the life-history of thought-material lies the source of the thinking that, the closer time t is to "right now", the less-qualified and more precise will be the mathematical rendering of whatever the conditions are being studied. Once a concept is detached from its historical anchor, it becomes easy to justify resorting to steady-state, equilibrium-based models.

This problem became especially acute for those who were striving to put the new social science of economics on a basis that would appear as rigorous as physics or any of the other natural sciences. In this respect, Alfred Marshall, in particular, as the modernizer of economic theory on the basis of what is known as "marginal analysis", shared with many of his peers in the other social sciences, a serious misconception of the thrust of Darwin's explanation of natural selection, his primary evolutionary mechanism. Darwin said only that the emergence of a species distinct in definite ways from its immediate predecessors and new to the surrounding natural environments, generally marked the final change in the sequence of steps in an evolutionary process. The essence of his argument concerned the non-linearity of the final step, the leap from which was formerly one species to distinctly another species. The *proximity* of that last step to the vantage-point of the observer might be centuries or millennia or longer, and therefore the length of that time interval was ***not*** the relevant element. Rather, the length of time that may have passed between the last observed change in a species-line and the point in time at which its immediate predecessor emerged – the characteristic time of the predecessor species – was the time period in which all the changes so significant for later on were prepared. This latter could be eons, spanning perhaps several geological eras. This idea of t_{NATURAL} as characteristic time was what the linearizers in the social sciences were especially bent upon obscuring. The idea was to marginalize any role for, or consideration of, t_{NATURAL}. Victory could then be, and was in fact, declared on all fronts for $\boldsymbol{t}_{\mathrm{LINEAR}}$.

6.4.6 First Interim "Time"-ly Conclusion: The Anti-Nature Essence of Linearized Time

The mathematical models used and applied to "get a handle" on engineered and/or natural phenomena are, first and foremost, *images* of an ideal form. It is not possible for t_{LINEAR} and $t_{NATURAL}$ to be or remain one and the same. A linear model applied to linearly-engineered phenomena may work well under circumstances where some operating limits have been experimentally verified. However, a linear model applied to phenomena that are themselves not linear is another matter. Nothing in nature is linear, so featuring linear independence in any model purporting to reflect the reality of a situation where everything affects everything else, certainly seems questionable. In Nature, there are literally dependencies upon dependencies. The notion of any system operating in isolation, or of modeling the solution of any problem presented in Nature by assuming the condition of some isolated system or sequence of such systems, is aphenomenal. Changes of state occur, appear or disappear in Nature both continuously as well as discontinuously, and they cannot be subjected to laboratory-type control. There is no such thing as a "steady state". Problems as found in their natural setting always appear "ill-posed". Up to now, however, there seems to have been a concerted effort not to attempt solutions to problems in this state. Instead a problem that looks like the actual problem but which can be posed in more or less linear form is solved instead. This result is declared to be something approximating the solution of the actual problem, given the addition of certain conditions and boundaries to the original problem's definition. This distorting technique starts very early, with high-school/first-year university instruction in how to solve problems associated with simple harmonic motion of a pendulum using Newton's Second Law of Motion in a linearized approximation as the governing equation. One convention widely adopted up to now, involves artificially and arbitrarily hedging the reality to be observed in nature with various time-constraints so that some relatively tractable mathematical model may be applied. This is not unrelated to the fact that the solution schema developed for such models have become ever more elaborate. These linearized images have served to sustain an illusion that nature's secrets are being discovered at a rate that is in lock-step with the advances taking place in the technology of electronic computation (Islam, 2005a).

6.4.7 Second Interim "Time"-ly Conclusion: Making Time Stand Still by Way of Linearized Visualization of Space

Before we can begin to learn and appreciate the form and content of nature and science, we have to unlearn much of what we thought we knew. Chief among this collection of problems is how we have already become trained by the society, culture and education system to conceive and accept the metaphors and correspondences of engineered space and time, represented, essentially, in two dimensions (2-D). It is indeed a considerable accomplishment that, utilizing perspective and the projective plane implicit in its geometry, what is actually a third spatial dimension, can be represented to us convincingly within a two-dimensional plane. However, the price at which this is achieved is something that's remarked upon far less. The fourth dimension; i.e., time itself, is made to disappear. In fact, whether the context is in the fine arts or engineered space and time, we have learned a certain visual "grammar", so to speak, with all spatial visualizations and representations. We know no other "language" but that in which either:

1. time is frozen - as in a snapshot - or
2. time is represented, not as the fourth dimension, but rather as something that varies independently of any phenomenon occurring within it.

The modern history of communications media and information transfer really begins with the famous Canaletto landscapes of 16th century Italy, incorporating perspective and immediately overthrowing in that same moment, the centuries-long authority of the Holy Roman Catholic Church message we were meant to receive from works of art. With the emergence of the new approach in art of the Renaissance, the principles underlying representational art works of the early and high Middle Ages were reversed. Any previously authorized message already vetted carefully as to the acceptability of its content and the morality of its purpose would hereafter become extraneous and secondary to the information gathered by the visual cortex of the individual observer.

The new approach made the visual arts accessible at all levels of society for the first time. Perspective in Renaissance paintings, and the findings of anatomy regarding the movement and distribution of weight in the human frame, manifested now in Renaissance sculptures, overthrew the centuries-long monopoly of Church authority with the bluntest directness. This was bracingly liberating and bound to provoke ever-deeper questioning of

Church authority in other fields. By enabling Humanity to reclaim from Nature something that Authority had denied, these transformations within mass communication media (turning art into a mass medium was itself the key to the transformation) unleashed a social and intellectual revolution. However, even as the new "grammar" of perspective-based representation of three-dimensional space, a space that now appeared to be living rather than representing a purely imaginary phantasm or idea, overwhelmed the previously accepted canons of visual arts, and overthrew with it the long-asserted timelessness of the Church's approved truths, the new visual canon served up another illusion of reality: the timeless snapshot-like image.

Over the next four centuries, expressed as a struggle to capture the moving image, and later the live image, further development of mass communications media, associated systems and technologies of information transfers, wrestled with just about every imaginable and practical aspect of how to engineer the appropriate representations of time and space. Interwoven throughout this development, are parts of the history of development of analog and then digital electronic media, of the individual or limited-edition static-image to the mass-marketed photographic static images, and of the illusion of the moving picture; i.e., an illusion created by overwhelming the visual cortex with 24 still frames per second, and then superimposing a sound track over this same moving picture (the talking motion picture.) Also interwoven are the stories of the unmodulated telegraphic signal whose information is contained in its sequencing to the modulated signal overlaid with an audio carrier (telephone and radio), the modulated signal overlaid with visual and audio carrier signals (television), the encoding of information in digitized sequences (computers), and the digital encoding of information on a transmitted carrier signal (cell phones, the Internet). All these technological aspects have been exhaustively discussed and examined by many people. Less cogently commented upon, but still mentioned, at least, are the political-economic transitions that also developed within this historical tapestry. Ranging from products produced by private individuals, or craft-oriented companies, in production prior to the Industrial Revolution, business was intended for finite, relatively small markets of certain individuals of for privately-owned companies. Then, in the 19th and early 20th centuries, products were mass produced for the output for social mass markets, which quickly grew to the readily-socialized mass production of our own time, conducted under increasingly narrowly monopolized ownership., I substantially rewrote the previous two sentences and I'm not completely sure I have the meaning correct?, Nevertheless, what remains unmentioned and uncommented upon anywhere in these historical recapitulations, is whatever happened to the tangible-intangible nexus involved at each stage of any of these developments. We cannot hope seriously to make headway towards, much less accomplish, serious nature-science of phenomena or an authentic science of the tangibles-intangibles nexus, without filling in that part of the tapestry as well. That which is natural can be neither defended nor sustained without first delimiting and then restricting the sphere of operation of everything that is anti-Nature.

This absence of discussion of whatever happened to the tangible-intangible nexus involved at each stage of any of these developments is no mere accidental or random fact in the world. It flows directly from a Eurocentric bias that pervades, well beyond Europe and North America, the gathering and summation of scientific knowledge everywhere. Certainly, it is by no means a property inherent, either in technology, or in the norms and demands of the scientific method *per se*, or even within historical development; that time is considered so intangible as to merit being either ignored as a fourth dimension, or conflated with tangible space as something varying independently of any process underway within any or all dimensions of three-dimensional space.

6.5 What is New Versus what is Permitted: Science and the Establishment?

This section discusses how New Science enabled settling contradictions between knowledge that was new and knowledge whose publication could not be permitted — by tossing Nature away

6.5.1 "Laws" of Motion, Natural "Law" & Questions of Mutability

From the individual's earliest exposure to science in the classroom, this point is repeatedly impressed on the learner's consciousness: "law" in science is an expression of some relationship among elements that holds true somewhere in time and space. What is only glimpsed later on in the development of a career in a scientific field, is that not all "scientific" laws are equal. Some may describe an empirical relationship. Others define fundamental features common to an entire category of processes. Newton's calculus held out a seductive promise of all relationships becoming in principle quantifiable, even computable for some unique solution or set of solutions. However like all "law", this promise was a double-edged sword.

Would such laws as those that Newton's calculus might describe and/or articulate relationships that captured the reality of change in the natural world, or would they be mere snapshots, freezing some relationship in an artificial bubble of permanence? Essentially what all of this illustrates is that the student of science is inured from the earliest exposure to the subject matter that "law" and "necessity" are not necessarily the same.

As shown by the history of the reality, on the one hand, of the effect of the Law of Gravity on the acceleration towards the earth of freely-falling objects and the insistence, on the other hand, from Aristotle's day until even after Galileo, that this rate was a function of the object's mass, it is entirely possible for quite fundamental laws to operate even as their very existence remains vehemently denied. The usual explanation is that somehow people eventually knew better and gave up the discredited notion. Meanwhile, this fails to address how a vast amount of data had accumulated over the millennia — data that seriously challenged the official orthodoxy long before that orthodoxy gave up defending the erroneous, discredited view, which was something that was done by those who pursued alternative explanations and who shut down their lines of research., Again, not sure if this sentence I rewrote is completely correct?, There's more to these stories than the happy ending of an enlightened understanding displacing a backward one.

In our own day, it has been possible to witness truly extraordinary efforts undertaken from the highest levels on down to ensure survival of the status quo by stifling any consciousness or source of consciousness without any alternatives. As one of the recent leaders of this campaign, former British prime minister,(now Baroness) Thatcher, used to intone: "There Is No Alternative" (Seatini, 2007). Using the initial letter of each word in the phrase to form an acronym, critics have labeled this ongoing campaign, the "TINA syndrome". Over the 25 years or so, this campaign has emerged in a wide range of manifestations, throughout all fields of study in politics, economics and policy. Throughout the social sciences and even in the natural sciences, assertion of the TINA syndrome and the struggle against its assertion have together spurred an intense and renewed interest in the meaning of "law" in general, and of how particular processes may be considered to be governed by some sort of law, pattern, or set of relationships. It is difficult enough to conceive anything more intangible than a "relate" or relationship, let alone one such as the TINA syndrome that has produced such wide and highly tangible impacts. There is indeed no alternative at this point but to take the plunge and examine what the *brouhaha* is all about.

Once again, the bellwether of what would eventually emerge in the natural sciences, came in the earliest attempts to establish some of the new social sciences on a more rigorous basis, complete with their own "laws of motion" à *la* Newton in physics. The industrial revolution was already underway for a generation in Britain when the political economist Adam Smith famously put forward his theory of the so-called "invisible hand":

> *...every individual necessarily labours to render the annual revenue of the society as great as he can. He generally, indeed, neither intends to promote the public interest, nor knows how much he is promoting it. By preferring the support of domestic to that of foreign industry, he intends only his own security; and by directing that industry in such a manner as its produce may be of the greatest value, he intends only his own gain,* ***and he is in this, as in many other cases, led by an invisible hand to promote an end which was no part of his intention****. Nor is it always the worse for the society that it was no part of it. By pursuing his own interest he frequently promotes that of the society more effectually than when he really intends to promote it. I have never known much good done by those who affected to trade for the public good. [Emphasis added – Ed.] (Smith, 1776).*

Implicit in Smith's invocation of the superiority of the individual pursuing his self-interest over the interests of society or the public, lies a notion of the shortest conceivable time-span, one in which $\Delta t \rightarrow 0$: "he intends only his own gain". Herein lurks the aphenomenal heart of Adam Smith's darkness: If what happens to "self-interest" is transplanted to a context in which $\Delta t \rightarrow \infty$ is considered, this aphenomenality becomes starkly evident. Self-interest in the long-term becomes the pursuit of the gain or benefit *for society as a whole*. Otherwise, it would be akin to dividing by zero, something that would cause the model to "blow up" (Zatzman and Islam, Ch. 5, *ibid*). Significantly, Smith does not say that money-capital wedded to the short-term immediate intentions of some individuals (or grouping of common interests) would not achieve its aims. He confines himself instead to observing that objectives which formed no part of the originating set of immediate short-term intentions, *viz.*, "an end which was no part of his intention", might also come to be realized, thanks to the intervention of the "invisible hand".

All the defenders of, and apologists for the status quo have pointed to Adam Smith's argument as their theoretical justification for opposing, in principle, any state intervention in the economy. Chanting their mantra of the invisible hand, policy-makers on the same wavelength, have been confining and restricting such intervention to those parts of economic space in which there operates no profitable production of goods or services with which such intervention would be competing. How sound is this chain of reasoning, however?

The Soviet bloc disappeared during 1989-1991. Since then, one rather profound and deeply disturbing truth about the TINA syndrome has begun to emerge more starkly than ever before. Neither the State as bogeyman, nor the State as employer-substitute, remains a viable option for Humanity. There is in fact no alternative but something *other than* either of these options. This is a subversive consciousness. It is very much resisted by both proponents of TINA and their detractors. Once Δt exceeds some characteristic time, and a measure of the change in economic space, s, exceeds the lone individual, *i.e.*, as $\Delta s > >1$, then the role of the human factor - social consciousness may become decisive. It increasingly must displace the previous need for a State "above society" to play all or any of its previously accustomed roles in economic life. At that point, intentions that serve society as a whole, no longer require the application of some power previously delegated to an external force (the State) in order to prevail as a norm. That is the stage in which the intangible - good intentions - can finally command the tangible. (Zatzman and Islam, Ch.6, *ibid*). Although this discourse relates to the social sciences and their relationship to the formulation of policy, there is one particularly notable knock-on, knock-on or knock-off?, effect in the natural sciences and technological development: mere knowledge of ingenious technical solutions to problems provides no hope, or—what is worse—false hope, if intentions are tied to self-interest in the short-term.

The significance of Smith's observation of the so-called "invisible hand" is that the outcome of normal operations of commodity production are achieved *independently* of the will of any individual participant or group, *viz.*, "an end which was no part of his intention". Clearly this contains an implicit rejection of external control of people's private economic choices — whether by government *or by any other group* (cartels or other monopoly giants, for example)—that would deign to interfere with individual private choices. Smith was looking meanwhile into a very different world, at a different time. He believed that "an end which was no part of his intention" came about as a byproduct of how competition operates to "regulate", in a rough and overall manner, both the supply of and demand for socially necessary goods and services. The will of any consumer(s) or producer(s) by itself would never suffice. For Smith, the secret to the "law of motion" of an industrial commodity economy, lies in how the marketplace under conditions of free competition allocated economic resources.

Smith aspired in this manner to render economics as scientific as physics. Underlying Smith's view was a set of philosophical assumptions, independent of his economic research, which formed a definitive perspective regarding the place of scientific reasoning of any kind within human thoughts, in general, which were actually very close to the outlook of Sir Isaac Newton's. This was broad 18th-century Deist philosophical outlook, already prevalent among a broad section of the European intelligentsia of his day. Anything could be examined as the outcome of a process comprising observable, definable stages and steps, and linked ultimately to some Prime Mover (or initiating force). During the 17th and 18th centuries, for most scientists, an analysis ascribing a process to some Prime Mover manifesting itself as a Newtonian "mechanism", was the best of all possible worlds. On the one hand, a natural occurrence could be accounted for on its own terms, without having to invoke any mystical forces, divine interventions or anything else not actually observed or observable. On the other hand, the divinity of Creation did need not be dispensed with or challenged. On the contrary: this divinity was being reaffirmed, albeit indirectly "at a certain remove" insofar as whatever was required to sustain or reproduce the process in question could now be attributed to some even more fundamental "law of motion". In any event, such "laws of motion" had more fundamental properties as being indispensable and necessary; without them, no investigation could be carried very far or penetrate nature's secrets.

Since the publication of Charles Darwin's *Origin of Species (1859),* a revolution was occasioned in the scientific outlook. It has become so complete and all-encompassing that it is hard to remember, from the vantage point of the start of the 21st century, that much of the support for, and embrace of, Newtonian mechanism (and the attendant penchant in many fields for "laws of motion") derived from the beliefs that it could be reconciled with a "creationist" assumption, not just about Man *within* Nature, but about the very existence of Nature itself.

Re-examined in this light, the impact of Smith's assertions about the "invisible hand" among his contemporaries can be better understood. In essence, he was declaring that:

1. Economic life comprised phenomena that could be analyzed and comprehended as scientifically and as objectively as Newton had analyzed and disclosed the laws of physical motion of all forms of matter in Nature and even the universe; and

2. Such investigations would provide yet another proof of the divinity of Man's existence within that natural universe.

Between the time of Sir Isaac Newton in the early 1700s and that of Charles Darwin in the middle third of the 1800s, these considerations were framed and understood by scientific investigators within a larger context, *viz.*, the conception of "natural law". Whereas "laws of motion" were valid and accounted for, everything essential about phenomena within definite boundary conditions, "natural law" was fundamental on yet another plane. Using the scientific method, Man could come to know, understand and make use of laws of motion or natural laws – laws operating within observable processes in Nature itself, and discoverable from systematic observation of these processes. However: **these natural laws in themselves were immutable.** This was the same as with any mathematical function whose "Newtonian" derivative yielded an instantaneous rate of change between points on its graph but which itself did not change. In fact, it was precisely this notion of the immutability of natural law that was assumed and implicit within the general and more widely-accepted view that some law (s) of motion, eventually connectible back to a Prime Mover, must account for any and every process observed in Nature. The conundrum was reduced to this: if natural laws were not immutable, science would be compelled to account for innumerable random divine interventions in any natural process, at any time. Such a course could drag science back into the swamp of the metaphysical idealism of Bishop Berkeley – Newton's great antagonist – who famously explained in *The Analyst (1734)* that objects in physical nature continued to exist beyond our perception because God exists to cognize them, whenever human beings are not available to cognize them. In the words of a limerick popularized widely in the 19th century specifically satirizing Berkeley:

There was a young man who said "God
Must think it exceedingly odd
If he finds that this tree
Continues to be
When there's no one about in the Quad."
"Dear Sir, your astonishment's odd;
I am always about in the Quad
And that's why this tree
Will continue to be
Since observed by

Yours faithfully,
God." (Berkeley, 2007)

No one would accept something so contrary to common sense; science and scientists would become laughing-stocks. If natural laws were not held to be immutable, how could logical reasoning guarantee that error could be detected and rejected?

The actual solution of this conundrum came into practice, in the course of further, deeper-level research into actual phenomena. Starting with Darwin in natural science in the middle of the 19th century and extending early in the 20th century to physics and chemistry with the elaboration of theories of quantum mechanics, it has become increasingly clear that the mutability or immutability of any natural law is actually a function of, and dependent on, the time-scale selected for observation and study. **The problem here in general is one of method.**

The particular source of the problem lies with how the methods of scientific investigation which are applied to comprehend the material actually deal with temporal factors — the *passage* of time, as well as the *role* of time. Consider here, for example, the well-known line of attack against appropriate time-consciousness launched in the field of geology. In the geological record, entire species appear in one epoch only to disappear in a later one; ludicrously, this has been adduced by so-called "Creationists" as evidence that Darwin's theory of evolution - which used such leaps and gaps precisely to explain speciation - must be untrue! Of course, evidence of this kind proved only that the notion that evolution should take place as a smooth process uninterrupted by quantum leaps - the very view that Darwin's analysis and evidence definitively refuted - was devoid of reality. The same issue of time-scale is now just beginning to be understood regarding some of the earliest states of matter in the first few picoseconds of the Big Bang.

In our own day, the exposure of these absurdities is helping to finish hammering the final nails into the coffin of the "TINA syndrome", at least for this generation. All phenomena or effects duly observed in any natural or social process arise from some verifiable cause, but in accordance with the operation of some body of law that remains constant and consistent, and always within some definite spatial-temporal boundaries. To argue immutability outside such boundaries is, at the least, open to serious question.

6.5.2 Scientific Disinformation

To deny or ignore the existence, role(s) and/or consequences of such boundaries is a source of scientific disinformation.

"Scientific disinformation" is a most apt description of the condition in which provision of scientific theory and researched data nevertheless leave the relevant authorities incapacitated when it comes to framing and/or selecting a course of action to carry out consciously programmed changes in the status quo. It explains very well why, for example, literally millions of people in our own time have become perfectly well aware that there is a problem posed by global warming, yet no research body or governmental agency has proven capable of getting a handle on the essential features of the problem and preparing interventions that could turn the situation around.

Assume for the moment that this societal condition is recognized — and no longer disputed — as a scientific and verifiable fact. For the wide range of researchers across the natural sciences, the matter of taking some social responsibility is immediately posed. It is important to establish causes and effects in order to sort out the dynamics of this condition. How to alleviate the negative consequences of such a condition in various areas - from the ability to access necessary resources to the health of the population, etc., - would accordingly preoccupy specialists in the relevant respective fields. However, there is indeed a way to present the evidence of this condition in its various aspects, and of the extremely negative consequences flowing from this condition, so that everything is to blame for the condition, and hence no one thing is to blame for any one part within the overall situation. One approach that fills this bill very nicely is the resort to statistical methods - especially those involving correlation.

One of the most important consequences of resorting to statistical methods was the finessing of the need to establish and distinguish cause from effect. To be able to assert that **A** and **B** are related by some correlation, coefficient χ appears highly suggestive of underlying reality even as it skirts at the same time, the entire issue of whether **A→B**, **B→A**, or actually **Q→A** and **R→B** while in fact no causal relationship whatsoever exists between **A** and **B**. Correlation is very useful where causal relations are already known and established. In the social sciences, however, in the absence of - or inability to gather - any other evidence from more direct or more thorough experimental observation, it has become *de rigeur* to employ correlation to imply or suggest a causal relationship. Is the publication of caveats about the distinction between demonstrating a correlation and suggesting some relationship of cause-and-effect sufficient to shield such activity from merited condemnation as a serious abuse of the requirements of scientific integrity?

One of the most fundamental requirements of science properly conducted, is that one's work at the end of the day draws some line of demarcation between what is known to be false and what may not yet be fully understood to be the truth. Detection of error and elimination of falsehood are absolutely fundamental to scientific enterprise at any level. In this respect, the "correlation" bucket has holes in it big enough for a veritable spotlight to coruscate. Consider the following example: If one were to correlate "intensity of religious faith", "presence of exact bus fare" and "frequency of arrival at a preset destination on public transit", any number of clearly nonsensical, as well as a number of apparently reasonable, correlations might be elaborated, *e.g.*, "faith and a two-dollar coin gets you downtown on the bus." However, regardless of how anyone might go about weighing the various possible renderings of the available evidence, the results would always be insufficient to rule out possibilities lying on the farthest margins and perhaps bordering on nonsense, *e.g.*, what happens if you have the two-dollar coin but have a lack of faith? This converts the likely acceptance of the apparently more reasonable seeming possibility (or possibilities) into a matter of purely personal prejudice. It is no longer guided by a procedure that meets the fundamental requirement of any scientific method, *viz.*, that a clearly erroneous result will be excluded by the weight of the evidence and not by the prejudice of the investigator.

The issue is not statistical modes of reasoning as such, but how they are deployed — and especially why rather self-evident abuses of these modes of reasoning are widely, even seemingly routinely, indulged.Speaking purely in terms of how the logic of an explanation for a phenomenon comes to be constructed when inputs are "probable" or "likely", but not actually known, if any of the steps on the path of reasoning toward an actually correct conclusion are themselves false, neither the Bayesian methods of inferring conditional probabilities (Jevons, 1871, nor Pearsonian methods of statistical correlation (Pearson, 1892) will assist the investigator in reasoning to the particular conclusion that will be demonstrably most consistent with known facts. Statistical modes of reasoning carefully employed, in a context where there exists some actual knowledge of definite causes and definite affects, can be subtly powerful. However, it is an entirely different story when reasoning proceeds from the grouping of data according to statistical procedures derived from the norms of some abstract probability distribution. No grouping of data, however well-fitted to some known probability distribution, can ever substitute for establishing actual causes and actual effects. Substitution of the "statistically likely" or "probable" in the absence of knowledge of what is actually the case, is a truly inexcusable breach of scientific integrity.

Consider this syllogism:

- Any collection of objects will fall freely from a higher level to a lower level on the earth at a rate proportional to the mass of each object-mass. This means each object-mass may be expected to arrive on the lower level at a different point in time [major premise].
- A waterfall is a collection of objects of different mass [minor premise].
- Therefore the molecules in a waterfall must reach the lower level at different times [conclusion-deduction].

If the information relayed above in either the major or minor premises is derived from a scenario of what is merely probable (as distinct from what is actually known), the conclusion which happens to be correct, would be not only acceptable as something independently knowable, but reinforced as something also statistically likely. This then, finesses determining the truth or falsehood of any of the premise, and, eventually, someone is bound to "reason backwards" to deduce the statistical likelihood of the premises from the conclusion. This latter version, in which eventually all premises are falsified, as a result of starting out with a false assumption asserted as a conclusion, is the most typical feature of aphenomenal models (Zatzman and Islam, Chapter 5, *ibid*).

Up until now, proponents of the arguments in favor of relying on statistical procedures and processes to rigourise social science have held an ace-in-the-hole. Apart from investigations of extremely limited phenomena, and recognizing that results cannot be reliably duplicated where input conditions cannot be fully or faithfully replicated, lab-controlled experimental reproducibility is really not an option in the social sciences, and most certainly not of the kind routinely utilized in the natural sciences. Does it follow from this, however, that, because the entire complex process cannot be reproduced in a controlled experiment, therefore that phenomena observed in society—its politics and its economics cannot be ascribed accurately to definite causes? Instead of addressing this meat of the matter, advocates of statistical methodology, at the heart and soul of rigorous social science, raise a diversion. The counter-argument advanced on this point is that, without a probability measure, there is a rather wide and unrestrained latitude for subjective biases to substitute themselves in the place of carefully-reasoned opinion and judgment. Instead of assuming that the status quo is all there is, however, why not instead, incorporate, properly and duly, all characteristic historical time-dependent conditions attending the emergence or disappearance of phenomena? Surely this would serve to rein in such arbitrariness.

That "properly and duly" caveat is particularly important in this connection. It is embarrassingly easy to edit the record and arrange historical data so that no single determinable cause or clearly-defined pathway of causation can be arrived at. The trick, like a good Agatha Christie detective thriller, is simply to make many causes equally suspect. What is the difference between someone from the social sciences who asserts that a phenomenon has so many causes that no one cause or pathway can be sorted out, and a student of the physical or natural sciences suggesting that the mass of an object and the acceleration due to gravity, could be equally responsible for how soon an object freely falling towards the earth will land? ***Improper and undue manipulation of the intangible aspects of temporal factors*** is one of the ways that such felonies can be committed against scientific integrity and the authority of authentic knowledge with impunity.

6.5.3 Galileo's Experimental Program: An Early Example of the Nature-Science Approach

Although science and knowledge are not possible without data measurements, neither science nor knowledge is reducible to data, techniques of measurement, or methods of data analysis. Before operationalizing any meaningful solution(s) to a research problem, there is a crucial prior step – the stage of "experimental design," of formulating the operational steps or the discrete sub-portions of the problematic, in which our knowledge is incomplete. For this, there is no magic formula or guaranteed road to success. This is where the true art of the research scientist comes into play. In adopting the standpoint and approach of nature-science, in which actual phenomena in nature or society provide the starting-point, the different roles of the investigator as participant on the one hand and as observer on the other can be clearly delineated as part of the research program.

Galileo's experiments with freely falling bodies are well known and widely discussed in the literature of theoretical physics, the his- tory of science and technology, the history of the conflicts between science and religion, and a number of other areas. There is a wide- spread consensus about his principled stance in defense of the findings in his own research, coupled with some ongoing disputes as to how consistently he could defend principles against the pressures of the Inquisition. In a departure from that path in the literature, rather than adding or passing judgment on Galileo's conduct vis-à-vis the Church authorities, this paper addresses the backbone of Galileo's stand, namely, his conviction that his method was a more reliable guide to finding the truth of the nature of freely falling bodies than any guesswork by Aristotle, a Greek who had been dead for two millennia. The standpoint adopted here is that Galileo's research program represents an early model of the nature-science approach — the first by a European, in any event. Its "correction" by those who came after him, on the other hand, represents a corruption of his method by the mandates of "New Science," mandates whereby subsequent investigators would become so preoccupied with tangible evidences to the point of excluding other considerations.

The following information about Galileo's approach to experimental methods and the controversy that continued around it, taken from the summary from the Wikipedia entry "Two New Sciences," summarizes the most significant developments on which the claim of the first European proponent of the nature- science approach is based:

The Discourses and Mathematical Demonstrations Relating to Two New Sciences (*Discorsi e dimostrazioni matematiche, intorno a due nuove scienze*, 1638) was Galileo's final book and a sort of scientific testament covering much of his work in physics over the preceding thirty years.

Unlike the Dialogue Concerning the Two Chief World Systems (1632), which led to Galileo's condemnation by the Inquisition following a heresy trial, it could not be pub- lished with a license from the Inquisition. After the failure of attempts to publish the work in France, Germany, or Poland, it was picked up by Lowys Elsevier in Leiden, The Netherlands, where the writ of the Inquisition was of little account.

The same three men as in the Dialogue carry on the discussion, but they have changed. Simplicio, in particular, is no longer the stubborn and rather dense Aristotelian; to some extent he represents the thinking of Galileo's early years, as Sagredo represents his middle period. Salviati remains the spokesman for Galileo.

Galileo was the first to formulate the equation for the dis- placement s of a falling object, which starts from rest, under the influence of gravity for a time t:

$$s = 1/2gt^2$$

He (Salviati speaks here) used a wood molding, "12 cubits long, half a cubit wide and three finger-breadths thick" as a ramp with a straight, smooth, polished groove to study rolling balls ("a hard, smooth and very round bronze ball"). He lined the groove with "parchment, also smooth and polished as possible". He inclined the ramp at various angles, effectively slowing down the acceleration enough so that he could measure the elapsed time.

He would let the ball roll a known distance down the ramp, and used a water clock to measure the time taken to move the known distance. This clock was "a large vessel of water placed in an elevated position; to the bottom of this vessel was soldered a pipe of small diameter giving a thin jet of water, which we collected in a small glass during the time of each descent, whether for the whole length of the channel or for a part of its length; the water thus collected was weighed, after each descent, on a very accurate balance; the differences and ratios of these weights gave us the differences and ratios of the times, and this with such accuracy that although the operation was repeated many, many times, there was no appreciable discrepancy in the results." (Website 7)

It is critical to add that, instead of clocking standardized "seconds" or minutes, this method of time measurement calibrates one natural motion by means of another natural duration.

The water clock mechanism described above was engineered to provide laminar flow of the water during the experiments, thus providing a constant flow of water for the durations of the experiments. In particular, Galileo ensured that the vat of water was large enough to provide a uniform jet of water.

Galileo's experimental setup to measure the literal flow of time, in order to describe the motion of a ball, was palpable enough and persuasive enough to found the sciences of mechanics and kinematics. (ibid.)

Although Galileo's procedure founded "time" in physics, in particular, on the basis of uniformity of flow in a given interval, this would later be generalized as the notion of a linear flow of time. Einstein would later overthrow this notion with regard to the vastnesses of space in the universe and what the nature-science approach proposes to correct in all investigations of processes unfolding in the natural environment of the earth.

The law of falling bodies was discovered in 1599. But in the 20th century some authorities challenged the reality of Galileo's experiments, in particular the distinguished French historian of science Alexandre Koyré. The experiments reported in Two New Sciences to determine the law of acceleration of falling bodies, for instance, required accurate measurements of time, which appeared to be impossible with the technology of 1600. According to Koyré, the law was arrived at deductively, and the experiments were merely illustrative thought experiments.

Later research, however, has validated the experiments. The experiments on falling bodies (actually rolling balls) were replicated using the methods described by Galileo, and the precision of the results was consistent with Galileo's report. Later research into Galileo's unpublished working papers from as early as 1604 clearly showed the reality of the experiments and even indicated the particular results that led to the time-squared law. (ibid.)

Of interest here is the substance of Koyré's challenge - that the time it would take objects to fall to the ground from the top of the Tower of Pisa could never have been measured precisely enough in Galileo's day to justify his conclusion. Of course, sub- sequent experimental verification of Galileo's conclusions settles the specific question, but Koyré's objection is important here for another reason.

What if, instead of following Galileo's carefully framed test, there was a series of increasingly precise measurements of exactly how long it took various masses in free fall to reach the ground from the same height? The greater the precision, the more these incredibly small differences would be magnified. One could hypothesize that air resistance accounted for the very small differences, but how could that assertion then be positively demonstrated? If mod- ern statistical methods had been strictly applied to analyzing the data generated by such research, magnificent correlations might be demonstrated. None of these correlations, however, would point conclusively to the uniformity of acceleration of the speed at which these freely falling objects descend over any other explanation. As long as the focus remained on increasing the precision of measure- ment, the necessity to drop Aristotle's explanation entirely (that object fall freely at speeds proportional to their mass) would never be established unambiguously.

If the model diagrammed in Figure 6.3 were true, the theory would be verified. What, however if the model were false from the outset, betrayed by retention of a first assumption that was not more carefully scrutinized? With more observations and data, this modeling exercise could carry on indefinitely, e.g., a 5-parameter univariate non- linear "function" like $y = \mathbf{a}x + \mathbf{b}x^2 + \mathbf{c}x^3 + \mathbf{p}x^4 + \mathbf{q}x^5$... that continues towards ever higher degrees of "precision."

The following possible representation of how the data resulting from such a falsified experimental approach — as partially illustrated in Fig 6.3 —reinforces this conclusion.

Galileo's act of publishing the *Discorsi* in 1638 as the main summary of the most important part of his life's work was no ordinary act of defiance. It was his affirmation before the entire world that he had stuck to his original research program and never took the easy way out. The nature-science approach settles the question of whether one is on the right path to begin with, and that is Galileo's primary accomplishment. Today the time has arrived to take matters to the next stage. In order to capture and distinguish real causes and real effects as a norm in all fields of scientific and engineering research, it is necessary to apply what Albert Einstein elaborated – from examining the extremities of space, namely that time is a fourth dimension, to the observation of all phenomena, including those that are natural/social and those that are more deliberately engineered.

Proponents of the "New Science" of tangibles have long insisted that Galileo is truly their founding father. In fact, their premise has been that the collection and sifting of the immediate evidence of nature and its processes is the sum total of reliable scientific knowledge.

Without conscious investigation of nature, and without critical questioning of conclusions that no longer fit the available evidence, how could science advance? This contribution is acknowledged by recognizing Galileo as the first European practitioner of nature-science.

Although science and knowledge are bound to give rise to data and measurement, the reverse is not necessarily true. In the scramble to produce models that generate information about phenomena with increasing specificity and precision, it is easy to lose sight of this elementary insight and identify advances in data gathering and management with advances in scientific understanding. The coherence of research efforts undertaken in all fields has come under threat from this quarter; conventional research methods and approaches have come into question; and novel alternatives excite growing interest. As the review of the "engineering approach" serves to reconfirm, what decides the usefulness of any mathematical modeling tool is not the precision of the tool itself, but rather the observer's standpoint of a real process unfolding in nature. In fact, this ultimately limits or determines its actual usefulness. As Galileo's example clearly shows, regardless of the known value of improved precision of measurement, greater precision cannot overcome problems that arise from having embarked on the wrong path to begin with. The same goes for how precisely, i.e., narrowly, one specifies the reference frame of an observer or what is being observed. Conventional methods of modeling have long since passed the point of being able to specify to the point of excluding the general context of the reality surrounding the observer or the phenomena of interest. The apparent short-term gain in specificity may end up being vastly overcompensated by such things as the impossibility of attaching any meaningful physical interpretation to elaborately computed mathematical results and, with that, the subsequent realization that a research effort may have begun by looking in the wrong direction in the first place.

6.5.4 Implications of Einstein's Theory of Relativity on Newtonian Mechanics

The underlying problems that philosophy of science addressed before World War II were linked to deciding what weight to give Einstein's disturbance of the Newtonian worldview of "mechanism." There was considerable concern over the principle of relativity and how far some of its new concepts might be taken, e.g., where the idea of time as a fourth dimension might be taken next.

Alongside this came pressing concerns after World War II to eliminate the powerful influence of the advances in science coming out of the Soviet Union. These concerns were based on a paradigm shift that was universally rejected in Western countries but nevertheless produced powerful results in many engineering fields throughout the Soviet system that could not be ignored. The notion on which all scientific research in the Soviet Union came to be based was that the core of scientific theory-building depended on sorting out the necessary and sufficient conditions that account for observed phenomena. In the West, this was allowed only for mathematics. Introducing this notion into investigations of, and practical engineering interventions in, natural processes, was dis- missed as "excessive determinism." Instead, an entire discussion of "simplicity" as a diversionary discourse was developed.

This discourse had an aim. First and foremost it was intended to avoid dealing with the entire matter of necessary and sufficient conditions. This approach reduced the process of selecting the more correct, or most correct, theoretical explanation accounting for one's data to the investigator's personal psychology of preference for "the simple" over "the complex." This was put forward in opposition to the notion that a scientific investigator might want to account for what is in line with his, her, or others' observations or counter others' observations of the same phenomenon (Ackermann 1961; Bunge 1962; Chalmers 1973b; Feuer 1957, 1959; Goodman 1961; Quine 1937; Rudner 1961; Schlesinger 1959, 1961).

As discussed in Chapter 2, the notion that nature is dynamic has been known for at least two and a half millennia. However, Einstein's revelation of the existence of the fourth dimension was a shock because all scientific and engineering models used in the post-Renaissance world were steady-state models. However, the acceptance of these models, without consideration of their existence in the real world, led to subsequent models that remain aphenomenal. In this, it is important to note that an aphenomenal hypothesis will entirely obscure the actual pathway of some natural phenomena that anyone with a correct starting point would set out to study.

Unless this is recognized, resorting to curve-fitting and/or other retrofitting of data to highly linearized preconceptions and assumptions concerning the elementary physics of the phenomenon in its natural environment will not validate the aphenomenal model.

With the revolutionary paper of Albert Einstein, the understanding of light was placed on a fully scientific basis for the first time as a form of matter that radiates as an energy wave (Website 8). To establish this it was necessary to breach the wall in human thought that was created as a result of following the misconceived starting-point that "light" was the opposite of "dark."

How had an everyday notion become such an obstacle to scientific understanding? Investigation unexpectedly disclosed the culprit, lurking in what was universally acknowledged for the three hundred and twenty-seven years preceding Einstein's paper, to represent one of the most revolutionary scientific advances of all time, namely, the affirmation by Isaac Newton that "motion" was the opposite of "rest" (Website 9). Nowhere does Newton's mechanism explicitly deny that motion is the very mode of existence of all matter. However, his system crucially fails to affirm this as its first principle. Instead, his First Law of Motion affirms that a body in motion remains in motion (dis- regarding drag effects due to friction) and an object at rest remains at rest unless acted upon by an external force. The corollary flowing immediately from this law is not that motion is the mode of existence of all matter, but only that motion is the opposite of rest.

In Newton's day, no one doubted that the earth and everything in the heavens had been created according to a single all-encompassing clock. Among Christians (mainly Catholics and Protestants during the Reformation), all that was disputed at this time was whether the clock was set in motion by a divine creator and then carried on heedless of human whims, or whether humans, by their individual moral choices, could influence certain operations of this clock. What are the implications, however, of affirming motion as the mode of matter's very existence? The most important implication is that any notions regarding the universe as a mechanism operating according to a single all-encompassing clock lose all coherence. Newton's great antagonist, Bishop George Berkeley, excoriated Newton's failure to close the door to such heretically anti-Christian views (Stock, 1776).

In the same moment that the unquestioning belief in a single all-encompassing clock is suspended or displaced, it becomes critical to affirm the existence of, and a role for, the observer's frame of reference. Newton, his contemporaries, and other men and women of science before Einstein's

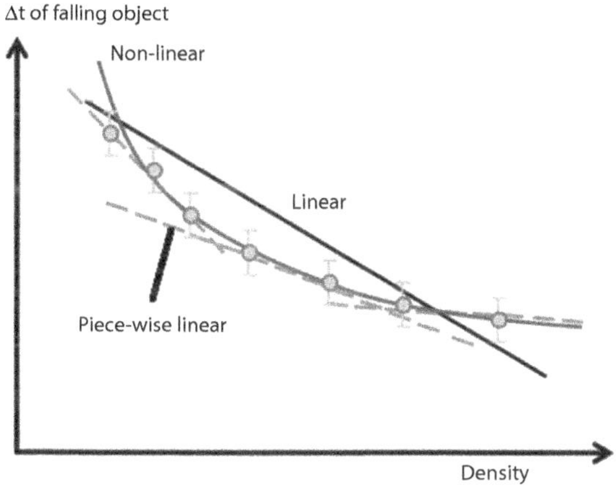

Figure 6.3 Linear and piecewise linear analyses of the Δt of a freely-falling object.

theory of relativity were certain that Newton's Laws of Motion defined the mechanism governing force of any kind, motion of any type, as well as any of their possible interactions. What we now understand is that, in fact, Newton's laws of motion actually defined only the possibilities for a given system of forces, one that would moreover appear "conservative" only because any observer of such a system was assumed to stand outside it.

To seventeenth-century European contemporaries of Isaac Newton, this matter of frame of reference was of no moment whatsoever. For them, the flora, humans, and non-human fauna of the known world had never been anywhere other than where they currently were. Nor was it generally known or understood that people lived in human social collectives other than those already known since "ancient times," an era beginning in western Asia some finite but unknown number of years before the birth of Jesus Christ. There was no reason to wrestle with the prospect of any other frame of reference either in historical time or in spaces elsewhere in the universe. Only by means of subsequent research, of a type and on a scale that could not have been undertaken in Newton's day, did it become possible to establish that in any space-time coordinates anywhere in the universe, from its outermost cosmic reaches to the innermost sub-atomic locale, mass, energy, and momentum would be and must be conserved regard- less of the observer's frame of reference.

In a universe defined by a single clock and common reference frame, three principal physical states of matter — vapor, solid, and liquid —

could be readily distinguished. (By driving externally applied energy, bounded by powerful electro-magnetic force fields, and still without having to re-adjust any other prevailing assumptions about a single clock and common reference-frame, a fourth highly transient plasma state could be further distinguished.) Overall, motion could and would still be distinguished as the opposite of rest. What, however, can be said to connect matter in a vapor, solid, or liquid state to what happens to a state of matter at either sub-atomic or cosmic spatial scales? These are the regions for which the conservation of matter, energy, and momentum must still be accounted. However, in these regions, matter cannot possibly be defined as being at rest without introducing more intractable para- doxes and contradictions. It is only when motion is recognized as the mode of existence of all matter in any state that these paradoxes become removable.

6.6 Deconstruction of Einstein's Concept of Reality, Mass, Time, and Energy

6.6.1 Origin of Aphenomenality

Zatzman and Islam (2007b) have offered an extensive review of Aristotle's philosophy and provided one with scientific explanation of why that philosophy is equivalent to launching the science of tangibles. In economic life, tangible goods and services and their circulation provide the vehicles whereby intentions become, and define, actions. Locked inside those tangible goods and services, inaccessible to direct observation or measurement, are intangible relations – among the producers of the goods and services and between the producer and Nature – whose extent, cooperativeness, antagonism and other characteristic features are also framed and bounded by intentions at another level, in which the differing interests of producers and their employers are mutually engaged. In economic terms, Zatzman and Islam (2007b) identified two sources of distortion in this process. They are: 1) linearization of complex societal non-linear dependencies (functions and relationships) through the introduction of the theories of marginal utility (MU); and 2) lines in the plane intersect as long as they are not parallel, *i.e.*, as long as the equation-relationships they are supposed to represent are not redundant. The first source removes very important information pertaining to social interactions and the second source enables the use of the "=" sign, where everything to its left is equated to everything to its right. Equated quantities cannot only be manipulated, but – especially – interchanged, according to the impeccable logic, as sound as Aristotle (who first propounded it), which says: two quantities each equal to a third quantity must themselves be equal to one another or, symbolically, that 'A = C' and 'B = C' implies that 'A = B'. Scientific implications of this logic will be discussed in the latter part of this section. Here, in philosophical sense, the introduction of this logic led to the development of a 'solution'. As further arguments will be built on this 'a solution', soon this 'a solution' will become 'the solution' as all relevant information are removed during the introduction of the aphenomenal process. This would lead to the emergence of 'equilibrium', 'steady state', and various other phenomena in all branches of New Science. It won't be noticeable to common people that these are not natural systems. If anyone questions the non-existence of such a process, s/he will be marginalized as 'conspiracy theorists', 'pseudo-scientists', and numerous other derogatory designations. This line of thinking would explain why practically all scientists up until Newton had tremendous difficulty with the Establishment in Europe. In the post-Renaissance world, the collision between scientists and the Establishment was erased — *not* because the Establishment became pro-science, but more likely because the New Scientists became equally obsessed with tangibles, devoid of time function as well as intention (Zatzman and Islam, 2007a). Theoretically, both of these groups subscribed to the same set of misconceptions or aphenomenal bases that launched the technology development in the post-renaissance era. Islam *et al.* (2010) identified these misconceptions as:

1) Chemicals are chemicals or energy is energy (meaning they are not function of the pathway and are measured as a function of a single dimension); 2) If you cannot see, it doesn't exist (only tangible expressions, as 'measurable by certain standard' are counted); 3) Simulation equals emulation (if there is agreement between reality and prediction at a given point in time, the entire process is being emulated). Zatzman and Islam (2007a) and Khan and Islam (2007a) attributed these misconceptions to the pragmatic approach (whatever works must be true), which can be traced back original Greek philosophy.

The immediate consequence of the science of tangibles is that every decision that emerges is scientifically false. The removal of a dimension and ignoring the consequences is inherent to this consequence (Islam *et al.*, 2010a). Because time is the most important dimension, the omission of this dimension has the severest consequences. Time also forms the pathway of the science of intangibles. Intention, on the other hand, forms the root or foundation of the science of intangibles. Ignoring any of these would render the process aphenomenal. Zatzman and Islam (2007b) cited a number of examples to establish this assertion. One of them is as follows:

Compare the circle to the sphere, in Figure 6.4. The shortest distance between two points A and B on the circle's circumference is a straight (secant) line joining them, whereas the curved arc of the circumference between the two points A and B is always longer than the secant. The shortest distance between two points on the surface of a sphere, on the other hand, is always a curve and can never be a straight line. Furthermore, between two points A and B lying on any great-circle of the sphere (*i.e.*, lying along any of an infinite number of circles that may be drawn around the surface of the sphere whose diameter is identical to that of the sphere itself) and between any other two points, say C and D, a curved line of the same length joining each pair of points will subtend a different amount of arc. For the points not on the great-circle, the same length-distance will subtend a greater arc than for points on a great-circle; curvature k in each case is uniform but different. On the other hand, what makes a circle circular rather than, say, elliptical, is precisely the condition whereby equal distances along its circumference subtend equal arcs, because curvature k anywhere along the circumference of a circle is the same.

Is anyone whose comprehension remains confined to cases of the circle in flat planar space likely to infer or extrapolate from such knowledge whatever they would need in order to grasp and use such differences found on the curved surface of the sphere? Of course not: indeed, any solution or solutions obtained for a problem formulated in two-dimensional space can often appear utterly aphenomenal when transferred or translated to three-dimensional space. In terms of the originating example of the fundamental metrics of the two kinds of surfaces: to those working in the environment of spherical surface-space, it becomes quickly obvious the shortest distance between two points that stand really close to one another on the surface of a sphere will approximate a straight line. In fact, of course, it is actually a

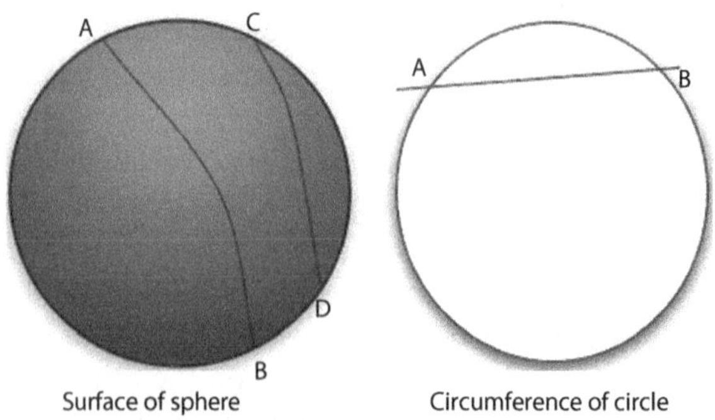

Figure 6.4 Sphere and Circle Compared (redrawn from Zatzman and Islam, 2007b).

curve and the idea that "the shortest distance between two points is a straight line" remains something that only approximates the situation in an extremely restricted subspace on the surface of a sphere. In a pragmatic sense, it is quite acceptable to propose that the shortest distance between two points is a straight line. However, in a real world which is anything but linear, this statement cannot form the basis for subsequent logical development as it would falsify all subsequent logical trains. Zatzman and Islam (2007b) called this attempt to simplify a model used to account for complex phenomena by chopping dimensions away at the outset (with the idea of adding them back in at the end) was "about as practicable as trying to make water run uphill."

The above point is further expounded by giving a simple, yet thought-provoking example (Zatzman and Islam, 2007). This example involves an experiment carried out without the knowledge of the findings of Galileo and Newton. Someone could observe and write up everything about an experiment involving dropping a weighty lead ball from the top of the Tower of Pisa. On another occasion the same could be done for an experiment entailing dropping a feather from the top of the Tower. Comparing the two, the key observation might well be that the lead ball fell to the ground faster and, from that fact, it might even be speculated that the difference in the time taken to reach the ground was a function of the difference in the weight of the feather and the lead ball. They argued that no matter how much more precise the measuring of the two experiments became, in the absence of any other knowledge or discovery it would be difficult to overthrow or reject this line of reasoning. Now, let's say the measuring devices available for subsequent repetition of this pair of experiments over the next 10, 20 or 30 years develop digital readouts, to three or four decimal places of accuracy. Quite apart from what engineering marvels these 'precision clocks' have led to, the fact would remain that this advancement in engineering does not help correcting the conclusion. In fact, the engineering advancement without regard to the basis of fundamental science would actually increase confidence toward the wrong conclusion. This is the situation for which engineering (making of precision clocks) only serves to strengthen the prejudice of the science of tangibles.

Another example involves the discussion of the speed of light. It is 'well recognized' in the New Science world that the speed of light is the maximum achievable speed. The argument is based on the famous Einstein's equation of $E=mc^2$. However, scientific scrutiny of this 'well recognized' statement in Physics shows that this statement is aphenomenal. When it is stated what is the speed of light, it is asserted that this is the speed of light (or radiation) within a vacuum. Then, when one asks what is a vacuum, it emerges that the vacuum in true sense[17] does not exist and one must resort to a definition, rather than a phenomenal state.

Historically, there were always a debate about the limit of speed of light, but only in the post-renaissance culture that speed of light became a 'constant'. For example note the following quote from Wikipedia (Website 10, 2007):

> "Many early Muslim philosophers initially agreed with Aristotle's view that light has an infinite speed. In the 1000s, however, the Iraqi Muslim scientist, Ibn al-Haytham (Alhacen), the "father of optics", using an early experimental scientific method in his *Book of Optics*, discovered that light has a finite speed. Some of his contemporaries, notably the Persian Muslim philosopher and physicist Avicenna, also agreed with Alhacen that light has a finite speed. Avicenna "observed that if the perception of light is due to the emission of some sort of particles by a luminous source, the speed of light must be finite."
>
> The 14th century scholar Sayana wrote in a comment on verse Rigveda 1.50.4 (1700–1100 BCE—the early Vedic period): "Thus it is remembered: [O Sun] you who traverse 2202 yojanas [ca. 14,000 to 30,000 km] in half a nimesa [ca. 0.1 to 0.2 s]", corresponding to between 65,000 and 300,000 km/s, for high values of *yojana* and low values of *nimesa* consistent with the actual speed of light."

In modern science, the justification of a constant *c* is given through yet another definition – that of vacuum itself. The official NIST site (on Fundamental Physics constant) lists this constant as having a value of 299 792 458 m/s, with a standard uncertainty of zero (stated as 'exact'). Now, all of a sudden, a definition, rather than a phenomenon (measurable quantity) becomes the basis for future measurements. It is so because the fundamental SI unit of length, the meter, has been defined since October 21, 1983[18],

---

[17]Merriam-Webster online dictionary lists the four meanings of vacuum, none phenomenal:

1 : emptiness of space

2 a : a space absolutely devoid of matter **b** : a space partially exhausted (as to the highest degree possible) by artificial means (as an air pump) **c** : a degree of rarefaction below atmospheric pressure

3 a : a state or condition resembling a vacuum : VOID <the power *vacuum* in Indochina after the departure of the French – Norman Cousins > **b** : a state of isolation from outside influences ["people who live in a *vacuum*...so that the world outside them is of no moment" — W.S.Maugham]

4 : a device creating or utilizing a partial vacuum.

one meter being the distance light travels in a vacuum in 1/299,792,458 of a second. This implies that any further increase in the precision of the measurement of the speed of light will actually change the length of the meter, the speed of light being maintained at 299,792,458 m/s. This is equivalent to changing the base of a logical train from reality to aphenomenality and effectively creating a process that penalizes any improvement in discovering the truth (the true speed of light in this case). Abou-Kassem *et al.* (2007) argued that this definition of meter is not scientifically any more precise than the original definition that was instituted by the French in 1770's in the following form: one meter is defined as 1/10,000,000 the distance from the North Pole to the Equator (going through Paris). They also discussed the spurious arrangement of introducing the unit of time as a second. It wasn't until 1832 that the concept of second was attached to the SI arrangement. The original definition was 1 second = 1 mean solar day/864,000. As late as 1960 the ephemeris second, defined as a fraction of the tropical year, officially became part of the new SI system. It was soon recognized that both mean solar day and mean tropical year both vary, albeit slightly, more 'precise' (apparent assertion being more precise means closer to the truth) unit was introduced in 1967. It was defined as 9,192,631,770 cycles of the vibration of the cesium 133 atom. The assumption here is that vibration of cesium 133 atom is exact, this assumption being the basis of Atomic clock. Only recently, it has been revealed that this assumption is not correct, creating an added source of error in the entire evaluation of the speed of light. On the other hand, if purely scientific approach is taken, one would realize that the true speed of light is neither constant nor the highest achievable speed. Clayton and Moffat (1999) discussed the phenomenon of variable light speed. Also, Schewe and Stein (1999) discussed the possibility of very low speed of light. In 1998, the research group of Lene Hau showed that the speed of light can be brought down to as low as 61 km/hour (17 m/s) by manipulating the energy level of the medium (Hau *et al.*, 1999). Two years later, the same research group reported near halting of light (Liu *et al.*, 2001). The work of Bajcsy *et al.* (2003) falls under the same category except that they identified the tiny mirror-like behavior of the media, rather than simply low energy level. More recent work on the subject deals with controlling light rather than

[18]Decided at the Seventh Conférence générale des poids et mesures (CGPM; the same acronym is used in English, standing for "General Conference on Weights and Measures". It is one of the three organizations established to maintain the International System of Units (SI) under the terms of the Convention du Mètre (Metre Convention) of 1875. It meets in Paris every four to six years.

observing its natural behavior (Ginsberg *et al.*, 2007). Abou-Kassem *et al.* (2007) used the arguments provided by previous physicists and constructed Figure 6.5. It is clear from Figure 6.5 that the assumption that 'speed of light', 'vacuum', and 'unit of time', 'unit of distance' are some arbitrarily set constants do not change the true nature of nature, which remains continuously dynamic. Note that media density can be converted into media energy, only if continuous transition between energy and mass is considered. Such transition, as will be clearer in the following section, is rarely talked about in the context of engineering (Khan *et al.*, 2007). This graph also reveals that that once definitions and assertions have been accepted in face values and are not subject to further scrutiny, the possibility of increasing knowledge (as in being closer to discover the truth about nature) is diminished.

Recently, a chemical engineer argued that the above graph is only a trivial manifestation of what Bose-Einstein theory would have predicted some 100 years ago. According to him, Bose-Einstein theory predicts the speed of light within an infinitely dense medium must be zero. Similar to the attitude expressed by the food scientist (in the Introduction of the present paper), this line of thinking unfortunately characterizes modern-day education in which scientists become so focused on obvious tangible expressions that they are incapable to thinking beyond the most tangible aspect of the research. Recently (2001), Eric Cornell, Wolfgang Ketterle, and Carl Wieman were awarded Nobel Prize for "for the achievement of Bose-Einstein condensation in dilute gases of alkali atoms, and for early fundamental studies of the properties of the condensates". Their work, however, far from vindicating science of tangibles, has in fact made it more necessary to

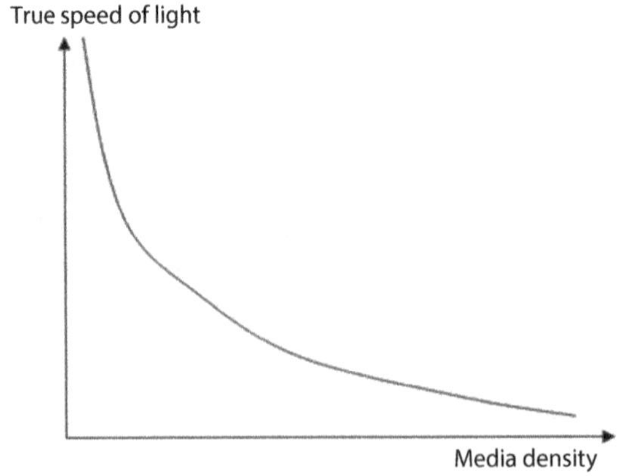

Figure 6.5 Speed of light as a function of media density.

take an alternate approach. Eric Cornell's most popular invited lecture is titled: Stone Cold Science: Things Get Weird Around Absolute Zero. This 'weird'-ness cannot be predicted with the science of tangibles. One further citation of Eric Cornell is: "What was God thinking? Science can't tell" (Cornell, 2005). Far from being content that all discoveries have been made and, therefore, there is nothing more to discover, a true researcher in pursuit of knowledge readily picks up on these shortcomings of conventional thinking, and as history frequently shows, it is within the inquisitive mind that answers reside. Ironically, this would also mean, discoveries are made because one is not afraid to make mistakes and, more importantly, theorize without fear of the Establishment. Satyendranath Bose, who did not have a PhD, himself did his part of the mistake. In fact, as the following quote from Wikipedia shows (Website 11), his findings were initially discarded by mainstream journals and he had to resort to sending Einstein his original work. It was only then that Einstein introduced the idea from light to mass (photon to atom) and the paper was finally published. If Bose did not have the tenacity and if Einstein did not have the decency to consider thinking "outside the box", one could not begin to think what would be the state of laser and all the gadgets that we take for granted today.

"Physics journals refused to publish Bose's paper. It was their contention that he had presented to them a simple mistake, and Bose's findings were ignored. Discouraged, he wrote to Albert Einstein, who immediately agreed with him. His theory finally achieved respect when Einstein sent his own paper in support of Bose's to Zeitschrift für Physik, asking that they be published together. This was done in 1924." (from Website 11)

In the core of any scientific research, the attitude must be such that there is no blind faith or automatic acceptance of an existing principle. In this particular case of Nobel Prize winning work of Eric Cornell, he would not be able to observe anomalies if he took Bose-Einstein theory as absolute true (Tung *et al.*, 2006). In fact, all experimental observations indicate, there has to be major adjustments, if not re-formulation, made to the Bose-Einstein theory, as evidenced by follow up research that won Nobel Prize in 2005 (discussion below).

As recently as 2005, Nobel Prize in Physics was awarded jointly to Roy Glauber 'for his contribution to the quantum theory of optical coherence' and to John Hall and Theodor Hänsch 'for their contributions to the development of laser-based precision spectroscopy, including the optical frequency comb technique'. Indeed these discoveries have something to with Bose-Einstein theory, but far from being a regurgitation or redemption of the theory, they open up opportunities for new discoveries.

Even after the above series of Nobel Prize quality research, following questions remain unanswered. They are:

Bose envisioned dissimilar characteristics of any two photons and Einstein extended that work to study with what degree of freedom two dissimilar atoms would behave within a balanced system. The work of the Nobel Physics laureate discovered the separation of certain atoms under very cold conditions, yet they concluded "all atoms are absolutely identical". The reason behind this conclusion? The difference between atoms could not be measured. Fleischhauer (2007) began to question this assumption by comparing milimetre distance in a lab in quantum scale with the distance between the moon and the earth in mega scale and calls 'indistinguishability' as applicable to 'far apart'. However, soon (in the same article) he yields to the pragmatic approach and concludes, "But it also shows that we are entering a state of unprecedented experimental control of coherent light and matter waves. That could bring very real technological benefits: applications that spring to mind include quantum information interfaces that allow the transfer of a quantum bit encoded in a photon to a single atom, as well as ultra-sensitive rotation sensors and gravity detectors". One must wonder What would happen if this pragmatic reasoning was eliminated?

With the above conclusion, how can we observe asymmetry in experimental observations (as reported by Tung *et al.*, 2006)? It must be noted that perfect symmetry is aphenomenal: Nature does not have any example of perfect symmetry. Such asymmetry is rarely taken in account in conventional theory (Dorogovtsev *et al.*, 2000).

Later work of 2005 comes up with similar conclusion about photons, with tremendous implications on light source development and monitoring (e.g. GPS). How would this explain the difference in various light sources, particularly between natural light and artificial light?

For both sets of discoveries, the focus is on 'engineering' phenomena rather than observing. What are the possible implications of this 'engineering' on quality of energy source or matter, *e.g.*, nano materials that are being engineered?

What do these theories say about how transition between mass and energy takes place? Satyendra Nath Bose wrote (Website 11) the real gas equation of state some 100 years ago, what can be said today about real fluids that are still modeled with aphenomenal (unreal) first premises?

What does variable speed of light mean in terms of the definition of the unit of time. What would a paradigm shift in standards and technology development entail if one were to consider truly natural unit of time (Zatzman, 2008). Should the shift be to more 'engineering' or more natural?

6.6.2 Reality of Einstein

In 1936, Einstein published an essay titled: Physics and reality. This publication summarizes Einstein's concept of reality, mass, time, and energy. In his work, Einstein (1936) uncritically accepts and, without skipping a beat, links perception, i.e., immediate and stored sense-experience and knowledge directly.

What is the issue?

The issue is that cognition is the stage in which thought material, comprising perception along with much else, becomes truly humanized. Even though cognitive processes are also influenced by distinct experiences accumulated by each individual, that humanization is what makes the results of our cognition useful to all of us collectively as human beings.

So, Einstein's "science" amounts to the conventional cataloguing|databasing exercise carried out for the purpose of systematizing patterns discovered with what we think we know or have uncovered. His main aim here is actually trying to raise a question as to whether we have to find a use for each new discovery. This is where he wants moral judgment to be more closely exercised. He has something quite specific in mind and we mislead ourselves if we rush to judgment before paying some attention to the matter of context.

This piece is written and published in 1936. This was a time of universal chaos as the Great Depression deepened and the appeal of fascist powers that "something must be done" to sweep away everything that interferes with the will of the Leader became palpable throughout western Europe.

Einstein's most revolutionary assumption was the connection of perception with time. Mathematically, it can be expressed as $t = f$ (perception). Perception is in itself function of the individual's past and the environment. Mathematically, it amounts to:

Perception = $g(t)$, $h(s)$,

Where $g(t)$ is individual history and conscious experience and h (s) is the environment which is a function of space at any given time period, Δt as long as Δt is very small.

This means, any time period becomes a function of infinite number of possibilities. This is the essence of quantum theory that asserts that there is numerous history of every particle, ranging from subatomic size to universe (Hawking, 2010). Apart from creating multiple history violating most important feature of creation (i.e., uniqueness), this manipulation of the time function distorts the definition of work, power as well as efficiency. As seen in earlier section, this is equivalent to creating different standard at the whim of the 'authority'.

In addition, if the perception is not in conformance with reality, the same data will not produce phenomenal thought material. Similar to matter of energy and material, an aphenomenal basis will produce aphenomenal outcome and the entire cognition process will be reversed.

The matter of perception cannot have any influence on reality that exist independent of the observer. For instance, a blind person doesn't see the difference between night and day but that perception doesn't change anything about the night and day. This fact demonstrates the absurdity of Quantum theory that makes reality a function of perception. As stated earlier, reality cannot be a function of perception, unless there is no such thing as reality. Aristotle called this a mistake of Zeno, who failed to distinguish between actual (present) and potential (future) motion, creating paradoxical description of time and motion. This also created a "god's eye" view, often referred to in modern day as 'Einstein's notion of time'. Aristotle 'remedied' it by differentiating between actual (present) and potential (future). Then asserted, "Everything is potential infinity and nothing is actual infinity." This in essence reversed the notion of reality concept of Plato. Einstein invoked the same conclusion by calling any 'perception' a 'reality'. Invoking perception as the root is equivalent to introducing different sources to a non-linear system, it causes the onset of chaos. Such has been the case of cognition process in modern science. As a chaotic system produces multiple solutions (depending on the origin), one of the outcomes is likely to be matched with the desired outcome, making it appealing to the pragmatic approach. This point is made in a cartoon in Picture 6.1.

Perception is a function of the environment, which might be "polluted" with disinformation and ignorance. Therefore, in absence of a criterion for truth, there is no guarantee (in fact, the opposite is guaranteed) there will be any beneficial outcome. In fact, if such cognition persists, long-term disability to align perception with knowledge can be hampered to the extent that mental disorder can set in. This can be best characterized as 'addiction' mode as shown in the downward line in the Figure 6.6.

It is also important to note that perception is a function of propaganda and can be effectively manipulated by the Authority in order to create fear that is beneficial to the Establishment. Gallop poll results (Figure 6.7) regarding the perception of terror attack in the post-9-11 era is a proof of how such perception has nothing to do with reality. This aspect of 'reality' or 'truth' cannot be explained away with dogmatic definitions and "pissing contest a la Bill Clinton", who infamously stated, "it depends on the meaning of the word 'is' ".

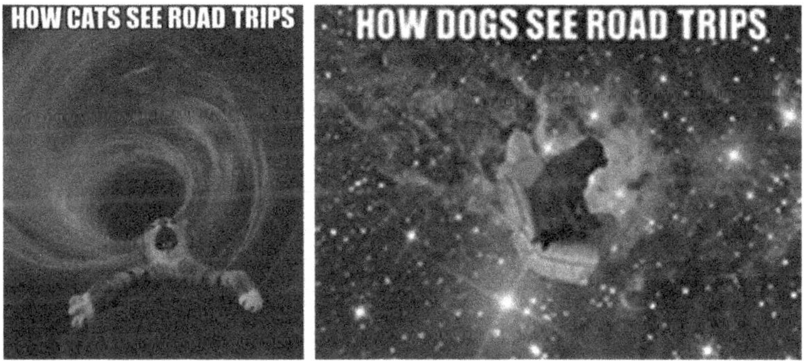

Picture 6.1 If reality is a function of perception, reality itself becomes a paradox.

Zatzman and Islam (2007b) identified the most significant contribution of Newton in mathematics as the famous definition of the derivative as the limit of a difference quotient involving changes in space or in time as small as anyone might like, but not zero, *viz.*

$$\frac{d}{dt}f(t) = \lim_{\Delta t \to \infty} \frac{f(t + \Delta t) - f(t)}{\Delta t} \qquad [1]$$

Without regards to further conditions being defined as to when and where differentiation would produce a meaningful result, it was entirely possible to arrive at "derivatives" that would generate values in the range of a function at points of the domain where the function was not defined or did not exist. Indeed: it took another century following Newton's death before mathematicians would work out the conditions – especially the requirements for continuity of the function to be differentiated within the domain of values – in which its derivative (the name given to the ratio-quotient generated by the limit formula) could be applied and yield reliable results. Kline (1972) detailed the problems involving this breakthrough formulation of Newton. However, no one in the past did propose an alternative to this differential formulation, at least not explicitly. The following figure (Figure 6.8) illustrates this difficulty.

In this Figure, economic index (it may be one of many indicators) is plotted as a function of time. In nature, all functions are very similar. They do have local trends as well as global trend (in time). One can imagine how the slope of this graph on a very small time frame would quite arbitrary and how devastating it would be to take that slope to a long-term. One can easily show the trend, emerging from Newton's differential quotient would be diametrically opposite to the real trend.

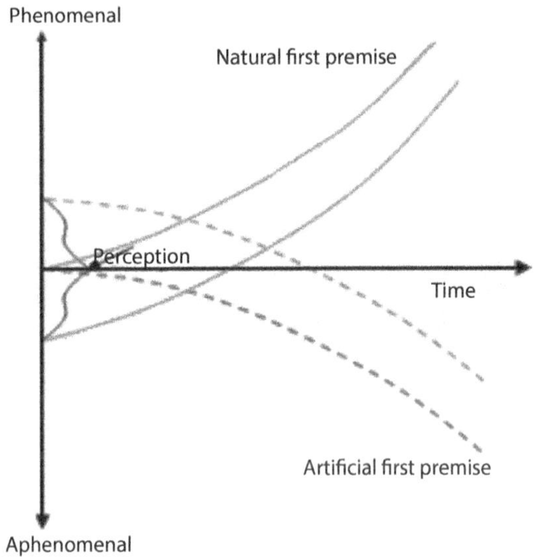

Figure 6.6 If the first premise is not real, perception becomes inherently false and leads to aphenomenal cognition (modified from Khan and Islam, 2012).

In brief, Dogma introduced an aphenomenal standard, namely "Jesus the son of God" and made multiple points exist at the same point in time (trinity). Newton invoked calculus to make more than 3 points exist simultaneous in time, whereas Einstein invoked quantum theory to make multiple history (the time function) to co-exist.

Einstein's energy equation was already deconstructed in a previous section (Section 6.3). In his photoelectric theory, Einstein took Maxwell's rigid sphere model as true, whereas Maxwell himself took that model from Newton. By continuing to build on the same first premise, they have accumulated error and traveled the downward path, depicted in Figure 6.6. By contrast, consider how Ibn Haitham took the model of Aristotle and deconstructed it based on simple logic. He discarded Aristotle's conclusion that light has infinite speed and reconstructed a model that, until today, serves as the only model that can distinguish sunlight from artificial light. While it is well known that sunlight is the essence of life and that artificial light is something that is used to torture people, Einstein's theory or any other optic theory cannot explain scientifically how this is possible. In addition, Ibn Haitham undid another one of the old theories, which is that something that comes out of our eyes makes it possible for us to see. Instead of that theory, he introduced the notion of something entering your eye that makes it possible to see. This "something" was later

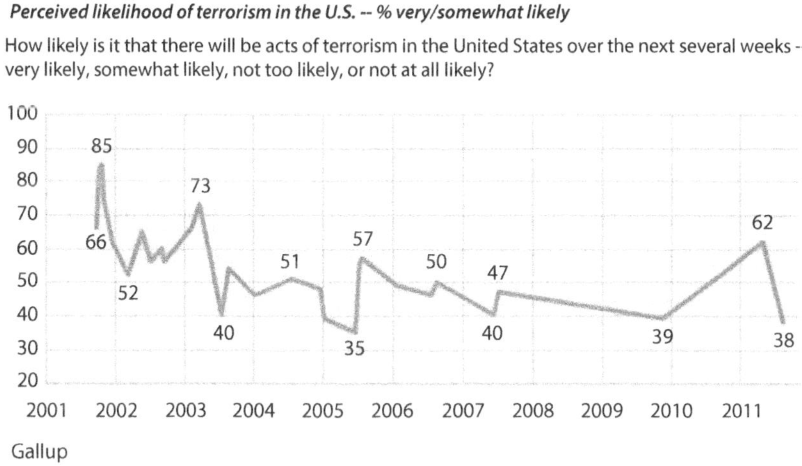

Figure 6.7 How perception is related to propaganda.

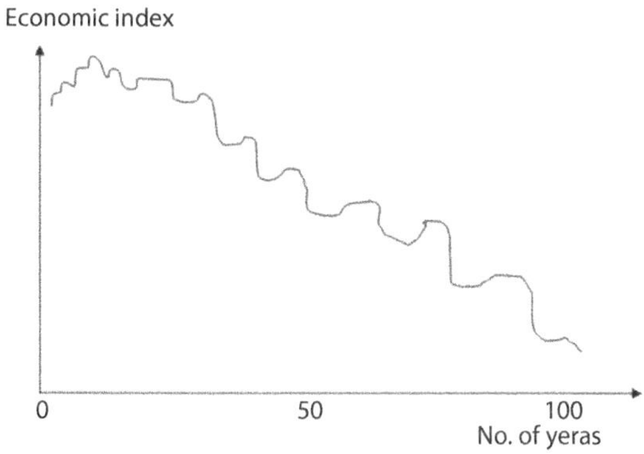

Figure 6.8 Economic wellbeing is known to fluctuate with time (adapted from Islam *et al.*, 2010a).

proclaimed to be a photon. This notion was correct, but the denomination as well as the attribution of various properties made further research on the topic of light characterization impossible. For instance, because this theory postulates that all photons are alike and do not have mass, the source of light cannot have an impact on the quality of light, leading to the same difficulty that made it impossible to discern between sunlight and artificial light. Whereas, if Ibn Haitham's theory was used correctly, one would be able to correlate the toxic nature of the light source (e.g. power-saving light) with long-term impacts, such as breast cancer, brain dysfunction, and numerous other reported correlations. This would also unravel the science behind skin-cancer causing chemicals that are often included in suntan or sun protection lotions. Another example is worth mentioning here, which is a recently touted new form of energy-saving light. This light was excellent in energy savings as well as producing the "white light" effect. However, it was also performing the so-called "belly dance". When the source was sent to the International Space Station for a probe, it was discovered that the "belly dance" subsided or disappeared. Gravity was found to be the reason behind the belly dance. Could this be explained with existing light theories? Of course not, because if photons have zero mass, how could gravity affect them? This paradoxical modus operandi continues when it comes to dark matters (infinite mass, but no energy) and dark energy (infinite energy but no mass) in the realm of the discussion in Cosmic physics.

So, where's the catch? Ibn Haitham didn't read Aristotle's work to believe in it. He read the theory, used the criterion on truth and falsehood and was able to decipher true elements from the volumes of work. Einstein didn't dare use the same logic about Maxwell's "laws" or he didn't have the criterion that Ibn Haitham was equipped with. Einstein was not a Christian, but he certainly was a believer of Eurocentric philosophy. This fundamental inability to discern truth from falsehood (called Furqan in Arabic), is missing from European new science or social science. As a result, what we see is constant confusion about everything that governs our daily lives.

6.7 Conclusions

All theories of mass and energy are critically reviewed in view of determining spurious basis. Based on the review, the following conclusions can be reached:

1. Unless illogical premises are removed, a comprehensive mass balance cannot emerge; and
2. Unless a different first premise is used, an experiment cannot prove anything and will fail to validate a theory;
3. Galileo was the only scientist to attempt an experiment with a first premise starkly different from previous scientists;

4. Nature of fallacy and illogical premises has degraded with time, starting from atomism, all the way down to quantum mechanics;
5. This finding has tremendous bearing on mass, energy and human thought characterization.

7

The Nature-Science Approach: Conclusions of Book I

7.1 Summary

Throughout history, matter, energy and human thought have been the primary foci of human civilization. This to the extent that, civilization itself is actually defined by the manner in which we humans' matter and energy requirements are planned with human thought material (HTM).

The current volume has discussed how matter and energy affect HTM, triggering physical and mental conditions. Earlier chapters were devoted to demonstrating — with numerous examples — the objective fact that, unfortunately contrary to what we have been led on to believe by existing approaches taken to solving existing problems, New Science — outside the conditions that define a conservative Newtonian framework — seems to lack any ability to analyze phenomena objectively let alone point to paths to curing illness and other unstable conditions of the living human body.

Ultimately having failed to produce even a credible scientific determination of cause and effect among many phenomena found throughout this part of the natural order that resist useful study in controlled laboratory conditions, most of New Science has to be set aside in the quest for meaningful and practical solutions to the challenges posed to the optimum functioning of the human person in an environment still beset by may kinds of disease, sickness and other sources of ill health.

As the final chapter of the current volume, this seems an excellent point at which to ask whether there is a hopeful trend to counter this doom-and-gloom, as well as to summarize the main points of the preceding chapters and render a conclusion. Here we will now proceed to draw a line under all that we have discussed so far by way of introducing an entirely new perspective — one that the authors have labeled 'the nature-science approach.'

The nature-science approach opens many pathways to overcoming the ravages of sickness and ill health. Many of its specific aspects are taken up in the following volumes of this series.[1] In this chapter we sketch its outlines and bring out where the nature-science approach can overcome serious deficiencies of Newtonian-based methods already discussed in the previous six chapters.

Previous chapters have demonstrated that the preeminent shortcoming of post-Newtonian 'New Science' is to be found in fundamental premises that are spurious or unnatural. As argued in earlier chapters, natural cognition and some other centuries-old pre-Newtonian premises have been allowed to stand… but for unjustifiable reasons. Once this subset of spurious assumptions is removed at the fundamental level, however, a new basis emerges. This new basis itself invokes a paradigm shift in characterization of matter, energy, and HTM.

A further major shortcoming of the Newtonian-based approach is the truncation or elimination of the time function. This distortion has caused serious discontinuities that have led to outcomes similar to what follows from a spurious first premise.

The natural order on this earth — not just in the outer-spatial regions of Einstein's and Bose's conception — is four-dimensional. However, invoking time as a 4th dimension cannot remedy such truncation. The nature science approach remedies this shortcoming on another path, by invoking a continuous time function at a fundamental level. This characterization can explain behavior of matter of all sizes, ranging from subatomic to galactic scale, without resorting to any discontinuity or dogmatic assertion. Most intriguing: such an approach can also explain phenomena, such as radiation, chemical poisoning, light pollution, pathways of carcinogens, and others. In addition, this

---

[1]This approach is something the authors have developed over the last dozen years or so. It connects matter, energy, time and HTM in the same breath.

approach explains ailments (both physical and mental) and addictions as a departure from natural balance in terms of matter, energy, and/or HTM. It, therefore, removes all paradoxes that are ubiquitous in modern medicine and environmental science. This is the first step toward developing a comprehensive diagnostic scheme, followed by cure.

7.2 Introduction

There are a number of obstacles inherent in the project to establish a science of intangibles based on Nature. The challenge is twofold:

i. to demonstrate where current "science" — social or natural — is wrong **as Science**; and
ii. to eliminate the division imposed between "Nature" and human conscience by pointing out what is necessary for human conscience

All previous chapters so far were dedicated to establishing the just-mentioned first point; this chapter addresses the second point.

7.2.1 The 'Origin-Pathway' Approach of Nature-Science versus the 'Input-Output' Approach of Engineering

One of the great achievements of engineering science has been its identification of the starting point and ending point of a process. This frames the application of the relevant governing equations which must be fulfilled if the proposed system arrangements are to give rise to usable outputs. This is fundamental to engineering designs, as these overall procedures effectively also define various other criteria that the processes must fulfill.

One of the lesser leaps of human imagination, on the other hand, has been the ready identification of what works for a human-engineered design with what is actually going on within any natural process that some human-engineered design purports to simulate. Can the actual pathway within a natural process be identified with the unique solutions generated by some relevant governing equations that must be fulfilled if some proposed system arrangements are to give rise to usable outputs? Does obtaining an output that is similar to what the natural process produced prove or verify the simulation in any way or to any degree? What about the actual pathway within the natural process? Even if the simulation may have verified the unique solution(s) predicted by applying certain governing equations, does that prove or demonstrate that the assumptions underlying those equations and their application are also true for the natural system which is being simulated? Logical deductions or the extrapolations of inferences from conclusions back to the first premises are not enough. Only actual research and investigation can establish the actual pathways and hence also the truth of what's really going on in such circumstances.

The input-output approach of engineering methods may corroborate the processing of certain calculations, according to certain assumptions. However, this approach in itself discloses nothing about theory or truth. For that, a model of the science of nature and how it actually works is needed. The appropriate path on which to emulate what is discovered about how nature actually works is to model the origins and pathways of the process and then engineer so as to emulate that model. This chapter lays down some markers and other guidelines of this "nature science" approach that highlight the importance of the delinearized history standpoint already elaborated.

7.2.2 Reference Frame and Dimensionality

There is an important matter in the natural processes of time and its disclosures and passages, according to differential frames-of-reference, especially for component processes of a larger integrated process. This is crucial for documenting and reconstructing the actual pathways and steps of an overall process. Conventional analysis encounters many difficulties when it comes to sorting out such pathways and their distinctive components. The standpoint implicit within conventional analysis is that of the neutral external observer, a standpoint that renders such reconstruction almost impossible.

The form in which this standpoint expresses itself is itself embedded in the conventions that come with the "usual" linearizations, *viz.*, viewing time as the independent variable, varying independently of the processes being observed, and "adding"/accounting" other additional effects by linear superposition. To the external observer, intermediate changes-of-state at the interface of successive sub-processes are "invisible". These changes and even the interface itself are "invisible" in much the same way that the third dimension is invisible at the interfaces of processes observed in two dimensions. This may account for why analysis based on comparing output to input "works" best of all precisely with the most linearized models. Methods of linear superposition finesse this problem of the invisibility of the interface by glossing over it. Just because an equation, or set of equations, describing the transformation of an overall process from input to output, can or may be decomposed into a set of linear superpositions, does not mean that any or each of these superpositions describes or represents any actual pathway, or portion thereof, unfolding within Nature.

There is a highly significant difference between that which tangibly exists for the five senses in some finite portion of time and space and that which exists in Nature independently of our perceptual functioning in some finite portion of time and space.

If we grasp the latter first, then the former can be located as a subset. However, errors will occur if we proceed from the opposite direction, according to the assumption that what is perceivable about a process or phenomenon in a given finite portion of time and space, contains everything typical and/or characteristic of the natural environment surrounding and sustaining the process or phenomenon as observed in some given finite portion of time and space. Proceeding according to this latter pattern, for example, mediaeval medical texts portrayed the human fetus as a "homunculus", a miniaturized version of an adult person. Proceeding according to the former pattern, on the other hand, if we take phase z as a complex variable, de Moivre's Theorem can be used to readily generate expressions for cos nz and sin nz, whereas (by comparison) if we struggle with constructions of right triangles in the two-dimensional plane, based on some angle x, it is a computationally intensive task just to derive cos 2x and sin 2x, and to, produce?, orders of magnitude more difficult to extend the procedure to derive cos nx and sin nx.

At the conceptual level, there is another factor that interferes with how scientists and researchers cognize what is observed of a natural process. It arises from that fact that time $\boldsymbol{t}$ is not orthogonal to the other three spatial dimensions. Of course, time is no less a dimension for not being orthogonal. Socially available knowledge is also not orthogonal with respect to time $\boldsymbol{t}$ (nor with respect to the other three spatial dimensions, for that matter). The point is: orthogonality is not in itself a relationship built into dimensionality. It applies only to the arrangements scientists and mathematicians have developed to render three spatial dimensions simultaneously visible, i.e., tangible. However, that orthogonalised model is implicit in how scientists and researchers attempt to get a handle on what is observed in any process that they have undertaken to study.

7.2.3 Can "Lumped Parameters" Address Phenomena of only Partial Tangibility?

One technique of engineering is to characterize the operational functioning of a process or phenomenon according to lumped parameters. The idea is that component phenomena are treated "as a class". Thus, for example, in electric circuit theory and actual electrical engineering design, for certain applications, it is useful to think of resistance in a circuit as one single resistor device of a certain rating that represents the magnitude of resistance / reactance across the entire circuit or circuit section, between the input and output of the given section. But again, this modeling is only as good as the initial underlying assumptions — which include (in this case) near-zero frequency, virtually direct current flow and very small potential differences.

Nature-science would want to address such questions as the true resistance, current flow and voltage at any actual point in the real circuit. All that the lumped parameter approach enables is to characterize the overall result/ difference/change at output, compared to the input, without worrying about the details of what happened at any point between the input and the output. So long as the result computed according to Ohm's Law matches the difference measured between the input and the output, this approach opens the door to impose any interpretation to account for whatever happened. How, for example, could the lumped parameter approach account for what happens when an electric eel discharges 1,000 volts and cleans its ambient marine environment?

7.2.4 Standardizing Criteria and the Intangible Aspects of Tangible Phenomena

Closely related to the technique of characterising the operation of a process by means of lumped parameters is the technique of assessing / describing overall performance of the process under study (or development) according to objective, external, uniform "standards" or norms. In the MKS system of SI units, for example, the meter is standardized as a unit of distance, according to the length of some rod of some special element maintained in a vacuum bell at a certain temperature and pressure in some location in Paris, France. Similarly the NIST in Washington DC standardizes the duration of the "second" as the fundamental unit of time, according to an atomic clock, etc. The problem with all such standards is that the question of the standard's applicability for measuring something about the process-of-interest is never asked beforehand. Consider, for example, the known very considerable physical differences between the way extremely high-frequency [tiny-wavelength] EM waves on the one hand, and much lower-frequency [much-greater wavelength] audible-sound waves, on the other hand that each propagate. The meter may be quite reasonable for the latter case. Does it follow that the nanometer —remember: it is based on subdividing the meter into one billion units —is equally reasonable for the former case? The physical reality is that the standard meter bar in Paris actually varies in length by a certain number of picometers or nanometers just within an Earth year. If the process-of-interest is EM radiation traversing light-years through space, however, a variation of the standard metre by 1 nanometer or even by 1000 picometers, will make nonsense of whatever measure we assign to something happening in the physical universe at this scale.

What the objectivity, externality and uniformity of standards enables is a comparison based on what the human observer can directly see, hear, smell, touch or taste—or more indirectly, measure, according to standards that can be tangibly grasped within ordinary human understanding. Is science reducible to that which may be tangibly grasped within ordinary human understanding? If it were, we could and should have spent the last several centuries since Galileo, fine-tuning our measurements of the speed of bodies falling freely toward the Earth so as to catalogue this feature for different classes of objects according to Aristotle's principle — a principle seemingly quite reasonable, perfectly tangible yet utterly erroneous — that the speed with which objects fall freely toward the Earth is a function of their mass.

This example hints at the solution to the conundrum: once the principle of gravity as a force — something that cannot be directly seen, heard, smelt, touched or tasted — acting everywhere on the Earth was grasped, measuring and comparing the free fall of objects, according to their mass had to be given up. It was the attraction due to gravity that was the relevant common and decisive feature characteristic to all these freely-falling objects, not their individual masses. So, standards of measurement applied to phenomena and processes in Nature should cognize features that are characteristic to those phenomena and processes, not externally applied, regardless of their appropriateness or inappropriateness.

Instead of measuring the overall performance of a process or phenomenon under study or development, according to criteria that are characteristic, statistical norms, should be?, applied that compare and benchmark performance relative to some absolute external standard. (Zatzman, 2008a)

7.2.5 Some Retrograde Consequences of Eurocentric Bias for New Science

The authors summarize the central deficiency of New Science, dealing with both the ph ysical aspect ("hard science") and intangible aspect ("social science") as follows:

i. aphenomenal first premise (source);
ii. aphenomenal function that connects events with the first premise. Most functionality involves the time function that is introduced as a fourth dimension (rather than an implicit continuous function) in order to render a 'steady state' into an 'unsteady steady state'. In addition to this spurious insertion of time function, it is expanded with a functionality that doesn't exist in nature.

As the authors detailed elsewhere recently[2], there was a definite stage-by-stage process under way here. Both Averröes' 'the creator' and Thomas Aquinas's 'God' were used to denominate monotheist faith. Then, the concepts of science and religion became a matter of conflicting paradox (Pickover, 2004). Averröes called the (only) creator as 'The Truth' (In Qu'ranic Arabic, the words 'Truth' and 'Creator' refer to the same entity). His first premise pertained to the book (*Qur'an*) that said, "Verily unto Us is the first and the last (of everything)"(89.13).

Contrast this to a "modern" view of a creator, illustrated by Carl Sagan's words quoted at (Hawking, 1988): "This is also a book about God…or perhaps about the absence of God. The word God fills these

[2]See M R Islam, G M Zatzman and J S Islam *Reconstituting the Curriculum* (Scrivener-Wiley 2014)

[3]As discussed more completely in the Appendix to Chapter 5 of the current volume, the essence of Creator and Creation was already obscured through introduction of the (Catholic) trinity and the concept of 'original sin' in the early Christian church, when — hundreds of years later — Thomas Aquinas would smuggle in his profound changes in the form of cognition. The special orders founded inside the Catholic church to combat the appeal of Lutheranism and other proto-Protestant doctrines before and after the Peasants' Wars in German-speaking Europe — the Benedictines, Franciscans and especially the Jesuits produced the effect most desired within the Papacy, viz., the identification of the Pope with Christ's vicar on earth and every feudal lord (no matter how

pages. Hawking embarks on a quest to answer Einstein's famous question about whether God had any choice in creating the universe. Hawking is attempting, as he explicitly states, to understand the mind of God. And this makes all the more unexpected the conclusion of the effort, at least so far: a universe with no edge in space, no beginning or end in time, and nothing for a Creator to do."[3]

This absence of discussion of whatever happened to the tangible-intangible nexus involved at each stage of any of these developments is neither accidental, random, nor arbitrary. It flows directly rather from the Eurocentric bias just discussed, a bias that today pervades the gathering and summation of scientific knowledge everywhere, well beyond Europe and North America. It is by no means a property inherent - either in technology as such, or in the norms and demands of the scientific method *per se*, or even within historical development - that time is considered so intangible as to merit being either ignored as a fourth dimension, or conflated with tangible space as something varying independently of any process underway within any or all dimensions of three-dimensional space.

Mustafiz *et al.* (2007) identified the need of including a continuous time function as starting point of acquiring knowledge. According to them, the knowledge dimension does not get launched unless time as a continuous function is introduced. They further show that the knowledge dimension is not only possible but necessary. The knowledge is conditioned not only by the quantity of information gathered in the process of conducting research, but also by the depth of that research, *i. e.*, the intensity of one's participation in finding things out. In and of themselves, the facts of nature's existence and of our existence within it neither guarantee nor demonstrate our consciousness of either, or the extent of that consciousness. Our perceptual apparatus enables us to

---

insubstantial) as the domestic version of Christ's vicar to his half-enslaved "tenant" farmers. In every aspect, the same doctrinal philosophy has been applied.

Today, the European mindset accepts the supremacy of the secular over the religious while striving to maintain a future for the religious by way of dogmatizing every narration of historical facts, even as these dogmas themselves are in their turn eventually falsified by scientific findings.

Thus — for example — As the entire world celebrate Christmas, the alleged birth day of Jesus, even the pope writes that it was a mistake that dates back to sixth century (Squires, 2012). Catholic Encyclopedia (1911 edition) states:

"Christmas was not among the earliest festivals of the Church... The first evidence of the feast is from Egypt." Further, "Pagan customs centering around the January calends gravitated to Christmas." Under "Natal Day," Origen, an early Catholic writer, admitted,

record a large number of discrete items of data about the surrounding environment. Much of this information we organize naturally and indeed unconsciously. The rest we organize according to the level to which we have trained, and-or come to use, our own brains. Hence, neither can it be affirmed that we arrive at knowledge directly or merely through perception, nor can we affirm being in possession at any point in time of a reliable proof or guarantee that our knowledge of anything in nature is complete.

These questions are not new. What is new in this book is the realization that their answers could be found hundred years ago and the environmental, technological, and moral disaster of today averted... had we not resorted to doctrinal philosophy. Instead of looking at our previous scientists with contempt as though they were somewhat inferior human beings, if we took their natural cognition process and continued with the theories that they had advanced, we would be living in a different world.

7.3 Struggle for Social Reform: Internal and External Factors

Social reform is an essential component of human cognition, which is the driver of overall welfare of the society. There are two sources for the impetus to renew or otherwise reform the educational process. Each source needs its own accounting.

One of these sources is physiologically and psychologically internal. It has to do with what we call human thought material (HTM). The other source appears and emerges external to any particular individual or individuals. It plays a major parallel role in the change, development and motion of education systems across entire social orders and

"...in the Scriptures, no one is recorded to have kept a feast or held a great banquet on his birthday. It is only sinners (like Pharaoh and Herod) who make great rejoicings over the day on which they were born into this world."

"Christmas..." the *Encyclopedia Americana*, 1956 edition states, "was not observed in the first centuries of the Christian church, since the Christian usage in general was to celebrate the death of remarkable persons rather than their birth...[A] feast was established in memory of this event [Christ's birth — *Edd.*] in the 4th century. In the 5th century the Western church ordered the feast to be celebrated on the day of the Mithraic rites of the birth of the sun and at the close of the Saturnalia, as no certain knowledge of the day of Christ's birth existed."

The Catholic encyclopedia (1990) states that the probable date of Jesus' birth is May or April. Contrast that with the books of *Hadith*, detailing every aspect of Prophet

civilizations. It appears sometimes as an ideological component, sometimes as a religious component, and sometimes as an expression of what is referred to as individual conscience.[4]

Historically, a struggle has developed between two very different conceptions regarding this external component and its significance. Eurocentrism is the dominant characteristic of one of these conceptions. Islamically-based belief forms the outstanding characteristic of the other conception.

All conflicts, including ideological ones, have their "casualties", so to speak, but the struggle to establish a truly human consciousness of Right is inevitable and continuous. Hence, bemoaning the conflict element and its casualties is beside the point. Damages and loss of various kinds inevitably emerge from deep ideologically-based disagreements. It is also worse than useless to adopt an "above-the-fray" posture and hold forth about educational reform without acknowledging and taking a stand within this struggle. The issue is to be aware of the conflict and take one's stand within it. More light than heat may then appear in whatever answers are put forward regarding the central question of what comes next in educational reform.

Eurocentric and Islamic views of this matter have frequently come into conflict in the past and still do so today. As a result, it has become impossible to render the story of the development of each of these outlooks in a straight line.

Muhammad's life with dates and times preserved from nearly 1200 years ago (counting from today).

This divergence in Christian and Muslim pathways caught the attention of Zatzman and Islam (2007a), where it was noted that, speaking historically, open challenges to the first premise — i.e., where the divergence is set — has become such a taboo that there is no documented case of anyone challenging it and surviving the wrath of the Establishment (Church alone in the past, Church and European colonialism and imperialism since the Renaissance). Even challenging some of the cursory premises has been hazardous, as demonstrated by Galileo. Today, we continue to avoid challenging the first premise; even now — in the 'teen years' of the Information Age so to speak — it continues to be hazardous, if not fatal, to challenge the first premise or secondary premises. It has been possible to maintain this *modus operandi* as a result of the imposition of

7.4 Consequences of Nature-Science for Classical Set Theory and Conventional Notions of Mensuration

What happens to certain long-assumed fundamental truths about set theory when applied to a set of elements from the real world of nature? Two fundamental theorems of set theory seem to fall by the wayside, viz., the notion that every set is its own subset, and the notion that every set contains/includes a null set.

The first proposition becomes questionable if we watch what happens to certain elements of the set as time passes; they undergo changes, so the totality of elements at time t+ t ["t plus delta-t"] will no longer be the same as the totality was at time t. At the latter point in time, it can no longer be affirmed or guaranteed that the set is still a subset of what it was at the earlier point in time. The passage of time itself has altered the applicability of an equivalence relation on the set. The second proposition becomes problematic when we identify that there is no such thing as an empty set anywhere in nature. So, no set of objects in nature can contain any such thing as a null set. However, these two propositions together are crucial axioms, fundamental to the conventional algebra of sets. Without major modifications to this algebra, the passage of time thus renders impossible its application to sets of objects in nature, and hence also the application of the entire apparatus of formal logic of which that algebra is but the symbolic representation.

This suggests that, because these quantitative methods are premised on assumptions that negate any role whatsoever for critical qualitative differences, whatever results we are obtaining by applying conventional mathematical logic to describing and predicting natural phenomena in their quantitative aspects cannot be said to confirm or refute any

---

new "laws" that purport to protect 'freedom of religion' and even 'freedom of speech'. For special-interest groups, this has opened a Pandora's box full of 'clash of civilizations' and every other aphenomenal model now current (Zatzman and Islam, 2007b).

Avoiding discussion of any theological nature, Zatzman and Islam (2007a) nevertheless managed to challenge the aphenomenal first premise. Rather than basing the first premise on the Truth à *la* Averröes, they mentioned the importance of individual acts. Each action would have three components: 1) origin (intention); 2) pathway; 3) consequence (end). Averröes talked about origin being the truth; they talked about intention that is real. How can an intention be real or false? They equate real with natural. Their work outlines fundamental features of nature and shows there can be only two options: natural (true) or artificial (false). It establishes Aristotle's logic of anything being 'either A or not-A' as useful only to discern between true (real) and false (artificial), and why

underlying natural laws operating overall. All modern mathematical notions of numbers and how a quantity is "composed", are based on these set-theoretic notions refined to their most advanced stages in the work of Georg Cantor at the end of the 19th century (Wallace, 2003). If the foundations of Cantor's set theory cannot be justified or maintained consistently when composing numbers of natural objects and collections of natural objects, then conclusions based on the quantitative results flowing from the mathematics based on this theory cannot be said to be fully warranted. They are at best conditional, and then only in very narrowly-defined time frames.

The truly striking implications of this problem only begin to become clear when we consider what the meanings can be of equality or identity when we are dealing with objects or sets of objects in nature.

The next stage of this argument is: what constitutes an equivalence relation when we are dealing with objects or sets of objects in nature? Following the notion of abstraction of form that has been the guideline since the time of the early mathematicians of ancient Greece, at some level of abstraction we assert that the result of an expression or operation on the left-hand side of an equals-sign is represented on the right-hand side of the equals-sign. So, 2 + 3 = 5. The character of this abstraction, known as a "number" is such that the very same equals-sign, and implicitly the same equivalence relation, applies to the expression "257 - 251 +1", i.e., 257 - 251 +1 = 5", as well. From the time of Euclid onward, this equivalence relation has been described in its abstract essence, thus: two quantities that are each equivalent to some third quantity must be equal to one another, or "A=B and C=B implies that A=C". In terms of the numerical expressions just introduced, this amounts to saying that we are permitted to state that 2 + 3 = 257 - 251 +1, because 2 + 3 = 5 and 257 - 251 +1 = 5 also. But the pathway on

this should be so. In order to ensure the end being real, this work reintroduced the criterion of Khan (2006) and Khan and Islam (2007b). If something were convergent when time was extended to infinity, the end would be assuredly real. In fact, if this criterion is used, we are spared questioning as to the 'intention' of an action. If in doubt, all that remains is to investigate where the activity will end up if time, *t* goes to infinity.

[4]This component is by and large intangible. In western societies, scientific formation and education in the various fields of the natural sciences, based on Newtonian notions of mechanism, tends largely to dismiss any role for, or even any existence of, intangible factors. Although classical Greek theorists of what was known as the dialectic, such as Heraclitus, acknowledged the continuous operation of a flux and contradiction between form and content, formal European philosophy also mostly downplayed any role(s) for intangible factors as such. Early in the 19th century, the German philosopher Georg

the LHS (viz., "2 + 3") is very different from that expressed on the LHS ("257 - 251 +1"). According to the mathematics of tangibles, all that matters is that these very different operations result in the same final quantity [i.e., the "5"]. We need a mathematics of intangibles that no longer dismisses, or behaves indifferently to the fact, that the quality inherent in the "5" produced by the pathway stipulated as "2 + 3" is not the same as the quality inherent in the "5" produced by the pathway stipulated as "257 - 251 +1".

This reasoning may at first hearing, seem somewhat child-like, if not childish. However, its consequences are anything but childish. For example, it is a fact widely observed since ancient times — across literally hundreds of social formations, from tribes and clans to so-called "civilized peoples" to white-Europeans and their descendants — that Nature seems to hoard no long-term surpluses. If the essential economics of modern advanced society is the hoarding of vast surpluses and their hypothecation (through lending at interest, investing for a return, etc.) into yet vaster surpluses, the economics of nature is that of sustainability, premised on neither hoarding nor hypothecating anything. In time, and over time, everything is used, reused, recycled etc.

7.5 Conclusions

7.5.1 Summary

This book presents a comprehensive theory that first explains causes of all ailments, both physical and mental. The theory is based on the premise that nature is perfect and deviation from natural environment, including human behavior causes all ailments. It follows that any disease can be cured by reversing the insult to the environment as well as

Wilhelm Friedrich Hegel overhauled the entire notion of dialectics, while continuing to maintain that the movement of contradictions and of the negation of the negation took place only in the realm of ideas, unconnected to anything in the physical world. Hegel did, nevertheless, update one long overlooked notion from classical Greek dialectic concerning the relationship between Form and Content, and it would be this update that would inadvertently open a path to applying the dialectical logic and mode of analysis especially to social reality. In Hegel's words, the key link between dialectical flux and actual subsequent change was "the overthrowing of the Form, the transformation of the Content" (Lenin, 1976). As the years tick by, Zatzman and Islam's *Economics of Intangibles* (Nova Science, 2005) continues to stand out as an original attempt —within our current collectivity of intellectual fashions — to deconstruct Newtonian mechanism and apply this same perspective to the natural sciences.

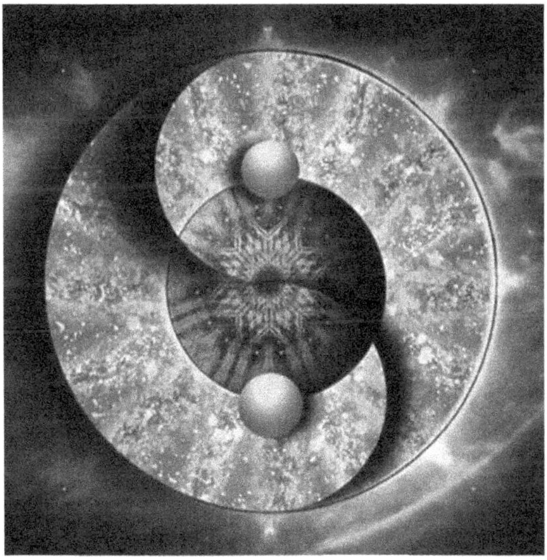

Picture 7.1 The biggest to the smallest entity can be described with a yin yang representation.

natural behavior. While this empowers a person and puts him in control of his health, it contradicts every modern theory of medicine during the New science era. The contradiction is explained through the difference in the first premise of New science that implies that nature can be and should be engineered for profiteering. This understanding explains why New science doesn't have a single cure to a disease, the term 'cure' meaning actually restoring health rather than 'managing' symptoms. It also explains why every medicine or chemical solution has been more toxic, with longer-term peril, than the original stimulant of the disease. These medicines are all artificial with varying degree of aphenomenality. Most importantly, the cause of none of the so-called 'incurable' diseases is known, while numerous vaccines as well as management medicines are pushed in the market. The book also explains how the current dismal state of socio-political status of the world is because of the preposterous nature of solutions offered as 'the solution' in every aspect of society. By using a different first premise, the book invokes a paradigm shift in chemical and pharmaceutical industry practices.

The book uses the ancient scientific description of yin yang (Picture 7.1) duality to describe both phenomenal and artificial objects, including matter, energy, and thought material. It is established that both artificial and natural appear to be similar only for a very small span of time. In the long run, artificial and natural follow diametrically opposite pathways. With time, artificial spirals down to continuous insult and decay of the natural environment, whereas the natural ones continuously improve the natural environment (Pictures 7.2 and 7.3). The former one is metaphorically called 'cancer'. Every ailment is then described as a 'cancer', of tangible (physical ailment) and intangible (mental ailment) category. The consequence of this observation is profound as modern era is characterized by a superflux of artificial objects. Artificial human thought material (HTM), on the other hand, is centuries old, coinciding with the installment of Roman Catholic church dogma and dogmatic thinking. Such dogmatic thinking has been motivated by short-term economic gains, which happen to be the driver of a culture of artificial in all aspects of life.

The book establishes link between addiction and artificial objects and HTM. Every addiction is rooted to an artificial object (mass or energy)

Picture 7.2 Artificial objects of thought materials spiral down to a condition later known as ailment.

Picture 7.3 Natural objects and thought materials improve the environment and remedy all insults and ailments to a state better than the original state

or HTM. For conscious adults, such HTM-linked addiction leads to cognitive addiction, in which a person becomes incapable of changing his original premise, no matter how illogical it is. Such fascination is best described as 'deliberate autism' (distinct from regular Autism that is disconnected from thought materials). There is also another genre of cognitive malfunction that is called 'deliberate schizophrenia' (as distinct from schizophrenia that is disconnected from thought materials). This one emerges from lustful adherence to a dogma because of the lucrative outcome that follows as an end result. These attitudes explain why New science has been unable to break out of dogmatic thinking and continued to feed corporations with theories that support the corporate agenda. On the tangible side, substance abuse is equivalent to 'deliberate schizophrenia' whereas continuous dependence on pharmaceutical drugs is equivalent to 'deliberate Autism'.

This book reconnects humanity with Conscience and establishes the role of humanity in a global society, bringing back previously known but forgotten in the post Roman Catholic Church period notion of purpose of life. It establishes dogma-free, delinearized history of the universe and defines the role of mass in producing energy. Thereby, it connects pollution (of both mass and energy) to human intervention with short-sighted policies (called 'aphenomenal intention'). The book then lays out logical discourses reconstructing mass-energy interaction leading to onset of diseases as well as environmental insult. The book then shows how every solution offered to a particular disease or environmental status differs from the ones proposed in last few centuries of New science, both in theory and practice. A second volume of this book will be entirely dedicated to addresses specific ailments with scientific cure.

In brief, the book revives very old concept of using nature to heal environmental and health insult. Such approach was once captured in the works of Avicenna, whose 'canons of medicine' remains the most widely used for the longest time in history of medicine and human health. However, Avicenna didn't have the 'luxury' of knowing the perils of artificial medicine or corporate-sponsored policies that govern all aspects of modern society. The book captures the development in all branches of science and social science in last 900 years and provides the reader with a deconstruction of dogma and reconstruction of truly scientific logic, empowering the readership with a powerful manual for cognition as well physical welfare.

7.5.2 Conclusions (in detail)

Based on the discussion put forward in previous chapters, the following conclusions can be made:

7.5.2.1 Dogmatic Cognition and Illogical/Aphenomenal Premises of New Science

All theories during the era of New science are based on fundamentally non-existent and illogical premises. This leads to corruption of scientific cognition, turning it into dogmatic cognition. Such cognition process makes it impossible to identify the root causes of corruption and leads to numerous contradictions in all aspects of science and social science, none of which can be resolved without questioning the logical validity of fundamental premises. Because such change in fundamental premises did not occur, all solutions offered in New science have no validity outside of the immediate short-term. The identification of such shortcomings of New science is crucial because the same flawed theories are often used to rectify the conclusion that emerges from any such cognition. Disconnection from truly scientific cognition is equivalent to the onset of first level of 'cancer' that has crippled modern age and has set the scientific community in motion of a spiral down mode, resulting in today's dismal state of the world.

7.5.2.2 Disconnection of Origins from Process

All science and social science theories have a common problem of disconnection between creation and creator as well as conflation between the traits of creator and creation. It then followed with disconnection of humanity from conscience, replacing conscience with a mechanical process, turning on the robotization process. Because engineering solutions were all based on this flawed science, they became extremely short-sighted without any long-term merit. Nonetheless, these solutions were adopted because of their economic appeal. Because of the fact that economic system itself is driven by the same premises of New science that led to the corruption of the entire process, a spiral down mode became the only outcome of the entire technology development process. However, it was also promoted to be the only possible solution, without any alternative pathway to follow.

Figure 7.1 shows how both artificial and natural products (mass, energy, or HTM) act the same way except that their direction of movement is opposite to each other. Without tracking the time function, the description of material behavior has no meaning. This also shows how any natural material that cures an ailment cannot be substituted with an artificial chemical. Such artificial chemicals include all modern medicines, vaccines, antibiotic, and therapies for which nothing can be substituted for natural antidotes (e.g. cow pox, horse serum, fungus, meditation).

7.5.3 Tangible/Intangible Conundrum or Yin Yang Cycle

The failure that led to the current technological disaster is caused by failure in policies that govern our current society, in all scales, ranging from municipality to UN. How policies affect society, thereby setting in motion changes in the environment that later affect individuals that then engage in altering lifestyle and food consumption, can be explained by the yin yang or tangible/intangible conundrum. Because this mode evolves from disconnection with humanity, the entire process spins out of control (Figure 7.2).

All policies are based on dogmatic cognition and false theories and motivated by short-term, leading to 'cancer' of the society. This causes

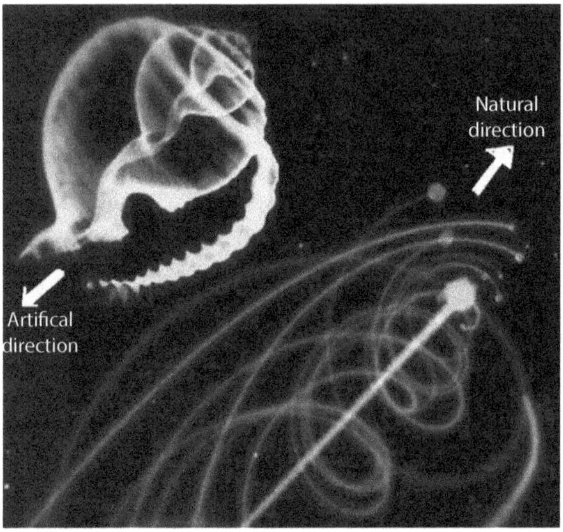

Figure 7.1 Natural artificial both act the same way, except for the time function.

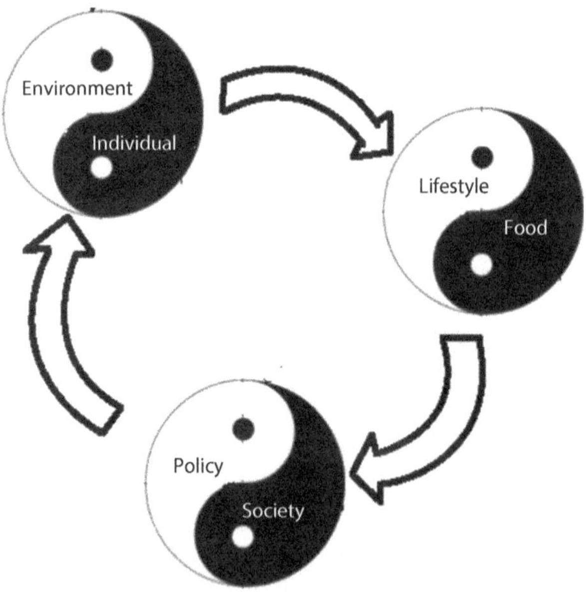

Figure 7.2 Spiral-down process of policy and societal yin yang that starts with disconnection of humans from conscience

collapse leading to environmental degradation that brings about human health degradation. Behavior of this negative yin yang cycle is akin to cancer that deviates the footprint of organic DNA of the universal order.

7.5.4 Honey-Sugar-Saccharine-Aspartame (HSSA) Degradation

In all aspects of society, the degradation, represented by the HSSA degradation, forms the basis for all maladies of the present era. HSSA degradation represents a movement from natural to artificial. This degradation is intricately linked to profiteering by the technology development companies that are linked to policy makers, completing the loop that spirals down both environmental and public health. As environmental degradation continued, causing havoc on human health, profit margin skyrocketed for chemical and pharmaceutical industries that introduced antidotes that are more toxic than the original offenders that they are supposed to remediate. In all these processes, natural solutions represent the Honey paradigm. For instance, original antibiotic, found in fungus, original vaccine found in cowpox, original vitamin C found in organic lemon juice, or original mantra 'food is medicine' they all represented sustainable solutions. When artificial substitutes are introduced, the artificial chemicals acted exactly opposite to the original remedy and created long-term havoc to the organic system, be it human body or the society. As for the cognition process, dogma-free cognition is considered to be the Honey paradigm. Departure from this paradigm causes irreversible harm to the organism with magnitude of harm increasing as a process becomes more artificial.

The degradation is true even for diagnostic tools. Newly developed diagnostic tools and imaging technologies all involve toxic radiation, capable of triggering various ailments, including cancer. Ironically, the new generation of diagnostic tools are touted for detection of cancer cells.

7.5.5 Origin of Addiction

Both chemical and cognitive addictions are linked to artificial. Similar to a cancer cell that mutates uncontrollably without regard to the main pathway followed by an organism, artificial products (energy, mass, or HTM) gain traction through an addictive process. In this, every dogmatic theory is addictive in the sense that one becomes illogically attached to it (cognitive dissonance) and becomes incapable of making a paradigm shift. This process then fuels all clinical or chemical research that continues to produce more and more artificial chemicals with increasingly greater addictive properties. Unfortunately, this process is linked to profit making and contributes to addiction to money or profit in the shortest possible term. Every solution to the problem of addiction, therefore, becomes more addictive and more insidious than the previous solutions.

Similar to artificial chemicals, artificial energy is also addictive. Because all forms of energy is related to a physical source and because every type of energy is related to motion of physical 'particles' of both the source and the medium, the nature of harm caused by artificial chemical is similar to artificial energy. There too, HSSA degradation applies. For instance, sunlight represents Honey, whereas flame from processed natural gas represents sugar, progressively taking a downward turn with tungsten light bulb, then fluorescent light, LED, etc. This addiction eventually turns into allergy to organic food or energy source. On the cognition side, correct dogma-free cognition becomes the subject of all out crusade.

7.5.6 Origin of Ailments

The theory presented in this book can explain such phenomena as light pollution, noise pollution, addiction, chemical poisoning, behavior of carcinogens, onset of cancer, and others. This theory explains ailments (both physical and mental) and addictions as a departure from natural balance in terms of matter, energy, and/or human thought material (HTM). For conscious adults, Conscience plays the most pivotal role. For children or others not in control of their intention, ailments start from environmental stimuli. In either case, it's the history of departure from natural lifestyle and the environment that causes both physical and mental ailment.

7.5.6.1 Role of Magnitude and Frequency of Stimuli

Environmental stimuli accumulate and pass the threshold of onset of a disease, depending on the magnitude of the insult as well as the natural frequency of the stimuli. Energy stimuli is the driver in this process. Pictures 7.4 and 7.5 demonstrate the role of magnitude and frequency

Picture 7.4 The role of frequency in passing threshold concentration that triggers an ailment

Picture 7.5 The role of magnitude of insults that trigger an ailment

in crossing the threshold. Consider the electric fan as the representative of the organic body and the guns the stimuli.

Let natural frequency of a material be f_m; the frequency of the energy source be f_e, and the ratio f_e / f_m be denoted by F_r:

*Scenario 1: Near resonance frequency, FrNoentity*1.

When the energy frequency is close to natural frequency of a matter of interest, resonance-like movement makes immediate impact on the matter. In extreme situation, this would lead to irreparable damage to materials. However, this damage would be at a short time span. This is equivalent to poisoning in physiological sense.

Scenario 2: Large F_r

This scenario deals with a situation for which the frequency difference is high. The probability of penetration of energy matter increases within the cellular structure of the body. With this probability also increases the probability of contamination and eventual alteration of a DNA structure. If indeed a DNA structure is 'hit', the probability of attracting energy matter increases dramatically. Two competing phenomena result from this onset. As the body is continuously exposed the artificial energy source, the size of the affected body mass increases as other cells get affected and become 'cancerous'. At the same time, when such 'dead' cell is recognized by the body, the immune system attempts to remove the cell from the onset site. If the structure of a cancer cell is inspected, it becomes clear that such structure is difficult to remove. However, at present, there is little evidence as to how a cancer cell is purged in an organic system. From the shape of the cancer cells, it appears that it would be very difficult for them to travel within an organic mass unless its molecular structure is broken down to the extent that it is recognized as a 'debris' in an organic system.

It is important to note that natural frequency of humans vary with age. This makes people at different group vulnerable to different types of stimuli. New science conflates this effect with genetic disorder. Figure 7.3 Shows how heart rate varies for different ages. Naturally, children are more dynamic and their body parts are renewed faster. This would necessitate faster replenishment of energy.

There is what is characteristic but there is also fundamental frequencies. For example, other natural frequencies in one's body are unrelated

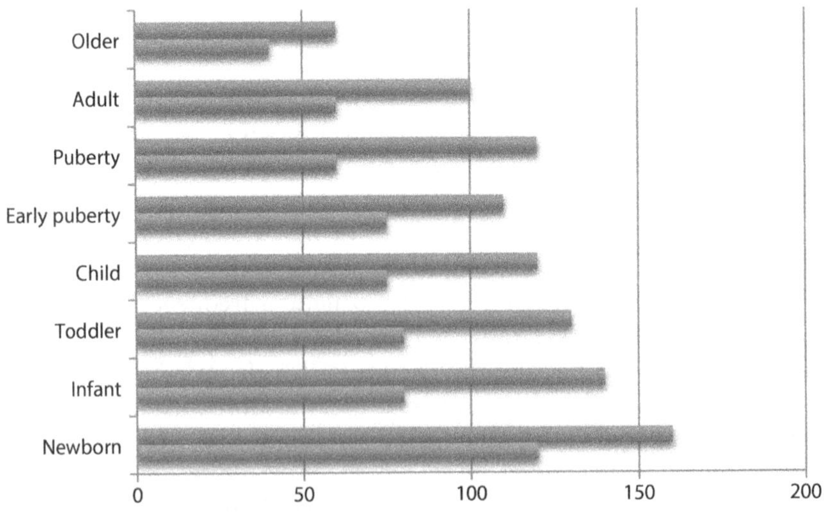

Figure 7.3 Maximum and minimum heart rate for different age groups

Table 7.1 Characteristic number and sizes of various human body parts.

Element	Characteristic number	Characteristic size, m
Limbs	4	10^{-2}
Organs	13	10^{-2}
Bones	300	10^{-2}
Hairs	200,000	10^{-4}
Veins and arteries	10^{6}	10^{-5}
Nerves	10^{7}	10^{-6}
Cells	$75.\ 10^{12}$	10^{-12}

to heartbeat frequency. Thus, it would be difficult to believe that an individual's brainwave frequency, for example, could be entirely accounted for from investigation of phenomena occurring within the cerebral cortex alone by itself. This book establishes that we have no control over these frequencies. What we have control over is our intention. Thankfully that doesn't affect the universal order but doubly thankfully does affect an individual's long-term future.

Table 7.1 shows characteristic number and sizes of various human body parts. Table 7.2 shows renewal times for various human body cells. Together, human body is a complex organic structure that is described as a "universe" in the galaxy model presented in the book.

Table 7.3 lists some symptoms of vibration exposure at low frequencies, along with the frequency ranges at which the symptoms are most predominant (data adapted from Griffin, 1990)

Table 7.4 shows natural frequency range of various body parts.

Table 7.5 shows pathological signs that are associated with different body parts and their natural frequency ranges.

As stated earlier, this frequency range is altered with age. This is another trait of nature.

7.5.7 The Role of Medical Intervention

Because every medication, vaccine, and antibiotic in modern age is artificial, any medical intervention automatically invokes conflict with natural cycles of a human body. The upper graph of Figure 7.4 shows the

Table 7.2 List showing the renewal time for various types of human body cells.

Cell Type	Characteristic life span
Red Cells	120 days
Platelets	10 days
Neutrophils	4-5 days
Fat Cells	8 years
Cardiomyocytes	1-.45% per year/7-32% per year
Crypts in colon	7 days
B Cells (Mouse)	4-7 weeks
Whole skin	1 month
Stomach Cells	2 days
Taste buds	10 days
Skeleton	10 Years
Nails	6-10 months
Hair	~1cm per month
Lung cell surface	2-3 weeks

natural cycles of a human. Typically, any ailment, if treated with natural remedy, leads to an improved immune system than the one prior to the ailment. This is reflected in the zigzag nature of the curve. The smooth curve shows the average of a human system. It is recognized that each person goes through an optimum age for which the immune system and the overall status of health is at its peak. The same human, if exposed to artificial treatment and medical intervention, would be subject to the lower graph. Every time a vaccine is administered, antibiotic given, or medicine consumed to 'cure' or prevent a disease, the body lapses into a state worse than the one that existed prior the ailment or the procedure administered. Because the body becomes busy fighting 'management' pills, it never has time to reach the optimum in terms of immune system or healthy status. Instead, the body goes through a spiraling down mode with continuous decline in immune system and

Table 7.3 Frequency ranges and ailment.

Symptoms	**Frequency (Hz)**
General feeling of discomfort	4-9
Head symptoms	13-20
Lower Jaw symptoms	6-8
Influence on speech	13-20
"Lump in throat"	12-16
Chest Pains	5-7
Abdominal pains	4-10
Urge to urinate	10-18
Increased muscle tone	13-20
Influence on breathing movements	4-8
Muscle contractions	4-9

dependence on artificial stop-gap measures or 'health management' regimen. This lucrative business of pharmaceutical industry comes with a catastrophic consequence on human health, each medication working like a toxic, albeit legal, drug.

Medicines are indeed planned according to the specific demands of a certain age group. For instance, hormonal treatment, birth control pills, heart medications, vitamin supplement, vaccines, flu shots, etc. are administered differently for different age groups, making sure every age group is most pre-occupied fighting artificial chemicals as the profit margin of the pharmaceutical industry skyrockets.

7.5.8 Hopelessness of Modern Medical Science and Medical Diagnosis

While Newton had assumed everything in an absurd state of 'steady state', the addition of time as a fourth dimension didn't improve the underlying science. Albert Einstein's most famous assumption linked perception with time as well as reality, taking humanity back beneath the time of Plato, who discarded everything in transition as non-existent

Table 7.4 Natural frequency range of various body parts (Adopted from Walsh, 1992; Jessel et al., 2013; and Ivancevic and Ivancevic, 2007). Because any form of energy would contain frequency range interfering with natural body frequency, every form of artificial energy would create some form of dissonance. This alone can explain the origin of ailments.

Body part	**Frequency range (Hz)**
Neuron	500-1000
Brain	20-30
Eyeball	20-90
Shoulder girdle	4-5
Lower arm	16-30
Arm	4-5
Abdominal mass	4-6
Legs	2 (flexible) – 20 (rigid)
Brain freq. Beta	14-30 Hz
Heart	If 120 beats/m 2 Hz
Stomach	58-65 Hz
Liver	55-60 Hz
Bone	38-43 MHz
Lungs	58-65 Hz
Pancreas	60-80 Hz
Visionary range	120 MHz

or unreal. The dilemma is depicted in Figure 7.5 Every diagnosis is a function of perception based on short-term history. At the end the option the physician is looking at is confined to selection of a drug, irrespective of the validity of the perception. Once a drug reduces the symptoms, it is considered to be 'successful'. If it doesn't reduce the symptoms, a new series of drugs are investigated. Diagnostic techniques are not any help because each of these techniques has assumptions embedded in it that makes it impossible to see the big picture. When it comes to more sophisticated diagnostic tools, each of them includes

Table 7.5 Experimentally determined natural frequencies (from Wieckowski, 2013).

Body part	**Natural frequency Hz**	**Possible pathological signs**
Head	4-5; 17-25	Pains, vertigo, desequilbrium, feeling of pressure in the larynx, nausea, psychological tiredness
Head with neck	20-30	
Shoulder and head	20-30	
Jaw	6-8	
Eyeballs	60-90; 40-90	
Abdominal organs	4.5-10	Pain, nasea, feeling of fullness, urge to urinate, weakness and tiredness
Liver	3-4	
Stomach	2-3	
Urinary bladder	10-18	
Kidneys	6-8	
Thorax	5-7, 4-11	Respiratory disorders, breathlessness, quickening of breath, anxiety, blood pressure change, quickened heart beat, speech disorder
Organs in the thorax	5-9	
Lungs	4-11	

(*Continued*)

Table 7.5 cont.

Body part	Natural frequency Hz	Possible pathological signs
Heart	4-6	
Trachea, bronchi	12-16	
Upper part of the torso: shoulders and head	4-5, 20-30	Joint and muscular pains, pains in the cervical spine, increased muscle tone, feeling of tiredness
Lower part of the torso	4-6	Joint and muscular pains, pains in the sacral spine and lumbar spine, increased muscle tone, feeling of tiredness
Pelvis	5-9	
Vertibral column	10-12	
Sacral spine	8-12	
Lumbar spine	8-12	
Lower limbs	5	Joint pains, increased muscle tone, numb sensation and muscle spasm
Hips	5	
Calves	20	
Feet	-	

(Continued)

Table 7.5 cont.

Body part	**Natural frequency Hz**	**Possible pathological signs**
Upper limbs	4-5	Joint pains, increased muscle tone
Arm	16-30	Muscular pains, involuntary muscular contraction, difficulty carrying out physical tasks
Forearm	4-6	
Hand	20-30	
Upper limbs	4-5	
Arm	16-30	
Forearm	4-6	
Hand	20-30	

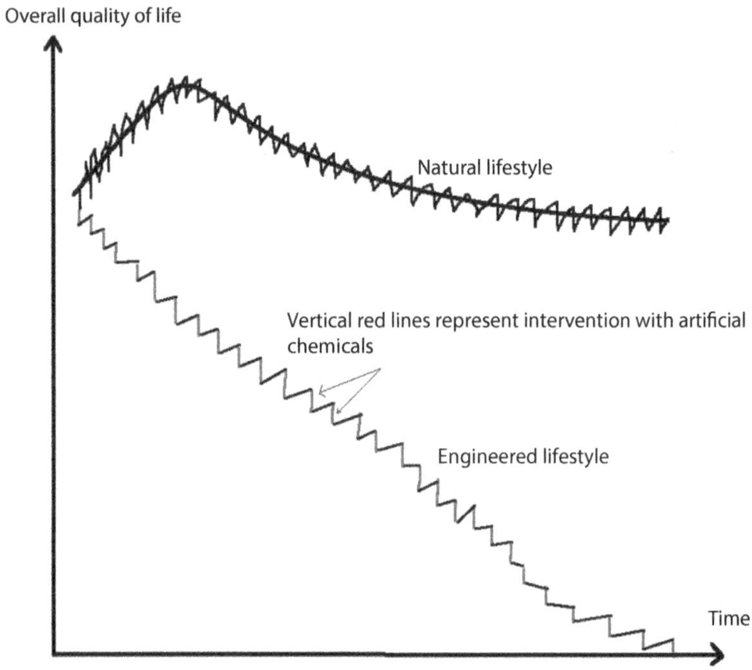

Figure 7.4 Role of ailment on natural and artificial lifestyle

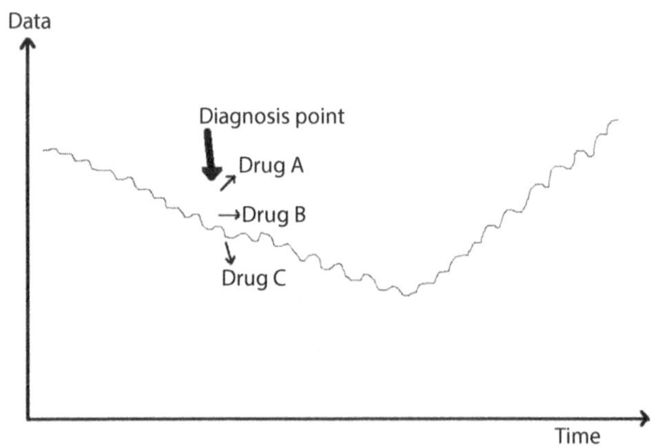

Figure 7.5 Diagnosis is a function of doctor's perception, which is based on myopic analysis or false premise.

artificial energy sources that are capable of inducing onset of debilitating imbalance in the body.

The assumption of 'defective gene' as well as consideration of 'family history' doesn't improve the predictive ability of a physician. The process remains obscure as long as the big picture is not seen. Seeing the big picture requires a holistic approach - something medical professionals are incapable of mastering.

7.5.9 Cause of Disease and Irreversible Environmental Damage

For conscious adults, any disease starts with intangible components of the aphenomenal domain, such as fear, anxiety, lust for short-term outcome, myopic outlook (focus on tangibles). The presence of such aphenomenal components triggers movement in the spiral down direction. This in itself is the product of the environment as well as personal history (experience) of a person. The downward motion of the intangible accompanies downward motion of tangible components of artificial mass and energy. Even though the motion in the spiraling down mode begins immediately, the impact is 'felt' later and only after the aphenomenal components have gained momentum with enough 'snowball' effect. The avalanche theory in the book predicts that each of these 'snow balls' is capable of maintaining spiral down movement and cause enough damage to trigger an ailment. Figure 7.6 shows how the 'cancerous' 'snow balls' move in a direction different from the main direction of the organic components. The term 'cancerous' is both metaphorical and literal. Even though, it is commonly believed that cancer cells are unique because they are not recognized by the body as 'enemies' and are not attacked, this belief itself is not logical. With a different premise that is discussed in this book, it is concluded that human body is infinitely in tuned with universal order and never oblivious to the fact that some 'cancer' cells are dragging down the entire organism toward death and destruction. It turns out that perpetual injection of artificial chemicals, including food, energy source, medications, etc. create a scenario that changes overall nature of the organism. A person becomes addicted to artificial and becomes allergic to natural products (e.g. sunlight, mother's milk). Such condition with overt symptoms was once recognized as Chronic fatigue syndrome (CFS). It has received different denominations and is now recognized as malfunction of the central nervous system, resulting in metabolic, infectious or post-infectious,

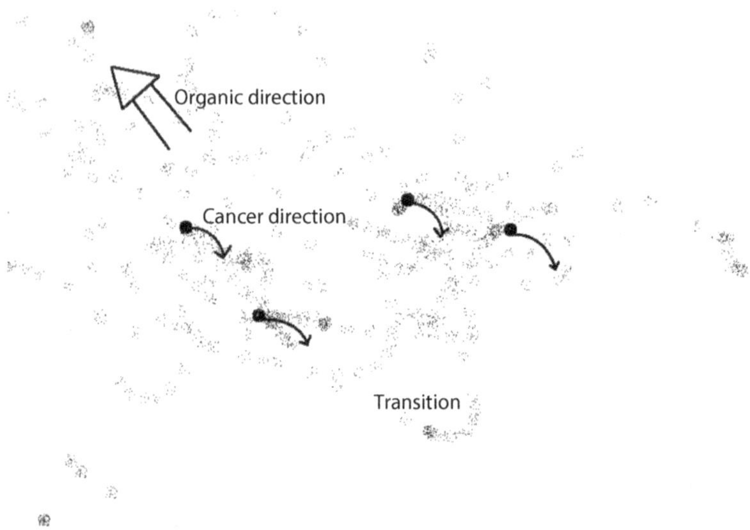

Figure 7.6 Onset of cancer cells.

cardiovascular, immune system or psychiatric disorder. At the core of this malady remains the key to onset of a disease. Figure 7.6 Shows how the entire organic process can be dragged down and the natural course be reversed if enough 'cancer' cells gain control of the body. Until the time when the organic side capitulates to the artificial side, transitions take place and no symptom develops. After the symptoms have emerged, it is already a sign of capitulation. If these symptoms are suppressed without treating with a natural cure, the problem intensifies. (Figure 7.6)

This process of onset and propagation of a disease can be depicted with the galaxy model, presented in this book. Figure 7.7 depicts the galaxy model. After a "cancerous" cell manages to survive and reproduce following an order different from the natural order, the entire organism develops a malady. Here, the quicker the cell regeneration the more vulnerable a cell is (Table 7.2). After a cancer cell takes hold, it is only a matter of time that it would spread, much like a drop of paint within a painting that is still not dry. The galaxy model portrays such phenomena much like what is shown in Figure 7.7. It is further concluded that the notion of arresting spread of a cancer is an absurd concept. Equally absurd is the notion of injecting artificial chemical or artificial energy for remedying cancer. It is equivalent to adding another coating of white paint and hoping to contain the white paint.

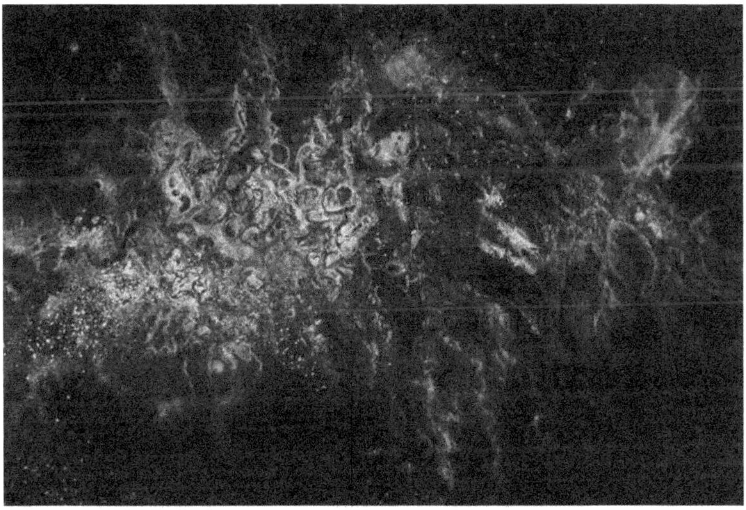

Figure 7.7 Galaxy portrayal of an organic system.

7.5.10 Greening of Pharmaceutical Industry

The book establishes that artificial mass and energy are the cause of physical illness whereby artificial (e.g. dogmatic) cognition is the source of mental ailment. It follows that these conditions can only be aggravated by using artificial medication or dogmatic therapy.

7.5.10.1 Phases of Life

The role of intangible should be considered in characterizing human life span. From infancy to adolescent, a child has no control over his/her intention and, therefore, all ailments arise from environmental stimuli. Similarly, people suffering from severe dementia, schizophrenia, dementia or other 'mental disease' would fall under the same category, meaning their ailments are the result of environmental stimuli of mass, energy or both.

The next characterization is in terms of mental and physical fitness. Mental health is more important than physical health because mental strength is directly related to immunity and ability to repel harm due to artificial mass and energy sources. Mental health can be improved by focusing on long-term and avoiding myopic vision. A myopic vision is facilitated by so-called 'deliberate autism', the lust for short-term outcome. Deliberate autism is adherence to status quo, no matter how illogical it is. Such focus on short-term gains is maintained with 'deliberate schizophrenia', in which a person insists on focusing on short-term gains so much that he behaves as though he is addicted to schizophrenia and unable to cognize with any logical approach. Every time, he conflates every issue so that the intended outcome is justified.

7.5.10.2 Recognizing the Stimulant

The environment as well as food and drink intake should be studied in order to identify artificial elements and ranked according to the level of artificiality as per the HSSA criterion. The ranking should be done for energy sources as well.

In terms of intangible, recognize the role of 'deliberate schizophrenia', 'deliberate autism'. In presence of these conditions, a person is incapable of recognizing the root causes of an ailment, let alone taking steps toward remedying them. The role of environment and social conditions should be understood as well. The next phase of intangibles are the supplements (e.g. vitamins, hormonal, minerals) that are consumed. If from artificial sources, they are negative stimulants and must be recognized as such. It will also include all chemical medications that are used to 'manage' certain health conditions.

7.5.10.3 Remove Negative Stimulant

As long as a person is still alive, there is hope for reversing the condition. This can be done by removing the artificial stimulants from all forms of food and energy. This must be followed by natural replacements. There is no evidence today that any condition cannot be reversed. Therefore, no one should be told, his/her condition is beyond repair. Of course, death is not something that can be avoided but there is no need to invite death by telling a patient there is no hope of recovery from an ailment.

7.5.10.4 Replacement of Artificial with Natural

This must be comprehensive, including changes in the environment, entourage, social setting, culture and overall lifestyle. On the intangible side, it should involve reconnecting with the purpose of life, the role of universal order, and the logic of karma. On the tangible side, all

Table 7.6 Essential chemicals, their functions, and their natural sources.

Chemical	Natural sources	Function
Calcium	Milk, soy milk, yogurt, hard cheeses, cereals, fish bones, bone marrow	bone growth and strength, blood clotting, muscle contraction, etc.
Choline	Milk, liver, eggs, peanuts	Overall cell replacement
Chromium	Broccoli, potatoes, meats, poultry, fish, some cereals	Helps control blood sugar levels
Copper	Seafood, nuts, seeds, wheat bran cereals, whole grains	Helps process iron and related vital functions
Fiber	Plant foods, including oatmeal, lentils, peas, beans, fruits, and vegetables	Helps with digestion, lowers LDL ("bad") cholesterol, helps feeling content, and helps maintain blood sugar levels
Fluoride	Natural spring water, some sea fish	Prevents cavities in teeth, helps with bone growth
Folic acid (folate)	Dark, leafy vegetables, citrus fruits, whole grain breads, beans, peas, lentils, avocado, okra, cereals, seeds and nuts	Helps prevent birth defects, important for heart health and for cell development
Iodine	Seaweed, seafood, dairy products, sea salt, mountain salt	Helps make thyroid hormones
Iron	Cereals, beans, lentils, beef, turkey (dark meat), soy beans, spinach	Needed for red blood cells and many enzymes
Magnesium	Green leafy vegetables, nuts, dairy, soybeans, potatoes, whole wheat, quinoa	Helps with heart rhythm, muscle and nerve function, bone strength
Manganese	Nuts, beans and other legumes, tea, whole grains	Helps form bones and make some enzymes
Molybdenum	Legumes, leafy vegetables, grains, nuts	Needed to make some enzymes

(*Continued*)

Table 7.6 cont.

Chemical	Natural sources	Function
Phosphorus	Milk and other dairy products, peas, meat, small fish, eggs, some cereals and breads	Normal cell functions, helps make energy, needed for bone growth.
Potassium	Potatoes, bananas, yogurt, milk, yellow fin tuna, soybeans	Helps control blood pressure, makes kidney stones less likely
Selenium	Organ meats, seafood, dairy, some plants (if grown in soil with selenium), Brazil nuts	Protects cells from damage; helps manage thyroid hormone.
Sodium	Sea salt, mountain salt, all vegetables	Important for fluid balance and related activities, e.g. blood pressure, vision, heartbeat; HCl in stomach, nervous system.
Vitamin A	Sweet potatoes, carrots, spinach,	Maintaining good vision, protect cells from radiation, and fight against cancer, infections, and radiation damage.
Vitamin B1 (thiamine)	Whole-grain, bread, cereals, Durian, Sweet Potatoes, Avocado, Bananas, Sunflower Seeds, Bell Peppers, chili peppers, Quinoa, Wild Rice	Helps the body process carbs and some protein
Vitamin B2 (riboflavin)	Milk, bread products, Quinoa, sweet potatoes, pecan, avocado	Helps convert food into energy. Also helps make red blood cells; useful for eye, skin, tendon, ligament

(*Continued*)

Table 7.6 cont.

Chemical	Natural sources	Function
Vitamin B3 (niacin)	Meat, fish, poultry, whole grain breads, wild rice, buckwheat, avocado, Parsley	Helps with digestion and with making cholesterol, important for maintaining healthy skin and blood circulation
Vitamin B5 (pantothenic acid)	Chicken, beef, potatoes, oats, cereals, tomatoes, Sweet Potatoes, Buckwheat, Jerusalem Artichokes	Helps with metabolism, the immune system, and babies' brain development; is used by the body to convert fats and carbohydrates into usable energy, and to produce healthy fats in the body. It also plays a role in supporting the adrenal glands and in the production of stress-related hormones.
Vitamin B6	chickpeas, potatoes, organ meats, Durian , Sweet Potatoes, Avocado, Bananas, Sunflower Seeds, Bell Peppers, Red bell peppers, Dandelion Greens, Mango, Brussels Sprouts, Buckwheat, Quinoa, Wild Rice	plays a critical role in new cell formation, which makes it especially important for regenerating tissue such as the skin. It is also involved in the metabolism of carbohydrates and in a broad range of nervous system activity, including the production of neurotransmitters.
Vitamin B7 (biotin)	Liver, fruits, meats	Helps body make fats, protein, and other products for proper functioning of cells.
Vitamin B12	Fish, poultry, meat, dairy products	Helps body make red blood cells

(*Continued*)

Table 7.6 cont.

Chemical	Natural sources	Function
Vitamin C	Red and green peppers, kiwis, oranges and other citrus fruits, strawberries, broccoli, tomatoes	Helps protect against cell damage, supports the immune system, and helps body make collagen
Vitamin D	Sunlight, Fish liver oils, fatty fish, milk products	Needed for bones, muscles, the immune system, and communication between the brain and the rest of the body
Vitamin E	sunflower seeds, almonds, peanut butter, vegetable oils	Helps protect cells against damage
Vitamin K	Green vegetables like spinach, collards, and broccoli; Brussels sprouts; cabbage	Important in blood clotting and bone health
Zinc	Red meats, some seafood	Supports the immune system and nerve function. Also important for reproduction.

minerals, supplements, vitamins, hormones should be replaced with natural sources. Here is a list of such chemicals and their natural sources.

7.5.10.5 Medicines and Therapies With Natural Substitutes

This book presents the argument that natural remedy is the only way to cure a disease. Once cured, not only the original state of health will be restored, the immune system will be improved from the original state and the person will be less vulnerable to the same ailment. While the discussion in the section above revolves around replacing trace chemicals with natural alternatives or food sources for overall good health, this section focuses on medicines and therapies after the onset of a disease.

Table 7.7 Lists a number of debilitating diseases with their conventional medicines and natural alternatives. Note that for each of these conditions, honey and blackseed are common antidotes and, therefore, omitted from the table. Listed is the most useful 'green solution' that targets a specific ailment. Volume 2 of this book will have a comprehensive list of all remedies with discussion of science behind the remedies.

7.5.10.6 Mental Conditioning and Staged Prevention

This book concludes that for conscious and matured adults, mental conditioning is the first stage of prevention. Mental conditioning includes reorientation for long-term vision, meditation, positive visualization, and organic lifestyle. Initial resistance must be treated like a disorder of addiction. Health education should involve taking ownership of the intention and be able to set both short-term and long-term goals with a focus on long-term, in conformance with the universal order. This so-called approach of obliquity is necessary for achieving inner peace. Once a person realizes that the intention must be in conformance with the universal order and long-term peace, he is prepared to face the world with an attitude that safeguards him from long-term harms (Figure 7.8).

Just like phenomenal intention is the source of a well balanced lifestyle, natural water is the source of well balanced matter and energy. In order to achieve progress in personal or societal health, all policies have to emerge from phenomenal intention and all matter has to emerge from naturally purified water. This is shown in Figure 7.9.

Table 7.7 Certain diseases, Modern medical solutions, and Green solutions.

Disease	Modern medical solution	Green solution
Cancer	Chemotherapy	H-86, mudbath
	Radiotherapy	sunbath
	Infrared therapy	Natural fire place, clay oven cooked food
	Gene therapy	Natural hormone induction
Diabetes	Actos, Avandia, Byetta, Metform	DBX-13, organic whole wheat, organic rice (with mineral coating)
Arthritis	Motrin, Bextra, Naproxen, Celebrex	Organic cinnamon, ginger, Chinese Thunder God Vine, willow bark extract, feverfew, cat's claw and stinging nettle. They have similar ingredients as naproxen (Aleve) and ibuprofen (Advil), but of natural source.
Alzheimer	Aricept, Rivastigmine (Exelon), Galantamine (Razadyne).	Amyloban in Chinese herbal mushroom. Amyloban contains DLPE and hericenone derivatives as active ingredients, it may play a pivotal role in protecting neuronal cells particularly from ER–stress exerted assault, ultimately preventing the onset of neurodegenerative diseases (AD).

(Continued)

Table 7.7 cont.

Disease	Modern medical solution	Green solution
High blood pressure	Beta blockers, cancium channel blockers and ACE inhibitors, Lipitor, Zocor, Baycol, Crestor	Policosanol (extract from sugar cane that Policosanol decreases cholesterol production in the liver and to increase the break down of LDL (low-density lipoprotein or "bad") cholesterol. It also decreases the stickiness of particles in the blood known as platelets, which might help reduce blood clots.
Insomnia	Ambien, Lunesta, Restoril, Sonata	Valerian; Natural cortisol from organic sources, such as coffee, grapefruit, liquorice

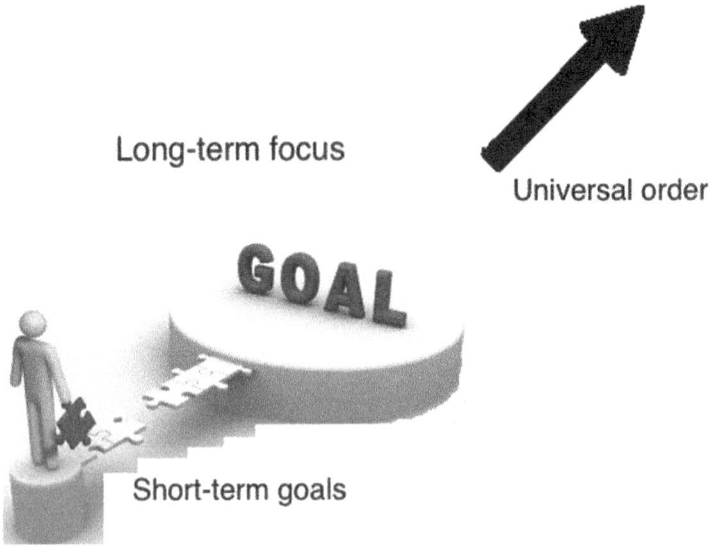

Figure 7.8 The approach of obliquity is necessary for grounding and achieving personal inner peace.

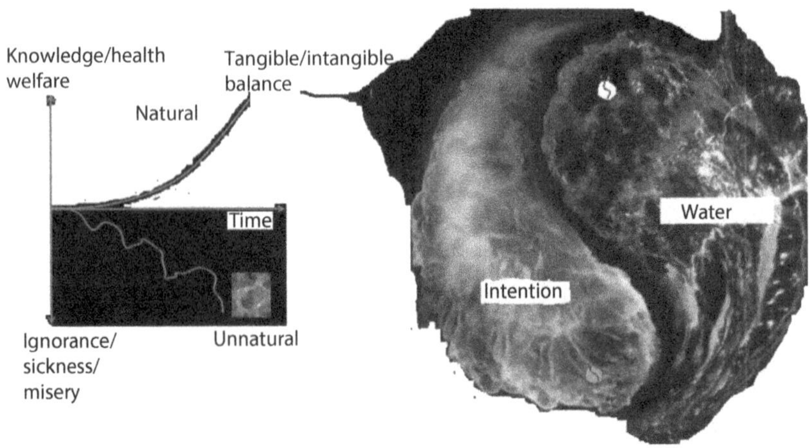

Figure 7.9 The Knowledge Model and the source of good health.

In order to prevent a disease, the entire history of the person, including family background must be included. It is important to note that family history cannot be used to create support for aphenomenal premises such as 'defective gene'. The history only creates a basis and determines how the individual reacts to artificial stimuli, including medicine, surgical treatments, chemical abuse, and lifestyle.

Primordial prevention consists of actions to minimize future hazards to health and hence inhibit the establishment factors (environmental, economic, social, behavioural, cultural) known to onset a disease. It addresses broad health determinants rather than preventing personal exposure to risk factors, which is the goal of primary prevention.

Cleanup must start with water, therefrom inserting natural remedies both in matter and energy. This book provides one with the science of water and redefines energy after eliminating dogmatic assertions that energy is disconnected from mass.

7.6 The Need for Change

Historical time in social development and characteristic time in natural processes each exist, and operate, objectively and independent of our wills or even our perceptions. They are certainly not perceived as such by us humans living in the present. We cognize these phenomena, and their objectivity, only in the process of summing up matters on the basis of looking back from the vantage point of the present.

We may idealize the arc of change, development and/or motion of a process. This idealization can be as tractable or as complex as we desire, with a view to being reproducible in experiments of various kinds. What weight is to be assigned, however, to any conclusions drawn from analysis of this idealization and how it works? Can those conclusions apply to what is actually happening in the objective social or natural processes? The nub of this problem is that the input-state and output-state of an actual process can be readily simulated in any such idealization or its experimental reproduction. The actual pathway — meaning how matters actually proceeded from input to output — is very likely, however, another matter entirely.

Essentially, that idealization takes the form of a First Assumption. The most dangerous such First Assumptions are the most innocent-seeming. Consider, for example, the notion of the speed of light taken as a constant — in a vacuum. Where in nature is there a vacuum? Since no such location is known to exist anywhere in nature, if the speed of light is observed to vary, *i.e.*, not be constant, does this mean any observed non-constant character can be ascribed to the absence of a vacuum? So, therefore, does the original definition remain valid? Or, does it mean rather that we need better measuring instruments? This notion of the speed of light being constant in a vacuum has been retro-fitted to make it possible to bridge various gaps in our knowledge of actually-observed phenomena. It is an example of an idealization. By fitting a "logically necessary" pathway of steps between input and output, however, on the basis of applying conclusions generated by an idealization of some social or natural process to the social or natural process itself, it becomes trivially easy to create the appearance of a smooth and gradual development or evolution from one intermediate state to another intermediate state. In such linearising and smoothing, some information-loss, perhaps even a great deal, necessarily occurs. Above all, however, what is being passed off as a scientific explanation of phenomena, is in fact an aphenomenal construction on the actual social or natural processes. This aphenomenal modeling of reality closes all loops and bridges all gaps with fictions of various kinds.

One necessary corrective action to this hopeless course should rely instead on the closest possible observation of input-state (*i.e.*, historical origin), pathway and output-state (i.e., the present condition, as distinct from a projection) of the actual social or natural process — starting with the present, i.e., meaning the current output-state. Whatever has been clearly established, and whatever still remains incompletely understood is then summed up. A process of elimination is launched. This is based on abstracting the absence to advance a hypothesis that might account for whatever gaps remain in the observer's knowledge, which is also possible to test. The observer plans out some intervention(s) that can establish in practice whether the hypothesized bridging of the gaps in knowledge indeed account for what's been "missing".

All processes explained up to now, rather simplistically, only insofar as their change, development and motion conforms to known laws of social or natural developments can be reviewed by these same methods and their conventional explanations replaced with these essentially "delinearized" histories.

7.7 The Nature Science Approach

We have thus far established, without fundamental shift in first premise, we cannot begin to cognize any process. This chapter shows how a new premise can lead to an entirely new and logical way to cognize. We call that the nature science approach and it has been called 'science of intangibles' before (e.g. Khan and Islam, 2012). This approach has the following essential components:

- Any study should involve the recognition of the pathways of a product, including the origin. By tracking the origin and recognizing the role of matters that are conventionally not accounted for in the science of tangibles, one can begin to see the full picture. This analysis is typical of the science of intangibles.
- A new material characterization methodology is proposed that would include the role of intangibles automatically. With this approach, the importance of sustainability will be inherently featured in any product quality.
- If products are characterized based on the methodology provided in this chapter, the devastating effect of unsustainable products that are ubiquitous today will become apparent.
- Any disease or ailment arises from departure from natural cognition, natural environment/lifestyle, and/or natural food. The ailment can be mental or physical.
- Repeated insult/injury with energy pollution, mass pollution or lifestyle corruption can lead to permanent irreversible damage, often manifested as "cancer" of tangible (cancer) or intangible (cognitive dissonance) kind.
- The cure to any disease is to go back to the root cause and replace the source with natural alternatives.

References and Bibliography

Abegunde D.O., Mathers C.D., Adam T., Ortegon M., Strong K. The burden and costs of chronic diseases in low-income and middle-income countries. *Lancet* 370. 2007. 1929:38.

Abou-Donia MB, El-Masry EM, Abdel-Rahman AA, McLendon RE, Schiffman SS. Splenda alters gut microflora and increases intestinal p-glycoprotein and cytochrome p-450 in male rats. *J Toxicol Environ Health A.* 71(21). 2008. 1415:29.

Abou-Kassem, J.H., 2007, "Engineering Approach vs the Mathematical Approach in Developing Reservoir Simulators", *Journal of Nature Science and Sustainable Technology 1 (1)* 35:68.

Abou-Kassem, J.H., Farouq Ali, S.M., Islam, M.R..A Basic Approach to Reservoir Simulation, (Houston TX: Gulf Publishing 2006), 455pp.

Abou-Kassem, J.H., Zatzman, G.M, and Islam, M.R., (2008) Some Implications of the Nature Science Standpoint for Mathematical Modeling Methods and Their Appications: Part 1: The True Meaning of Einstein's Fourth Dimension", *J. Nature Sci. and Sust. Tech.*, 2 (1/2) 1:20.

Abramson, C. I. And Boyd, B. J. An automated apparatus for conditioning proboscis extension in honey bees, Apis mellifera L. *J. Entomol. Sci.* 36. 2001. 78:92.

Abramson, C. I., Stone, S. M., Ortez, R. A., Luccardi, A., Vann, K. L., Hanig, K. D., and Rice, J The development of an ethanol model using social insects — I: Behavior studies of the honey bee (Apis mellifera L.). *Alcohol Clin. Exp. Res.* 24. 2000. 1153:1166.

Adams, MJ.; Lefkowitz, EJ.; King, AM.; Carstens, EB. "Recently agreed changes to the International Code of Virus Classification and Nomenclature", *Archives of Virology* 158 (12). Dec 2013. 2633:39.

Adler, L. S. 2000. The ecological significance of toxic nectar. *Oikos* 91 409:420.

Afik, O. Dag, A., Kerem, Z. & Shafir, S., 2006, Analyses of Avocado (Persea americana) Nectar Properties and their Perception by Honey bees (Apis mellifera), *J Chem Ecol*, 32 1949:63.

Afik, O., Dag, A., And Shafir, S., The effect of avocado (*Persea americana*) nectar composition on its attractiveness to honey bees (*Apis mellifera*). *Apidologie* 37. 2006. 317:325.

Ahmad O.B., Boschi-Pinto C., Lopez A.D., Murray C., Lozano R., Inoue M. World Health Organization *Age standardisation of rates: a new WHO standard*. 2001 (GPE Discussion Paper Series: no.31 EIP/GPE/EBD).

Ahmed, R.L., K. H. Schmitz, K. E. Anderson, W. D. Rosamond, and A. R. Folsom, "The metabolic syndrome and risk of incident colorectal cancer," *Cancer* 107 (1) 2006. 28:36,.

Ahn, J., A. Schatzkin, J. V. Lacey Jr. et al., "Adiposity, adult weight change, and postmenopausal breast cancer risk," *Archives of Internal Medicine 167 (19)* 2007. 2091:2102.

Al Waili NS, Saloom KY, "Effects of topical honey on post-operative wound infections due to gram positive and gram negative bacteria following caesarean sections and hysterectomies," *Eur J Med Res* 4. 1999. 126:30.

Al-Jabri, A.A., S. A. Al-Hosni, B. C. Nzeako, Z. H. Al- Mahrooqi, and H. Nsanze, "Antibacterial activity of Omani honey alone and in combination with gentamicin," *Saudi Medical Journal* 26 (5) 2005. 767:771.

Al-Waili, N.S., "Clinical and mycological benefits of topical application of honey, olive oil and beeswax in diaper dermatitis," *Clinical Microbiology and Infection* 11 (2) 2005. 160:163.

Al-Waili, N.S., "Investigating the antimicrobial activity of natural honey and its effects on the pathogenic bacterial infections of surgical wounds and conjunctiva," *Journal of Medicinal Food* 7(2) 2004. 210–222

Al-Waili, N.S., "Topical application of natural honey, beeswax and olive oil mixture for atopic dermatitis or psoriasis: partially controlled, single-blinded study," *Complementary Therapies in Medicine* 11 (4) 2003. 226:234.

Al-Waili, N.S., "Topical honey application versus acyclovir for the treatment of recurrent herpes simplex lesions," *Medical Science Monitor* 10 (8) 2004. MT94:MT98.

Al-Waili, N.S., and A. Haq, "Effect of honey on antibody production against thymus-dependent and thymus-independent antigens in primary and secondary immune responses," *Journal of Medicinal Food* 7 (4) 2004. 491:494,

Al-Waili, N.S., M. Akmal, F. S. Al-Waili, K. Y. Saloom, and A. Ali, "The antimicrobial potential of honey from United Arab Emirates on some microbial isolates," *Medical Science Monitor* 11 (12) 2005. BR433:BR438.

Al-Waili, N.S., T. N. Al-Waili, A. N. Al-Waili, and K. S. Saloom, "Influence of natural honey on biochemical and hematological variables in AIDS: a case study," *The Scientific World Journal* 6 2006. 1985:1989.

Alandejani, T., J. Marsan, W. Ferris, R. Slinger, and F. Chan, "Effectiveness of honey on Staphylococcus aureus and Pseudomonas aeruginosa biofilms," *Otolaryngology—Head and Neck Surgery* 141 (1) 2009. 114:118.

Alden, J.R. 1996. George Washington: A Biography, (Baton Rouge LA: LSU Press).

Aljadi, A.M., and M. Y. Kamaruddin, "Evaluation of the phe- nolic contents and antioxidant capacities of two Malaysian floral honeys," *Food Chemistry* 85 (4) 2004. 513:518.

Aljazeera. 2013. "Report: Annual NYC inmate cost exceeds four years at Harvard", Sept. 30.

Allen KL, Molan PC, Reid GM. "A survey of the antibacterial activity of some New Zealand honeys" *J Pharm Pharmacol* 43. 1991. 817:22.

Allender S., Scarborough P., Peto V. 2008. European cardiovascular disease statistics (Brussels: European Heart Network).

Alnaqdy, A., A. Al-Jabri, Z. Al Mahrooqi, B. Nzeako, and H. Nsanze, "Inhibition effect of honey on the adherence of Salmonella to intestinal epithelial cells in vitro," *International Journal of Food Microbiology* 103 (3) 2005. 347:351.

Alwine J.C. "Modulation of host cell stress responses by human cytomegalovirus," *Curr.* Top. Microbiol. Immunol. 2008. 325 263:79. (doi:10.1007/978-3-540-77349-8).

Amemori, S., A. Ootani, S. Aoki, et al., "Adipocytes and preadipocytes promote the proliferation of colon cancer cells in vitro," *American Journal of Physiology* 292 (3) 2007. G923:G929..

American Psychiatric Association (2013). *Diagnostic and Statistical Manual of Mental Disorders (Fifth* ed.) (Arlington VA: American Psychiatric Publishing 797:798. ISBN 978-0-89042-555-8.

American Psychiatric Association. (1994). *Fact Sheet: Violence and Mental Illness*. Washington, DC: American Psychiatric Association.

Anderson C, Franks NR, McShea DW. The complexity and hierarchical structure of tasks in insect societies. *Anim Behav (200)* 62 643-651.

Anderson C, Ratnieks FLW. Worker allocation in insect societies: coordination of nectar foragers and nectar receivers in honey bee (*Apis mellifera*) colonies. *Behav Ecol Sociobiol* (1999) 46 73-81.

Anklam, E. 1998. A review of the analytical methods to determine the geographical and botanical origin of honey. *Food Chem.* 63 549–562.

Appelbaum PC, MRSA—the tip of the iceberg. *Clin Microbiol Infect* (2006) 12 (Suppl 2):3-10.

Appleby, L., Mortensen, P. B., Dunn, G., & Hiroeh, U. (2001). Death by homicide, suicide, and other unnatural causes in people with mental illness: a population-based study. *The Lancet* 358 2110-2112.

Armstrong, D.G., "Manuka honey improved wound healing in patients with sloughy venous leg ulcers," *Evidence-Based Medicine* 14 (5) [2009] p.148,.

Aronson, E., & Mills, J. (1959). The effect of severity of initiation on liking for a group. *The Journal of Abnormal and Social Psychology*, 59(2), 177.

Arslan D, Legendre M, Seltzer V, Abergel C, Claverie JM. Distant Mimivirus relative with a larger genome highlights the fundamental features of Megaviridae. Proceedings of the National Academy of Sciences of the United States of America. 2011;108(42):17486–91. doi:10.1073/pnas.1110889108. PMID 21987820. Bibcode:2011PNAS.10817486A.

Arvin AM, Greenberg HB. New viral vaccines. *Virology*. 2006;344(1):240–9. doi:10.1016/j.virol.2005.09.057.

Asadi Pooya, A.A., M. R. Pnjehshahin, and S. Beheshti, "The antimycobacterial effect of honey: an in vitro Study," *Rivista di Biologia*, vol. 96, no. 3, pp. 491–495, 2003.

Asaria P, MacMahon E. Measles in the United Kingdom: can we eradicate it by 2010?. *BMJ*. 2006;333(7574):890–5. doi:10.1136/bmj.38989.445845.7C. PMID 17068034.

Asaria P., Chisholm D., Mathers C., Ezzati M., Beaglehole R. Chronic disease prevention: health effects and financial costs of strategies to reduce salt intake and control tobacco use. *Lancet*. 2007;370(9604):2044–2053.

Associated Press, The (2013), "Washington couple weep as they are sentenced to nearly 30 years in prison for death of Hana Williams", October 30.

Asuquo, M., G. Ugare, G. Ebughe, and P. Jibril, "Marjolin's ulcer: the importance of surgical management of chronic Evidence-Based Complementary and Alternative Medicine cutaneous ulcers," *International Journal of Dermatology*, vol. 46, supplement 2, pp. 29–32, 2007.

Asuquo, M.E., A. M. Udosen, I. A. Ikpeme, et al., "Cutaneous squamous cell carcinoma in Calabar, Southern Nigeria," *Clinical and Experimental Dermatology*, vol. 34, no. 8, pp. 870–873, 2009.

Aumann, R.J., 1964, "Markets with a Continuum of Traders", *Econometrica* (1964).

Aumann, R.J., 1974, "Subjectivity and Correlation in Randomized Strategies", *Journal of Mathematical Economics* (1974).

Aumann, R.J., and Branderburger, A., 1987, "Correlated Equilibrium as an Expression of Bayesian Rationality", *Econometrica* (1987).

Aumann, R.J., and Hart, S., *Handbook of Game Theory with Economic Applications*. Elsevier (Vol. 1: 1992, Vol. 2: 1994, Vol. 3: 2002)

Aune, E., Røislien, J., Mathisen, M., Thelle, D.S., and Otterstad, J.E., 2011, "The "smoker's paradox" in patients with acute coronary syndrome: a systematic review", *BMC Medicine*, 9:97 doi:10.1186/1741-7015-9-97.

Azoulay A, Garzon P, Eisenberg MJ (2001). "Comparison of the mineral content of tap water and bottled waters". *J Gen Intern Med* 16 (3): 168–75. doi:10.1111/j.1525-1497.2001.04189.x. PMC 1495189. PMID 11318912.

Bachman, W. W. And Waller, G. D. 1977. Honeybee responses to sugar solutions of different compositions. *J. Apic. Res.* 16:165–169.

Baggesen DL, Sørensen G, Nielsen EM, Wegener HC. Phage typing of Salmonella Typhimurium – is it still a useful tool for surveillance and outbreak investigation?. *Eurosurveillance*. 2010;15(4):19471.

Baker, H. G. 1977. Non-sugar chemical constituents of nectar. *Apidologie* 8:349–356.

Baker, J.P., 2008, "Mercury, Vaccines, and Autism: One Controversy, Three Histories", *Am J Public Health*. 2008 February; 98(2): 244–253.

Barman S, Ali A, Hui EK, Adhikary L, Nayak DP. Transport of viral proteins to the apical membranes and interaction of matrix protein with glycoproteins in the assembly of influenza viruses. *Virus Res*. 2001;77 (1):61–9.

Barnes, H. E. (1932) Prohibition Versus Civilization: Analyzing the Dry Psychosis. New York: Viking

Barnett ED, Wilder-Smith A, Wilson ME. Yellow fever vaccines and international travelers. *Expert Rev Vaccines.* 2008;7(5):579–87. doi:10.1586/14760584.7.5.579. PMID 18564013.

Baroni-Urbani C, Buser MW, Schilliger E. Substrate vibrations during recruitment in ant social organization. *Insect Soc* 1988;35:241-250.

Barozzi P, Potenza L, Riva G, Vallerini D, Quadrelli C, Bosco R, Forghieri F, Torelli G, Luppi M. B cells and herpesviruses: a model of lymphoproliferation. *Autoimmun Rev.* 2007;7(2):132–6.

Barrangou R, Fremaux C, Deveau H, et al. CRISPR provides acquired resistance against viruses in prokaryotes. *Science.* 2007;315(5819):1709–12.

Barré-Sinoussi, F. *et al.* Isolation of a T-lymphotropic retrovirus from a patient at risk for acquired immune deficiency syndrome (AIDS). Science. 1983;220(4599):868–871. doi:10.1126/science.6189183. PMID 6189183. Bibcode:1983Sci..220.868B.

Barth FG, Hrncir M, Jarau S., 2008, Signals and cues in the recruitment behavior of stingless bees (Meliponini). *J Comp Physiol A* 2008;194:313-327.

Bates, C., 2011, Global cancer cases rise 20 per cent in a decade to hit 12million a year, Mail online, Sept. 8.

Beaglehole R., Bonita R., Horton R. Priority actions for the non-communicable disease crisis. *Lancet.* 2011;377(9775):1438–1447.

Beard, K. W.; Wolf, E. M. (2001). "Modification in the Proposed Diagnostic Criteria for Internet Addiction". *Cyber Psychology & Behavior* 4 (3): 377 doi:10.1089/109493101300210286.

Becker, A., 2015, "Why does time always run forward and never in reverse", *BBC,* March 9.

Beer, S. and Waisel, Y., 1982, "Effects of light and pressure on photosynthesis in two seagrasses", *Aquatic Botany,* Volume 13, 1982, Pages 331–337

Bejamin, M. and Ramchndran, N., 2015, "10 frightening facts about Tom Cotton", *Salon,* March 14.

Bell, J., 2013,One child in three is overweight or obese, UAE study shows, The National (UAE), June 12.

Bellon M, Nicot C. Telomerase: a crucial player in HTLV-I-induced human T-cell leukemia. *Cancer genomics & proteomics.* 2007;4(1):21–5.

Belpomme, D., P. Irigaray, A. J. Sasco, et al., "The growing incidence of cancer: role of lifestyle and screening detection (Review)," *International Journal of Oncology,* vol. 30, no. 5, pp. 1037–1049, 2007.

Bener, A., H. Ayub, R. Kakil, and W. Ibrahim, "Patterns of cancer incidence among the population of Qatar: a worldwide comparative study," *Asian Pacific Journal of Cancer Prevention,* vol. 9, no. 1, pp. 19–24, 2008.

Bergh O, Børsheim KY, Bratbak G, Heldal M. High abundance of viruses found in aquatic environments. *Nature.* 1989;340(6233):467–8.

Berman DM, Nicklas BJ, Ryan AS, Rogus EM, Dennis KE, Goldberg AP: Regulation of lipolysis and lipoprotein lipase after weight loss in obese, postmenopausal women, *Obes Res* 2004, 12:32-39.

Beshers SN, Fewell JH. Models of division of labor in social insects. *Annu Rev Entomol* 2001;46:413-440.

Beshers SN, Traniello JFA. The adaptiveness of worker demography in the attine ant Trachymrymex-septentrionalis. *Ecology* 1994;75:763-775.

Bhattacharjee, Y., 2013, Earliest Known Galaxy Formed Stars at a Breakneck Pace, *Science* Oct. 25., vol. 342, no. 6157, p. 411

Bianchine PJ, Russo TA. The role of epidemic infectious diseases in the discovery of America. *Allergy Proc.* 1992 [Retrieved 16 September 2008];13(5):225–32. doi:10.2500/108854192778817040.

Bickle TA, Krüger DH. Biology of DNA restriction. *Microbiol.* Rev. 1 June 1993;57(2):434–50. PMID 8336674.

Biello, D., 2006, Social Honeybee Shares Genetic Secrets, *Scientific American*, Oct. 25.

Biesmeijer JC, Slaa EJ. Information flow and organization of stingless bee foraging. *Apidologie* 2004;35:143-157.

Biesmeijer JC. The occurrence and context of tremble dancing in free-foraging honey bees (Apis mellifera). *Behav Ecol Sociobiol* 2003;53:411-416.

Bitterman, M. E., Menzel, R., Fietz, A., And Schafer, S. 1983. Classical conditioning of proboscis extension in honey bees (Apis mellifera). *J. Comp. Psychol.* 97:107–119.

Bjorge, T., A. Engeland, S. Tretli, and E. Weiderpass, "Body size in relation to cancer of the uterine corpus in 1 million Norwegian women," *International Journal of Cancer*, vol. 120, no. 2, pp. 378–383, 2007.

Bland JM, Altman DG, 1986, Statistical methods for assessing agreement between two methods of clinical measurement. *Lancet* 1986;1:307–10.

Block, J. J. (2008). "Issues for DSM-V: Internet addiction" *American Journal of Psychiatry* 165, 306–307.

Blum AS, Soto CM, Wilson CD, et al. An Engineered Virus as a Scaffold for Three-Dimensional Self-Assembly on the Nanoscale. Small. 2005;7:702. doi:10.1002/smll.200500021.

Bodeker GC, Ryan TJ, Ong CK, "Traditional approaches to wound healing". *Clin Dermatol* 1999;17:93-8.

Bodewes, R., Fraaij, P.L., Geelhoed-Mieras M.M., van Baalen CA, Tiddens HA, van Rossum AM, van der Klis FR, Fouchier RA, Osterhaus AD, Rimmelzwaan GF., "Annual vaccination against influenza virus hampers development of virus-specific CD8$^+$ T cell immunity in children.", *J Virol.* 2011 Nov;85(22):11995-2000. doi: 10.1128/JVI.05213-11. Epub 2011 Aug 31.

Boevink P, Oparka KJ. Virus-host interactions during movement processes. *Plant Physiol.* 2005;138(4):1815–21.

Bogdanov S, Nature and origin of the antibacterial substances in honey. *Lebensm Wiss Technol* 1997;30:748-53.

Bohannon, J. 2015, New study questions tropc that conservatives are happier than liberals Science, March 12, DOI: 10.1126/science.aab0311.

Bolino CM, Bercik P. Pathogenic factors involved in the development of irritable bowel syndrome: focus on a microbial role; *Infect Dis Clin North Am.* 2010 Dec;24(4):961-75.

Bolts, W., 1772, *Considerations on India Affairs*, London, 1772-5, I, passim; Harry Verelst, A View of the Rise, Progress and Present State of the English Government in Bengal, London, 1772, passim; L.S. Sutherland, The East India Company in Eighteenth-Century Politics, Oxford, 1952, pp.219-22,255-8.

Bonabeau E, Theraulaz G, Deneubourg JL, Aron S, Camazine S. Self-organization in social insects. *Trends Ecol Evol* 1997;12:188-193.

Bonn D, Sweet solution to superbug infections? *Lancet Infect Dis* 2003;3:608.

Bonod, I., Sandoz, J. C., Loublier, Y., And Pham-Delegue, M. H. 2003. Learning and discrimination of honey odours by the honey bee. *Apidologie* 34:147–159.

Bosy-Westphal A, Eichhorn C, Kutzner D, Illner K, Heller M, Müller MJ. The age-related decline in resting energy expenditure in humans is due to the loss of fat-free mass and to alterations in its metabolically active components. *J Nutr* 2003;133:2356–62.

Bosy-Westphal A, Reinecke U, Schlorke T, et al. Effect of organ and tissue masses on resting energy expenditure in underweight, normal weight and obese adults. *Int J Obes Relat Metab Disord* 2004;28:72–9.

Bouchard, S., Vonhof, M. J., Fenton, M. B., And Monette, G. 2000. Nutrient preferences of Brazilian hummingbirds. *Wilson Bull.* 112:558–562.

Bouchardy, C., G. Fioretta, E. Rapiti, et al., "Recent trends in prostate cancer mortality show a continuous decrease in several countries," *International Journal of Cancer*, vol. 123, no. 2, pp. 421–429, 2008.

Boudreau, K.J., Guinan, E.C., Lakhani, K.R., and Ried, C., 2013, "The Novelty Paradox & Bias for Normal Science: Evidence from Randomized Medical Grant Proposal Evaluations", *Harvard Business School, Working Knowledge*, Jan. 10, paper no. 7173.

Bowe, W.P., Alan C Logan. Acne vulgaris, probiotics and the gut-brain-skin axis - back to the future? *Gut Pathogens* 2011, 3:1.

Boyle, P., and B. Levin, World Cancer Report, WHO-IARC, 2008.

Bradbury JW, Vehrencamp SL. Principles of animal communication. Sunderland (MA): Sinauer Associates; 1998.

Bray, F., and B. Moller, "Predicting the future burden of cancer," *Nature Reviews Cancer*, vol. 6, no. 1, pp. 63–74, 2006.

Brehm, J. W. (1956). Postdecision changes in the desirability of alternatives. *The Journal of Abnormal and Social Psychology*, 52(3), 384.

Brian L. Gilmore, B.L., Shannon P. Showalter, S.P., Madeline J. Dukes, M.J., b Justin R. Tanner, J.R., a Andrew C. Demmert, A.C., a Sarah M. McDonalda, S.M., and Deborah F. Kelly, D.F., 2013, "Visualizing viral assemblies in a nanoscale biosphere", *Lab Chip*, 2013,13, 216-219, DOI: 10.1039/C2LC41008G

Briffault, R., 1919, *The making of humanity*, Publisher London, G. Allen & Unwin ltd. London, 388 pp.

Brock, F. et al. Reliability of nitrogen content (%N) and carbon:nitrogen atomic ratios (C:N) as indicators of collagen preservation suitable for radiocarbon dating. *Radiocarbon* 54, 879–886 (2012)

Bronk Ramsey, C., Scott, M. & van der Plicht, H. Calibration for archaeological and environmental terrestrial samples in the time range 26–50 ka cal BP. *Radiocarbon.* 55, 2021–2027 (2013)

Brouns SJ, Jore MM, Lundgren M, *et al.* Small CRISPR RNAs guide antiviral defense in prokaryotes. *Science.* 2008;321(5891):960–4. doi:10.1126/ science.1159689. PMID 18703739. Bibcode:2008Sci..321.960B.

Brumfiel, G. (March 16, 2012), "Neutrinos not faster than light", *Nature News*, doi:10.1038/ nature2012.10249, retrieved March 16, 2012.

Brumfiel, G. 2011, "Particles break light-speed limit", *Nature News*, doi:10.1038/news 2011.554, September 23.

Burd M, Howard JJ. Optimality in a partitioned task performed by social insects. *Biol Lett* 2008;4:627-629.

Burnett M.E., Wang S.Q., 2011, "Current sunscreen controversies: a critical review". *Photodermatology, Photoimmunology & Photomedicine* 27 (2): 58–67.

Byrne, 2006, The Secret, Atria Books, 198 pp.

Calderone NW, Johnson BR. The within-nest behaviour of honey bee pollen foragers in colonies with a high or low need for pollen. *Anim Behav* 2002;63:749-758.

Calderone NW, Page RE. Genotypic variability in age polyethism and task specialization in the honey bee, Apis mellifera (Hymenoptera, Apidae). *Behav Ecol Sociobiol* 1988;22:17-25.

Cammaerts MC, Cammaerts R. Food recruitment strategies of the ants Myrmica-sabuleti and Myrmica-ruginodis. *Behav Process* 1980;5:251-270.

Cammaerts MC. The organization of food gathering in Myrmica-scabrinodis. *Insect Soc* 1980;27:328-342.

Can-Alonzo, C., Quezada-Euan, J. J. G., Xiu-Ancona, P., Moo-Valle, H., Valdovinos-Nunez, G. R., And Medina-Peralta, S. 2005. Pollination of 'criollo' avocados (Persea americana) and the behavior of associated bees in subtropical Mexico. *J. Apic. Res.* 44:3–8.

Cani PD, Delzenne NM. Involvement of the gut microbiota in the development of low grade inflammation associated with obesity: focus on this neglected partner; *Acta Gastroenterol Belg.* 2010 Apr-Jun;73(2):267-9.

Cani PD, Delzenne NM. The role of the gut microbiota in energy metabolism and metabolic disease; *Curr Pharm Des.* 2009;15(13):1546-58.

Cannata D., Fierz Y, Vijayakumar A, LeRoith D., 2010, Type 2 diabetes and cancer: what is the connection?, Mt Sinai J Med. *2010 Mar-Apr*;77(2):197-213, doi: 10.1002/msj.20167.

Caplan, S. E.; High, A. (2006). "Beyond excessive use: The interaction between cognitive and behavioral symptoms of problematic internet use". *Communication Research Reports* 23 (4): 265. doi:10.1080/08824090600962516.

Carels RA, Darby LA, Cacciapaglia HM, Douglass OM: Reducing cardiovascular risk factors in postmenopausal women through a lifestyle change intervention, *J Womens Health (Larchmt)* 2004, 13:412-426.

Carlborg, F.W., 1981, Dose-response function in carcinogenesis and the Weibull model, Food Cosmet Toxicol., April 19 (2) 255-263.

Carlborg, F.W., 1982, The threshold and the virtually safe dose, Fd. *Chem. Toxic.* Vol. 20, 219.

Carlborg, F.W., 1985, , A cancer risk assessment for saccharin, Fd Chem. Toxic., vol. 23, no. 45, 499-506.

Carmichael L. An annotated historical account of canine parvovirus. *J. Vet. Med. B Infect. Dis. Vet. Public Health.* 2005;52(7–8):303–11. doi:10.1111/j.1439-0450.2005.00868.x

Carrillo-Infante, C., G. Abbadessa, L. Bagella, and A. Gior- dano, "Viral infections as a cause of cancer (Review)," *International Journal of Oncology*, vol. 30, no. 6, pp. 1521–1528, 2007.

Carroll, J. S., Padilla-Walker, L., Nelson, L. J., Olson, C. D., Barry, C. M., & Madsen, S. D. (2008). Generation XXX. *Journal of Adolescent Research*, 23, 6–30.

Carter, C., Shafir, S., Yehonatan, L., Palmer, R. G., And Thornburg, R. , . 2006. A novel role for proline in plant floral nectars. *Naturwissenschaften* 93:72–79.

Carter, J., 2014, *A Call to Action: Women, Religion, Violence, and Power*, Simon & Schuster, 211 pp.

CAS

Cascalho M, Platt JL. Novel functions of B cells. Crit. Rev. Immunol. 2007;27 (2):141–51. PMID 17725500.

Casens, S. Desk Encyclopedia of General Virology. Boston: Academic Press; 2009. ISBN 0-12-375146-2. p. 167–174.

Cassill D. Rules of supply and demand regulate recruitment to food in an ant society. *Behav Ecol Sociobiol* 2003;54:441-450.

Cássio C. Ibiapina, Álvaro A. Cruz, Paulo A. M. Camargos, 2004, Hydrofluoroalkane as a propellant for pressurized metered-dose inhalers: history, pulmonary deposition, pharmacokinetics, efficacy and safety, *Journal de Pediatria, J Pediatr (Rio J).* 2004;80(6):441-6.

Casswall TH, Fischler B. Vaccination of the immunocompromised child. *Expert review of vaccines.* 2005;4(5):725–38. doi:10.1586/14760584.4.5.725.

Caulfield, T., 2012, *The Cure for Everything: Untangling the Twisted Messages about Health, Fitness and Happiness*, Viking Canada, 320 pp.

Caulfield, T., 2013, "The paradoxes of pop science", *Policy Options*, September-October, vol. 34, no. 7.

CBS News, 2015, Ultrasounds show fetuses react to mothers' smoking, March 24.

CDC, 2008,"Americans Carry Body Burden of Toxic Sunscreen Chemical". *Environmental Working Group Report*, March 25, 2008.

Cello J, Paul AV, Wimmer E. Chemical synthesis of poliovirus cDNA: generation of infectious virus in the absence of natural template. *Science.* 2002;297(5583):1016–8. doi:10.1126/science.1072266. PMID 12114528. Bibcode:2002Sci..297.1016C.

Chakraborty, R., 2014, "The Bengal Famine: How the British engineered the worst genocide in human history for profit", *Yourstory.com*, August 15.

Chan, A.L., 2014, "What You Need To Know Before You Reuse That Plastic Water Bottle", August 14.

Chao, L.,1997, "Evolution of sex and the molecular clock in RNA viruses", *Gene.* Dec. 31: 205(1-2): 301-308.

Chen YP, Zhao Y, Hammond J, Hsu H, Evans JD, Feldlaufer MF. Multiple virus infections in the honey bee and genome divergence of honey bee viruses. *Journal of Invertebrate Pathology.* October–November 2004;87(2–3):84–93. doi:10.1016/j.jip.2004.07.005. PMID 15579317.

Chhetri, A.B., and Islam, M.R., 2008, *Inherently Sustainable Technology Developments*, Nova Science Publishers, New York, 452 pp.

Chirife J, Scarmato G, Herszage L, Scientific basis for use of granulated sugar in treatment of infected wounds. *Lancet* 1982;1:560-1.

Choi, C. Q., 2011, "Leading light: What would faster-than-light neutrinos mean for physics", *Scientific American*, Oct. 13.

Chou, J., Y. C. Lin, J. Kim, et al., "Nasopharyngeal carcinoma—review of the molecular mechanisms of tumori- genesis," *Head and Neck*, vol. 30, no. 7, pp. 946–963, 2008.

Chugani HT, Phelps ME, Mazziotta JC., 1987, Positron emission tomography study of human brain functional development. *Ann Neurol* 1987;22:487–97.

Circulation 2004, 109:672-693.

Clayton, M.A. and Moffat, J.W., (1999) Dynamical Mechanism for Varying Light Velocity as a Solution to Cosmological Problems, *Physics Letters B*, vol. 460, No.3-4, pp. 263-270.

Codoner-Franch, P., V. Valls-Belles, A. Arilla-Codoner, and E. Alonso-Iglesias, "Oxidant mechanisms in childhood obesity: the link between inflammation and oxidative stress," *Translational Research*, vol. 158, no. 6, pp. 369–384, 2011.

Coghlan, A., 2005, Autism rises despite MMR ban in Japan, New Scientists, March 3.

Cogni R, Oliveira PS. Recruitment behavior during foraging in the neotropical ant Gnamptogenys moelleri (Formicidae: Ponerinae): does the type of food matter? *J Insect Behav* 2004;17:443-458.

Cohen, J., 2011, BREAKTHROUGH OF THE YEAR:HIV Treatment as Prevention, *Science*, 23, December, Vol. 334 no. 6063 p. 1628.

Cohen, M.S., Chen, Y.Q., Marybeth McCauley, M., *et al.*, 2011, "Prevention of HIV-1 Infection with Early Antiretroviral Therapy", *N Engl J Med*, Aug. 11, 365:493-505.

Cohen, S.M., D. T. Purtilo, and L. B. Ellwein, "Ideas in pathology. Pivotal role of increased cell proliferation in human carcinogenesis," *Modern Pathology*, vol. 4, no. 3, pp. 371–382, 1991.

Coleman JR, Papamichail D, Skiena S, Futcher B, Wimmer E, Mueller S. Virus attenuation by genome-scale changes in codon pair bias. *Science*. 2008;320 (5884):1784–7. doi:10.1126/science.1155761. PMID 18583614. Bibcode:2008Sci..320.1784C.

Collett TS, Graham P, Durier V. Route learning by insects. *Curr Opin Neurobiol* 2003;6:718-725.

Collier, Leslie; Balows, Albert; Sussman, Max (1998) Topley and Wilson's Microbiology and Microbial Infections ninth edition, Volume 1, Virology, volume editors: Mahy, Brian and Collier, Leslie. Arnold. ISBN 0-340-66316-2.

Collins R., Peto R., MacMahon S. Blood pressure, stroke, and coronary heart disease. Part 2. Short-term reductions in blood pressure: overview of randomised drug trials in their epidemiological context. *Lancet*. 1990;335:827–838.

Comprehensive Material Balance Equation on Detection of the Causes of Medical Disorders, Part 1: Elimination of Paradoxes", *Journal of Characterization and Development of Novel Materials*, vol. 7, no. 1, 17-43.

Contents in Edible Parts of Broccoli Inflorescences after Domestic Cooking. *J. of the Science of Food and Agriculture*, vol. 83, pp. 1511–1516.

Cooper RA, Halas E, Molan PC, The efficacy of honey in inhibiting strains ofPseudomonas aeruginosa from infected burns. *J Burn Care Rehabil* 2002;23:366-70.

Cooper RA, Molan PC, Harding KG, The sensitivity to honey of gram-positive cocci of clinical significance isolated from wounds. *J Appl Microbiol* 2002;93:857-63.

Cooper RA, Wigley P, Burton NF, Susceptibility of multiresistant strains ofBurkholderia cepacia to honey. *Lett Appl Microbiol* 2000;31:20-4.

Cooper, R.A., P. C. Molan, L. Krishnamoorthy, and K. G. Harding, "Manuka honey used to heal a recalcitrant surgical wound," *European Journal of Clinical Microbiology and Infectious Diseases*, vol. 20, no. 10, pp. 758–759, 2001.

Cordero-Macintyre ZR, Peters W, Libanati CR, Espana RC, Howell WH, Lohman TG: Effect of a weight-reduction program on total and regional body composition in obese postmenopausal women, *Ann N Y Acad Sci* 2000, 904:526-535.

Corrigan, P.W., Rowan, D., Green, A., et al. (2002). Challenging two mental illness stigmas: Personal responsibility and dangerousness. *Schizophrenia Bulletin*, 28, 293-309.

Couture P, Hulbert AJ., 1995, Relationship between body mass, tissue metabolic rate, and sodium pump activity in mammalian liver and kidney. *Am J Physiol* 1995;268:R641–50.

Couzin ID, Krause J, Franks NR, Levin SA. Effective leadership and decision-making in animal groups on the move. *Nature* 2005;433:513-516.

Couzin ID. Collective cognition in animal groups. *Trends Cogn Sci* 2009;13:36-43.

Crane, L.M., M. Schaapveld, O. Visser, M. W. Louwman, J. T. Plukker, and G. M. van Dam, "Oesophageal cancer in The Netherlands: increasing incidence and mortality but improving survival," *European Journal of Cancer*, vol. 43, no. 9, pp. 1445–1451, 2007.

Crawford, Dorothy H. Viruses: A Very Short Introduction. Oxford University Press, USA; 2011. ISBN 0-19-957485-5. p. 16.

Creager AN, Morgan GJ. After the double helix: Rosalind Franklin's research on Tobacco mosaic virus. *Isis*. 2008;99(2):239–72. doi:10.1086/588626. PMID 18702397.

Curr Treat Options Cardiovasc Med 2005, 7:61-74.

Curry, A., 2013, "Archaeology: The milk revolution", *Nature*, July 13.

Cutting, K.F., "Honey and contemporary wound care: an overview," *Ostomy Wound Management*, vol. 53, no. 11, pp. 49–54, 2007.

D'Herelle F. On an invisible microbe antagonistic toward dysenteric bacilli: brief note by Mr. F. D'Herelle, presented by Mr. Roux. *Research in Microbiology*. 2007;158(7):553–4. doi:10.1016/j.resmic.2007.07.005. PMID 17855060.

Dag, A., Afik, O., Yeselson, Y., Schaffer, A., And Shafir, S. 2006. Physical, chemical and palynological characterization of avocado (Persea americana Mill.) honey in Israel. *Int. J. Food Sci. Technol.* 41:387–394.

Dag, A., Fetscher, A. E., Afik, O., Yeselson, Y., Schaffer, A., Kamer, Y., Waser, N. M., Madore, M. A., Arpaia, M. L., Hofshi, R., And Shafir, S. 2003. Honey bee (Apis mellifera) strains differ in avocado (Persea americana) nectar foraging preference. *Apidologie* 34:299–309.

Dargan, G., 2006, Intelligent Design, *Encounter*, ABC, Nov. 26.

Darwin, C. 1871. The descent of man and selection in relation to sex. London, Murray.

Dawkins, R. (1976). *The Selfish Gene*. New York City: Oxford University Press. ISBN 0-19-286092-5.

DCP2, "Controlling Cancer in Developing Countries; pre- vention and treatment strategies merit further study," 2007, http://www.dcp2.org/.

De Biseau JC, Pasteels JM. Regulated food recruitment through individual behavior of scouts in the ant, Myrmica-sabuleti (Hymenoptera, Formicidae). *J Insect Behav* 1994;7:767-777.

de Onis, Blossner, M., and M. Borghi E., 2010, Global prevalence and trends of overweight and obesity among preschool children, *Am. J. Clin. Nutr.*, vol. 92, 1257-64.

Deedwania PC, Volkova N: Current Treatment Options for the Metabolic Syndrome.

Deibert P, König D, Schmidt-Trucksaess A, Zaenker KS, Frey I, Landmann U, Berg A: Weight loss without losing muscle mass in pre-obese and obese subjects induced by a high-soy-protein diet, *Int J Obes Relat Metab Disord* 2004, 28:1349-1352.

Deibert, P., König, D., Mara Z Vitolins, M.Z., Ulrike Landmann, U., Frey, I., Zahradnik, H.P., and Berg, A., 2007, Effect of a weight loss intervention on anthropometric measures and metabolic risk factors in pre- versus postmenopausal women, *Nutrition Journal* 2007, 6:31 doi:10.1186/1475-2891-6-31.

Del Piano M, Morelli L, Strozzi GP, Allesina S, Barba M, Deidda F, Lorenzini P, Ballaré M, Montino F, Orsello M, Sartori M, Garello E, Carmagnola S, Pagliarulo M, Capurso L. Probiotics: from research to consumer; *Dig Liver Dis.* 2006 Dec;38 Suppl 2:S248-55.

Demirel, Y., 2012, Energy, Green Energy and Technology, DOI: 10.1007/978-1-4471-2372-9_2, Springer-Verlag London Limited, 39-45.

Deneubourg JL, Goss S. Collective patterns and decision making. *Ethol Ecol Evol* 1989;1:295-311.

Derbyshire D., 2011, "How humans are 97% the same as orangutans: New research shows how DNA matches", *The Daily Mail UK*, Jan. 27.

Deters AM, Schröder KR, Hensel A. Kiwi fruit (Actinidia chinensis L.) polysaccharides exert stimulating effects on cell proliferation via enhanced growth factor receptors, energy production, and collagen synthesis of human keratinocytes, fibroblasts, and skin equivalents; *J Cell Physiol.* 2005 Mar;202(3):717-22.

Detrain C, Deneubourg JL. Collective decision-making and foraging patterns in ants and honey bees. *Adv Insect Physiol* 2008;35:123-173.

DHHS. Mental Health: A Report of the Surgeon General. Rockville, MD: U.S. Department of Health and Human Services, Substance Abuse and Mental health Services Administration, Center for Mental Health Services, National Institutes of Health, National Institute of Mental Health, 1999. http://www.surgeongeneral.gov/library/mentalhealth/toc.html

Diamond, J., 1999. Guns, Germs, and Steel: The Fates of Human Societies (New York: W.W. Norton)

DiBaise JK, Zhang H, Crowell MD, Krajmalnik-Brown R, Decker GA, Rittmann BE. Gut microbiota and its possible relationship with obesity. *Mayo Clin Proc.* 2008 Apr;83(4):460-9.

DiLorenzo, T., 2002, The Real Lincoln: A New Look at Abraham Lincoln, His Agenda, and an Unnecessary War, Random House LLC, ISBN 9780307559388.

Dimmock, N.J; Easton, Andrew J; Leppard, Keith (2007) Introduction to Modern Virology sixth edition, Blackwell Publishing, ISBN 1-4051-3645-6.

Dinesh-Kumar SP, Tham Wai-Hong, Baker BJ. Structure—function analysis of the tobacco mosaic virus resistance gene N. *PNAS.* 2000;97(26):14789–94.

Dixon B., Bacteria can't resist honey. *Lancet Infect Dis* 2003;3:116.

Dobyns, F. (1940) The Amazing Story of Repeal: An Expose of the Power of Propaganda. Chicago: Willett, Clark & Co.

Dogan, S., X. Hu, Y. Zhang, N. J. Maihle, J. P. Grande, and M. P. Cleary, "Effects of high-fat diet and/or body weight on mammary tumor leptin and apoptosis signaling pathways in MMTV-TGF-alpha mice," *Breast Cancer Research*, vol. 9, no. 6, p. R91, 2007.

Doucet E, St Pierre S, Almeras N, Mauriege P, Despres JP, Richard D, Bouchard C, Tremblay A: Fasting Insulin Levels Influence Plasma Leptin Levels Independently from the Contribution of Adiposity: Evidence from Both a Cross-Sectional and an Intervention Study, *J Clin Endocrinol Metab* 2000, 85:4231-4237.

Dreier, M., Wölfling, K. Beutel, M.E. (2014). Internet Addiction in Youth. Internetsucht bei Jugendlichen. *Monatsschrift Kinderheilkunde.* doi: 10.1007/s00112-013-3069-2. 1-6.

Dudley, K.W., *et al.*, 1976, U.S. Patent 3,933,323, Solid state solar to microwave energy converter system and apparatus.

Duesberg, P.H., 1996, *Inventing the AIDS Virus*, Regnery USA, 720 pp.

Duffy S, Holmes EC. Validation of high rates of nucleotide substitution in geminiviruses: phylogenetic evidence from East African cassava mosaic viruses. *The Journal of General Virology.* 2009;90(Pt 6):1539–47.

Dvash, L., Afik, O., Shafir, S., Schaffer, A., Yeselson, Y., Dag, A., And Landau, S. 2002. Determination by near-infrared spectroscopy of perseitol used as a marker for the botanical origin of avocado (Persea americana Mill.) honey. *J. Agric. Food Chem.* 50:5283–5287.

Eddy, J.J., M. D. Gideonsen, and G. P. Mack, "Practical considerations of using topical honey for neuropathic diabetic foot ulcers: a review," *Wisconsin Medical Journal*, vol. 107, no. 4, pp. 187–190, 2008.

Editorial, 2013, "Recent 6 'Big Pharma' frauds", *Biospectrum*, Aug. 6.

Edwin A.M., and Gale, 2002, "The Rise of Childhood Type 1 Diabetes in the 20th Century", *Diabetes*, 51:3353–3361.

Efem SEE, Clinical observations on the wound-healing properties of honey. *Br J Surg* 1988;75:679-81.

Einstein MH, Schiller JT, Viscidi RP, Strickler HD, Coursaget P, Tan T, Halsey N, Jenkins D. Clinician's guide to human papillomavirus immunology: knowns and unknowns. *The Lancet Infectious Diseases.* 2009;9(6):347–56. doi:10.1016/S1473-3099(09)70108-2. PMID 19467474.

Einstein, A., 1936, Physics and Reality (translation by Jean Picard), available at www.kostic.niu.edu

Elia M., 1992, Organ and tissue contribution to metabolic rate. Kinney JM, Tucker HN, editors., Energy metabolism: tissue determinants and cellular corollaries. New York, NY: Raven Press, 1992:61–80.

Engdahl, F.W., 2007, Seeds of Destruction: Hidden Agenda of Genetic Manipulation', 341 pp., ISBN Number: 978-0-937147-2-2

Engel, C., 2006. How Much Collusion? A Meta-Analysis on Oligopoly Experiments, in *MPI Collective Goods Preprint No. 2006/27* (Bonn: Max Planck Institute, December).

English, H.K., A. R. Pack, and P. C. Molan, "The effects of manuka honey on plaque and gingivitis: a pilot study," *Journal of the International Academy of Periodontology*, vol. 6, no. 2, pp. 63–67, 2004.

EPA, 1996, SW-846, 6010B: Inductively coupled plasma-atomic emission spectrometry. Revision 2.

Ernst ND, Cleeman JI: National cholesterol education program keeps a priority on lifestyle modification to decrease cardiovascular disease risk. *Curr Opin Lipidol* 2002, 13:69-73.

Estevinho, L., A. P. Pereira, L. Moreira, L. G. Dias, and E. Pereira, "Antioxidant and antimicrobial effects of phenolic compounds extracts of Northeast Portugal honey," *Food and Chemical Toxicology*, vol. 46, no. 12, pp. 3774–3779, 2008.

Evangelista, M., 2013, "Law, Ethics, and the War on Terror", *Wiley*, 176 pp.

Evans TA. Foraging and building in subterranean termites: task switchers or reserve labourers? *Insect Soc* 2006;53:56-64.

Executive Summary of The Third Report of The National Cholesterol Education Program (NCEP) Expert Panel on Detection, Evaluation, And Treatment of High Blood Cholesterol In Adults (Adult Treatment Panel III) *JAMA* 2001, 285:2486-2497.

Ezzati M., Hoorn S.V., Rodgers A. Estimates of global and regional potential health gains from reducing multiple major risk factors. *Lancet.* 2003;362 (9380):271–280.

FAO/WHO/UNU Human energy requirements. Rome, Italy: Joint FAO/WHO/UNU Expert Consultation, 2004:35–50.

Faria AN, Ribeiro Filho FF, Gouveia F Sr, Zanella MT: Impact of visceral fat on blood pressure and insulin sensitivity in hypertensive obese women, *Obes Res* 2002, 10:1203-1206.

Fauzi, A.N., M. N. Norazmi, and N. S. Yaacob, "Tualang honey induces apoptosis and disrupts the mitochondrial membrane potential of human breast and cervical cancer cell lines," *Food and Chemical Toxicology*, vol. 49, no. 4, pp. 871- 878, 2011.

Felsenstein, J. (1974). The evolutionary advantage of recombination. *Genetics* 78, 737-756.

Fenner, J. N. Cross-cultural estimation of the human generation interval for use in genetics-based population divergence studies. *Am. J. Phys. Anthropol.* 128, 415–423 (2005)

Fenstermacher, J.D., Nagaraja, T., Davies, K.R., 2001, Overview of the structure and function of the blood-brain barrier in vivo, in 2001, edited by Kobiler, Lustig and Shapiro, Springer, New York.

Festinger, L. (1957). A Theory of cognitive dissonance. Stanford, CA: Stanford University Press.

Festinger, L. (1959). Some attitudinal consequences of forced decisions. *Acta Psychologica*, 15, 389-390.

Festinger, L. (1962). "Cognitive dissonance". *Scientific American* 207 (4): 93–107.

Festinger, L. (Ed.). (1964). Conflict, decision, and dissonance (Vol. 3). Stanford University Press.

Festinger, L., & Carlsmith, J. M. (1959). Cognitive consequences of forced compliance. *The Journal of Abnormal and Social Psychology*, 58(2), 203.

Finegold, JA, Asaria, P, Francis, DP., 2013, "Mortality from ischaemic heart disease by country, region, and age: statistics from World Health Organisation and United Nations.", *Int J Cardiol.* 2013 Sep 30;168(2):934-45. doi: 10.1016/j.ijcard.2012.10.046. Epub 2012 Dec 4.

Fink RC, Roschek B Jr, Alberte RS, 2009, HIV type-1 entry inhibitors with a new mode of action, *Antivir Chem Chemother.* 2009;19(6):243-55.

Fiorillo, V., 2014, "Q&A: Rutgers Law Prof Who Says Pedophilia Is Not a Crime

Firstenberg, O., Peronel, T., Liang, Q., Gorshkov, A.V., Lukin, M.D., and Vuletic, V., 2013, Attractive photons in a quantum nonlinear medium, *Nature*, vol. 502, Oct., 71-75.

Fischlechner M, Donath E. Viruses as Building Blocks for Materials and Devices. *Angewandte Chemie International Edition.* 2007;46(18):3184–93.

Fitzpatrick, M., 2005, "Review: Evidence of Harm. Mercury in Vaccines and the Autism Epidemic". *British Medical Journal*, May, 330 (7500): 1154. doi:10.1136/bmj.330.7500.1154. Retrieved 2008-10-14.

Follow us: @MailOnline on Twitter | DailyMail on Facebook

Forno E, Onderdonk AB, McCracken J, Litonjua AA, Laskey D, Delaney ML, Dubois AM, Gold DR, Ryan LM, Weiss ST, Celedón JC. Diversity of the gut microbiota and eczema in early life; *Clin Mol Allergy.* 2008 Sep 22;6:11.

Forterre P, Philippe H. The last universal common ancestor (LUCA), simple or complex?. *The Biological Bulletin.* 1999;196(3):373–5; discussion 375–7.

Fortuine R. Smallpox decimates the Tlingit (1787). *Alaska Med.* 1988 [Retrieved 16 September 2008];30(3):109.

Forzano, F., Borry, P., Cambon-Thomsen, A., Hodgson. S.,V., Tibben, A., de Vries, P., van El, C., and Cornel, M., 2010, Italian appeal court: a genetic predisposition to commit murder? *Eur J Hum Genet. 2010 May*; 18(5): 519–521.

Fowler, S.P. 65th Annual Scientific Sessions, American Diabetes Association, San Diego, June 10-14, 2005; Abstract 1058-P. Sharon P. Fowler, MPH, University of Texas Health Science Center School of Medicine, San Antonio.

Franceschi, S., and C. La Vecchia, "Cancer epidemiology in the elderly," *Critical Reviews in Oncology/Hematology*, vol. 39, no. 3, pp. 219–226, 2001.

Franks NR, Deneubourg JL. Self-organizing construction in ants: individual worker behavior and the nest's dynamics. *Anim Behav* 1997;54:779-796.

Frazier TH, DiBaise JK, McClain CJ. Gut microbiota, intestinal permeability, obesity-induced inflammation, and liver injury; *JPEN J Parenter Enteral Nutr.* 2011 Sep;35(5 Suppl):14S-20S.

French, V.M., R. A. Cooper, and P. C. Molan, "The antibacterial activity of honey against coagulase-negative staphylococci," *Journal of Antimicrobial Chemotherapy*, vol. 56, no. 1, pp. 228–231, 2005.

Frisch, K. V., 1967, The Dance Language and Orientation of Bees. Cambridge, Mass.: The Belknap Press of Harvard University Press.

Fry, PS, 2003, "Perceived self-efficacy domains as predictors of fear of the unknown and fear of dying among older adults.". *Psychol Aging* 18 (3): 474–86.

Frydman-Marom A1, Levin A, Farfara D, Benromano T, Scherzer-Attali R, Peled S, Vassar R, Segal D, Gazit E, Frenkel D, Ovadia M., 2011, Orally administrated cinnamon extract reduces β-amyloid oligomerization and corrects cognitive impairment in Alzheimer's disease animal models, *PLoS One*. 2011 Jan 28;6(1):e16564. doi: 10.1371/journal.pone.0016564.

Fu, Q., Li, H., Moorjani, P., *et al.*, 2014, Genome sequence of a 45,000-year-old modern human from western Siberia, *Nature* 514, 445–449 (23 October 2014) doi:10.1038/nature13810

Fu, Q. et al. A revised timescale for human evolution based on ancient mitochondrial genomes. *Curr. Biol.* 23, 553–559 (2013)

Fukuda, M., K. Kobayashi, Y. Hirono, et al., "Jungle honey enhances immune function and antitumor activity," *Evidence-Based Complementary and Alternative Medicine*, vol. 2011, Article ID 908743, 2011.

Gallagher D, Allen A, Wang ZM, Heymsfield SB, Krasnow N. Smaller organ tissue mass in the elderly fails to explain lower resting metabolic rate. *Ann N Y Acad Sci* 2000;904:449–55.

Gallagher D, Belmonte D, Deurenberg P, et al. Organ-tissue mass measurement allows modeling of REE and metabolically active tissue mass. *Am J Physiol* 1998;275:E249–58.

Gallagher, J., 2015, Being overweight 'reduces dementia risk', BBC, April 10.

Gallup Poll, 2005, "What Frightens America's Youth", gallup.com (29 March 2005).

Gao F, Bailes E, Robertson DL, et al. Origin of HIV-1 in the Chimpanzee Pan troglodytes troglodytes. *Nature.* 1999;397(6718):436–441. doi:10.1038/17130.

Garcia, M., A. Jemal, E. Ward, et al., Global Cancer Facts & Figures 2007, American Cancer Society, *Atlanta*, Ga, USA, 2007.

Gardener, M. C. And Gillman, M. P. 2002. The taste of nectar—A neglected area of pollination ecology. *Oikos* 98:552–557.

Gardner, A.G., 2013, How Did Washington Make His Millions? *CW Journal*, Winter.

Gazit, S. and Degani, C. 2002. Reproductive biology, pp. 101–133, in A. W. Whiley, B. Schaffer, and B. N. Wolstenholme (eds.). The Avocado: Botany, Production and Uses. CAB International, Wallingford.

Gethin, G., and S. Cowman, "Bacteriological changes in sloughy venous leg ulcers treated with manuka honey or hydrogel: an RCT," *Journal of Wound Care*, vol. 17, no. 6, pp. 241–247, 2008.

Gethin, G.T., S. Cowman, and R. M. Conroy, "The impact of manuka honey dressings on the surface pH of chronic wounds," *International Wound Journal*, vol. 5, no. 2, pp. 185– 194, 2008.

Ghashm, A.A., N. H. Othman, M. N. Khattak, N. M. Ismail, and R. Saini, "Antiproliferative effect of Tualang honey on oral squamous cell carcinoma and osteosarcoma cell lines," *BMC Complementary and Alternative Medicine*, vol. 10, article 49, 2010.

Gheldof, N., and N. J. Engeseth, "Antioxidant capacity of honeys from various floral sources based on the determination of oxygen radical absorbance capacity and inhibition of in vitro lipoprotein oxidation in human serum samples," *Journal of Agricultural and Food Chemistry*, vol. 50, no. 10, pp. 3050– 3055, 2002.

Gibbons, R. D.; Brown, C. H.; Hur, K.; Marcus, S. M.; Bhaumik, D. K.; Erkens, J. A.; Herings, R. M.C.; Mann, J. J. (2007). "Early Evidence on the Effects of Regulators' Suicidality Warnings on SSRI Prescriptions and Suicide in Children and Adolescents". *American Journal of Psychiatry* 164 (9): 1356–63.

Gillespie, D., 2010, Sweet Poison, *e-Penguin*, 208 pp.

Ginsberg, N.S., Garner, S.R., Hau, L.V., (2007) Coherent control of optical information with matter wave dynamics, *Nature*, vol. 445, Feb. 8, pp. 623-626.

Giovannucci, E., Harlan, D.M., and Douglas Yee, D., 2010, "Diabetes and Cancer: A consensus report", *Diabetes Care*, July, 1674-1685.

Giurfa, M., Nunez, J., Chittka, L., And Menzel, R. 1995. Color preferences of flower-naive honey bees. *J. Comp. Physiol. A* 177:247–259.

Glassner, B. (1999). The Culture of Fear: Why Americans are Afraid of the Wrong Things, Basic Books, New York.

Gleba, YY; Giritch, A. Recent Advances in Plant Virology. Caister Academic Press; 2011. ISBN 978-1-904455-75-2. Plant Viral Vectors for Protein Expression.

Goebel, T. A., Derevianko, A. P. & Petrin, V. T. Dating the Middle to Upper Paleolithic transition at Kara-Bom. *Curr. Anthropol.* 34, 452–458 (1993)

Goldberg, H., 2014, "China Now Has the Second-Most Millionaires in the World", *Time*, June 10.

Goldman L. The decline in coronary heart disease: determining the paternity of success. *Am J Med.* 2004;117:274–276.

Goldwert, L., 2012, "Sugar is as addictive as cocaine, and causes obesity, diabetes, cancer and heart disease: Researchers", *New York Daily News*, April 2.

Gomez-Caravaca, A.M., M. Gomez-Romero, D. Arraez- Roman, A. Segura-Carretero, and A. Fernandez-Gutierrez, "Advances in the analysis of phenolic compounds in products derived from bees," *Journal of Pharmaceutical and Biomedical Analysis*, vol. 41, no. 4, pp. 1220–1234, 2006.

Goodpasture EW, Woodruff AM, Buddingh GJ. The cultivation of vaccine and other viruses in the chorioallantoic membrane of chick embryos. *Science.* 1931;74(1919):371–372. doi:10.1126/science.74.1919.371. PMID 17810781. Bibcode:1931Sci..74.371G.

Gopalakrishna, A. and Kumar, T.V.P., 2008, ""Intravenous injection of elemental mercury: A report of two cases", *Indian J Plast Surg.* 2008 Jul-Dec; 41(2): 214–218.

Gordon DM. The organization of work in social insect colonies. *Nature* 1996;380:121-124.

Goris N, Vandenbussche F, De Clercq K. Potential of antiviral therapy and prophylaxis for controlling RNA viral infections of livestock. *Antiviral Res.* 2008;78(1):170–8. doi:10.1016/j.antiviral.2007.10.003.

Goudsmit, Jaap. Viral Sex. Oxford Univ Press, 1998.ISBN 978-0-19-512496-5 ISBN 0-19-512496-0

Goyret J, Farina WM. Non-random nectar unloading interactions between foragers and their receivers in the honey bee hive. *Naturwissenschaften* 2005;92:440-443.

Graham P, Cheng K. Ants use the panoramic skyline as a visual cue during navigation. *Curr Biol* 2009;19:R935-R937.

Graham, D.W., 2006, *Explaining the Cosmos: The Ionian Tradition of Scientific Philosophy*, Princeton: Princeton University Press.

Granberg, D. & Brown, T. A., (1995), "The Monty Hall Dilemma", *Personality and Social Psychology Bulletin* 21 (7): 711–729.

Grandone, A. Amato, C. Luongo, N. Santoro, L. Perrone, and E. M. del Giudice, "High-normal fasting glucose levels are associated with increased prevalence of impaired glucose tolerance in obese children," *Journal of Endocrinological Investigation*, vol. 31, no. 12, pp. 1098–1102, 2008.

Gratz, S.W., Hannu Mykkanen, and Hani S El-Nezami. Probiotics and gut health: A special focus on liver diseases; *World J Gastroenterol.* 2010 January 28; 16(4): 403-410.

Greene MJ, Gordon DM. Interaction rate informs harvester ant task decisions. *Behav Ecol* 2007;18:451-455.

Greene MJ, Gordon DM. Social insects—cuticular hydrocarbons inform task decisions. *Nature* 2003;423:32-32.

Greenfacts, 2012, http://www.greenfacts.org/en/digests/state-of-the-environment.htm

Greer S, Alexander GJ. Viral serology and detection. *Baillieres Clin. Gastroenterol.* 1995;9(4):689–721. doi:10.1016/0950-3528(95)90057-8.

Gregory, J. Leyer, Shuguang Li, Mohamed E. Mubasher, Cheryl Reifer, Arthur C. Ouwehand, Probiotic Effects on Cold and Influenza-Like Symptom Incidence and Duration in Children; *Pediatrics* 2009;124;e172.

Grenoble, R., 2015, Colorblind People See Certain Colors For The First Time Ever, Are Visibly Moved, The Huffington Post, March 23.

Griffin, M.J., 1990, *Handbook of Human Vibration*, Academic Press, London, 988 pp.

Griffiths, M. 2000. Internet addiction-time to be taken seriously?. *Addiction Research & Theory*, 8 (5), pp. 413-418.

Griffiths, M. D. (1998) 'Internet addiction: does it really exist?' in Gackenbach, J. (ed), Psychology and the Internet. New York: Academic Press, pp. 61–75.

Griffiths, M. D. (2000) p.414-415. 'Internet Addiction - Time To Be Taken Seriously? (Editorial)' in Addiction Research, 2000, Vol. 8, No. 5, pp 413-418. OPA (Overseas Publishers Association).

Grohol, J. M. (1999). Internet Addiction Guide. Psych Central available online.

Gruter C, Farina WM. Past experiences affect interaction patterns among foragers and hive-mates in honey bees. *Ethology* 2009;115:790-797.

Guerrini, R., Belmonte, A., and Genton, P., 1998, "Antiepileptic Drug-Induced Worsening of Seizures in Children", *Epilepsia*, 39(Suppl. 3):SZ-SIO.

Haarbo J, Marslew U, Gotfredsen A, Christiansen C: Postmenopausal hormone replacement therapy prevents central distribution of body fat after menopause.

Hackman RM, Havel PJ, Schwartz HJ, Rutledge JC, Watnik MR, Noceti EM, Stohs SJ, Stern JS, Keen CL: Multinutrient supplement containing ephedra and caffeine causes weight loss and improves metabolic risk factors in obese women: a randomized controlled trial, *Int J Obes (Lond)* 2006.

Hagler, J. R. And Buchmann, S. L. 1993. Honey-bee (Hymenoptera, Apidae) foraging responses to phenolic-rich nectars. *J. Kans. Entomol. Soc.* 66:223–230.

Hagler, J. R., Cohen, A. C., And Loper, G. M. 1990. Production and composition of onion nectar and honey-bee (Hymenoptera, Apidae) foraging activity in Arizona. *Environ. Entomol.* 19:327–331. HARBORNE, J. B. 1993. Introduction to Ecological Biochemistry. Academic Press, San Diego.

Hall, A. J., Jepson, P. D., Goodman, S. J. & Harkonen, T. "Phocine distemper virus in the North and European Seas — data and models, nature and nurture". *Biol. Conserv.* 131, 221–229 (2006)

Hall, S., 2000, The last of the Neanderthals, National Geography, Oct.

Hamer DH, Hu S, Magnuson VL, Hu N, Pattatucci AM (July 1993). "A linkage between DNA markers on the X chromosome and male sexual orientation". *Science* 261 (5119): 321–7. doi:10.1126/science.8332896. PMID 8332896.

Hamer, D. H., 1994, The Science of Desire: The Search for the Gay Gene and the Biology of Behavior (Simon and Schuster, 1994) ISBN 0-684-80446-8

Hamer, D.H., 1999, Living with Our Genes: Why They Matter More Than You Think with Peter Copeland (Anchor, 1999) ISBN 0-385-48584-0

Hamming, R.W., 1984, "Numerical Analysis for Scientists and Engineers", McGraw-Hill, New York, 2nd Edition, pp. 721.

Hampson AW, Mackenzie JS. The influenza viruses. *Med. J. Aust.* 2006;185(10 Suppl):S39–43.

Hansen S (2002). "Excessive internet usage or 'internet addiction'? The implications of diagnostic categories for student users". *Journal of Computer Assisted Learning* 18: 232–236.

Hau, L.V., Harris, S.E., Dutton, Z., and Behroozi, C.H., (1999) Light Speed Reduction to 17 Meters Per Second in an Ultra Cold Atomic Gas, *Nature*, vol. 397, pp. 594–598

Hauptman LM. Smallpox and American Indian; Depopulation in Colonial New York. *N Y State J Med.* 1979;79(12):1945–9.

Hawking, S., 1973-2010, *The Large Scale Structure of Space-Time* (1973) *A Brief History of Time* (1988) *Black Holes and Baby Universes and Other Essays* (1994) *The Universe in a Nutshell* (2001) *On the Shoulders of Giants* (2002) *A Briefer History of Time* (2005) *God Created the Integers* (2005) *The Grand Design* (2010).

Healy D, and Aldred G (2005). Antidepressant drug use and the risk of suicide. *International Review of Psychiatry* 17, 163-172.

Hegazi, A.G., and F. K. Abd El-Hady, "Influence of honey on the suppression of human low density lipoprotein (LDL) peroxidation (In vitro)," *Evidence-Based Complementary and Alternative Medicine*, vol. 6, no. 1, pp. 113–121, 2009.

Hempstead, K.A., and Phillips, J.A., 2015, "Rising Suicide Among Adults Aged 40–64 Years: The Role of Job and Financial Circumstances," *the American Journal of Preventive Medicine*, Volume 48, Issue 5 May (online on Feb. 27).

Henning, J. A., Peng, Y. S., Montague, M. A., And Teuber, L. R. 1992. Honey-bee (Hymenoptera, Apidae) behavioral-response to primary alfalfa (Rosales, Fabaceae) floral volatiles. *J. Econ. Entomol.* 85:233–239.

Henriques, A.F., RR.E.Jenkins,N.F.Burton,andRR.A.Cooper, "The intracellular effects of manuka honey on Staphylococ- cus aureus," *European Journal of Clinical Microbiology and Infectious Diseases*, vol. 29, no. 1, pp. 45–50, 2010.

Heppermann, B., "Towards evidence based emergency medicine: Best BETs from the Manchester Royal Infirmary. Bet 3. Honey for the symptomatic relief of cough in children with upper respiratory tract infections," *Emergency Medicine Journal*, vol. 26, no. 7, pp. 522–523, 2009.

Hernandez-Ono A, Monter-Carreola G, Zamora-Gonzalez J, Cardoso-Saldana G, Posadas-Sanchez R, Torres-Tamayo M, Posadas-Romero C: Association of visceral fat with coronary risk factors in a population-based sample of postmenopausal women.

Herper, M., 2010, The World's Most Expensive Drugs, *Forbes*, Feb. 22.

Heshka, S. and D.B. Allison, 2001, Is obesity a disease? *International Journal of Obesity*, 2001, vol. 25, 1401-1404.

Heuson, J.C., N. Legros, and R. Heimann, "Influ- ence of insulin administration on growth of the 7,12- dimethylbenz(a)anthracene-induced mammary carcinoma in intact, oophorectomized, and hypophysectomized rats," *Cancer Research*, vol. 32, no. 2, pp. 233–238, 1972.

Hicks, J., 2010, The pursuit of Sweet: A history of saccharin, Chemical Heritage Magazine, Spring.

Hiday, V. A. (2006). Putting Community Risk in Perspective: a Look at Correlations, Causes and Controls. *International Journal of Law and Psychiatry*, 29, 316-331.

Hiebert, S. M. And Calder, W. A. 1983. Sodium, potassium, and chloride in floral nectars-energy-free contributions to refractive-index and salt balance. *Ecology* 64:399– 402.

Hilleman MR. Strategies and mechanisms for host and pathogen survival in acute and persistent viral infections. *Proc. Natl. Acad. Sci. U.S.A.* 2004;101 Suppl 2:14560–6. doi:10.1073/pnas.0404758101. PMID 15297608. Bibcode:2004PNAS.10114560H.

Ho, M.W., 2011, DNA Sequence Reconstituted from Water Memory? *ISIS Report* 20/07/11.

Hogenboom, M., 2013, Selfish traits not favoured by evolution, study shows, BBC report, Oct.

Holland, J., 2010, The Fifteen Biggest Lies about the Economy: And Everything Else the Right Doesn't Want You to Know about Taxes, Jobs, and Corporate America, Wiley

Hölldobler B, Wilson EO. Multiple recruitment systems of African weaver ant Oecophylla-longinoda (Latreille) (Hymenoptera, Formicidae). *Behav Ecol Sociobiol* 1978;3:19-60.

Hölldobler B, Wilson EO. The ants. Cambridge (MA): Harvard University Press; 1990.

Hölldobler B. Recruitment behavior in Camponotus-socius (Hymenoptera, Formicidae). *J Comp Physiol A* 1971;75:123-142.

Hölldobler B. The chemistry of social regulation: multicomponent signals in ant societies. *Proc Natl Acad Sci U S A* 1995;92:19-22.

Hooper, S.J., M. J. Wilson, and S. J. Crean, "Exploring the link between microorganisms and oral cancer: a systematic review of the literature," *Head and Neck*, vol. 31, no. 9, pp. 1228–1239, 2009.

Hope, J., 2015, Why Japan banned MMR vaccine, Daily Mail (UK), March 1.

Horstman, J. (2009). The Scientific American Day in the Life of Your Brain. Jossey-Bass / Wiley. ISBN 978-0470376232.

Hossain, M.E., and M.R. Islam, 2009, *An Advanced Analysis Technique for Sustainable Petroleum Operations*, VDM Publishing Ltd., Germany, 750 pp.

Hossain, M.E., Rahman, M.S., Ketata, C. and Islam, M.R. (2009). Molecular Structure Evaluation of Beeswax and Paraffin Wax by Solid-State 13C CP/MAS NMR. *Journal of Characterization and Development of Novel Materials*, 1(2), 101 - 110.

Howard BV, Manson JE, Stefanick ML, Beresford SA, Frank G, Jones B, Rodabough RJ, Snetselaar L, Thomson C, Tinker L, Vitolins M, Prentice R: Low-fat dietary pattern and weight change over 7 years: the Women's Health Initiative Dietary Modification Trial.

Howard RW, Blomquist GJ. Ecological, behavioral, and biochemical aspects of insect hydrocarbons. *Annu Rev Entomol* 2005;50:371-393.

HR.com, 2005, Hangover at work, July 1.

Hrbacek, J., M. Urban, E. Hamsikova, et al., "Serum antibodies against genitourinary infectious agents in prostate cancer and benign prostate hyperplasia patients: a case-control study," *BMC Cancer*, vol. 11, article 53, 2011.

Hrncir M, Schmidt VM, Schorkopf DLP, Jarau S, Zucci R, Barth FG. Vibrating the food receivers: a direct way of signal transmission in stingless bees (Melipona seminigra). *J Comp Physiol A* 2006;192:879-887.

Hsu A, Heshka S, Janumala I, et al. Larger mass of high-metabolic-rate organs does not explain higher resting energy expenditure in children. *Am J Clin Nutr* 2003;77:1506–11.

http://lilt.ilstu.edu/critique/Fall%202004%20Docs/Singh_Ajit.pdf

Hu J, Ludgate L. HIV-HBV and HIV-HCV coinfection and liver cancer development. *Cancer Treat. Res.* 2007;133:241–52. doi:10.1007/978-0-387-46816-7_9.

Hublin, J. J. The earliest modern human colonization of Europe. *Proc. Natl Acad. Sci. USA* 109, 13471–13472 (2012)

Huffington Post, 2014, The Most Used Drugs According To 2014's Global Drug Survey, April 14.

Hugot, J.P., Inflammatory bowel disease: a complex group of genetic disorders; *Best Practice & Research Clinical Gastroenterology*, Volume 18, Issue 3, Pages 451-462, June 2004.

Hulsegge G, Susan H, Picavet J, et al., 2013, Today's adult generations are less healthy than their predecessors: Generation shifts in metabolic risk factors: the Doetinchem Cohort Study. *Eur J Prevent Cardiol*, April 10.

Hutchinson, J.S., DiNicolantonio, R., Veroni, M, Cleverdon, M., 1983, Evidence for a functional central dopaminergic insufficiency in the spontaneously hypertensive rat, Clin Exp Pharmacol Physiol. *May-Jun*;10(3):311-4.

Huxley, R., "Theroleoflifestyleriskfactorsonmortalityfrom colorectal cancer in populations of the Asia-Pacific region," *Asian Pacific Journal of Cancer Prevention*, vol. 8, no. 2, pp. 191–198, 2007.

Hyland KM, Cao TT, Malechuk AM, Lewis LA, Schneider SS. Vibration signal behaviour and the use of modulatory communication in established and newly founded honey bee colonies. *Anim Behav* 2007;73:541-551.

Illner K, Brinkmann G, Heller M, Bosy-Westphal A, Müller MJ. Metabolically active components of fat free mass and resting energy expenditure in non obese adults. *Am J Physiol* 2000;278:E308–15.

Ingle, R., J. Levin, and K. Polinder, "Wound healing with honey—a randomised controlled trial," *South African Med- ical Journal*, vol. 96, no. 9, pp. 831–835, 2006.

Inouye, D. W. And Waller, G. D., 1984. Responses of honey bees (Apis-mellifera) to amino-acid solutions mimicking floral nectars. *Ecology* 65:618–625.

Institute of Medicine, Improving the Quality of Health Care for Mental and Substance-Use Conditions. Washington, DC: Institute of Medicine, 2006.

Int J Obes Relat Metab Disord 2002, 26:33-39.

Ionisio, D., Khanna, A.K., Nicolaou, Raghavan, V., Juneja, S., Sengupta, A., Messeri, D., 2008, "For-profit policies and equitable access to antiretroviral drugs in resource-limited countries", Future HIV Ther., vol. 2, no.1, 25-36.

Irvine, W. 1955. Apes, Angels, and Victorians: The Story of Darwin, Huxley, and Evolution. New York: McGraw-Hill.

Ish-Am, G. 1998, Improving avocado pollination with bumblebees: 3 seasons summary. *Calif. Avoc. Soc.* 82:119–135.

Ish-Am, G. And Eisikowitch, D. 1998. Low attractiveness of avocado (Persea americana Mill.) flowers to honey bees (Apis mellifera L.) limits fruit set in Israel. *J. Hortic. Sci. Biotechnol.* 73:195–204.

Ish-Am, G., 1994. Interrelationship between avocado flowering and honey bees and its implication on the avocado fruitfulness in Israel. PhD Dissertation. Tel-Aviv University, Tel-Aviv.

Ish-Am, G., Barrientos-Priego, A. F., Castaneda-Vildozola, A., and GAZIT, S. 1999. Avocado (Persea americana Mill.) pollinators in its region of origin. *Rev. Chapingo, Ser.: Hortic.* 5:137–143.

Ishizu, T., Tsujino, E., Winarno, H., Ohashi, K., And Shibuya, H. 2001. A complex of perseitol and K+ ion from Scurrula fusca (Loranthaceae). *Tetrahedron* 42:6887–6889.

Islam, J.S., 2015, Greening of an Isolated Desert Land, *JNSST*, vol. 9, issue 1.

Islam, J.S., 2015, "Delinearized history of African people in North America", book chapter in *Delinearized Global History*, Nova Science Publishers, New York, to be published.

Islam, J.S., G.M. Zatzman, M.A.H. Mughal, and M.R. Islam, 2014a, "A Knowledge-Based Cognition Model", *Journal of Information*, Intelligence, *and Knowledge*, vol. 6, no. 1.

Islam, J.S., G.M. Zatzman, M.A.H. Mughal, and M.R. Islam, 2014b, "A comprehensive theory of mass, energy, time, and human thought material (HTM), *Journal of Nature Science and Sustainable Technology*, vol. 8, no. 1, 181-215.

Islam, J.S., Zatzman, G.M., Mughal, M.A.H., and Islam, M.R., 2015, "Implications of a Comprehensive Material Balance Equation on Detection of the Causes of Medical Disorders, Part 1: Elimination of Paradoxes", *Journal of Characterization and Development of Novel Materials*, vol. 7, no. 1, 17-44.

Islam, M.R. 2005. Unraveling the Mysteries of Chaos and Change: The Knowledge-Based Technology Development Keynote Address to Joint Session of 5th International Conference on Composite Science & Technology and 1st *International Conference on Modeling*, Simulation and Applied Optimization Sharjah UAE 1-3 February 2005.

Islam, M.R., 2015, *Unconventional Gas Reservoirs*, Elsevier, 655 pp.

Islam, M.R., A.B. Chhetri, and M.M. Khan, 2012, *Green Petroleum*, Scrivener-Wiley, 605 pp.

Islam, M.R., A.B. Chhetri, M.M. Khan, 2010, *Greening of Petroleum Operations*, Scrivener Wilcy, 852 pp.

Islam, M.R., H. Mousavizadeghan, S. Mustafiz, and J.H. Abou-kassem, 2010a, *Reservoir Simulation: Advanced Approach*", Scrivener-Wiley, 468 pp.

Islam, M.R., Islam, J.S., Zatzman, G.M., and Mughal, M.A.H., 2013, Paradigm Shift in Education: A New Curriculum Fit for the Information Age, *Journal of Information, Intelligence and Knowledge*, Nova Science Publishers, vol. 5, no. 4, 1-17.

Islam, M.R., Mousavizadeghan, H., Mustafiz, S., and Abou-kassem, J.H., 2010, *Reservoir Simulation: Advanced Approach"*, Scrivener-Wiley, 468 pp.

Islam, M.R., Zatzman, G.M., Islam, J.S., 2013, *Reconstituting the curriculum*, Scrivener-Wiley, 503 pp.

Islam, Mohammad Ariful, Cheol-Heui Yun, Yun-Jaie Choi, and Chong-Su Cho. Microencapsulation of Live Probiotic Bacteria; *J. Microbiol. Biotechnol.* (2010), 20(10), 1367-1377

Iurlina, M.O., A. I. Saiz, R. Fritz, and G. D. Manrique, "Major flavonoids of Argentinean honeys. Optimisation of the extraction method and analysis of their content in relationship to the geographical source of honeys," *Food Chemistry*, vol. 115, no. 3, pp. 1141–1149, 2009.

Iwamoto T, Suzuki N, Tanabe K, Takeshita T, Hirofuji T. Effects of probiotic Lactobacillus salivarius WB21 on halitosis and oral health: an open-label pilot trial; *Oral Surg Oral Med Oral Pathol Oral Radiol Endod.* 2010 Aug;110(2):201-8.

Iyer LM, Balaji S, Koonin EV, Aravind L. Evolutionary genomics of nucleo-cytoplasmic large DNA viruses. *Virus Res.* 2006;117(1):156–84. doi:10.1016/j.virusres.2006.01.009. PMID 16494962.

Jackson, A., C. Cole, J. Esquiro, and M. Edwards, "Obesity in primary care patients in Kelantan, Malaysia: prevalence, and patients' knowledge and attitudes," *Southeast Asian Journal of Tropical Medicine and Public Health*, vol. 27, no. 4, pp. 776– 779, 1996.

Jaganathan, S.K., and M. Mandal, "Antiproliferative effects of honey and of its polyphenols: a review," *Journal of Biomedicine and Biotechnology*, vol. 2009, Article ID 830616, 13 pages, 2009.

Jaganathan, S.K., and M. Mandal, "Involvement of non- protein thiols, mitochondrial dysfunction, reactive oxygen species and p53 in honey-induced apoptosis," *Investigational New Drugs*, vol. 28, no. 5, pp. 624–633, 2010.

Jaganathan, S.K., S. M. Mandal, S. K. Jana, S. Das, and M. Mandal, "Studies on the phenolic profiling, anti-oxidant and cytotoxic activity of Indian honey: in vitro evaluation," *Natural Product Research*, vol. 24, no. 14, pp. 1295–1306, 2010.

Janiak, A., 2008, *Newton as Philosopher*, Cambridge University Press, 208 pp.

Jay, S. C. 1986. Spatial management of honey-bees on crops. *Annu. Rev. Entomol.* 31:49–65.

Jayaram H, Estes MK, Prasad BV. Emerging themes in rotavirus cell entry, genome organization, transcription and replication. *Virus Res.* 2004;101 (1):67–81. doi:10.1016/j.virusres.2003.12.007.

Jeanne RL. Group size, productivity, and information flow in social wasps. In: Detrain C, Deneubourg JL, Pasteels JM, editors. Information processing in social insects. Basel (Switzerland): Birkhauser Press; 1999. p. 3-30.

Jeanne RL. Regulation of nest construction behaviour in Polybia occidentalis. *Anim Behav* 1996;52:473-488.

Jeanne RL. Social complexity in the Hymenoptera, with special attention to the wasps. In: Kikuchi T, Azuma N, Higashi S, editors. Genes, behaviors, and evolution of social insects. Sapporo (Japan): Hokkaido University Press; 2003. p. 81-130.

Jeanne RL. The organization of work in Polybia occidentalis: costs and benefits of specialization in a social wasp. *Behav Ecol Sociobiol* 1986;19:333-341.

Jha, A., 2011, Obesity in middle age increases risk of dementia, *The Guardian, UK*, May 2.

Jha, A., 2013, "Stephen Hawking: physics would be 'more interesting' if Higgs boson hadn't been found", *The Guardian*, Tuesday 12 November 2013.

Johnson BR. A self-organizing model for task allocation via frequent task quitting and random walks in the honey bee. *Am Nat* 2009;174:537-547.

Johnson BR. Division of labor in honey bees: form, function, and proximate mechanisms. *Behav Ecol Sociobiol* 2010;64:305-316.

Johnson BR. Organization of work in the honey bee: a compromise between division of labour and behavioural flexibility. *Proc R Soc Lond B Biol Sci* 2003;270:147-152.

Johnson BR. Reallocation of labor in honey bee colonies during heat stress: the relativeroles of task switching and the activation of reserve labor. *Behav Ecol Sociobiol* 2002;51:188-196.

Johnson BR. Within-nest temporal polyethism in the honey bee. *Behav Ecol Sociobiol* 2008;62:777-784.

Johnson DW, van Eps C, Mudge DW, et al., Randomized, controlled trial of topical exit-site application of honey (Medihoney) versus mupirocin for the prevention of catheter-associated infections in hemodialysis patients. *J Am Soc Nephrol* 2005;16:1456-62.

Johnson NP, Mueller J. Updating the accounts: global mortality of the 1918–1920 "Spanish" influenza pandemic. *Bull Hist Med.* 2002;76(1):105–15.

Johnson, B.R., 2010, "Task partitioning in honey bees: the roles of signals and cues in group-level coordination of action", *Behavioral Ecology*, Volume 21, Issue 6Pp. 1373-1379.

Jones, M., 2014, The Madness of Sir Isaac Newton, *Know Your Scientist*, March 27.

Jordan MC, Jordan GW, Stevens JG, Miller G. Latent herpesviruses of humans. *Annals of Internal Medicine.* 1984;100(6):866–80.

Jordan, D. and Walsh, M., 2008, "White Cargo: The Forgotten History of Britain's White Slaves in America", March, NYU Press, 320 pp.

Joseph P. Mailman School of Public Health, Columbia University, 2000. Available: http://www.indiana.edu/~icmhsr/amerview1.pdf

Jung, R.T., 1997, "Obesity as a disease," British Medical Bulletin, vol. 53, *no. 2, pp. 307*–321, 1997.

Kabir Z., and L. Clancy, "Lifestyle-related cancer death rates in Ireland: decreasing or increasing?" *Irish Medical Journal*, vol. 99, no. 2, pp. 40–42, 2006.

Kadooka Y, Sato M, Imaizumi K, Ogawa A, Ikuyama K, Akai Y, Okano M, Kagoshima M, Tsuchida T. Regulation of abdominal adiposity by probiotics (Lactobacillus gasseri SBT2055) in adults with obese tendencies in a randomized controlled trial; *Eur J Clin Nutr.* 2010 Jun;64(6):636-43.

Kaiser W, Steinerkaiser J. Neuronal correlates of sleep, wakefulness and arousal in a diurnal insect. *Nature* 1983;301:707-709.

Kaiser, J., 2015, Why a powerful cancer drug only helps some patients, *Science*, March 12, DOI: 10.1126/science.aab0310.

Kaku, M., 1994, *Hyperspace*, Doubleday, New York.

Kearns, C. A. And Inouye, D. W. 1993. Techniques for Pollination Biologists. University Press of Colorado, Niwot.

Keenan, J. 2004 'New Imperialism' for Saharan peoples' Cambridge Review of International Affairs. Publisher:Routledge, part of the Taylor & Francis Group Issue:Volume 17, Number 3 / October, 2004, pp. 421-436.

Ketata, C., Satish, M.G., Islam, M.R., 2007, "Chaos Laws of Motion", Proc. International Conference of Computational and Experimental Engineering and Sciences (ICCES-2007), Miami, Florida, Jan.

Khan, M.I. and M.R. Islam, 2007, *The Petroleum Engineering Handbook: Sustainable Operations*, Gulf Publishing Company, Houston, TX, 461 pp.

Khan, M.I. and M.R. Islam, 2007, *True Sustainability in Technological Development and Natural Resource Management*, Nova Science Publishers, NY, 381 pp.,

Khan, M.M. and Islam, M.R., 2012, *Zero-Waste Engineering*, Scrivener-Wiley, 465 pp.

Khan, M.M., Zatzman, G.M. and Islam, M.R. (2008) The Formulation of a Comprehensive Mass and Energy Balance Equation, Proc. ASME International Mechanical Engineering Congress and Exposition, Nov. 2-6, Boston, MA.

Kim J, Wang ZM, Heymsfield ZM, Baumgartner RN, Gallagher D.,2002, Total-body skeletal muscle mass: estimation by new dual-energy X-ray absorptiometry method. *Am J Clin Nutr*, 76:378–83.

Kim S.J., & Byrne, S. (2011). Conceptualizing personal web usage in work contexts: A preliminary framework. *Computers in Human Behavior.* 27 (6), 2271-2283.

Kim, Y. S. And Smith, B. H. 2000. Effect of an amino acid on feeding preferences and learning behavior in the honey bee, Apis mellifera. *J. Insect Physiol.* 46:793–801.

Kirby, D., 2006 *The Evidence of Harm*, St. Martin's Press, 460 pp.

Kircher, M., Stenzel, U. & Kelso, J. Improved base calling for the Illumina Genome Analyzer using machine learning strategies. *Genome Biol.* 10, R83 (2009)

Kirchheimer, S., 2004, Supplement Speeds Wound Healing, *WebMD News*, July 8.

Kirchner WH, Lindauer M. The causes of the tremble dance of the honey bee, Apis mellifera. *Behav Ecol Sociobiol* 1994;35:303-308.

Klein BA, Olzsowy KM, Klein A, Saunders KM, Seeley TD. Caste-dependent sleep of worker honey bees. *J Exp Biol* 2008;211:3028-3040.

Klein E, Kis LL, Klein G. Epstein-Barr virus infection in humans: from harmless to life endangering virus-lymphocyte interactions. *Oncogene.* 2007;26(9):1297–305.

Klein, S., 2010, Fatty food may cause cocaine like addiction, March 31, *Health*, http://cnn.com.

Kline, M., (1972) Mathematical Thought from Ancient to Modern Times, Oxford University Press, New York, USA, pp. 1238

Kloppenberg, J.T., 2004, Pragmatism and the practice of history: from Turner and Du Bois to Today, *Metaphilosophy*, vol. 35, nos. 1&2, Jan.,

Knight, R. J. 2002. History, distribution and uses, pp. 1–14, in A. W. Whiley, B. Schaffer, and B. N. Wolstenholme (eds.). The Avocado: Botany, Production and Uses. CAB International, Wallingford.

Knip, M., Veijola, R., Virtanen, S.M., Hyöty, H., Vaarala, O., and Åkerblom, H.K., 2005, "Environmental Triggers and Determinants of Type 1 Diabetes", *Diabetes*, December 2005 vol. 54 no. suppl 2 S125-S136.

Knipe, David M; Howley, Peter M; Griffin, Diane E; Lamb, Robert A; Martin, Malcolm A; Roizman, Bernard; Straus Stephen E. (2007) Fields Virology, Lippincott Williams & Wilkins. ISBN 0-7817-6060-7.

Kohn, G. C. 1999 *Dictionary of War* Fitzroy Dearborn, Chicago, London, UK.

Koike K. Hepatitis C virus contributes to hepatocarcinogenesis by modulating metabolic and intracellular signalling pathways. *J. Gastroenterol. Hepatol.* 2007;22 Suppl 1:S108–11. doi:10.1111/j.1440-1746.2006.04669.x.

Kong, A. et al. Rate of de novo mutations and the importance of father's age to disease risk. *Nature* 488, 471–475 (2012)

Kotowska M, Albrecht P, Szajewska H (2005). Saccharomyces boulardii in the prevention of antibiotic-associated diarrhoea in children: a randomized double-blind placebo-controlled trial. *Alimentary Pharmacology & Therapeutics.* 21 (5): 583-90.

Kotseva K., Wood D., De Backer G. Cardiovascular prevention guidelines in daily practice: a comparison of EUROASPIRE I, II, and III surveys in eight European countries. *Lancet.* Mar 14, 2009;373(9667):929–940.

Kozisek F., 2005, "Health risks from drinking demineralised water", Nutrients in Drinking Water. *World Health Organization.* pp. 148–63. ISBN 92-4-159398-9.

Krauss, L, 2012, A Universe from Nothing: Why there is something rather than nothing?, Atria Books.

Krauss, L.M., Dawkins, R., 2013, *A Universe from Nothing: Why there is something rather than Nothing*, Atria Books, 240 pp.

Kruis, W., G Forstmaier, C Scheurlen, F Stellaard. Effect of diets low and high in refined sugars on gut transit, bile acid metabolism, and bacterial fermentation; *Gut*, 1991, 32, 367-371

Kuhn, S. L. & Zwyns, N. Rethinking the initial Upper Paleolithic. *Quat. Int.* http://dx.doi.org/10.1016/j.quaint.2014.05.040 (2014)

Kuk JL, Lee S, Heymsfield SB, Ross R: Waist circumference and abdominal adipose tissue distribution: influence of age and sex, *Am J Clin Nutr* 2005, 81:1330-1334.

Kumar, P., 2004, "Facing Competition: The History of Indigo Experiments in Colonial India, 1897-1920)", PhD Dissertation, Georgia Institute of Technology, August, 401 pp.

Kuniyasu, H., Y. Kitadai, H. Mieno, and W. Yasui, "Heli- cobactor pylori infection is closely associated with telomere reduction in gastric mucosa," *Oncology*, vol. 65, no. 3, pp. 275–282, 2003.

Kuper, H., H. O. Adami, and D. Trichopoulos, "Infections as a major preventable cause of human cancer," *Journal of Internal Medicine*, vol. 249, no. 741, pp. 61–74, 2001.

Kuroki, Y. et al. Comparative analysis of chimpanzee and human Y chromosomes unveils complex evolutionary pathway. *Nature Genet.* 38, 158–167 (2006)

Kutner, J., 2015, "Genetic sexual attraction is normal, and very real": A woman describes the reality of parent-child incest", *Salon*, Feb. 17.

Kuulasmaa K., Tunstall-Pedoe H., Dobson A. Estimation of contribution of changes in classic risk factors to trends in coronary event rates across the WHO MONICA project populations. *Lancet*. 2000;355:675–687.

Kwakman, P. H. S., Johannes P. C. Van den Akker, Ahmet Güçlü, Hamid Aslami, Jan M. Binnekade, Leonie de Boer, Laura Boszhard, Frederique Paulus, Pauline Middelhoek, Anje A. te Velde, Christina M. J. E. Vandenbroucke-Grauls, Marcus J. Schultz, and Sebastian A. J. Zaat, 2008, Medical-Grade Honey Kills Antibiotic-Resistant Bacteria In Vitro and Eradicates Skin Colonization, Clin Infect Dis. (2008) 46 (11): 1677-1682. doi: 10.1086/587892

Lahav, E. And Whiley, A. W., 2002. Irrigation and mineral nutrition, pp. 259–298, in A. W. Whiley, B. Schaffer, and B. N. Wolstenholme (eds.). The Avocado: Botany, Production and Uses. CAB International, Wallingford.

Lando, H.A., B. Borrelli, L. C. Klein, et al., "The landscape in global tobacco control research: a guide to gaining a foothold," *American Journal of Public Health*, vol. 95, no. 6, pp. 939–945, 2005.

Lane JM. Mass vaccination and surveillance/containment in the eradication of smallpox. *Curr. Top. Microbiol. Immunol.* 2006;304:17–29.

Langergraber, K. E. et al. Generation times in wild chimpanzees and gorillas suggest earlier divergence times in great ape and human evolution. *Proc. Natl Acad. Sci. USA* 109, 15716–15721 (2012)

Lapidus L, Bengtsson C, Larsson B, Pennert K, Rybo E, Sjöström L: Distribution of adipose tissue and risk of cardiovascular disease and death: a 12 year follow up of participants in the population study of women in Gothenburg, Sweden, *Br Med J* 1984, 289:1257-1261.

Larsson B, Svärdsudd K, Welin L, Wilhelmsen L, Björntorp P, Tibblin G: Abdominal adipose tissue distribution, obesity, and risk of cardiovascular disease and death: 13 year follow up of participants in the study of men born in 1913, *Br Med J* 1984, 288:1401-1404.

Laura R. Marks, Emily A. Clementi, Anders P. Hakansson, 2013, Sensitization of Staphylococcus aureus to Methicillin and Other Antibiotics In Vitro and In Vivo in the Presence of HAMLET, *PLoS One*, May 1, 2013DOI:10.1371/journal.pone.0063158

Lazaridis, I. et al. Ancient human genomes suggest three ancestral populations for present-day Europeans. *Nature* 513, 409–413 (2014)

Le Conte Y, Hefetz A. Primer pheromones in social hymenoptera. *Annu Rev Entomol* 2008;53:523-542.

Le Marchand, L., J. H. Hankin, L. R. Wilkens, et al., "Com- bined effects of well-done red meat, smoking, and rapid N- acetyltransferase 2 and CYP1A2 phenotypes in increasing Evidence-Based Complementary and Alternative Medicine colorectal cancer risk," *Cancer Epidemiology Biomarkers and Prevention*, vol. 10, no. 12, pp. 1259–1266, 2001.

Le Page C, Génin P, Baines MG, Hiscott J. Interferon activation and innate immunity. *Rev Immunogenet.* 2000;2(3):374–86.

Leaf, A., 1973, "Search for the Oldest People". *National Geographic*, Jan., pp. 93-118.

Lee, S.H., 2012, Detection of human papillomavirus L1 gene DNA fragments in postmortem blood and spleen after Gardasil® vaccination—A case report, Advances in Bioscience and Biotechnology, vol. 3, 1214-1224.

Lehrer RI, Barton A, Daher KA, Harwig SS, Ganz T, Selsted ME, Interaction of human defensins with Escherichia coli: mechanism of bactericidal activity. *J Clin Invest* 1989;84:553-61.

Lenzer, J. (2006). "Antidepressants double suicidality in children, says FDA". *BMJ* 332 (7542): 626. doi:10.1136/bmj.332.7542.626-c. PMC 1403257. PMID 16543316.

Lepage, C., L. Remontet, G. Launoy, et al., "Trends in incidence of digestive cancers in France," *European Journal of Cancer Prevention*, vol. 17, no. 1, pp. 13–17, 2008.

Lerner L (2000) Good and bad science in US schools. *Nature* 407: 287–290.

Lerner, L., 2005, "Creationism Evolves: The Advent of the ID-zoic Era", available in website: http://www.younginquirers.com/newsletter/0504/lerner.html

Levi F., Chatenoud L., Bertuccio P., Lucchini F., Negri E., La Vecchia C. Mortality from cardiovascular and cerebrovascular diseases in Europe and other areas of the world: an update. *Eur J Cardiovasc Prev Rehabil.* 2009;16(3):333–350.

Levi F., Lucchini F., Negri E., La Vecchia C. Trends in mortality from cardiovascular and cerebrovascular diseases in Europe and other areas of the world. *Heart.* 2002;88(2):119–124.

Levy SB, 2002, The antibiotic paradox: how misuse of antibiotics destroys their curative powers. Cambridge: Perseus Publishing.

Levy. SB, Marshall B, Antibacterial resistance worldwide: causes, challenges and responses. *Nat Med* 2004;10:122-9.

Lewis, C., 2014, *935* Lies: The Future of Truth and the Decline of America's *Moral Integrity*, Perseus Books, 364 pp.

Lewis LA, Schneider SS, Degrandi-Hoffman G. Factors influencing the selection of recipients by workers performing vibration signals in colonies of the honey bee, Apis mellifera. *Anim Behav* 2002;63:361-367.

Lewis, T., 2013, Incredible Technology: How to Explore the Microscopic World, *Livescience*, Aug. 3

Ley CJ, Lees B, Stevenson JC: Sex- and menopause-associated changes in body-fat distribution, *Am J Clin Nutr* 1992, 55:950-954.

Li, H. & Durbin, R. Fast and accurate short read alignment with Burrows–Wheeler transform. *Bioinformatics* 25, 1754–1760 (2009)

Li, H. & Durbin, R. Inference of human population history from individual whole-genome sequences. *Nature* 475, 493–496 (2011)

Liefke C, Holldobler B, Maschwitz U. Recruitment behavior in the ant genus Polyrhachis (Hymenoptera, Formicidae). *J Insect Behav* 2001;14:637-657.

Lim S.S., Gaziano T.A., Gakidou E. Prevention of cardiovascular disease in high-risk individuals in low-income and middle-income countries: health effects and costs. *Lancet.* 2007;370(9604):2054–2062.

Lim, G.C., "OverviewofcancerinMalaysia,"*JapaneseJournal of Clinical Oncology*, vol. 32, pp. S37–S42, 2002.

Lim, M.L., Land, M.F., 2007, "Sex-specific UV and Fluorescence Signals in Jumping Spiders", Jan., vol. 26, No. 315 (5811), pp. 481.

Linares J, The VISA/GISA problem: therapeutic implications. *Clin Microbiol Infect*2001;7(Suppl 4):8-15.

Linde B, Hjemdahl P, Freyschuss U, Juhlin-Dunndelt A., 1989, Adipose tissue and skeletal muscle blood flow during mental stress. *Am J Physiol* 1989;256:E12–8.

Liu, C., Dutton, Z., Behroozi, C.H. and Hau, L.V., (2001) Observation of Coherent Optical Information Storage in an Atomic Medium Using Halted Light Pulses, *Nature*, vol. 409, pp. 490–493

Liu, F. L., Fu, W. J., Yang, D. R., Peng, Y. Q., Zhang, X. W., And He, J. Z. 2004. Reinforcement of bee-plant interaction by phenolics in food. *J. Apic. Res.* 43:155–157.

Liu, X., Sherman, G., Robinson P., Witney, G., And Arpaia, M. L. 1995. Nectar sugar composition of selected avocado cultivars and related species. *Subtrop. Fruit News* 3:8–9.

Livio, M., 2005, The Equation That Couldn't Be Solved: How mathematical genius discovered the language of symmetry, Simon & Schuster, New York, 353 pp.

Loewe L (2006) "Quantifying the genomic decay paradox due to Muller's ratchet in human mitochondrial DNA" *Genetics Research* 87:133-159.

Lofgren IE, Herron KL, West KL, Zern TL, Brownbill RA, Ilich JZ, Koo SI, Fernandez ML: Weight loss favorably modifies anthropometrics and reverses the metabolic syndrome in premenopausal women, *J Am Coll Nutr* 2005, 24:486-493.

Lohman TC, Roche AF, Martorell R, 1988, editors., Anthropometric standardization reference manual. Champaign, IL: Human Kinetics.

Lohman TG, Chen Z. Dual energy X-ray absorptiometry, 2005, Heymsfield SB, Lohman TG, Wang ZM, Going SB, editors., Human body composition. 2nd ed Champaign, IL: Human Kinetics, 2005:63–78.

Lomonossoff, GP. Recent Advances in Plant Virology. Caister Academic Press; 2011. ISBN 978-1-904455-75-2. Virus Particles and the Uses of Such Particles in Bio- and Nanotechnology.

London-Shafir, I., Shafir, S., And Eisikowitch, D. 2003. Amygdalin in almond nectar and pollen—facts and possible roles. *Plant Syst. Evol.* 238:87–95.

Long GW, Nobel J, Murphy FA, Herrmann KL, Lourie B. Experience with electron microscopy in the differential diagnosis of smallpox. *Appl Microbiol.* 1970;20(3):497–504.

Lopez A.D., Mathers C.D., Ezzati M., Jamison D.T., Murray C. The burden of disease and mortality by condition: data, methods, and results for 2001. *2001. Global burden of disease and risk factors.* Chapter 3.

Low, S., M. C. Chin, and M. Deurenberg-Yap, "Review on epidemic of obesity," *Annals of the Academy of Medicine, Singapore*, vol. 38, no. 1, pp. 57–65, 2009.

Ludwig, D.S. and Willett, W.C., 2013, "Three Daily Servings of Reduced-Fat Milk: An Evidence-Based Recommendation?", *JAMA Pediatr.* 2013;167 (9):788-789

Lukashev AN. Role of recombination in evolution of enteroviruses. *Rev. Med. Virol.* 2005;15(3):157–67. doi:10.1002/rmv.457.

Lusby PE, Coombes A, Wilkinson JM, Honey: a potent agent for wound healing?*J Wound Ostomy Continence Nurs* 2002;29:295-300.

Luttge, U. 1977. Nectar composition and membrane transport of sugars and amino acids: A review on the present state of nectar research. *Apidologie* 8:305–319.

Lwoff A, Horne R, Tournier P. A system of viruses. Cold Spring Harb. *Symp. Quant. Biol.* 1962;27:51–5.

Lwoff A, Horne RW, Tournier P. A virus system. *C. R. Hebd. Seances Acad. Sci.* 1962;254:4225–7.

Lye, H.S., Chiu-Yin Kuan, Joo-Ann Ewe, Wai-Yee Fung and Min-Tze Liong. The Improvement of Hypertension by Probiotics: Effects on Cholesterol, Diabetes, Renin, and Phytoestrogens; *Int. J. Mol. Sci.* 2009, *10*, 3755-3775.

Lynch, M. & Gabriel, W. (1990). Mutation load and the survival of small populations. *Evolution* 44, 1725-1737.

Lynch, M., Butcher, R. B. D. & Gabriel, W. (1993). The mutational meltdown in asexual populations. *J. Hered.* 84, 339-344.

M. R. Islam, J. S. Islam, G. M. Zatzman and M. A. H. Mughal, 2013a, Paradigm Shift in Education: A New Curriculum Fit for the Information Age, *Journal of Information, Intelligence and Knowledge*, Nova Science Publishers, vol. 5, no. 4, 1-17.

M. R. Islam, J. S. Islam, G. M. Zatzman and M. A. H. Mughal, 2013, Paradigm Shift in Education: A New Curriculum Fit for the Information Age, *Journal of Information, Intelligence and Knowledge*, Nova Science Publishers, vol. 5, no. 4, 1-17.

M.R. Islam, G.M. Zatzman, I. Ashour, and J.S. Islam, 2011, "Comparison Between Conventional and Knowledge-Based Teaching Methodologies", *Journal of Nature Science and Sustainable Technology*, Nova Science Publishers, vol. 4, issue 4, 21-33.

M.R. Islam, J.S. Islam and G.M. Zatzman, 2013, *Reconstituting the curriculum*, 2013 Wiley Scrivener, USA, 536 pp.

M.R. Islam, J.S. Islam, G.M. Zatzman, and M.S. Rahman, *Greening of Chemical and Pharmaceutical Industries*, to be published in 2015, Wiley Scrivener.

M.R. Islam, J.S. Islam, G.M. Zatzman, M.S. Rahman, and Khan, M.M., 2015b, "God, God's particle, God's wrath and force of nature: Delinearized history of mass, energy and time", *Advances in Sustainable Petroleum Engineering Science*, vol. 6, issue 4.

MacMahon S., Peto R., Cutler J. Blood pressure, stroke, and coronary heart disease. Part 1. Prolonged differences in blood pressure: prospective observational studies corrected for the regression dilution bias. *Lancet.* 1990;335:765–774.

Mafauzy, M., 2006, "Diabetes mellitus in Malaysia," *Medical Journal of Malaysia*, vol. 61, no. 4, pp. 397–398, 2006.

Mafauzy, M., N. Mokhtar, W. B. Wan Mohamad, and M. Musalmah, "Diabetes mellitus and associated cardiovascular risk factors in North-East Malaysia," *Asia-Pacific Journal of Public Health*, vol. 11, no. 1, pp. 16–19, 1999.

Magden J, Kääriäinen L, Ahola T. Inhibitors of virus replication: recent developments and prospects. *Appl. Microbiol. Biotechnol.* 2005;66(6):612–21. doi:10.1007/s00253-004-1783-3.

Mahabir S, Baer D, Johnson LL, Roth M, Campbell W, Clevidence B, Taylor PR: Body Mass Index, percent body fat, and regional body fat distribution in relation to leptin concentrations in healthy, non-smoking postmenopausal women in a feeding study, *Nutr J* 2007, 6:3.

Mailleux AC, Buffin A, Detrain C, Deneubourg JL. Recruiter or recruit: who boosts the recruitment in starved nests in mass foraging ants? *Anim Behav* 2010;79:31-35.

Majamaa H, Isolauri E. Probiotics: a novel approach in the management of food allergy; *J Allergy Clin Immunol.* 1997 Feb;99(2):179-85.

Makarova KS, Grishin NV, Shabalina SA, Wolf YI, Koonin EV. A putative RNA-interference-based immune system in prokaryotes: computational analysis of the predicted enzymatic machinery, functional analogies with eukaryotic RNAi, and hypothetical mechanisms of action. *Biol. Direct.* 2006;1:7. doi:10.1186/1745-6150-1-7.

Mallery DL, McEwan WA, Bidgood SR, Towers GJ, Johnson CM, James LC; McEwan; Bidgood; Towers; Johnson; James (November 2010). "Antibodies mediate intracellular immunity through tripartite motif-containing 21 (TRIM21)". Proceedings of the National Academy of Sciences of the United States of America 107 (46): 19985–90.

Manago A. M., Graham M. B., Greenfield P. M., Salimkhan G. (2008). "Self-presentation and gender on MySpace". *Journal of Applied Developmental Psychology*, vol. 29, pp. 446-458.

Mang B, Wolters M, Schmitt B, Kelb K, Lichtinghagen R, Stichtenoth DO, Hahn A., 2006, Effects of a cinnamon extract on plasma glucose, HbA, and serum lipids in diabetes mellitus type 2., *Eur J Clin Invest.* May;36(5): 340-4.

Manhattan, A., 1983, The Vatican Billions, ISBN-13: 978-0937958162

Manning, J. C. (2006). The impact of Internet pornography on marriage and the family: A review of the research. *Sexual Addiction & Compulsivity*, 13, 131–165. March 12, 2015.

Markl H, Fuchs S. Alarm by rapping in carpenter ants (Camponotus, Formicidae, Hymenoptera). *Z Vergl Physiol* 1972;76:204-225.

Markl H, Hölldobler B. Recruitment and food-retrieving behavior in Novomessor (Formicidae, Hymenoptera). 2. Vibration signals. *Behav Ecol Sociobiol* 1978;4:183-216.

Marshall JAR, Bogacz R, Dornhaus A, Planque R, Kovacs T, Franks NR. On optimal decision-making in brains and social insect colonies. *J R Soc Interface* 2009;40:1065-1074.

Martin, J.E., *et al.*, 1990, U.S. Patent 4,955,562, Microwave powered aircraft.

Mato, I., Huidobro, J. F., Simal-Lozano, J., And Sancho, M. T. 2003. Significance of nonaromatic organic acids in honey. *J. Food Prot.* 66:2371–2376.

Matsuzaki S, Rashel M, Uchiyama J, Sakurai S, Ujihara T, Kuroda M, Ikeuchi M, Tani T, Fujieda M, Wakiguchi H, Imai S. Bacteriophage therapy: a revitalized therapy against bacterial infectious diseases. *Journal of Infection and Chemotherapy: Official Journal of the Japan Society of*

Chemotherapy. 2005;11(5):211–9. doi:10.1007/s10156-005-0408-9. PMID 16258815.

Matter L, Kogelschatz K, Germann D. Serum levels of rubella virus antibodies indicating immunity: response to vaccination of subjects with low or undetectable antibody concentrations. *J. Infect. Dis.* 1997;175(4):749–55. doi:10.1086/513967. PMID 9086126.

Maughan, T., 2015, The dystopian lake filled by the world's tech lust, *BBC*, April 5, 2015.

Mauron, A., 2003, "Biology and ethics: the paradoxes of the natural", *9th Postgraduate Course for Training in Reproductive Medicine and Reproductive Biology, Geneva Foundation for Medical Education and Research.*

Mawar N, Saha S, Pandit A, Mahajan U. The third phase of HIV pandemic: social consequences of HIV/AIDS stigma & discrimination & future needs [PDF]. *Indian J. Med. Res.* 2005 [Retrieved 13 September 2008];122 (6):471–84. PMID 16517997.

Mayanagi G, Kimura M, Nakaya S, Hirata H, Sakamoto M, Benno Y, Shimauchi H. Probiotic effects of orally administered Lactobacillus salivarius WB21-containing tablets on periodontopathic bacteria: a double-blinded, placebo-controlled, randomized clinical trial; *J Clin Periodontol.* 2009 Jun;36(6):506-13. Epub 2009 Apr 22.

Maynard Smith, J. (1958). The Theory of Evolution. London, Penguin Books. ISBN 0-14-020433-4.

Maynard Smith, J. (1968) Mathematical Ideas in Biology. Cambridge University Press. ISBN 0-521-07335-9.

Maynard Smith, J. (1972) On Evolution. Edinburgh University Press. ISBN 0-85224-223-9.

Maynard Smith, J. (1974b) Models in Ecology, Cambridge University Press, ISBN 0-521-20262-0.

Maynard Smith, J. (1978d) The Evolution of Sex. Cambridge University Press. ISBN 0-521-29302-2.

Maynard Smith, J. (1982d) Evolution and the Theory of Games. Cambridge University Press. ISBN 0-521-28884-3.

Maynard Smith, J. (1986b) The Problems of Biology. Oxford: Oxford University Press. ISBN 0-19-289198-7.

Maynard Smith, J. (1988a) Did Darwin Get it Right?: Essays on Games, Sex and Evolution. London, Chapman & Hall. ISBN 0-412-03821-8.

Maynard Smith, J. (1989a) Evolutionary Genetics. Oxford: Oxford University Press. ISBN 0-19-850231-1.

Maynard Smith, J. (ed.) (1981d) Evolution Now. London, Macmillan. ISBN 0-7167-1426 4.

Maynard Smith, J. and Szathmáry, E. (1997) The Major Transitions in Evolution. New York: Oxford University Press. ISBN 0-19-850294-X.

Maynard Smith, J. and Szathmáry, E. (1999) The Origins of Life: From the Birth of Life to the Origin of Language. Oxford: Oxford University Press. ISBN 0-19-286209-X.

Maynard Smith, J.; Price, G.R. (1973). "The logic of animal conflict". *Nature* 246 (5427): 15–18. Bibcode:1973Natur.246..15S. doi:10.1038/246015a0.

McCourt, C.K., D. G. Mutch, R. K. Gibb, et al., "Body mass index: relationship to clinical, pathologic and features of microsatellite instability in endometrial cancer," *Gynecologic Oncology*, vol. 104, no. 3, pp. 535–539, 2007.

McCown, T. D. & Keith, A. The Stone Age of Mount Carmel Vol. 2 (Clarendon, Oxford, 1939)

McCrory MA, Gomez TD, Bernauer EM, Mole PA: Evaluation of a new air displacement plethysmograph for measuring human body composition, *Med Sci Sports Exerc* 1995, 27:1686-1691.

McCully KK, Posner JD., 1995, The application of blood flow measurements to the study of aging muscle. *J Gerontol* 1995;50:130–6.

McDonald, K.T., 2013, Spin-Orbit Coupling in the Earth-Moon System, Princeton University Report, updated July 12, 2013.

McFarland L, Bernasconi P (1993). Saccharomyces boulardii: a review of an innovative biotherapeutic agent; *Microbial Ecology in Health and Disease* 6: 157-71.

McHenry, H.M., 2009, "Human Evolution", in Michael Ruse & Joseph Travis (eds), Evolution: The First Four Billion Years, pp. 256–280,

McInnes, M., 2014, The Honey Diet, Hodder & Stoughton, UK, 151 pp.

McLaughlin-Drubin, M.E., and K. Munger, "Viruses associ- ated with human cancer," *Biochimica et Biophysica Acta*, vol. 1782, no. 3, pp. 127–150, 2008.

McLean AA. Development of vaccines against hepatitis A and hepatitis B. *Rev. Infect. Dis.* 1986;8(4):591–8.

Mead, M.M, 2008, Contaminants in Human Milk: Weighing the Risks against the Benefits of Breastfeeding, *Environ Health Perspect*, Oct; 116(10): A426–A434.

Menocal, M.R., 2003, *The Ornament of the World: How Muslims, Jews, and Christians Created a Culture of Tolerance in Medieval Spain*, Little, Brown, and Company, 303 pp.

Mental Health America. American Opinions on Mental Health Issues. Alexandria: NMHA, 1999.

Mertl AL, Traniello JFA. Behavioral evolution in the major worker subcaste of twig-nesting Pheidole (Hymenoptera: Formicidae): does morphological specialization influence task plasticity? *Behav Ecol Sociobiol* 2009;63:1411-1426.

Metabolism 1991, 40:1323-1326.

Metzner KJ. Detection and significance of minority quasispecies of drug-resistant HIV-1. *J HIV Ther.* 2006;11(4):74–81.

Meyer, M. et al. A high-coverage genome sequence from an archaic Denisovan individual. *Science* 338, 222–226 (2012)

Michener CH. The comparative social behavior of bees. *Annu Rev Entomol* 1969;14:299-342.

Michener CH. The social behavior of the bees. Cambridge (MA): Harvard University Press; 1974.

Mindel A, Sutherland S. Genital herpes — the disease and its treatment including intravenous acyclovir. *J. Antimicrob. Chemother.* 1983;12 Suppl B:51–9.

Miorin, P.L., N. C. Levy Jr., A. R. Custodio, W. A. Bretz, and M. C. Marcucci, "Antibacterial activity of honey and propolis from Apis mellifera and Tetragonisca angustula against Staphylococcus aureus," *Journal of Applied Microbiology*, vol. 95, no. 5, pp. 913–920, 2003.

MIT (Slice of), 2013, "MIT Researchers Develop a Shocking Treatment for Facebook Addiction".

Moghaddam, A.A., M. Woodward, and R. Huxley, "Obesity and risk of colorectal cancer: a meta-analysis of 31 studies with 70,000 events," *Cancer Epidemiology Biomarkers and Prevention*, vol. 16, no. 12, pp. 2533–2547, 2007.

Moghul, S.N., 2013, "20 Health benefits of honey", The Times of India, Sept. 13.

Moglich M, Holldobler B. Communication and orientation during foraging and emigration in the ant Formica-fusca. *J Cell Comp Physiol* 1975;101:275-288.

Mojica FJ, Díez-Villaseñor C, García-Martínez J, Soria E. Intervening sequences of regularly spaced prokaryotic repeats derive from foreign genetic elements. *J. Mol. Evol.* 2005;60(2):174–82. doi:10.1007/s00239-004-0046-3.

Molan PC, Betts JA, Clinical usage of honey as a wound dressing: an update. *J Wound Care* 2004;13:353-6.

Molan PC, The antibacterial activity of honey. *1. The nature of the antibacterial activity. Bee World* 1992;73:5-28.

Molan PC, The antibacterial activity of honey. *2. Variation in the potency of the antibacterial activity. Bee World* 1992;73:59-76.

Molan, P.C., "Potential of honey in the treatment of wounds and burns," *American Journal of Clinical Dermatology*, vol. 2, no. 1, pp. 13–19, 2001.

Molan, P.C., "The potential of honey to promote oral wellness," *General dentistry*, vol. 49, no. 6, pp. 584–589, 2001. [74] J. Hon, "Using honey to heal a chronic wound in a patient with epidermolysis bullosa," British Journal of Nursing, vol. 14, no. 19, pp. S4–S12, 2005.

Monbiot, G., 2013, Farming subsidies: this is the most blatant transfer of cash to the rich, The Guardian, July 1.

Mondal, S. and Pahan, K., 2015, Cinnamon Ameliorates Experimental Allergic Encephalomyelitis in Mice via Regulatory T Cells: Implications for

Multiple Sclerosis Therapy', *PLoS One.*, January 8, 2015DOI: 10.1371/journal.pone.0116566

Moore OA, Smith LA, Campbell F, Seers K, McQuay HJ, Moore RA, Systematic review of the use of honey as a wound dressing. *BMC Complement Altern Med*2001;1:2.

Moorehead, B., 1995, *Words Aptly Spoken*, Overlake Christian Bookstore, 198 pp.

Moreno MA, et al. (2011), Problematic internet use among US youth: A systematic review" *Archives of Pediatrics and Adolescent Medicine* 165, 797–805.

Moreno MA, Jelenchick LA, Christakis DA (2013). Problematic internet use among older adolescents: A conceptual framework, *Computers and Human Behavior,* 29, 1879–1887, http://dx.doi.org/10.1016/j.chb.2013.01.053.

Mosca L, Appel LJ, Benjamin EJ, Berra K, Chandra-Strobos N, Fabunmi RP, Grady D, Haan CK, Hayes SN, Judelson DR, Keenan NL, McBride P, Oparil S, Ouyang P, Oz MC, Mendelsohn ME, Pasternak RC, Pinn VW, Robertson RM, Schenck-Gustafsson K, Sila CA, Smith SC Jr, Sopko G, Taylor AL, Walsh BW, Wenger NK, Williams CL: Evidence-based guidelines for cardiovascular disease prevention in women.

Moss RB, Davey RT, Steigbigel RT, Fang F. Targeting pandemic influenza: a primer on influenza antivirals and drug resistance. *The Journal of Antimicrobial Chemotherapy*. 2010;65(6):1086–93. doi:10.1093/jac/dkq100.

Mota, P.G., 2010, "Darwin's sexual selection theory – a forgotten idea", *Antropologia Portuguesa* 26/27, 2009/2010: 149-161.

Mougios V, Kazaki M, Christoulas K, Ziogas G, Petridou A: Does the intensity of an exercise programme modulate body composition changes? *Int J Sports Med* 2006, 27:178-181.

Mousavizadegan,,S.H., Mustafiz, S., Rahman, M., 2007, "The Adomian Decomposition Method on Solution of Non-Linear Partial Differential Equations", *Journal of Nature Science and Sustainable Technology*, vol. 1, No. 1, pp.115-131.

Moutsatsou, P., "The spectrum of phytoestrogens in nature: our knowledge is expanding," *Hormones*, vol. 6, no. 3, pp. 173–193, 2007.

Müller MJ, Bosy-Westphal A, Kutzner D, Heller M. Metabolically active components of fat-free mass and resting energy expenditure in humans: recent lessons from imaging technologies. *Obesity (Silver Spring)* 2002;3:113–22.

Muller, H. J. (1963). The need for recombination to prevent genetic deterioration. *Genetics* 48, 903-903.

Muller, H. J. (1964). The relation of recombination to mutational advance. *Mutat. Res.* 1, 2-9.

Muller, R.J., 2008, "Neurotheology: Are We Hardwired for God?", Psyhiatric Times, May 1.

Müller, U. C. et al. The role of climate in the spread of modern humans into Europe. *Quat. Sci. Rev.* 30, 273–279 (2011)

Mulvey, E. P. (1994). Assessing the evidence of a link between mental illness and violence. *Hospital and Community Psychiatry*, 45, 663-668.

Murray CJ, Lopez AD, Chin B, Feehan D, Hill KH (December 2006). "Estimation of potential global pandemic influenza mortality on the basis of vital registry data from the 1918–20 pandemic: a quantitative analysis". *Lancet* 368 (9554): 2211–8.

Mykkanen L, Laakso M, Pyorala K: High plasma insulin level associated with coronary heart disease in the elderly, *Am J Epidemiol* 1993, 137:1190-1202.

Nadège Philippe, et al. "Pandoraviruses: amoeba viruses with genomes up to *2.5 Mb reaching that of parasitic eukaryotes", Science, 19 July* 2013: 341 (6143), 281-286.

Naiemi NA, Duim B, Savelkoul PH, et al., Widespread transfer of resistance genes between bacterial species in an intensive care unit: implications for hospital epidemiology. *J Clin Microbiol* 2005;43:4862-4.

NAMI, 2014, website: http://www.nami.org/Content/NavigationMenu/Inform_Yourself/About_Public_Policy/Policy_Research_Institute/Policymakers_Toolkit/Spending_Money_in_all_the_Wrong_Places_Jails.pdf, last accessed October 15, 2014.

Naoki, S., 2014, *Wireless Power Transfer via Radiowaves*. John Wiley & Sons. ISBN 1118862961.

NASA, 2013, website: http://map.gsfc.nasa.gov/universe/uni_shape.html, last accessed Nov., 2013.

Navon-Venezia S, Ben Ami R, Carmeli Y, Update on Pseudomonas aeruginosaand Acinetobacter baumannii infections in the healthcare setting. *Curr Opin Infect Dis* 2005;18:306-13.

Nazri, S.M., M. K. Imran, I. M. Ismail, and A. A. Faris, "Prevalence of overweight and self-reported chronic diseases among residents in Pulau Kundur, Kelantan, Malaysia," *Southeast Asian Journal of Tropical Medicine and Public Health*, vol. 39, no. 1, pp. 162–167, 2008.

New Freedom Commission on Mental Health, Achieving the Promise: Transforming Mental Health Care in America. Final Report. DHHS Pub. No. SMA-03-3832. Rockville, MD: 2003.

Ng, M., Fleming, T., Robinson, M., Thomson, B., *et al.*, 2013, Global, Regional, and National prevalence of overweight and obesity in children and adults during 1980-2013: systematic analysis for the global burden of disease study, *Lanset*, 384 (9945): 766-81.

Nicolson, S. W. And W.-Worswick, P. V. 1990. Sodium and potassium concentrations in floral nectars in relation to foraging by honey bees. *S. Afr. J. Zool.* 25:93–96.

Nieh JC, Tautz J. Behaviour-locked signal analysis reveals weak 200–300 Hz comb vibrations during the honey bee waggle dance. *J Exp Biol* 2000;203:1573-1579.

Nieh JC. Recruitment communication in stingless bees (Hymenoptera, Apidae, Meliponini). *Apidologie* 2004;35:159-182.

Nieh JC. Stingless-bee communication. *Am Sci* 1999;87:428-435.

Nieh JC. The honey bee shaking signal: function and design of a modulatory communication signal. *Behav Ecol Sociobiol* 1998;42:23-36.

Nobel Lectures, Physics 1981–1990, (1993) Editor-in-Charge Tore Frängsmyr, Editor Gösta Ekspång, World Scientific Publishing Co., Singapore.

Norrby E. Nobel Prizes and the emerging virus concept. *Arch. Virol.* 2008;153 (6):1109–23. doi:10.1007/s00705-008-0088-8. PMID 18446425.

Norton, A., 2015, Teens' Heavy Pot Smoking Tied to Memory Problems, WebMD.com, March 15.

Norval, M; Wulf, HC, 2009, "Does chronic sunscreen use reduce vitamin D production to insufficient levels?". *The British journal of dermatology* 161 (4): 732–6.

O'Donnell S, Bulova SJ. Worker connectivity: a review of the design of worker communication systems and their effects on task performance in insect societies. *Insect Soc* 2007;54:203-210.

O'Donnell S. Polybia wasp biting interactions recruit foragers following experimental worker removals. *Anim Behav* 2006;71:709-715.

Oddo, L.P., T. A. Heard, A. Rodriguez-Malaver, et al., "Composition and antioxidant activity of Trigona carbonaria honey from Australia," *Journal of Medicinal Food*, vol. 11, no. 4, pp. 789–794, 2008.

Olalde, I. et al. Derived immune and ancestral pigmentation alleles in a 7,000-year-old Mesolithic European. *Nature* 507, 225–228 (2014)

Olney, J.W. Excitotoxins in foods. *Neurotoxicology* 1994;15:535-544.

Ong, A., 2008, "The Human and Ethical Living," In *Globalizing the Research Imagination*, ed. Jane Kenway and Johannah Fahey, pp.87-100. London: Routledge.

Othman, N.H., and A. A. Zin, "Association of colorectal carcinoma with metabolic diseases; experience with 138 cases from Kelantan, Malaysia," *Asian Pacific Journal of Cancer Prevention*, vol. 9, no. 4, pp. 747–751, 2008.

Othman, N.H., 2012, Honey and Cancer: Sustainable Inverse Relationship Particularly for Developing Nations—A Review, *Evidence-Based Complementary and Alternative Medicine* Volume 2012, Article ID 410406, 10 pages doi:10.1155/2012/410406

Othman, N.H., Z. M. Nor, and B. M. Biswal, "Is Kelan tan joining the global cancer epidemic?—Experience from hospital Universiti Sains Malaysia; 1987–2007," *Asian Pacific Journal of Cancer Prevention*, vol. 9, no. 3, pp. 473–478, 2008.

Ouyang P, Sung J, Kelemen MD, Hees PS, DeRegis JR, Turner KL, Bacher AC, Stewart KJ: Relationships of insulin sensitivity with fatness and fitness and in older men and women, *J Womens Health (Larchmt)* 2004, 13:177-185.

Owens, E. W., Behun, R. J., Manning, J. C., & Reid, R. C. (2012). The impact of Internet pornography on adolescents: A review of the research. *Sexual Addiction & Compulsivity*, 19, 99–122.

Oxfam, 2015, 'Wealth: Having It All and Wanting More,' Report released at World Economic Forum, Davos, Switzerland.

Padilla-Walker, L.M., Nelson, L.J., Carroll, J.S. & Jensen, A.C. (2009). More than just a game: Video game and internet use during emerging adulthood. *Journal of Youth and Adolescence*, 39, 103 – 113. DOI 10.1007/s10964-008-9390-8.

Page RE, Mitchell SR. Self organization and the evolution of division of labor. *Apidologie* 1998;29:101-120.

Page RE, Robinson GE. The genetics of division of labor in honey bee colonies. *Adv Insect Physiol* 1991;23:117-169.

Page, R. E., Erber, J., And Fondrk, M. K. 1998. The effect of genotype on response thresholds to sucrose and foraging behavior of honey bees (Apis mellifera L.). *J. Comp. Physiol. A* 182:489–500.

Palese P. Making better influenza virus vaccines?. *Emerging Infect. Dis.* 2006;12 (1):61–5. doi:10.3201/eid1201.051043.

Pankiw T. Cued in: honey bee pheromones as information flow and collective decision-making. *Apidologie* 2004;35:217-226.

Parkin, D.M., F. I. Bray, and S. S. Devesa, "Cancer burden in the year 2000. The global picture," *European Journal of Cancer*, vol. 37, supplement 8, pp. S4–S66, 2001.

Parkin, D.M., M. Almonte, L. Bruni, G. Clifford, M. P. Curado, and M. Piñeros, 2008, "Burden and trends of type-specific human papillomavirus infections and related diseases in the Latin America and Caribbean Region," *Vaccine*, vol. 26, supplement 11, pp. L1–L15, 2008.

Parrish, C; Kawaoka Y (2005). "The origins of new pandemic viruses: the acquisition of new host ranges by canine parvovirus and influenza A viruses". *Annual Rev Microbiol* 59: 553–86.

Pastoret PP, Schudel AA, Lombard M. Conclusions—future trends in veterinary vaccinology. *Rev. – Off. Int. Epizoot.* 2007;26(2):489–94, 495–501, 503–9.

Patterson KB, Runge T. Smallpox and the Native American. *Am. J. Med. Sci.* 2002;323(4):216–22. doi:10.1097/00000441-200204000-00009. PMID 12003378.

Patterson KD, Pyle GF. The geography and mortality of the 1918 influenza pandemic. *Bull Hist Med.* 1991;65(1):4–21.

Patterson, N. et al. Ancient admixture in human history. *Genetics* 192, 1065–1093 (2012)

Patton JT, Vasquez-Del Carpio R, Spencer E. Replication and transcription of the rotavirus genome. *Curr. Pharm. Des.* 2004;10(30):3769–77.

Peele, S., 2010, God, Lincoln, and Depression, *Addiction in Society*, Jan. 2

Pearl, J., 2009, Causality: Models, Reasoning, and Inference, Cambridge University Press (2nd edition), ISBN 0-521-77362-8.

Perez, A.R., M. T. Iglesias, E. Pueyo, M. Gonzalez, and C. de Lorenzo, "Amino acid composition and antioxidant capacity of Spanish honeys," *Journal of Agricultural and Food Chemistry*, vol. 55, no. 2, pp. 360–365, 2007.

Pescosolido, B.A., Martin, J.K., Link, B.G., et al. Americans' Views of Mental Health and Illness at Century's End: Continuity and Change. Public Report on the MacArthur Mental health Module, 1996General Social Survey. Bloomington: Indiana Consortium for Mental Health Services Research and

Pescosolido, B.A., Monahan, J. Link, B.G. Stueve, A., & Kikuzawa, S. (1999). The public's view of the competence, dangerousness, and need for legal coercion of persons with mental health problems. *American Journal of Public Health*, 89, 1339-1345.

Petersen KF, Beftoy D, Dufour S, et al., 2003, Mitochondrial dysfunction in the elderly: possible role in insulin resistance. *Science* 2003;300:1140–2.

Petrov, V. 1970. Mineral constituents of some Australian honeys as determined by atomic absorption spectrophotometry. *J. Apic. Res.* 9:95–101.

Pi-Sunyer FX, Aronne LJ, Heshmati HM, Devin J, Rosenstock J: Effect of rimonabant, a cannabinoid-1 receptor blocker, on weight and cardiometabolic risk factors in overweight or obese patients: RIO-North America: a randomized controlled trial, *JAMA* 2006, 295:761-775.

Pichichero, E., R. Cicconi, M. Mattei, M. G. Muzi, and A. Canini, "Acacia honey and chrysin reduce proliferation of melanoma cells through alterations in cell cycle progression," *International Journal of Oncology*, vol. 37, no. 4, pp. 973–981, 2010.

Pirez N, Farina WM. Nectar-receiver behavior in relation to the reward rate experienced by foraging honey bees. *Behav Ecol Sociobiol* 2004;55:574-582.

Pischon, T., U. Nothlings, and H. Boeing, "Obesity and cancer," *Proceedings of the Nutrition Society*, vol. 67, no. 2, pp. 128–145, 2008.

Planck, M., 1901, "On the Law of the Energy Distribution in the Normal Spectrum", *Ann.* Phys., 4, 553.

Pocock, J G A. 2004. The Ancient Constitution and the Feudal Law Cambridge

Potera C (Mar–Apr 1998). "Trapped in the Web?". *Psychology Today* 31 (2): 66–70.

Prangishvili D, Garrett RA. Exceptionally diverse morphotypes and genomes of crenarchaeal hyperthermophilic viruses. *Biochem. Soc. Trans.* 2004;32(Pt 2):204–8. doi:10.1042/BST0320204. PMID 15046572.

Pressing J, Reanney DC. Divided genomes and intrinsic noise. *J Mol Evol.* 1984; 20(2):135–46. doi:10.1007/BF02257374.

Proceedings of the Nutrition Society of New Zealand, 2008, Volume 32, pages 20-25. *Forty Second Annual Conference held in combination with the Australian Nutrition Society at Massey University*, Albany Campus, Auckland, New Zealand (December 2007).

Projan SJ, Why is big Pharma getting out of antibacterial drug discovery? *Curr Opin Microbiol* 2003;6:427-30.

Prostitution, Harm, and Gender Inequality: Theory, Research and PolicyAuthor: Maddy Coy, M., (ed.), 2012, ProstitutionProstitution, harm, and gender inequality: Theory, Research and Policy, Ashgate Publishing, 207p.

Prüfer, K. et al. The bonobo genome compared with the chimpanzee and human genomes. *Nature* 486, 527–531 (2012)

Prüfer, K. et al. The complete genome sequence of a Neanderthal from the Altai Mountains. *Nature* 505, 43–49 (2014)

Pulitzer MP, Amin BD, Busam KJ. Merkel cell carcinoma: review. *Advances in Anatomic Pathology*. 2009;16(3):135–44.

Pyrzynska K., and M. Biesaga, "Analysis of phenolic acids and flavonoids in honey," *TrAC Trends in Analytical Chemistry*, vol. 28, no. 7, pp. 893–902, 2009.

Raghavan, M. et al. Upper Palaeolithic Siberian genome reveals dual ancestry of Native Americans. *Nature* 505, 87–91 (2014)

Ragland K.E., Selvin S., Merrill D.W. The onset of decline in ischemic heart disease mortality in the United States. *Am J Epidemiol.* 1988;127:516–531.

Raguso, R. A. 2004. Why are some floral nectars scented? *Ecology* 85:1486–1494.

Ramstedt M. Fluctuations in male ischaemic heart disease mortality in Russia 1959–1998: assessing the importance of alcohol. *Drug Alcohol Rev.* 2009;28(4):390–395.

Ranc, K., Jørgensen, M.E., Søren Friis, S., and Carstensen, B., 2014, "Mortality after cancer among patients with diabetes mellitus: effect of diabetes duration and treatment", *Diabetologia* (2014) 57:927–934.

Ranlet P. The British, the Indians, and smallpox: what actually happened at Fort Pitt in 1763?. *Pa Hist.* 2000 [Retrieved 16 September 2008];67 (3):427–41. PMID 17216901.

Rao, A.V., Alison C Bested, Tracey M Beaulne, Martin A Katzman, Christina Iorio5, John M Berardi6 and Alan C Logan. A randomized, double-blind, placebo-controlled pilot study of a probiotic in emotional symptoms of chronic fatigue syndrome; Gut Pathogens 2009, 1:6.

Rapp, K., J. Klenk, H. Ulmer *et al.*, "Weight change and cancer risk in a cohort of more than 65,000 adults in Austria," *Annals of Oncology*, vol. 19, no. 4, pp. 641–648, 2008.

Rasmussen, M. et al. An Aboriginal Australian genome reveals separate human dispersals into Asia. *Science* 334, 94–98 (2011)

Ratnieks FLW, Anderson C. Task partitioning in insect societies. *II. Use of queueing delay information in recruitment. Am Nat* 1999;154:536-548.

Read more at http://www.phillymag.com/news/2014/10/06/pedophilia-not-a-crime-rutgers-margo-kaplan/#CDcbB5KlhxfHed64.99

Read more: http://www.dailymail.co.uk/health/article-17509/Why-Japan-banned-MMR-vaccine.html#ixzz3TBJhCHbO

Reeves, G.K., K. Pirie, V. Beral, J. Green, E. Spencer, and D. Bull, "Cancer incidence and mortality in relation to body mass index in the Million Women Study: cohort study," *British Medical Journal*, vol. 335, no. 7630, p. 1134, 2007.

Reich, D. et al. Denisova admixture and the first modern human dispersals into Southeast Asia and Oceania. *Am. J. Hum. Genet.* 89, 516–528 (2011)

Reich, D. et al. Genetic history of an archaic hominin group from Denisova Cave in Siberia. *Nature* 468, 1053–1060 (2010)

Reimer, P. J. et al. IntCal13 and Marine13 radiocarbon age calibration curves 0–50,000 Years cal BP. *Radiocarbon* 55, 1869–1887 (2009)

Reinhard J, Srinivasan MV, Guez D, Zhang SW. Floral scents induce recall of navigation and visual memories in honey bees. *J Exp Biol* 2004;207:4371-4381.

Relman DA, Schmidt TM, MacDermott RP, Falkow S, Identification of the uncultured bacillus of Whipple's disease. *N Engl J Med* 1992;327:293-301.

Reuters, 2015, Wireless electricity: Scientists send energy 55 meters away through air,

Reuters, 2015a, US troops, contractors sexually abused Colombian girls with impunity –report", March 25.

Reynolds, G., 2015, One Twin Exercises, the Other Doesn't, *The New York Times*, March 4.

Rhoades, D. F. And Bergdahl, J. C. 1981. Adaptive significance of toxic nectar. *Am. Nat.* 117:798–803.

Richards, M. P. & Trinkaus, E. Out of Africa: modern human origins special feature: isotopic evidence for the diets of European Neanderthals and early modern humans. *Proc. Natl Acad. Sci. USA* 106, 16034–16039 (2009)

Riedel, S., 2005, "Edward Jenner and the history of smallpox and vaccination", *Proc (Bayl Univ Med Cent)*. 2005 Jan; 18(1): 21–25.

Roberts B., Gilmore A., Stickley A. Changes in smoking prevalence in 8 countries of the former Soviet Union between 2001 and 2010. *Am J Public Health*. 2012;102(7):1320–1328.

Roberts B., Stickley A., Balabanova D., Haerpfer C., McKee M. The persistence of irregular treatment of hypertension in the former Soviet Union. *J Epidemiol Community Health*. 2012;66(11):1079–1082.

Roberts, H.J.- Aspartame Disease: An Ignored Epidemic Dr. Harold Waisman – 52 week oral toxicity study – http://dorway.com/raoreport.pdf

Robin, C., 2004, Fear: The History of a Political Idea, Barnes and Noble, 314 pp.

Robson, D., 2015, The surprising downsides of being clever, *BBC*, April 15

Roces F, Hölldobler B. Vibrational communication between hitchhikers and foragers in leaf cutting ants (Atta cephalotes). *Behav Ecol Sociobiol* 1995;37:297-302.

Roces F, Tautz J, Hölldobler B. Stridulation in leaf cutting ants: short range recruitment through plant-borne vibrations. *Naturwissenschaften* 1993;80:521-524.

Rodríguez T., Malvezzi M., Chatenoud L. Trends in mortality from coronary heart and cerebrovascular diseases in the Americas: 1970–2000. *Heart.* 2006;92(4):453–460.

Roger V.L. Epidemiology of myocardial infarction. *Med Clin North Am.* 2007;91(4):537–552.

Rosamond W., Flegal K., Furie K. Heart disease and stroke statistics—2008 update: a report from the American Heart Association Statistics Committee and Stroke Statistics Subcommittee. *Circulation.* 2008;117(4): e25–e146.

Rosamond W.D., Chambless L.E., Folsom A.R. Trends in the incidence of myocardial infarction and in mortality due to coronary heart disease, 1987 to 1994. *N Engl J Med.* 1998;339:861–867.

Rose G., "Strategy of prevention: lessons from cardiovascular disease", *Brit Med J* 1981; 282: 1847-51.

Rosen, FS. Isolation of poliovirus—John Enders and the Nobel Prize. *New England Journal of Medicine.* 2004;351(15):1481–83. doi:10.1056/NEJMp048202. PMID 15470207.

Rosen, L. D. *et al.* (2012), iDisorder: Understanding Our Obsession with Technology and Overcoming Its Hold On Us, New York: Palgrave Macmillan, ISBN 9780230117570

Ross R, Janssen I. Computed tomography and magnetic resonance imaging. Heymsfield SB, Lohman TG, Wang ZM, Going SB, editors, 2005, Human body composition. 2nd ed Champaign, IL: Human Kinetics, 2005:89–108.

Rossmann MG, Mesyanzhinov VV, Arisaka F, Leiman PG. The bacteriophage T4 DNA injection machine. *Current Opinion in Structural Biology.* 2004;14(2):171–80. doi:10.1016/j.sbi.2004.02.001. PMID 15093831.

Roughgarden, J., 2005, *Evolution's Rainbow*: Diversity, *Gender, and Sexuality in Nature and People,* University of California Press, 474 pp.

Roulston A, Marcellus RC, Branton PE. Viruses and apoptosis. *Annu. Rev. Microbiol.* 1999;53:577–628. doi:10.1146/annurev.micro.53.1.577.

Rubin, G.S., 2011, A Gayle Rubin Reader, John Hope Franklin Center, 504 pp.

Rudin D, Shah SM, Kiss A, Wetz RV, Sottile VM. Interferon and lamivudine vs. interferon for hepatitis B e antigen-positive hepatitis B treatment: meta-analysis of randomized controlled trials. *Liver Int.* 2007;27(9):1185–93. doi:10.1111/j.1478-3231.2007.01580.x.

Rushton, J.P. and Ankney, C.D., 2009, "Whole brain size and general mental ability: A review", *International Journal of Neuroscience*, 119:691–731.

Rushton, J.P., Jensen, A.R., 2005, "Thirty years of research on race differences in cognitive ability, "*Psychology, Public Policy, and Law*, vol. 11, no. 2, 235.

Rushton, J.P., Jensen, A.R., 2010, "Race and IQ: A Theory-based review of the research in Richard Nisbett's intelligence and how to get it", *The Open Psychology Journal*, vol. 3, 9-35.

Safdar N, Kluger DM, Maki DG, A review of risk factors for catheter-related bloodstream infection caused by percutaneously inserted, noncuffed central venous catheters: implications for preventive strategies. *Medicine (Baltimore)*2002;81:466-79.

Sandbulte MR, Westgeest KB, Gao J, Xu X, Klimov AI, Russell CA, Burke DF, Smith DJ, Fouchier RA, Eichelberger MC. Discordant antigenic drift of neuraminidase and hemagglutinin in H1N1 and H3N2 influenza viruses. Proceedings of the National Academy of Sciences of the United States of America. 2011;108(51):20748–53.

Sandeman DC, Tautz J, Lindauer M. Transmission of vibration across honeycombs and its detection by bee leg receptors. *J Exp Biol* 1996;199:2585-2594.

Sanjuán R, Nebot MR, Chirico N, Mansky LM, Belshaw R. Viral mutation rates. *Journal of Virology*. 2010;84(19):9733–48. doi:10.1128/JVI.00694-10. PMID 20660197.

Sankararaman, S., Patterson, N., Li, H., Pääbo, S. & Reich, D. The date of interbreeding between Neandertals and modern humans. *PLoS Genet.* 8, e1002947 (2012)

Saunders, Venetia A.; Carter, John. Virology: principles and applications. Chichester: John Wiley & Sons; 2007. ISBN 0-470-02387-2. p. 72.

Scally, A. & Durbin, R. Revising the human mutation rate: implications for understanding human evolution. *Nature Rev. Genet.* 13, 745–753 (2012)

Scarborough P., Bhatnagar P., Wickramasinghe K., Smolina K., Mitchell C., Rayner M. British Heart Foundation; 2010. Coronary heart disease statistics.

Scarborough P., Wickramasinghe K., Bhatnagar P., Rayner M. British Heart Foundation; London: 2011. Trends in coronary heart disease 1961–2011.

Schewe, P.F. and Stein, B., (1999) Light has been Slowed to a Speed of 17 m/s, American Institute of Physics, Bulletin of Physics News, No. 415, February 18.

Schiffman M, Castle PE, Jeronimo J, Rodriguez AC, Wacholder S. Human papillomavirus and cervical cancer. *Lancet*. 2007;370 (9590):890–907.

Schmickl T, Crailsheim K. TaskSelSim: a model of the self-organization of the division of labour in honey bees. *Math Comp Model Dyn Syst* 2008;14:101-125.

Schmidz-Nielsen K., 1984, Scaling: why is animal size so important? Cambridge, United Kingdom: Cambridge University Press, 1984.

Schmitt, J., Warner, K., and Gupta, S. 2010, The High Budgetary Cost of Incarceration, Center for Economic and Policy Research Report, 17 pp.

Schneider SS, Lewis LA. The vibration signal, modulatory communication and the organization of labor in honey bees, Apis mellifera. *Apidologie* 2004;35:117-131.

Schneider SS, Painter-Kurt S, Degrandi-Hoffman G. The role of the vibration signal during queen competition in colonies of the honey bee, Apis mellifera. *Anim Behav* 2001;61:1173-1180.

Schneider SS, Stamps JA, Gary NE. The vibration dance of the honey bee 1. Communication regulating foraging on 2 time scales. *Anim Behav* 1986a;34:377-385.

Schneider SS, Stamps JA, Gary NE. The vibration dance of the honey bee 2. The effects of foraging success on daily patterns of vibration activity. *Anim Behav* 1986b;34:386-391.

Schneider SS. The modulation of worker activity by the vibration dance of the honey bee, Apis mellifera. *Ethology* 1987;74:211-218.

Schneider SS. The vibration dance activity of successful foragers of the honey bee, Apis mellifera (Hymenoptera, Apidae). *J Kansas Entomol Soc* 1986;59:699-705.

Schröder FH, Hugosson J, Roobol MJ, et al. Prostate-cancer mortality at 11 years of follow-up. *N Engl J Med.* 2012; 366(11):981-990.

Seeley T.D., Visscher P.K., Passino K.M. (2006). "Group decision making in honey bee swarms". *American Scientist* 94: 220–229. doi:10.1511/2006.3.220.

Seeley TD, Kühnholz S, Weidenmüller A. The honey bee's tremble dance stimulates additional bees to function as nectar receivers. *Behav Ecol Sociobiol* 1996;39:419-427.

Seeley TD, Morse RA. The nest of the honey bee (Apis mellifera L.). *Insectes Soc* 1976;23:495-512.

Seeley TD, Weidenmüller A, Kuhnholz S. The shaking signal of the honey bee informs workers to prepare for greater activity. *Ethology* 1998;104:10-26.

Seeley TD. Division of labor betweens scouts and recruits in honey bee foraging. *Behav Ecol Sociobiol* 1983;12:253-259.

Seeley TD. Social foraging in honey bee: how nectar foragers assess their colony nutritional status. *Behav Ecol Sociobiol* 1989;24:181-199.

Seeley TD. The tremble dance of the honey bee: message and meanings. *Behav Ecol Sociobiol* 1992;31:375-383.

Seeley TD. The wisdom of the hive. Cambridge (MA): Harvard University Press; 1995.

Seeley TD. Thoughts on information and integration in honey bee colonies. *Apidologie* 1998;29:67-80.

Seid, M. A.; Castillo, A.; Wcislo, W. T. (2011). "The Allometry of Brain Miniaturization in Ants". *Brain, Behavior and Evolution* 77 (1): 5–13. doi:10.1159/000322530. PMID 21252471.

Selvam R, Maheswari P, Kavitha P, Ravichandran M, Sas B, Ramchand CN. Effect of Bacillus subtilis PB6, a natural probiotic on colon mucosal inflammation and plasma cytokines levels in inflammatory bowel disease; *Indian J Biochem Biophys*. 2009 Feb;46(1):79-85.

Seow, A., J.M.Yuan,W.P.Koh,H.P.Lee,andM.C. Yu, 2006, "Diabetes mellitus and risk of colorectal cancer in the Singapore Chinese Health Study," *Journal of the National Cancer Institute*, vol. 98, no. 2, pp. 135–138, 2006.

Seow, A., S. W. Duffy, M. A. McGee, J. Lee, and H. P. Lee, "Breast cancer in Singapore: trends in incidence 1968–1992," *International Journal of Epidemiology*, vol. 25, no. 1, pp. 40– 45, 1996.

Sessa R, Palagiano C, Scifoni MG, di Pietro M, Del Piano M. The major epidemic infections: a gift from the Old World to the New?. *Panminerva Med*. 1999;41(1):78–84. PMID 10230264.

Shafir, S., Wiegmann, D. D., Smith, B. H., And Real, L. A. 1999. Risk-sensitive foraging: Choice behaviour of honey bees in response to variability in volume of reward. *Anim. Behav*. 57:1055–1061.

Shakespeare, W., 1608, "King Lear I,i: 72 in Evans, G Blakemore, et al., edd., 1974, The Riverside Shakespeare, Houghton-Mifflin", Boston, pp. 1923

Shapiro, B. et al. A Bayesian phylogenetic method to estimate unknown sequence ages. *Mol. Biol. Evol*. 28, 879–887 (2011)

Sharma, V., V. S. Chauhan, G. Nath, A. Kumar, and V. K. Shukla, "Role of bile bacteria in gallbladder carcinoma," *Hepato-Gastroenterology*, vol. 54, no. 78, pp. 1622–1625, 2007.

Sharp, A., "Beneficial effects of honey dressings in wound management," *Nursing Standard*, vol. 24, no. 7, pp. 66–72, 2009.

Shenk, J.W., 2005, Lincoln's Great Depression, *The Atlantic*, Oct. 1

Shors, Teri (2008). Understanding Viruses. Jones and Bartlett Publishers. ISBN 0-7637-2932-9.

Shu, X.O., Linet, M.S., Steinbuch, M., Wen, W.Q., Buckley, J.D., Neglia, J.P., Potter, J.D., Reaman, G.H., and Robison, L.L., 1999, Breast-Feeding and Risk of Childhood Acute Leukemia. *JNCI J Natl Cancer Inst (1999)* 91 (20): 1765-1772.

Shuda M, Feng H, Kwun HJ, Rosen ST, Gjoerup O, Moore PS, Chang Y. T antigen mutations are a human tumor-specific signature for Merkel cell polyomavirus. *Proceedings of the National Academy of Sciences of the United States of America*. 2008;105(42):16272–7. doi:10.1073/pnas.0806526105. PMID 18812503. Bibcode:2008PNAS.10516272S.

Siddique, K., S. Bhandari, and G. Harinath, "Epsteln-Barr virus (EBV) posittie anal B cell lymphoma: a case report and review of literature," *Annals of the Royal College of Surgeons of England*, vol. 92, no. 3, pp. W7–W9, 2010.

Sinclair, A. (1962) *Prohibition: The Era of Excess*. Boston: Little, Brown

Sinclair J. Human cytomegalovirus: Latency and reactivation in the myeloid lineage. *J. Clin. Virol.* 2008;41(3):180–5.

Singh, A., 2005, "University of Winnipeg, The Bush Doctrine:*Creating Discord in International Security Critique: A worldwide journal of politics*,

Singleton, V. L., Orthofer, R., And Lamuela-Raventos, R. M. 1999. Analysis of total phenols and other oxidation substrates and antioxidants by means of Folin–Ciocalteu reagent. *Meth. Enzymol.* 299:152–178.

Sissons JG, Bain M, Wills MR. Latency and reactivation of human cytomegalovirus. *J. Infect.* 2002;44(2):73–7.

Skinner, S.J., 2013, The Costly Paradox of Health-Care Technology", *MIT Technology Review*, Sept. 5.

Skoglund, P. & Jakobsson, M. Archaic human ancestry in East Asia. *Proc. Natl Acad. Sci. USA* 108, 18301–18306 (2011)

Slessor KN, Winston ML, Le Conte Y. Pheromone communication in the honey bee (Apis mellifera L.). *J Chem Ecol* 2005;31:2731-2745.

Smith, M. J. and Harper, D. (2003) *Animal Signals*. Oxford University Press. ISBN 0-19-852685-7.

Smolina K., Wright F.L., Rayner M., Goldacre M.J. Determinants of the decline in mortality from acute myocardial infarction in England between 2002 and 2010: linked national database study. *BMJ*. 2012;344:d8059.

Snyder WS, Cook MJ, Nasset ES, Karhausen LR, Howells GP, Tipton IH., 1975, Report of the Task Group on Reference Man. Oxford, United Kingdom: Pergamon Press.

Sokal, R. And Rohlf, J. 1995. Biometry. W. H. Freeman and Company, New York.

Sookeun Byun, et al. (2009). "Internet Addiction: Metasynthesis of 1996–2006 quantitative research". *Cyberpsychology & Behavior* 12 (2): 203–7.

Soosaar JL, Burch-Smith TM, Dinesh-Kumar SP. Mechanisms of plant resistance to viruses. *Nature Reviews Microbiology*. 2005;3(10):789–98.

Soto CM, Blum AS, Vora GJ, et al. Fluorescent signal amplification of carbocyanine dyes using engineered viral nanoparticles. *J. Am. Chem. Soc.* 2006;128(15):5184–9. doi:10.1021/ja058574x.

Southwick, E. E., Loper, G. M., And Sadwick, S. E. 1981. Nectar production, composition, energetics and pollinator attractiveness in spring flowers of western New York. *Am. J. Bot.* 68:994–1002.

Spurling, M.C., and Vinson, D. C., 2003, "Alcohol-Related Injuries: Evidence for the Prevention Paradox", report presented at the *North American Primary Care Research Group annual meeting, Banff, Alberta*, Canada; October 27, 2003.

Stabentheiner A, Kovac H, Brodschneider R. Honey bee colony thermoregulation—regulatory mechanisms and contribution of individuals in dependence on age, location and thermal stress. *PloS ONE* 2010;5:e8967.

Stanley WM, Lauffer MA. Disintegration of tobacco mosaic virus in urea solutions. *Science.* 1939;89(2311):345–347. doi:10.1126/science.89.2311.345. PMID 17788438. Bibcode:1939Sci..89.345S.

Stanley WM, Loring HS. The isolation of crystalline tobacco mosaic virus protein from diseased tomato plants. *Science.* 1936;83(2143):85. doi:10.1126/science.83.2143.85. PMID 17756690. Bibcode:1936Sci..83..85S.

Steinhardt E, Israeli C, Lambert R.A. Studies on the cultivation of the virus of vaccinia. *J. Inf Dis.* 1913;13(2):294–300. doi:10.1093/infdis/13.2.294.

Steinhoff, U. (June 2005). "Who controls the crowd? New findings and old questions about the intestinal microflora". *Immunol. Lett.*99 (1): 12-6.

Stelmakiene, A., Ramanauskiene, K.,Vitalis Briedis, V., and Daiva Leskauskaite, D., 2012, "Examination of rheological and physicochemical characteristics in Lithuanian honey", *African Journal of Biotechnology* Vol. 11(60), pp. 12406-12414, 26 July.

Stroh, M, 2004, "A Contrarian View Of Cancer: Once shunned for his AIDS theories, Peter Duesberg is back in the spotlight", April 1.

Strong K., Mathers C., Leeder S., Beaglehole R. Preventing chronic diseases: how many lives can we save? *Lancet.* 2005;366(9496):1578–1582.

Stuart RJ, Moffett MW. Recruitment communication and pheromone trails in the neotropical ants, Leptothorax (Nesomyrmex) spininodis and L (N) Echinatinodis. *Experientia* 1994;50:850-852.

Stuckler D., McKee M., Ebrahim S., Basu S. Manufacturing epidemics: the role of global producers in increased consumption of unhealthy commodities including processed foods, alcohol, and tobacco. *PLoS Med.* 2012;9(6): e1001235.

Stuebe, A., 2009, The Risks of Not Breastfeeding for Mothers and Infants, Rev Obstet Gynecol. *2009 Fall*; 2(4): 222–231.

Subrahmanyam M, Honey dressing versus boiled potato peel in the treatment of burns: a prospective randomized study. *Burns* 1996;22:491-3.

Subrahmanyam M, Topical application of honey in treatment of burns. *Br J Surg*1991;78:497-8.

Subrahmanyam M., A prospective randomised clinical and histological study of superficial burn wound healing with honey and silver sulfadiazine. *Burns*1998;24:157-61.

Subrahmanyam M., Honey impregnated gauze versus polyurethane film (OpSite) in the treatment of burns—a prospective randomised study. *Br J Plast Surg*1993;46:322-3.

Subrahmanyam M., Honey-impregnated gauze versus amniotic membrane in the treatment of burns. *Burns* 1994;20:331-3.

Subramanya D, Grivas PD. HPV and cervical cancer: updates on an established relationship. *Postgrad Med.* 2008;120(4):7–13.

Suez J, Korem T, Zeevi D, et al. (September 2014). "Artificial sweeteners induce glucose intolerance by altering the gut microbiota". *Nature* 514 (7521): 181–6.

Suler, J. 2004. The psychology of text relationships. Online counseling: A handbook for mental health professionals, pp. 19-50.

Sullivan, A., Carl E Nord and Birgitta Evengård. Effect of supplement with lactic-acid producing bacteria on fatigue and physical activity in patients with chronic fatigue syndrome;*Nutrition Journal*2009, 8:4.

Sumpter DJT. The principles of collective animal behaviour. *Philos Trans R Soc B Biol Sci* 2006;361:5-22.

Surratt, Carla G (1999). Netaholics?: The creation of a pathology Commack, NY: Nova Science Publishers.

Suzan-Monti M, La Scola B, Raoult D. Genomic and evolutionary aspects of Mimivirus. *Virus Research*. 2006;117(1):145–155. doi:10.1016/j.virusres.2005.07.011.

Swanson, 2010, God, Science, and the Universe, Eloquent Books, Durban, CT, 444 pp.

Swellam, T., N. Miyanaga, M. Onozawa, et al., "Antineoplastic activity of honey in an experimental bladder cancer implan- tation model: in vivo and in vitro studies," *International Journal of Urology*, vol. 10, no. 4, pp. 213–219, 2003.

Tanner CJ. Aggressive group behavior in the ant Formica xerophila is coordinated by direct nestmate contact. *Anim Behav* 2008;76:1335-1341.

Tautz J, Rohrseitz K. What attracts honey bees to a waggle dancer? *J Comp Physiol A* 1998;183:661-667.

Telis, G., 2013, Why That Bee Sting Might Be Good for You, *Science*, 24 October, published on line.

Temin, HM, Baltimore D. RNA-directed DNA synthesis and RNA tumor viruses. *Adv. Virus Res.* 1972 [Retrieved 16 September 2008];17:129–86. doi:10.1016/S0065-3527(08)60749-6. PMID 4348509.

Terrab, A. And Heredia, F. J. 2004. Characterization of avocado (Persea americana Mill) honeys by their physicochemical characteristics. *J. Sci. Food Agric.* 84:1801–1805.

The Guardian, 2015, "No convictions over 500 black and Asian deaths in custody", March 21, 2015.

The Y Chromosome Consortium A nomenclature system for the tree of human Y-chromosomal binary haplogroups. *Genome Res.* 12, 339–348 (2002)

Thenius R, Schmickl T, Crailsheim K. Optimisation of a honey bee-colony's energetics via social learning based on queuing delays. *Connect Sci* 2008;20:193-210.

Theraulaz G, Gautrais J, Camazine S, Deneubourg JL. The formation of spatial patterns in social insects: from simple behaviours to complex structures. *Philos Trans R Soc A* 2003;361:1263-1282.

Thom C. The tremble dance of honey bees can be caused by hive-external foraging experience. *J Exp Biol* 2003;206:2111-2116.

Thomas, P., 2005, The Ecologist report, September, http://www.wnho.net/the_ecologist_aspartame_report.htm

Thomssen R. Live attenuated versus killed virus vaccines. *Monographs in allergy*. 1975;9:155–76.

Thorp, R. W., Briggs, D. L., Esters, J. R., And Erickson, E. H. 1975. Nectar fluorescence under ultra-violet irradiation. *Science* 189:476–478.

Toh, C.K., "The changing epidemiology of lung cancer," *Methods in Molecular Biology*, vol. 472, pp. 397–411, 2009.

Tonks, A., R. A. Cooper, A. J. Price, P. C. Molan, and K. P. Jones, "Stimulation of TNF-α release in monocytes by honey," *Cytokine*, vol. 14, no. 4, pp. 240–242, 2001.

Tonks, A.J., E. Dudley, N. G. Porter, et al., "A 5.8-kDa component of manuka honey stimulates immune cells via TLR4," *Journal of Leukocyte Biology*, vol. 82, no. 5, pp. 1147- 1155, 2007.

Tonks, A.J., R. A. Cooper, K. P. Jones, S. Blair, J. Parton, and A. Tonks, "Honey stimulates inflammatory cytokine production from monocytes," *Cytokine*, vol. 21, no. 5, pp. 242–247, 2003.

Torrey, E.F.,Kennard, A.D., Eslinger, D., Lamb, R., Pavle, J., 2010, More Mentally Ill Persons Are in Jails and Prisons Than Hospitals: A Survey of the States, Joint TAC and TSA report, http://www.treatmentadvocacycenter.org/storage/documents/final_jails_v_hospitals_study.pdf, last accessed October 15, 2014.

Towner JS, Khristova ML, Sealy TK, *et al.* Marburgvirus genomics and association with a large hemorrhagic fever outbreak in Angola. *J. Virol.* 2006;80(13):6497–516. doi:10.1128/JVI.00069-06. PMID 16775337.

Traniello JFA. Recruitment behavior, orientation, and organization of foraging in carpenter ant Camponotus-pennsylvanicus degeer (Hymenoptera, Formicidae). *Behav Ecol Sociobiol* 1977;2:61-79.

Trinkaus, E. & Ruff, C. B. Diaphyseal cross-sectional geometry of Near Eastern Middle Paleolithic humans: the femur. *J. Archaeol. Sci.* 26, 409–424 (1999)

Trochim, W., & Kane, M. (2005). Concept mapping: an introduction to structured conceptualization in health care. *International Journal for Quality in Health Care*, 17, 187–191.

Tschinkel WR. The fire ants. Cambridge (MA): Harvard University Press; 2006.

Tsui, B., 2012, "What's the secret to learning a second language?", *Salon*, October 27.

Tsutsui ND. Scents of self: the expression component of self/nonself recognition systems. *Ann Zool Fenn* 2004;41:713-727.

Tunstall-Pedoe H for the WHO MONICA Project. WHO; 2003. MONICA monograph and multimedia sourcebook: world's largest study of heart disease, stroke, risk factors, and population trends 1979–2002.

Tunstall-Pedoe H., Kuulasmaa K., Mähönen M., Tolonen H., Ruokokoski E., Amouyel P. Contribution of trends in survival and coronary-event rates to changes in coronary heart disease mortality: 10-year results from 37 WHO MONICA project populations. Monitoring trends and determinants in cardiovascular disease. *Lancet.* 1999; 353(9164):1547–1557.

Tunstall-Pedoe H., Vanuzzo D., Hobbs M. Estimation of contribution of changes in coronary care to improving survival, event rates, and coronary heart disease mortality across the WHO MONICA project populations. *Lancet.* 2000;355:688–700.

Turel, O. & Serenko, A. (2010). "Is mobile email addiction overlooked?". *Communications of the ACM* 53 (5): 41–43. doi:10.1145/1735223.1735237.

Turel, O., Serenko, A. & Bontis, N. (2011). Family and Work-Related Consequences of Addiction to Organizational Pervasive Technologies. *Information & Management*, 48(2–3): 88–95.

U.S. Department of Agriculture, Office of Inspector General; *FSIS National Residue Program for Cattle*, Audit Report 24601-08-KC, March 2010.

Uemura K., Pisa Z. Trends in cardiovascular disease mortality in industrialized countries since 1950. *World Health Stat Q.* 1988;41:155–178.

Umene K. Mechanism and application of genetic recombination in herpesviruses. *Rev. Med. Virol.* 1999;9(3):171–82.

Underwood, E., 2015, "Can Sound open the brain for therapies? ", *Science*, March 12, Vol. 347 no. 6227 pp. 1186-1187, DOI:10.1126/science.347.6227.1186.

Underwood, E., 2015, Magnetic 'rust' controls brain activity, *Science*, March 12, DOI: 10.1126/science.aab0309.

University Press [digital imprint of reissued 1987 update of 1957 1st Edition]

Valko, M., D. Leibfritz, J. Moncol, M. T. D. Cronin, M. Mazur, and J. Telser, "Free radicals and antioxidants in normal physiological functions and human disease," *The International Journal of Biochemistry & Cell Biology*, vol. 39, no. 1, pp. 44–84, 2007.

Vallejo, F., Tomas-Barberan. F.A. and Garcıa-Viguera, C., (2003) Phenolic Compound

Van Rijn K. "Lo! The poor Indian!" colonial responses to the 1862–63 smallpox epidemic in British Columbia and Vancouver Island. *Can Bull Med Hist.* 2006 [Retrieved 16 September 2008];23(2):541–60. PMID 17214129.

Vaziri, H., Zatzman, G.M. and Islam, M.R., (2007) "Preface", Handbook of Sustainable Petroleum Engineering Operations, Gulf Publishing Co., Houston, 452 pp.

Velazquez, N. and Best, R., (2002) Methodology for the Energy Analysis of an Air Cooled GAX Absorption Heat Pump Operated by Natural Gas and Solar Energy, *Applied Thermal Engineering*, vol. 22, pp. 1089–1103

Viegas, J., 2015, "Female Chimps Seen Making, Wielding Spears", Discovery.com, April 14.

Vieira-Neto EHM, Mundim FM, Vasconcelos HL. Hitchhiking behaviour in leaf-cutter ants: an experimental evaluation of three hypotheses. *Insect Soc* 2006;53:326-332.

Visscher PK, Camazine S. Collective decisions and cognition in bees. *Nature* 1999;397:400-400.

Visscher PK, Seeley TD. Foraging strategies of honey bee colonies in a temperate deciduous forest. *Ecology* 1982;63:1790-1801.

Visscher PK. Group decision making in nest-site selection among social insects. *Ann Rev Entomol* 2007;52:255-275.

Vithanage, V. 1990. The role of the European honeybee (Apis mellifera L.) in avocado pollination. *J. Hortic. Sci.* 65:81–86.

Vogel, G., 2015, Don't edit embryos, researchers warn, *Science*, March 13, DOI: 10.1126/science.aab0314.

von Frisch K. The dance language and orientation of bees. Cambridge (MA): Harvard University Press; 1967.

Von Frisch, K. 1950. Bees. Cornell University Press, Ithaca, NY.

Von Frisch, K. 1967. The Dance Language and Orientation of Bees. Harvard University Press, Cambridge. WALLER, G. D. 1972. Evaluating responses of honey bees to sugar solutions using an artificial flower feeder. Ann. Entomol. Soc. Am. 65:857–862.

VonLow, E.C., F.G.Perabo,R.Siener,andS.C.Muller,"Facts and fiction of phytotherapy for prostate cancer: a critical assessment of preclinical and clinical data," *In Vivo*, vol. 21, no. 2, pp. 189–204, 2007.

Wahl, O. (1995). Media Madness: Public Images of Mental Illness. New Brunswick, NJ: Rutgers University Press.

Wahl, O.F., et al. (2002). Newspaper coverage of mental illness: is it changing? *Psychiatric Rehabilitation Skills*, 6, 9-31.

Wakabayashi H, Nishiyama Y, Ushivama T, Maeba T, Maeta H. Evaluation of the effect of age on functioning hepatocyte mass and liver blood flow using liver scintigraphy in preoperative estimations for surgical patients: comparison with CT volumetry. *J Surg Res* 2002;106:246–53.

Waller, G. D., Carpenter, E. W., And Ziehl, O. A. 1972. Potassium in onion nectar and its probable effect on attractiveness of onion flowers to honey bees. *J. Am. Soc. Hortic. Sci.* 97:535–539.

Wang Z, Heshka S, Heymsfield SB, Shen W, Gallagher D., 2005, A cellular level approach to predicting resting energy expenditure across the adult years. *Am J Clin Nutr* 2005;81:799–806.

Wang Z, Heshka S, Wang J, et al. Metabolically active portion of fat-free mass: a cellular body composition level modeling analysis. *Am J Physiol Endocrinol Metab* 2007;292:E49–53.

Wang Z, Heshka S, Zhang K, Boozer C, Heymsfield SB. Resting energy expenditure: systematic organization and critique of prediction methods. *Obes Res* 2001;9:331–6.

Wang Z, Heymsfield SB, Ying Z, Pierson RN, Jr, Gallagher D, Gidwani S. A cellular level approach to predicting resting energy expenditure: Evaluation of applicability in adolescents. *Am J Hum Biol* 2010;22:476–83.

Wang, J. C., Hu, S. H., Wang, J. T., Chen, K. S., Chia, Y. C., 2005, "Hypoglycemic effect of extract of Hericium erinaceus". *Journal of the Science of Food and Agriculture* 85 (4): 641–646, doi:10.1002/jsfa.1928.

Wang, J., Fan, H. C., Behr, B. & Quake, S. R. Genome-wide single-cell analysis of recombination activity and de novo mutation rates in human sperm. *Cell* 150, 402–412 (2012)

Wang, Y.C., McPherson, K. Marsh, T., Gortmaker, SS.L., and Brown, M., 2011, Health and Economic burden of the projected obesity trends in the USSA and the UK, *Lanset*, 378: 815-25.

Wang, Z., Ying, Z., Bosy-Westphal, A., Zhang, J., Schautz, B., Later, W., Heymsfield, S.B., and Müller, M.J., 2014, Specific metabolic rates of major organs and tissues across adulthood: evaluation by mechanistic model of resting energy expenditure, *Am J Clin Nutr.* 2010 Dec; 92(6): 1369–1377.

Warren, C.W., N. R. Jones, A. Peruga, et al., "Global youth tobacco surveillance, 2000–2007," *MMWR Surveillance Summaries*, vol. 57, no. 1, pp. 1–28, 2008.

Warren, R., 2002, The Purpose Driven Life: What on Earth Am I here for? Zondervan, Grand Rapids, MI, 330 pp.

Weast, R. C. 1988. CRC Handbook of Chemistry and Physics. CRC Press, Boca Raton, FL.

Webb, J., 2015, Repeated remembering 'wipes similar memories', *BBC*, March 17.

Weisberg S. Applied linear regression. 3rd ed Hoboken, NJ: John Wileys & Sons, 2005.

Weissert, W., 2013, "Federal judge accused of making racial comments", NECN.com, June 4.

Weissman MM, Bland RC, Canino GJ, et al., 1999, Prevalence of suicide ideation and suicide attempts in nine countries. *Psychological Medicine*, 1999; 29(1): 9-17.

Weller, C., 2013, 'Biological Immortality'? Gabby Williams' Genetic Condition Prevents The 8-Year-Old From Aging, *Medical Daily*, Aug. 19.

Westlake, S., and N. Cooper, "Cancer incidence and mortality: trends in the United Kingdom and constituent countries,1993 to 2004," *Health Statistics Quarterly*, no. 38, pp. 33–46, 2008.

White JW Jr, Subers MH, Studies on honey inhibine. 2. A chemical assay. *J Apic Res* 1963;2:93-100.

WHO Monica project Ecological analysis of the association between mortality and major risk factors of cardiovascular disease. *Int J Epidemiol.* 1994;23:505–516.

Wijesinghe, M. Weatherall, K. Perrin, and R. Beasley, "Honey in the treatment of burns: a systematic review and meta-analysis of its efficacy," *New Zealand Medical Journal*, vol. 122, no. 1295, pp. 47–60, 2009.

Williams, S., 2015, "Humans may harbor more than 100 genes from other organisms", *Science*, 12 March, DOI: 10.1126/science.aab0307

Willix DJ, Molan PC, Harfoot CG, A comparison of the sensitivity of wound-infecting species of bacteria to the antibacterial activity of manuka honey and other honey. *J Appl Bacteriol* 1992;73:388-94.

Wilson EO. Chemical communication in the social insects. *Science* 1965;149:1064-1071.

Wilson EO. The insect societies. Cambridge (MA): Harvard University Press; 1971.

Winkler, A.; Dörsing, B.; Rief, W.; Shen, Y.; Glombiewski, J. A. (2013). "Treatment of internet addiction: A meta-analysis". *Clinical Psychology Review* 33 (2): 317–329. doi:10.1016/j.cpr.2012.12.005. PMID 23354007.

Winston, M. L. 1987. The Biology of the Honey Bee. Harvard University Press, Cambridge.

Winston, M. L. And Slessor, K. N. 1993. Applications of queen honey-bee mandibular pheromone for beekeeping and crop pollination. *Bee World* 74:111–128.

Witthöft T, Möller B, Wiedmann KH, *et al.* Safety, tolerability and efficacy of peginterferon alpha-2a and ribavirin in chronic hepatitis C in clinical practice: The German Open Safety Trial. *J. Viral Hepat.* 2007;14(11):788–96. doi:10.1111/j.1365-2893.2007.00871.x. PMID 17927615.

Wolstenholme, B. N. 2002. Ecology: climate and the edaphic environment, pp. 71–100, in A. W. Whiley, B. Schaffer, and B. N. Wolstenholme (eds.). The Avocado: Botany, Production and Uses. CAB International, Wallingford.

Wommack KE, Colwell RR. Virioplankton: viruses in aquatic ecosystems. *Microbiol. Mol. Biol. Rev.* 2000;64(1):69–114. doi:10.1128/MMBR.64.1.69-114.2000.

Wong, M., 2011, "Harmful chemicals in 60pc of plastic toys", *South China Morning Post*, May 19.

Woo, K.J., Y. J. Jeong, J. W. Park, and T. K. Kwon, "Chrysin- induced apoptosis is mediated through caspase activation and Akt inactivation in U937 leukemia cells," *Biochemical and Biophysical Research Communications*, vol. 325, no. 4, pp. 1215–1222, 2004.

World Health Organisation. World Health Organisation; 1992. International statistical classification of disease and related health problems. 10th revision.

World Health Organisation statistical Information System WHO mortality database. http://www.who.int/healthinfo/statistics/mortality

Worobey M, Holmes EC. Evolutionary aspects of recombination in RNA viruses. *J. Gen. Virol.* 1999;80 (Pt 10):2535–43.

Wykes, G. R. 1952. The preferences of honey bees for solutions of various sugars which occur in nectar, *J. Exp. Biol.* 29:511–518.

Xu J.Q., Kochanek K.D., Murphy S.L., Tejada-Vera B. vol. 58 no. 19. National Center for Health Statistics; Hyattsville, MD: 2010. Deaths: final data for 2007. (National vital statistics reports).

Xue, Y. et al. Human Y chromosome base-substitution mutation rate measured by direct sequencing in a deep-rooting pedigree. *Curr. Biol.* 19, 1453–1457 (2009)

Yaghoobi, N., N. Al-Waili, M. Ghayour-Mobarhan, et al., "Natural honey and cardiovascular risk factors; effects on blood glucose, cholesterol, triacylglycerole, CRP, and body weight compared with sucrose," *The Scientific World Journal*, vol. 8, pp. 463–469, 2008.

Yang, Y.X., S. Hennessy, and J. D. Lewis, "Type 2 diabetes mellitus and the risk of colorectal cancer," *Clinical Gastroenterology and Hepatology*, vol. 3, no. 6, pp. 587–594, 2005.

Yao, L., N. Datta, F. A. Tomás-Barberán, F. Ferreres, I. Martos, and R. Singanusong, "Flavonoids, phenolic acids and abscisic acid in Australian and New Zealand Leptospermum honeys," *Food Chemistry*, vol. 81, no. 2, pp. 159–168, 2003.

Yeole, B.B., "Trends in cancer incidence in female breast, cervix uteri, corpus uteri, and ovary in India," *Asian Pacific Journal of Cancer Prevention*, vol. 9, no. 1, pp. 119–122, 2008.

You T, Berman DM, Ryan AS, Nicklas BJ: Effects of hypocaloric diet and exercise training on inflammation and adipocyte lipolysis in obese postmenopausal women, *J Clin Endocrinol Metab* 2004, 89:1739-1746.

You T, Ryan AS, Nicklas BJ: The metabolic syndrome in obese postmenopausal women: relationship to body composition, visceral fat, and inflammation, *J Clin Endocrinol Metab* 2004, 89:5517-5522.

Young, K. 2009. Understanding online gaming addiction and treatment issues for adolescents. *The American Journal of Family Therapy*, 37 (5), pp. 355-356.

Young, K. S. (1998). Caught in the Net. New York: John Wiley. ISBN 0471191590.

Young, K. S. 1998. Caught in the net. New York: J. Wiley.

Young, K., S., and Nabuco de Abreu, C. (ed.) (2011) Internet addiction, A handbook and guide to evaluation and treatment. Hoboken: John Wiley & Sons, Inc.

Yuen, M.F., J.L.Hou,andA.Chutaputti,"Hepatocellularcar- cinoma in the Asia Pacific region," *Journal of Gastroenterology and Hepatology*, vol. 24, no. 3, pp. 346–353, 2009.

Yusuf S., Hawken S., Ounpuu S. Effect of potentially modifiable risk factors associated with myocardial infarction in 52 countries (the INTERHEART study): case–control study. *Lancet.* 2004;364:937–952.

Yusuf S., Islam S., Chow C.K. Use of secondary prevention drugs for cardiovascular disease in the community in high-income, middle-income, and low-income countries (the PURE Study): a prospective epidemiological survey. *Lancet.* 2011;378(9798):1231-1243.

Zaid, S.S., S. A. Sulaiman, K. N. Sirajudeen, and N. H. Othman, "The effects of tualang honey on female reproductive organs, tibia bone and hormonal profile in ovariectomised rats—animal model for menopause," *BMC Complementary and Alternative Medicine*, vol. 10, article 82, 2010.

Zaini, A., 2000, "Where is Malaysia in the midst of the Asian epidemic of diabetes mellitus?" *Diabetes Research and Clinical Practice*, vol. 50, supplement 2, pp. S23–S28, 2000.

Zatzman, G.M., 2008, The Honey, Sugar, Saccharin, Aspartame, or HSSA Syndrome: A Note;pp. 369-373, Nature Science and Sustainable Technology Compendium, Volume 1, Nova Science Publishers, NY.

Zatzman, G.M., and M.R. Islam, 2007, *Economics of Intangibles*, Nova Science Publishers, NY, 407 pp.

Zatzman, G.M., Chhetri, A.B., Khan, M.M., Al-Maamari R., Islam, M.R., 2008, "Colony Collapse Disorder- The Case for a Science of Intangibles", *Journal of Nature Science and Sustainable Technology*, vol. 2, No.2.

Zatzman, G.M., and Islam, M.R., 2007a, "Truth, Consequences and Intentions: The Study of Natural and Anti-Natural Starting Points and Their Implications", *J. Nature Science and Sustainable Technology*, vol. 1, no. 2, 169-174.

Zhang T, Breitbart M, Lee WH, Run J, Wei, CL, *et al.* (2006) RNA viral community in human feces: Prevalence of plant pathogenic viruses. *PLoS Biol* 4(1): e15. doi:10.1371/journal.pbio.0040015

Zhou, Y. et al. (2009). Gray matter abnormalities in Internet addiction: A voxel-based morphometry study. *European Journal of Radiology.* doi:10.1016/j.ejrad.2009.10.025

Zimmer, C., 2004, "Faith-Boosting Genes: A search for the genetic basis of spirituality", *Scientific American*, Oct.

zur Hausen H. Novel human polyomaviruses—re-emergence of a well known virus family as possible human carcinogens. International Journal of Cancer. *Journal International Du Cancer.* 2008;123(2):247–50. doi:10.1002/ijc.23620.

WEBSITES

Website 1: Screening for prostate cancer: U.S. Preventive Services Task Force Recommendation Statement. US Preventive Services Task Force Website. http://www.uspreventiveservicestaskforce.org/uspstf12/prostate/draftrecprostate.htm. Updated October 2011. Accessed April 10, 2013.

Website 2: http://www.illovosugar.co.za /World_of_sugar/ Sugar_Statistics/International.aspx, last accessed Sept. 29, 2013.

Website 3: NY government report, New York's oil and natural gas history – a long story, but not the final chapter. http://www.dec.ny.gov/docs/materials_minerals_pdf/nyserda2.pdf.

Website 4 from Chapter 7 energy book

Website 4: http://www.sacredlotus.com/theory/yinyang.cfm, last accessed November 7, 2013.

Website 5: http://en.wikipedia.org/wiki/Satyendranath_Bose, last accessed August 5, 2014

Website 6: http://www-groups.dcs.st-and.ac.uk/history/Biographies/Pythagoras.html

Website 7: http://en.wikipedia.org/wiki/Two_New_Sciences

Website 8: Einstein, Albert. 1905. "On The Electrodynamics Of Moving Bodies" (1923 English translation of the first published version of the theory of special relativity)

Website 9: Newton, Isaac.1687. Philosophiae naturalis principia mathematica [1729 translation by Andrew Motte], Ch 1. "Of the method of first and last ratios of quantities, by the help whereof we demonstrate the propositions that follow".

Website 10: http://en.wikipedia.org/wiki/Speed_of_light

Website 11: http://en.wikipedia.org/wiki/Satyendranath_Bose

Index

Also of Interest

Check out these published volumes in the Advanced Materials Series

By the Same Authors

Reconstituting the Curriculum, by M.R. Islam, ISBN 9781118472897. This book presents a new definition of curriculum and what it should consist of, with a view toward creating a more ethical, educated, and thinking person.

Zero-Waste Engineering, by M. R. Islam et al., ISBN 9780470626047. In this controversial new volume, the author explores the question of zero-waste engineering and how it can be done, efficiently and profitably. *NOW AVAILABLE!*

Sustainable Energy Pricing, by Gary Zatzman, ISBN 9780470901632. In this controversial new volume, the author explores a new science of energy pricing and how it can be done in a way that is sustainable for the world's economy and environment. *NOW AVAILABLE!*

Advanced Petroleum Reservoir Simulation, by M.R. Islam, S.H. Mousavizadegan, Shabbir Mustafiz, and Jamal H. Abou-Kassem, ISBN 9780470625811. The state of the art in petroleum reservoir simulation. *NOW AVAILABLE!*

Sustainable Resource Development, by Gary Zatzman, ISBN 9781118290392. Taking a new, fresh look at how the energy industry and we, as a planet, are developing our energy resources, this book looks at what is right and wrong about energy resource development. This book aids engineers and scientists in achieving a true sustainability in this field, both from an economic and environmental perspective. *NOW AVAILABLE!*

The Greening of Petroleum Operations, by M. R. Islam et al., ISBN 9780470625903. The state of the art in petroleum operations, from a "green" perspective. *NOW AVAILABLE!*

Other books published by Scrivener Publishing

Ethics in Engineering, by James Speight and Russell Foote, ISBN 9780470626023. Covers the most thought-provoking ethical questions in engineering.

Emergency Response Management for Offshore Oil Spills, by Nicholas P. Cheremisinoff, PhD, and Anton Davletshin, ISBN 9780470927120. The first book to examine the Deepwater Horizon disaster and offer processes for safety and environmental protection. *NOW AVAILABLE!*

Energy Storage: A New Approach, by Ralph Zito, ISBN 9780470625910. Exploring the potential of reversible concentrations cells, the author of this groundbreaking volume reveals new technologies to solve the global crisis of energy storage. *NOW AVAILABLE!*

Biogas Production, Edited by Ackmez Mudhoo, ISBN 9781118062852. This volume covers the most cutting-edge pretreatment processes being used and studied today for the production of biogas during anaerobic digestion processes using different feedstocks, in the most efficient and economical methods possible. *NOW AVAILABLE!*

Bioremediation and Sustainability: Research and Applications, Edited by Romeela Mohee and Ackmez Mudhoo, ISBN 9781118062845. Bioremediation and Sustainability is an up-to-date and comprehensive treatment of research and applications for some of the most important low-cost, "green," emerging technologies in chemical and environmental engineering. *NOW AVAILABLE!*

Green Chemistry and Environmental Remediation, Edited by Rashmi Sanghi and Vandana Singh, ISBN 9780470943083. Presents high quality research papers as well as in depth review articles on the new emerging green face of multidimensional environmental chemistry. *NOW AVAILABLE!*

Bioremediation of Petroleum and Petroleum Products, by James Speight and Karuna Arjoon, ISBN 9780470938492. With petroleum-related spills, explosions, and health issues in the headlines almost every day, the issue of remediation of petroleum and petroleum products is taking on increasing importance, for the survival of our environment, our planet, and our future. This book is the first of its kind to explore this difficult issue from an engineering and scientific point of view and offer solutions and reasonable courses of action. *NOW AVAILABLE!*